TECHNICAL CORRECTIONS

A Policy on Geometric Design of Highways and Streets

1990

Prepared by the

American Association of State Highway and Transportation Officials

TECHNICAL CORRECTIONS

A Policy on Geometric Design of Highways and Streets

1990

American Association of State Highway and Transportation Officials
444 North Capitol Street, N.W. Suite 225
Washington, D.C. 20001
(202) 624-5800

Preface to Technical Corrections

Following the publication of *A Policy on Geometric Design of Highways and Streets, 1990*, some technical and editorial corrections were identified by the AASHTO Task Force on Geometric Design that were not found by the Task Force in its final prepublication review. For the convenience of users, this booklet contains corrected reprints of the pages on which major corrections occur (see Appendix) and lists those corrections which are less significant. It is recommended that users of this book clearly mark the pages on which corrections have been made, and refer to this booklet for the information contained therein.

Chapter II Design Controls and Criteria

Page 21 **Table II-1**

Correction: Change reference "19872 STAA" to "1982 STAA"

Pages 36-37 **Figures II-12 and II-13**

Correction: Reverse drawings (See *Items A and B* in Appendix)

Page 55 **Figure II-20**

Correction: Insert values on x-axis and y-axis (See *Item C* in Appendix)

Page 72 **Second Paragraph**

Correction: Change reference to "Figure II-21" to "Figure II-24"

Page 73 **Top Paragraph**

Correction: Change reference to "Figure II-22" to "Figure II-25"

Chapter III Elements of Design

Page 154 **Table III-6**

Correction: Delete "Maximum" in seventh column heading but retain superscript "a" (See *Item D* in Appendix)

Page 249 **Table III-27**

Correction: Below table, superscript "a" appearing in table caption should be at end of Line 1 beginning with "Ratio of flow rate. . ."

Page 264 **Table III-35**

Correction: Change description in upper left of table to read "Distance from Traveled-Way Pavement Edge[a]" instead of "Distance Traveled Pavement[a]"

Page 346 **Figure IV-4 (f)**

Correction: Show 48° (See *Item E* in Appendix)

Page 370 **End of first paragraph**

Correction: Change "flushed" to "flush"

Page 417 **Reference 26**

Correction: Change "Struc tures" to "Structures"

Chapter V Local Roads and Streets

Page 425 **Table V-7**

Correction: Change five values of runoff length (See *Item F* in Appendix)

Pages 454-455 **Figures V-7 and V-8**

Correction: Reverse drawings (See *Items G and H* in Appendix)

Page 466 **Reference 4**

Correction: Add date "1981"

Page 466 **Reference 5**

Correction: Change date "1976" to "1984"

Chapter VI Collector Roads and Streets

Page 487 **Section "Border Area"**

Correction: At bottom of paragraph, add: "back of the curb as practical for safety reasons. Breakaway features should be built when feasible and as an aid to safety considerations."

Page 488 **Section "Intersection Design", first paragraph**

Correction: Delete fourth and fifth lines of paragraph (text identical to correction made on Page 487).

Page 489 **Section "Street and Roadway Lighting", third paragraph**

Correction: Change "Information" to "Informational"

Page 490 **Section "References**

Correction: In Reference 5 change date "1988" to "1981"

Page 490 **Section "References"**

Correction: Add to list "6. *An Informational Guide for Roadway Lighting*, Washington, D.C.: AASHTO, 1984"

Page 491 **Section "Additional References"**

Correction: Delete first reference "*An Informational Guide. . .*"

Chapter VII Rural and Urban Arterials

Page 515 **Section "Cross Section and Right-of-Way Widths", third paragraph**

Correction: Change "Figure VII-5C" to "Figure VII-5G"

Chapter VIII Freeways

Page 591 **Figure VIII-1**

Correction: Revise A/C* line to show change at slip-ramp. Change line type for ROW and A/C to be consistent. (See *Item I* in Appendix)

Chapter IX At-Grade Intersections

Page 693 **First sentence**

Correction: Insert "not" after "since these vehicles are. . ." to read "since these vehicles are not generally selected. . ."

Page 711 **Table IX-3**

Correction: Delete "s" from "Notes". Capitalize "p" at the beginning of the Note.

Page 718 **Section "Channelized Islands," first paragraph**

Correction: Change reference to islands "c and d" to "b and d"

Page 726 **Last Paragraph**

Correction: Change reference number for "MUTCD" from "2" to "5"

Page 749 **Figure IX-34**

Correction: In first graph, change reference "16" to reference "7". In second graph, change reference "1" to reference "2".

Page 791 **Table IX-15**

Correction: Under column head "10% Left Turns" and subsection "40-mph Operating Speed", change third value from "305" to "330"

Page 801 **Figure IX-57A**

Correction: Change "WB-65 (Interstate Truck)" to "WB-62"

Page 827 **Figure IX-69**

Correction: Change figure caption to read "Multilane and two-lane marking."

Page 833 **Figure IX-71**

Correction: Change figure caption to read "14 ft to 18 ft median width left-turn design."

Page 834 **Figure IX-72**

Correction: Change figure caption to read "Median left-turn design for median width in excess of 18 ft."

Note: Where corrections have been made to titles of tables or figures, as on pages 827, 833, and 834, similar corrections should be made in the Table of Contents "List of Figures" on page xl.

Appendix

Item A	Figure II-12
Item B	Figure II-13
Item C	Figure II-20
Item D	Table III-6
Item E	Figure IV-4 (f)
Item F	Table V-7
Item G	Figure V-7
Item H	Figure V-8
Item I	Figure VIII-1

THIS TURNING TEMPLATE SHOWS THE TURNING PATHS OF THE AASHTO DESIGN VEHICLES. THE PATHS SHOWN ARE FOR THE LEFT FRONT OVERHANG AND THE OUTSIDE REAR WHEEL. THE LEFT FRONT WHEEL FOLLOWS THE CIRCULAR CURVE, HOWEVER, ITS PATH IS NOT SHOWN.

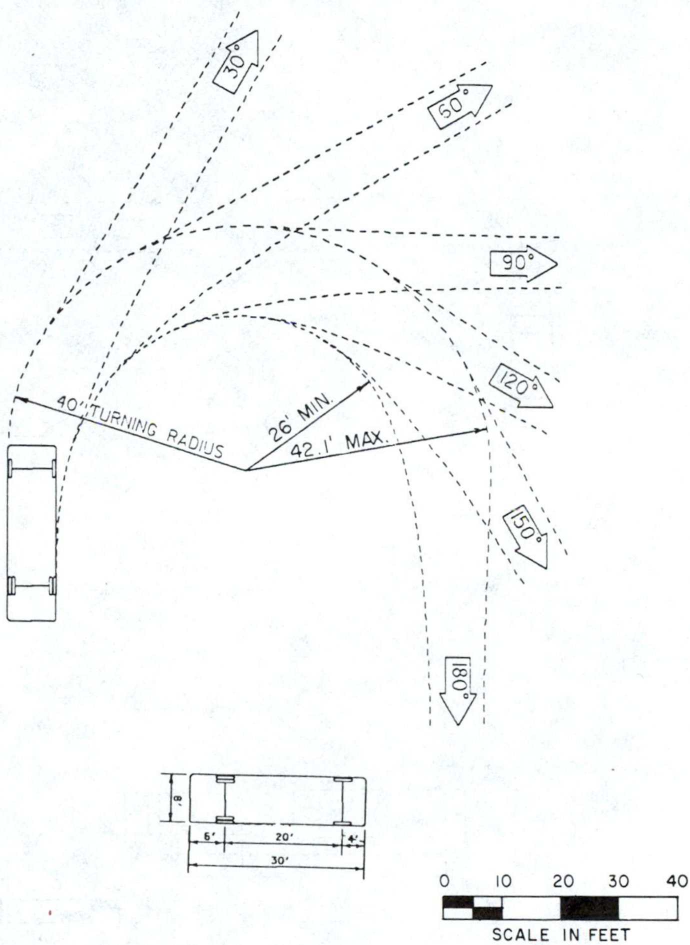

Source: Texas State Department of Highways and Public Transportation

Figure II-12. Minimum turning path for MH design vehicle.

Item A

THIS TURNING TEMPLATE SHOWS THE TURNING PATHS OF THE AASHTO DESIGN VEHICLES. THE PATHS SHOWN ARE FOR THE LEFT FRONT OVERHANG AND THE OUTSIDE REAR WHEEL. THE LEFT FRONT WHEEL FOLLOWS THE CIRCULAR CURVE, HOWEVER, ITS PATH IS NOT SHOWN.

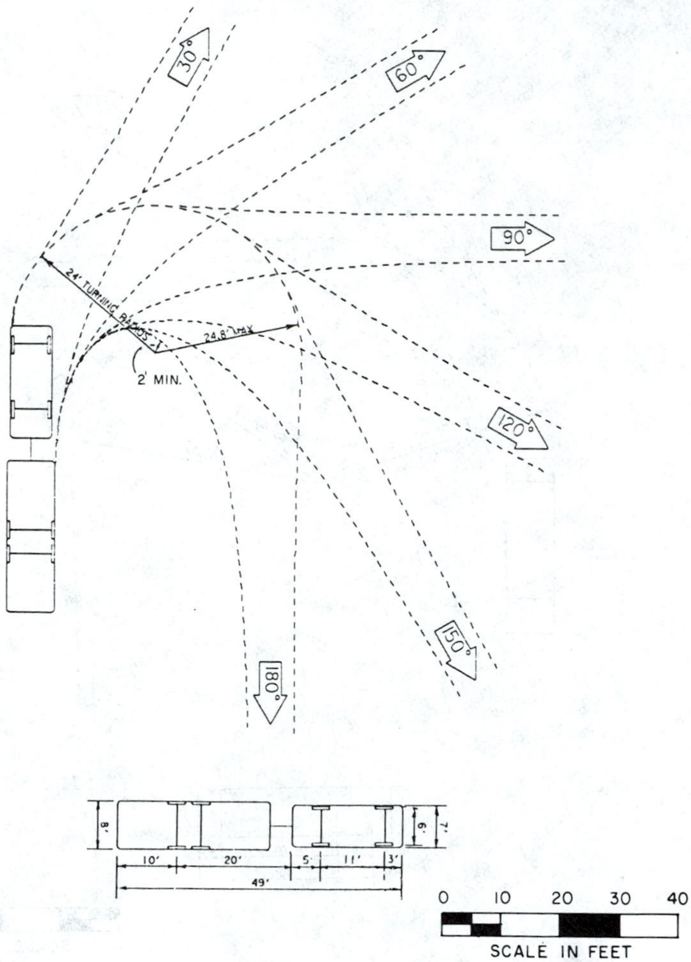

Source: Texas State Department of Highways and Public Transportation

Figure II-13. Minimum turning path for P/T design vehicle.

Item B

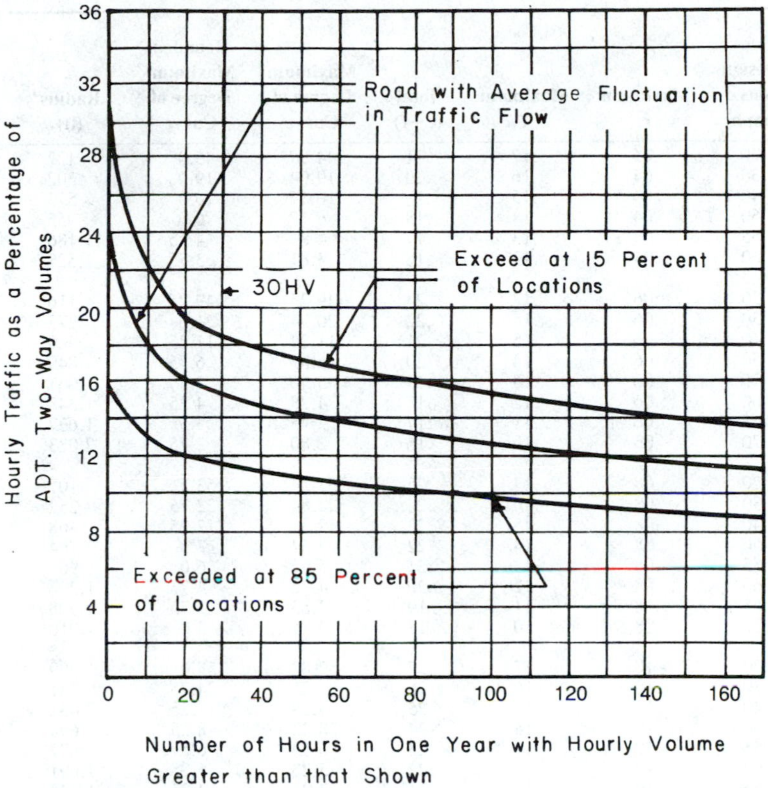

Figure II-20. Relation between peak-hour and average daily traffic volumes on rural arterials.

quacy in design can be judged by the 29 times during the year it is exceeded. The maximum hourly volume, which is about 25 percent of the ADT, exceeds 30 HV by about 67 percent. Whether or not 30 HV is too high for practical economy in design can be judged by the trend in the hourly traffic volumes lower than the 30th highest. The middle curve in Figure II-20 indicates that during 170 hours of the year the traffic volume exceeds 11.5 percent of the ADT. The lowest of this range of hourly volumes is about 23 percent less than 30 HV.

Another fortunate characteristic of 30 HV is that, as a percentage of ADT, it varies only slightly from year to year in spite of significant changes in ADT.

Item C

Design Speed (mph)	Maximum e	Maximum f	Total (e+f)	Maximum Degree of Curve	Rounded Maximum Degree of Curve	Radius[a] (ft)
20	.04	.17	.21	44.97	45.0	127
30	.04	.16	.20	19.04	19.0	302
40	.04	.15	.19	10.17	10.0	573
50	.04	.14	.18	6.17	6.0	955
55	.04	.13	.17	4.83	4.75	1,186
60	.04	.12	.16	3.81	3.75	1,528
20	.06	.17	.23	49.25	49.25	116
30	.06	.16	.22	20.94	21.0	273
40	.06	.15	.21	11.24	11.25	509
50	.06	.14	.20	6.85	6.75	849
55	.06	.13	.19	5.40	5.5	1,061
60	.06	.12	.18	4.28	4.25	1,348
65	.06	.11	.17	3.45	3.5	1,637
70	.06	.10	.16	2.80	2.75	2,083
20	.08	.17	.25	53.54	53.5	107
30	.08	.16	.24	22.84	22.75	252
40	.08	.15	.23	12.31	12.25	468
50	.08	.14	.22	7.54	7.5	764
55	.08	.13	.21	5.97	6.0	960
60	.08	.12	.20	4.76	4.75	1,206
65	.08	.11	.19	3.85	3.75	1,528
70	.08	.10	.18	3.15	3.0	1,910
20	.10	.17	.27	57.82	58.0	99
30	.10	.16	.26	24.75	24.75	231
40	.10	.15	.25	13.38	13.25	432
50	.10	.14	.24	8.22	8.25	694
55	.10	.13	.23	6.53	6.5	877
60	.10	.12	.22	5.23	5.25	1,091
65	.10	.11	.21	4.26	4.25	1,348
70	.10	.10	.20	3.50	3.5	1,637
20	.12	.17	.29	62.10	62.0	92
30	.12	.16	.28	26.65	26.75	214
40	.12	.15	.27	14.46	14.5	395
50	.12	.14	.26	8.91	9.0	637
55	.12	.13	.25	7.10	7.0	807
60	.12	.12	.24	5.71	5.75	996
65	.12	.11	.23	4.66	4.75	1206
70	.12	.10	.22	3.85	3.75	1528

NOTE: In recognition of safety considerations, use of e_{max} = 0.04 should be limited to urban conditions.
[a]Calculated using rounded maximum degree of curve.

Table III-6. Maximum degree of curve and minimum radius determined for limiting values of e and f, rural highways and high-speed urban streets.

Item D

islands in intersection areas or sometimes at the outer edge of the shoulder. Curbs should not be placed adjacent to high-speed lanes unless they serve a definite purpose. Any of the mountable sections in Figure IV-4 might be used for a median curb. Where drainage curbs are used in conjunction with metal median barriers, they should be installed directly under or behind the barrier. Drainage curbs should not be used with concrete median barriers. Improperly placed curbs may cause errant vehicles to vault a barrier or to strike it so that the vehicle overturns.

Shoulder curbs are mountable curbs placed at the outer edge of the shoulder to control drainage, improve delineation, and reduce erosion. They may be part of the longitudinal drainage system being combined with a gutter sec-

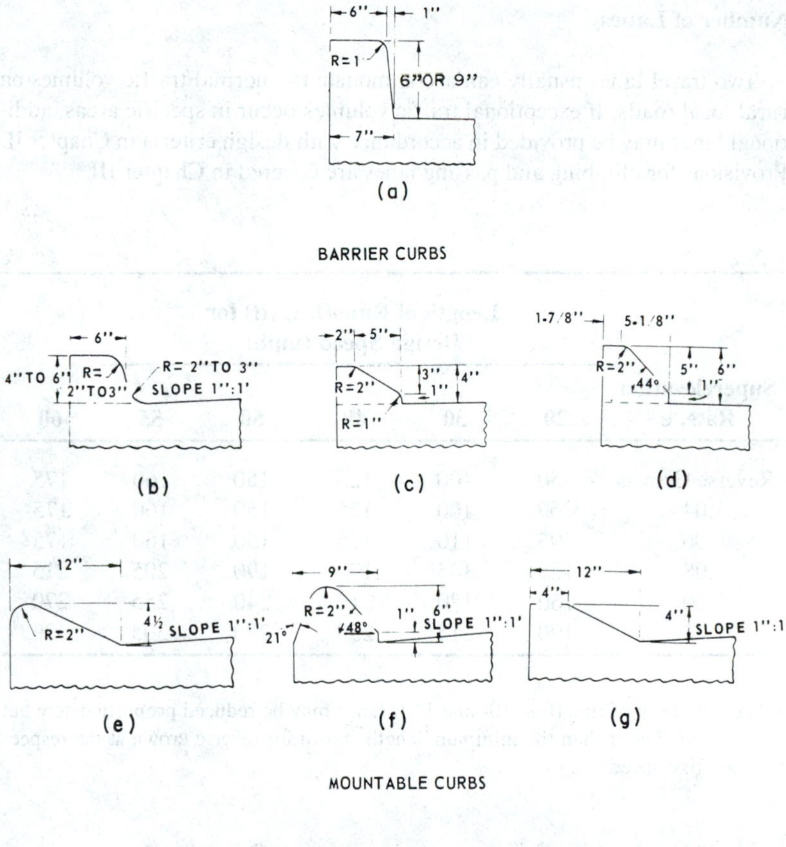

Figure IV-4. Typical highway curbs.

Item E

which case the superelevation should be not more than 0.08. For aggregate roads superelevation should be not more than 0.12.

Depending on the maximum superelevation value, the maximum degree of curvature to be used for different design speeds should be as given in Table V-6.

Superelevation runoff is the length of highway needed to accomplish the change in cross slope from a section with the adverse crown removed to a fully superelevated section. Minimum lengths of runoff are given in Table V-7. Adjustments in design runoff lengths may be necessary for smooth riding, surface drainage, and good appearance. General discussion on this topic appears in Chapter III.

Number of Lanes

Two travel lanes usually can accommodate the normal traffic volumes on rural local roads. If exceptional traffic volumes occur in specific areas, additional lanes may be provided in accordance with design criteria in Chapter II. Provisions for climbing and passing lanes are covered in Chapter III.

Superelevation Rate, e	Length of Runoff, L (ft) for Design Speed (mph)					
	20	30	40	50	55	60
Reverse Crown	50	100	125	150	160	175
.04	60	100	125	150	160	175
.06	95	110	125	150	160	175
.08	125	145	170	190	205	215
.10	160	180	210	240	255	270
.12	190	215	250	290	305	320

NOTE: Length of runoff on 10- and 11-ft lanes may be reduced proportionately but no shorter than the minimum length shown for reverse crown at the respective speed.

Table V-7. Minimum length for superelevation runoff for 2-12 ft lanes of pavement.

Item F

THIS TURNING TEMPLATE SHOWS THE TURNING PATHS OF THE AASHTO DESIGN VEHICLES. THE PATHS SHOWN ARE FOR THE LEFT FRONT OVERHANG AND THE OUTSIDE REAR WHEEL. THE LEFT FRONT WHEEL FOLLOWS THE CIRCULAR CURVE, HOWEVER, ITS PATH IS NOT SHOWN.

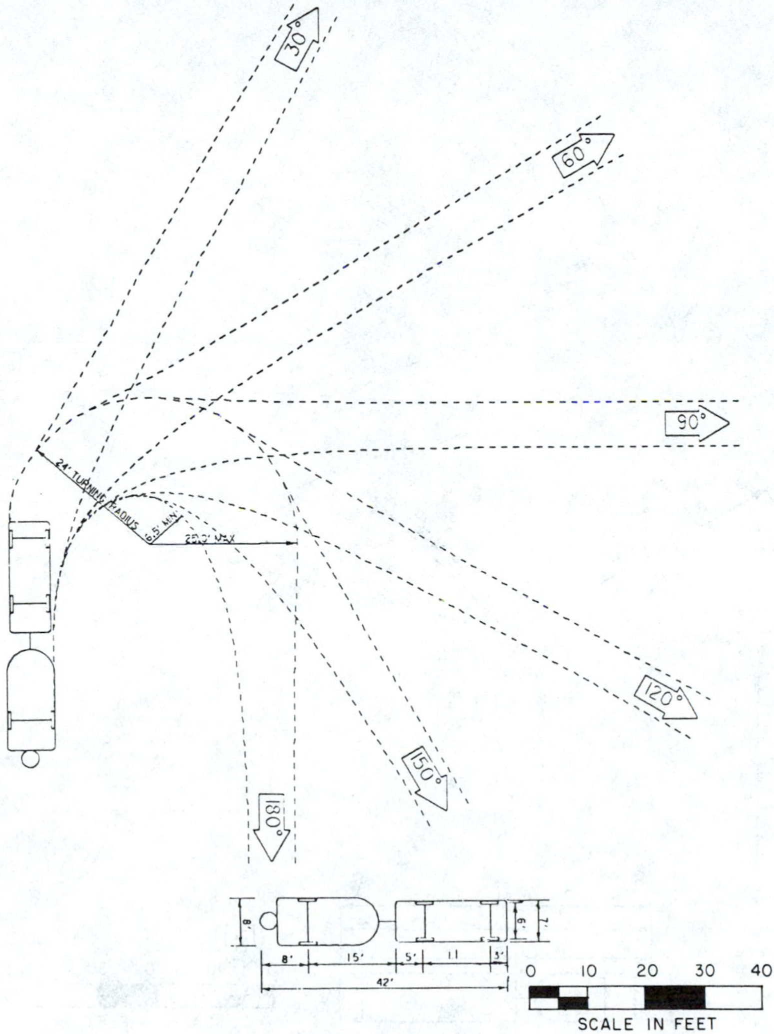

Source: Texas State Department of Highways and Public Transportation

Figure V-7. Minimum turning path for PB design vehicle.

Item G

THIS TURNING TEMPLATE SHOWS THE TURNING PATHS OF THE AASHTO DESIGN VEHICLES. THE PATHS SHOWN ARE FOR THE LEFT FRONT OVERHANG AND THE OUTSIDE REAR WHEEL. THE LEFT FRONT WHEEL FOLLOWS THE CIRCULAR CURVE, HOWEVER, ITS PATH IS NOT SHOWN.

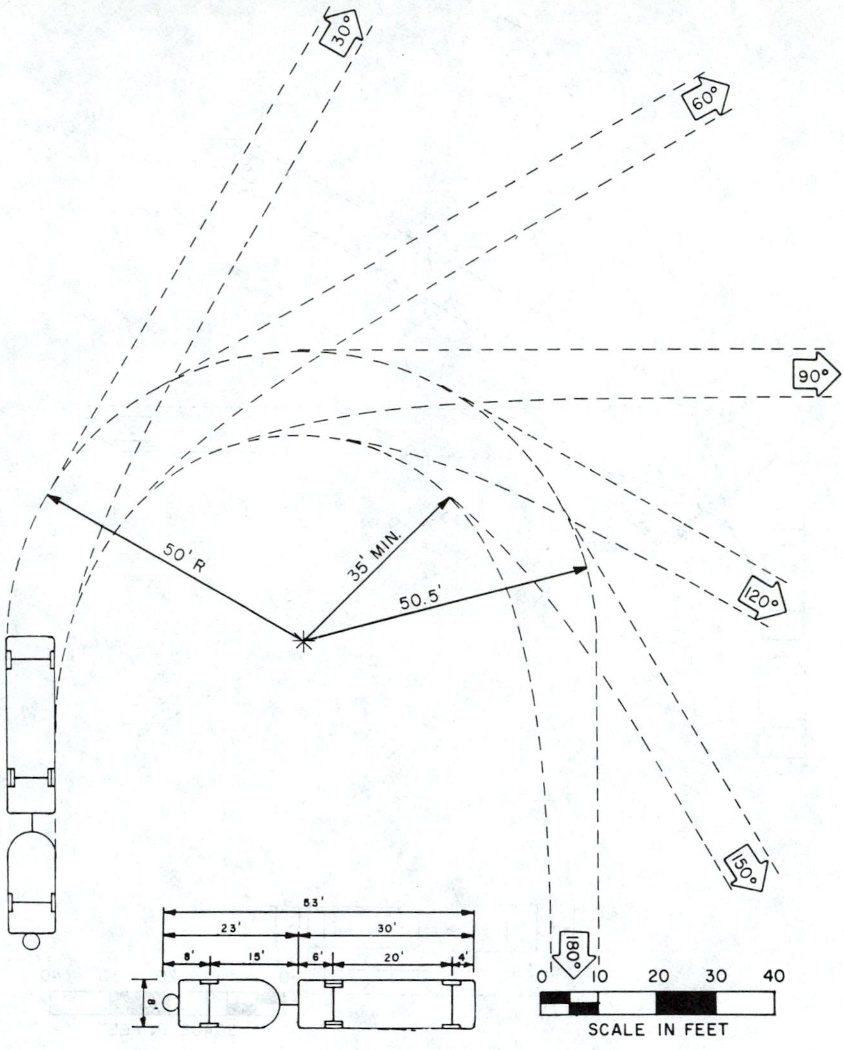

Source: Cal Trans

Figure V-8. Minimum turning path for MH-B design vehicle.

Item H

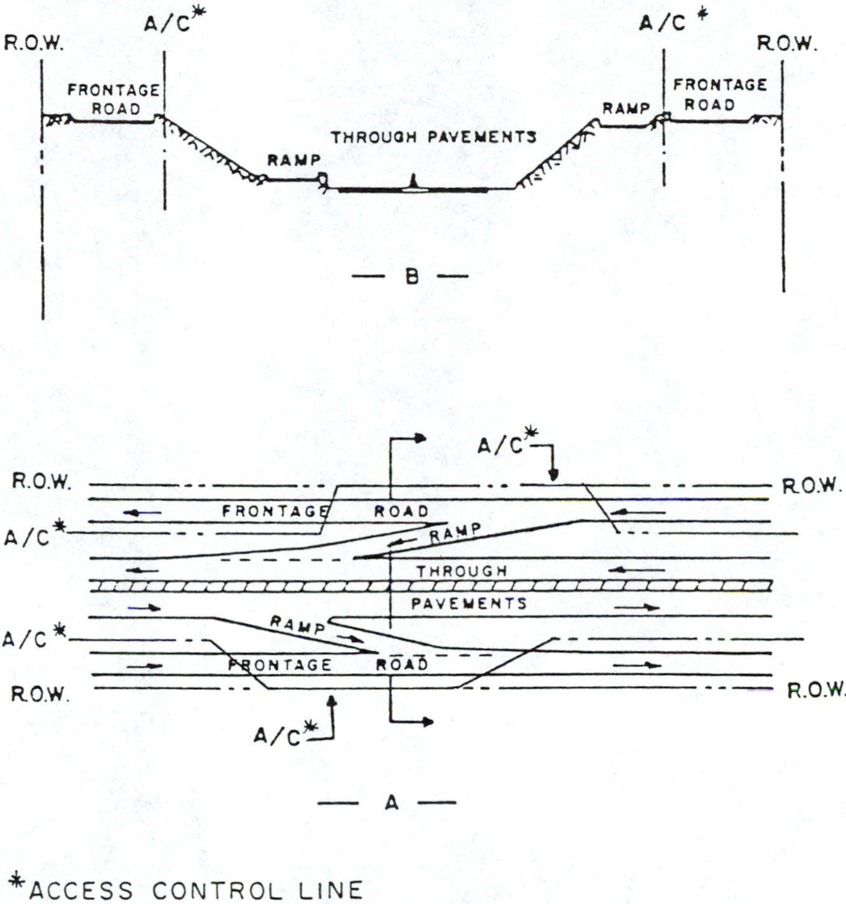

Figure VIII-1. Typical cross sections and right-of-way for depressed freeways.

plus border. The minimum median width of 10 to 22 ft is based on the assumption that for depressed freeways the ultimate section is constructed initially. However, where additional width is provided in the median for stage construction, the median should be widened in multiples of 12 ft (width of a traffic lane). Where ramps are not required, the uniform-width section should be graded to provide slopes as flat as possible within the available right-of-

NOTES

NOTES

A POLICY
on
GEOMETRIC DESIGN
of
HIGHWAYS
and
STREETS

1990

American Association of State
Highway and Transportation Officials
Suite 225, 444 North Capitol Street, N.W.
Washington, D.C. 20001
(202) 624-5800

ISBN 1-56051-001-3

AMERICAN ASSOCIATION OF STATE HIGHWAY AND TRANSPORTATION OFFICIALS

EXECUTIVE COMMITTEE—1989

President:	James P. Pitz, Michigan
Vice President:	Kermit Justice, Delaware

Elected Regional Members:

Region I	Howard Yerusalim, Pennsylvania
	Dana Connors, Maine
Region II	James Harrington, North Carolina
	Neil Wagoner, Louisiana
Region III	Darrel Rensink, Iowa
	Bernard Hurst, Ohio
Region IV	Charles L. Miller, Arizona
	Garth F. Dull, Nevada

Past Presidents:

Leno Menghini, Wyoming
John R. Tabb, Mississippi

Secretary of Transportation:

Samuel K. Skinner (Ex Officio)

Secretary-Treasurer:

Clyde E. Pyers, Maryland

Chairpersons of Standing Committees:

Duane Berentson, Washington – Administration
Frederick P. Salvucci, Massachusetts – Planning
Raymond E. Stotzer, Texas – Highways
Ronald R. Fiedler, Wisconsin – Highway Traffic Safety
Franklin E. White, New York – Water
Hal Rives, Georgia – Aviation
Robert N. Bothman, Oregon – Research
Ray D. Pethtel, Virginia – Public Transportation
Horace B. Edwards, Kansas – Railway (NCSRO)
Arthur J. Rock, Jr. – Special Committee of Commissioners and Boards

Executive Director:

Francis B. Francois, AASHTO (ex officio)

TASK FORCE ON GEOMETRIC DESIGN

Officers:

Hugh G. Downs	Maryland	Chairman	1975–1977
* Donald B. Jackson	Mississippi	Chairman	1978–1981
		Member	1975–1978
Brooks O. Nichols	Arkansas	Chairman	1982–1986
		Vice Chairman	1978–1982
		Member	1975–1978
George F. Sisson	Iowa	Chairman	1986–Present
		Member	1979–1986
Malcom D. Graham	New York	Vice Chairman	1975–1977
Thomas R. Bright	Illinois	Vice Chairman	1982–1983
		Member	1975–1982
Wilson B. Harkins	FHWA	Secretary	1975–1986
C. Larry King	FHWA	Secretary	1986–Present

Members and Members Representatives:

Clifford M. Albright	APWA	1976–1979
Ronald E. Bockstruck	Washington	1976–1977
James O. Brewer	Kansas	1986–Present
Anthony M. Capizzi	Maryland	1986–Present
Blaise M. Carriere	APWA	1976–1982
Harold D. Cooner	Texas	1979–1986
Wayne F. Cobine	Oregon	1986–Present
Edwin W. Dayton	New Jersey	1979–1986
**Donald G. Diller	Wyoming	1975–Present
Roger H. Edwards	New York	1978–1986
W. M. Foster	Washington	1975–1977
Andrew J. Gazda	Illinois	1976–Present
Charles A. Goessel	New Jersey	1986–Present
William A. Hartig	West Virginia	1975–1979
Ronald E. Heinz	FHWA	1982–1987
Frank Holzmann	Texas	1986–Present
* Irvin C. Hughes	Maryland	1976–1982
Milton L. Johnson	NACE	1976–Present
Leland J. Kissinger	FHWA	1978–1982
**Stephen C. Kologi	Montana	1975–Present
John N. LaPlante	NLC	1989–Present
David K. O. Leer	North Dakota	1986–Present
Robert L. Lewis	Texas	1978–1983
Mark A. Marek	Texas	1986–Present
James E. McCarty	NLC	1976–1986
Kenneth P. McLaughlin	Iowa	1975–1979
James P. Oxley	Arizona	1988–Present
Robert Radliff	New York	1986–Present
Norman H. Roush	West Virginia	1979–Present
E. Lamar Smith	Mississippi	1982–Present
Harry A. Smith	TRB	1979–1986
Marshall Sulloway	NLC/APWA	1976–Present
Donald C. Tillman	APWA	1976–1982
Robert L. Walters	Arkansas	1982–Present
Thomas O. Willett	FHWA	1987–Present
Wilfred Wolf	North Dakota	1975–1982

* Deceased
**Original Task Force Members

HIGHWAY SUBCOMMITTEE ON DESIGN—1989

Chairman: W. E. Schaefer, California
Vice Chairman: Robert W. Gubala, Connecticut
Secretary: Thomas Willett, FHWA

Members:
- Alabama, Don Arkle, Ray D. Bass
- Alaska, Steve Sisk, Timothy Mitchell
- Arizona, Robert P. Mickelson, James P. Oxley
- Arkansas, Bob Walters, Paul DeBusk
- California, Walter P. Smith, Karl Kampe, W. E. Schaefer
- Colorado, James E. Siebels, Kenneth F. Mauro
- Connecticut, Daniel Coffey, Earle R. Munroe
- Delaware, William S. Scarborough, Raymond Harbeson
- D.C., Gary Burch
- Florida, Kenneth N. Morefield
- Georgia, Walker Scott, Hoyt J. Lively, Charles Lewis
- Hawaii, Tit Mun Chun
- Idaho, Richard Sorenson, Larry Van Over
- Illinois, Andrew Gazda, Don Wolaver
- Indiana, Gregory L. Henneke
- Iowa, George F. Sisson
- Kansas, Bert Stratmann, James Brewer
- Kentucky, Glen Kelly, Larry Blevins
- Louisiana, Charles M. Higgins
- Maine, Alan L. Smith
- Maryland, Anthony M. Capizzi, Robert Douglass
- Massachusetts, Mario Tocci
- Michigan, Ho-Lum Wong
- Minnesota, Roger M. Hill
- Mississippi, E. Lamar Smith, Wendel T. Ruff
- Missouri, Frank Carroll, J. F. Roberts
- Montana, Steve Kologi, Ken Skoog
- Nebraska, Lyman D. Freemon
- Nevada, Michael McFall, Ronald W. Hill
- New Hampshire, Gilbert S. Rogers
- New Jersey, Kenneth Afferton, Charles A. Goessel
- New Mexico, J. L. Pacheco
- New York, Michael Cuddy, Robert Radliff
- North Carolina, D. R. (Don) Morton, J. T. Peacock, Jr.
- North Dakota, David K. O. Leer, Ken Birst
- Ohio, Everett J. Schaefer, Theodore J. Stitt
- Oklahoma, Bruce E. Taylor, Richard Hankins
- Oregon, Duane O. Christensen, John W. Sullivan
- Pennsylvania, Fred W. Bowser
- Puerto Rico, Efrain Irizarry
- Rhode Island, Edmond Parker
- South Carolina, Robert L. White, William M. DuBose
- South Dakota, James Kastner, K. C. Wilson
- Tennessee, Ken Wynne, Clellon Loveall
- Texas, Frank D. Holzmann, Roger G. Welsch
- U.S. DOT, Robert Bates (FAA), Thomas Willett (FHWA)
- Utah, Les Jester, Stephen Noble
- Vermont, Robert M. Murphy, Arthur J. Goss
- Virginia, E. C. Cochran, Jr., R. E. Atherton
- Washington, E. R. (Skip) Burch
- West Virginia, Norman Roush, Randolph Epperly
- Wisconsin, Thomas E. Carlsen
- Wyoming, Donald Diller

Affiliate Members
- Alberta, P. F. (Peter) Tajcnar
- Guam, Nonato C. Hallera
- Hong Kong, H. C. Beaton
- Manitoba, A. Boychuk
- Mariana Islands, John Pangelinan
- New Brunswick, C. Herbert Page
- Newfoundland, Terry McCarthy
- Northwest Territories, A. Gamble
- Nova Scotia, J. W. Pertus
- Ontario, Frank DeVisser
- Saskatchewan, Al Popoff

Associate Members-State
- N.J. Turnpike Authority, Allen J. Lewis
- Mass. Metro. Dist. Comm., E. Leo Lydon
- Port Auth. of NY & NJ, Alfred F. Grimaldi

Associate Members-Federal
- Bureau of Indian Affairs, Neil N. Rousseau
- U.S. Department of Agriculture-Forest Service, John Holt, Tim Rogan

PREFACE

This Policy was developed as part of the continuing work of the Standing Committee on Highways. The Committee, then titled the Committee on Planning and Design Policies, was established in 1937 to formulate and recommend highway engineering policies. It has developed *A Policy on Geometric Design of Rural Highways*, 1954 and 1965 editions, *A Policy on Arterial Highways in Urban Areas*, 1957, *A Policy on Design of Urban Highways and Arterial Streets*, 1973, *Geometric Design Standards for the National System of Interstate and Defense Highways*, 1956 and 1967, *Geometric Design Standards for Highways Other Than Freeways*, 1969, *A Policy on Geometric Design of Highways and Streets*, 1984, and a number of other AASHO and AASHTO policy and "guide" publications.

An AASHTO publication is typically developed through the following steps: (1) The Committee selects subjects and broad outlines of material to be covered. (2) The appropriate subcommittee and its task forces, in this case the Subcommittee on Design and its Task Force on Geometric Design, assemble and analyze relevant data and prepare a tentative draft. Working meetings are held and revised drafts are prepared, as necessary, and reviewed by the Subcommittee, until agreement is reached. (3) The manuscript is then submitted for Standing Committee on Highways approval, and then Executive Committee approval. Standards and policies must be adopted by a two-thirds vote by the Member Departments before publication. During the developmental process, comments were sought and considered from all the States, the Federal Highway Administration, and representatives of the American Public Works Association, the National Association of County Engineers, the National League of Cities, and other interested parties.

Changes in allowable maximum dimensions for truck-trailer combinations, the revision of the *Highway Capacity Manual* in 1985 and other developments subsequent to the 1984 edition prompted AASHTO to again update this policy.

Following the procedures outlined above, the Task Force on Geometric Design met in 1987 to begin the work that has resulted in this current edition. It is anticipated that a continuing effort will be maintained by the Subcommittee on Design to keep the policy up to date with future developments in geometric design.

CONTENTS

CHAPTER TITLES

Chapter I
HIGHWAY FUNCTIONS

Page

Chapter II

DESIGN CONTROLS AND CRITERIA

Chapter III
ELEMENTS OF DESIGN

Chapter IV
CROSS SECTION ELEMENTS

Chapter V

LOCAL ROADS AND STREETS

Chapter VI
COLLECTOR ROADS AND STREETS

Chapter VII
RURAL AND URBAN ARTERIALS

Chapter VIII
FREEWAYS

Chapter IX
AT-GRADE INTERSECTIONS

Chapter X
GRADE SEPARATIONS AND INTERCHANGES

LIST OF TABLES

LIST OF FIGURES

FOREWORD

As highway designers, highway engineers strive to provide for the needs of highway users while maintaining the integrity of the environment. Unique combinations of requirements that are always conflicting result in unique solutions to the design problems. The guidance supplied by this text, *A Policy on Geometric Design of Highways and Streets*, is based on established practices and is supplemented by recent research. This text is also intended to form a comprehensive reference manual for assistance in administrative, planning, and educational efforts pertaining to design formulation.

The fact that new design values are presented herein does not imply that existing streets and highways are unsafe, nor does it mandate the initiation of improvement projects. This publication is not intended as a policy for resurfacing, restoration or rehabilitation (R.R.R.) projects. For projects of this type, where major revisions to horizontal or vertical curvature are not necessary or practical, existing design values may be retained. Specific site investigations and accident history analysis often indicate that the existing design features are performing in a satisfactory manner. The cost of full reconstruction for these facilities, particularly where major realignment is not required, will often not be justified. Resurfacing, restoration and rehabilitation projects enable highway agencies to improve highway safety by selectively upgrading existing highway and roadside features without the cost of full reconstruction. When designing 3R projects the designer should refer to TRB Special Report 214, *Designing Safer Roads: Practices for Resurfacing, Restoration and Rehabilitation* and related publications for guidance.

The intent of this policy is to provide guidance to the designer by referencing a recommended range of values for critical dimensions. Sufficient flexibility is permitted to encourage independent designs tailored to particular situations. Minimum values are either given or implied by the lower value in a given range of values. The larger values within the ranges will normally be used where the social, economic and environmental (S.E.E.) impacts are not critical.

Emphasis has been placed on the joint use of transportation corridors by pedestrians, cyclists and public transit vehicles. Designers should recognize the implications of this sharing of the transportation corridors and are encouraged to consider not only vehicular movement, but also movement of people, distribution of goods, and provision of essential services. A more comprehensive transportation program is thereby emphasized.

Cost-effective design is also emphasized. The traditional procedure of comparing highway-user benefits with costs has been expanded to reflect the needs of non-users and the environment. Although adding complexity to the analysis, this broader approach also takes into account both the need for a given project and the relative priorities among various projects. The results of this approach may need to be modified to meet the needs-versus-funds problems that highway administrators face. The goal of cost-effective design is not merely to give priority to the most beneficial individual projects but to provide the most benefits to the highway system of which each project is a part.

Most of the technical material that follows is detailed or descriptive design information. Design guidelines are included for freeways, arterials, collectors, and local roads, in both urban and rural locations, paralleling the functional classification used in highway planning. The book is organized into functional chapters to stress the relationship between highway design and highway function. An explanation of functional classification is included in Chapter I.

These guidelines are intended to provide operational efficiency, comfort, safety, and convenience for the motorist. The design concepts presented herein were also developed with consideration for environmental quality. The effects of the various environmental impacts can and should be mitigated by thoughtful design processes. This principle, coupled with that of esthetic consistency with the surrounding terrain and urban setting, is intended to produce highways that are safe and efficient for users, acceptable to non-users and in harmony with the environment.

This publication supersedes the 1984 AASHTO publication of the same name. Because the concepts presented could not possibly be completely covered in one book, references to additional literature are given at the end of each chapter.

Chapter I
HIGHWAY FUNCTIONS

SYSTEMS AND CLASSIFICATIONS

The classification of highways into different operational systems, functional classes, or geometric types is necessary for communication among engineers, administrators, and the general public. Different classification schemes have been applied for different purposes in different rural and urban regions. Classification of highways by design types based on the major geometric features (e.g., freeways and conventional streets and highways) is the most helpful one for highway location and design procedures. Classification by route numbering (e.g., U.S., State primary, State secondary, and county numbering systems) is the most helpful for traffic operations. Administrative classification (e.g., State Federal aid primary, State Federal aid secondary, State primary and State secondary) is used to denote the levels of government responsible for, and the method of financing, highway facilities. Functional classification, the grouping of highways by the character of service they provide, was developed for transportation planning purposes. Comprehensive transportation planning, an integral part of total economic and social development, uses functional classification as an important planning tool. The emergence of functional classification as the predominant method of grouping highways is consistent with the policies contained in this publication.

THE CONCEPT OF FUNCTIONAL CLASSIFICATION

This section introduces the basic concepts required for understanding the functional classification of highway facilities and systems.

Hierarchies of Movements and Components

A complete functional design system provides a series of distinct travel movements. The six recognizable stages in most trips include main movement, transition, distribution, collection, access, and termination. For example, Figure I-1 shows a hypothetical highway trip using a freeway, where the main movement of vehicles is uninterrupted, high-speed flow. When

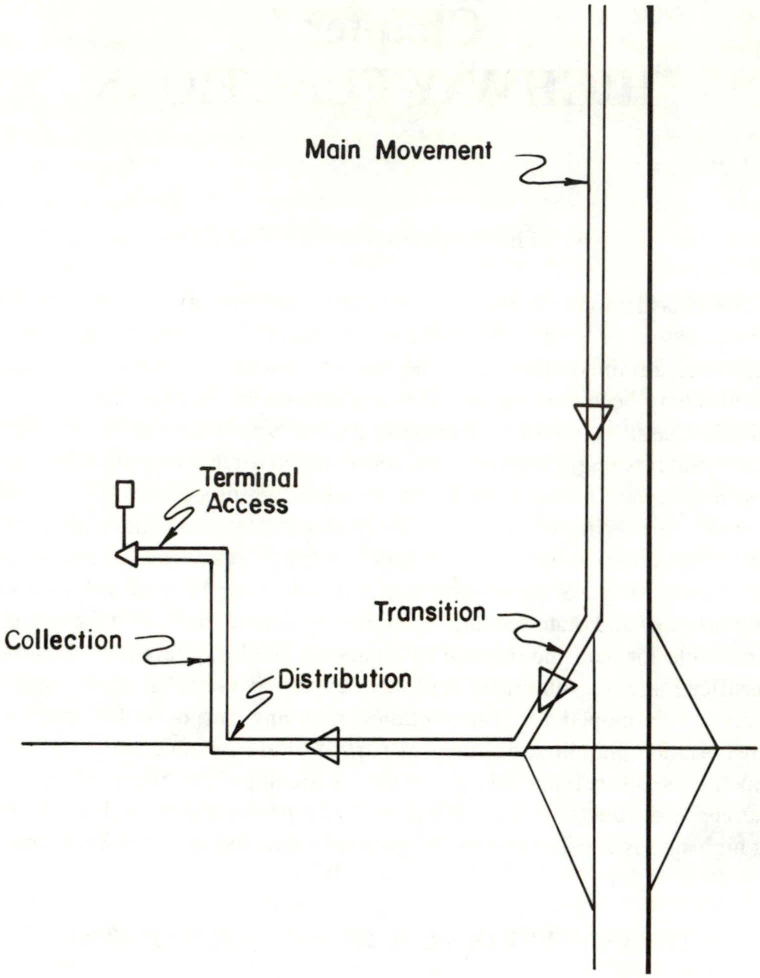

Figure I-1. Hierarchy of movement.

approaching destinations from the freeway, vehicles reduce speed on freeway ramps, which act as transition roadways. The vehicles then enter moderate-speed arterial (distributor facilities) that bring them nearer to the vicinity of their destination neighborhoods. They next enter collector roads that penetrate neighborhoods. The vehicles finally enter local access roads that provide direct approaches to individual residences or other terminations. At their destinations the vehicles are parked at an appropriate terminal facility.

Each of the six stages of a typical trip is handled by a separate facility designed specifically for its function. Because the movement hierarchy is based on the total amount of traffic volume, freeway travel is generally highest in the movement hierarchy, followed by distributor arterial travel, which is in turn higher in the movement hierarchy than travel on collectors and local access routes.

Although many trips can be subdivided into all of the six recognizable stages, intermediate facilities are not always needed. The complete hierarchy of circulation facilities relates especially to conditions of low-density suburban development, where traffic flows are cumulative on successive elements of the system. However, it sometimes is desirable to reduce the number of components in the chain. For instance, a large single traffic generator may fill one or more lanes of a freeway during certain periods. In this situation, it is expedient to lead traffic directly onto a freeway ramp without introducing arterial facilities that unnecessarily mix already-concentrated traffic flows with additional vehicles. This deletion of intermediate facilities does not eliminate the functional need for the remaining parts of the flow hierarchy or the functional design components, although it may change their physical characters. The order of movement is still identifiable.

The failure to recognize and accommodate by suitable design each of the different trip stages of the movement hierarchy is a prominent cause of highway obsolescence. Conflicts and congestion occur at interfaces between public highways and private traffic-generating facilities when the functional transitions are inadequate. Examples are commercial driveways that lead directly from a relatively high-speed arterial into a parking aisle without intermediate provisions for transition deceleration and arterial distribution or, more seriously, freeway ramps that lead directly into or from large traffic generators such as major shopping centers.

Inadequate acceptance capacity of the distributor arterial or internal circulation deficiencies within the traffic absorber create the danger that traffic may back up onto the freeway. Successful internal design that provides facilities to accommodate all the intermediate functions between the high-speed freeway and the terminal parking facility will alleviate such a situation.

In the case of the freeway leading to a large traffic generator, deceleration from rapid movement on the freeway occurs on the exit ramp. Distribution to various parking areas is then accomplished by primary distribution-type roads or lanes within the parking facility. These roads or lanes supplant the distributor arterial function. Collector-type roads or lanes within the parking facility may then deliver segments of the entering flow to the parking bays. The parking aisle, in leading to individual parking space terminals, then becomes the equivalent of an access street. Thus, the principal functions within the hierarchal movement system are recognizable. It can be pointed out that each functional category also is related to a range of vehicle speeds.

The same principles of design are also relevant to terminal facilities that adjoin distributor arterials or collectors. The functional design of the facility includes each movement stage, with internal circulation in the terminal design to accommodate the order of movement. The necessity of designing for all stages of the movement hierarchy varies with the size of the traffic generator. For relatively small generators, two or more stages may be accommodated on the same internal facility. For larger traffic generators, each movement stage must have a separate functional facility.

To determine the number of design components necessary, the customary volumes of traffic handled by public streets of different functional categories can be compared. The volume range on private internal facilities can be related to the comparable range on public streets. These volumes may not be directly comparable, inasmuch as the physical space available within a private facility is smaller and the standards of operation are necessarily quite different. However, the same principles of flow specialization and movement hierarchy can be applied.

Some further examples may demonstrate how the principles of movement hierarchy are related to a logical system of classification of traffic generation intensity. At the highest practical level of traffic generation a single generator fills an entire freeway, and for this condition, intermediate public streets could not be inserted between the generator and the freeway, so the various movement stages must be accommodated internally with appropriate design features. At the next level of traffic generation a single traffic generator could fill a single freeway lane; it is then appropriate to construct a freeway ramp for the exclusive use of the generator without intervening public streets. At still smaller volumes it becomes desirable to combine the traffic from several generators with additional traffic before the flow arrives at a freeway entrance ramp. The road performing this function then becomes a collector facility, accumulating these small flows until a traffic volume that will fill the freeway ramp is reached.

Similar principles can be applied at the distributor arterial level of service. If a given traffic generator is of sufficient size, an exclusive intersection driveway for that generator is justified. In other cases an intermediate collector street should combine smaller traffic flows until they reach a volume that warrants an intersection along the distributor. The same theory can be applied with regard to the criteria for direct access to the collector street. A moderately sized traffic generator usually warrants a direct connection to the collector without an intermediate access street; however, in a district of single-family residences a local access street should assemble the traffic from a group of residences and lead it in to a collector street at a single point of access. In practice, direct access to arterials and collectors must be provided from commercial and residential properties, particularly in established neighborhoods.

In short, each element of the functional hierarchy can serve as a collecting facility for the next higher element, but an element must be present only where the intermediate collection is necessary to satisfy the spacing and traffic volume requirements of the next higher facility. By defining the spacing and volume requirements for a system element, it is possible to determine in which cases it is necessary to use the full system and in which cases intermediate elements may be bypassed.

Functional Relationships

Functional classification thus groups streets and highways according to the character of service they are intended to provide. This classification recognizes that individual roads and streets do not serve travel independently. Rather, most travel involves movement through networks of roads and can be categorized relative to such networks in a logical and efficient manner. Thus, functional classification of roads and streets is also consistent with categorization of travel.

A schematic illustration of this basic idea is shown in Figure I-2. In Figure I-2A, lines of travel desire are straight lines connecting trip origins and destinations (circles). The relative widths of the lines indicate the relative amounts of travel desire. The relative sizes of the circles indicate the relative trip generating and attracting power of the places shown. Because it is impractical to provide direct-line connections for every desire line, trips must be channelized on a limited road network in the manner shown in Figure I-2B. Heavy travel movements are directly served or nearly so; the smaller movements are channeled into somewhat indirect paths. The facilities in Figure I-2 are labeled local access, collector, and arterial, which are terms that describe their functional relationships. In this scheme the functional hierarchy is also seen to be related to the hierarchy of trip distances served by the network.

A more complete illustration of a functionally classified rural network is shown in Figure I-3. The arterial highways generally provide direct service between cities and larger towns, which generate and attract a large proportion of the relatively longer trips. Roads of the intermediate functional category (collectors) serve small towns directly, connecting them to the arterial network. Roads of this category collect traffic from the local roads, which serve individual farms and other rural land uses or distribute traffic to these local roads from the arterials.

Although this example has a rural setting, the same basic concepts also apply in urban and suburban areas. A similar hierarchy of systems can be defined; however, because of the high intensity of land use and travel, specific travel generation centers are more difficult to identify. In urban and suburban areas additional considerations, such as the spacing of intersections, become

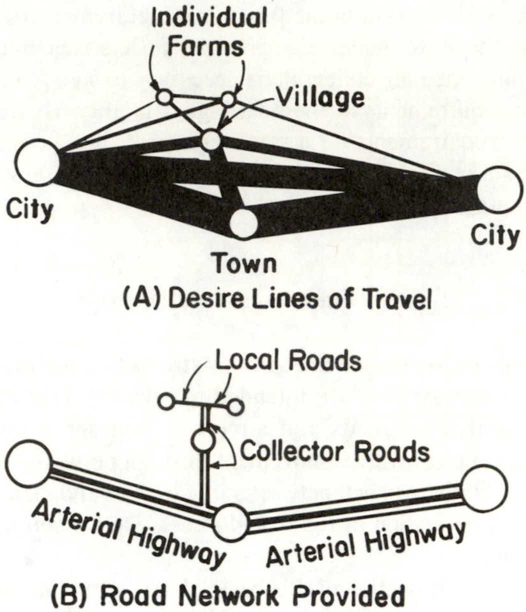

Figure I-2. Channelization of trips.

more important in defining a logical and efficient network. A schematic illustration of a functionally classified suburban street network is shown in Figure I-4.

Allied to the idea of traffic categorization is the dual role that the highway and street network plays in providing (1) access to property and (2) travel mobility. Access is a fixed requirement of the defined area. Mobility is provided at varying levels of service. Mobility can incorporate several qualitative elements, such as riding comfort and absence of speed changes, but the most basic factor is operating speed or trip travel time.

Figure I-2 shows that the concept of traffic categorization leads logically not only to a functional hierarchy of road classes but also to a similar hierarchy of relative travel distances served by these road classes. The hierarchy of travel distances can be related logically to functional specialization in meeting the property access and travel mobility requirements. Local rural facilities emphasize the land access function. Arterials for main movement or distribution emphasize the high level of mobility for through movement. Collectors offer approximately balanced service for both functions. This scheme is illustrated conceptually in Figure I-5.

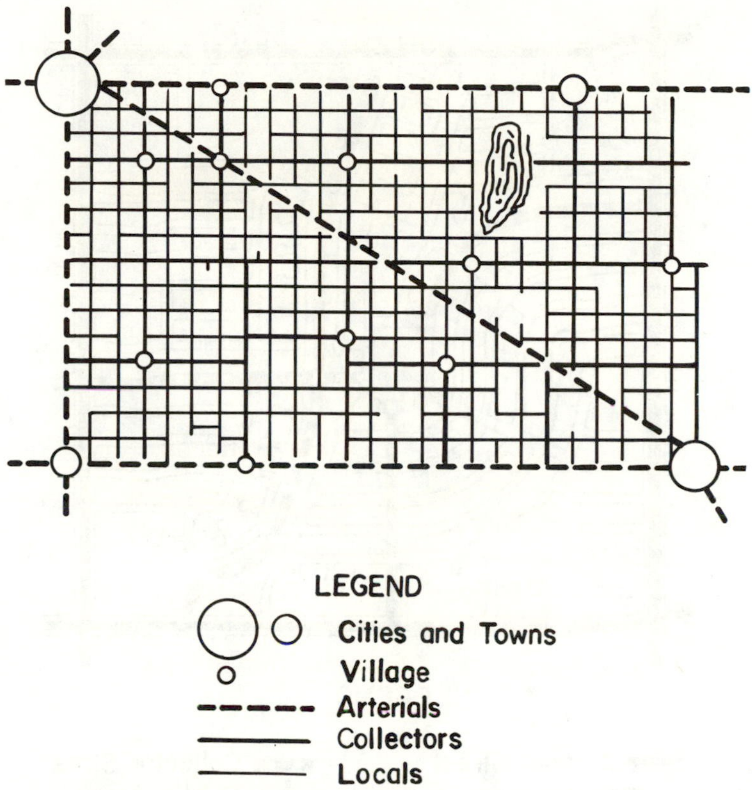

Figure I-3. Schematic illustration of a functionally classified rural highway network.

Access Needs and Controls

The two major considerations in classifying highway and street networks functionally are access and mobility. The conflict between serving through movement and providing access to a dispersed pattern of trip origins and destinations necessitates the differences and gradations in the various functional types. Regulated limitation of access is necessary on arterials to enhance their primary function of mobility. Conversely, the primary function of local roads and streets is to provide access (implementation of which causes a limitation of mobility). The extent and degree of access control is thus a significant factor in defining the functional category of a street or highway.

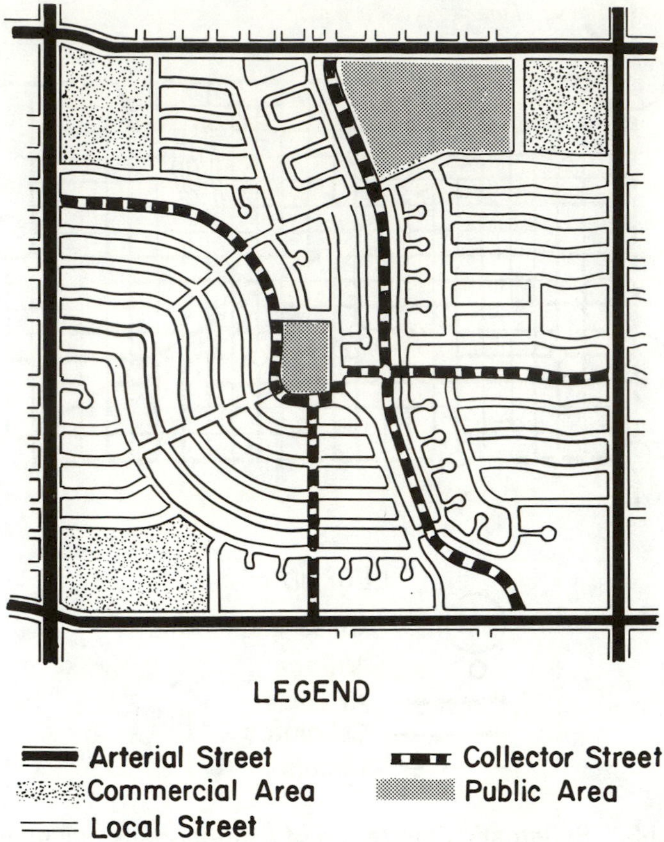

Figure I-4. Schematic illustration of a portion of a suburban street network.

Further discussion of the various degrees of access control appropriate to street and highway development is provided in the section "Access Control" in Chapter II.

FUNCTIONAL SYSTEM CHARACTERISTICS

This section contains definitions and characteristics of highway facilities in urban and rural settings based on their functional classifications. It presents information, in revised form, from the Federal Highway Administration publication *Highway Functional Classification: Concepts, Criteria and Procedures* (**1**).

Definitions of Urban and Rural Areas

Urban and rural areas have fundamentally different characteristics with regard to density and types of land use, density of street and highway networks, nature of travel patterns, and the way in which these elements are related. Consequently, urban and rural functional systems are classified separately.

Urban areas as discussed herein are considered those places within boundaries set by the responsible State and local officials having a population of

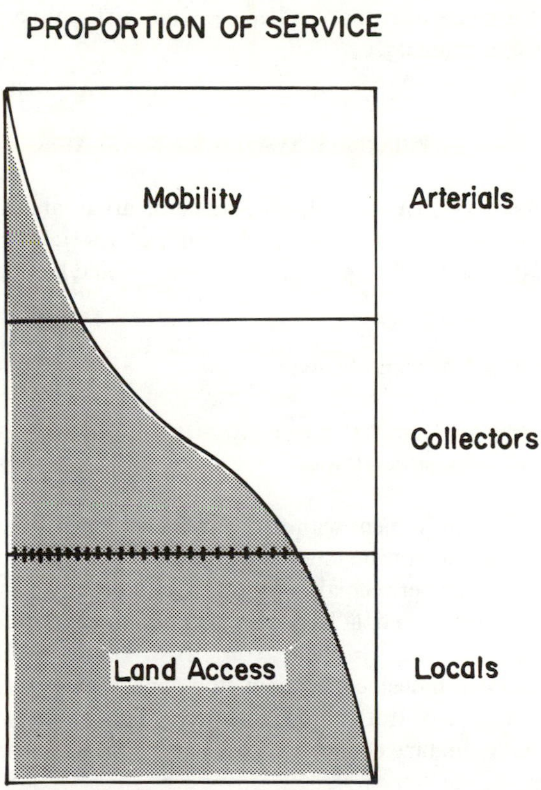

Figure I-5. Relationship of functionally classified systems in service traffic mobility and land access.

5,000 or more. Urban areas are further subdivided into *urbanized areas* (population of 50,000 and over) and *small urban areas* (population between 5,000 and 50,000). For design purposes, the population forecast for the design year should be used. (For legal definition of *Urban* areas, see Section 101 of Title 23, U.S. Code.)

Rural areas are those areas outside the boundaries of *urban areas.*

Functional Categories

The roads making up the functional systems differ for urban and rural areas. The hierarchy of the functional systems consists of principal arterials (for main movement), minor arterials (distributors), collectors, and local roads and streets; however, in urban areas there are relatively more arterials with further functional subdivisions of the arterial category; whereas in rural areas there are relatively more collectors with further functional subdivisions of the collector category.

Functional Systems for Rural Areas

Rural roads consist of facilities outside of urban areas. The names provided for the recognizable systems are principal arterials (roads), minor arterials (roads), major and minor collectors (roads), and local roads.

Rural Principal Arterial System

The rural principal arterial system consists of a network of routes with the following service characteristics:

1. Corridor movement with trip length and density suitable for substantial statewide or interstate travel.
2. Movements between all, or virtually all, urban areas with populations over 50,000 and a large majority of those with populations over 25,000.
3. Integrated movement without stub connections except where unusual geographic or traffic flow conditions dictate otherwise (e.g., international boundary connections or connections to coastal cities).

In the more densely populated states, this class of highway includes most (but not all) heavily traveled routes that might warrant multilane improvements; in the majority of states, the principal arterial system includes most (if not all) existing rural freeways.

The principal arterial system is stratified into the following two design types: (1) freeways and (2) other principal arterials.

Rural Minor Arterial System

The rural minor arterial road system, in conjunction with the rural principal arterial system, forms a network with the following service characteristics:

1. Linkage of cities, larger towns, and other traffic generators (such as major resort areas) that are capable of attracting travel over similarly long distances.
2. Integrated interstate and intercounty service.
3. Internal spacing consistent with population density, so that all developed areas of the state are within reasonable distances of arterial highways.
4. Corridor movements consistent with items (1) through (3) with trip lengths and travel densities greater than those predominantly served by rural collector or local systems.

Minor arterials therefore constitute routes, the design of which should be expected to provide for relatively high travel speeds and minimum interference to through movement.

Rural Collector System

The rural collector routes generally serve travel of primarily intracounty rather than statewide importance and constitute those routes on which (regardless of traffic volume) predominant travel distances are shorter than on arterial routes. Consequently, more moderate speeds may be typical. To define rural collectors more clearly, this system is subclassified according to the following criteria:

- **Major Collector Roads.** These routes (1) serve county seats not on arterial routes, larger towns not directly served by the higher systems, and other traffic generators of equivalent intracounty importance, such as consolidated schools, shipping points, county parks, and important mining and agricultural areas: (2) link these places with nearby larger towns or cities, or with routes of higher classifications; and (3) serve the more important intracounty travel corridors.

- **Minor Collector Roads.** These routes should (1) be spaced at intervals consistent with population density to accumulate traffic from local roads and bring all developed areas within reasonable distances of collector roads, (2) provide service to the remaining smaller communities, and (3) link the locally important traffic generators with their rural hinterland.

Rural Local Road System

The rural local road system, in comparison to collectors and arterial systems, primarily provides access to land adjacent to the collector network and serves travel over relatively short distances. Local road mileage, of course, constitutes all rural mileage not classified as principal arterial, minor arterial, or collector road mileage.

Extent of Rural Systems

The functional criteria for road systems have been expressed herein primarily in qualitative rather than quantitative terms. Because of varying geographic conditions (e.g., population densities, spacing between and sizes of cities, and densities and patterns of road networks), criteria on sizes of population centers, trip lengths, traffic volumes, and route spacings do not apply to all systems in all States. However, the results of classification studies conducted in many States show considerable consistency (when expressed in percentages of total rural road mileage) in the relative extents of the functional systems.

Highway systems developed by using these criteria are generally expected, in all states except Alaska and Hawaii, to fall within the percentage ranges shown in Table I-1. The higher values of the ranges given in Table I-1 apply to states having less extensive total road networks relative to the population density. In states having more extensive total road networks relative to the population density, the lower values are applicable. The range of percentages of rural collectors represents the total mileage of both major and minor collector roads and applies to the statewide rural mileage totals; the percentages in particular counties may vary considerably from the statewide average. Areas having an extensive regular grid pattern of roads usually have a smaller percentage of collectors than areas within which geographic conditions have imposed a restricted or less regular pattern of road development.

Systems	Percentage of Total Rural Miles
Principal arterial system	2–4
Principal arterial plus minor arterial system	6–12, with most states falling in 7–10 percent range
Collector road	20–25
Local road system	65–75

Table I-1. Typical distribution of rural functional systems.

Functional Highway Systems in Urbanized Areas

The four functional highway systems for urbanized areas are urban principal arterials (streets), minor arterials (streets), collectors (streets), and local streets. The differences in the nature and intensity of development in rural and urban areas warrant corresponding differences in urban system characteristics relative to the correspondingly named rural systems.

Urban Principal Arterial System

In every urban environment, one system of streets and highways can be identified as unusually significant in terms of the nature and composition of travel it serves. In small urban areas (population under 50,000), these facilities may be very limited in number and extent, and their importance may be derived primarily from the service provided to through travel. In urbanized areas, their importance also derives from service to rurally oriented traffic, but equally or even more importantly, from service for major circulation movements within these urbanized areas.

The urban principal arterial system serves the major centers of activity of urbanized areas, the highest traffic volume corridors, and the longest trip desires and carries a high proportion of the total urban area travel on a minimum of mileage. The system should be integrated both internally and between major rural connections.

The principal arterial system carries most of the trips entering and leaving the urban area, as well as most of the through movements bypassing the central city. In addition, significant intra-area travel, such as between central business districts and outlying residential areas, between major inner-city communities, and between major suburban centers, is served by this class of facility. Frequently, the principal arterial system carries important intra-urban as well as inter-city bus routes. Finally, in urbanized areas, this system provides continuity for all rural arterials that intercept the urban boundary.

Because of the nature of the travel served by the principal arterial system, almost all fully and partially controlled access facilities are usually part of this functional class. However, this system is not restricted to controlled-access routes. To preserve the identification of controlled-access facilities, the principal arterial system should be stratified as follows: (1) interstate, (2) other freeways and (3) other principal arterials (with partial or no control of access).

The spacing of urban principal arterials is closely related to the trip-end density characteristics of particular portions of the urban areas. Although no firm spacing rule applies in all or even in most circumstances, the spacing between principal arterials (in larger urban areas) may vary from less than 1 mile in the highly developed central business areas to 5 miles or more in the sparsely developed urban fringes.

For principal arterials, service to abutting land is subordinate to travel service to major traffic movements. Only facilities within the subclass of other principal arterials are capable of providing any direct access to land, and such service should be purely incidental to the primary functional responsibility of this class of roads.

Urban Minor Arterial Street System

The minor arterial street system interconnects with and augments the urban principal arterial system. It accommodates trips of moderate length at a somewhat lower level of travel mobility than principal arterials do. This system distributes travel to geographic areas smaller than those identified with the higher system.

The minor arterial street system includes all arterials not classified as principal. This system places more emphasis on land access than the higher system does and offers lower traffic mobility. Such a facility may carry local bus routes and provide intracommunity continuity but ideally does not penetrate identifiable neighborhoods. This system includes urban connections to rural collector roads where such connections have not been classified as urban principal arterials for internal reasons.

The spacing of minor arterial streets may vary from ⅛ to ½ mile in the central business district to 2 to 3 miles in the suburban fringes but is normally not more than 1 mile in fully developed areas.

Urban Collector Street System

The collector street system provides both land access service and traffic circulation within residential neighborhoods and commercial and industrial areas. It differs from the arterial system in that facilities on the collector system may penetrate residential neighborhoods, distributing trips from the arterials through the area to their ultimate destinations. Conversely, the collector street also collects traffic from local streets in residential neighborhoods and channels it into the arterial system. In the central business district, and in other areas of similar development and traffic density, the collector system may include the entire street grid. The collector street system may also carry local bus routes.

Urban Local Street System

The local street system comprises all facilities not in one of the higher systems. It primarily permits direct access to abutting lands and connections to the higher order systems. It offers the lowest level of mobility and usually contains no bus routes. Service to through-traffic movement usually is deliberately discouraged.

Extent of Mileage and Travel on Urban Systems

Table I-2 contains the typical distribution of travel volume and mileage of the functional systems for urbanized areas. Systems developed for urbanized areas using the criteria herein usually fall within the percentage ranges shown.

Functional Classification as a Design Type

This text has utilized the functional classification system as a design type of highway. Two major difficulties exist in this effort. The first major problem involves the freeway. The freeway is not a functional class in itself but is normally classified as a principal arterial. It does, however, have unique geometric criteria that demand a separate design designation apart from other arterials. Therefore, a separate chapter of freeways has been included along with chapters on arterials, collectors, and local roads and streets. The addition of the universally familiar term "freeway" to the basic functional classes seems preferable to the adoption of a completely separate system of design types.

Systems	Range	
	Travel Volume (percent)	Mileage (percent)
Principal arterial system	40–65	5–10
Principal arterial plus minor arterial street systems	65–80	15–25
Collector street system	5–10	5–10
Local street system	10–30	65–80

Table I-2. Typical distribution of urban functional systems.

The second major difficulty is that in the past geometric design criteria and capacity levels have traditionally been based on a classification of traffic volume ranges. Under such a system, highways with comparable traffic volumes are constructed to the same standards and provide identical levels of service, although there may be considerable difference in the function they serve.

Under a functional classification system, standards and level of service vary according to the function of the highway facility. Volumes serve to further refine the standards for each class.

Arterials are expected to provide a high degree of mobility for the longer trip length. Therefore, they should provide a high operating speed and level of service. Since access to abutting property is not their major function, some degree of access control is desirable to enhance mobility. The collectors serve a dual function in accommodating the shorter trip and feeding the arterials. They must provide some degree of mobility and also serve abutting property. Thus, an intermediate design speed and level of service is appropriate. Local roads and streets have relative short trip lengths, and because property access is their main function, there is little need for mobility or high operating speeds. This function is reflected by use of a lower design speed and level of service.

The functional concept is important to the designer. Even though many of the geometric standards could be determined without reference to the functional classification, the designer must keep in mind the overall purpose that the street or highway is intended to serve. This concept is consistent with a systematic approach to highway planning and design.

The first step in the design process is to define the function that the facility is to serve. The level of service required to fulfill this function for the anticipated volume and composition of traffic provides a rational and cost-effective basis for the selection of design speed and geometric criteria within the ranges of values available to the designer. The use of functional classification as a design type should appropriately integrate the highway planning and design process.

REFERENCES

1. *Highway Functional Classification: Concepts, Criteria and Procedures.* U.S. Department of Transportation, Federal Highway Administration. Washington, D.C.: U.S. Government Printing Office, 1974, pp II-8 through II-14.

Chapter II
DESIGN CONTROLS AND CRITERIA

INTRODUCTION

This chapter discusses those characteristics of vehicles, pedestrians, and traffic that act as criteria for the optimization or improvement in design of the various highway and street functional classes.

DESIGN VEHICLES

General Characteristics

The physical characteristics of vehicles and the proportions of variously sized vehicles using the highways are positive controls in geometric design. Therefore, it is necessary to examine all vehicle types, select general class groupings, and establish representatively sized vehicles within each class for design use. Design vehicles are selected motor vehicles with the weight, dimensions, and operating characteristics used to establish highway design controls for accommodating vehicles of designated classes. For purposes of geometric design, each design vehicle has larger physical dimensions and larger minimum turning radius than those of almost all vehicles in its class. The largest of all the several design vehicles are usually accommodated in the design of freeways.

Three general classes of vehicles have been selected, namely, passenger cars, trucks, and buses/recreational vehicles. The passenger car class includes compacts and subcompacts plus all light vehicles and light delivery trucks (vans and pickups). The truck class includes single-unit trucks, truck tractor-semitrailer combinations, and trucks or truck tractors with semitrailers in combination with full trailers. Buses/recreational vehicles include single unit buses, articulated buses, motor homes, and passenger cars or motor homes pulling trailers or boats.

The dimensions for 15 design vehicles representing vehicles within these general classes are given in Table II-1. In the design of any highway facility the largest design vehicle likely to use that facility with considerable frequency or a design vehicle with special characteristics that must be taken into account in dimensioning the facility is used to determine the design of such critical features as radii at intersections and radii of turning roadways. Design vehicle profiles can be determined from the dimensions shown in Table II-1.

Minimum Turning Paths of Design Vehicles

Figures II-1 through II-15 present the minimum turning paths for the 15 critical design vehicles. The principal dimensions affecting design are the minimum turning radius, the tread width, the wheelbase, and the path of the inner rear tire. Effects of driver characteristics (such as the rate at which the driver approaches centripetal acceleration) and of the slip angles of wheels are minimized by assuming that the speed of the vehicle for the minimum radius (sharpest) turn is less than 10 mph.

The boundaries of the turning paths of the several design vehicles when making the sharpest turns are established by the outer trace of the front overhang and the path of the inner rear wheel. This turn assumes that the outer front wheel follows the circular arc defining the minimum turning radius as determined by the vehicle steering mechanism. The minimum radii of the outside and inside wheel paths are given in Table II-2.

Geometric design requirements for trucks and buses are much more severe than they are for passenger vehicles. Trucks and buses are wider and have longer wheelbases and greater mimimum turning radii. These are the principal characteristic dimensions affecting horizontal highway design. The longer single-unit trucks and buses require greater minimum turning radii than most vehicle combinations, but because of their greater offtracking, the longer vehicle combinations also require greater widths of turning paths.

A semitrailer combination is a truck tractor with a semitrailer, either with or without a full trailer, or a truck with one or more full trailers. Because semitrailer-combination sizes and turning characteristics vary widely, there are several semitrailer design vehicles: (1) a design vehicle representative of medium tractor-semitrailer combinations (WB-40), (2) a design vehicle representative of larger tractor-semitrailer combinations commonly in use (WB-50), (3) a design vehicle representative of a larger tractor-semitrailer full trailer combinations commonly in use (WB-60), (4) a design vehicle representative of a larger tractor- semitrailer combination allowed on selected highways by the Surface Transportation Assistance Act of 1982 (WB-62), (5) a design vehicle representative of a larger tractor-semitrailer grandfathered on selected highways by the Surface Transportation Assistance Act of 1982 (WB-67), (6) a design vehicle representative of tractor-semitrailer full trailer-

Design Vehicle Type	Symbol	Dimension (ft)										
		Overall			Overhang							
		Height	Width	Length	Front	REAR	WB_1	WB_2	S	T	$WB_3$3	WB_4
Passenger car	P	4.25	7	19	3	5	11					
Single unit truck	SU	13.5	8.5	30	4	6	20					
Single unit bus	BUS	13.5	8.5	40	7	8	25					
Articulated bus	A-BUS	10.5	8.5	60	8.5	9.5	18		4[a]	20[a]		
Combination trucks												
Intermediate semitrailer	WB-40	13.5	8.5	50	4	6	13	27				
Large semitrailer	WB-50	13.5	8.5	55	3	2	20	30				
"Double Bottom" semi-trailer—full-trailer	WB-60	13.5	8.5	65	2	3	9.7	20	4[b]	5.4[b]	20.9	
Interstate Semitrailer	WB-62*	13.5	8.5	69	3	3	20	40–42				
Interstate Semitrailer	WB-67**	13.5	8.5	74	3	3	20	45–47				
Triple Semitrailer	WB-96	13.5	8.5	102	2.5	3.3	13.5	20.7	3.3[d]	6[d]	21.7	21.7
Turnpike Double Semitrailer	WB-114	13.5	8.5	118	2	2	22	40	2[c]	6[c]	44	
Recreation vehicle												
Motor home	MH		8	30	4	6	20					
Car and camper trailer	P/T		8	49	3	10	11	18	5			
Car and boat trailer	P/B		8	42	3	8	11	15	5			
Motor Home and Boat Trailer	MH/B		8	53	4	8	20	21	6			

* = Design vehicle with 48′ trailer as adopted in 19872 STAA (Surface Transportation Assistance Act)
** = Design vehicle with 53′ trailer as grandfathered in 1982 STAA (Surface Transportation Assistance Act)
a = Combined dimesion 24, split is estimated.
b = Combined dimesion 9.4, split is estimated.
c = Combined dimension 8, split is estimated.
d = Combined dimension 9.3, split is estimated.
WB_1, WB_2, WB_3, WB_4 are effective vehicle wheelbases.
S is the distance from the rear effective axle to the hitch point.
T is the distance from the hitch point to the lead effective axle of the following unit.

Table II-1. Design vehicle dimensions.

Design Vehicle Type	Passenger Car	Single Unit Truck	Single Unit Bus	Articulated Bus	Semitrailer Intermediate	Semitrailer Combination Large	Semitrailer Full Trailer Combination	Inter-State Semi-Trailer	Inter-State Semi-Trailer	Triple Semi-Trailer	Turnpike Double Semi-Trailer	Motor Home	Passenger Car with Travel Trailer	Passenger Car with Boat and Trailer	Motor Home and Boat Trailer
Symbol	P	SU	BUS	A-BUS	WB-40	WB-50	WB-60	WB-62*	WB-67**	WB-96	WB-114	MH	P/T	P/B	MH/B
Minimum design turning radius (ft)	24	42	42	38	40	45	45	45	45	50	60	40	24	24	50
Minimum inside radius (ft)	13.8	27.8	24.4	14.0	18.9	19.2	22.2	9.1	00	20.7	17	26.0	2.0	6.5	35

* Design vehicle with 48′ trailer as adopted in 1982 STAA (Surface Transportation Assistance Act)

** Design vehicle with 53′ trailer as grandfathered in 1982 STAA (Surface Transportation Assistance Act)

Table II-2. Minimum turning radii of design vehicles.

full trailer combinations (triples) selectively in use (WB-96), and (7) a design vehicle representative of larger tractor-semitrailer-full trailer (turnpike double) selectively in use (WB-114). Although turnpike doubles and triple trailers are not permitted on many highways, their occurrence does warrant inclusion in this publication.

The minimum turning radii and the transition lengths shown are for turns at less than 10 mph. Higher speeds lengthen the transition curves and require larger radii than the minimums. The radii are considered minimum for this application, although skilled drivers might be able to reduce them.

The dimensions of design vehicles take into account dimensional trends in motor vehicle manufacture and represent a composite of the vehicles currently in operation; however, the design vehicle dimensions must represent the values critical to geometric design and are thus greater than nearly all vehicles belonging to the corresponding vehicle classes.

The turning dimensions shown in Figures II-1 through II-15 were derived by both the use of scale models and computer plots for combination or articulated units. Alternate methods can be used to compute the dimensions that will give slightly different values.

The P design vehicle, with the dimensions and turning characteristics shown in Figure II-1, satisfies the requirements for representing the passenger car class.

The SU design vehicle characteristics are suitable for all single-unit trucks and small buses; the control dimensions for its minimum turning path suffice for a number of buses and truck combinations now in operation. (See Figure 11-2.) On most facilities serving truck traffic or large buses, however, the design vehicle either for semitrailer combinations or large buses should be considered in design.

A separate bus design vehicle is required because of the trend toward intercity and transit buses with longer wheelbases. A design vehicle designated "BUS" with a 25-ft wheelbase and an overall length of 40 ft, as shown in Figure II-3, has been selected. These dimensions are also applicable to rubber-tired electric trolley buses. Although the trolley buses are only 11.5 ft high, they require approximately 18.5 ft of overhead clearance.

The buses serving a particular urban area may not conform to the dimensions shown in Figure II-3. An example is the articulated bus now serving certain cities. Longer than conventional buses, its permanent hinge near the center allows maneuverability. Figure II-4 displays the critical dimensions for the A-BUS design vehicle.

The highway designer must also be aware that for certain buses the combination of ground clearance, overhang, and roadway vertical curvature may present problems in hilly areas.

THIS TURNING TEMPLATE SHOWS THE TURNING PATHS OF THE AASHTO DESIGN VEHICLES. THE PATHS SHOWN ARE FOR THE LEFT FRONT OVERHANG AND THE OUTSIDE REAR WHEEL. THE LEFT FRONT WHEEL FOLLOWS THE CIRCULAR CURVE, HOWEVER, ITS PATH IS NOT SHOWN.

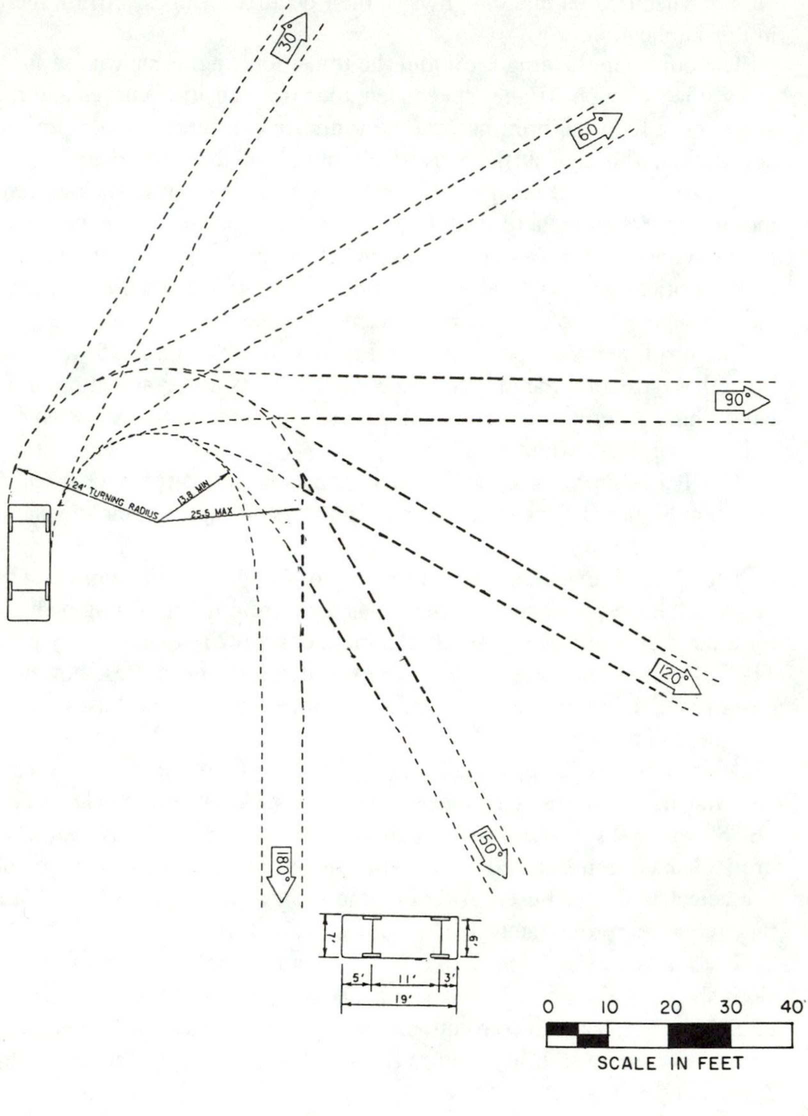

Source: Texas State Department of Highways and Public Transportation

Figure II-1. Minimum turning path for P design vehicle.

THIS TURNING TEMPLATE SHOWS THE TURNING PATHS OF THE AASHTO DESIGN VEHICLES. THE PATHS SHOWN ARE FOR THE LEFT FRONT OVERHANG AND THE OUTSIDE REAR WHEEL. THE LEFT FRONT WHEEL FOLLOWS THE CIRCULAR CURVE, HOWEVER, ITS PATH IS NOT SHOWN.

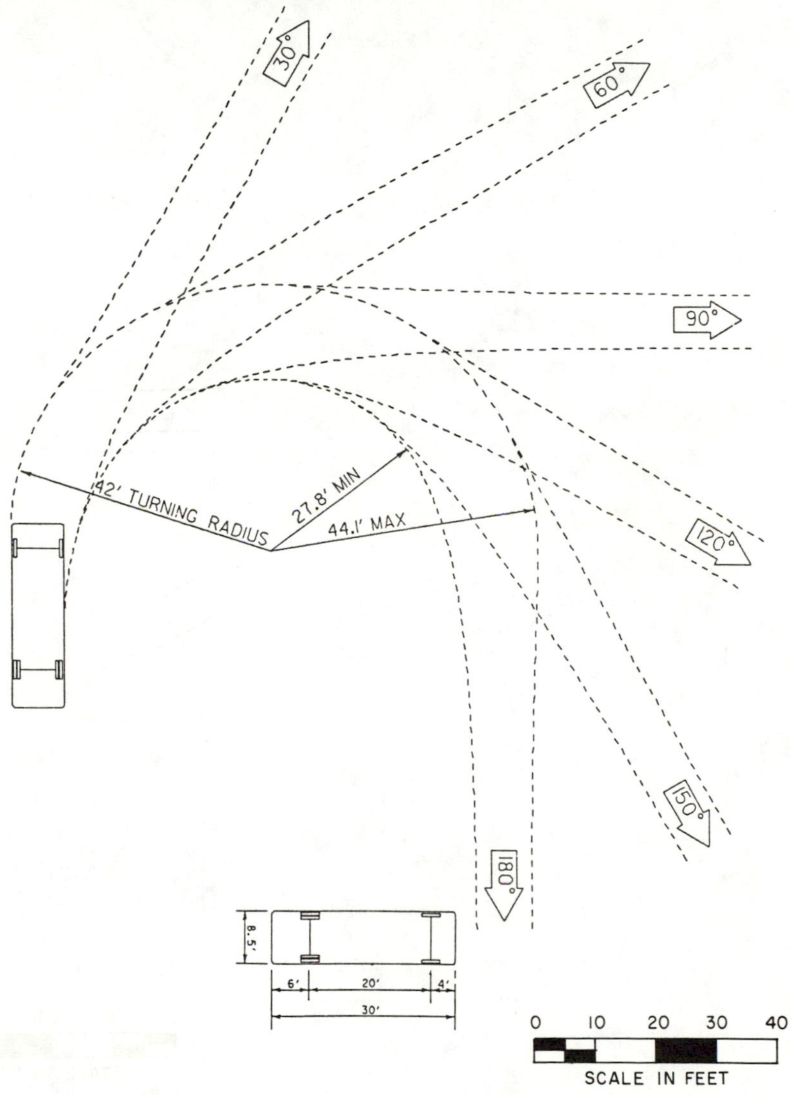

Source: Texas State Department of Highways and Public Transportation

Figure II-2. Minimum turning path for SU design vehicle.

THIS TURNING TEMPLATE SHOWS THE TURNING PATHS OF THE AASHTO DESIGN VEHICLES. THE PATHS SHOWN ARE FOR THE LEFT FRONT OVERHANG AND THE OUTSIDE REAR WHEEL. THE LEFT FRONT WHEEL FOLLOWS THE CIRCULAR CURVE, HOWEVER, ITS PATH IS NOT SHOWN.

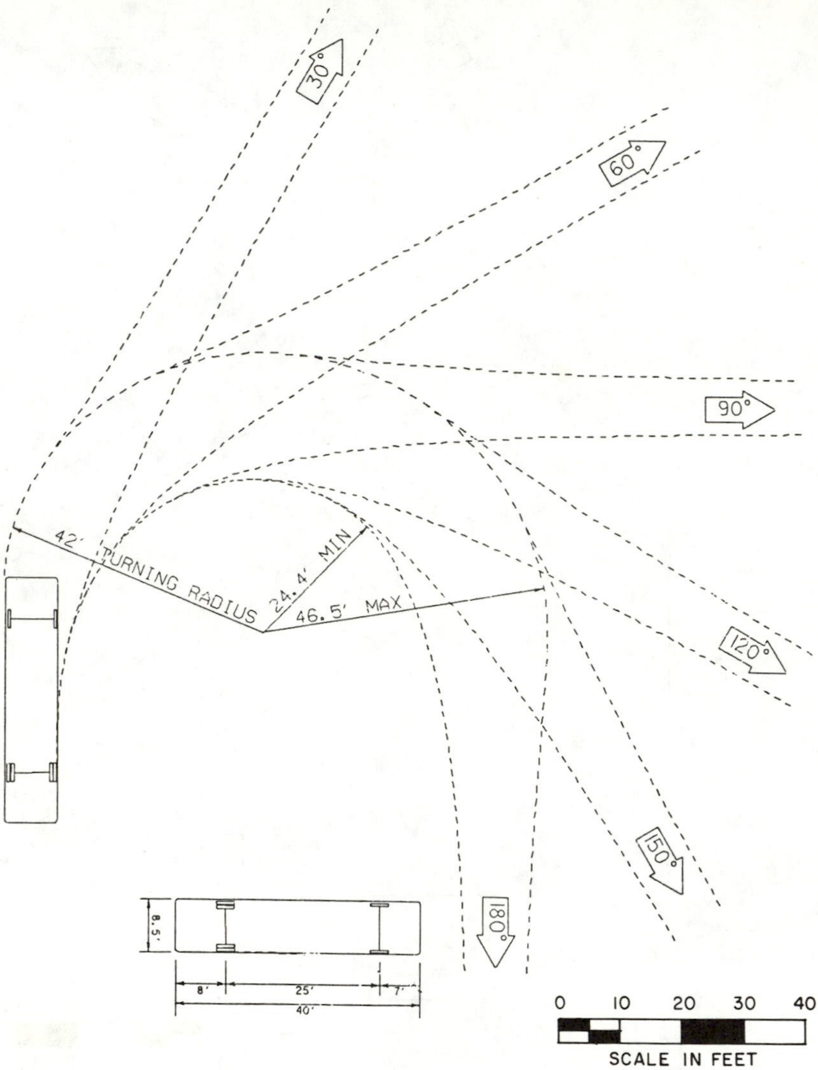

Source: Texas State Department of Highways and Public Transportation

Figure II-3. Minimum turning path for BUS design vehicle.

THIS TURNING TEMPLATE SHOWS THE TURNING PATHS OF THE AASHTO DESIGN VEHICLES. THE PATHS SHOWN ARE FOR THE LEFT FRONT OVERHANG AND THE OUTSIDE REAR WHEEL. THE LEFT FRONT WHEEL FOLLOWS THE CIRCULAR CURVE, HOWEVER, ITS PATH IS NOT SHOWN.

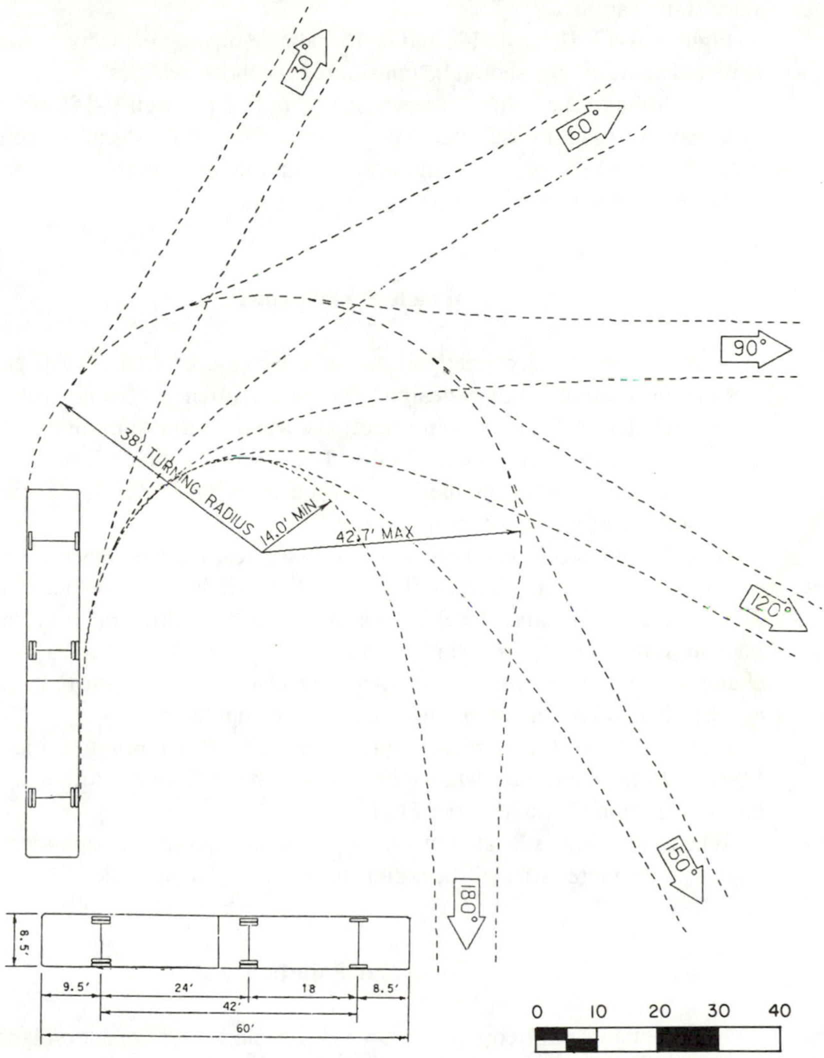

Source: Texas State Department of Highways and Public Transportation

Figure II-4. Minimum turning path for A-BUS design vehicle.

Of the truck tractor-semitrailer combinations shown in Figure II-5 through II-11, the WB-50 is critical or many design purposes. Figure II-6 shows that the inside radius of 19.2 ft. and the outside radius of 46.3 ft. should be considered in design.

Figures II-12, II-13, II-14, and II-15 indicate turning paths for typical recreational vehicles that should be considered for those vehicles.

In addition to the vehicles shown in Figure II-1 through II-15, other vehicles may be used for selected applications. With the advent of computer derived offtracking plots (1), the designer can determine the characteristics of the selected vehicles if it differs from those shown.

Vehicle Performance

Acceleration and deceleration rates of vehicles are often critical parameters in determining highway design. These rates often govern the dimensioning of such design features as intersections, freeway ramps, climbing or passing lanes, and turnout bays for buses. The following data are not meant to depict average vehicles for the design types, but rather the low-horsepower (compact) car or the loaded truck or bus.

Based on its acceleration and deceleration rates, the passenger car seldom controls design. From Figures II-16 and II-17, it is obvious that relatively rapid accelerations and decelerations are possible, although they may be uncomfortable for the passengers and unnecessary. Also, due to the rapid changes being made in vehicle operating characteristics, current data on acceleration and deceleration may soon become outdated.

Figure II-16 is the original figure from the 1984 edition of the Green Book, with new research data superimposed over it from the Michigan Vehicle Acceleration Characteristics Study. (**33**)

When a highway is located in a recreational area, compensation for the performance characteristics of recreation vehicles should be made.

Vehicular Pollution

The pollutants emitted from motor vehicles and their impact on land uses adjacent to highways are factors affecting the highway design process. As each vehicle travels along the highway, its emits pollutants into the atmosphere and transmits noise to the surrounding area. The highway designer must recognize these impacts and evaluate them before selection of transportation alternatives. There are many factors that affect the rate of pollutant emission from vehicles. Among these are vehicle mix, vehicle speed, ambient

THIS TURNING TEMPLATE SHOWS THE TURNING PATHS OF THE AASHTO DESIGN VEHICLES. THE PATHS SHOWN ARE FOR THE LEFT FRONT OVERHANG AND THE OUTSIDE REAR WHEEL. THE LEFT FRONT WHEEL FOLLOWS THE CIRCULAR CURVE, HOWEVER, ITS PATH IS NOT SHOWN.

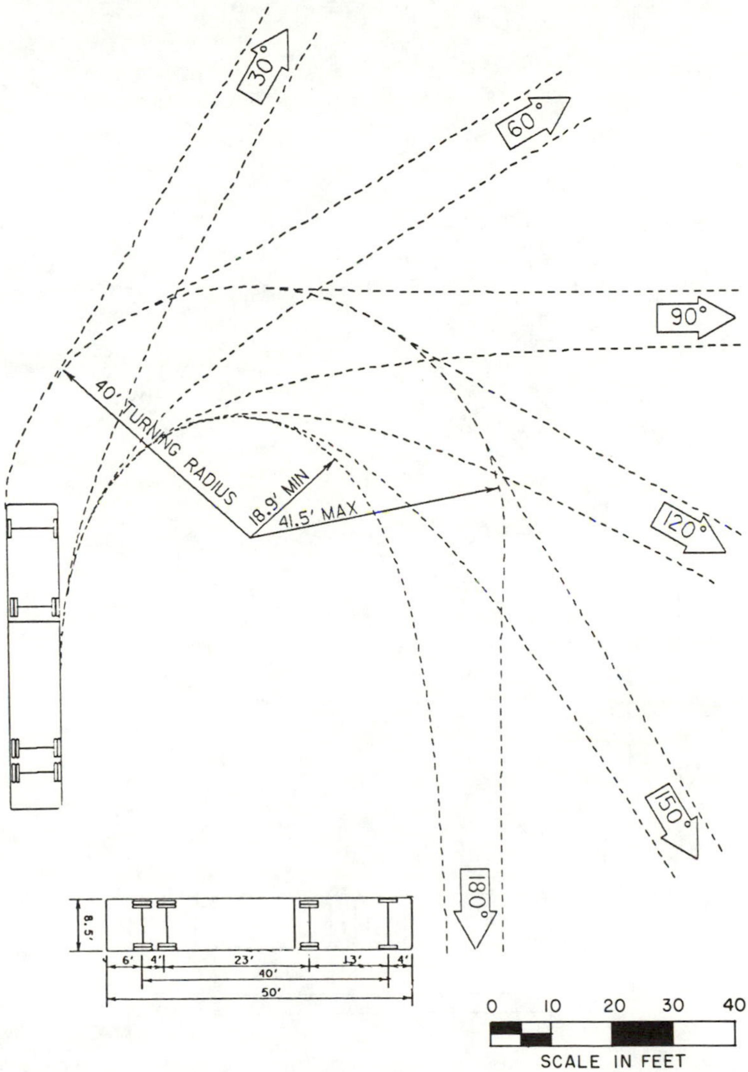

Source: Texas State Department of Highways and Public Transportation

Figure II-5. Minimum turning path for WB-40 design vehicle.

THIS TURNING TEMPLATE SHOWS THE TURNING PATHS OF THE AASHTO DESIGN VEHICLES. THE PATHS SHOWN ARE FOR THE LEFT FRONT OVERHANG AND THE OUTSIDE REAR WHEEL. THE LEFT FRONT WHEEL FOLLOWS THE CIRCULAR CURVE, HOWEVER, ITS PATH IS NOT SHOWN.

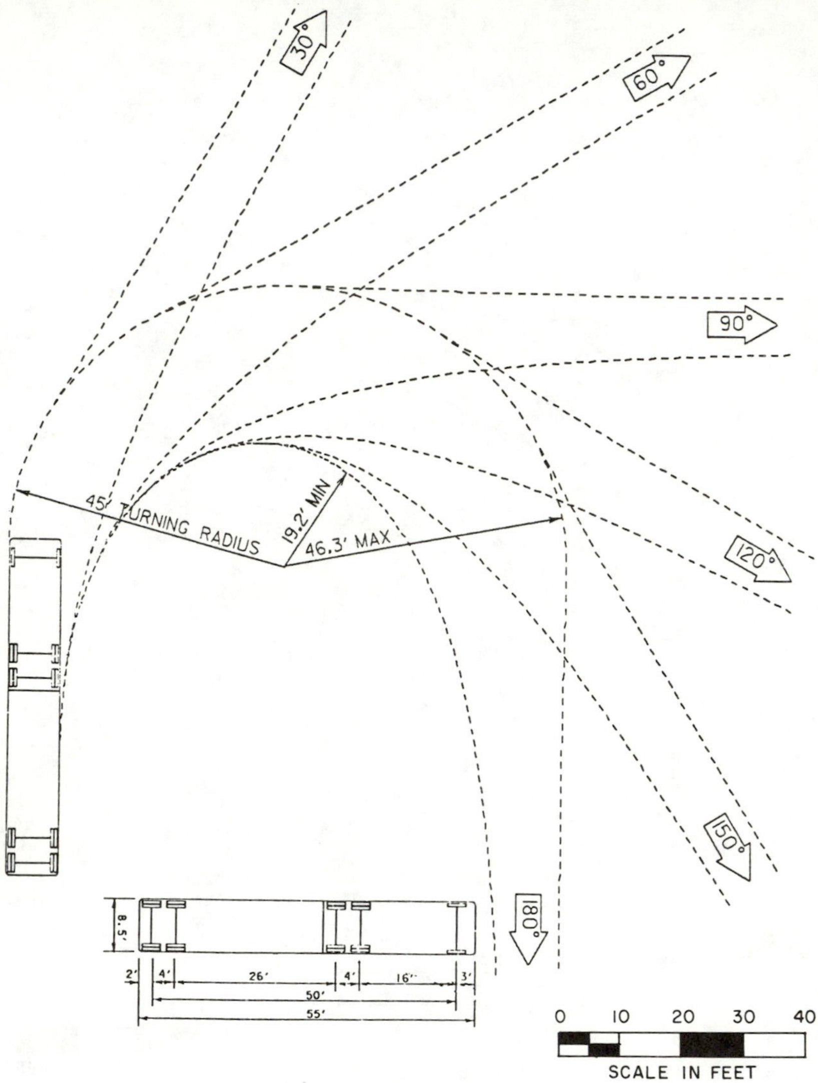

Source: Texas State Department of Highways and Public Transportation

Figure II-6. Minimum turning path for WB-50 design vehicle.

THIS TURNING TEMPLATE SHOWS THE TURNING PATHS OF THE AASHTO DESIGN VEHICLES. THE PATHS SHOWN ARE FOR THE LEFT FRONT OVERHANG AND THE OUTSIDE REAR WHEEL. THE LEFT FRONT WHEEL FOLLOWS THE CIRCULAR CURVE, HOWEVER, ITS PATH IS NOT SHOWN.

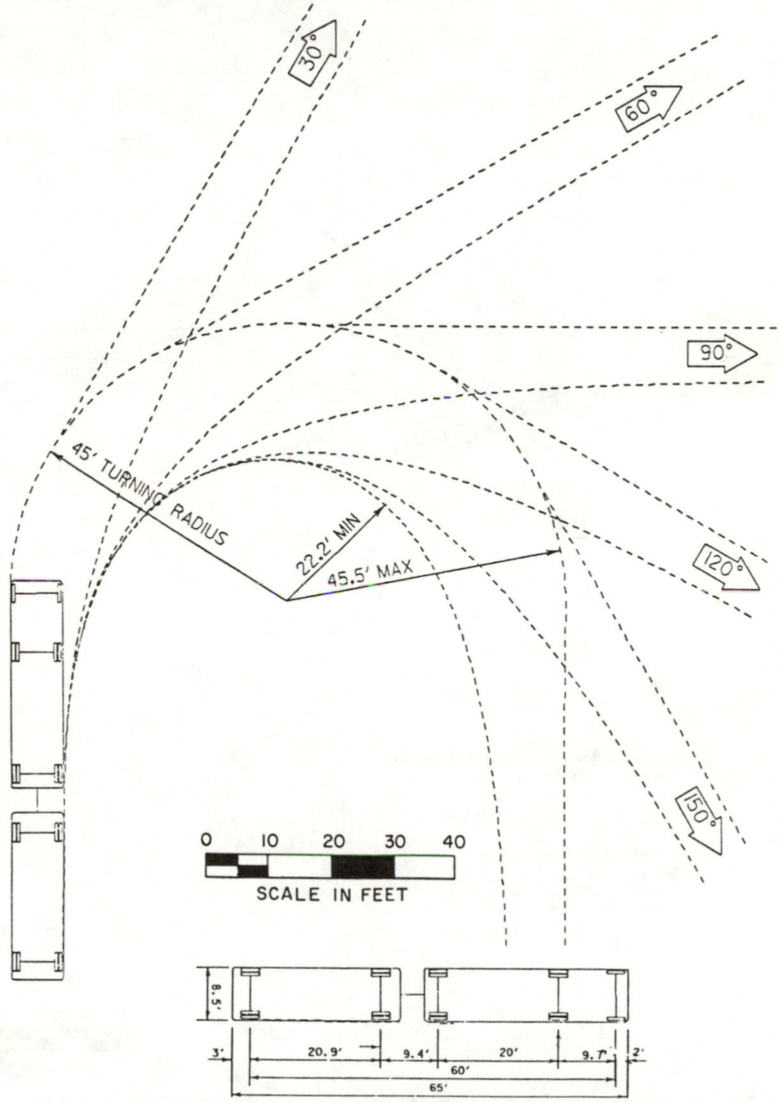

Source: Texas State Department of Highways and Public Transportation

Figure II-7. Minimum turning path for WB-60 design vehicle.

THIS TURNING TEMPLATE SHOWS THE TURNING PATHS OF THE AASHTO DESIGN VEHICLES. THE PATHS SHOWN ARE FOR THE LEFT FRONT OVERHANG AND THE OUTSIDE REAR WHEEL. THE LEFT FRONT WHEEL FOLLOWS THE CIRCULAR CURVE, HOWEVER, ITS PATH IS NOT SHOWN.

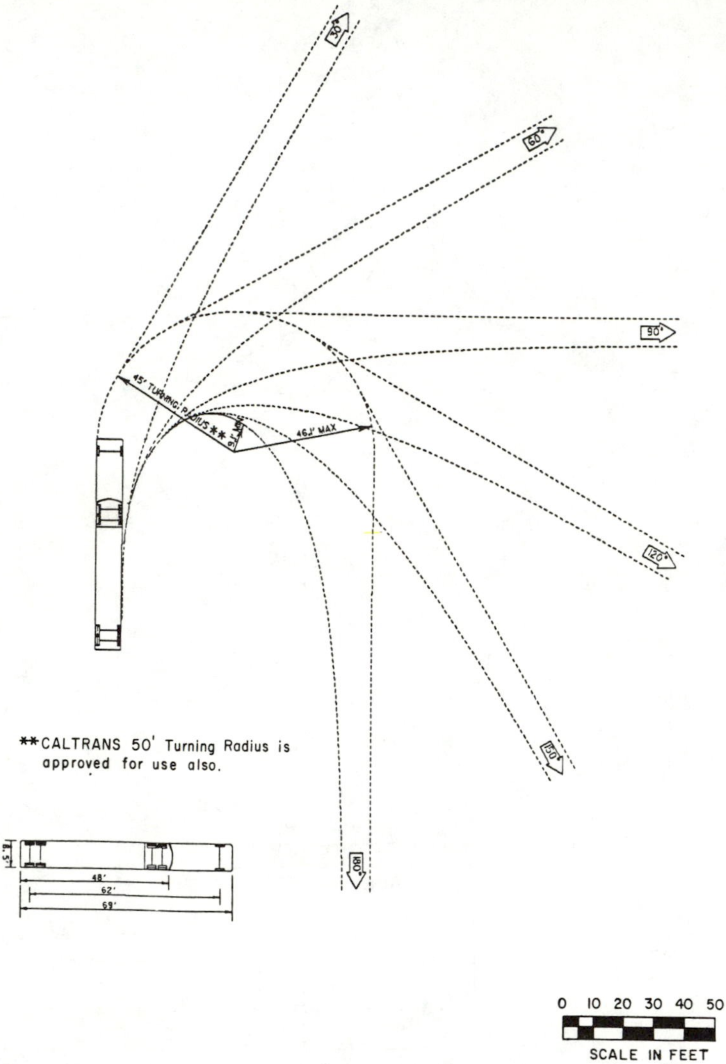

*Design vehicle with 48' trailer as adopted in 1982 Surface Transportation Assistance Act (STAA)

Source: Texas State Department of Highways and Public Transportation

Figure II-8. Minimum turning path for WB-62 (Interstate Semitrailer)*

THIS TURNING TEMPLATE SHOWS THE TURNING PATHS OF THE AASHTO DESIGN VEHICLES. THE PATHS SHOWN ARE FOR THE LEFT FRONT OVERHANG AND THE OUTSIDE REAR WHEEL. THE LEFT FRONT WHEEL FOLLOWS THE CIRCULAR CURVE, HOWEVER, ITS PATH IS NOT SHOWN.

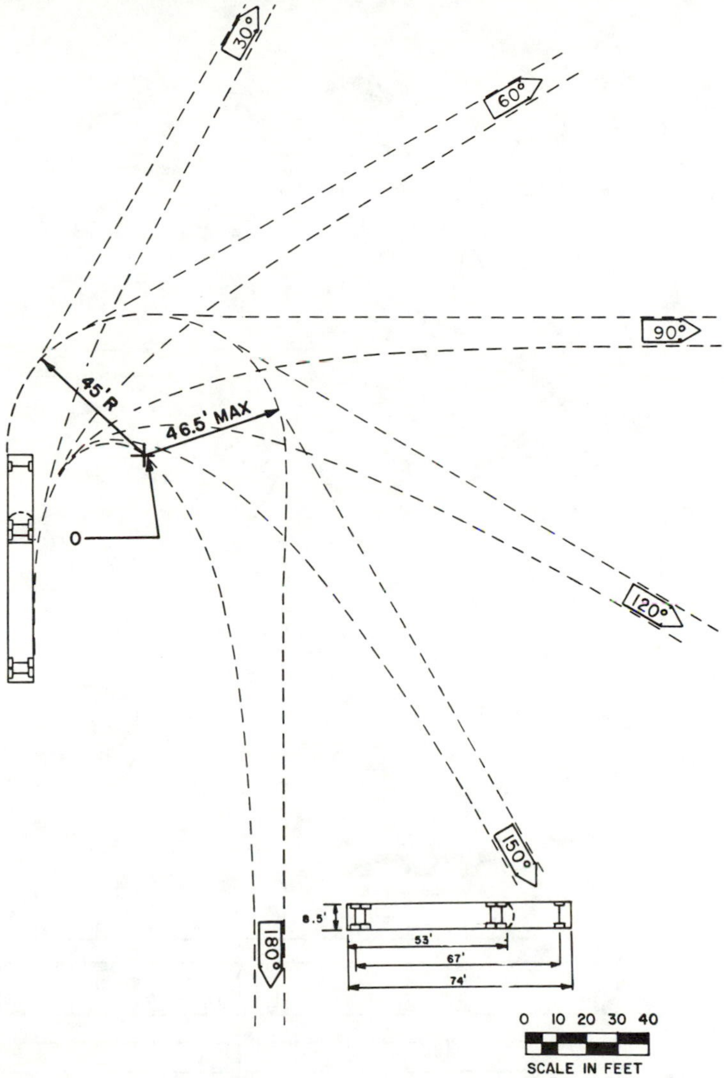

*Design vehicle with 53' trailer as grandfathered in 1982 Surface Transportation Assistance Act (STAA)
Source: Texas State Department of Highways and Public Transportation

Figure II-9. Minimum turning path for WB-67 (Interstate Semitrailer)*

THIS TURNING TEMPLATE SHOWS THE TURNING PATHS OF THE AASHTO DESIGN VEHICLES. THE PATHS SHOWN ARE FOR THE LEFT FRONT OVERHANG AND THE OUTSIDE REAR WHEEL. THE LEFT FRONT WHEEL FOLLOWS THE CIRCULAR CURVE, HOWEVER, ITS PATH IS NOT SHOWN.

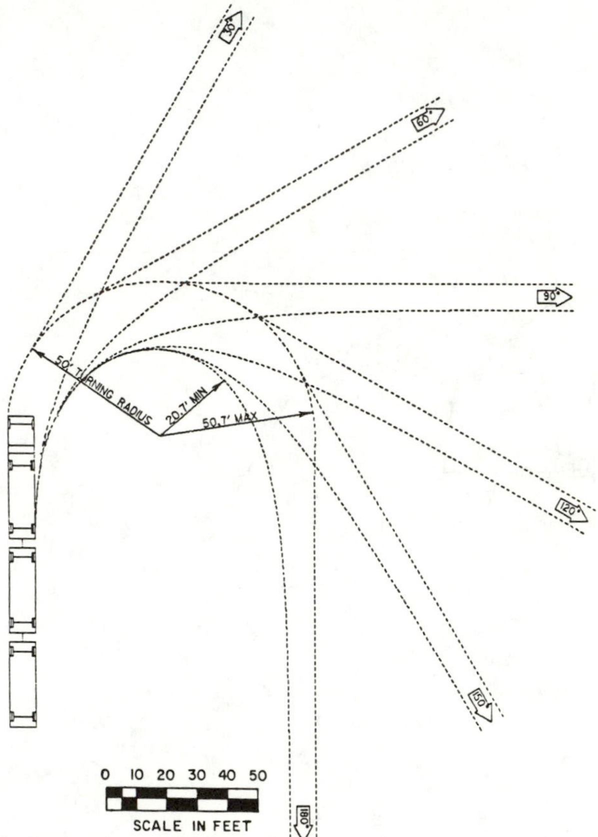

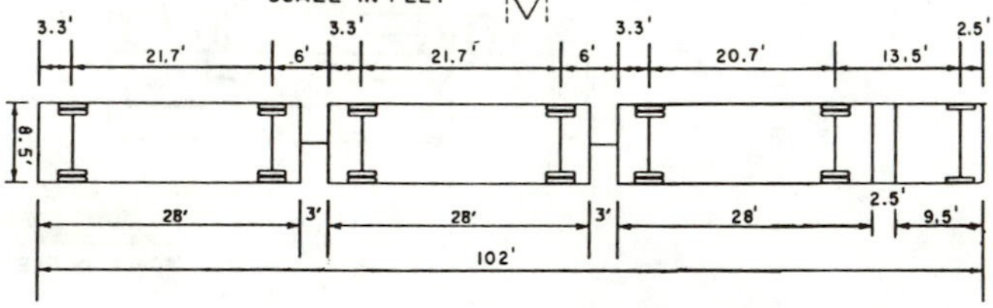

Source: Texas State Department of Highways and Public Transportation

Figure II-10. Minimum turning path for WB-96 design vehicle. (Triple Trailer)

THIS TURNING TEMPLATE SHOWS THE TURNING PATHS OF THE AASHTO DESIGN VEHICLES. THE PATHS SHOWN ARE FOR THE LEFT FRONT OVERHANG AND THE OUTSIDE REAR WHEEL. THE LEFT FRONT WHEEL FOLLOWS THE CIRCULAR CURVE, HOWEVER, ITS PATH IS NOT SHOWN.

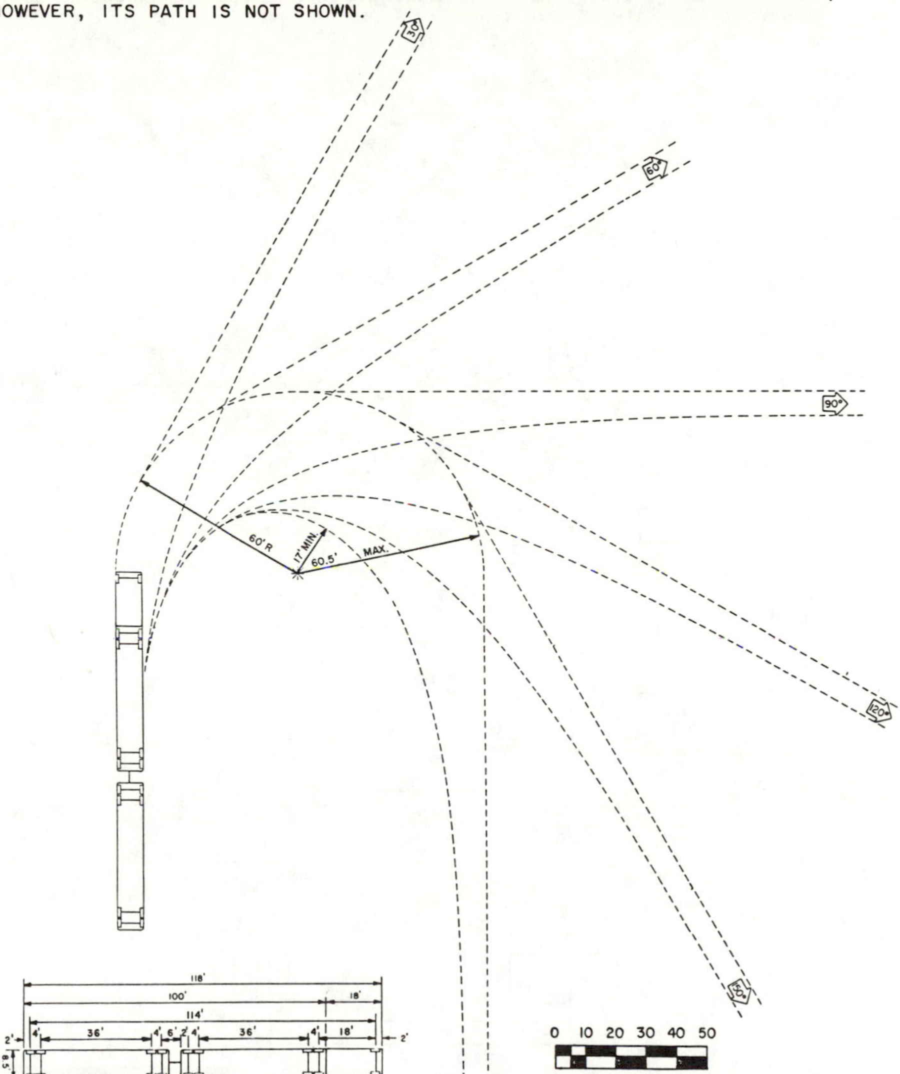

Source: Texas State Department of Highways and Public Transportation

Figure II-11. Minimum turning path for WB-114 design vehicle. (Turnpike Double Trailer)

THIS TURNING TEMPLATE SHOWS THE TURNING PATHS OF THE AASHTO DESIGN VEHICLES. THE PATHS SHOWN ARE FOR THE LEFT FRONT OVERHANG AND THE OUTSIDE REAR WHEEL. THE LEFT FRONT WHEEL FOLLOWS THE CIRCULAR CURVE, HOWEVER, ITS PATH IS NOT SHOWN.

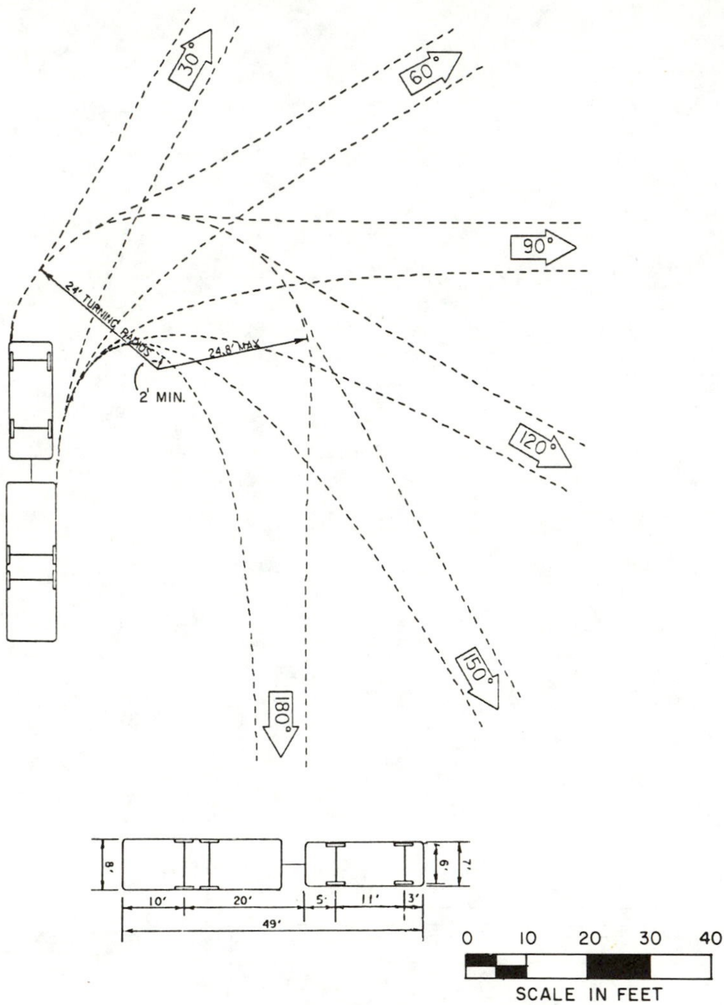

Source: Texas State Department of Highways and Public Transportation

Figure II-12. Minimum turning path for MH design vehicle.

THIS TURNING TEMPLATE SHOWS THE TURNING PATHS OF THE AASHTO DESIGN VEHICLES. THE PATHS SHOWN ARE FOR THE LEFT FRONT OVERHANG AND THE OUTSIDE REAR WHEEL. THE LEFT FRONT WHEEL FOLLOWS THE CIRCULAR CURVE, HOWEVER, ITS PATH IS NOT SHOWN.

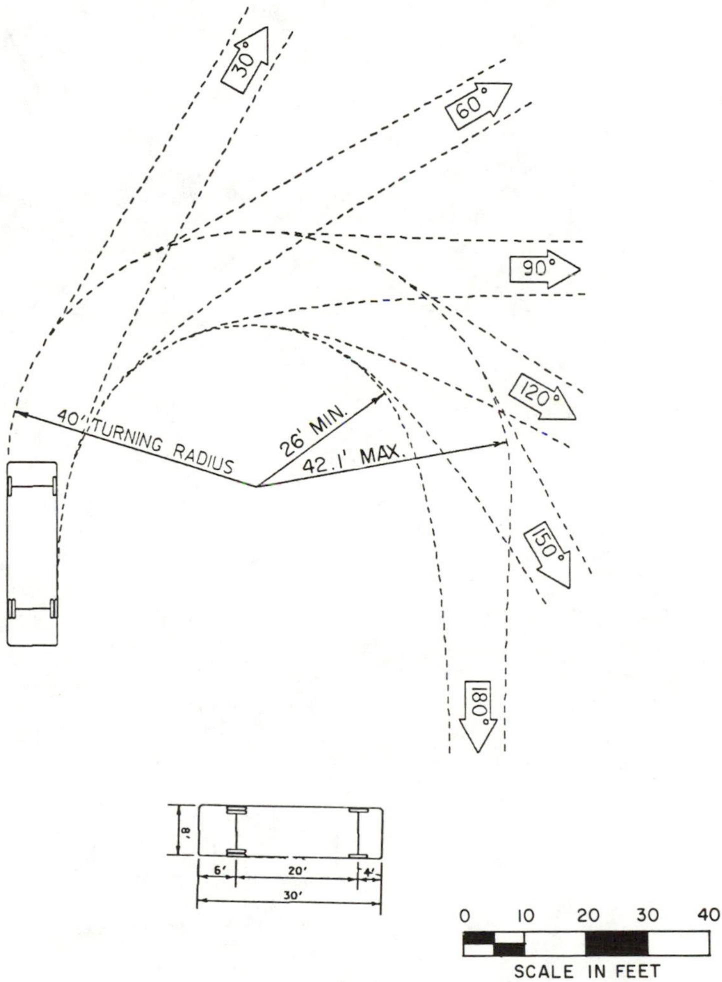

Source: Texas State Department of Highways and Public Transportation

Figure II-13. Minimum turning path for P/T design vehicle.

THIS TURNING TEMPLATE SHOWS THE TURNING PATHS OF THE AASHTO DESIGN VEHICLES. THE PATHS SHOWN ARE FOR THE LEFT FRONT OVERHANG AND THE OUTSIDE REAR WHEEL. THE LEFT FRONT WHEEL FOLLOWS THE CIRCULAR CURVE, HOWEVER, ITS PATH IS NOT SHOWN.

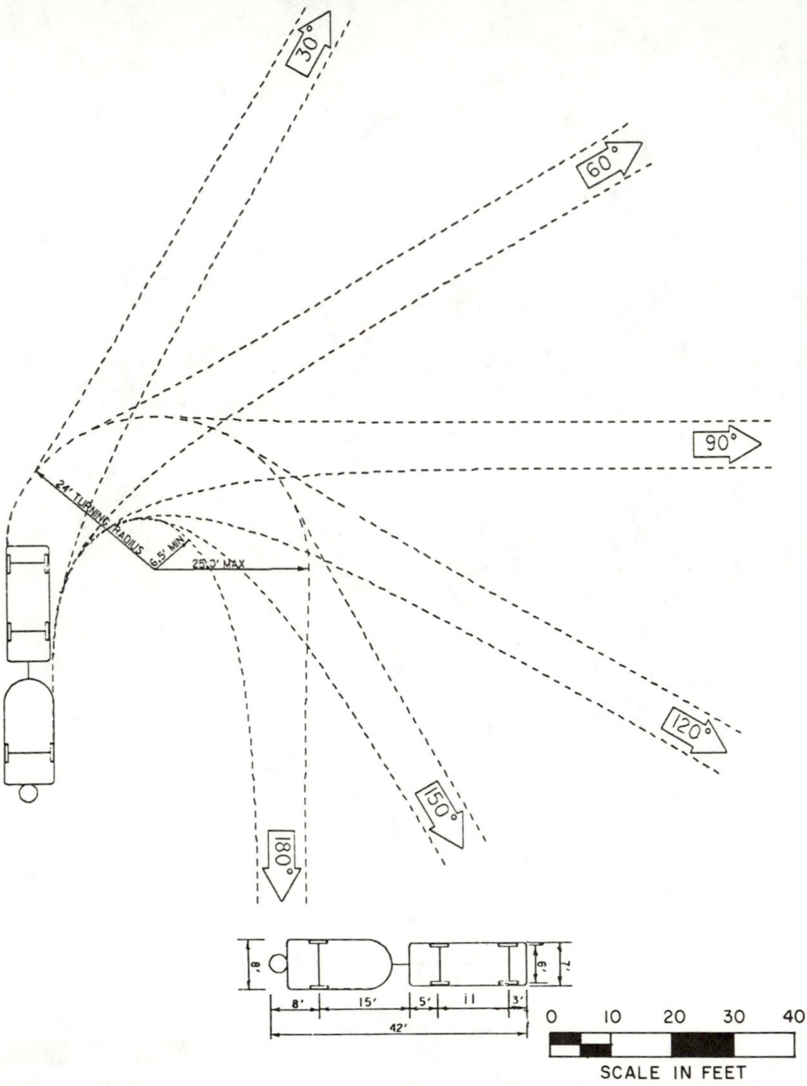

Source: Texas State Department of Highways and Public Transportation

Figure II-14. Minimum turning path for P/B design vehicle.

THIS TURNING TEMPLATE SHOWS THE TURNING PATHS OF THE AASHTO DESIGN VEHICLES. THE PATHS SHOWN ARE FOR THE LEFT FRONT OVERHANG AND THE OUTSIDE REAR WHEEL. THE LEFT FRONT WHEEL FOLLOWS THE CIRCULAR CURVE, HOWEVER, ITS PATH IS NOT SHOWN.

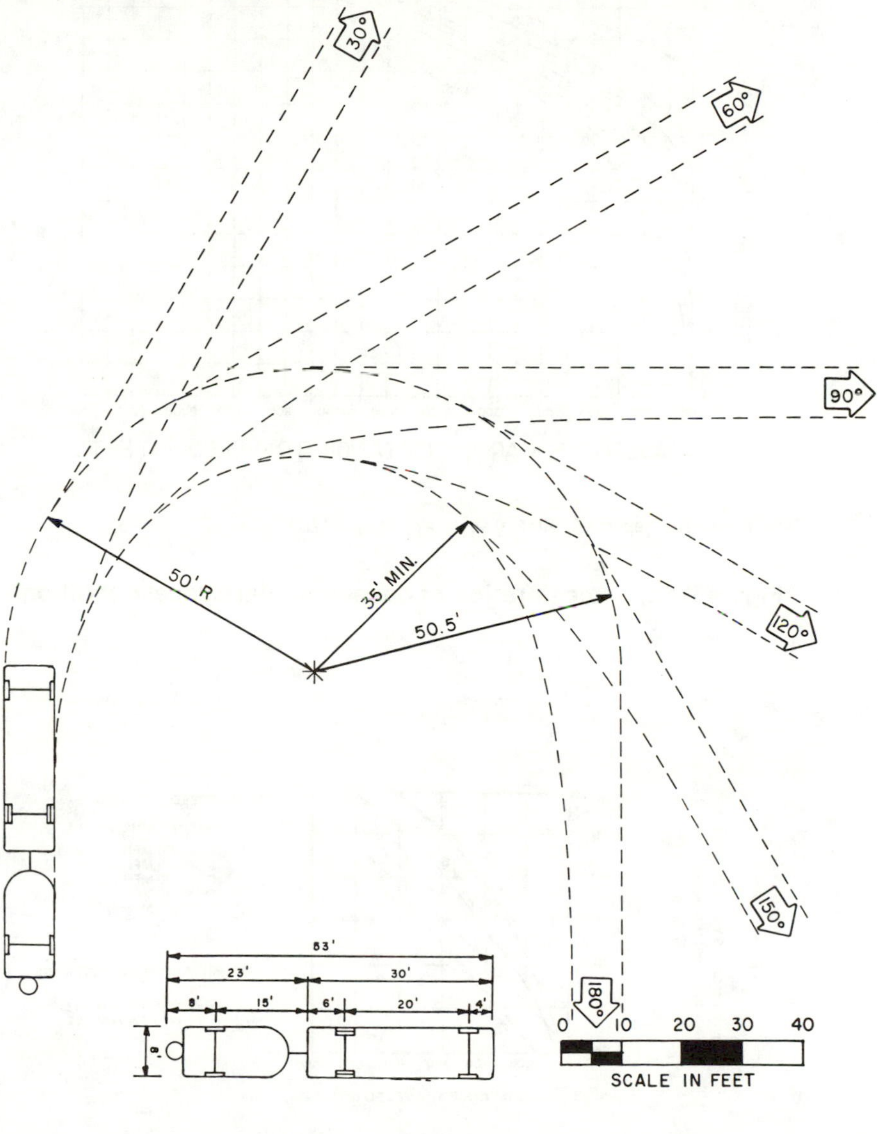

Source: CalTrans

Figure II-15. Minimum turning path for MH/B design vehicle.

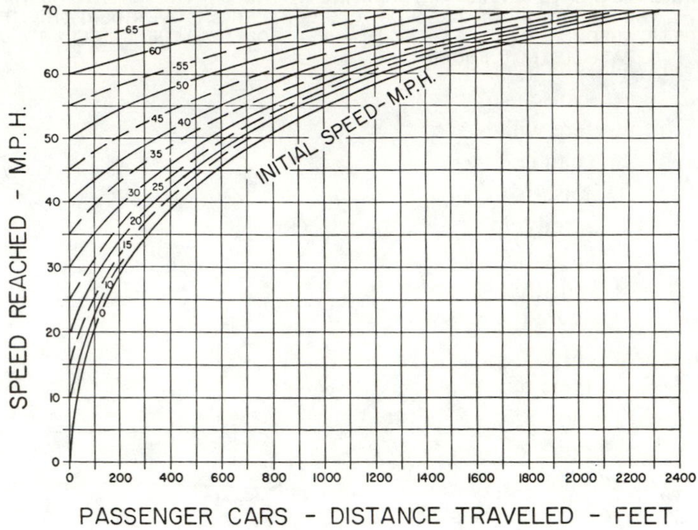

Source: Michigan Report NCHRP 270 as developed by NY D.O.T. (33)

Figure II-16. Acceleration of passenger cars, level conditions.

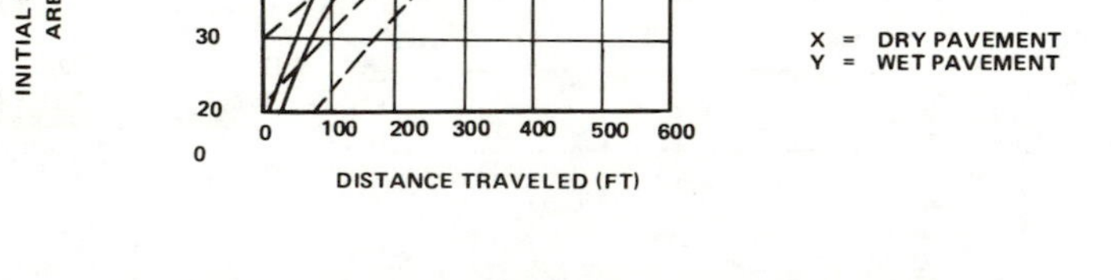

Figure II-17. Deceleration distances for passenger vehicles approaching intersections.

air temperature, ages of vehicles on the road, and percentage of vehicles operating in a cold mode.

Pollution from vehicles in the form of noise must also be recognized by the highway designer. Noise is unwanted sound, a subjective result of sounds that intrude on or interfere with activities such as conversation, thinking, reading, or sleeping. Sound can exist without people—noise cannot.

Motor vehicle noise is generated by the functioning of equipment within the vehicle, by its aerodynamics, by the action of tires on the roadway, and, in metropolitan areas, by the short-duration sounds of braking squeal, exhaust backfires, horns, and sirens of emergency vehicles.

Trucks and automobiles are the major noise-producing vehicles on the nation's highways. Motorcycles are a factor to be considered because of the rapid increase in their numbers in recent years. Trucks, particularly heavy diesel-powered trucks, present the most difficult noise problem on the highway. The development of more powerful engines has generally increased noise. Modern automobiles are relatively quiet, particularly at the lower cruising speeds, but exist in such numbers as to make their total noise contribution significant. A steep grade has little influence on the noise of automobiles but can cause an increase in noise levels for large trucks. Noise produced by automobiles increases dramatically with speed. Unlike the automobiles, truck noise levels are less influenced by speed because the factors (including acceleration noise) that are not affected by speed directly usually make-up a major part of the total noise.

For passenger cars, noise produced under normal conditions is primarily from the engine exhaust system and the tire-roadway interaction. Under high-speed cruise conditions the dominant source is from the tire-roadway interaction. For conditions of maximum acceleration the engine system noise predominates. Cars at constant highway speeds give much the same noise reading whether or not the engine is operating, because the noise is principally produced by the tire-roadway interaction with some added wind noise.

Truck noise has several principal components originating from such sources as exhaust, engine gears, fans, and air intack. At higher speeds, tire-roadway interaction and wind noise add to the problem. As in passenger cars, the noise produced by large diesel trucks is primarily from the engine exhaust system and the tire-roadway interaction. For trucks, however, engine exhaust noise tends to dominate tire-roadway interaction for most operating conditions. This is particularly true during acceleration. A major source of the noise of large trucks is the exhaust stack.

The quality of the noise varies with the number and operating conditions of the vehicles; the directionality and amplitude of the noise vary with highway design features. The highway designer must therefore be concerned with how highway locations and design influence the vehicle noise perceived by person residing or working nearby.

DRIVER PERFORMANCE

Introduction

An appreciation of driver performance is essential to proper highway design and operation. Design suitability rests as much on the ability of the highway to be used safely and efficiently as on any other criterion. When drivers use a highway designed to be compatible with their capabilities and limitations, their performance is aided. When a design is incompatible with the attributes of drivers, the chances for driver errors increase, and accidents and inefficient operation often result.

At the start of the 20th century, approximately 4 percent of America's population was 65 years of age or older. Persons 65 years of age or older accounted for 15 percent of the driving age population in 1986, and will increase to 22 percent by the year 2030.

Elderly drivers and pedestrians are a significant and rapidly growing segment of the traffic stream with a variety of age-related sensory-motor impairments. As a group, they have the potential to adversely affect the highway system's safety and efficiency. Data are lacking on the extent of these impairments and on design and operational standards. Basic policy decisions are also lacking. However, there is agreement that elderly road users require mobility, and that they should be accommodated by the highway's design and operational characteristics to the greatest extent practicable. Thus, designers and engineers should be aware of the problems and requirements of the elderly, and consider applying applicable measures to aid their performance.

This section provides information about driver performance useful to highway engineers in the design and operation of highways. It describes drivers in terms of performance—how they interact with the the highway and its information system and why they make errors.

The material draws extensively from *Positive Guidance in Traffic Control* (**2**), which contains information on the attributes of drivers, on the tasks of driving, and on information handling by the driver. When positive guidance is applied to design, competent drivers, using well-designed highways with appropriate information displays, can perform safely and efficiently. Properly designed and operated highways in turn provide positive guidance to drivers.

The Driving Task

The driving task depends on receiving and using information. The information received in transit is compared with the information already possessed by drivers. Decisions are then made and actions of control performed. Driving encompasses a number of discrete and interrelated activities. When grouped by performance, the activities fall into three levels: control, guidance, and navigation. These activities are ordered on scales of complexity of task and importance for safety.

Simple overlearned steering and speed control are at one end of the scale (control). Road-following and safe-path maintenance in response to road and traffic conditions are at midlevel (guidance). Trip planning and route following are at the other end of the scale (navigation.)

The driving task may be complex and demanding, several individual activities may be performed simultanenously, requiring smooth and efficient handling and integration of information. Driving often occurs at high speeds, under time pressures, in unfamiliar locations, and under adverse environmental conditions. The driving task may at other times be so simple and undemanding that a driver becomes inattentive. The key to safe, efficient performance is error-free information handling.

Driver errors occur because drivers do not know what is required, because situations may lead to task overload or inattentiveness, and because deficient or inconsistent design or information display may cause confusion. Driver errors also result from pressures of time, complexity of decisions, or profusion of information. Control and guidance errors may directly contribute to accidents. Navigational errors resulting in delay contribute to inefficient operations and may indirectly lead to accidents.

The Guidance Task

Highway design and traffic operations have the greatest effect on guidance. An appreciation of this task level is needed by the highway designer to aid driver performance.

Lane Placement and Road Following

Lane placement and road-following decisions are basic to vehicle guidance, with steering and speed control judgements representing major elements. A feed-back process is used to follow alinement and grade in keeping with road and environmental conditions. Obstacle-avoidance decisions are integrated into lane placement and road-following activities. This portion of the guidance task level is continually performed — when no other traffic is present (singularly), or when it is shared with other activities (integrated).

Car Following

Car-following decisions are more complex than road-following decisions because they involve speed control modifications. In car following, drivers constantly modify their speed to maintain safe gaps. They have to assess the speed of the lead vehicle and the speed and position of the other vehicles in the traffic stream and continually detect, assess, and respond to changes.

Overtaking and Passing

Overtaking and passing decisions are still more complex, requiring modifications in road and car-following and in speed control. In passing, drivers must judge the speed and acceleration potential of their own vehicle, the speed of the lead vehicle, the speed and rate of closure of the approached vehicle, and the presence of an acceptable gap in the traffic stream.

Other Guidance Activities

Other guidance task-level activities include merging, lane changing, avoidance of pedestrians, and response to traffic control devices. Most of these activities require complex decisions, judgments, and predictions.

The Information System

Each element that provides information is part of the information system of the highway. Formal sources of information are the traffic control devices specifically designed to display information. Informal sources include such elements as design features, pavement joints, tree lines, and traffic. Together, the formal and informal sources display the information drivers need to perform the task of driving safely and efficiently. Formal and informal sources of information are interrelated and must reinforce and augment each other to be maximally useful.

For every decade after age 25, drivers need twice the brightness at night to receive visual information. Hence, by age 75, some drivers may need 32 times the brightness they did at age 25.

Traffic Control Devices

Traffic control devices provide guidance and navigation information that often is not otherwise available or apparent. These devices include regulatory, warning, and route guidance information. Traffic control devices, such as markings and delineation, also display additional information that augments some roadway or environmental feature. These devices aid in the reception of information that might otherwise be overlooked or difficult to receive. Information relative to the proper traffic control devices is available in the *Manual on Uniform Traffic Control Devices* (**3**).

The Roadway and its Environment

Selection of speeds and paths is dependent on the drivers being able to view the road. Drivers must see the road directly in front of their vehicles and far enough in advance to predict with a high degree of accuracy the alinement, grade, width, and related aspects of the roadway. The view of the road includes its immediate environment. Such appurtenances and obstacles as shoulders, sign supports, bridge piers, abutments, guardrail, and median barriers that affect the driver must be clearly visible.

Information Handling

Drivers use many of their senses to gather information. They detect changes in vehicle handling through instinct. They feel road surface texture through vibrations in the steering wheel. They hear the sirens of emergency vehicles and see alinement, markings, and signs. Most information is received visually.

Throughout the driving task, drivers perform several functions almost simultaneously. They look at information sources, make numerous decisions, and perform necessary control actions. Sources of information (some needed, others not) compete for their attention. Needed information must be in the field of view, be available when needed, be in a usuable form, and be capable of getting the drivers' attention.

Because drivers can only attend to one visual information source at a time, they integrate the various information inputs and maintain an awareness of the changing environment through an attention-sharing process. Drivers sample visual information obtained in short-duration glances, shifting their attention from one source to another. They make some decisions immediately and delay others; they rely on judgment, estimation, and prediction to fill in gaps.

Reaction Time

Information takes time to process. Drivers' reaction times increase as a function of increased decision complexity and information content. The longer the reaction time, the greater the chance for error. Johannson and Rumar (**4**) measured brake reaction time for expected and unexpected signals. Their results show that when the signal is expected, reaction time averages about 2/3 sec. with a few drivers taking as long as 2 sec. With unexpected signals, reaction times increased by 35 percent. Thus, for a simple, unexpected decision and action, some drivers may take as long as 2.7 sec. to respond. A complex decision with several alternatives may take several seconds longer than a simple decision. Figure II-18 shows this relationship for median-case drivers, while Figure II-19 shows this relationship for 85th-percentile drivers. The figures show information in bits. A bit is a term used to quantify the amount of information needed to make a decision. Long processing time decreases the time available to attend to other information and increases chances for error.

Designs should take reaction times into account. It should be recognized that drivers vary in their response and take longer to respond when decisions are complex or features are unexpected. Clear sightlines and adequate decision sight distance provide margins for error.

Primacy

Primacy relates the relative importance to safety of competing information. Control and guidance information is importance because the related errors contribute directly to accidents. Navigation information assumes a lower primacy because errors lead to inefficient traffic flow. Accordingly, the design should focus the drivers' attention on the safety-critical elements or high-priority information sources. Providing clear sightlines and good visual quality achieves this goal.

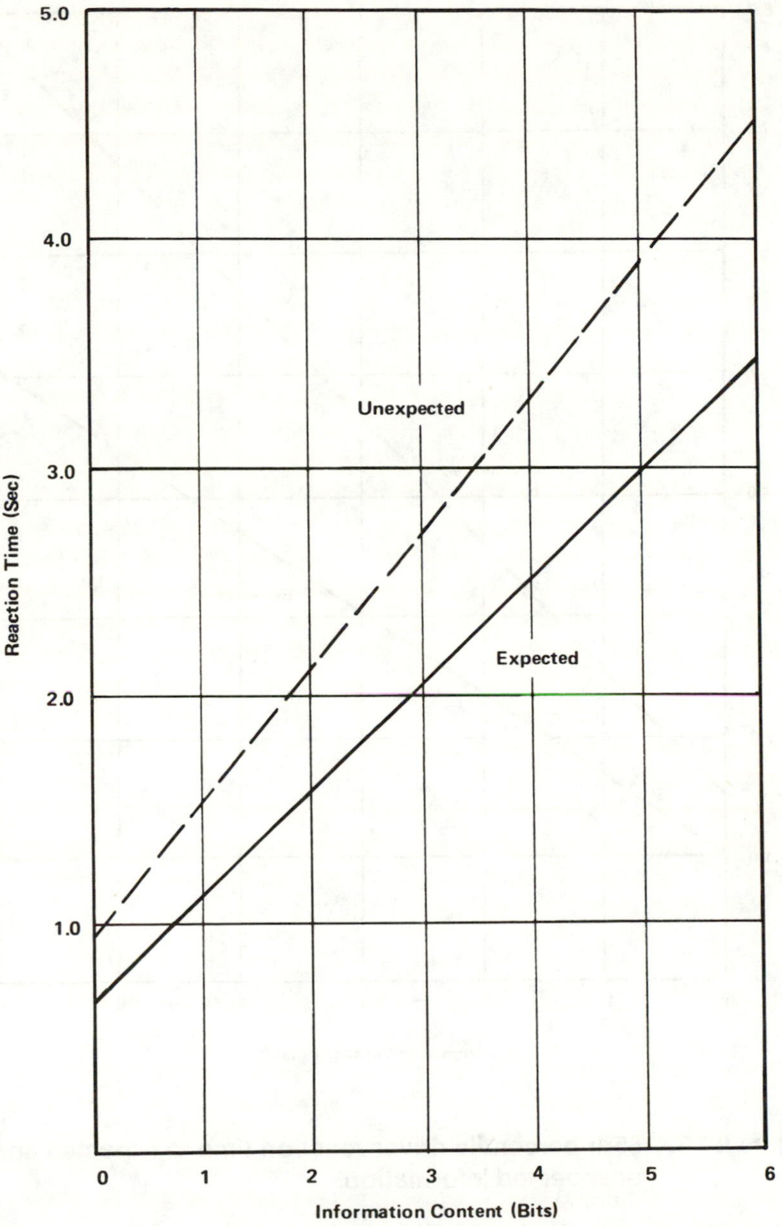

Figure II-18. Median driver reaction time to expected and unexpected information.

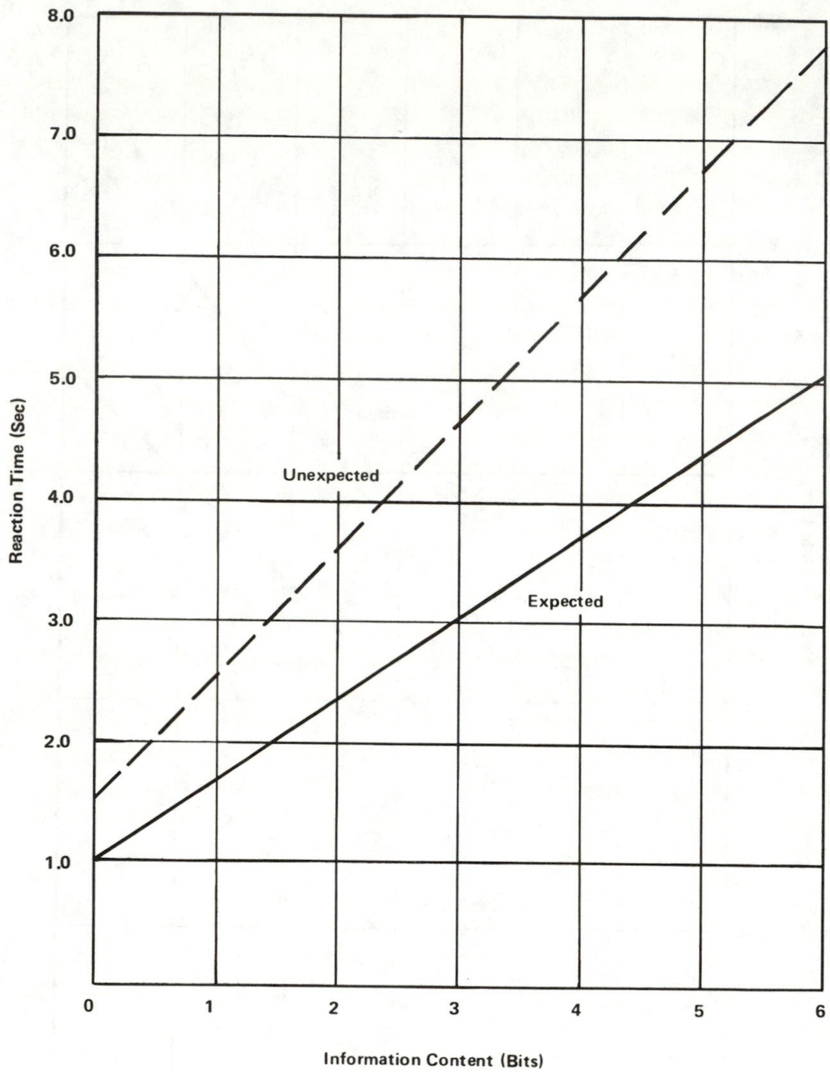

Figure II-19. 85th-percentile driver reaction time to expected and unexpected information.

Expectancy

Expectancies are formed by the drivers' experience and training. Situations that generally occur in the same way, and successful responses to these situations, are incorporated into the drivers' stores of knowledge. Expectancy relates to the drivers' readiness to respond to common situations in predictable and successful ways. It affects how drivers perceive and handle information, and modify the speed and accuracy of their responses.

Reinforced expectancies help drivers respond rapidly and correctly. Unusual, unique, or uncommon situations that violate expectancies may cause longer response times, inappropriate responses, or errors.

Most design features are sufficiently similar to create expectancies related to common geometric, operational, and route characteristics. For example, because most freeway interchanges have right exits, drivers generally expect to exit from the right. This aids performance by enabling rapid and correct responses when right exits are to be negotiated. There are, however, instances where expectancies are violated. For example, if an exit is on the left, then the right-exit expectancy is incorrect, and response times may be lengthened or errors committed.

The production of designs in accordance with prevalent expectancies is one of the most important ways to aid performance. Unusual or nonstandard design should be avoided, and design elements should be applied consistently throughout a highway segment. Care should be taken to maintain consistency from one segment to another. When drivers get the information they expect from the highway and its traffic control devices, their performance tends to be error free. When they do not get what they expect, or get what they do not expect, errors may result.

Driver Error

A common characteristic of may high-accident locations is that they place large or unusual demands on the information-processing capabilities of drivers. Inefficient operation and accidents usually occur where the chances for information-handling errors are high. At locations when the design is deficient the possibility of error and inappropriate driver performance increases.

Error Due to Driver Deficiencies

Many errors are caused by deficiences in a driver's capabilities or temporary states which, in conjunction with an inadequate design or difficult situation, produce the failure. For example, insufficient experience and training often contribute to the inability to recover from a skid. Similarly, inappropriate risk taking may lead to errors in gap acceptance while passing (**5**). Poor glare recovery may cause elderly drivers to miss information at night (**6**).

Adverse psychophysiological states also lead to driver failures. These include decreased performance caused by alcohol and drugs, in which the link to accidents has been clearly established. The effects of fatigue, caused by sleep deprivation from extended periods of driving without rest, or prolonged exposure to monotonous environments, or both, also contribute to accidents involvement (**7**).

It is generally not possible for a design or operational procedure to reduce errors caused by innate driver deficiencies. However, designs should be as forgiving as possible to lessen the consequences of these kinds of failure. Errors committed by competent drivers can be reduced by proper design and operation. Most individuals possess the attributes and skills to drive properly and are neither drunk, drugged, nor fatigued at the start of their trips. When drivers overextend themselves, fail to take proper rest breaks, or drive for prolonged periods, they ultimately reach a less-than-competent state. Fatigued drivers represent a sizable portion of the long-trip driving population and should be considered in freeway design.

Because the state-of-the-art of knowledge about the elderly's impairments is incomplete, and given the lack of agreement on policy matters, it is not feasible to make specific recommendations on measures to aid the elderly. However, the following have been suggested as design and traffic control measuers with the potential to aid the elderly driver:

- Assess all standards to determine the feasibility of designing for the 95th or 99th percentile driver (i.e., the elderly).
- Improve sight distance by modifying designs and removing obstructions, particularly at intersections and interchanges.
- Assess sight triangles for adequacy.
- Provide decision sight distances.
- Simplify and redesign intersections and interchanges that require multiple information reception and processing.
- Consider alternate designs to reduce conflicts.
- Provide protected movements, particularly for left turns.
- Eliminate yield situations.

Error Due to Situation Demands

Drivers often commit errors when they have to perform several highly complex tasks at the same time under extreme time pressure (**8**). This type of error usually happens in urban locations with closely spaced decision points, intensive land use, complex design features, and heavy traffic. Resultant information-processing demands beyond the drivers' capabilities may cause confused or inadequate understanding or overloaded thought processes.

Other locations present the opposite situations and are associated with different types of errors. These typically rural locations may have widely spaced decision points, sparse land use, smooth alinement, and light traffic. Information demands are thus minimal, and rather than being overloaded, drivers' thought processes may be underloaded. Errors are caused by a state of decreased vigilance in which drivers fail to detect, recognize, or respond to new, infrequently encountered, or unexpected elements or information sources.

Speed and Design

Speed reduces the visual field, restricts peripheral vision, and limits the time available to receive and process information. Highways built to high design standards help compensate for these limitations by simplifying control and guidance activities, by aiding drivers with appropriate information, by placing this information within the cone of clear vision, by eliminating much of the need for peripheral vision, and by simplifying the decisions required and spacing them further apart to decrease information-processing demands.

Current freeway design has approached the goal of allowing drivers to operate at high speeds in comfort and safety. Design standards utilizing control of access have reduced the potential for conflict. Drivers are given a clear path. Obstructions have been eliminated or made forgiving, and a clear roadside has been provided. The modern freeway has provided features of alinement and profile that, when taken together with other factors, encourage high operating speeds.

Although improved design has produced significant benefits, it has also created potential problems. For example, driving at night at high speeds may lead to reduced forward vision because of the inability of headlights to illuminate hazards in sufficient time for some drivers to respond (**9**). In addition, the severity of accidents is generally greater with increased speed.

Finally, the very fact that freeways succeed in providing safe, efficient transportation can lead to difficulties. The Institute of Traffic Engineers (**10**) indicated that, "Freeways encourage drivers to extend the customary length and duration of their trips. This results in driver fatigue and slower reaction as well as a reduction in attention and vigilance."

Thus, extended periods of high-speed driving on highways with low demand for information processing may not always be conducive to proper information handling by drivers and may therefore lead to driver fatigue. Design should take these possible adverse effects into account and lessen their consequences. These problems imply that long sections of flat, tangent roadway shold be avoided and flat, curving alinement that follows the natural contours of the terrain should be used whenever possible. Well-spaced rest areas also prove helpful.

Design Assessment

This section has described the way drivers use information provided by the highway and its appurtenances and has shown the interdependence between design and information display. Both should be assessed in the design stage. Because drivers "read" the road and the environment and believe what they appear to be indicating (even if traffic control devices making up the formal information system are indicating something contrary), a highway segment that is inadequately designed may not operate safely and efficiently even with the correctly compensating controls. Conversely, an adequately designed highway may not operate properly without an appropriate complement of traffic control devices.

Designers should consider their design of a particular segment as it will be when constructed. They should determine how the highway will interact with its environment, what its information system will be like, and the extent to which its information system will complement and augment the segment. The design should be considered from the point of view of drivers unfamiliar with the highway to determine how the segment will appear to them and what effects it will have on their performances. Consideration should be given to the visual quality of the road. Locations with potential for overload or underload of information should be identified and the situation rectified. The adequacy of the sightlines and distances should be assessed. Finally, a determination should be made whether unusual vehicle maneuvers will be required and whether prevalent driver expectancies may be violated.

Potential driver problems can be anticipated before a facility is built by using information about the driving tasks and possible driver errors in an assessment of design. When trade-offs are necessary they can be made with the attributes of drivers in mind to ensure that the resultant design is compatible with them. Properly designed and operated highways that provide positive guidance to the drivers operate at an unprecedented level of safety and efficiency. One of the primary efforts of the designer should be to ensure that this standard is always maintained.

TRAFFIC CHARACTERISTICS

General Considerations

The design of a highway or any part thereof should be based upon factual data, among which are those relating to traffic. All data should be considered jointly. Financing, quality of foundations, availability of materials, cost of right-of-way, and other factors have important bearing on the design, but traffic indicates the service for which the improvement is being made and directly affects the geometric features of design, such as widths, alinements, and grades. It is no more rational to design a highway without traffic information than it is to design a bridge without knowledge of the weights and numbers of vehicles it is intended to support. Traffic information serves to establish the loads for geometric highway design.

Traffic data for a road or section of road generally are available. The data collected by State or local agencies include traffic volumes for days of the year and times of the day, as well as the distribution of vehicles by types and by weights. The data also include information on trends from which the designer may estimate the traffic to be expected in the future.

Volume

Average Daily Traffic

The general unit of measure for traffic on a highway is the average daily traffic (ADT), defined as the total volume during a given time period (in whole days), greater than one day and less than one year, divided by the number of days in that time period. The current ADT volume for a highway can be determined readily where continuous traffic counts are available. Where only periodic counts are taken, ADT volume can also be established by correcting the results according to such factors as the season, month, or day of week.

Knowledge of ADT volume is important for many purposes, e.g., determining annual usage as justification for proposed expenditures or for design of structural elements of a highway, but its direct use in the geometric design of highways is not appropriate because it does not indicate the variation in the traffic occurring during the various months of the year, days of the week, and hours of the day. The amount by which the volume of an average day is exceeded on certain days is appreciable and varied. At the average rural location the volume on certain days may be double the ADT. Thus, a highway designed for the traffic on an average day would be required to carry a volume greater than the design volume a considerable portion of the year, and on many days the volume carried would be much greater than the design volume.

Peak-Hour Traffic

Traffic volume during an interval of time shorter than a day more appropriately reflects the operating conditions that should be used for design if traffic is to be properly served. The brief but frequently repeated rush-hour periods are significant in this regard. In nearly all cases a practical and adequate time period is one hour.

The traffic pattern on any highway shows considerable variation in traffic volumes during the different hours of the day and in hourly volumes throughout the year. It must be determined which of these hourly traffic volumes should be used in design. It would be wasteful to predicate the design on the (maximum) peak-hour traffic of the year, yet the use of the average hourly traffic would result in an inadequate design. The hourly traffic volume used in design should not be exceeded very often or by very much. On the other hand, it should not be so high that traffic would rarely be great enough to make full use of the resulting facility. One guide in determining the hourly traffic best fitted for use in design is a curve showing variation in hourly traffic volumes during the year.

Figure II-20 shows the relation between the highest hourly volumes and ADT on rural arterials taken from an analysis of traffic count data, covering a wide range of volumes and geographic conditions. The curves in the chart were prepared by arranging all of the hourly volumes of one year expressed as a percentage of ADT in a descending order of magnitude. The middle curve is the average for all locations studied and represents a highway with average fluctuation in traffic flow.

These curves lead to the conclusion that the hourly traffic used in design should be the 30th highest hourly volume of the year, abbreviated as 30 HV. The reasonableness of 30 HV as a design control is indicated by the change that results from choosing a somewhat higher or lower volume. The curve in Figure II-16 steepens quickly to the left of the point showing the 30th highest hour, indicating much higher volumes for the inclusion of only a few more of the higher hourly volumes. The curve flattens to the right, indicating many hours in which the volume is not much less than 30 HV.

On rural roads with average fluctuation in traffic flow, 30 HV is about 15 percent of the ADT. Whether or not this hourly volume is too low for ade-

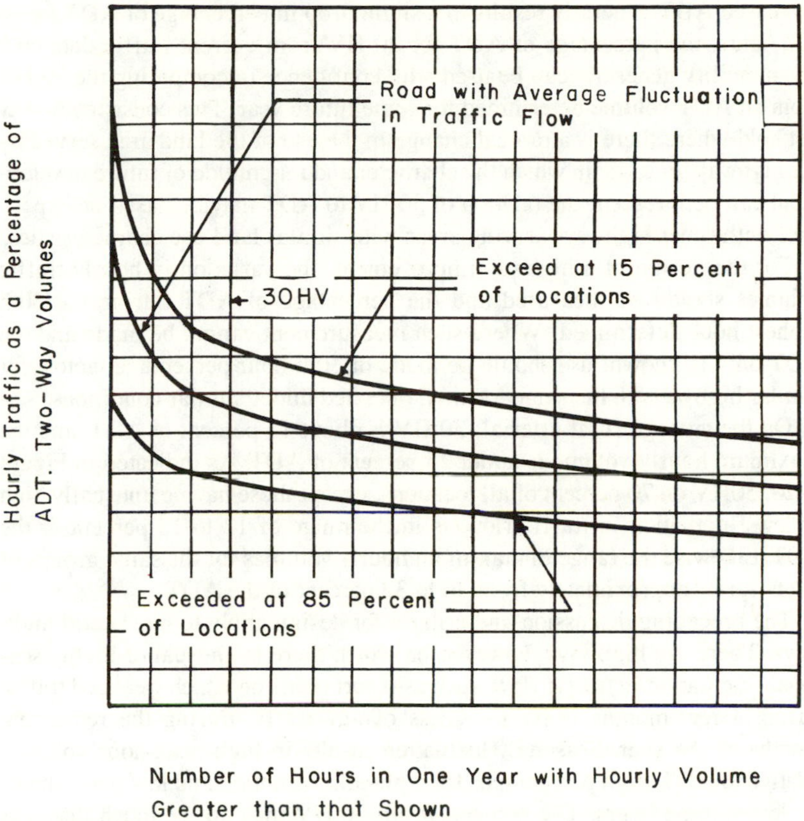

Figure II-20. Relation between peak-hour and average daily traffic volumes on rural arterials.

quacy in design can be judged by the 29 times during the year it is exceeded. The maximum hourly volume, which is about 25 percent of the ADT, exceeds 30 HV by about 67 percent. Whether or not 30 HV is too high for practical economy in design can be judged by the trend in the hourly traffic volumes lower than the 30th highest. The middle curve in Figure II-20 indicates that during 170 hours of the year the traffic volume exceeds 11.5 percent of the ADT. The lowest of this range of hourly volumes is about 23 percent less than 30 HV.

Another fortunate characteristic of 30 HV is that, as a percentage of ADT, it varies only slightly from year to year in spite of significant changes in ADT.

Increased ADT generally results in a slight drop in percentage of ADT for 30 HV. Thus, the percentage of ADT for 30 HV from current traffic data on a given facility generally can be used with confidence in computing the 30 HV from an ADT volume determined for some future year. This consistency may not hold where there is a radical change in the use of the land area served by the highway. In cases in which the character and magnitude of future development can be foreseen, the relation of 30 HV to ADT may be based on experience with other highways serving areas with similar land-use characteristics.

For the design of a highway improvement, the variation in hourly traffic volumes should be measured and the percentage of ADT during the 30th highest hour determined. Where such measurement cannot be made and the ADT only is known, use should be made of 30th-hour percentage factors for similar highways in the same locality, operated under similar conditions.

On the average rural arterial, 30 HV is about 15 percent of ADT, and the maximum hourly volume is about 25 percent of ADT. As indicated in Figure II-20, 30HV on 70 percent of all locations, except those having unusually high or low fluctuation in traffic flow, is in the range of 12 to 18 percent of the ADT. Likewise the range in maximum hourly volumes for the same groups of roads varies approximately from 16 to 32 percent of the ADT.

The preceding discussion and criteria for design apply to most rural highways. There are highways, however, on which there is unusual or highly seasonal fluctuation in traffic flow, such as resort roads on which weekend traffic during a few months is far in excess of any traffic during the remaining months of the year. Seasonal fluctuation results in high peak-hour volumes relative to ADT, high percentages for high-volume hours, and low percentages for low-volume hours. The percentage for 30 HV may not be much different from the percentage for 30 HV on most rural roads, but the percentage criterion may not have application. Economy dictates a design that results in somewhat less satisfactory traffic operation during seasonal peaks than on rural roads with normal fluctuations, and the public generally will accept such conditions. On the other hand, design should not be so economical that severe congestion results during peak hours. It may be desirable, therefore, to choose an hourly volume for design, which is, say, about 50 percent of the volumes expected to occur during a very few maximum hours of the design year whether or not it is equal to 30 HV. Some congestion would be experienced by traffic during peak hours but the capacity would not be exceeded. A check should be made to ensure that the expected maximum hourly traffic does not exceed the capacity.

The design hourly volume (DHV) therefore should be 30 HV of the future year chosen for design. Exceptions may be made on roads with high seasonal traffic fluctuation, where a different volume may need to be used. The 30-HV criterion also applies in general to urban areas, but where the fluctuation in traffic flow may be radically different from that on rural highways, other relations may have to be considered.

In urban areas an appropriate hourly volume may be determined from the study of traffic during the normal daily peak periods. Because of the recurring morning and afternoon peak traffic flow, there is usually little difference between the 30th highest hourly volume and the 200th highest hourly volume. For the usual urban condition the highest hourly volume is found during the afternoon work-to-home travel peaks. One approach for determining a suitable hourly volume for use as a basis for design is to select the highest afternoon peak traffic flow for each week and then average these values for 52 weeks of the year as the hourly volume for use in design. If all the morning peak-hour volumes are less than the afternoon peaks, the average of the 52 weekly afternoon peak-hour volumes would have about the same value as the 26th highest hourly volume of the year. If the morning peaks are equal to the afternoon peaks, the average of the afternoon peaks would be about equal to the 50th highest hourly volume. The volumes represented by the 26th and 50th hour are not sufficiently different from the 30 HV value to affect design. Therefore, for use in urban design the 30th highest hourly volume can be accepted because it is a reasonable representation of daily peak hours during the year. Exception may be necessary in those areas or locations where concentrated recreational or other travel during some seasons of the year results in a distribution of traffic volume of such nature that a sufficient number of hourly volumes are so much greater than the 30 HV that they cannot be tolerated and a higher value must be considered in design. Specific measurement of such volumes should be made and analyzed.

Traffic estimates used for the design of urban streets and highways are usually expressed as ADT volumes derived from the urban transportation planning process. Recently, however, consideration has been given to the development of DHVs by making peak-hour traffic assignments in lieu of ADT assignments. The availability of the 1980 census journey-to-work information is undoubtedly having a major influence on this latter approach.

In the usual case future travel demand is arrived at through the urban transportation planning process in terms of total daily trips that are assigned to the transportation system. Consideration of the split between public and private transportation is also incorporated into the process. These assigned trips then become volumes on links of the future street and highway network.

In some instances these volumes (ADT) are provided to the highway designers directly; in others they are converted to design hour directional volumes by the operational transportation study staff. From a practical standpoint the latter approach may be the more desirable because that transportation study staff is in a better position to evaluate the effects that the assumptions inherent in the planning process would have on the resultant design volumes.

Design hourly volumes (two-way), i.e., the 30 HV or its equivalent, may be determined by applying a representative percentage (usually 8 to 12 percent in urban areas) to the ADT. In many cases this percentage, based on data obtained in a traffic count program, is developed and applied systemwide; in other cases factors may be developed for different classes of facility or areas of the urban region or both. This is probably a more desirable approach. At least one highway agency has developed regression equations representing the relationships between peak flow and ADT. Different equations are applied, depending on the number of lanes and the range of the ADT volumes.

Directional Distribution

For two-lane rural highways the DHV is the total traffic in both directions of travel. On highways with more than two lanes and on two-lane roads where important intersections are encountered or where additional lanes are to be provided later, knowledge of the hourly traffic volume in each direction of travel is essential for design.

For the same ADT a multilane highway with a high percentage of traffic in one direction during the peak hours may require more lanes than a highway having the same ADT but with a lesser percentage. During peak hours on most rural highways, from 55 to 70 percent of the traffic is in one direction, being as much as 80 percent occasionally. For two multilane highways carrying equal traffic, one may have a one-way traffic load 60 percent greater than the other during the peak hours. As an example, consider a rural road designed for 4,000 vehicles per hour (vph) total for both directions. If during the design hour the directional distribution is equally split, or 2,000 vph is one direction, two lanes in each direction may be adequate. If 80 percent of the DHV is in one direction, at least three lanes in each direction would be required for the 3,200 vph; and if the 1,000-vehicles-per-lane criterion is rigidly applied, four lanes in each direction would be required.

Traffic distribution by directions during peak hours is generally consistent from year to year and from day to day on a given rural road, except on some highways serving recreational areas. The measured directional distribution may be assumed to apply to the DHV for the future year for which the facility is designed, except for urban highways.

The directional distribution of traffic on multilane facilities during the design hour (DDHV) should be determined by field measurements on the facility under consideration or on parallel and similiar facilities. In the latter case the parallel facilities should preferably be those from which traffic, for the most part, would be diverted to the new highway. The DDHV applicable for use on multilane facilities may be computed by multiplying the ADT by the percentage that 30 HV is of the ADT, and then by the percentage of traffic in the predominant direction during the design hour. Thus, if the DHV is 15 percent of the ADT and the directional distribution at that hour is 60:40, the DDHV is $0.15 \times 0.60 \times$ ADT, or 9 percent of the ADT. If the directional ADT is known for only one direction, the ADT nearly always is twice the directional ADT.

In designing at-grade intersections and interchanges it is necessary to know the volumes of all movements occurring during the design hour. This information is required for both the morning and evening peaks because the traffic pattern may change significantly from one peak hour to the other. Normally, a design is based on the DHV, which is to be accommodated during the morning rush hour in one direction and during the evening rush hour in the other direction. Total (two-way) volumes may be the same during both of these peaks, but the percentage of traffic in the predominant direction is reversed. At intersections the percentage of approaching traffic that turns to the right and to the left on each intersection leg must be determined separately for the morning and evening peaks. This information should be determined from on-the-ground counts, from origin and destination data, or both.

Composition of Traffic

Vehicles of different sizes and weights have different operating characteristics, which must be considered in highway design. Besides being heavier, trucks generally are slower and occupy more roadway space and consequently impose a greater traffic effect on the highway than passenger vehicles do. The overall effect on traffic operation of one truck is often equivalent to several passenger cars, the number of equivalent passenger cars depending on the gradient and the passing sight distance available. Thus the larger the proportion of trucks in a traffic stream, the greater the traffic load and the highway capacity required.

For uninterrupted traffic flow, as typically found in rural areas, the various sizes and weights of vehicles as they affect traffic operation can be grouped into two general classes:

1. Passenger cars —all passenger cars including light delivery trucks.
2. Trucks —all buses, single-unit trucks, and truck combinations except the light delivery trucks.

A light delivery truck is a single-unit truck, such as a van or pickup, with size and operating characteristics similar to those of a passenger car and commonly used for short-haul, light delivery service.

Vehicles in the truck class are normally those having 9,000 lb. or greater gross vehicle weight (GVW) rating of the manufacturer and vehicles having dual tires on the rear axle.

These definitions apply in the use herein of the terms "passenger cars" or "passenger car traffic" and "trucks" or "truck traffic."

In the passenger car class, most of the vehicles have similar operating characteristics. In the truck class, characteristics vary considerably, particularly in size and weight/horsepower ratio. Although this variation may be significant between any two trucks, the effect of all trucks in a traffic stream can be evaluated and is similar on most highways under comparable conditions. Accordingly, for the geometric design of a highway it is essential to have traffic data on vehicles in the truck class. These data generally indicate the major types of trucks and buses as percentages of all traffic expected to use the highway.

For design purposes the percentage of truck traffic during the peak hours should be known. In rural areas, comprehensive data usually are not available on the distribution of traffic by vehicle types during the peak hours; however, the percentage of truck traffic during the peak hours is known to be less than the percentage for the 24-hour periods. As the peak hour is approached, the passenger car traffic generally increases at a greater rate than truck traffic does. Most trucks operate steadily throughout the day, and much of the over-the-road hauling is done at night and during early morning hours. In the vicinity of major truck and bus terminals the scheduling of regular truck and bus runs may result in concentration of trucks during certain hours of the day. However, because of the delays caused by other traffic during peak hours, the schedules generally are made to avoid these hours.

For design of a particular highway, data on composition of traffic should be determined by traffic studies. Truck traffic (exclusive of light delivery trucks) should be expressed as a percentage of total traffic during the design hour, of total two-way traffic in the case of a two-lane highway; and of total traffic in the predominant direction of travel, in the case of a multilane highway.

Under urban interrupted-flow conditions the criteria differ somewhat. At important intersections, percentages of trucks during both the morning and evening peaks should be determined. Variations in truck traffic between the different movements at intersections may be substantial and may appreciably affect the geometric layout. The percentage of trucks during a particular hour of the day may vary considerably; it is advisable to count trucks for several peak hours that are considered representative of the 30th highest or design hour. A convenient value that appears reliable is the average of the truck traffic percentages for a number of weekly peak hours. For highway capacity purposes, local city-transit buses must be considered separately from other trucks and buses.

Future Projection of Traffic Demands

New highways or improvements of existing highways usually should not be based on current traffic volumes alone, but consideration should be given to the future traffic expected to use the facilities. A highway should be designed to accommodate the traffic that might occur within the life of the facility under reasonable maintenance.

It is difficult to define the life of a highway because major segments may have different lengths of physical life. Each segment is subject to variations in estimated life expectancy because of influences not readily subject to analysis such as obsolescence, unexpected radical changes in land use, and resulting changes in traffic volumes, pattern, and load. Right-of-way and grading might be considered to have a physical life expectancy of 100 years, minor drainage structures and base courses 50 years, bridges 25 to 100 years, and surfacing 10 to 30 years, all assuming adequate maintenance and no allowance for obsolescence. Bridge life may vary depending on the number of heavy, repetitious loads. Surfacing life varies widely, depending largely on initial expenditures and the repetition of heavy axle loads.

The assumption of no allowance for functional obsolescence is open to serious question. Two of the principal causes of obsolescence are reduction in the level of service provided and increase in accidents due to roadside interference, increase in number of at-grade intersections, and increase in traffic volumes thereon. Fortunately, these principal causes are minimized on freeways, which have full access control and no grade crossings. On highways other than freeways these sources of obsolescence are much less common in rural areas than in urban and suburban areas.

It is a moot question whether design should be based on life expectancy. The decision is greatly influenced by economics. A highway might be designed for traffic 50 years hence with the expectation that the surfacing would be restored in 25 years. However, if the added cost of such a design over one for a 25-year life expectancy is appreciable, it might be imprudent to make the added investment. The savings could earn interest for 25 years before the more costly highway is really needed, to say nothing of the cost of increased maintenance for the larger highway for 25 years. Also, most highways are adequate for traffic expected to develop much later than the design year because traffic volumes higher than design volumes can be accommodated with some inconvenience, such as reduction in speed, and because a highway is designed with an even number of lanes and without fractional lanes. For example a design ADT of 10,000 or 15,000 will require a four-lane highway that can handle two or three times that volume depending on several factors as later discussed. Thus, this highway would be adequate for traffic long after the design year and in many cases indefinitely.

It is well if the year for which traffic is used in design is no further ahead than that for which traffic can be estimated with reasonable accuracy. Many highway engineers believe the maximum period is in the range of 15 to 24 years. A period of 20 years is widely used as a basis for design. Estimating traffic beyond this period on a specific facility usually is not justified because of probable changes in the general regional economy, population, and land development along the highway, which cannot be predicted with any degree of assurance.

Even estimating traffic for a 20-year design period may not be appropriate for many reconstruction or rehabilitation projects. The projects may be developed on the basis of a shorter design period (5 to 10 years) because of the uncertainties of predicting traffic as well as funding constraints.

Speed

Speed is one of the most important factors to the traveler in selecting alternate routes or transportation modes. The value of a transportation facility in carrying people and goods is judged by its convenience and economy, which are directly related to its speed. The attractiveness of a public transportation system or a new highway are each weighed by the traveler in terms of time, convenience, and money saved. Hence, the desirability of rapid transit may well rest with how rapid it actually is. The speed of vehicles on a road or highway depends, in addition to capabilities of the drivers and their vehicles, upon four general conditions: the physical characteristics of the highway and its roadsides, the weather, the presence of other vehicles, and the speed limitations (either legal or because of control devices). Although any one of these may govern, the effects of these conditions are usually combined.

The objective in design of any engineered facility to be used by the public is to satisfy the demands for service in the safest and most economical manner. The facility should therefore accommodate nearly all demands with reasonable adequacy and also not fail completely under the severe or extreme load. In applying this principle to the design of highways, with particular reference to speed demands, provision should be made for a speed that satisfies nearly all drivers. Only a small percentage of drivers travel at extremely high speed, and it is not economically feasible to design for them. They can use the highway, of course, but must travel at speeds somewhat less than they consider desirable. On the other hand, the speed chosen for design should not be that used by drivers under unfavorable conditions, such as inclement weather, because the highway then would be unsafe for drivers under favorable conditions, and would not satisfy reasonable demands.

Operating Speed

Operating speed is the highest overall speed at which a driver can travel on a given highway under favorable weather conditions and under prevailing traffic conditions without at any time exceeding the safe speed as determined by the design speed on a section-by-section basis.

Design Speed

Design speed is the maximum safe speed that can be maintained over a specified section of highway when conditions are so favorable that the design features of the highway govern. The assumed design speed should be a logical one with respect to the topography, the adjacent land use, and the functional classification of highway. Except for local streets where speed controls are frequently included intentionally, every effort should be made to use as high a design speed as practicable to attain a desired degree of safety, mobility, and efficiency while under the contstraints of environmental quality, economics, aesthetics, and social or political impacts. Once selected, all of the pertinent features of the highway should be related to the design speed to obtain a balanced design. Above-minimum design values should be used where feasible, but in view of the numerous constraints often encountered, practical values should be recognized and used. Some features, such as curvature, superelevation, and sight distance, are directly related to, and vary appreciably with, design speed. Other features, such as widths of pavements and shoulders and clearances to walls and rails, are not directly related to design speed, but they affect vehicle speed, and higher standards should be accorded these features for the higher design speeds. Thus, when a change is made in design speed, many design elements of the highway are subject to change.

The design speed chosen should be consistent with the speed a driver is likely to expect. Where a difficult condition is obvious, drivers are more apt to accept lower speed operation than where there is no apparent reason for it. A highway of higher functional classification may justify a higher design speed than a less important facility in similar topography, particularly where the savings in vehicle operation and other operating costs are sufficient to offset the increased costs of right-of-way and construction. A low design speed, however, should not be assumed for a secondary road where the topography is such that drivers are likely to travel at high speeds. Drivers do not adjust their speeds to the importance of the highway, but to their perception of the physical limitations and traffic thereon.

The speed selected for design should fit the travel desires and habits of nearly all drivers. Where traffic and roadway conditions are such that drivers can travel at their chosen speed, there is always a wide range in the speeds at which various individuals operate their vehicles. A cumulative distribution of vehicles speeds has the typical S pattern when plotted as percent of vehicles versus observed speeds. The design speed chosen should be a high-percentile value in this speed distribution curve, i.e., nearly all inclusive of the typically desired speeds of drivers, wherever this is feasible.

The speed distribution curves in Figure II-21 illustrate the range in speed that should be considered in a determination of assumed design speed. A design speed of 70 mph should be maintained on freeways, expressways, and other major highways. This speed will ensure an adequate design if the speed restriction is removed. Also, a larger percent of the vehicles traveling at the faster speeds will be safely accommodated.

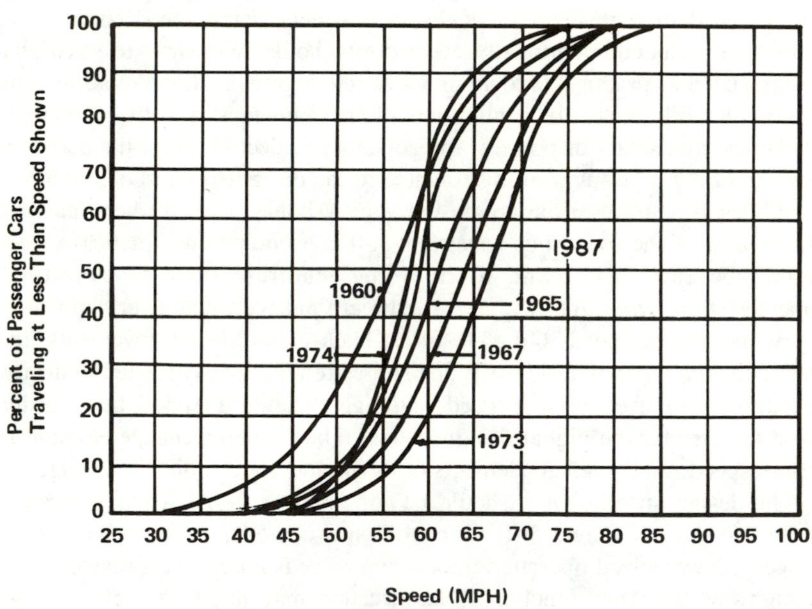

Figure II-21. Distribution of representative passenger car speeds on rural Interstate Highways.

These data lead to the conclusion that where physical features of the highway are the principal speed controls and where most drivers have been conditioned to operate near the speed limit, a top design speed of 70 mph would fit a very high-percentile speed. On a highway designed for this speed, a small percent of drivers might still operate at higher speed when volume is low and all other conditions are favorable. However, for a design speed of 50 mph, satisfactory performance could be expected only on certain highways. When the minimum design speed is used as the criterion, it is important to have the speed limit enforced during offpeak hours.

On many freeways, particularly in suburban and rural areas, a design speed of 60 mph or higher can be provided with little additional cost above that required for a design speed of 50 mph. The corridor of the main line may be relatively straight and the character and location of interchanges permit high-speed design. Under these conditions a design speed of 70 mph is desirable. Flat curvature and ample sight distance usually result in safer highways.

Generally, there is no design speed distinction between a ground-level, an elevated, or a depressed freeway. However, the operating characteristics on elevated freeways differ somewhat from those on depressed freeways. On an elevated highway, traffic exits the facility on downgrade ramps and enters on upgrade ramps. This condition is less desirable than the opposite one on a depressed highway because vehicles, particularly loaded trucks, entering the elevated freeway on an ascending grade require long distances to reach the running speed on the freeway (see section on "Running Speed"). Moreover, vehicles leaving the elevated freeway on a descending grade require additional braking to reach the running speed of the arterial street and consequently may tend to slow down on the through-traffic lanes in advance of the ramp terminal. Parallel deceleration lanes or longer ramp lengths and lesser grades are frequently used to reduce the problem of vehicles slowing on the main lanes. Nevertheless, running speeds on elevated freeways are apt to be slightly lower than those on depressed freeways of the same standards, especially when access points are closely spaced. In northern climates, elevated structures are subject to rapid freezing of precipitation as a result of their exposure and may require the use of lesser superelevation rates, which affect both running and design speed. Although speeds on viaducts are less than those on comparable depressed sections, the difference probably is small. Therefore, the appropriate design speeds of 50 to 70 mph apply to both elevated and depressed freeways.

With an overall range in design speeds of 20 to 70 mph, it has been found desirable for design speeds below 60 mph to use increments of 10 mph. Smaller increments show little distinction in design elements between one design speed and the next higher design speed, and larger increments of 15 to 20 mph cause too large a difference in design dimensions of features between any two design speeds. In some instances there may be an advantage in using intermediate values of 5-mph increments as 35, 45, or 55 mph for reasons of decreasing or increasing gradually from one design speed to another. Increments of 5 mph also may be pertinent to design of turning roadways, ramps, and low-speed roads. The use of 10-mph increments in design speed below 60 mph does not preclude the use of 5-mph increments for traffic control purposes, such as speed-zone signs. It has been found desirable to select 5-mph increments in design speeds above 60 mph. A 5-mph increment is more significant at high speeds because many of the important design elements related to design speed vary exponentially with the speed.

Although the selected design speed establishes the maximum degree of curvature and minimum sight distance necessary for safe operation, there should be no restriction on the use of flatter horizontal curves or greater sight distances where such improvements can be provided as a part of economic design. Even in rugged terrain an occasional tangent or flat curve may be fitting. These would not necessarily encourage drivers to speed up, but if a succession of them is introduced, drivers will naturally resort to higher speeds, and that section of highway should be designed for a higher speed. A substantial length of tangent between sections of curved alinement also is apt to encourage high-speed operation. In such cases a higher speed should be assumed and all geometric features, particularly that of sight distance on crest vertical curves, should be related to it.

A pertinent consideration in selecting design speeds is the average trip length. The longer the trip, the greater the desire for expeditious movement. In design of a substantial length of highway it is desirable where feasible to assume a constant design speed. Changes in terrain and other physical controls may dictate a change in design speed on certain sections. If so, the introduction of a lower design speed should not be affected abruptly but should be affected over sufficient distance to permit drivers to change speed gradually before reaching the section of highway with the lower design speed.

Where it is necessary to reduce design speed, many drivers may not perceive the lower speed condition ahead, and it is important that they be warned well in advance. The changing condition should be indicated by such controls as speed-zone signs and curve-speed signs.

On arterial streets the design speed control applies to a lesser degree than on other high-type highways. On rural highways or on high-type urban facilities, a certain percentage of vehicles are able to travel at near the safe speed determined by geometric design elements, but on arterial streets the top speeds for several hours of the day are limited or regulated to that at which the recurring peak volumes can be handled. Speeds are governed by the presence of other vehicles traveling en masse both in and across the through lanes and by traffic control devices rather than by the physical characteristics of the street. During periods of low-to-moderate volume, speeds are governed by such factors as speed limits, midblock frictions, and intersectional frictions. When arterial street improvements are being planned, the design speed should be considered with factors such as speed limits, physical and economic constraints, and the likely running speeds that can be attained during offpeak hours, which would influence the selection of the speed design.

Horizontal alinement generally is not the governing factor in restricting speeds on arterial streets. Proposed improvements generally are patterned to the existing street system, and minor horizontal alinement changes are commonly made at intersections. The effect of these alinement changes is usually minor because operation through the intersection is regulated by the type of traffic controls needed to handle the volume of cross and turning traffic. Superelevation may be provided at curves on arterial streets but in a different manner or degree than on the open highway. The wide pavement areas, proximity of adjacent development, control of cross slope and profile for drainage, and the frequency of cross streets and entrances all combine to result in pavements with low superelevation. Likewise, the width of lanes, offset to curbs, proximity of poles and trees to the traveled way, presence of pedestrians within the right-of-way, and nearness of business or residential buildings, singly and in combination, often nullify speed characteristics of a highway with good alinement and flat profiles. Yet, good alinement and flat profiles should always be strived for in the design of arterial streets, because safety and operating characteristics are improved, particularly during offpeak periods.

Topography can materially affect the choice of design speed on arterial streets. Many of our cities were developed along watercourses and include areas varying from gently rolling to mountainous terrain. The streets originally were constructed with minor grading to fit the topography. Because the arterial street usually is developed to fit an existing street, both through business and residential areas, it generally follows a varying vertical alinement. Once the design speed is determined, the proper sight distance should be assured at all crests. Profile conditions with long, continuous grades should also be designed with proper consideration of speeds of operation of mass transit and commercial vehicles. Extra lanes on the upgrades may be needed so that this portion of the route can match other portions in capacity and enable vehicles that can proceed at reasonable speed to pass slower moving vehicles.

Arterial streets should be designed and control devices regulated, where feasible, to permit running speeds of 20 to 45 mph. Lower speeds in this range are applicable for local and collector streets through residential areas and for the arterial streets through the more crowded business areas, while the higher speeds apply to the high-type arterials in the outlying suburban areas. For the arterial streets through the crowded business areas, coordinated signal control through successive intersections generally is necessary to permit even the lower speeds. Many cities have substantial lengths of streets controlled so as to operate at running speeds of 15 to 25 mph. At the other extreme, in suburban areas it is common experience on preferred streets to adopt some form of speed zoning or speed control to prevent hazardously high speeds. In these areas the infrequent pedestrian or occasional vehicle on a cross street may be unduly exposed to accident hazard caused by through drivers gradually gaining speed as the frequency of urban restrictions are left behind or by drivers retaining the speed of the open road as they enter the city. Thus, although through traffic should be expedited to the extent feasible, it may be equally important to establish a ceiling speed to reduce hazard and serve local traffic. Speed zones of 25 to 35 mph are common, increasing in 5-mph increments as the route extends outward to the edge of urban development.

A posted maximum speed, as a matter of practicability in determination and enforcement, is not the highest speed that might be used by relatively few drivers. Instead, it usually is about the 85-percentile speed value determined by observation of a sizable sample of vehicles. Such a value is within the "pace" or 10-mph speed range used by most drivers. Speed zones cannot be made to operate properly if arbitrarily determined or selected. They must be consistent with conditions along the street and subject to reasonable enforcement.

With running speeds of 20 to 45 mph it follows that pertinent design speeds for arterial streets range from 30 to 60 mph. Design speed for a major urban arterial highway depends largely on the frequency of at-grade intersections and the amount of cross and turning traffic but should generally be 40 to 60 mph. For the less important urban arterials the design speeds are somewhat lower. All elements of an urban arterial highway should be designed for safe operation at a speed of at least 30 mph.

The preceding paragraphs describe the basis for, and various considerations that need to be examined when selecting a design speed. It is evident from this discussion that there are meaningful differences between the design criteria applicable to low and high design speeds. Because of these distinct differences, it is necessary to establish certain limits. For application in this publication, the maximum limit for low or lower design speed is 40 mph, and the minimum limit for high speed is 50 mph. The resultant intermediate design speed range (approximately 10 mph) could have conditions approximating either those of high or low design speed. Such conditions would govern in the selection of the appropriate criteria. The specific basis for selection in the intermediate range is discussed in Chapters III through VII.

Running Speed

In design it is necessary to know actual vehicle speeds for traffic en masse to be expected on highways of different design speeds and various volume conditions. Speed of operation is one measure of the service that a highway renders, and it affords a means of evaluating road-user costs and benefits. The running speed is the speed of a vehicle over a specified section of highway, being the distance traveled divided by the running time (the time the vehicle is in motion).

One means of obtaining an equivalent average running speed on an existing facility where flow is reasonably continuous is to measure the spot speed. The average spot speed is the arithmetic mean of the speeds of all traffic at a specified point. For short sections of highway on which speed characteristics do not vary materially, the average spot speed may be considered as being representative of the average running speed. On longer stretches of rural highway, spot speeds measured at several points, where each represents the speed characteristics pertinent to a selected segment of highway, may be averaged (taking relative lengths into account) to represent the average running speed.

Average spot speeds, which generally are indicative of average running speeds, have been measured over a period of years in many States on highway sections of favorable alinement. The average speed, slowly increased over the years, then leveled out, and subsequently dropped with the advent of the 55-mph speed limit. Since then it has decreased very slightly on highways where the 55-mph speed limit is still in effect. On interstate highways with 65 mph speed limits the increase has been greater.

Experience on horizontal curves shows that speeds are lower than those on tangent alinement and that the difference between average spot speed and calculated design speed on such curves becomes less as the degree of curvature becomes greater. In this regard, it is generally accepted that a greater proportion of drivers operate near or at the design speed on highways with low design speed than on highways with high design speed. It is also known that some sections of low-design-speed highways are frequently overdriven, with an appreciable number of drivers exceeding the design speed.

Observed speeds of free-moving vehicles on horizontal curves indicate that low-design-speed curves yield an average spot speed close to the design speed; on high-design-speed curves the average speed is substantially below the design speed and approaches the average spot speed found on long stretches of tangent alinement. Because horizontal curvature is the principal factor related to design speed on open sections of rural highways and average spot speed approximates the average running speed for such conditions, a useful relation between the highway design speed and the average running speed (for low-volume conditions) may be established from these data. Comparing the observed average speeds with calculated design speeds, it is found that on sections having a 30-mph design speed, the average running speed is approximately 90 to 95 percent of the design speed.

The general relation between design speed and average running speed (the average for all traffic or component of traffic, being the summation of distances divided by the summation of running times; it is approximately equal to the average of the running speeds of all vehicles being considered) is illustrated in Figure II-22. The upper curve represents the conditions for low traffic volume as just described. As traffic volume increases on any highway, the average running speed decreases because of interference among vehicles. The curve labeled "Intermediate Volume" represents the relation between design speed and average running speed when the volume approximates the design service volume for rural highways. Should the volume exceed the intermediate level, the average running speed would be further lowered, and in the extreme case, where the volume is approaching the capacity of the highway, the speed of traffic is influenced more by congestion than by the design speed, especially where the design speed is above 50 mph. The relation between design speed and average running speed for very high traffic volumes is illustrated by the lower curve in Figure II-22. This curve is of academic interest only. It establishes a limiting condition for average running speeds but it is of little value in design. Highways should usually be designed to accommodate their traffic volumes without being subjected to the high degree of congestion represented by this curve.

A design that satisfies the requirements for average running speed at low volume is adequate for traffic using the highway when the volumes are higher and the speeds are lower. At low volumes about 50 percent of all vehicles travel at speeds within 5 mph of the average running speeds, as shown by the speed distribution curves in Figure II-21. For volumes in the intermediate range about 90 percent of all vehicles travel at or less than the average running speed representative of low volumes. For this reason low volumes control certain highway elements, such as lane and shoulder widths, treatment of intersection curves, and speed-change lanes.

Average running speed on a given highway varies somewhat during the day, depending primarily on the volume of traffic. Therefore, when reference is made to running speed it should be clear whether this speed is for peak hours or offpeak hours or whether it is an average for the day. The first two are of concern in design and operation; the latter is of importance in economic analyses.

Figure II-23 depicts the relationship between average travel speed and rate of flow under ideal conditions. Figure II-23 show minor differences between

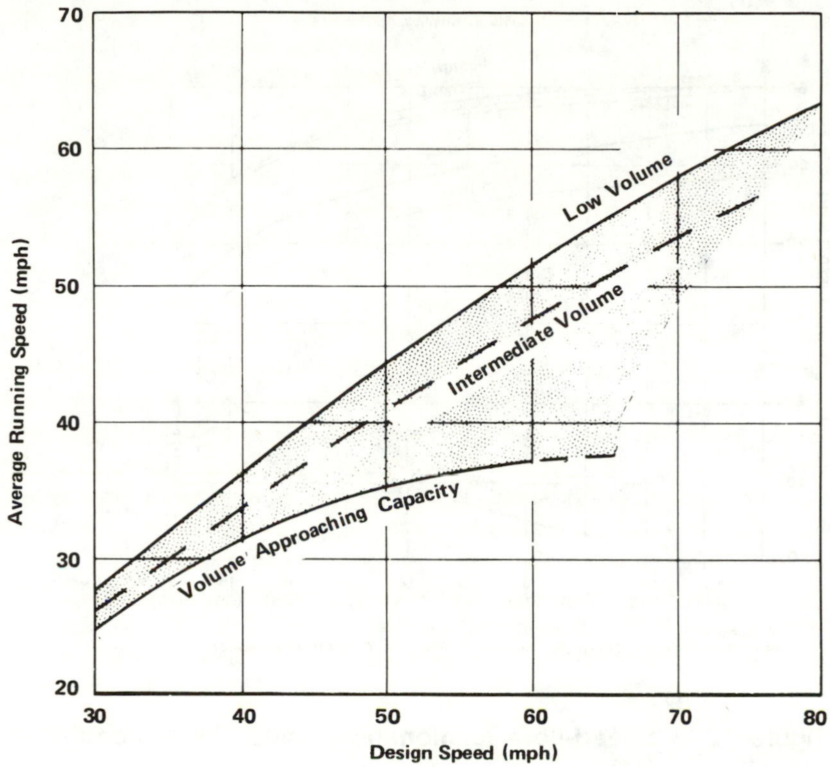

Figure II-22. Relation of average running speed and volume conditions.

four-, six-, and eight-lane freeways for a 70-mph design speed. Figure II-23 depicts two important characteristics:

1. There is a substantial range of flow over which speed is relatively insensitive to flow; this range extends to fairly high flow rates.
2. As flow approaches capacity, speed drops off at an extremely sharp rate.

The data for Figure 11-23 are taken from the *Highway Capacity Manual* (HCM) (**11**).

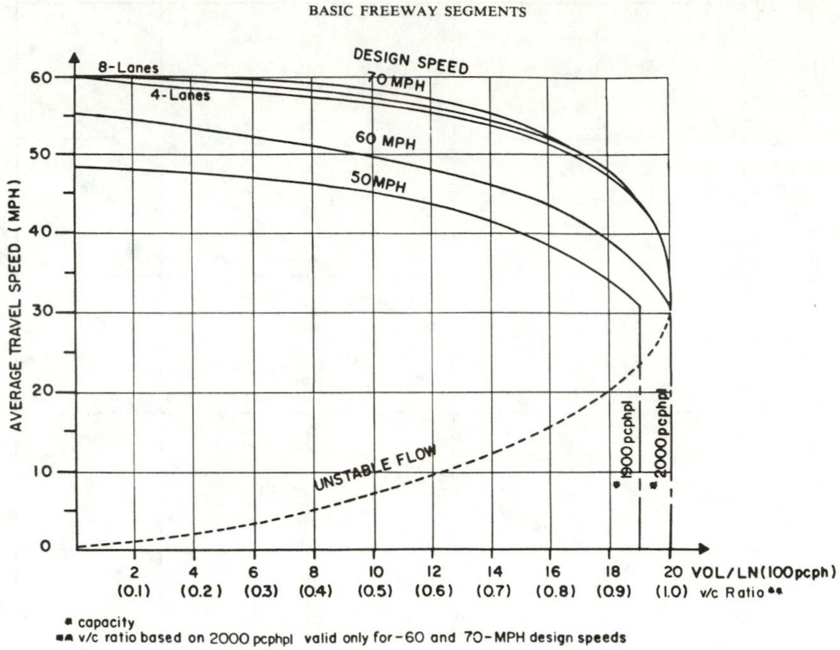

Figure II-23. Speed-flow relationships under ideal conditions. (11)

Traffic-Flow Relationships

Traffic flow on roadways is measured in terms of number of vehicles per hour, of average speed in miles per hour, and of traffic density in vehicles per mile. These three variables, volume, speed, and density, are all interrelated and are used to describe the predictable traffic-flow relationships.

Density, the number of vehicles per length of roadway, increases as vehicles crowd closer and closer together. As Figure II-21 shows, this increased crowding takes place when speed decreases and drivers can comfortably follow closer behind other vehicles. The solid line is an indication of what densities are possible for a given speed. As this solid line changes into the dashed line, flow approaches a condition where there is very little forward movement.

Traffic volumes also vary with density from points of zero flow to a maximum flow at a certain density. The points of zero flow represent either no vehicles at all or so many vehicles on the roadway that flow has jammed to a stop. The maximum flow is at the peak of the hyperbolic-shaped curve shown in Figure II-22. Again, the dashed line depicts drop in flow as the capacity of the roadway is exceeded and the corresponding rise in density as the vehicles are jammed together.

Interferences to traffic flow cause speeds to be reduced, vehicles to travel closer together, and density to increase. Interference may be caused by weather conditions, cross traffic, a disabled vehicle, or other marginal conditions. As one or more of these conditions cause more interference, the volume within certain limits can still be mantained but with further reduction in speed, closer vehicle spacing, and greater density. When interference becomes so great that despite closer vehicle spacing and greater density the average speed drops below that necessary to maintain stable flow, there is a rapid decrease in speed and traffic flow, and severe congestion sets in.

When traffic on a highway encounters interference that limits or reduces the roadway capacity in a single area, the result is a bottleneck. If the flow

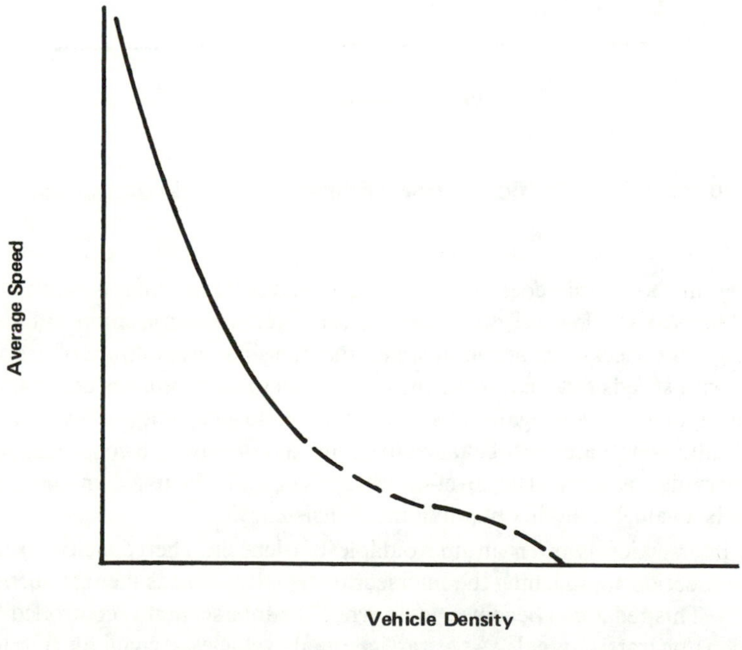

Figure II-24. Average speeds related to vehicle densities.

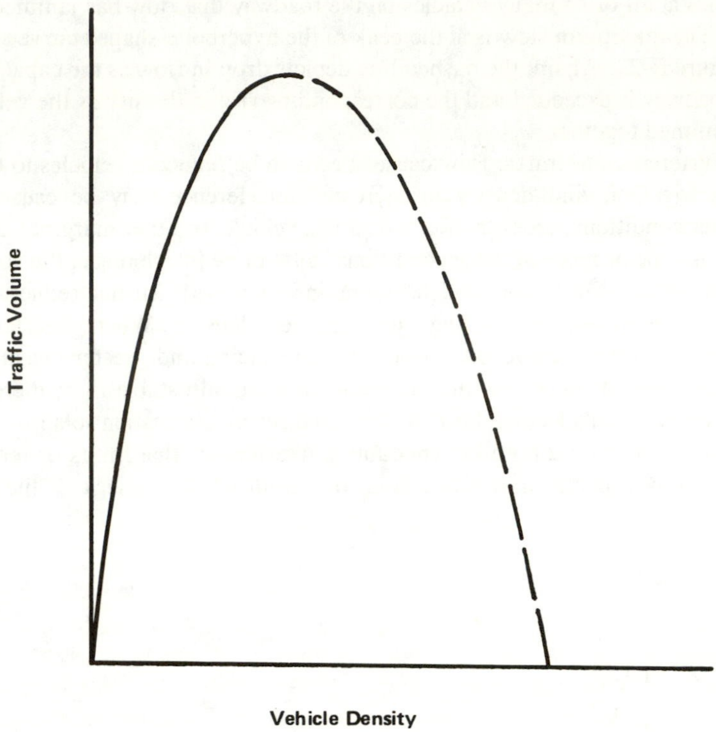

Figure II-25. Traffic volumes related to vehicle densities.

entering this bottleneck does not exceed its capacity, flow remains stable and no problems arise. However, when the upstream section carries more vehicles than the bottleneck can accommodate, the familiar breakdown of traffic results with speeds reduced to crawling and a queuing of vehicles accumulating until incoming flow again falls below the outflow capacity. To avoid bottleneck situations, care must be taken to design a roadway with consistent volume-carrying capacity. The level-of-service concept discussed in the next section is a valuable aid in obtaining this consistency.

An intersection is often an unavoidable bottleneck. The capacity on the roadway section approaching the intersection usually exceeds the intersection capacity. This reduction becomes acute when the intersection is controlled by stop signs or traffic signals. At a traffic signal, vehicles arriving during the red phase encounter a zero-capacity bottleneck. These vehicles form a queue until the green phase when the restraint is removed and the queue is discharged. If the incoming volume is too high, not all the vehicles in the queue can be discharged during the green phase, and there is a continuing build up of the queue.

Arrivals at the intersection are somewhat predictable in urban areas where the approaching vehicles are platooned by upstream signals. In suburban or rural locations, vehicle arrivals are often random. This randomness must be recognized in design of adequate cycle times, turn-lane storage lengths, and approach capacity.

At all bottlenecks where the traffic must slow down or stop, each vehicle and its occupants suffer a certain delay. Delay results in time loss (hence monetary loss) to the vehicle's occupants, increased fuel consumption, and additional air pollution.

HIGHWAY CAPACITY

General Characteristics

The term "capacity" is used to express the maximum hourly rate at which persons or vehicles can reasonably be expected to traverse a point or uniform section of a lane or a roadway during a given time period under prevailing roadway and traffic conditions. In the generic sense the term encompasses broader relations between highway characteristics and conditions, traffic composition and flow patterns, and the relative degree of congestion at various traffic volumes throughout the range from very light volumes to those equaling the capacity of the facility as defined above. The subject is discussed in the generic sense herein.

The following sections are a brief review of the principles and major factors concerning highway design capacity and summarize design values for the several basic rural and urban highway types. To determine the capacity for a particular highway design, the designer is referred for guidance to the *Highway Capacity Manual* (HCM) (**11**), which has been used as the basic reference for this section. The current edition, published in 1985, modifies some procedures contained in earlier editions, based on more recent research.

Application

Highway capacity information serves three general purposes:

1. It is used in transportation planning studies to assess the adequacy or sufficiency of existing highway networks to service current traffic and

to estimate the time in the future when traffic growth may overtake the capacity of the highways or perhaps may reach a level somewhat below capacity but one that would nevertheless result in an undesirable degree of congestion.

2. Highway capacity information is of vital concern in the design of highways. A knowledge of highway capacity is essential to the proper fitting of a planned highway to the requirements of traffic, both in the selection of the highway type and in determining dimensional needs such as number of lanes and minimum lengths for weaving sections.
3. Highway capacity information is utilized in traffic operation analyses for many purposes, but especially for isolating bottleneck locations (either existing or potential) and preparing estimates of operational improvements that may be expected to accrue from prospective traffic control measures or from spot alterations in the highway geometry.

The degree of exactness varies in the capacity data required for these several uses. For traffic operation analyses, in which the success of minor improvements may be measured in terms of a few vehicles per hour, a high degree of precision is desirable. For highway design a much lower order of precision suffices because the traffic data are frequently estimated for a period 10 to 20 years in the future and involve not only approximations of traffic volumes but also approximations of such factors as traffic composition and movement patterns. The discussion herein is intended to provide information in sufficient detail to ensure a reasonable balance between the standards for the highway being designed and the estimated future traffic so that the future operating conditions will not fall below an acceptable level. Those who need a greater accuracy than that obtainable from the contents herein should refer to the several reports on the subject, especially the HCM (**11**).

Capacity as a Design Control

Design Service Flow Rate Versus Design Volume

Design volume is the volume of traffic estimated to use a certain type of facility during the design year, which is a year usually 10 to 20 years ahead. Design volume is a product of the planning process. The derivation of the design hour volume (DHV) is discussed earlier in this chapter in the section "Traffic Characteristics and Flow."

Design service flow rate is the maximum hourly flow rate of traffic that a projected highway of designed dimensions would be able to serve without the degree of congestion falling below a preselected level as described hereinafter.

The objective in highway design is to create a highway of appropriate type with dimensional values and alinement characteristics such that the resulting design service flow rate is at least as great as the traffic flow rate during the peak 15-minute period of the design hour, but not enough greater as to represent extravagance or waste. Where this objective is accomplished, a well-balanced, economical highway system will result.

Measures of Congestion

In the tripartite system involving the roadway, the traffic using it, and the degree of congestion, the first two variables are measurable in exact units. For example, the roadway either is or is not a freeway with full control of access; its cross-section dimensions can be expressed in feet or other units of measure; the steepnesses of the grades are measurable slopes expressed in percent or feet per foot. Likewise, traffic flow can be expressed as numbers of vehicles per unit of time; traffic composition can be expressed in terms of percentages of vehicles of each class; its peaking characteristics and directional distribution can also be described. A scale of values for expressing the degree of congestion is, however, a much more elusive measure. Numerous indicators of overall service have been suggested, including safety, freedom to maneuver, the ratio of traffic volume to capacity (v/c), the highest overall speed at which a driver can travel under prevailing traffic conditions without exceeding the design speed for individual sections of the highway (i.e., operating speed), average running speed, and others. In the case of signalized intersections the stopped delay encountered by motorists is a commonly used measure of congestion.

For uninterrupted traffic flow, i.e., flow not influenced by signalized intersections, traffic operational conditions are defined by three primary measures: speed, volume (or rate of flow), and density. Density is the critical parameter describing traffic operations. It describes the proximity of vehicles to one another, and reflects the freedom to maneuver within the traffic stream. As density increases from zero, rate of flow also increases because more vehicles are on the roadway. While this is happening, speed begins to decline (due to the interaction of vehicles). This decline is virtually negligible at low densities and rates of flow. As density continues to increase, however, a point is reached at which speed declines precipitously. The maximum rate of flow is reached when the product of increasing density and decreasing speed results in reduced flow.

The maximum rate of flow for any given facility is its capacity. The density at which this occurs is referred to as *critical density*, and the speed at which it occurs is called *critical speed*. As capacity is approached, flow becomes more unstable because available gaps in the traffic stream are fewer. At capacity, there are no usable gaps in the traffic stream, and any perturbation from vehicles entering or leaving the facility, or from internal lane changing maneuvers, creates a disturbance that cannot be effectively damped or dissipated. Thus, operation at or near capacity is difficult to maintain for long periods of time without the formation of upstream queues, and forced or breakdown flow becomes almost unavoidable. For this reason, most facilities are designed to operate at volumes less than capacity.

For interrupted flow, such as that occurring on streets where traffic is controlled by signals, the highway user is not so much concerned with attaining a high travel speed as he is with avoiding lengthy stops or a succession of stops at intersection after intersection. Average stopped-time delay is the principal measure of effectiveness used in evaluating signalized intersections. Stopped-time delay is used because it is reasonably easy to measure, and is conceptually simple. It is a characteristic of intersection operations which is closely related to motorist perceptions of quality of traffic flow.

Relation Between Congestion and Traffic Flow Rate

Congestion does not mean a complete stoppage of traffic flow. Rather it can be thought of as a restriction or interference to normal free flow. For any given class of highway, be it freeway, highway with free access, or local street, congestion increases with an increase in rate of flow until the rate of flow is almost equal to the capacity of the facility, at which point congestion becomes acute. The gradual increase in congestion with increase in rate of flow is apparent no matter what measure is used as an index of congestion.

The relationship between average travel speed and traffic flow rate per lane on a freeway of high-type design is illustrated in Figure II-23. The average travel speed decreases to approximately 30 mph as the traffic flow rate increases to approximately 2,000 passenger cars per lane per hour. At this flow rate any minor disruption in the free flow of traffic causes traffic on the freeway to operate on a stop-and-go basis, with a resulting decrease in traffic flow rate that can be served. The data for Figure II-23 are taken from the HCM (**11**).

Highway sections where the paths of substantial volumes of traffic must cross within comparatively short distances are called weaving sections. Average running speed, and hence the degree of congestion, is a function not only of the volume of traffic involved in the weaving (crossing) movements but also of the distance within which the weaving maneuvers must be completed.

Weaving is covered under a separate subsection later in this chapter.

On arterial streets within the urban environment, average running speed varies only slightly with changes in traffic flow rate. However, delay at signalized intersections may increase dramatically as flow rates approach capacity. Therefore, greater degrees of congestion result in reduced overall travel speeds and higher average travel times.

Acceptable Degrees of Congestion

From the standpoint of the highway user it would be preferable for each individual user to have exclusive right to the highway at the time he finds occasion or need to use it. Moreover, he would prefer that all highways be of types that would permit speeds far in excess of those normally afforded by urban surface streets. However, users recognize that if others are to share the costs of transportation facilities, they are also entitled to share in their use. They will readily accept a moderate amount of congestion. Just what degree of congestion the motoring public is willing to accept as reasonable remains a matter of conjecture, but it is known to vary with a number of factors.

The average motorist understands in a general way that corrective measures to alleviate congestion may be more costly in some instances than in others and will generally accept a higher degree of congestion in those areas where improvements can be made only at a substantial cost. Motorists also accept more willingly a high degree of restraint if trips are short rather than long but are not satisfied with the type of operations that occur when the volume of traffic approaches the capacity of the facility.

From a highway administrator's point of view, the degree of congestion that highway users must be called upon to endure is geared to the availability of resources. Historically, funds have never been sufficient to meet all needs, and it has been a severe strain to advance the highway improvement program rapidly enough to prevent many facilities from becoming loaded beyond their capacity.

The matter of deciding on the degree of congestion that should be used as a goal in planning and designing highway improvements is resolved by weighing the desires of the motorists against the resources available for satisfying these desires. The degree of congestion that should not be exceeded during the design year on a proposed highway is realistically determined by the following: (1) estimates of the operating conditions that the majority of motorists will accept as satisfactory, (2) estimates of the highest standard of highway improvement that the governmental jurisdiction can support, and (3) reconciliation of the demands of the motorist and the general public with the means available to meet these demands.

This reconciliation of desires with available resources is an administrative process of high importance. First, the decision must be made as to the degree of congestion that should not be exceeded during the period when the facility is called on to carry its design traffic load. Then the design requirements for that facility (such as number of lanes) can be estimated from the relations discussed in the foregoing sections.

Principles for Acceptable Degrees of Congestion

No scientific method exists for deciding the maximum degree of congestion to be accepted as a basis for design. This decision lends itself neither to a modeling technique nor to the insertion of coefficients into a computer program. Nevertheless, some principles or guidelines will aid in arriving at a decision. These are itemized and discussed below.

1. *The highway should be so designed that when it is carrying the design volume, the traffic demand will not exceed the capacity of the facility even during short intervals of time.*

Conditions become intolerable for the motorist when the traffic demand exceeds the capacity of the street or highway. Moreover, when stoppages occur on highways (other than those controlled by signals), the flow rate is drastically reduced below the volume that can be served with traffic flowing at the optimum speed of 30 to 35 mph.

Stoppages will occur if the capacity is exceeded even for short intervals of time. Because traffic does not flow uniformly throughout a full hour, allowance should be made for peaking within the hour.

For freeways the traffic flow rate during short peak intervals is not likely to exceed the capacity of the freeway (usually about 2,000 passenger cars per lane per hour) unless the demand for the total hour exceeds about 1,600 to 1,800 cars per lane per hour. For highways without access control, under uninterrupted flow the total hourly volume generally will not exceed 85 or 90 percent of capacity. Capacity may vary from about 1,200 to 2,000 passenger cars per lane per hour, depending on a variety of factors.

Where traffic is controlled by signals at intersections, the relationship between delay and capacity may be extremely complex. It is possible to have unacceptably large delays and long queues where traffic demand approaches 75 to 85 percent of capacity. The reverse is also possible: an intersection approach where traffic demand equals capacity may have low delays if the signal cycle is short and/or signal progression is favorable.

2. *The design finds volumes per lane should not exceed the rate at which traffic can dissipate from a standing queue.*

This principle finds application primarily on freeways and other high-type multilane highways. If traffic on a freeway lane is stopped for any cause even momentarily, it cannot recover at a rate of 2,000 passenger cars per hour, which is the capacity of a lane flowing freely. If the traffic demand exceeds the rate at which cars can depart from the head of a standing queue, the queue will increase in length rather than be dissipated, even after the cause of the stoppage is removed.

The rate at which vehicles can depart from a standing queue is estimated by various authorities as being within the range of 1,500 to 1,800 passenger cars per lane per hour.

3. *Drivers should be afforded some choice of speed. The latitude in choice of speed should be related to the length of trip.*

This principle is applicable to all classes of streets and highways. The degree of freedom that should be afforded is a subjective determination.

If, for example, it would be determined that an average travel speed of about 50 mph is satisfactory for short freeway trips, the maximum flow rate that can be served by a high-type six-lane freeway is approximately 1,700 passenger cars per lane per hour.

The range of speed between the slowest and fastest driver would be about 15 mph. For longer trips, higher average travel speeds may be warranted, perhaps 5 mph higher than for short trips in densely developed areas. An average travel speed of, say, 55 mph can be achieved on freeways with traffic volumes up to about 1,500 passenger cars per lane per hour. The high cost of construction of urban freeways and the impact on the surrounding neighborhoods usually militates against an operating speed this high except in suburban areas as discussed subsequently under principle 6.

The foregoing relationships between speed and volume are approximate and apply to typical situations where a high design speed is used and where traffic lanes are 12 ft. wide. They will vary somewhat with the characteristics of the highway, such as design speed, number of lanes, and width of lane.

4. *Operating conditions should be such that they will provide a degree of freedom from driver tension that is related to or consistent with the length and duration of the trip.*

This principle may appear to be a corollary of the previous principle. However, principle 3 had to do with tensions growing out of impatience, whereas this one deals with tensions that develop from driving in a compact traffic stream at speeds that are recognizably too fast for safety but over which the individual is powerless to exercise control. If the individual reduces speed, the individual induces others to pass and cut in front, thereby reducing the gap that the individual was seeking to enlarge. Freeway travel at speeds of 40 to 50 mph under very high density conditions is a rather tense experience to many and is one that should not be endured if avoidable. There are no research data to support any recommendations as to the maximum length of time that drivers can or should endure travel under high density conditions, but it is commonly accepted that tensions build up with continued exposure.

Driver tensions associated with freeway densities of 42 passenger cars per mile per lane are generally considered acceptable for trips within most metropolitan areas. For long trips the mental concentration that is required and the tensions that develop while driving in such heavy traffic are excessive; consequently, lower volumes should be used for designing freeways that serve relatively long trips.

5. *Recognition should be taken of practical limitations that preclude the design of an ideal freeway.*

An ideal section of freeway is capable of carrying 2,000 passenger cars per lane per hour. More often than not it is necessary to compromise on design features to fit the freeway (or other arterial) within attainable right-of-way, or to economize on certain features such as curvature or lengths of speed-change lanes, or to locate interchanges closer to each other than would be desirable. In other words, it is usually impossible to design a section of freeway with uniform capacity throughout its length. Moreover, certain features that are required in order that the highway fulfill the definition of the term "ideal" are still under investigation. This much is known: Few freeways have a demonstrated capability for carrying as many as 2,000 passenger cars per lane per hour. On the other hand, there are few freeways that do not have the capability to handle as many as 1,700 passenger cars per lane per hour. Adverse effects of deficiencies in design of the types discussed above are not apparent unless the traffic volume exceeds about 1,500 passenger cars per lane per hour.

6. *The attitude of motorists toward adverse operating conditions is influenced by their awareness of the construction and rights-of-way costs necessary to provide better service.*

Highway users will accept poor operating conditions if they feel that the highway is the best that can be reasonably provided at the particular location. They recognize in a general way that highways are extremely costly in densely developed areas of high land values, in difficult terrain, and across major obstacles, such as navigable streams or harbors. Consequently, they will accept poorer operating conditions where highway costs are high than where there is no apparent reason for deficiencies that can be corrected at moderate expense. Because construction costs are frequently much higher in large cities than in small cities, the net result is that this principle tends to offset principle 3 insofar as the effect of length of trip within densely developed areas is concerned.

Reconciliation of Principles for Acceptable Degrees of Congestion

Freeways, under ideal design and traffic conditions, have a capacity of 2,000 passenger cars per lane per hour. Two-lane highways serving traffic in two directions have a capacity of 2,800 passenger cars per hour in both directions. Ideal conditions consist of uninterrupted flow, no marginal interference from vehicles or pedestrians, no commercial vehicles, 12-ft. lanes, adequate shoulders, high design speed, and no sight restrictions to overtaking and passing.

The capacities mentioned above may not be obtainable or desirable, depending on the design and desired use of the highway as explained in the foregoing principles. These principles point to the broad general conclusions that are summarized in the following paragraphs.

Freeways

For short trips, tolerance to congestion is governed to a considerable extent by driving tensions, with loss of travel time being of secondary importance—except that complete stoppages, or stop-and-go driving, are intolerable. These considerations suggest that the density of traffic on urban freeways preferably should not exceed 42 passenger cars per mile, per lane. Furthermore, if density does not exceed this level, there will be little difficulty from momentary stoppages, and minor substandard design features will have no noticeable adverse effect on operating conditions.

For longer metropolitan trips, travel time becomes more important to the user. Driver tensions associated with densities of 42 passenger cars per mile per lane, while not unbearable, are decidedly unpleasant. There are no criteria for fixing upon any definite value, but indications point to 30 passenger cars per mile per lane as resulting in an acceptable degree of congestion. This density will afford an average travel speed of about 55 mph with a flow rate of 1,500 passenger cars per lane per hour on a typical six-lane urban freeway.

For rural freeways, travel speed is the dominant consideration. On the basis of past experience, a density of 20 passenger cars per mile per lane will permit an average travel speed of about 57 mph on a typical four-lane freeway with a flow rate of 1,100 passenger cars per lane per hour.

Other Multilane Highways

Except where traffic is controlled by signals, measures of congestion are similar to those for freeways. Where the interference with traffic from marginal development is slight, the traffic densities that result in acceptable degrees of congestion on freeways may also be served by other multilane highways. This situation is notably true in rural areas. In urban areas the traffic volumes that can be served on other multilane highways at acceptable levels of congestion are generally somewhat lower than those for freeways, as will be discussed subsequently in this chapter.

Factors Other Than Traffic Volume That Affect Operating Conditions

The ability of a highway to serve traffic efficiently and effectively is influenced by the characteristics of the traffic and by the features of the highway.

Highway Factors

Few highways possess all the requirements for the ultimate in design characteristics. Although most modern freeways have adequate cross-section dimensions, many fall below the ideal with respect to design speed, weaving section design, and ramp terminal design. Inadequacies in these features will result in inefficient use of the remaining portions of the freeway.

On other classes of multilane highways the intersections at grade, even though unsignalized, often interfere with the free operation of traffic. Ribbon development with attendant driveways and interference from traffic entering and leaving the through-traffic lanes cause a loss in efficiency and lead to congestion and safety problems at relatively low volumes. The adverse effect, although readily apparent, is particularly difficult to quantify. Horizontal and vertical alinements are not always of a high standard, and it is sometimes necessary to compromise on cross-sectional dimensions. All of these conditions combine to cause the effects of congestion to be felt at lower traffic volumes than would be the case for highways designed with ideal features and protected by full control of access.

For urban streets with signalized intersections at relatively close intervals, the volumes of traffic that could otherwise be served are reduced as a result of loss of "go," or "green," time.

For a highway that is deficient in some of its characteristics and where the traffic stream is composed of a mixture of traffic rather than passenger cars only, compensatory adjustment factors need to be applied to the traffic flow rates acceptable as design values for ideal highway conditions. These adjustments are necessary to obtain volumes of mixed traffic that would result in the minimum acceptable operating conditions on the highway under consideration. The HCM (**11**) identifies significant highway features that may have an adverse effect on operating conditions. It provides factors and outlines procedures for determining the traffic volumes that can be served by highways that are not ideal in all respects. Features that could result in a highway being less than ideal in its characteristics include lanes and shoulders of substandard width, steep grades, low design speed, and the presence of interchange facilities including intersections, ramp terminals, and weaving sections. The HCM (**11**) should be referred to for a discussion of these features and their effects on operating conditions. However, the discussion therein relative to horizontal alinement, weaving sections, and ramp terminals is supplemented and amplified below.

Alinement

For traffic traveling at any given speed, the better the roadway alinement, the more traffic it can carry. It follows that congestion will generally be felt at lower volumes if the design speed is low than if the design speed is high. The highway must be subdivided into sections of consistant geometric design characteristics for analysis using the HCM (**11**) techniques. A single limiting curve or steep grade in an otherwise gentle alinement will thus be identified as the critical feature limiting the capacity of the roadway.

Weaving Sections

Weaving sections are highway segments where the pattern of traffic entering and leaving at contiguous points of access results in vehicle paths crossing each other. Where the distance in which the crossing is accomplished is relatively short in relation to the volume of weaving traffic, operations within the highway section will be congested. Some reduction in operating efficiency through weaving sections can be tolerated by highway users if the reduction is minor and the frequency of occurrence is not high. It is generally accepted that a reduction in operating speed of about 5 mph below that for the highway as a whole can be considered a tolerable degree of congestion for weaving sections.

Operating conditions within weaving sections are affected by both the length and width of the section as well as by the volume of traffic in the several movements. These relationships are discussed later in this chapter and in the HCM (**11**).

Ramp Terminals

Ramps and ramp terminals are features that can adversely influence operating conditions on freeways if the demand for their use is excessive or if their design is deficient. When congestion develops at freeway ramp junctions, some through vehicles avoid the outside lane (lane 1) of the freeway, thereby adding to the congestion in the remaining lanes. Thus, if there are only two lanes in one direction, the efficiency per lane is not as high on the average as that for three or more lanes in one direction.

The loss in efficiency is a function of the volume of traffic entering or leaving at the ramps, the distance between points of entry and exit, and the geometric layout at the terminals. Too little is known of these separate variables to permit a quantitative assessment of their effect when taken individually. Their combined effect is accounted for by levying a uniform assessment against lane 1, regardless of the causes or extent of interference at individual locations.

Apart from the effect on through traffic, the traffic that uses the ramps is exposed to a different form of congestion, which does not lend itself to measurement in terms of travel speed, delay, or driver tension. The degree of congestion in these movements is related to the total volume of traffic in lane 1 of the freeway in the vicinity of the ramp junction; that is, the combined volume of through traffic using lane 1 and the volume of traffic using the ramp. There is as yet no scientific basis for determining the maximum flow rate of traffic that can be served at a ramp terminal without producing excessive congestion. Such a determination must be arrived at intuitively. Experience indicates that the flow rates given in Table II-3 are acceptable for the conditions indicated.

The HCM (**11**) provides procedures for estimating volumes of through traffic in lane 1 of a freeway just upstream of an entrance or an exit ramp for various combinations of highway and traffic conditions. At an entrance, the

Type of Terminal	Type of Location	Maximum Acceptable Flow Rate in Lane 1 of Freeway
Entrance	Urban	1,450
	Rural	1,000
Exit	Urban	1,500
	Rural	1,050

Table II-3. Service flow rate in lane 1 of freeway in vicinity of ramp terminals.

total of the lane 1 flow rate and the ramp flow rate can be compared to the maximum acceptable flow rate in lane 1 of the freeway given for entrances in Table II-3.

For the exit, the lane 1 flow rate determined from the HCM (**11**) procedures is compared directly to the appropriate flow rates in Table II-3. When lane 1 flow rates are less than the flow rates obtained from Table II-3, the desired operating conditions are met. However, if lane 1 flow rates exceed Table II-3 flow rates, the junction will operate, but at a lower level of service, and a congested condition may result.

Traffic Factors

The usual traffic stream is composed of a mixture of vehicles: passenger cars, trucks, buses, and occasionally recreational vehicles. Traffic also does not flow at a uniform rate throughtout the hour, day, season, and year. Consideration must be given to these two variables, composition of traffic and fluctuations in flow, in deciding upon volumes of traffic that will result in acceptable degrees of congestion (service) and also upon the period of time over which the flow should extend.

The effect that trucks and buses have in contributing to congestion on highways is discussed in the HCM (**11**). Detailed procedures are outlined for converting volumes of mixed traffic to equivalent volumes of passenger cars. Factors for converting trucks, buses, and recreational vehicles to equivalent passenger car traffic on multilane highways are given in Table II-4. These values are applicable to long sections of highway. For operation on specific individual grades, reference should be made to the HCM (**11**).

Equivalency Factors, E			
Type of	Terrain		
Vehicle	Level	Rolling	Mountainous
Truck	1.7	4	8
Bus[a]	1.5	3	5
Recreational vehicles	1.6	3	4

[a] Separate consideration not warranted in most problems; use only where bus volumes are significant.

Table II-4. Factors for converting trucks, buses, and recreational vehicles to equivalent passenger cars (multilane highways)

The accepted unit of time for expressing flow rate is a 1-hour period. It is customary to design highways with a sufficient number of lanes and with other features that will enable the highway to accommodate the forecasted DHV for the design year, which is frequently 20 years from the date of construction.

Because flow is not uniform throughout an hour, there are certain periods within an hour during which congestion is worse than at other times. The HCM (**11**) considers operating conditions prevailing during the most congested 15-min period of the hour to establish the service level for the hour as a whole. Accordingly, the total hourly volume that can be served without exceeding a specified degree of congestion is equal to or less than 4 times the maximum 15-min count. The factor used to convert the rate of flow during the highest 15-min period to the total hourly volume is the peak hour factor (PHF). The PHF may be described as the ratio of the total hourly volume to the number of vehicles during the highest 15-min period multiplied by 4. It is never greater than 1.00 and is normally within the range of 0.75 to 0.95. Thus, for example, if the maximum flow rate that can be served by a certain freeway without excessive congestion is 4,200 vehicles per hour during the peak 15-min period, and further, if the PHF is 0.80, the total hourly volume that can be accommodated at that service level is 3,360 vehicles, or 80 percent of the traffic flow rate during the worst congested 15-min period.

Levels of Service

Techniques and procedures for adjusting operational and highway factors to compensate for conditions that are other than ideal are found in the HCM (**11**). It is desirable that the results of these procedures be made adaptable to highway design.

For anyone not intimately familiar with the level-of-service concept, it is difficult to visualize the operating conditions that characterize levels-of-service A through F. Table II-5 presents a brief description of the operating characteristics for each level of service and type of highway. The description of each of the several levels of service in terms of density, average travel speeds, and volume-to-capacity ratios (v/c), or in the case of intersections at grade, in terms of stopped delay, should aid the designer and the decision-maker in determining which level of service should be provided. The HCM (**11**) presents a more thorough discussion on the level-of-servicc concept. The division points between levels-of-service A through F were determined subjectively. The HCM (**11**) contains no recommendations for the applicability of the levels of service in the design of highways. Choice of level of service is properly left to the user of the HCM (**11**). The guidance in the foregoing discussion will enable the user of the HCM (**11**) to link the appropriate degrees of congestion to specific levels of service. The relationship between highway type and location and the level of service appropriate for design is summarized in Table II-6. This relationship is derived from the descriptions of the level of service, the accompanying traffic volumes, and the controls and criteria for acceptable degrees of congestion as outlined earlier in this discussion. As may be fitting to the conditions, the highway agency should strive to provide the highest level of service feasible. In heavily developed sections of metropolitan areas, conditions may necessitate the use of level-of-service D for freeways and arterials, but such use should be rare and at least level-of-service C should be strived for. For some urban and suburban highways, conditions may necessitate the use of level-of-service D.

Design Service Flow Rates

The traffic flow rates that can be served at each level of service are termed service flow rates. Once a level of service has been identified as applicable for design, the accompanying service flow rate logically becomes the design service flow rate, implying that if the traffic flow rate using the facility exceeds that value, operating conditions will fall below the level of service for which the facility was designed.

Level Service	Controlled Access Highways	Multilane Rural without Access Control	Two Lanes	Urban and Suburban Arterials
A	Free flow. Average travel speeds at or greater than 60 mph. Service flow rate of 700 passenger cars per hour per lane.	Average travel speed 57 mph or greater. Under ideal conditions, flow rate is limited to 700 passenger cars per lane per hour or 36 percent of capacity.	Average travel speeds of 50 mph or higher. Most passing maneuvers can be made with little or no delay. Under ideal conditions, a service flow rate of 420 passenger car per hour, total two-way, can be achieved.	Average travel speed of about 90 percent of free flow speed. Stopped delay at signalized intersections is minimal.
B	Reasonably free flow conditions. Average travel speed at greater than 57 mph. Service flow rate not greater than 1,100 passenger cars per hour per lane.	Reasonably free flow. Volume at which actions of preceding vehicle will have some influence on following vehicles. Flow rates will not exceed 54 percent of capacity of 1,100 passenger vehicles per lane per hour at a 53 mph average travel speed under ideal conditions.	Average travel speeds of 55 mph or higher. Flow rates may reach 27 percent of capacity with continuous passing sight distance. Flow rates of 750 passenger cars per hour, total two-way, can be carried under ideal conditions.	Average travel speeds drop due to intersection delay and inter-vehicular conflicts, but remain at 70 percent of free flow speed. Delay is not unreasonable.
C	Operation stable, but becoming more critical. Average travel speed of 54 mph, service flow at 77 percent of capacity or not more than flow rate of 1,550 passengar cars per hour per lane.	Stable flow to a flow rate not exceeding 71 percent of capacity of 1,400 passenger cars per lane per hour, under ideal conditions, maintainiing at least a 50-mph average travel speed.	Flow still stable. Average travel speeds of 52 mph or above with total flow rate under ideal conditions equal to 43 percent of capacity with continuous passing sight distance or 1,200 passenger cars per hour total two-way.	Stable operations. Longer queues at signals result in average travel speeds of about 50 percent of free flow speeds. Motorists will experience appreciable tension.

Table II-5. Level-of-service characteristics by highway type.

Level Service	Controlled Access Highways	Multilane Rural without Access Control	Two Lanes	Urban and Suburban Arterials
D	Lower speed range of stable flow. Operation approaches instability and is susceptible to changing conditions. Average travel speeds approx. 46 mph. Service flow rates at 93 percent of capacity. Flow rate cannot exceed 1,850 passenger cars per hour per lane.	Approaching unstable flow at flow rates up to 87 percent of capacity or 1,750 passenger cars per hour at an average travel speed of about 40 mph under ideal conditions.	Approaching unstable flow. Average travel speeds approximately 50 mph. Flow rates, two-direction, at 64 percent of capacity with continuous passing opportunity, or 1,800 passenger cars per hour total two-way under ideal conditions.	Approaching unstable flow. Average travel speeds down to 40 percent of free flow speed. Delays at intersections may become extensive.
E	Unstable flow. Average travel speeds of 30–35 mph. Flow rate at capacity or 2,000 passenger cars per hour per lane under ideal conditions. Traffic stream cannot dissipate even minor disruptions. Any incident may produce a serious breakdown.	Flow at 100 percent of capacity or 2,000 passenger cars per lane per hour under ideal conditions. Average travel speeds of about 30 mph.	Average travel speeds in neighborhood of 45 mph. Flow rate under ideal conditions, total two-way, equal to 2,800 passenger cars per hour. Level E may never be attained. Operation may go directly from Level D to Level F.	Average travel speeds 33 percent of free flow speed. Unstable flow. Continuous backup on approaches to intersections.
F	Forced flow. Freeway acts as a storage for vehicles backed up from downstream bottleneck. Average travel speeds range from near 30 mph to stop-and-go operation.	Forced flow, congested condition with widely varying volume characteristics. Average travel speeds of less than 30 mph.	Forced, congested flow with unpredictable characteristics. Operating speeds less than 45 mph.	Average travel speed between 25 and 33 percent of free flow speed. Vehicular backups, and high approach delays at signalized intersections.

Table II-5. Level-of-service characteristics by highway type. (Con't.)

Once a level of service has been selected, it is desirable that all elements of the roadway are consistently designed to this level. This consistency of design service flow rate results in near-constant freedom of traffic movement and operating speed, and flow interruptions from bottlenecks can be avoided.

The HCM (**11**) supplies the analytical base for design calculations and decisions, but the designer must use his or her judgment to select the proper level of service. The designer may possibly select a design service flow rate less than the anticipated demand for certain recreational routes or for environmental or land use planning reasons. Or, the guidance given in Table II-6 may be chosen for the appropriate values in particular locations.

Whether designing at intersection, an interchange, an arterial, or a freeway, the selection of the desired level of service must be carefully weighed because the adequacy of the roadway is dependent on this choice.

	Type of Area and Appropriate Level of Service			
Highway Type	**Rural Level**	**Rural Rolling**	**Rural Mountainous**	**Urban and Suburban**
Freeway	B	B	C	C
Arterial	B	B	C	C
Collector	C	C	D	D
Local	D	D	D	D

NOTE: General operating conditions for levels of service (source: Ref. 11):

A – free flow, with low volumes and high speeds.

B – reasonably free flow, but speeds beginning to be restricted by traffic conditions.

C – in stable flow zone, but most drivers restricted in freedom to select their own speed.

D – approaching unstable flow, drivers have little freedom to maneuver.

E – unstable flow, may be short stoppages.

Table II-6. Guide for selection of design levels of service.

Weaving Sections

Weaving sections occur where one-way traffic streams cross by merging and diverging maneuvers. The principal types of weaving sections are illustrated in Figure II-26. Weaving sections are designed, checked, and adjusted so that the level of service is consistent with the remaining highway. The design level of service of a weaving section is dependent on its length, number of lanes, acceptable degree of congestion, and relative volumes of individual movements. Large-volume weaving movements usually result in considerable friction and reduction in speed of all traffic. Further, there is a definite limit to the amount of traffic that can be handled on a given weaving section without undue congestion. This limiting volume is a function of the distribution of traffic between the weaving movements, the length of weaving section, and the number of lanes therein.

Weaving sections may be considered as simple or multiple. Figure II-27 shows a simple weaving section in which a single-entrance junction is followed by a single-exit junction. A multiple-weaving section consists of two or more overlapping weaving sections. A multiple weave may also be defined as that portion of a one-way roadway that has two consecutive entrance junctions followed closely by one or more exit junctions, or one entrance junction followed closely by two or more exit junctions, as shown in Figure II-28. Multiple weaving sections occur frequently in urban areas where there is need for collection and distribution of high concentrations of traffic. For further information concerning the operation and analysis of simple and multiple weaving sections refer to the HCM (**11**).

The weaving section should have a length and number of lanes based on the appropriate level of service as given in Table II-6. The HCM presents an equation for predicting the average running speed of weaving and non-weaving traffic based on roadway and traffic conditions. Level of service criteria for weaving sections are based on these average running speeds.

Multilane Highways Without Control of Access

These highways may be treated as freeways if major crossroads are infrequent or are separated in grade and if marginal development is so sparse as to generate little interference. Even on those highways where the interference is currently only marginal, the designer should be alert to the fact that by the design year the interference may be extensive unless the control of access is regulated. In most cases the designer should assume liberal crossroad and business improvements over the design life of the facility.

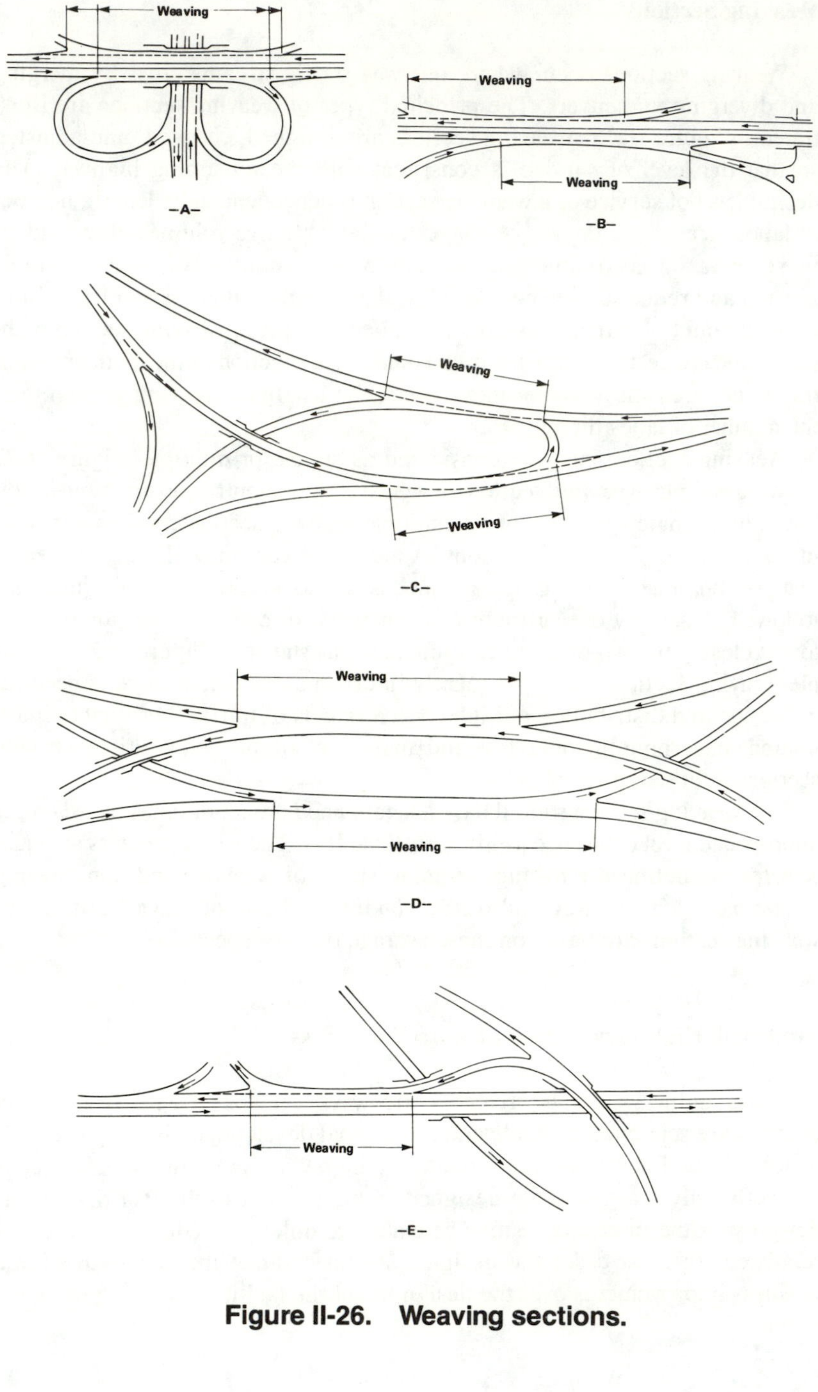

Figure II-26. Weaving sections.

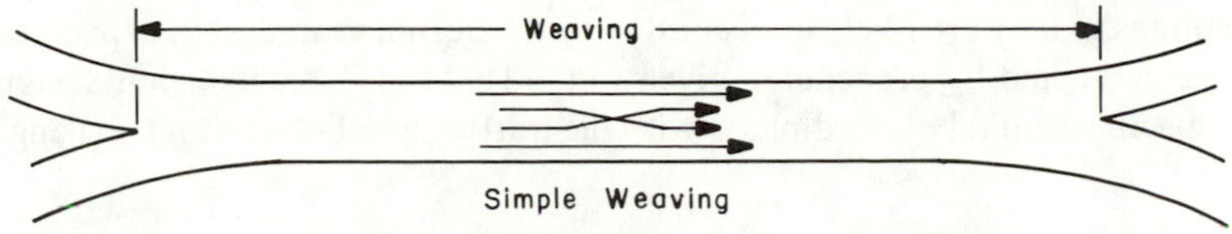

Figure II-27. Simple weaving section.

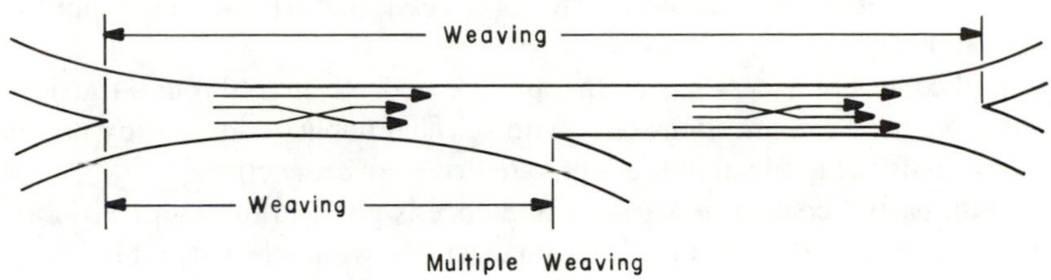

Figure II-28. Multiple weaving.

Where there are major crossroads or where marginal development results in more than slight interference, the facility should be treated as a multilane highway without access control.

Arterial Streets and Urban Highways

Arterial streets and urban highways are a troublesome class of highway from the standpoint of establishing design service flow rates. The reason for this is that the level of service provided by such facilities does not remain stable with the passage of time and tend to deteriorate in an unpredictable manner. The capacity of an arterial is generally dominated by the capacity of its individual signalized intersections. Level of service of an arterial is a function of average overall travel speed over a section of the arterial.

Signalized Intersections

Design capacities of signalized intersections are subject to a very large number of variables. To the extent that these variables can be predicted for conditions as they are likely to occur during the design year, design capacities can be estimated by procedures given in the HCM (**11**). Intersection design and spacing should be coordinated with the traffic signal design and phasing.

ACCESS CONTROL

The regulated limitation of access is called access control, which is achieved through the regulation of public access rights to and from properties abutting the highway facilities. These regulations generally are categorized as full control of access, partial control of access, and driveway and approach regulations.

Full control of access means that preference is given to through traffic by providing access connections only with selected public roads and by prohibiting crossing at grade and direct private driveway connections.

With partial control of access, preference is given to through traffic to a degree that, in addition to access connections with selected public roads, there may be some crossing at grade and some private driveway connections. Full or partial access control generally is accomplished by legally obtaining right-of-access from the abutting property owners (usually at the time of purchase of the right-of-way) or by the use of frontage roads.

Driveway or approach regulations may be applied even though no control of access is obtained. Each abutting property is permitted access to the street or highway; however, the location, number, and geometrics of the access points may be governed by the regulations.

The principal advantages of control of access are the preservation or upgrading service and safety. Fully controlled access highways or streets have no at-grade crossings and have carefully designed access connections. A highway or street with partial control of access has similar characteristics to one with full control but includes some crossroad intersections at grade and some carefully selected and predetermined land service connections.

The principal operational or functional difference between a street or highway with or without control of access is the degree of interference with through traffic by other vehicles or pedestrians entering, leaving, and crossing the highway. With control of access, entrances and exits are located at points best suited to fit traffic and land-use needs and are designed to enable vehicles to enter and leave safely with a minimum of interference with through traffic. Vehicles are prevented from entering or leaving elsewhere so that, regardless of the type and intensity of development of the roadside areas, a high quality of service is preserved and the accident potential is lessened. On streets or highways where there is no control of access and roadside businesses develop, interference from the roadside can become a major factor in reducing the capacity, increasing the accident potential, and eroding the mobility function that the facility was designed to provide.

Control of driveways and roadside development is an integral part of access control. If access points are adequately spaced and entering and exiting volumes are light, the street or highway functions efficiently. If access points are numerous and entering and exiting volumes are heavy, the capacity and safety of the facility are reduced. However, trade-offs must be made to serve through traffic development and land-use needs, particularly those involving industrial, commercial, or other high-use establishments.

A new rural highway without full control of access generally has few intersections or roadside businesses and the accident rate approximates that of a freeway. With time, businesses and intersections increase and the accident rate may double or triple as shown in Figure II-29. Over time, the freeway accident rate remains the same or may even decrease slightly. Full control of access is the most important single safety factor that may be designed into new highways. It is the key factor in the outstanding safety record of freeways (**12**).

For highways without full control of access, techniques are available to provide additional safety (**13**). The most obvious technique limits the number of driveways and intersections. Other techniques suggest changes in the design and operational measures both within and outside the highway right-of-way. The techniques aim to meet these functional objectives: (1) limit the number of conflict points, (2) separate basic conflict areas, (3) reduce maximum deceleration requirements, and (4) remove turning vehicles or queues for certain portions of the through lanes.

Some degree of access control should be included in the development of any street or highway, particularly a new facility where the likelihood of commercial development exists. The type of street or highway to be built should be coordinated with the local land use plan to ensure that the desired degree of access control can be maintained through local zoning ordinances or subdivision regulations. The access control may range from minimum driveway regulations to full control. Thus the extent and degree of access control that are feasible or ultimately possible are significant factors in defining the type of street or highway.

THE PEDESTRIAN

General Considerations

A pedestrian is any person afoot, and involvement of pedestrians in traffic is a major consideration in highway planning and design. Pedestrians are a part of every roadway environment, and attention must be paid to their presence in rural as well as urban areas. The urban pedestrian, being far more prevalent, more often influences roadway design features than the rural pedestrian does. Because of the demands of vehicular traffic in congested urban areas, it is often extremely difficult to make adequate provisions for pedestrians. Yet this must be done, because pedestrians are the lifeblood of

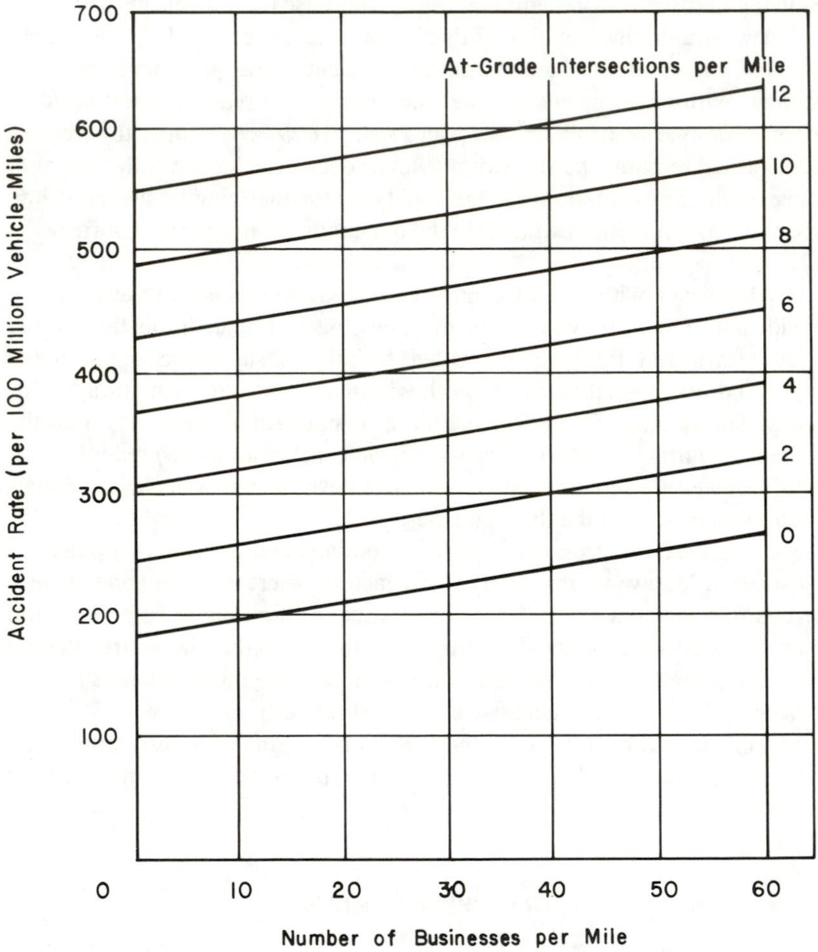

Figure II-29. Accident rate on 4-lane divided non-Interstate highways by number of at-grade intersections per mile and number of businesses per mile.

our urban areas, especially in the downtown and other retail areas. In general, the most successful shopping sections are those that provide the most comfort and pleasure for pedestrians. Pedestrian facilities include sidewalks, crosswalks, traffic control features, special walkways found on some portions of freeway right-of-way, and curb cuts (depressions) and ramps for the handicapped. They are also parts of bus stops or other loading areas, grade separations, and the stairs or escalators related to these facilities.

General Characteristics

To effectively plan and design pedestrian facilities, it is necessary to describe the typical pedestrian. The pedestrian will most likely not walk over a mile to work or over 0.5 mi to catch a bus, and about 80 percent of the distances traveled will be less than 3,000 ft (**14**). The typical pedestrian is a shopper about 50 percent of the time that he or she is a pedestrian and a commuter only about 11 percent of the time. As a consequence, a plot of pedestrian volumes peaks at about noon rather than at the commuter peak times. A typical plot of the hourly fluctuations in pedestrian volumes on a city street is shown in Figure II-30.

The pedestrian volumes are influenced by such transient conditions as weather or, in specific locations, advertised sales.

Pedestrian actions are less predictable than those of motorists. Many pedestrians consider themselves outside the law in traffic matters, and in many cases pedestrian regulations are not fully enforced. This makes it difficult to design a facility for safe and orderly pedestrian movement.

Pedestrians tend to walk in a path that represents the shortest distance between two points; thus along streets they often cross at midblock and fail to stay in crosswalks.

Pedestrians also have a basic resistance to changes in grade or elevation when crossing roadways and tend to avoid using special underpass or overpass pedestrian facilities. Also, pedestrian underpasses may be potential crime areas.

An important factor in relation to pedestrian accidents is age. Very young pedestrians are often careless in traffic from ignorance and exuberance, whereas the elderly appear inattentive or defiant toward motor vehicles and drivers. Pedestrian accidents can also be related to the lack of adequate sidewalks, which forces pedestrians to share the pavement with motorists.

The following have been suggested as measures with potential to aid the elderly pedestrian:

- Assess alternate designs at pedestrian crossings to protect elderly pedestrians.

- Lower walking speed criterion, particularly at wide signalized intersections.
- Provide refuge islands at wide intersections.
- Provide lighting at locations which require multiple information gathering and processing, and eliminate glare sources.
- Consider the traffic control system in the context of the geometric design to assure compatibility and to provide adequate advance warning of situations that could surprise or adversely affect the safety of elderly drivers and pedestrians.
- Use enhanced standard traffic control devices.
- Provide oversized, retroreflective signs with suitable legibility.

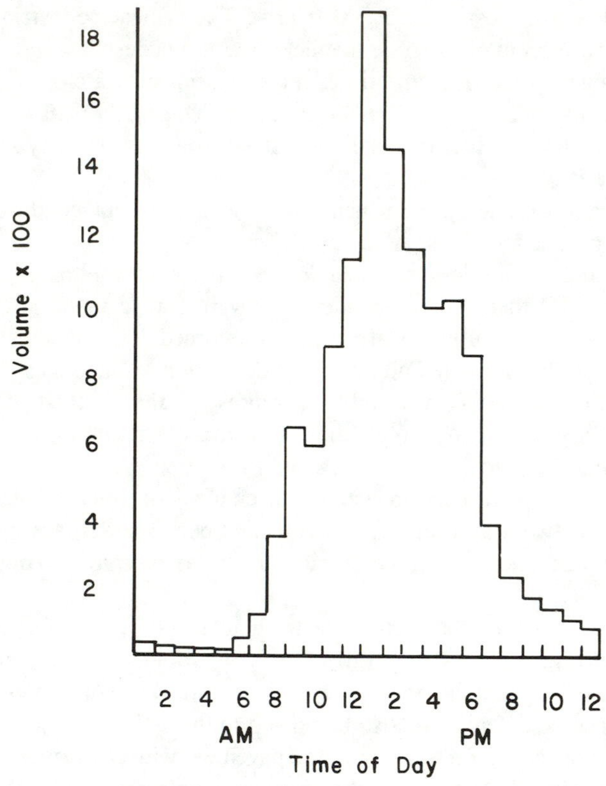

Figure II-30. Weekday average hourly pedestrian volumes on city street. (15)

- Consider increasing sign letter size to accommodate individuals with decreased visual acuity.
- Use properly located signals with large signal indications.
- Provide enhanced markings and delineation.
- Use repetition and redundancy.

Physical Characteristics

Body Area

The physical dimensions of the human body are reflected in the design of pedestrian facilities. For the design of sidewalks, stairs, or transit-loading areas, a knowledge of the width and depth of the body is most useful. Studies have shown that nearly all adult males have a shoulder width less than 20.7 in. and a depth of less than 13 in. For design purposes the area of a body is approximated by an ellipse 24 in. wide and 18 in. deep (**16**). These minimum dimensions apply only to situations where individuals are forced into close proximity. If a greater degree of comfort or mobility is required, a larger body area per person must be assumed.

As desired pedestrian mobility increases, space to go around a slower moving person or to avoid an oncoming or crossing person must be available. As curbs, stairs, or other irregularities to the ground surface are encountered, at least one pace or one stair-step separation is required for normal walking.

Walking Rate

There is a broad range of walking speeds among pedestrians. The rates when crossing a street are significant in design. Average walking speeds range from approximately 2.5 to 6.0 ft/sec with the *Manual on Uniform Traffic Control Devices* (MUTCD) (**3**) assuming a normal walking rate of 4.0 ft/sec. Older people will generally be in the slower part of this range.

Walking rates are faster at midblock than at intersections, are faster for men than women, and are affected by steep grades. Air temperature, time of day, trip purpose, and ice and snow all affect the pedestrian walking rate. Age is the best identified cause for slower walking rates, and in areas where there are many older people, a rate of 3 ft/sec should be considered.

Walkway Capacities

Figure II-31 shows that walking speeds decrease as the pedestrian density of the walkway increases. Figures II-32 and II-33 show that at an optimum speed and density, the walkway will carry the largest volume. The effective width used for walkway calculations must be reduced where parking meters, hydrants, newsstands, litter barrels, utility poles, or similar obstructions preclude the use of the full walkway. Walkway capacity calculations for sidewalks, for stairs, and for the effect of traffic signals involve different procedures as discussed in the following material.

Sidewalks

To quantify the relative mobility of the pedestrian and his conflicts with other pedestrians that influence his walking speed, maneuvering room, and feeling of comfort, degrees of level of service have been introduced (**17**). Similar to the level-of-service concept discussed earlier in this chapter and in the HCM (**11**), levels of service (A to F) reflect increasing crowding and decreasing freedom of movement. These are based on the available area per person and are defined in the following way (**18**):

Level-of-service A provides an average of 130 ft^2 per person or more, allowing each person to choose his desired walking speed and to avoid conflicts with other pedestrians.

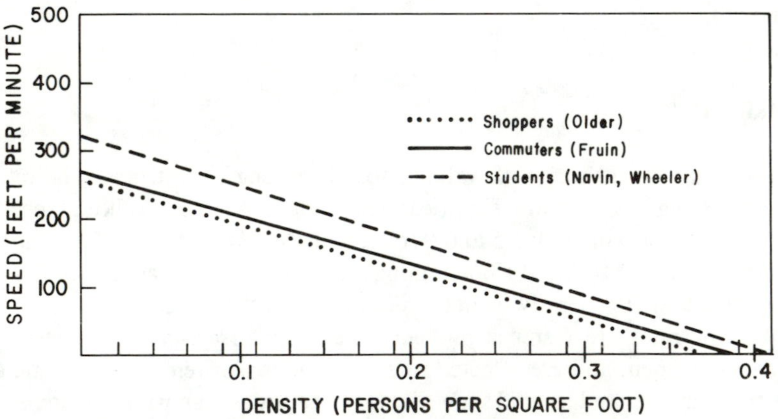

Figure II-31. Relationships between pedestrian speed and density.

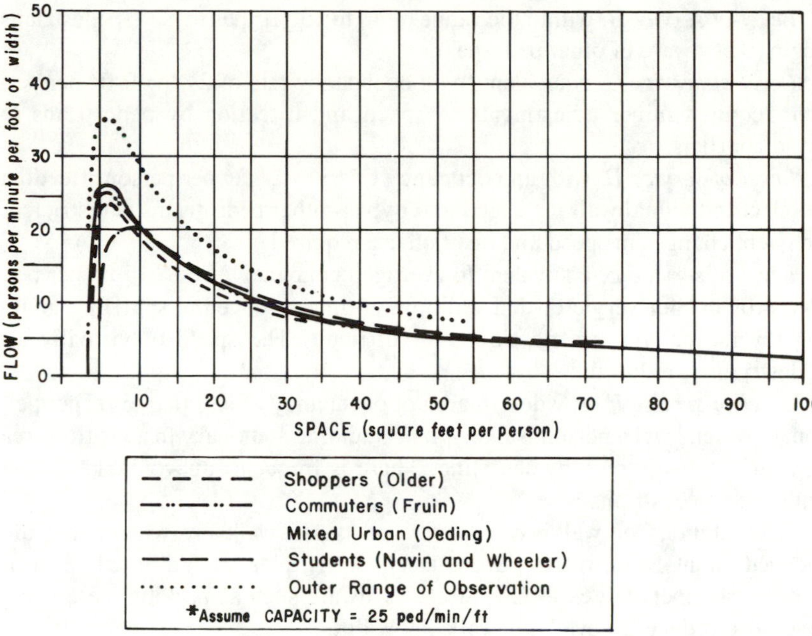

Figure II-32. Relationships between pedestrian flow and space.

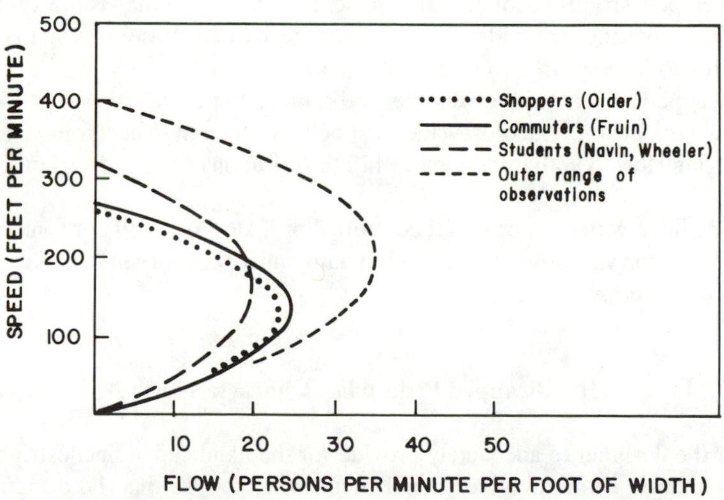

Figure II-33. Relationships between pedestrian speed and flow.

Level-of-service B, within the range of 40 to 130 ft^2 per person, pedestrians begin to be aware of other pedestrians.

Level-of-service C, equivalent to an occupancy rate of 24 to 40 ft^2 per person, requires minor adjustments to speed and direction by pedestrians to avoid conflicts.

Level-of-service D, with an occupancy of 15 to 24 ft^2 per person, freedom to select individual walking speed and bypass other pedestrians is restricted. Frequent changes in speed and position are required.

Level-of-service E, equivalent to average occupancy of 6 to 15 ft^2 per person, provides for very crowded walking, at times reduced to shuffling, making reverse or cross-traffic flow very difficult. The speed of virtually all pedestrians is reduced.

Level-of-service F provides an area of occupancy of less than 6 ft^2 per person, at which level a person is likely to be standing stationary in a waiting area or is able to walk only by shuffling. There is frequent, unavoidable contact with other pedestrians.

Computations of walkway capacity must use walkway widths that are reduced about 18 in. if there are adjacent walls, plus an additional 18 in. if window shoppers are expected. Street hardware such as parking meters and poles also reduce the available walkway width.

Intersections

When pedestrians encounter an intersection, there is a major interruption. The sidewalk should provide sufficient storage area for those waiting to cross plus area for cross traffic to pass.

Once pedestrians are given the walk indication, the crosswalk width becomes important. The crosswalk must be wide enough to accommodate the pedestrian flow in both directions within the duration of the pedestrian signal phase.

If the intersection is not signal controlled or if stop signs do not control the through motor vehicular traffic, pedestrians must wait for sufficient gaps in the traffic to cross.

Handicapped Pedestrian Characteristics

For the designer to adequately provide for the handicapped pedestrian, he must be aware of the range of impairments to expect so that the design can provide for them. In this way the mobility of this sector of our society may be greatly enhanced. The designer is cautioned to adequately review all local and national requirements to assume proper compliance with all rules and regulations. For further details see Chapter IV section "Curb Cut Ramps for the Physically Handicapped."

Walking Impairment

Ambulation difficulties range from persons who walk unassisted but with difficulty, to persons who require aid from braces, canes, or crutches, to persons confined to wheelchairs. Stairs, curbs, and raised channelizing islands are the major roadway obstructions to these pedestrians. Design modifications must provide ramps rather than stairs or curbs. The front wheels of a wheelchair are very sensitive to obstacles, and any bump may impair the progress of a wheelchair or increase the possibility of the overturning of a wheelchair moving forward.

Vision Impairment

Pedestrians with very limited vision require special consideration. Intersections are the major threat to their safety. Complicated crossings such as those at channelized intersections can be assisted by textured guide strips. Curb depressions for wheelchairs make locating the curb line difficult for the blind and a texturing or other indication of this area may be of assistance. Because crossings are often made by the sound of traffic, caution must be used when exclusive turn phases or other unusual traffic movements may confuse the blind pedestrian.

Mental Impairment

People with mental impairments are unable to drive and are therefore often captive pedestrians. To help ensure the correct response from these pedestrians, including young children, the pedestrian signals or other pedestrian-related facilities must be simple, straightforward, and consistent in their meaning.

Conclusions

Design of urban highways should include consideration of methods of increasing efficiency of operations in areas of heavy pedestrian-vehicular conflict. Major forms are the following: (1) provision of pedestrian subways or overcrossings, (2) the conversion from two-way to one-way street operation, (3) elimination of turns, (4) provision of separate signal phases for pedestrians, and (5) elimination of some crosswalks. These and other pedestrian considerations are covered in subsequent chapters.

BICYCLE FACILITIES

The bicycle has become an important element for consideration in the highway design process. Fortunately, it is practical for most of the mileage needed for bicycle travel to be comprised of the street and highway system much as it presently exists. At the same time, measures such as the following, which are generally of low capital intensity, can enhance considerably a route's safety and capacity for bicycle traffic:

- Paved shoulders.
- Wide outside traffic lane (14-ft minimum) if no shoulder.
- Bicycle-safe drainage grates.
- Adjusting manhole covers to the grade.
- Maintaining a smooth, clean riding surface.

At certain locations or in certain corridors, it is appropriate to supplement further the existing highway system by providing specifically designated bikeways (for either exclusive or nonexclusive bicycle use). To provide adequately for bicycle traffic, it is necessary to be familiar with bicycle dimensions, operating characteristics, and requirements. These factors determine acceptable turning radii, grades, and sight distance. In many instances, design features of separate bike facilities are controlled by the adjoining roadway, so that even then consideration of bicycles is an essential element of the design of the highway itself. For further guidance, the designer is referred to the latest edition of the AASHTO *Guide for Development of New Bicycle Facilities* (**20**).

SAFETY

Attention to highway safety has been emphasized by the Congress of the United States as well as other national committees concerned with safety. In July of 1973, after hearings on highway safety design and operation were conducted by subcommittees of the House Committee on Public Works, the following mandate was published by the Committee:

"Whose responsibility is it to see that maximum safety is incorporated into our motor vehicle transportation system? On this, the subcommittee is adamant. It is the responsibility of Government and specifically those agencies that, by law, have been given that mandate. This responsibility begins with the Congress and flows through the Department of Transportation, its Federal Highway Administration, the State Highway Departments and safety agencies, and the street and highway units of counties, townships, cities and towns. There is no retreating from this mandate, either in letter or in spirit." (**21**)

The national significance of vehicular accidents is shown in Table II-7 (**21, 22**).

These statistics have not been ignored. The design values given within this text have safety as their primary objective. Emphasis on safety has also come from Congress with passage of the Highway Safety Act of 1966, from the Federal Highway Administration (FHWA) by adoption of the AASHTO publi-

	1967	1972	1975	1986
Accidents				
Total accidents (rounded)	13,700,000	17,000,000	15,600,000	17,700,000*
Fatal accidents	44,500	48,800	39,900	41,100**
Nonfatal disabling injury accidents	1,200,000	1,400,000	1,200,000	1,940,476
Property damage accidents (including nondisabling injury accidents)	12,500,000	15,600,000	15,300,000	15,700,000
Deaths				
Total	52,900	52,000	45,600	46,100
Motor Vehicle Traffic				
Vehicle registered	96,887,000	119,384,000	139,000,000	176,191,339
Motor vehicle miles (millions)	965,132	1,268,343	1,315,000	1,838,240
Rates				
Motor vehicle deaths per 100 million vehicle miles	5.5	4.5	3.5	2.51
Motor vehicle accidents per 100 million vehicle miles	1,420	1,340	1,190	963

* NATIONAL SAFETY COUNCIL
** U.S. DEPT. OF TRANSPORTATION

Table II-7. Highway accident statistics for four typical years.

cation, *Highway Design and Operational Practices Related to Highway Safety* (**21**), from the report Enhancing Highway Safety in an Age of Limited Resources resulting from the TRB-conducted symposium sponsored by AASHTO and others in 1981 (**23**) and from many state and local governments. Emphasis on fuel conservation has had an additional effect. At least a portion of the drop in highway fatalities has resulted directly from the nationwide 55-mph speed limit.

It is seldom that an accident results from a single cause. There are usually several influences affecting the situation at any given time. These influences can be separated into three groups: the human element, the vehicle element, and the highway element. Although this policy is primarily concerned with highway characteristics and design, the psychological inferences are ever present. An error in perception or judgment or a faulty action on the part of the driver can easily lead to an accident. Highways should be designed so that a driver needs to make only one decision at a time and is never surprised by an unexpected situation where he must make a decision but does not have enough time. The number of accidents increases with an increase in the number of decisions required by the driver. Standardization in highway design features and traffic control devices plays an important role in reducing the number of required decisions; by this means the driver becomes aware of what to expect on a certain type of highway.

The most significant design factor contributing to safety is the provision of full access control. The beneficial effect of this element has been documented in reports of the cooperative research study (**24**) between the FHWA and 39 State highway agencies. One of the principal findings of this study is that the absence of access control invariably increased the accident rate.

The study (**24**) shows that accidents, injury, and fatality rates on Interstate highways are between 30 and 76 percent of comparable rates of conventional highways existing before the Interstate highways were opened to traffic. No other single design element can claim comparable savings.

A study (**25**) of rural roads shows the relationship of accidents to the number of access points. Traffic conflicts were defined in the study as the number of minor road intersections and principal access driveways to abutting property between major road intersections which were used as the break points between study sections. (See Figure II-34.)

The principle of full control of access is invaluable as a means for preserving the capacity of arterial highways and of minimizing accident hazards, but the principle does not have universal application. Highways without control of access are essential as land service facilities, and the design features and operating characteristics of these highways need to be carefully planned so that they will reduce conflicts and minimize the interference between vehicles to the maximum extent and still meet the needs of the highway users.

Speed is often a contributing factor in accidents, but it must be related to conditions. It is improper to say that any given speed is safer than another for all combinations of the many kinds of drivers, vehicles, highways, and local conditions. For a highway with particularly adverse roadway conditions, a relatively low speed may result in fewer accidents than a high speed, but this fact does not necessarily mean that all accident hazards can be compensated by

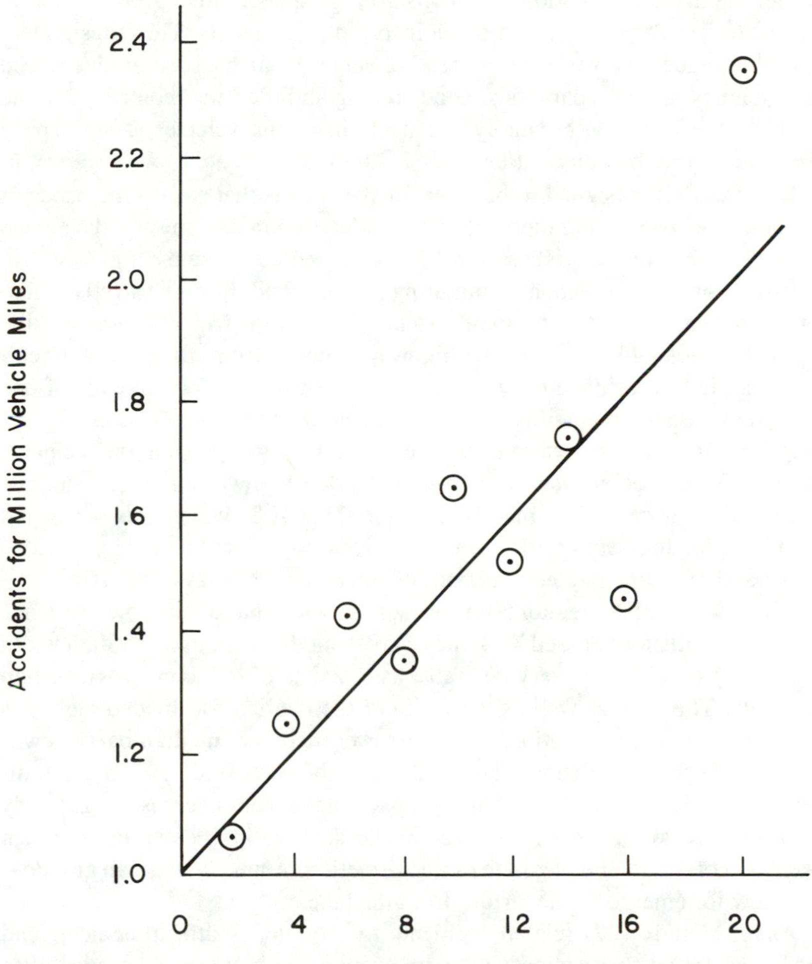

Figure II-34. Accident rate versus traffic conflicts.

low speeds. Likewise, on good roads vehicles traveling at relatively high speed may have lower accident involvement rates than vehicles traveling at lower speeds, but it does not necessarily follow that yet a higher speed would be even safer. The safest speed for any highway depends on design features, road conditions, traffic volumes, weather conditions, roadside development, spacing of intersecting roads, cross-traffic volumes, and other factors. Accidents are not related as much to speed as to the range in speeds from the highest to the lowest. Regardless of the average speed on a main rural highway, the greater the drivers' deviations from this average speed, either above or below, the greater are their chances of being involved in accidents. Thus, design features that reduce the variance in speed of vehicles (such as flat grades, speed change lanes, grade separations, good signing and marking) contribute to the safety of the highway. Normally, accidents involving vehicles at high speed are more severe than those at low speed.

Highways are designed to be safer for the speeds that satisfy the needs of most of the drivers using the facility. Consideration in design should be given to the type and characteristics of the drivers expected to use the highway. Trip purposes (such as recreation, commuting to work, and through travel) are factors affecting the design to some extent. Allied with trip purposes are the types of vehicles likely to use the highway, ranging from all passenger vehicles to a high percentage of heavy commercial vehicles. Where trips of one type predominate, the facility can be designed to fit the specific needs.

A study (**26**) on the effect of the Interstate Highway System on accidents shows a lower accident rate on four-lane divided highways than on four-lane undivided highways. This may be seen in Table II-8, which was developed from data for highways within Interstate Highway corridors during periods before and after opening new sections of Interstate Highways to traffic.

A highway with a median 50 ft. or more in width has a very low incidence of head-on collisions caused by vehicles crossing the median. A median width of up to 75 or 100 ft. is very desirable as a means of reducing cross-median accidents. The type as well as the width of the median for divided highways should be given consideration. With narrower medians, median barriers will eliminate head-on collisions, but at the cost of some increase in same-direction accidents because recovery space has been decreased. Properly designed median barriers minimize vehicle damage and lessen the accident likelihood of traffic moving in the same direction. A narrow median also does not allow for emergency departure from the lane.

Another study (**27**) related accidents to shoulder width, alinement, and grade and found that accident rates on sections with curves or grades were much higher than on level tangent highway sections and that the rates were highest on roads having combinations of sharp curves and steep grades.

This study which was limited to two-lane rural roads, lends strong support

Type and Class of Highway	Accident Rates for Highways With Average Daily Traffic Volumes					
	Under 2,000	2,000 to 3,999	4,000 to 7,999	8,000 to 15,999	16,000 to 31,999	32,000 and Over
	Accidents per 100 Million Vehicle Miles					
Before opening section of Interstate System						
Four-lane undivided	—	294	472	353	—	—
Four-lane divided	—	157	111	—	—	—
After opening sections of Interstate System						
Four-lane undivided	316	392	354	284	241	—
Four-lane divided	176	136	188	244	—	—
Interstate System	99	103	153	104	161	253

Table II-8. Accident rates on four-lane undivided and four-lane divided highways before and after opening sections of Interstate System in the same traffic corridor.

to the postulate that straight, level rural roads without intersections or significant numbers of private driveways are the safest highways within their general class. The few accidents that occur on straight level rural roads without intersections do not represent a stable source of information, however. There are wide variations between similar sections and between different years for the same section. An apparent correlation between one element such as shoulder width, for example, and accident rates is almost certain to be clouded by random variations of the accident pattern.

Accidents are likely to occur where drivers are called on to make decisions under circumstances where their vehicles are unable to respond properly as, for example, where a truck is descending a grade. It would be logical to expect more accidents on grades and curves than on level tangent highways where decisions are called for less frequently and vehicles are fully responsive. Design with tangent alinement can be overdone, however. On extremely long tangents, drivers have a tendency to completely relax, especially after driving on a congested highway before entering a freeway. On some freeways there has been concern over the number of accidents that occur when the driver apparently goes to sleep. It is considered highly desirable to provide gentle curvature and avoid a fixed cross section for long tangent sections. This can be handled by varying the median width, using independent roadway designs, and taking advantage of the terrain wherever feasible.

With the improvements in the elements of design for alinement, grade, and traveled way, the general area of roadside design has become of increasing importance. Accidents involving single vehicles running off the road constitute more than one-half of all fatal accidents on freeways.

When a vehicle leaves the roadway, the driver no longer has the ability to fully control the vehicle. Any object in or near the path of the vehicle becomes a contributing factor to the severity of the accident. The concept of the safer or forgiving roadside should not be viewed as a byproduct of the application of safety criteria to each element but as a planned segment of the total engineering for the highway. Chapters V and VI of reference **21** present an overview of the AASHTO policy in this area, which is reflected throughout this book in the geometric design criteria and elements.

Basic to the concept of the forgiving roadside is the provision of a clear recovery area. Studies have indicated that on high-speed highways, a relative level traversable width of 30 ft from the edge of the traveled way permits about 80 percent of the vehicles leaving the highway to be safely stopped or returned to the roadway. Even though the 30 ft is not a magic number and the application of engineering judgment is necessary, this figure has been used extensively as a guide for recovery zones. Within this area, two major items must be controlled by the designer: roadside slopes and unyielding obstacles. NCHRP Report 158 (**28**) summarizes research in the selection of safe roadside cross sections. The AASHTO's *Roadside Design Guide* (**29**) also discussed the effects slope and other topographic features have on the effectiveness of recovery areas. The AASHTO recommendations for treatment of roadside obstacles on existing highways consider them in the following priority:

1. Elimination of the hazard.
2. Relocation of the hazard to a point where it is less likely to be struck.
3. Use of breakaway devices to reduce the hazard.
4. Selection of a cost-effective traffic barrier (longitudinal barrier or crash cushion) to reduce accident severity.

The same priorities could be followed in design, particularly in upgrading and rehabilitation projects.

The design of guardrails and barrier systems (a hazard in themselves) has become a subject of considerable research. NCHRP Report 118 (**30**) and 129

(**31**), and AASHTO's *Roadside Design Guide* (**29**), are some of many published reports that deal with this subject. Of particular concern is the treatment of the end section of the guardrail or barrier.

Highway design engineers must recognize the dynamic development currently underway in the entire area of roadside design. Although this publication has attempted to deal with the subject in the most current possible manner, the ongoing research and implementation projects will undoubtedly offer newer and better results in the future. Highway design engineers should endeavor to use the most current acceptable information in their designs.

Communication with the motorist is probably one of the most complex problems of the design engineer; one of the best tools that he has for this purpose is the MUTCD (**3**), which depicts the national standards developed for all signing, signalization, channelization, and marking for all highways in the United States. The main message of the MUTCD (**3**) is the importance of uniformity.

Highway users are dependent on traffic control devices (signs, markings, and signals) for information, warning, and guidance. So great is the dependence that uniform high-quality devices are necessary for safe, efficient use and public acceptance of any highway regardless of its excellence in width, alinement, and structural design.

Any traffic control device should have the following characteristics: (1) fulfill an important need, (2) command attention, (3) convey a clear, simple meaning, (4) command respect of road users, and (5) give adequate time for response. In addition, devices that control or regulate traffic must be sanctioned by law.

Four basic principles are employed to ensure that these requirements are met: design, placement, maintenance, and uniformity. Consideration should be given to these principles of employment of traffic control devices during the design stage to ensure that the required number of devices can be held to a minimum and that those that are necessary can be properly placed.

The operation of a motor vehicle is a full-time occupation, particularly in congested areas. A driver should be able to operate his vehicle with minimum distractions. Advertising or other roadside signs should not be placed where they would interfere with or confuse the meaning of standard traffic control devices. Advertising signs with bright colors or flashing lights are especially objectionable in this respect. Lights shining toward a driver can blind him partially or fully for various periods depending on individual eye capability. Bright lights, in effect, can form a curtain hiding what is ahead and thus endanger motorists and pedestrians.

A large number of accidents on rural highways occur at intersections. Several studies have been made at intersections with varying conditions, and the results vary according to conditions studied. Factors to be considered are total traffic volume, percent of cross traffic, turning movements, type of traffic control, and type of intersection, including the design of the crossroad and the utilization of islands and channelization in the design.

Various studies indicate improvements in safety by channelization of intersections, introduction of safety refuge islands and sidewalks for pedestrians, lighting, signing, and traffic control devices. These have been taken into account in the development of this policy and resulting guides for geometric design.

A viable safety evaluation and improvement program, as part of the overall highway improvement program, is a necessity. The identification of safety hazards, the evaluation of the effectiveness of alternative solutions, and the programming of available funds to the most effective uses are of primary importance. The safety of the traveling public must be reflected throughout the highway program: in the spot safety projects, in the rehabilitation projects, in the construction of new highways, and elsewhere. Reference **(24)** provides a number of important recommendations on safety as it relates to the total highway program.

ENVIRONMENT

A highway necessarily has wide-ranging effects beyond that of providing traffic service to users. It is essential that the highway be considered as an element of the total environment. Environment as used herein refers to the totality of man's surroundings: social, physical, natural, and manmade. It includes human, plant, and animal communities and the forces that act on all three. The highway can and should be located and designed to complement its environment and serve as a catalyst to environmental improvement.

The area surrounding a proposed highway is an interrelated system of natural, manmade, and sociologic variables. Changes in one variable within this system cannot be made without some effect on other variables. Some of these consequences may be negligible, but others may have strong and lasting impact on the environment, including the sustenance and quality of human life. Because highway location and design decisions have an effect on adjacent area developments, it is important that environmental variables be given full consideration.

ECONOMIC ANALYSIS

Highway economics is concerned with the cost of a proposed improvement and the general value resulting from it. Proposed improvements should be analyzed in accordance with AASHTO's *A Manual on User Benefit Analysis of Highway and Bus-Transit Improvements* (**32**).

REFERENCES

1. Pilkington, G.B. and Howell, P.D. "A Simplified Procedure for Computing Vehicle Offtracking on Curves." *FHWA* RD-74-8, December 1983.
2. Alexander, G.J. and Lunenfeld, H. *Positive Guidance in Traffic Control.* Washington, D.C.: U.S. Department of Transportation, Federal Highway Administration, 1975.
3. U.S. Department of Transportation, Federal Highway Administration. *Manual on Uniform Traffic Control Devices for Streets and Highways.* Washington, D.C.: U.S. Government Printing Office, 1988.
4. Johannson, C. and Rumar, K. "Driver's Brake Reaction Time," *Human Factors*, Vol. 13, No. 1, 1971, pp. 22–27.
5. Fell, J.C. "A Motor Vehicle Accident Causal System. The Human Element." *Report No. DOT-HS-801-214*. Washington, D.C.: National Highway Traffic Safety Administration, July 1974.
6. Schmidt, I. and Connolly, P.D. "Visual Considerations of Man, the Vehicle and the Highways." *SP-279-SAE*. New York: Society of Automotive Engineers, 1966.
7. Tilley, D.H., Erwin, C.W., and Gianturco, D.T. "Drowsiness and Driving; Preliminary Report of a Population Survey." *Paper No. 730121-SAE*. New York: Society of Automotive Engineers, 1973.
8. King, G.F. and Lunenfeld, H. "Development of Information Requirements and Transmission Techniques for Highway Users." *NCHRP Report 123*, 1971, 239 pp.
9. Adler, B. and Lunenfeld, H. "Three Beam Headlight Evaluation." *Report No. HS-238-2-411-1*.
10. *Freeway Operations.* Washington, D.C.: Institute of Traffic Engineers, 1961.
11. *Highway Capacity Manual, Special Report No. 209*, Washington, D.C.: Highway Research Board, 1985. (p. 132)
12. Fee, Julie Anna, et al. "Interstate System Accident Research Study 1." U.S. Department of Transportation, Federal Highway Administration, October 1970.
13. Glennon, J.C., Valenta, J.J., Thorson, B.A., and Azzeh, J.A. "Technical Guidelines for the Control of Direct Access to Arterial Highways." Volumes 1 and 2, Midwest Research Institute, August 1975.
14. Maring, G.E. "Pedestrian Travel Characteristics." *Highway Research Record 406*, pp. 14–20.

15. Cameron, R.M. "Mechanical Measurement of Pedestrian Volumes." *TRB 498*, pp. 13–19.
16. Fruin, J.J. *Pedestrian Planning and Design*. Metropolitan Association of Urban Designers and Environmental Planners, Inc., 1971.
17. Older, S.J. "Movement of Pedestrians on Footways in Shopping Streets." *Traffic Engineering and Control*, August 1968, pp. 160–163.
18. Fruin, J.J. "Designing for Pedestrians: A Level-of-Service Concept." *HRB 355*, pp. 1–15.
19. Di Pietro, C.M. and King, L.E. "Pedestrian Gap-Acceptance." *HRB 308*, pp. 80–91.
20. *Guide for Development of New Bicycle Facilities*, Washington, D.C.: AASHTO, 1981.
21. *Highway Design and Operational Practice Related to Highway Safety*. Washington, D.C.: AASHTO, 1974.
22. *1976 Accident Facts*, Washington, D.C.: National Safety Council.
23. American Association of State Highway and Transportation Officials, et al. "*Enhancing Highway Safety in an Era of Limited Resources*." A report resulting from a symposium conducted by the Transportation Research Board, November 1981.
24. "Interstate System Accident Research." October 1970.
25. "Dart, O.K., Jr. and Mann, L., Jr. "Relationship of Rural Highway Geometry to Accident Rates in Louisiana." *Highway Research Record 312*, 1970.
26. Byington, S.R. "Interstate System Accident Research." *Public Roads, Vol. 32, No. 11, December 1963*.
27. Billion, C.E. and Stohner, W.R. "A Detailed Study of Accidents as Related to Highway Shoulders in New York State." *Proc. HRB*, 1957, p. 497.
28. Weaver, G.D., Marquis, E.L., and Olson, R.M. "Selection of Safe Roadside Cross Sections." *NCHRP Report 158*, 1975, 57 pp.
29. *Roadside Design Guide*. Washington, D.C.: AASHTO, 1989.
30. Michie, J.D. and Bronstad, M.E. "Location, Selection, and Maintenance of Highway Traffic Barriers." *NCHRP Report 118*, 1971, 96 pp.
31. Michie, J.D. and Bronstad, M.E. "Guardrail Crash Test Evaluation—New Concepts and End Designs." *NCHRP Report 129*, 1972, 89 pp.
32. *A Manual on User Benefit Analysis of Highway and Bus-Transit Improvements*. Washington, D.C.: AASHTO, 1977.
33. Olson, P.L., Cleveland, D.E. Fancher, P.S., Kostynuik, L.P. and Schneider, L.W. "Parameters Affecting Stopping Sight Distance." *NCHRP Report 270*, 1984 (developed by New York D.O.T. from report).

Chapter III
ELEMENTS OF DESIGN

SIGHT DISTANCE

General Considerations

The ability to see ahead is of the utmost importance in the safe and efficient operation of a vehicle on a highway. On a railroad, trains are confined to a fixed path, yet a block signal system and trained operators are necessary for safe operation. On the other hand, the path and speed of motor vehicles on highways and streets are subject to the control of drivers whose ability, training and experience are quite varied. For safety on highways the designer must provide sight distance of sufficient length that drivers can control the operation of their vehicles to avoid striking an unexpected object on the traveled way. Certain two-lane highways should also have sufficient sight distance to enable drivers to occupy the opposing traffic lane for passing overtaken vehicles without hazard. Two-lane rural highways should generally provide such passing sight distance at frequent intervals and for substantial portions of their length. Conversely, it normally is of little practical value to provide passing sight distance on two-lane urban streets or arterials. The length and interval of passing sight distance should be compatible with the criteria established in the chapter pertaining to that specific highway or street classification.

Sight distance is discussed in four steps: (1) the distances required for stopping, applicable on all highways; (2) the distances required for the passing of overtaken vehicles, applicable only on two-lane highways; (3) the distances needed for decisions at complex locations; and (4) the criteria for measuring these distances for use in design. The design of alinement and profile to provide these distances and to meet these criteria are described later in this chapter. The special conditions related to sight distances at intersections are discussed in Chapter IX.

Stopping Sight Distance

Sight distance is the length of roadway ahead visible to the driver. The minimum sight distance available on a roadway should be sufficiently long to enable a vehicle traveling at or near the design speed to stop before reaching a stationary object in its path. Although greater length is desirable, sight distance at every point along the highway should be at least that required for a below-average operator or vehicle to stop in this distance.

Stopping sight distance is the sum of two distances: the distance traversed by the vehicle from the instant the driver sights an object necessitating a stop to the instant the brakes are applied and the distance required to stop the vehicle from the instant brake application begins. These are referred to as brake reaction distance and braking distance, respectively.

Brake Reaction Time

Brake reaction time is the interval between the instant that the driver recognizes the existence of an object or hazard on the roadway ahead and the instant that the driver actually applies the brakes. This interval includes the time required to make the decision that a stop is necessary. Under certain conditions, such as emergency conditions denoted by flares or flashing lights, operators accomplish these tasks almost instantly. Under most other conditions the operator must subconsciously associate the object ahead with stationary objects adjacent to the roadway, such as walls, fences, trees, poles, or bridges, to determine that the object is also stationary or moving at a slow speed. These determinations take time, the amount of which varies considerably depending on the distance to the object, the acuity of the operator, the natural rapidity with which the driver reacts, atmospheric visibility, the type and the condition of the roadway, and the type, color, and condition on the hazard. Vehicle speed and the roadway environment probably also influence reaction time. Normally, an operator traveling at or near the design speed is more alert than one traveling at a lesser speed. An operator on an urban facility confronted by innumerable possibilities for conflicts from parked vehicles, driveways, and cross streets is also likely to be more alert than the same operator on a limited-access facility where such conditions should be almost nonexistent.

The study (**1**) referred to in Chapter II was based on data from 321 drivers who expected to apply their brakes. The median reaction-time value for these drivers was 0.66 sec with 10 percent requiring 1.5 sec or longer. These findings correlate with those of earlier studies in which alerted drivers were also used. Another study (**2**) gives 0.64 sec as the average value; 5 percent of the drivers required over 1 sec. In a third study (**3**) reaction-time values ranged from 0.4 to 1.7 sec. In the Johansson and Rumar study (**1**), when the signal was unexpected, the drivers' responses were found to increase by approximately 1 sec or longer; some reaction times being 1.5 sec or more. This increase substantiated earlier laboratory and road tests in which the conclusion was drawn that the driver who required 0.2 to 0.3 sec under alert conditions required 1.5 sec under normal conditions.

Minimum reaction times thus could be at least 1.64 sec; 0.64 sec for alerted drivers plus 1 sec for the unexpected signal. Because the studies used simple prearranged signals, they represent the least complex of roadway conditions. Even under these simple conditions it was found that some operators may take over 3.5 sec to respond. Because actual conditions on the highway are generally more complex than those of the studies and because there is wide diversity in the reaction times required, it is evident that the value adopted should be greater than 1.64 sec. In determination of sight distance for design, the reaction time should be larger than the average for all drivers under normal conditions. It should be large enough to include the reaction time required for nearly all drivers under most highway conditions. For approximately 90 percent of the drivers in the first study mentioned, a reaction time of 2.5 sec was found to be adequate. A reaction time of 2.5 sec has thus been assumed in the development of Table III-1.

A reaction time of 2.5 sec is considered adequate for more complex conditions that those of the various studies, but it is not adequate for the most complex conditions encountered by the driver. Additional consideration of the most complex conditions such as those found at multiphase at-grade intersections and ramp termini at through roadways can be found later in this chapter in the section "Decision Sight Distance."

Braking Distance

The approximate braking distance of a vehicle on a level roadway may be determined by the use of the standard formula:

$$d = \frac{V^2}{30f}$$

where:

d = braking distance, ft;
V = initial speed, mph; and
f = coefficient of friction between tires and roadways.

In this formula for braking distance the f factor is used as an overall or a single value that is representative for the whole of the speed change. Measurements show that f is not the same for all speeds. It decreases as the initial speed increases. It varies considerably because of many physical elements such as air pressure of tires, composition of tires, tire tread pattern and depth

Design Speed (mph)	Assumed Speed for Condition (mph)	Brake Reaction Time (sec)	Brake Reaction Distance (ft)	Coefficient of Friction f	Braking Distance on Level (ft)	Stopping Sight Distance Computed (ft)	Stopping Sight Distance Rounded for Design (ft)
20	20-20	2.5	73.3-73.3	0.40	33.3-33.3	106.7-106.7	125-125
25	24-25	2.5	88.0-91.7	0.38	50.5-54.8	138.5-146.5	150-150
30	28-30	2.5	102-7-110.0	0.35	74.7-85.7	177.3-195.7	200-200
35	32-35	2.5	117.3-128.3	0.34	100.4-120.1	217.7-248.4	225-250
40	36-40	2.5	132.0-146.7	0.32	135.0-166.7	267.0-313.3	275-325
45	40-45	2.5	146.7-165.0	0.31	172.0-217.7	318.7-382.7	325-400
50	44-50	2.5	161.3-183.3	0.30	215.1-277.8	376.4-461.1	400-475
55	48-55	2.5	176.0-201.7	0.30	256.0-336.1	432.0-537.8	450-550
60	52-60	2.5	190.7-220.0	0.29	310.8-413.8	501.5-633.8	525-650
65	55-65	2.5	201.7-238.3	0.29	347.7-485.6	549.4-724.0	550-725
70	58-70	2.5	212.7-256.7	0.28	400.5-583.3	613.1-840.0	625-850

Table III-1. Stopping sight distance (wet pavements).

of tread, type and condition of the pavement surface, and the presence of moisture, mud, snow, or ice. The braking distance also depends on the braking system of the vehicle. The several variables are allowed for if f is computed for each test from the standard formula. It thus represents the equivalent constant friction factor. The values of f in Figure III-1 (A and B) were calculated on this basis for some of the curves that represent tests in which speed and distance only were recorded.

Figure III-1A illustrates friction coefficients found by different investigators. Curves 1 to 6 are from a study (**4**) in which more than 1,000 measurements of forward skidding were made on 32 pavements in both wet and dry conditions. Several types of tires were being used. Coefficients of friction were computed by using actual stopping distances in the standard stopping formula. Curves 7 and 8 are representative of several curves of a study (**5**) in which over 50 surfaces were tested in the dry condition by three different methods using three types of tires. Curves 9 and 10 from the same study are representative of wet pavements. Curve 11 is the calculated equivalent f value for stopping distances measured (**3**) on high-type pavement; these were the only tests that included stops from speeds of 60 to 70 mph. This curve is the average of all stops measured, but comparison with on-the-road samples (at low speeds) shows that the vehicles and drivers were somewhat better than the average on the highways.

Because of the lower coefficients of friction on wet pavements as compared with dry, the wet condition governs in determining stopping distances for use in design. The coefficients of friction used for design criteria should represent not only wet pavements in good condition but also surfaces approaching the end of their useful lives. The values should encompass nearly all significant pavement surface types and the likely field conditions. Figure III-1B summarizes the results of a series of 600 measurements made on modern pavements in Germany (**6**). Each road section was tested at 20, 40, 60 and 80 km/hr (12.5, 25, 37.5, and 50 mph) by a locked-wheel trailer. Because the test methods differ, there is no direct correlation between these curves and those of Figure III-1A or Table III-1. However, Figure III-1B does indicate that there is a wide variation between the coefficients of friction for various pavements, which reflects the effect that surface texture has on stopping sight distance and, therefore, the importance of this factor.

The friction factor values used in calculating safe stopping distances should allow for worn tires as well as for new tires and for nearly all types of treads and tire compositions. The friction factor values used should also encompass the differences in vehicle and driver braking from various speeds. On the other hand, the values used need not be so low as to be suitable for pavements under icy conditions. Preferably, the f values used for design should be nearly all-inclusive, rather than average; however, available data are

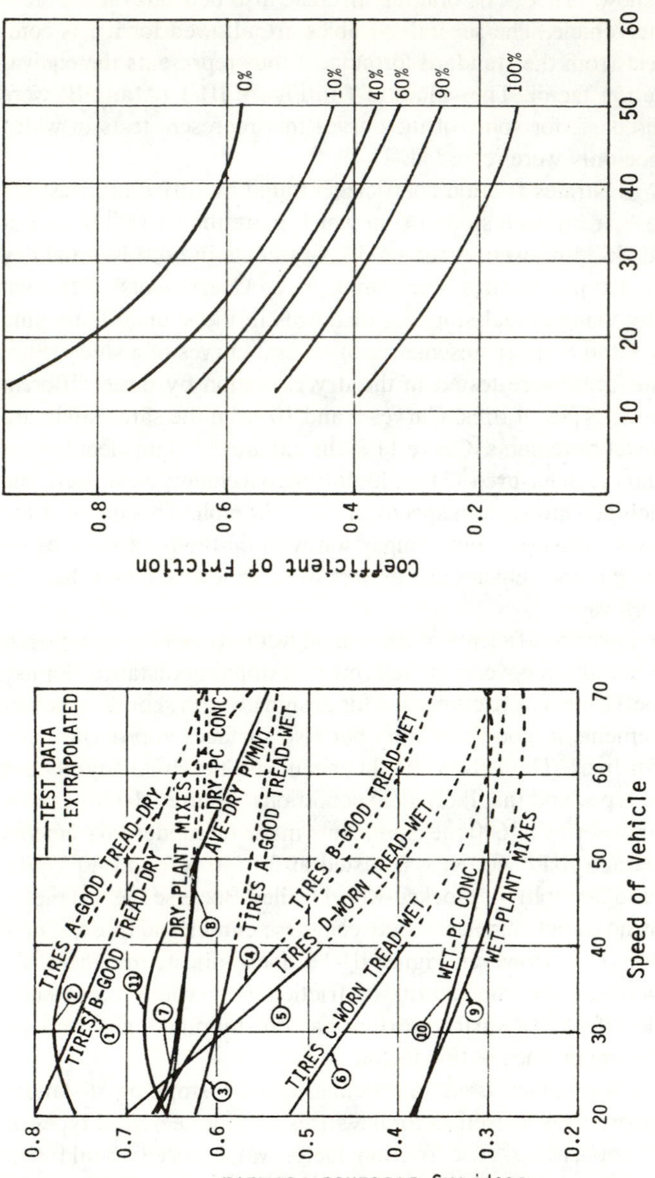

A. Skid resistance for various tire and pavement conditions

B. Skid resistance of pavement

Figure III-1. Variation in coefficient of friction with vehicular speed.

not fully detailed over the range for all these variables, and conclusions must be made in terms of the safest reported average values. The coefficients of friction in Table III-1 have been selected based on these criteria as being appropriate for the calculation of stopping sight distances. Comparison with Figure III-1A shows that the range of values in Table III-1 is generally conservative; the upper limit reflects the concern for the more recent measurements such as those of Figure III-1B, which show a few coefficients for wet pavements with values near or less than the values used for developing the lower limits for stopping sight distance.

The average running speed for low-volume conditions rather than design speed is used in formulating the limiting values for minimum stopping distance. This speed is the initial value given in the second column of Table III-1. Studies show that many operators drive just as fast on wet pavements as they do on dry. To account for this factor, design speed in place of average running speed is used to formulate stopping distance values, as shown by the higher values in the second column of Table III-1.

Design Values

The sum of the distance traversed during the brake reaction time and the distance to stop the vehicle is the minimum stopping sight distance. The computed distances for wet pavements and for various speeds at the assumed conditions are developed in Table III-1. A range of rounded values for direct use in design is shown in the last column of this table.

Any length of stopping sight distance within the range of values established on Table III-1 is acceptable for a specific speed. However, values approaching or exceeding the upper limit of the range should be used as the basis for design wherever conditions permit. Use of the upper limit values increases the margin of safety by providing for the drivers who operate at or near the design speed during wet weather. To ensure that new pavements will have initially, and will retain, coefficients of friction comparable to those given in the table, designs should meet the criteria established in the AASHTO *Guidelines for Skid Resistant Pavement Design* (**7**). Although research data may demonstrate that proposed pavements will have initially, and will retain coefficients of friction greater than those given in Table III-1, this finding should not be considered justification for using stopping sight distance values less than the minimums contained in this table. Variations in results (particularly because of the methods of determining coefficients of frictions) preclude such direct comparisons. If portions of existing pavements are to be retained, skid tests should be made. Although direct comparisons cannot be made, the need for increased lengths of stopping sight distance should be considered when tests yield coefficients of friction significantly below those in Table III-1.

Table III-1 includes data for the higher design speeds for wet conditions where no actual test data are available. Values for design speeds above 60 mph are extrapolated from the curves for lower speeds in Figure III-1A.

Effect of Grade on Stopping

When a highway is on a grade, the standard formula for braking distance is the following:

$$d = \frac{V^2}{30(f \pm G)}$$

in which G is the percent of grade divided by 100, and the other terms are as previously stated. The safe stopping distances on upgrades are shorter; those on downgrades are longer. The extent of grade corrections is indicated in Table III-2. These corrections are computed for wet conditions, the assumed design criterion used in Table III-1. Design speed is used in calculating downgrade corrections, average running speed in calculating upgrade corrections. The different criteria for descending and ascending grades are based on the effect grades have on the speed of individual vehicles, particularly trucks; the effect these vehicles have on the overall speed of the traffic stream; and the premise that many drivers, particularly those in automobiles, do not compensate completely for the changes in speed caused by grades.

On nearly all roads and streets the grade is traversed by traffic in both directions, but the sight distance at any point on the highway generally is different in each direction, particularly on straight roads in rolling terrain. As a general rule, the sight distance available on downgrades is larger than on upgrades, more or less automatically providing the necessary corrections for grade. This may explain why some design offices do not make corrections in stopping sight distance because of grade. Exceptions are one-way roads or streets, as on divided highways with independent design profiles for the two roadways for which the separate grade corrections are in order and the refinement in design is in keeping with the overall standards used.

Variation for Trucks

The derived minimum stopping sight distances directly reflect passenger car operation and might be questioned for use in design for truck operation. Trucks as a whole, especially the larger and heavier units, require longer stop-

Design Speed (mph)	Increase for Downgrades Correction in Stopping Distance (ft)			Assumed Speed for Condition (mph)	Decrease for Upgrades Correction in Stopping Distance (ft)		
	3%	6%	9%		3%	6%	9%
30	10	20	30	28	—	10	20
40	20	40	70	36	10	20	30
50	30	70	—	44	20	30	—
60	50	110	—	52	30	50	—
65	60	130	—	55	30	60	—
70	70	160	—	58	40	70	—

Table III-2. Effect of grade on stopping sight distance—wet conditions.

ping distances from a given speed than passenger vehicles do. However, there is one factor that tends to balance the additional braking lengths for trucks for given speeds with those for passenger cars. The truck operator is able to see the vertical features of the obstruction substantially farther because of the higher position of the seat in the vehicle. Separate stopping sight distances for trucks and passenger cars, therefore, are not used in highway design standards.

There is one situation that should be treated with caution, in which every effort should be made to provide stopping sight distances greater than the minimum design value. When horizontal sight restrictions occur on downgrades, particularly at the ends of long downgrades, the greater height of eye of the truck operator is of little value, even when the horizontal sight obstruction is a cut slope, when (on long downgrades) truck speeds may closely approach or exceed those of passenger cars. Although the average truck operator tends to be more experienced than the average passenger car operator and quicker to recognize hazards, it is best under such conditions to supply a stopping sight distance that meets or exceeds the values in Table III-1.

Decision Sight Distance

Stopping sight distances are usually sufficient to allow reasonably competent and alert drivers to come to a hurried stop under ordinary circumstances. However, these distances are often inadequate when drivers must make complex or instantaneous decisions, when information is difficult to perceive, or when unexpected or unusual maneuvers are required. Limiting sight distances to those provided for stopping may also preclude drivers from performing evasive maneuvers, which are often less hazardous and otherwise preferable to stopping. Even with an appropriate complement of standard traffic control devices in accordance with the MUTCD (**8**), stopping sight distances may not provide sufficient visibility distances for drivers to corroborate advance warning and to perform the necessary maneuvers. It is evident that there are many locations where it would be prudent to provide longer sight distances. In these circumstances, decision sight distance provides the greater length that drivers need.

Decision sight distance is the distance required for a driver to detect an unexpected or otherwise difficult-to-perceive information source or hazard in a roadway environment that may be visually cluttered, recognize the hazard or its threat potential, select an appropriate speed and path, and initiate and complete the required safety maneuver safely and efficiently (**51**). Because decision sight distance gives drivers additional margin for error and affords them sufficient length to maneuver their vehicles at the same or reduced speed rather than to just stop, its values are substantially greater than stopping sight distance.

Drivers need decision sight distances whenever there is a likelihood for error in either information reception, decision-making, or control actions (**9**). The following are examples of critical locations where these kinds of errors are likely to occur, and where it is desirable to provide decision sight distance: interchange and intersection locations where unusual or unexpected maneuvers are required; changes in cross section such as toll plazas and lane drops; and areas of concentrated demand where there is apt to be "visual noise" whenever sources of information compete, as those from roadway elements, traffic, traffic control devices, and advertising signs.

The decision sight distances in Table III-3 (1) provide values to be used by designers for appropriate sight distances at critical locations and (2) serve as criteria in evaluating the suitability of the sight lengths at these locations. Because of the additional safety and maneuverability these lengths yield, it is recommended that decision sight distances be provided at critical locations or that these points be relocated to locations where decision sight distance lengths are available. If it is not feasible to provide these distances because of horizontal or vertical curvature or if relocation is not possible, special attention should be given to the use of suitable traffic control devices for providing advance warning of the conditions that are likely to be encountered.

Decision sight distance values that will be applicable to most situations have been developed from empirical data. The decision sight distances vary depending on whether the location is on a rural or urban road, and on the type of maneuver required to avoid hazards and negotiate the location properly. Table III-3 shows decision sight distance values for various situations rounded for design. As can be seen in Table III-3, generally shorter distances are required for rural roads and when a stop is the maneuver involved.

The following avoidance maneuvers are covered in Table III-3:

- Avoidance Maneuver A: Stop on rural road.
- Avoidance Maneuver B: Stop on urban road.
- Avoidance Maneuver C: Speed/path/direction change on rural road.
- Avoidance Maneuver D: Speed/path/direction change on suburban road.
- Avoidance Maneuver E: Speed/path/direction change on urban road.

In computing and measuring decision sight distances, the 3.5-ft eye height and 6-in. object height criteria used for stopping sight distance have been adopted. Although drivers may have to be able to see the entire roadway situation, including the road surface, the rationalc for the 6-in object height is as applicable for decision sight distance as it is for stopping sight distance.

Design Speed (MPH)	Decision Sight Distance for Avoidance Maneuver (ft)				
	A	B	C	D	E
30	220	500	450	500	625
40	345	725	600	725	825
50	500	975	750	900	1025
60	680	1300	1000	1150	1275
70	900	1525	1100	1300	1450

Table III-3. Decision sight distance. (Source: Ref. 10)

Passing Sight Distance for Two-Lane Highways

Criteria for Design

Most roads and numerous streets are considered to qualify as two-lane, two-way highways on which vehicles frequently overtake slower moving vehicles, the passing of which must be accomplished on lanes regularly used by opposing traffic. If passing is to be accomplished with safety, the driver should be able to see a sufficient distance ahead, clear of traffic, to complete the passing maneuver without cutting off the passed vehicle in advance of meeting an opposing vehicle appearing during the maneuver. When required, a driver can return to the right lane without passing if he sees opposing traffic is too close when the maneuver is only partially completed. Many passings are accomplished without the driver seeing a safe passing section ahead, but design based on such maneuvers does not have the desired factor of safety. Because many cautious drivers would not attempt to pass under such conditions, design on this basis would reduce the usefulness of the highway. An alternative to providing passing sight distance will be found later in this chapter in the section "Passing Lane Sections on Two-Lane Roads".

Passing sight distance for use in design should be determined on the basis of the length needed to safely complete normal passing maneuvers. While there may be occasions to consider multiple passings, where two or more vehicles pass or are passed, it is not practical to assume such conditions in developing minimum design criteria. Instead, sight distance is determined for a single vehicle passing a single vehicle. Longer sight distances occur in design and these locations can accommodate an occasional multiple passing.

When computing minimum passing sight distances on two-lane highways for design use, certain assumptions for traffic behavior are necessary, some of which offer a wide choice. The assumed control for driver behavior should be that practiced by a high percentage of drivers, rather than the average driver. Such assumptions follow:

1. The overtaken vehicle travels at uniform speed.
2. The passing vehicle has reduced speed and trails the overtaken vehicle as it enters a passing section.
3. When the passing section is reached, the driver requires a short period of time to perceive the clear passing section and to react to start his maneuver.
4. Passing is accomplished under what may be termed a delayed start and a hurried return in the face of opposing traffic. The passing vehicle accelerates during the maneuver, and its average speed during the occupancy of the left lane is 10 mph higher than of the overtaken vehicle.
5. When the passing vehicle returns to its lane, there is a suitable clearance length between it and an oncoming vehicle in the other lane.

Some drivers accelerate at the beginning of a passing maneuver to an appreciably higher speed and then continue at a uniform speed until the passing is completed. Many drivers accelerate at a fairly high rate until just beyond the vehicle being passed and then complete the maneuver either without further acceleration or at reduced speed. For simplicity, extraordinary maneuvers are ignored and passing distances are developed with the use of observed speeds and times that fit the practices of a high percentage of drivers.

The minimum passing sight distance for two-lane highways is determined as the sum of the four distances (shown in the diagram at the top of Figure III-2):

d_1—Distance traversed during perception and reaction time and during the initial acceleration to the point of encroachment on the left lane.

d_2—Distance traveled while the passing vehicle occupies the left lane.

d_3—Distance between the passing vehicle at the end of its maneuver and the opposing vehicle.

d_4—Distance traversed by an opposing vehicle for two-thirds of the time the passing vehicle occupies the left lane, or ⅔ of d_2 above.

Extensive field observations of driver behavior during passing maneuvers were made by Prisk (**12**) during the years 1938 to 1941. Data from this study were grouped into three passing-speed groups, 30 to 40 mph, 40 to 50 mph, and 50 to 60 mph, and are given in Table III-4. A fourth speed group, 60 to 70 mph, has since been added based on extrapolated data obtained from Prisk's summary report. Time and distance values were determined in relation to the average speed of the passing vehicle. Speeds of overtaken vehicles were approximately 10 mph less than speeds of passing vehicles.

Three of the locations where observations were made in the Prisk study were restudied in 1957. Very little change was noted in the passing practices of drivers on the study sections despite increased vehicle performance capabilities. However, the drivers appeared to become more conservative in passing on acceptable, though relatively short, tangents; i.e., fewer passings occurred at any given volume level in 1957 than in 1938. A 1971 study (**53**) of vehicle passing performance on two-lane highways produced a different set of passing sight distance values, which values were subsequently reviewed in a 1978 study (**55**) which evaluated minimum passing sight distances. The 1978 study reported that Prisk's total safe passing sight distances, given in Table III-4, are greater than those determined in the 1971 study for all speeds except 70 mph. Thus, the minimum passing sight distances given in Table III-5 are generally conservative for modern vehicles and will be used herein.

Initial maneuver distance (d_1). The initial maneuver period has two components, a time for perception and reaction, and an interval during which the driver brings his vehicle from the trailing speed to the point of encroachment on the left or passing lane. To a great extent the two overlap. As a passing section of highway comes into view of a driver desiring to pass, he may begin to accelerate and maneuver his vehicle toward the centerline of the highway while deciding whether or not to pass. Studies show that the average passing

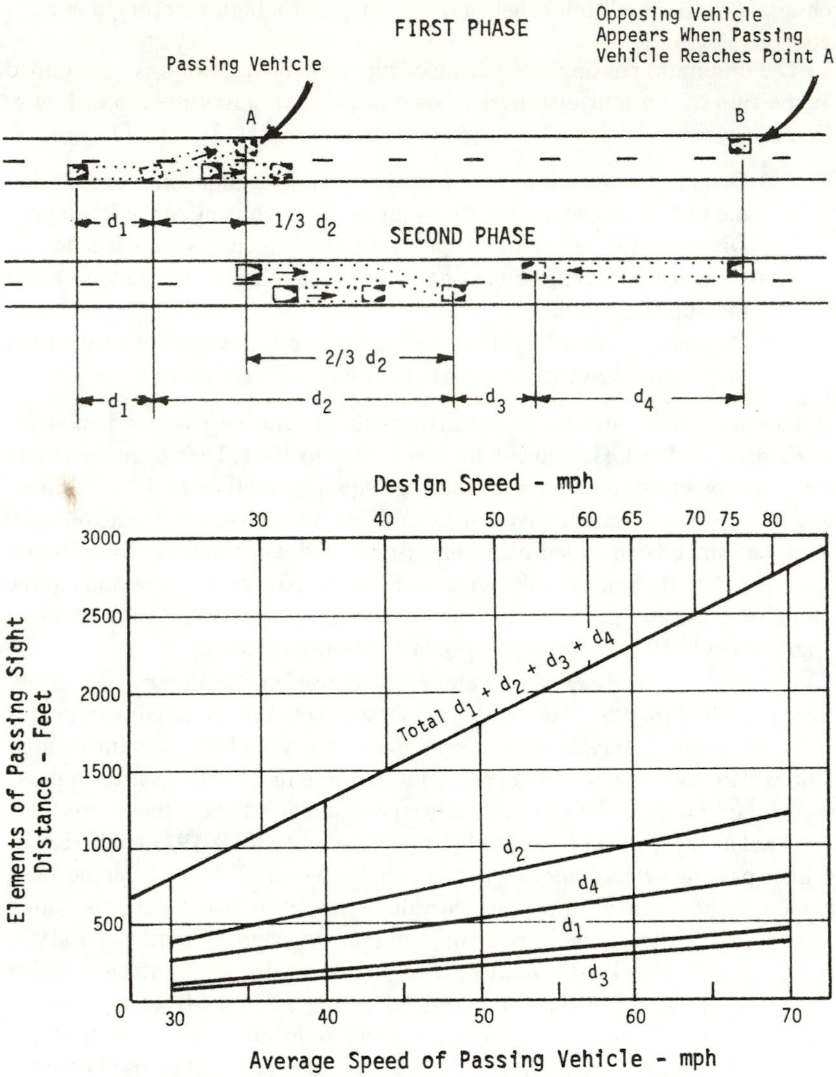

Figure III-2. Elements of and total passing sight distance— two-lane highways.

Speed Group (mph)	30-40	40-50	50-60	60-70
Average Passing Speed (mph)	**34.9**	**43.8**	**52.6**	**62.0**
Initial maneuver:				
a = average acceleration (mph/sec)[a]	1.40	1.43	1.47	1.50
t_1 = time (sec)[a]	3.6	4.0	4.3	4.5
d_1 = distance traveled (ft)	145	215	290	370
Occupation of left lane:				
t_2 = time (sec)[a]	9.3	10.0	10.7	11.3
d_2 = distance traveled (ft)	475	640	825	1,030
Clearance length:				
d_3 = distance traveled (ft)[a]	100	180	250	300
Opposing vehicle:				
d_4 = distance traveled (ft)	315	425	550	680
Total distance, $d_1 + d_2 + d_3 + d_4$ (ft)	1,035	1,460	1,915	2,380

[a]For consistent speed relation, observed values adjusted slightly.

Table III-4. Elements of safe passing sight distance—two-lane highways.

vehicle accelerates at less than its maximum potential, indicating that the initial maneuver period contains an element of time for perception and reaction. However, some drivers may remain in normal lane position while deciding to pass. The exact position of the vehicle during initial maneuver is unimportant because differences in resulting passing distances are insignificant.

The acceleration rate obtained from the passing study data in the three speed groups during the initial maneuver period varied from 1.41 to 1.47 mph/sec; the average time varied from 3.7 to 4.3 sec; and the average passing speeds were 34.9, 43.8 and 52.6 mph. For the 60-to 70-mph group, on the basis of extrapolated data, the average acceleration was assumed to be 1.50 mph/sec, the maneuver time 4.5 sec, and the average speed 62 mph.

The distance d_1 traveled during the initial maneuver period is computed from the following formula:

$$d_1 = 1.47t_1\left(v - m + \frac{at_1}{2}\right)$$

where:
- t_1 = time of initial maneuver, sec;
- a = average acceleration, mph/sec;
- v = average speed of passing vehicle, mph; and
- m = difference in speed of passed vehicle and passing vehicle, mph.

The acceleration, time, and distance traveled during the initial maneuver periods in passing are given in Table III-4. The d_1 line in Figure III-2 shows the distance plotted against the average passing speed.

Distance while passing vehicle occupies left lane (d_2). Passing vehicles were found in the study to occupy the left lane from 9.3 to 10.4 sec. The distance d_2 traveled in the left lane by the passing vehicle is computed by the following formula:

$$d_2 = 1.47vt_2$$

where: t_2 = time passing vehicle occupies the left lane, sec; and
v = average speed of passing vehicle, mph.

The time and distance traveled while the passing vehicle occupies the left lane are given in Table III-4. Distances are plotted against average passing speeds as curve d_2 in Figure III- 2.

Clearance length (d_3). The clearance length between the opposing and passing vehicles at the end of the maneuvers found in the passing study varied from 110 to 300 ft. This length, adjusted somewhat for practical consistency is shown as the clearance length d_3 in Table III-4 and on Figure III-2.

Distance traversed by an opposing vehicle (d_4). Passing sight distance includes the distance traversed by an opposing vehicle during the passing maneuver to minimize the chance of a passing vehicle meeting an opposing vehicle while in the left lane. Conservatively, this distance should be the distance traversed by an opposing vehicle during the entire time it takes to pass or during the time the passing vehicle is in the left lane, but such distance is questionably long. During the first phase of the passing maneuver the passing vehicle has not yet pulled abreast of the vehicle being passed, and even though the passing vehicle occupies the left lane, its driver can return to the right lane if he sees an opposing vehicle. It is unnecessary to include this trailing time interval in computing the distance traversed by an opposing vehicle. This time interval, which can be computed from the relative positions of passing and passed vehicle, is about one-third the time the passing vehicle occupies the left lane, so that the passing sight distance element for the opposing vehicle is the distance it traverses during two-thirds of the time the passing vehicle occupies the left lane. The opposing vehicle is assumed to be traveling at the same speed as the passing vehicle, so $d_4 = 2d_2/3$. The distance d_4 is given in Table III-4 and shown on Figure III-2.

Design Values

The "total" curve in Figure III-2 is determined by the sum of the named elements. For each passing speed, it indicates the minimum passing sight distance for a vehicle passing another vehicle traveling 10 mph slower, in the face of an opposing vehicle traveling at the same speed as the passing vehicle. On determination of a likely and logical relation between average passing speed and the highway design speed, these distances can be used to express the minimum passing sight distance needed for design purposes.

The ranges of speeds of the passed and passing vehicles are affected by traffic volume. When traffic volume is low (level of service A), there are few vehicles that need to be passed, but as volume increases (level-of-service D or lower) there are few, if any, passing opportunities. The speed of the passed vehicle has been assumed to be the average running speed at a traffic volume near capacity as represented by the curve for intermediate volumes in Figure II-22. The speed of the passing vehicle is assumed to be 10 mph greater. The assumed speeds for passing vehicles in Table III-5 represent the likely passing speeds on two-lane highways. Passing sight distances for these passing speeds would accommodate the majority of the desired passing maneuvers. They correspond to the "total" curve of Figure III-2. For convenience, a design speed grid has been superimposed on the upper half of the figure, permitting direct reading of the passing sight distance in terms of the design speed. The rounded values in the last column of Table III-5 are design values for minimum passing sight distance. In designing a highway these distances should be exceeded as much as practicable and such sections provided as often as can be done with reasonable costs so as to present as many passing opportunities as possible.

Design Speed (mph)	Assumed speeds: Passed Vehicle (mph)	Assumed speeds: Passing Vehicle (mph)	Minimum Passing Sight Distance (ft): Figure III-2	Minimum Passing Sight Distance (ft): Rounded
20	20	30	810	800
30	26	36	1,090	1,100
40	34	44	1,480	1,500
50	41	51	1,840	1,800
60	47	57	2,140	2,100
65	50	60	2,310	2,300
70	54	64	2,490	2,500

Table III-5. Minimum passing sight distance for design of two-lane highways.

These minimum passing sight distances for design should not be confused with other distances used as the warrants for placing no-passing zone pavement stripes on completed highways. Such values as shown in section 3B-5 of the MUTCD (**8**) are substantially less than design distances and are derived for traffic operating-control needs that are based on different assumptions from those for highway design.

Effect of Grade on Passing Sight Distance

Appreciable grades increase the sight distance required for safe passing. Passing is easier for the vehicle traveling downgrade because the overtaking vehicle can accelerate more rapidly than on the level and thus can reduce the time of passing, but the overtaken vehicle can also accelerate easily so that a situation akin to a racing contest may result.

The sight distances required to permit vehicles traveling upgrade to pass with safety are greater than those required on level roads because of reduced acceleration of the passing vehicle (which increases the time of passing) and the likelihood of opposing traffic speeding up (which increases the distance traveled by it). Compensating for this somewhat are the factors that the passed vehicle frequently is a truck that usually loses some speed on appreciable upgrades and that many drivers are aware of the greater distances needed for passing upgrade compared with level conditions.

If passings are to be performed safely on upgrades, the passing sight distance should be greater than the derived minimum. Specific adjustments for design use are unavailable, but the designer should recognize the desirability of increasing the minimums shown in Table III- 5.

Frequency and Length of Passing Sections

Sight distance adequate for passing should be encountered frequently on two-lane highways, and at each passing section the length of roadway with sight distance ahead equal to or greater than the minimum passing sight distance should be as long as feasible. Frequency and length of passing sections for highways depend principally on the topography, the design speed of highway, and the cost; for streets, the spacing of intersections is the principal consideration. On roadways with high volumes that approach capacity frequent and long passing sections are essential. On roadways with intermediate to low volumes the need is not as great, but passing sections are still an important adjunct for efficiency and safety.

It is not possible to directly indicate the frequency with which passing sections should be provided on two-lane highways due to the physical and cost limitations. On almost all roads and selected streets some passing sections are provided in the normal course of design, but the designer's appreciation of their importance and his studied attempt to provide them usually can ensure others at little or no additional cost. On some two-lane roads in steep mountainous terrain, it may be more economical to build intermittent four-lane sections with stopping sight distance in lieu of two-lane sections with passing sight distance.

The minimum passing sight distance is sufficient for a single or isolated passing only. Design with only minimum sight distance will not assure that safe passings can be made. Even on low-volume roadways a driver desiring to pass may, on reaching the passing section, find vehicles in the opposing lane and thus be unable to use the section or at least not be able to begin to pass at once.

The importance of frequent passing sections is illustrated by their effect on the service volume of a two-lane, two-way highway. Table 8.1 of the HCM (1985) shows, for example, that, for an average travel speed of 55 mph over level terrain the service flow rate is reduced from approximately 760 passenger cars per hour where there are no sight restrictions to about 530 where passing sight distance is available on only 40% of the highway. The effect of restricted sight distance on service volume is even more severe where a design speed lower than 70 mph is assumed as a basis for design.

These data indicate another possible criterion for passing sight distance design on two-lane highways. On sections of roadway that are several miles or more in length, the available sight distances along this length can be summarized to show the percentage of length with sight distance greater than the passing minimum. Analysis of capacity related to this percentage would indicate whether or not alinement and profile adjustments are necessary to accommodate the DHV's. When highway sight distances are analyzed over the whole range of lengths within which passings are made, a new design criterion may be evaluated. Where high traffic volumes are expected on a highway and a high level of capacity is to be maintained, it is virtually mandatory that frequent or nearly continuous passing sight distances are provided.

Sight Distance for Multilane Highways

It is not necessary to consider passing sight distance on highways or streets that have two or more traffic lanes in each direction of travel. Passing maneuvers on multilane roadways are expected to occur within the limits of each one-way traveled way. Thus passing maneuvers that require crossing the centerline of four-lane undivided roadways or crossing the median of four-lane roadways are reckless and should be prohibited.

It is imperative that multilane roadways have continuously adequate stopping sight distance, with greater than minimum distance being preferred. Lengths of stopping sight distance vary with vehicle speed and are discussed in detail at the beginning of this chapter.

Criteria for Measuring Sight Distance

Sight distance is the distance along a roadway that an object of specified height is continuously visible to the driver. This distance is dependent on the height of the driver's eye above the road surface, the specified object height above the road surface, and the height of sight obstructions within the line of sight.

Height of Driver's Eye

For sight distance calculations for passenger vehicles, the height of the driver's eye is considered to be 3.50 ft above the road surface. This value is based on studies (**14, 15, 16, 17**) which show that average vehicle heights decreased since 1960 to 4.25 ft with a comparable decrease in average eye heights to 3.50 ft. The average vehicle heights decreased 2.6 in. in this period, which correlates well with the 2.1-in. reduction in average eye heights. In the same time period the minimum height of eye decreased 2.5 in. to 3.31 ft. Because of this significant change in the minimum eye heights, the design eye height has been reduced from 3.75 ft to 3.50 ft. This change in eye height has the effect of lengthening minimum crest vertical curves by approximately 5 percent, thereby providing about 2.5 percent more sight distance. Because of various factors that appear to place practical limits on any further decreases in passenger car heights and the relatively small increases that further change would mandate in lengths of vertical curves, 3.50 ft is considered to be the height of driver's eye for measuring both stopping and passing sight distances. For large trucks the driver eye height ranges from 6 to 8 ft with the average height being 8 ft. Eight ft is assumed for design.

Height of Object

For stopping sight distance calculations, the height of object is considered to be 6 in. above the road surface. For passing sight distance calculations, the height of object is considered to be 4.25 ft above the road surface.

Stopping sight distance object. The object height of 6 in. was adopted for stopping sight distance calculation purposes in 1965. The basis for its selection was largely an arbitrary rationalization of possible hazardous object size and a driver's ability to perceive and react to a hazardous situation. If other vehicles were the only likely hazard to be encountered, the height of vehicle taillights, 1.5 to 2.0 ft, would be a sufficient object height. Such a height, however, would preclude a driver's seeing small animals, rocks, or other debris that are likely to be encountered in the roadway. It is considered that a 6-in.-high object is representative of the lowest object that can create a hazardous condition and be perceived as a hazard by a driver in time to stop before reaching it. Using object heights of less than 6 in. for stopping sight distance calculations results in considerably longer crest vertical curves. For example, if the roadway surface is used as the sighted object, crest vertical curves would have to be about 85 percent longer than when 6 in. is used as the object height. The object height of less than 6 in. could substantially increase construction costs because additional excavation would be required to provide the longer crest vertical curves. It is also doubtful that the driver's ability to perceive a hazardous situation would be increased.

Passing sight distance object. The object height of 4.25 ft is adopted for passing sight distance calculations, superseding the 4.5-ft object height, which had been used since 1940. Because vehicles are the objects that must be seen when passing and because the height of the average passenger vehicle body has been reduced to its current 4.25-ft height above the pavement, this height will be used for calculation purposes. Passing sight distances calculated on this basis are also considered adequate for night conditions because the beams of the headlights of an opposing vehicle generally are seen from a greater distance than its top could be seen in the daytime.

Sight Obstructions

On tangents the obstruction that limits the driver's sight distance is the road surface at some point on a crest vertical curve. On horizontal curves the obstruction that limits the driver's sight distance may be the road surface at some point on a crest vertical curve, or it may be some physical feature outside of the traveled way, such as a longitudinal barrier, a bridge-approach fill slope, a tree, foliage, or the backslope of a cut section. Accordingly, all highway construction plans should be checked in both the vertical and horizontal plane for sight distance obstructions.

Measuring and Recording Sight Distance on Plans

The design of horizontal alinement and vertical profile using sight distance and other criteria is covered later in this chapter, particularly the detail design of horizontal and vertical curves. Sight distance, however, should be considered in the preliminary stages of design when both the horizontal and vertical alinement are still subject to adjustment. By determining graphically the sight distances on the plans and recording them at frequent intervals, the designer can appraise the overall layout and effect a more balanced design by minor adjustments in the plan or profile. Methods for scaling sight distances are demonstrated in Figure III-3. The figure also shows a typical sight distance record that would be shown on the final plans.

Because the view of the highway ahead may change rapidly in a short distance, it is desirable to measure and record sight distance for both directions of travel at each station. Both horizontal and vertical sight distances should be measured and the shorter lengths recorded. In the case of two-lane highways, passing sight distance in addition to stopping sight distance should be measured and recorded.

Sight distance charts such as those in Figures III-41 through III-44 may be used to establish minimum lengths of vertical curves. Charts similar to Figures III-26A and III-26B are useful for determining the degree of horizontal curve or the lateral offset therefrom needed to provide the required sight distance. Once the horizontal and vertical alinements are tentatively established, the practical means of examining sight distances along the proposed highway is by direct scaling on the plans.

Horizontal sight distance on the inside of a curve is limited by obstructions such as buildings, hedges, wooded areas, high ground, or other topographic features. These generally are plotted on the plans. Horizontal sight is measured with a straightedge, as indicated at the upper left in Figure III-3. The cut slope obstruction is shown on the worksheets by a line representing the proposed excavation slope at a point 2.0 ft (average of 3.50 and 0.5 ft) above the road surface for stopping sight distance and at a point about 3.75 ft above the road surface for passing sight distance. The position of this line with respect to the centerline may be scaled from the plotted highway cross sections. Preferably, the stopping sight distance should be measured between points on the one traffic lane, and passing sight distance from the middle of one lane to the middle of the other lane. Such refinement on two-lane highways generally is not necessary and measurement to the centerline or pavement edge is suitable. Where there are changes of grade coincident with horizontal curves that have sight-limiting cut slopes on the inside, the line-of-sight intercepts the slope at a level either lower or higher than the assumed average height. In measuring sight distance the error in the use of the assumed 2.0- or 3.75-ft height usually can be ignored.

Vertical sight distance may be scaled from a plotted profile by the method illustrated at the right center of Figure III-3. A transparent strip with parallel edges 4.25 ft apart and with scratched lines 6 in. and 3.50 ft from the upper edge, in accordance with the vertical scale, is a useful tool. The 3.50-ft line is

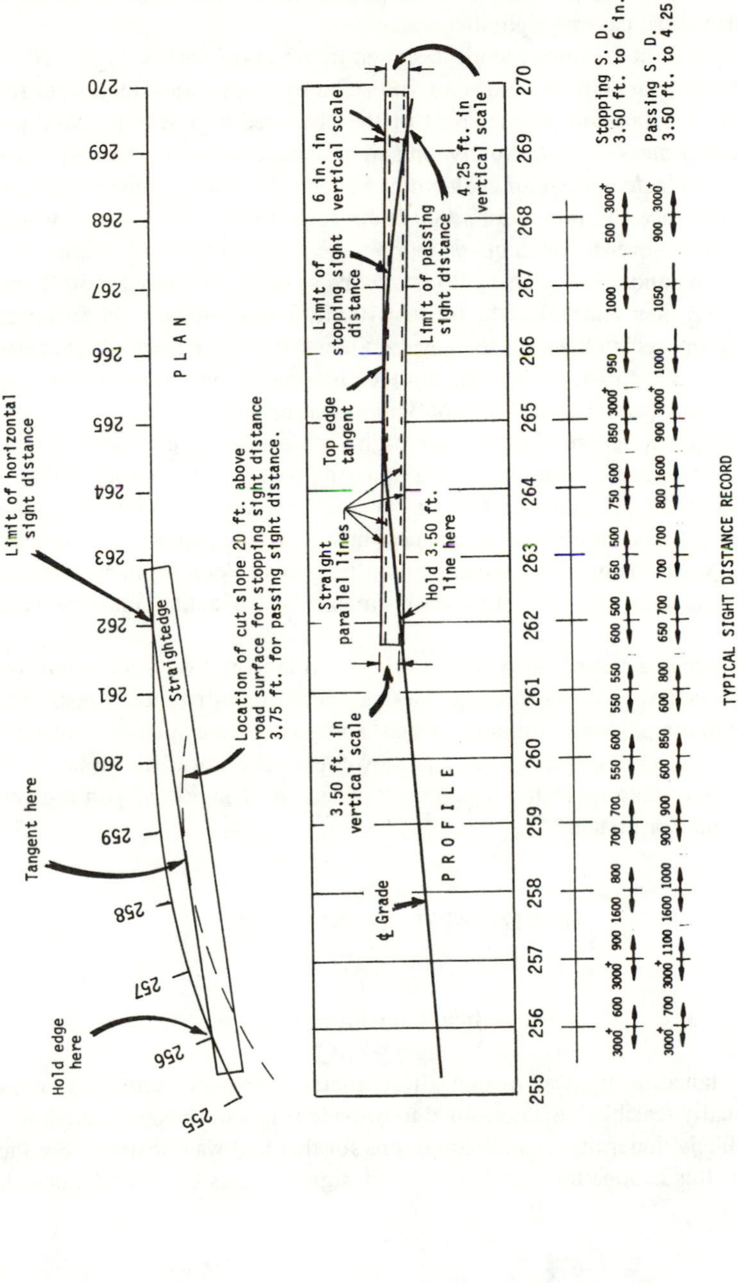

Figure III-3. Scaling and recording sight distances on plans.

placed on the station from which the vertical sight distance is desired, and the strip is pivoted about this point until the upper edge is tangent to the profile. The distance between the initial station and the station on the profile intersected by the 6-in. line is the stopping sight distance. The distance between the initial station and the station on the profile intersected by the lower edge of the strip is the passing sight distance.

A simple sight distance record is shown in the lower part of Figure III-3. Sight distances in both directions are indicated by arrows and figures at each station on the plan and profile sheet of the proposed highway. To avoid the extra work of measuring unusually long sight distances that may occasionally be found, a selected maximum value may be recorded. In the example shown, all sight distances of more than 3,000 ft are recorded as 3,000+, and where this occurs for several consecutive stations, the intermediate values are omitted. Sight distances less than 1,000 ft may be scaled to the nearest 50 ft and those greater than 1,000 ft to the nearest 100 ft. The available sight distances along a proposed highway also may be shown by other methods. Several States use a sight distance graph, plotted in conjunction with the plan and profile of the highway, as a means of demonstrating sight distances.

Sight distance records for two-lane highways may be used to advantage to tentatively determine the marking of no-passing zones in accordance with criteria given in the MUTCD (**8**). Marking of such zones is an operational rather than a design problem. No-passing zones thus established serve as a guide for markings when the highway is completed; the zones so determined should be checked and adjusted by field measurements before actual markings are placed.

Sight distance records also are useful on two-lane highways for determining the percentage of length of highway on which sight distance is restricted to less than the passing minimum, which is important in evaluating capacity. With recorded sight distances, as in the lower part of Figure III-3, it is a simple process to determine the percentage of length of highway with a given sight distance or greater.

HORIZONTAL ALINEMENT

Theoretical Considerations

For balance in highway design all geometric elements should, as far as economically feasible, be determined to provide safe, continuous operation at a speed likely under the general conditions for that highway or street. For the most part this is done through the use of design speed as the overall control.

In the design of highway curves it is necessary to establish the proper relation between design speed and curvature and also their joint relations with superelevation and side friction. Although these relations stem from the laws of mechanics, the actual values for use in design depend on practical limits and factors determined more or less empirically over the range of variables involved. These limits and factors are explained below with determination of logical controls for highway curve design.

When a vehicle moves in a circular path, it is forced radially outward by centrifugal force. The centrifugal force is counterbalanced by the vehicle weight component related to the roadway superelevation or the side friction developed between tires and surface or by a combination of the two. From the laws of mechanics, the basic point mass (curve) formula for vehicle operation on a curve is the following.

$$\frac{e + f}{1 - ef} = \frac{0.067V^2}{R} = \frac{V^2}{15R}$$

where: e = rate of roadway superelevation, ft/ft;
f = side friction factor;
V = vehicle speed, mph; and
R = radius of curve, ft.

The value of the product ef in this equation is always small. As a result, the (1 – ef) term is normally omitted in highway and street designs, thus providing slightly more conservative values.

When a vehicle travels at constant speed on a curve superelevated so that the f value is zero, the centrifugal force is balanced by the weight component of the vehicle and, theoretically, no steering force is required. A vehicle traveling faster or slower than the balance speed develops tire friction as steering effort is applied to prevent movement to the outside or to the inside of the curve. On nonsuperelevated curves, travel at different speeds is also possible by utilizing side friction in appropriate amounts to resist the varying centrifugal force.

General Considerations

From accumulated research and experience, limiting values have been established for e and f. Using the e_{max} value with a conservative f value in the basic formula permits determination of minimum curve radii for various design speeds. For a given design speed, use of curves with radii longer than the minimum calls for balance in the factors involved to determine the desirable superelevation rates below the maximum. These factors are discussed separately in the following sections.

Superelevation

There is a practical limit to the rate of superelevation. In areas subject to ice and snow the rate of superelevation cannot be greater than that on which vehicles standing or traveling slowly would slide down the cross slope when the pavement is icy. At higher speeds the phenomenon of partial hydroplaning can occur on curves with poor surface drainage that allows water buildup. Skidding occurs, usually at the rear wheels, when the lubricating effect of the water film reduces available lateral friction below the demand being made by cornering. When traveling slowly around a curve with high superelevation, negative lateral forces develop and the car is held in the proper path only when the driver steers up the slope or against the direction of the horizontal curve. This direction of steering is an unnatural movement on the part of the driver and possibly explains the difficulty of driving on roads where the superelevation is in excess of that required for normal speeds. Such a high rate of superelevation is undesirable on high-volume roads, as in urban and suburban areas, where there are numerous occasions when vehicle speeds are reduced considerably because of highway traffic volume or other conditions. Also some trucks have high centers of gravity and some cars are loosely suspended on the axles. When these vehicles travel slowly on steep cross slopes, a high percentage of the weight is carried by the inner tires.

Side Friction Factor

With wide variation in vehicle speeds on curves, there usually is an unbalanced force whether the curve is superelevated or not. This force results in tire side thrust which is counterbalanced by friction between tires and surface. The counterforce of friction is developed by distortion of the contact patch area of the tire.

The coefficient of friction f is the friction force divided by the weight perpendicular to the pavement and is expressed as the following simplified curve formula:

$$f = \frac{V^2}{15R} - e$$

This coefficient has been called lateral ratio, cornering ratio, unbalanced centrifugal ratio, friction factor, and side friction factor. Because of its widespread use, the last term is used herein. The upper limit of this factor is that at which the tire is skidding or at the point of impending skid. Because highway curves are designed to avoid skidding conditions with a margin of safety, the f values should be substantially less than the coefficient of friction of impending skid.

The side friction factor at which side skidding is imminent depends on a number of factors, among which the most important are the speed of the vehicle, the type and condition of the roadway surface, and the type and condition of the tires. Different observers have recorded different maximum rates at the same speeds for similar composition pavements, and logically so, because of the inherent differences in pavement texture, weather conditions, and tire condition. In general, studies show that the maximum side friction factors developed between new tires and wet concrete pavements range from about 0.5 at 20 mph to approximately 0.35 at 60 mph. For normal wet concrete pavement and smooth tires the value is about 0.35 at 45 mph. In all cases the studies show a decrease in friction values for an increase in speed (**18, 19, 20**).

Curves should not be designed directly on the basis of the maximum available side friction factor. The portion of the side friction factor that can be used with comfort and safety by the vast majority of drivers should be the maximum allowable value for design. Values that relate to pavements that are glazed, bleeding, or otherwise lacking in reasonable skid-resistant properties should not control design, because these conditions are avoidable and geometric design should be based on acceptable surface conditions attainable at reasonable cost.

In selecting maximum allowable side friction factors for use in design, one criterion is the point at which the centrifugal force is sufficient to cause the driver to experience a feeling of discomfort and cause him to react instinctively to avoid higher speed. The speed on a curve, at which discomfort due to the centrifugal force is evident to the driver, can be accepted as a design control for the maximum allowable amount of side friction. At lower nonuniform running speeds, which are typical in urban areas, drivers are more tolerant of discomfort, thus permitting employment of an increased amount of side friction for use in design of horizontal curves.

The ball-bank indicator has been widely used by research groups, local agencies, and highway departments as a uniform measure for the point of discomfort to set safe speeds on curves. It consists of a steel ball in a sealed glass tube. The ball is free to roll except for the damping effect of the liquid in the tube. Its simplicity of construction and operation has led to widespread acceptance as a guide for determination of safe speeds. With such a device mounted in a vehicle in motion, the ball-bank reading at any time is indicative of the combined effect of the body roll angle, the centrifugal force angle, and the superelevation angle as shown in Figure III-4.

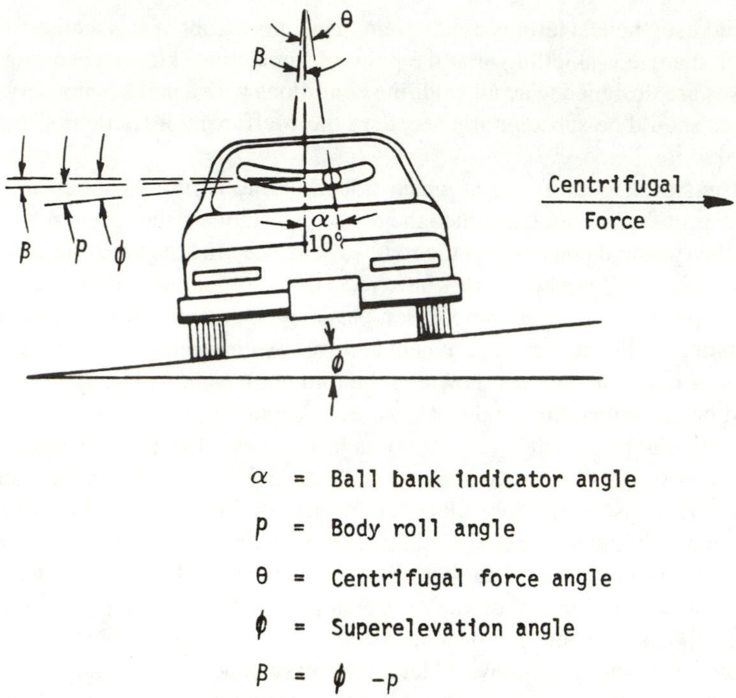

Figure III-4. Geometry for ball-bank indicator.

The centrifugal force developed as a vehicle travels at uniform speed on a curve causes the ball to roll out to a fixed angle position as shown in Figure III-4. A correction must be made for that portion of the force taken up in the small body roll angle. The indicated side force perceived by the vehicle occupants is thus on the order of $F \approx \tan(\alpha\text{-}p)$.

In a series of definitive tests (**20**) it was concluded that safe speeds on curves were indicated by ball-bank readings of 14° for speeds of 20 mph or less, 12° for speeds of 25 and 30 mph, and 10° for speeds of 35 through 50 mph. These ball-bank readings are indicative of side friction factors of 0.21, 0.18, and 0.15, respectively, for the test body roll angles and provide ample margin of safety against skidding.

From other tests (**21**), a maximum side friction factor of 0.16 for speeds up to 60 mph was recommended. For higher speeds this factor was to be reduced on an incremental basis. Speed studies on the Pennsylvania Turnpike (**19**) led to a conclusion that the side friction factor should not exceed 0.10 for design speeds of 70 mph and higher.

It should be recognized that other factors affect and act to control driver speed at conditions of high friction demand. Swerving becomes perceptible, drift angle increases, and increased steering effort is required to avoid involuntary lane line violation. Under these conditions the cone of vision narrows and is accompanied by an increasing sense of concentration and intensity considered undesirable by most drivers. These factors are more apparent to a driver under open road conditions.

When practical, the maximum factors selected should be conservative for dry pavements and provide a margin of safety for operating on pavements that are wet as well as ice or snow covered. The need for providing skid-resistant pavement surfacing for these conditions cannot be overemphasized because superimposed on the frictional demands dictated by roadway geometry are those often made by driving maneuvers such as braking, sudden lane changes, and minor changes in direction within the lane. In these short-term maneuvers the discomfort threshold is not perceived immediately and, consequently, high friction demand can exist but not be perceived in time for compensation by a comfortable speed reduction.

Figure III- 5 summarizes the finding of the cited tests relating to side friction factors recommended for curve design. Although some variation is noted, all are in agreement that the side friction factor should be lower for high-speed design than for low-speed design. Refer to subsequent sections in this chapter ("Design for All Rural Highways and High Speed Urban Streets" and "Design for Low Speed Urban Streets") for the recommended values of the side friction factor.

Distribution of e and f Over a Range of Curves

For a given design speed there are five methods for counteracting centrifugal force on curves by use of e or f, or both. These methods follow, and the resulting relation are illustrated in Figure III-6:

1. Superelevation and side friction are directly proportional to the degree of curve, i.e., a straight-line relation exists between D = O and D = D_{max}.
2. Side friction is such that a vehicle traveling at design speed has all centrifugal force counteracted in direct proportion by side friction on curves up to those requiring f_{max}. For sharper curves, f remains at f_{max} and e is then used in direct proportion to the continued increase in curvature until e reaches e_{max}.
3. Superelevation is such that a vehicle traveling at design speed has all centrifugal force counteracted in direct proportion by superelevation

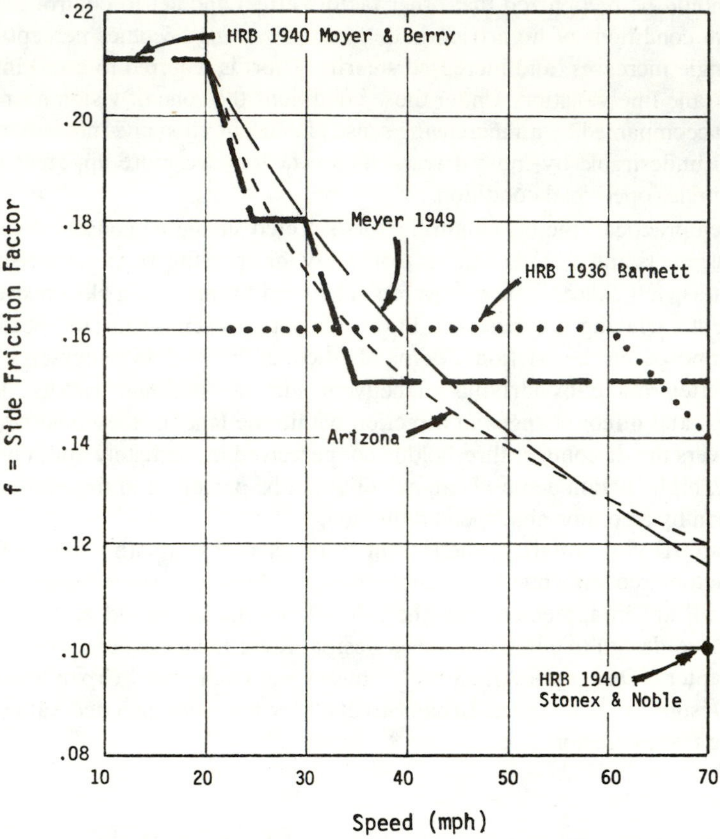

Figure III-5. Side friction factors.

on curves up to that requiring e_{max}. For sharper curves, e remains at e_{max} and f is then used in direct proportion to the continued increase in curvature until f reaches f_{max}.

4. Method 4 is the same as method 3, except that it is based on average running speed instead of design speed.
5. Superelevation and side friction are in a curvilinear relation with degree of curve, with values between those of methods 1 and 3.

Figure III-6A shows the comparative relations of superelevation versus degree of curve for these methods. Figure III-6B shows the corresponding value of side friction for a vehicle traveling at design speed, and Figure III-6C for one traveling at the corresponding average running speed.

The straight-line relation between superelevation and degree of curve of method 1 results in a similar relation between side friction and degree of curve for vehicles traveling at either design or running speed. This method has considerable merit and logic other than its simplicity. On any particular highway, the horizontal alinement consists of tangents and curves of varying sharpness up to the maximum for the design speed. Application of superelevation in amounts directly proportional to the degree of curve would, for vehicles traveling at uniform speed, result in side friction factors with a straight-line variation from zero on tangents (ignoring crown) to the maximum at the sharpest permissible curvature. This method might appear to be an ideal means of distributing the side friction factor, but its success depends on travel at a constant speed by each vehicle in the traffic stream, regardless of whether travel is on a tangent, a curve of intermediate degree, or a curve with the maximum curvature for that design speed. While uniform speed is the aim of most drivers and can be obtained on well-designed highways when volumes are not heavy, there is a tendency for some drivers to travel faster on tangents and the flatter curves than on the sharper curves, particularly after being delayed by inability to pass slower moving vehicles. This tendency points to the desirability of providing superelevation rates for intermediate curves somewhat in excess of those that result from use of method 1.

Method 2 uses side friction to counteract all centrifugal force up to the curvature corresponding to the maximum allowable side friction, and this maximum allowable side friction is available on all sharper curves. In this method, superelevation is introduced only after the maximum allowable side friction has been used. Therefore, no superelevation is required on flatter curves that require less than maximum allowable side friction for vehicles traveling at the design speed. (See curve 2 in Figure III-6A.) When superelevation is needed, it increases rapidly as curves with maximum allowable side friction grow sharper. Because this method is completely dependent on available side friction, its use is generally limited to locations where travel speed is not uniform, such as on urban streets. This method is particularly advantageous on low-speed urban streets where, because of various constraints, superelevation frequently cannot be provided.

In method 3, which was practiced many years ago, superelevation to counteract all centrifugal force when traveling at the design speed is provided on all curves up to that requiring maximum practical superelevation, and this maximum superelevation is provided on all sharper curves. Under this practice no side friction is required on flat curves with less than maximum superelevation for vehicles traveling at the design speed, curve 3 in Figure III-6B, and side friction required increases rapidly as curves with maximum superelevation grow sharper. Further, as shown by curve 3 in Figure III-6C, for vehicles traveling at average running speed, this superelevation method

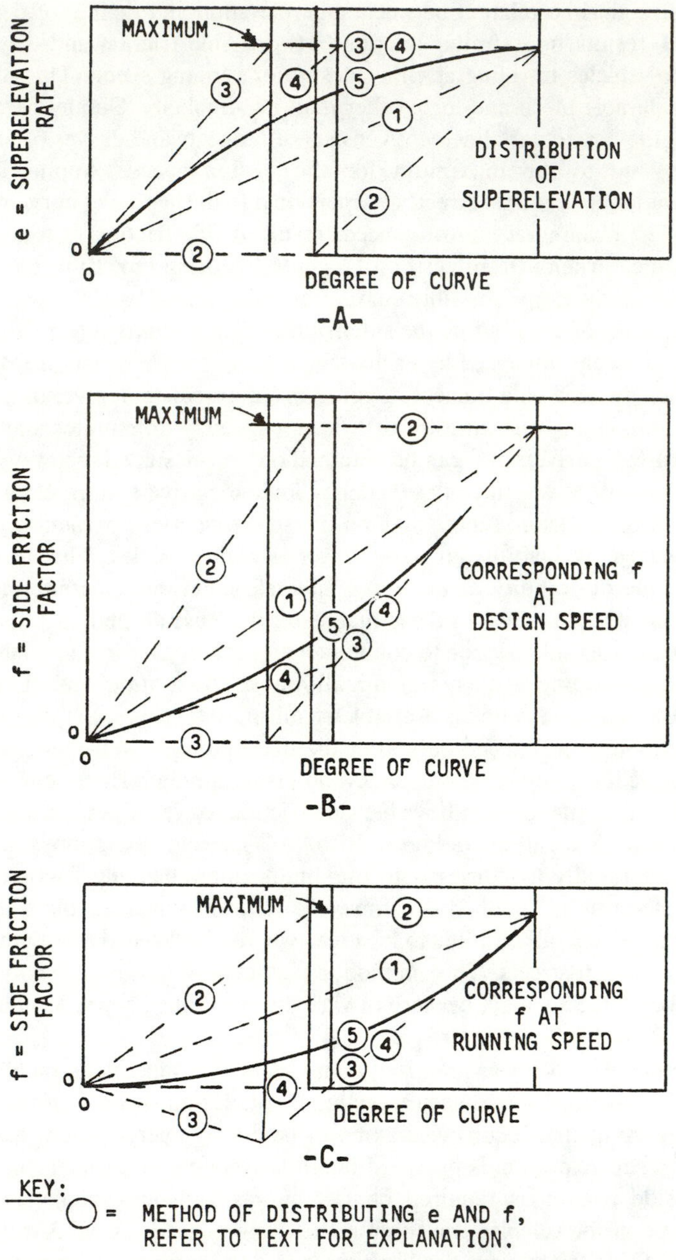

Figure III-6. Methods of distributing superelevation and side friction.

results in negative friction for curves from very flat up to about the middle of the range, above which curvature the side friction increases rapidly up to the maximum for the maximum degree of curve. This marked difference in side friction required for different curves is not logical and results in erratic driving, either at design or average running speed.

Method 4 is an attempt to overcome the deficiencies of method 3 by using superelevation on a speed lower than design speed. This method has been widely used with an average running speed for which all centrifugal force is compensated by superelevation of curves flatter than that requiring the maximum rate of superelevation. This average running speed was an approximation that, as presented in Table III-13, varies from 83 to 100 percent of design speed. Curve 4 in Figure III-6A shows that in using this method the maximum superelevation is reached near the middle of the curvature range. Figure III-6C shows that at average running speed no side friction is required up to this curvature, and side friction increases rapidly and in direct proportion for sharper curves. This method has the same disadvantages as method 3, but they apply to a smaller degree.

To favor the overdriving characteristics that occur on flat to intermediate curves, it is desirable that the superelevation approximate that obtained by method 4. Overdriving on such curves is not dangerous because superelevation counteracts nearly all centrifugal force at average running speed, and considerable side friction is available for greater speed. On the other hand, it also is desirable to favor method 1, which avoids a substantial part of the range of curves with maximum superelevation. Using method 5, a curved line (curve 5, as shown within the triangular working range between curves 1 and 4 in Figure III-6A) represents a superelevation and side friction and distribution reasonably satisfying both aspects. Curve 5, of parabolic form, represents a practical distribution over the range of curvature.

Design Considerations

In highway curve design it is necessary to determine superelevation rates which are applicable over the range of curvature for each design speed. One extreme of this range is the maximum superelevation established by practical considerations and used to determine the maximum curvature for each speed. The maximum superelevation is different for different highway conditions. At the other extreme, no superelevation is needed for tangent highways or highways with extremely long-radius curves. For curvature between these extremes and for a given design speed, the superelevation should be distributed in such manner that there is a logical relation between the side friction factor and the applied superelevation rate.

Maximum Superelevation Rates

For a given speed the maximum superelevation rate and the assumed value for maximum side friction factor in combination determine the maximum curvature. The maximum rates of superelevation usable on highways are controlled by several factors: climate conditions, i.e., frequency and amount of snow and ice; terrain conditions, i.e., flat or mountainous; type of area, i.e., rural or urban; and frequency of very slow-moving vehicles that would be subject to uncertain operation. Consideration of these factors jointly leads to the conclusion that no single maximum superelevation rate is universally applicable and that a range of values must be used.

The maximum superelevation rate for open highways in common use is of the order of 0.100 or sometimes a maximum rate of 0.120 is used. Use of rates above 0.08 invariably are in areas without snow and ice. Although higher superelevation rates offer an advantage to the group of drivers traveling at high speeds, current practice demonstrates that rates above 0.120 are beyond practical limits for open highways. This practice is in recognition of the combined controls of construction processes, maintenance difficulties, and operation of vehicles at low speeds.

Thus, a superelevation rate of 0.120 appears to represent a maximum practical value where snow and ice do not exist. A superelevation rate of 0.120 may be used on low-volume gravel-surfaced roads to facilitate cross drainage, however, rates of this magnitude can cause higher speeds, conducive to rutting and displacement of gravel. Generally 0.08 is recognized as a reasonable value.

Where snow and ice are factors, tests and experience show that a superelevation rate of about 0.080 is a logical maximum to minimize slipping across a highway when stopped or attempting to slowly gain momentum from a stopped position. One series of tests (**18**) yielded coefficient of friction values for ice ranging from 0.050 to 0.200 depending on the condition of the ice, i.e., wet, dry, clean, smooth, or rough. Other tests (**22**) corroborated these values. The lower value probably occurs only under thin film "quick freeze" conditions at a temperature of about 30°F in the presence of water on the pavement. Similar low values may prevail with thin layers of mud on the pavement surface, with oil or flushed spots, and with high speeds and a sufficient depth of water on the pavement surface to permit hydro-planing. Tests on loose or packed snow show coefficients of friction ranging from 0.200 to 0.400. Some agencies have adopted a maximum rate of 0.080. Some agencies believe too, that 0.080 represents a logical maximum, regardless of snow or ice conditions, to avoid excessive effort and erratic operation in driving slowly around the curve.

Where traffic congestion or extensive marginal development acts to curb top speeds, it is common practice to utilize a low maximum rate of superelevation, usually 0.040 to 0.060. Similarly, either a low maximum rate of superelevation or no superelevation is employed within important intersection areas or where there is a tendency to drive slowly because of turning and crossing movements, warning devices, and signals. In these areas it is difficult to warp crossing pavements for drainage without negative superelevation for some turning movements.

In summary of the above, it may be concluded that (1) several rates, rather than a single rate, of maximum superelevation should be recognized in establishing design controls for highway curves, (2) a rate of 0.120 should not be exceeded, and (3) a rate of 0.040 or 0.060 is applicable for urban design in areas with little or no constraints, or (4) superelevation may be omitted on low speed urban streets subjected to severe constraints. Accordingly, five maximum superelevation rates—0.040, 0.060, 0.080, 0.100 and 0.120—are used herein. For each of these rates the maximum curvature and actual superelevation rates for flatter curves are determined. For actual design use an agency will generally use different values within the range of rates for different road systems.

Maximum Degree of Curvature and Minimum Radius

The maximum degree of curvature (or the minimum radius) is a limiting value of curvature for a given design speed and is determined from the maximum rate of superelevation and the maximum allowable side friction factor. Use of sharper curvature for that design speed would call for superelevation beyond the limit considered practical or for operation with tire friction beyond the safe limit, or both. Thus, the maximum degree of curvature is a significant value in alinement design. The maximum degree of curvature is also a necessary and important control value for determination of superelevation rates for flatter curves.

The minimum safe radius R_{min} can be calculated directly from the simplified curve formula introduced earlier under the section, "Side Friction Factor."

$$R_{min} = \frac{V^2}{15\,(e + f)}$$

The degree of curve of a given circular curve is the angle (or number of degrees) subtended at the center by an arc of 100-ft length. By using D as the degree of circular curve (arc definition), D = 5,729.6/R, the simplified curve formula becomes the following:

$$D_{max} = \frac{85{,}660\,(e + f)}{V^2}$$

Design for All Rural Highways and High-Speed Urban Streets

On all rural highways, on urban freeways, and on urban streets where speed is relatively high and relatively uniform, horizontal curves are generally superelevated and successive curves are generally balanced to provide a smooth-riding transition from one curve to the next. A balanced design for a series of curves of varying radii is provided by the distribution of e and f values, as previously discussed, from maximum superelevation conditions to the crowned pavement section.

Figure III-7 shows the recommended values of the side friction factor for all rural highways and high-speed urban streets as a solid line superimposed on the analysis curves from Figure III-5. These recommended values provide a reasonable margin of safety at high speeds and give somewhat lower rates for the low design speeds than do some of the other curves. The lower rates at the low speeds provide a greater margin of safety to offset the tendency of many motorists to overdrive highways of low design speed.

From this discussion and from the study data previously discussed, it is concluded that safe, maximum allowable side friction factors for use in curve design for all rural highways and all high-speed urban streets should be as shown by the solid line in Figure III-7, varying directly with the design speed from 0.17 at 20 mph to 0.14 at 50 mph, thence varying directly to 0.10 at 70 mph. The recent research report "Side Friction For Superelevation on Horizontal Curves" confirms the validity of these design values (**59**).

On the basis of the maximum allowable side friction factors from Figure III-7, Table III-6 gives the maximum degree of curve and the minimum radius for each of the five maximum superelevation rates for design speeds from 20 mph to 70 mph.

Method 5, described previously, is recommended for the distribution of e and f for all curves less than the maximum degree of curve for all rural highways and high-speed urban streets. Use of method 5 is discussed in the following text, tables, and figures.

Procedure for Development of Finalized e Distribution

The side friction factors shown as the solid line on Figure III-7 represent the maximum allowable f values for each speed. When these values are used in conjunction with the recommended method 5, they determine the f distribution curves for the various speeds. Subtracting these computed f values from the computed value of e + f at design speed, the finalized e distribution is thus obtained. (See Figure III-8.)

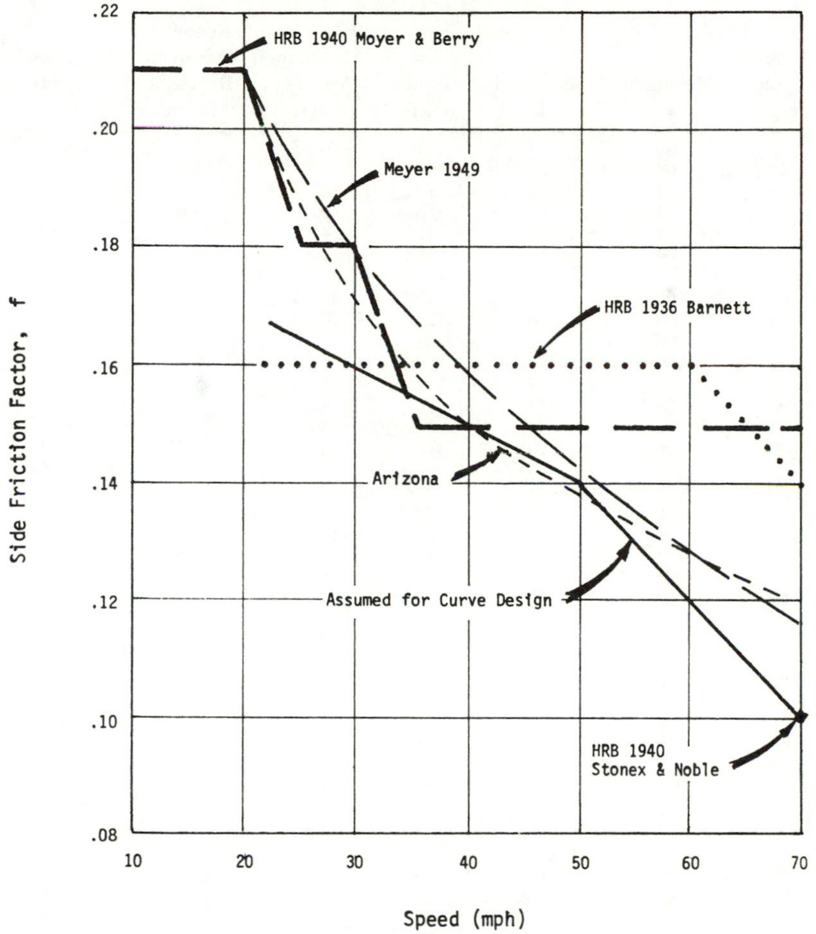

Figure III-7. Side friction factors for rural highways and high-speed urban streets.

The resulting finalized e distribution curves, based on method 5 and used herein, are shown in Figures III-9 to III-14.

As Figure III-8 illustrates, the f distribution curve at the design speed, using method 5, results in an unsymmetrical parabolic curve with legs 1 and 2. These legs correspond to curves 4 and 3-4, respectively, in Figure III-6B. The derivation of the equations necessary to compute the f and finalized e distributions are given on Figure III-8.

Design Speed (mph)	Maximum e	Maximum f	Total (e+f)	Maximum Degree of Curve	Rounded Maximum Degree of Curve	Maximum[a] Radius (ft)
20	.04	.17	.21	44.97	45.0	127
30	.04	.16	.20	19.04	19.0	302
40	.04	.15	.19	10.17	10.0	573
50	.04	.14	.18	6.17	6.0	955
55	.04	.13	.17	4.83	4.75	1,186
60	.04	.12	.16	3.81	3.75	1,528
20	.06	.17	.23	49.25	49.25	116
30	.06	.16	.22	20.94	21.0	273
40	.06	.15	.21	11.24	11.25	509
50	.06	.14	.20	6.85	6.75	849
55	.06	.13	.19	5.40	5.5	1,061
60	.06	.12	.18	4.28	4.25	1,348
65	.06	.11	.17	3.45	3.5	1,637
70	.06	.10	.16	2.80	2.75	2,083
20	.08	.17	.25	53.54	53.5	107
30	.08	.16	.24	22.84	22.75	252
40	.08	.15	.23	12.31	12.25	468
50	.08	.14	.22	7.54	7.5	764
55	.08	.13	.21	5.97	6.0	960
60	.08	.12	.20	4.76	4.75	1,206
65	.08	.11	.19	3.85	3.75	1,528
70	.08	.10	.18	3.15	3.0	1,910
20	.10	.17	.27	57.82	58.0	99
30	.10	.16	.26	24.75	24.75	231
40	.10	.15	.25	13.38	13.25	432
50	.10	.14	.24	8.22	8.25	694
55	.10	.13	.23	6.53	6.5	877
60	.10	.12	.22	5.23	5.25	1,091
65	.10	.11	.21	4.26	4.25	1,348
70	.10	.10	.20	3.50	3.5	1,637
20	.12	.17	.29	62.10	62.0	92
30	.12	.16	.28	26.65	26.75	214
40	.12	.15	.27	14.46	14.5	395
50	.12	.14	.26	8.91	9.0	637
55	.12	.13	.25	7.10	7.0	807
60	.12	.12	.24	5.71	5.75	996
65	.12	.11	.23	4.66	4.75	1206
70	.12	.10	.22	3.85	3.75	1528

NOTE: In recognition of safety considerations, use of e_{max} = 0.04 should be limited to urban conditions.
[a]Calculated using rounded maximum degree of curve.

Table III-6. Maximum degree of curve and minimum radius determined for limiting values of e and f, rural highways and high-speed urban streets.

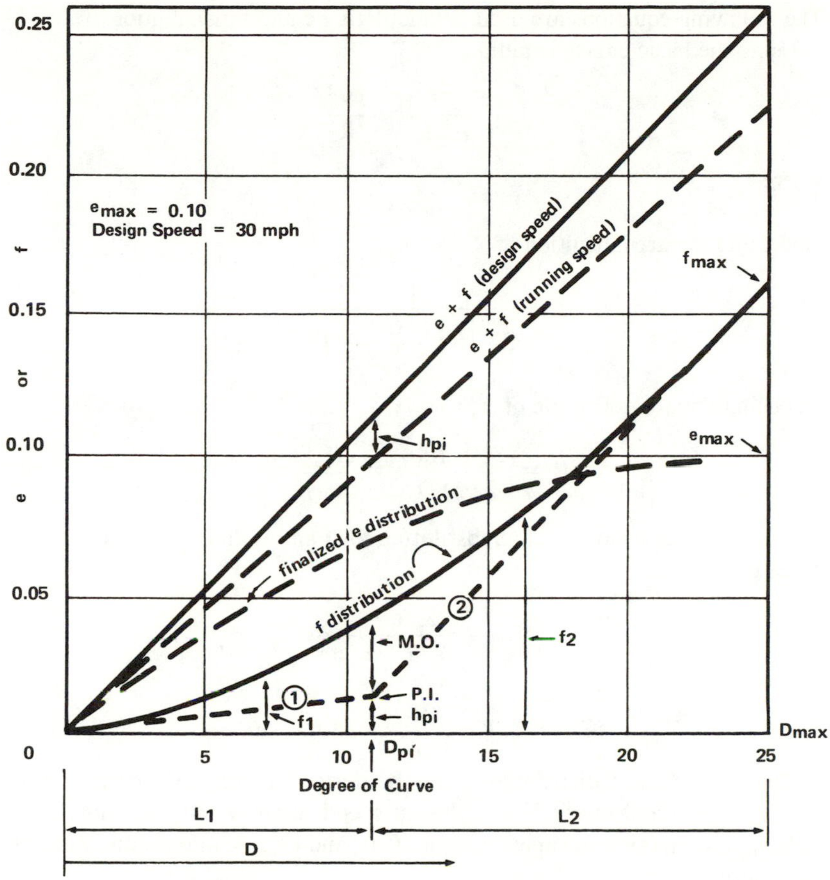

Figure III-8. Method 5 procedure for development of the finalized e distribution.

The following equations are used in calculating e and f distributions:

Using the basic curve formula

$$\frac{e + f}{1 - ef} = \frac{JV^2}{R}$$

where

land using the are definition of R.

$$J = \frac{1}{32.16\left(\frac{60^2}{5{,}280}\right)^2}$$

and using the arc definition of R,

$$R = \frac{360\,(100)}{2\,(\pi)\,D_c} = \frac{18{,}000}{(\pi)D_c}$$

where D_c = degree of curve. Substituting for J and R and solving for D_c produces the result,

$$D_c = \frac{K\,(e + f)}{V^2\,(1 - ef)}$$

where $K = \frac{RD_c}{J}$. If $D_c = 1$. $K = \frac{R}{J} = 85{,}659.56$.

Thus, the e and f distributions may be derived as follows, neglecting the $(1 - ef)$ term and given: V_D = design speed, mph; V_R = running speed, mph; e_{max} = maximum superelevation, ft/ft; and f_{max} = maximum allowable side friction factor.

$$D_{max} = \frac{K\,(e_{max} + f_{max})}{{V_D}^2}$$

where D_{max} = maximum D_c and

$$D_{PI} = \frac{Ke_{max}}{{V_R}^2}$$

where $D_{PI} = D_c$ at the point of intersection, PI, of legs (1) and (2) of the f distribution parabolic curve (= D_c at the point of intersection of e_{max} and $(e + f)_R$).

Because $(e + f)_D - (e + f)_R = h$, at point D_{PI} the equations reduce to the following:

$$h_{PI} = e_{max}\left(\frac{{V_D}^2}{{V_R}^2}\right) - e_{max}$$

where h_{PI} = PI offset from the D axis.
Also,

$$S_1 = \frac{h_{PI}}{D_{PI}}$$

where S_1 = slope of leg 1 and

$$S_2 = \frac{f_{max} - h_{PI}}{D_{max} - D_{PI}}$$

where S_2 = slope of leg 2.

The equation for the middle ordinate of an unsymmetrical vertical curve is the following:

$$MO = \frac{L_1 L_2 (S_2 - S_1)}{2(L_1 + L_2)}$$

where $L_1 = D_{PI}$ and $L_2 = D_{max} - D_{PI}$. It follows that

$$MO = D_{PI} (D_{max} - D_{PI}) \frac{S_2 - S_1}{2 D_{max}}$$

where MO = middle ordinate of the f distribution curve, and

$$(e + f)_D = \frac{(e_{max} + f_{max})D}{D_{max}}$$

in which $D = D_c$ at any point

Use the general vertical curve equation

$$\frac{Y}{MO} = \left(\frac{x}{L} \right)^2$$

with D measured from the vertical axis.

For $D \leq D_{PI}$,

$$f_1 = MO \frac{D}{D_{PI}}^2 + S_1 D$$

where f_1 = f distribution at any point $D \leq D_{PI}$ and $e_1 = (e + f)_D - f_1$, where e_1 = e distribution at any point $D \leq D_{PI}$.

For $D \geqq D_{PI}$,

$$f_2 = MO \left(\frac{D_{max} - D}{D_{max} - D_{PI}} \right)^2 + h_{PI} + S_2 (D - D_{PI})$$

where f_2 = f distribution at any point $D > D_{pi}$ and $e_2 = (e + f)_D - f_2$, where e_2 = e distribution at any point $D > D_{PI}$.

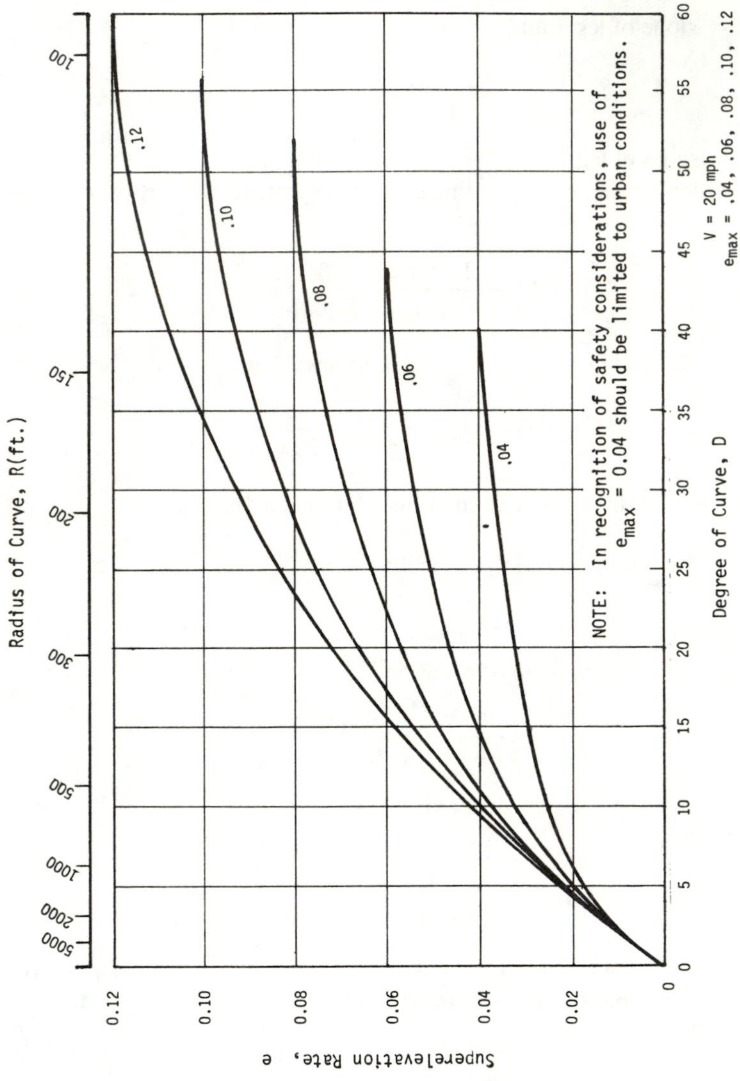

Figure III-9. Design superelevation rates.

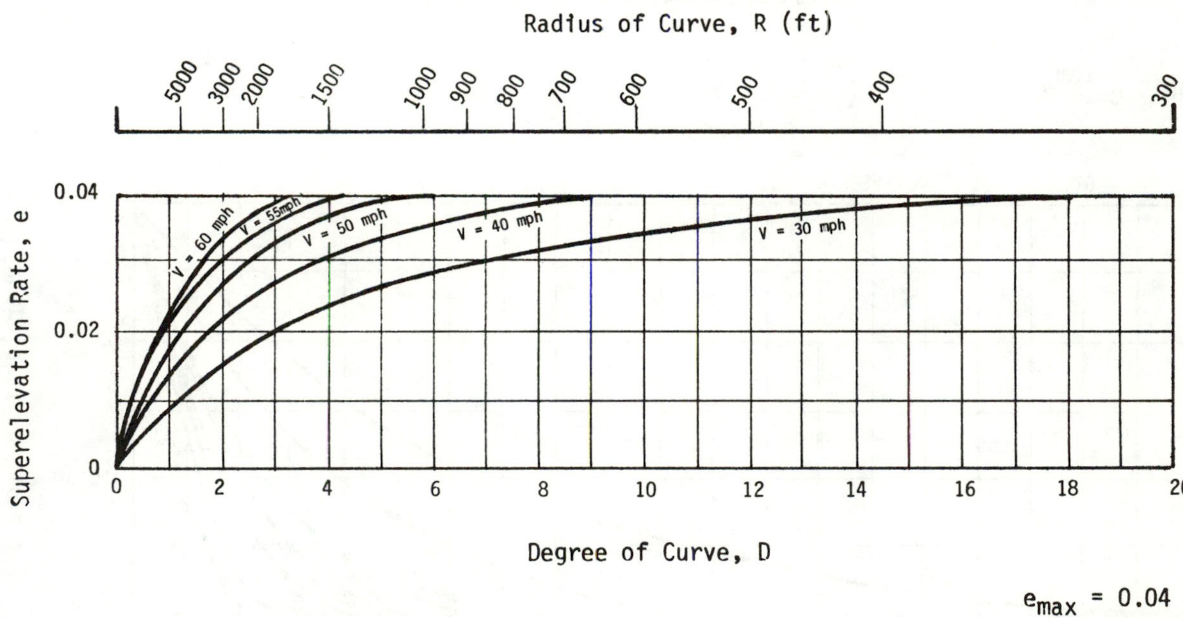

NOTE: In recognition of safety considerations, use of e_{max} = 0.04 should be limited to urban conditions.

Figure III-10. Design superelevation rates.

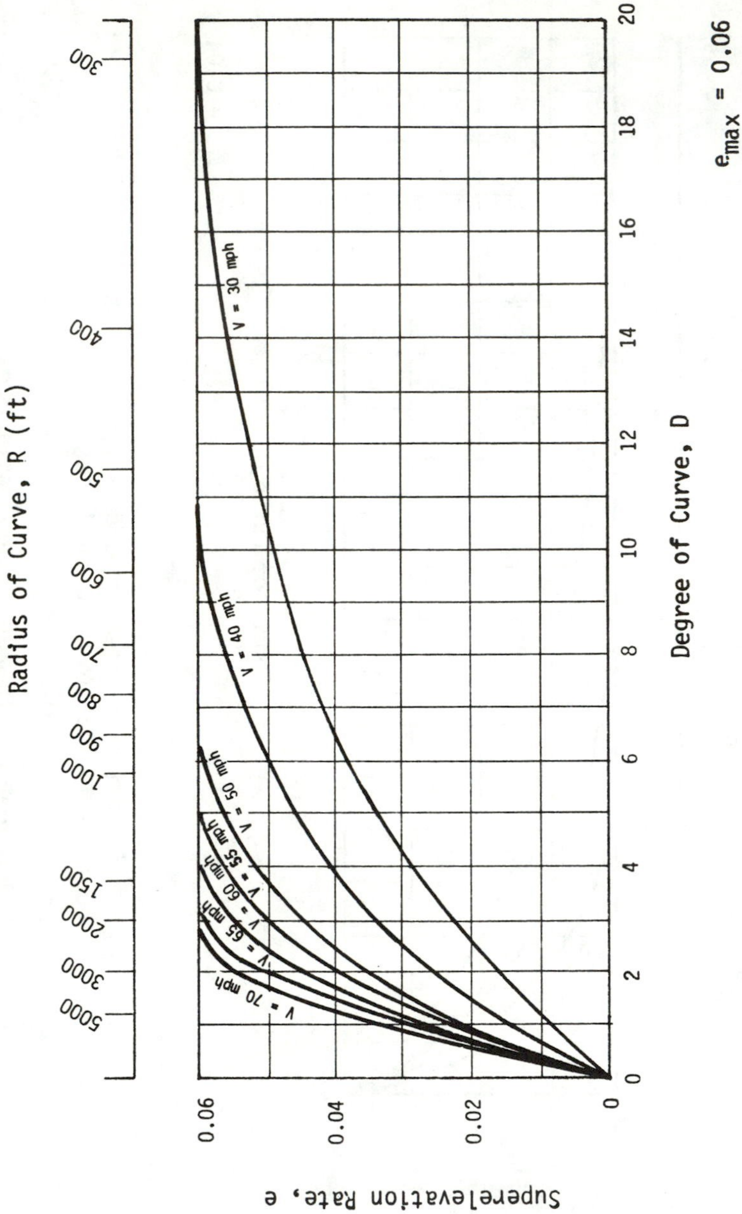

Figure III-11. Design superelevation rates.

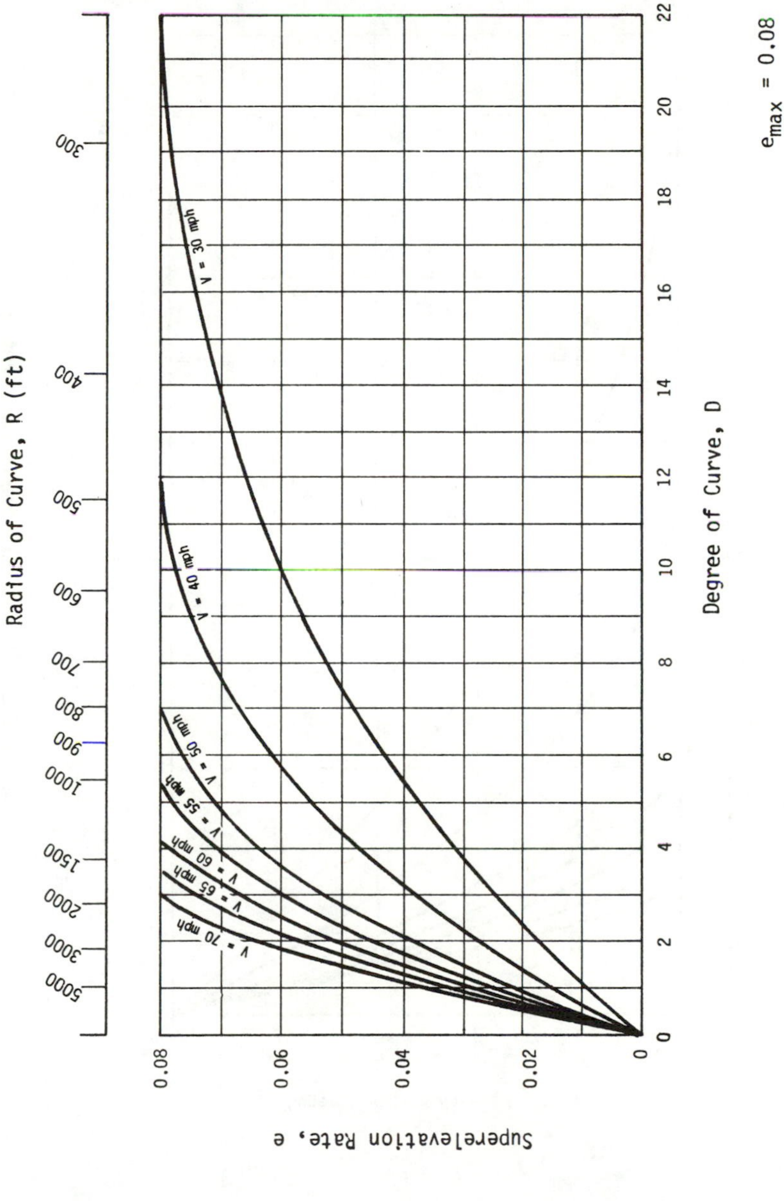

Figure III-12. Design superelevation rates.

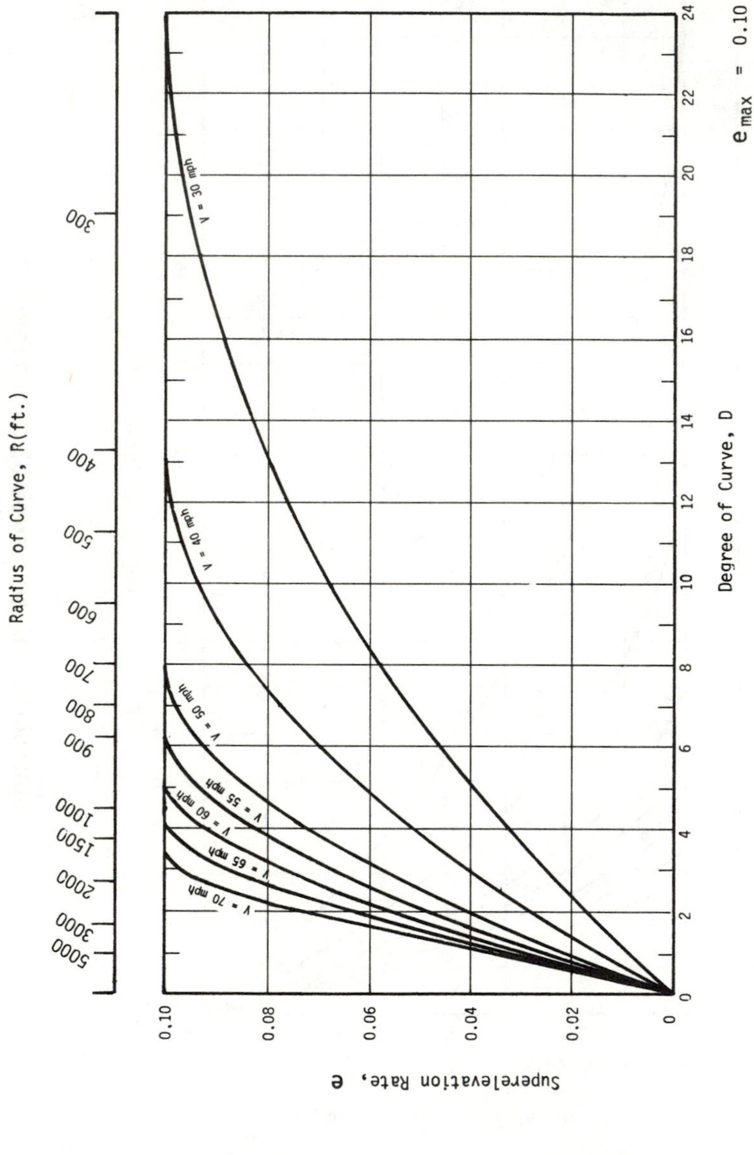

Figure III-13. Design superelevation rates.

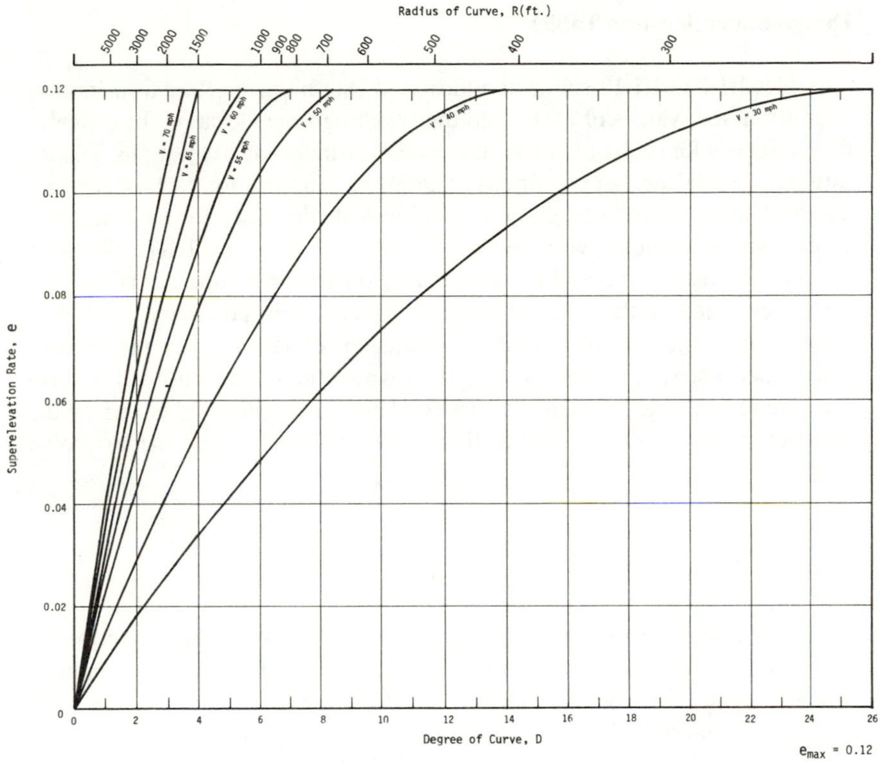

Figure III-14. Design superelevation rates.

Figure III-8 is a typical layout illustrating the method 5 procedure for development of the finalized e distribution. Here, the layout is for a given design speed of 30 mph and an e_{max} of 0.10. By using Tables III-13 and III-6 the running speed and maximum allowable side friction factor are determined to be 28 mph and 0.16, respectively. By using these given values in the equations shown, $D_{max} = 24.75°$, $D_{PI} = 10.93°$, and $h_{PI} = 0.015$. The slopes S_1 and S_2 of legs 1 and 2 are 0.0014 and 0.0105, respectively. Substituting, the middle ordinate becomes 0.028. The $(e + f)_{PI}$ value at point D_{PI} is 0.115. The f_1 $(=f_2)$ value at point D_{PI} is then h_{PI} + MO, or 0.043. The e distribution value is thus $(e + f)_{PI}$ at design speed for point D_{PI} minus the f_1 value for point D_{PI}, or 0.072. This corresponds to the e value shown for 11° at the 30-mph design speed in Table III-11.

Design Superelevation Tables

Tables III-7 to III-12 show, in addition to length of runoff or transition as later discussed, values of D, R, and the resulting superelevation for different design speeds for each of the five assumed maximum superelevations, i.e., for different general design conditions. The minimum radii for each of the five assumed maximum superelevations were calculated from the standard superelevation formula, with the use of f values from Figure III-7. Method 5 was used to distribute e and f in calculating the required superelevation rates for the remainder of the range of curves. The term "normal crown" (NC) designates curves that are so flat that the elimination of adverse crown is not considered necessary, and thus the normal crown sections can be used. The term "remove crown" (RC) designates curves where it is adequate to eliminate the adverse crown by superelevating the entire roadway at the normal crown slope.

Sharpest Curve Without Superelevation

The minimum rate of cross slope applicable to traveled ways is determined by drainage requirements. Consistent with the type of highway and amount of rainfall, snow, and ice, the usually accepted minimum values range from 0.015 for high-type surfaces to approximately 0.020 for low-type surfaces; see the section "Cross Slope" in Chapter IV. For discussion purposes, a value of 0.015 is used herein as a single value representative of high-type uncurbed pavements. Steeper cross slopes are generally needed on curbed pavements to minimize ponding on the outside through lane.

The shape or form of the normal pavement cross slope varies. Some States and many municipalities use a curved crown section for two-lane roadways, usually parabolic in form. Others employ a straight-line section for each lane.

Very flat horizontal curves require no superelevation. Traffic entering a curve to the right has some superelevation in the normal cross slope. Traffic entering a curve to the left has an adverse or negative superelevation in the normal cross slope, but with flat curves the side friction required to counteract both centrifugal force and negative superelevation is small. However, on successively sharper curves for the same speed, a point is reached where the combination of centrifugal force and negative superelevation overcomes the allowable side friction, and a positive slope across the entire pavement is desirable to help counteract the centrifugal force. A part of any superelevation standard is an expression for this condition, the minimum curvature for which superelevation is required, or stated otherwise, the maximum curvature for which a crowned pavement cross section is appropriate.

D	R (ft)	e_{max} = 0.04, V = 20 mph: e	e_{max} = 0.04, V = 20 mph: L(ft) Two Lanes	e_{max} = 0.06, V = 20 mph: e	e_{max} = 0.06, V = 20 mph: L(ft) Two Lanes	e_{max} = 0.08, V = 20 mph: e	e_{max} = 0.08, V = 20 mph: L(ft) Two Lanes	e_{max} = 0.10, V = 20 mph: e	e_{max} = 0.10, V = 20 mph: L(ft) Two Lanes	e_{max} = 0.12, V = 20 mph: e	e_{max} = 0.12, V = 20 mph: L(ft) Two Lanes
1°00′	5730	NC	0	NC	0	NC	0	NC	0	NC	0
2°00′	2865	NC	0	NC	0	NC	0	NC	0	NC	0
3°00′	1910	NC	0	NC	0	NC	0	NC	0	NC	0
4°00′	1432	RC	50	RC	50	RC	50	RC	50	RC	50
5°00′	1146	RC	50	.020	50	.021	50	.022	50	.022	50
6°00′	955	.020	50	.023	50	.025	50	.026	50	.026	50
7°00′	819	.022	50	.026	50	.028	50	.029	50	.030	50
8°00′	716	.023	50	.029	50	.031	50	.033	50	.034	50
9°00′	637	.024	50	.031	50	.035	50	.036	60	.038	60
10°00′	573	.025	50	.033	50	.037	60	.040	60	.041	70
12°00′	477	.027	50	.037	60	.043	60	.046	70	.048	80
14°00′	409	.028	50	.039	60	.047	70	.052	70	.055	90
16°00′	358	.030	50	.041	60	.051	70	.057	80	.061	100
18°00′	318	.031	50	.044	70	.054	80	.062	90	.067	110
20°00′	286	.032	50	.046	70	.057	90	.066	100	.072	120
24°00′	239	.035	60	.049	70	.062	100	.073	110	.081	130
28°00′	205	.036	60	.052	80	.067	100	.079	120	.089	140
32°00′	179	.038	60	.055	80	.070	110	.084	130	.096	150
36°00′	159	.039	60	.057	90	.074	110	.089	140	.102	160
40°00′	143	.040	60	.059	90	.076	120	.092	140	.107	170
44°00′	130	.040	60	.060	90	.078	120	.095	150	.111	180
48°00′	119			.060	90	.079	120	.098	150	.115	180
52°00′	110	D_{max} = 45°00′				.080	120	.099	150	.117	190
56°00′	102			D_{max} = 49°15′				.100	150	.119	190
60°00′	95					D_{max} = 53°30′				.120	190
								D_{max} = 58°00′			
										D_{max} = 62°00′	

e_{max} = 0.04, 0.06, 0.08, 0.10, 0.12

V = 20 mph

D = degree of curve
R = radius of curve
V = assumed design speed
e = rate of superelevation
L = minimum length of runoff (does not include tangent runout)
NC = normal crown section
RC = remove adverse crown, superelevate at normal crown slope

NOTE: In recognition of safety considerations, use of e_{max} = 0.04 should be limited to urban conditions.

Table III-7. Values for design elements related to design speed and horizontal curvature.

D	R (ft)	V = 30 mph e	V = 30 mph L(ft) Two Lanes	V = 30 mph L(ft) Four Lanes	V = 40 mph e	V = 40 mph L(ft) Two Lanes	V = 40 mph L(ft) Four Lanes	V = 50 mph e	V = 50 mph L(ft) Two Lanes	V = 50 mph L(ft) Four Lanes	V = 55 mph e	V = 55 mph L(ft) Two Lanes	V = 55 mph L(ft) Four Lanes	V = 60 mph e	V = 60 mph L(ft) Two Lanes	V = 60 mph L(ft) Four Lanes
0°15′	22,918	NC	0	0	NC	0	0	NC	0	0	NC	0	0	NC	0	0
0°30′	11,459	NC	0	0	NC	0	0	NC	0	0	NC	0	0	NC	0	0
0°45′	7,639	NC	0	0	NC	0	0	NC	0	0	RC	160	160	RC	175	265
1°00′	5,730	NC	0	0	NC	0	0	RC	150	225	.021	160	160	.023	175	265
1°30′	3,820	NC	0	0	RC	125	190	.024	150	225	.026	160	160	.029	175	265
2°00′	2,865	RC	100	150	.022	125	190	.027	150	225	.030	160	160	.033	175	265
2°30′	2,292	RC	100	150	.025	125	190	.030	150	225	.033	160	160	.037	175	265
3°00′	1,910	.020	100	150	.027	125	190	.033	150	225	.036	160	160	.039	175	265
3°30′	1,637	.022	100	150	.028	125	190	.035	150	225	.038	160	160	.040	175	265
4°00′	1,432	.024	100	150	.030	125	190	.037	150	225	.039	160	160			
5°00′	1,146	.026	100	150	.033	125	190	.039	150	225				Dmax = 3°45′		
6°00′	955	.028	100	150	.035	125	190	.040	150	225	Dmax = 4′45′					
7°00′	819	.030	100	150	.037	125	190									
8°00′	716	.031	100	150	.039	125	190	Dmax = 6°00′								
9°00′	637	.033	100	150	.040	125	190									
10°00′	573	.034	100	150	.040	125	190									
11°00′	521	.035	100	150												
12°00′	477	.036	100	150	Dmax = 10°00′											
13°00′	441	.037	100	150												
14°00′	409	.038	100	150												
16°00′	358	.039	100	150												
18°00′	318	.040	100	150												
19°00′	302	.040	100	150												
		Dmax = 19°00′														

$e_{max} = 0.04$

D = degree of curve
R = radius of curve
V = assumed design speed
e = rate of superelevation
L = minimum length of runoff (does not include tangent runout)

NC = normal crown section
RC = remove adverse crown, superelevate at normal crown slope

Note: Lengths rounded in multiples of 25 or 50 ft permit simpler calculations
In recognition of safety considerations, use of $e_{max} = 0.04$ should be limited to urban conditions

Table III-8. Values for design elements related to design speed and horizontal curvature.

D	R (ft)	V = 30 mph e	V = 30 mph L(ft) Two Lanes	V = 30 mph L(ft) Four Lanes	V = 40 mph e	V = 40 mph L(ft) Two Lanes	V = 40 mph L(ft) Four Lanes	V = 50 mph e	V = 50 mph L(ft) Two Lanes	V = 50 mph L(ft) Four Lanes	V = 55 mph e	V = 55 mph L(ft) Two Lanes	V = 55 mph L(ft) Four Lanes	V = 60 mph e	V = 60 mph L(ft) Two Lanes	V = 60 mph L(ft) Four Lanes	V = 65 mph e	V = 65 mph L(ft) Two Lanes	V = 65 mph L(ft) Four Lanes	V = 70 mph e	V = 70 mph L(ft) Two Lanes	V = 70 mph L(ft) Four Lanes
0°15′	22,918	NC	0	0	NC	0	0	NC	0	0	NC	0	0	NC	0	0	NC	0	0	NC	0	0
0°30′	11,459	NC	0	0	NC	0	0	NC	0	0	NC	0	0	NC	0	0	RC	190	190	RC	200	200
0°45′	7,639	NC	0	0	NC	0	0	RC	0	0	RC	160	160	.021	175	175	.024	190	190	.026	200	200
1°00′	5,730	NC	0	0	NC	0	0	.020	150	150	.023	160	160	.027	175	175	.030	190	190	.033	200	200
1°30′	3,820	NC	0	0	.020	125	125	.028	150	150	.032	160	160	.037	175	175	.041	190	190	.046	200	200
2°00′	2,865	RC	100	100	.025	125	125	.035	150	150	.040	160	160	.045	175	180	.050	190	210	.055	200	230
2°30′	2,292	.020	100	100	.030	125	125	.040	150	150	.045	160	170	.051	175	200	.056	190	230	.059	200	260
3°00′	1,910	.023	100	100	.034	125	125	.045	150	160	.050	160	190	.055	175	220	.059	190	250			
3°30′	1,637	.026	100	100	.038	125	125	.048	150	170	.054	160	210	.058	175	230	.060	190	250	Dmax = 2°45′		
4°00′	1,432	.029	100	100	.041	125	130	.052	150	180	.057	160	220	.060	175	240						
5°00′	1,146	.034	100	100	.046	125	140	.056	150	200	.060	160	230				D_{max} = 3°30′					
6°00′	955	.038	100	100	.050	125	160	.059	150	210				D_{max} = 4°15′								
7°00′	819	.041	100	110	.053	125	170				D_{max} = 5°15′											
8°00′	716	.043	100	120	.056	125	180	D_{max} = 6°45′														
9°00′	637	.046	100	120	.058	125	180															
10°00′	573	.048	100	130	.059	125	190															
11°00′	521	.050	100	140	.060	130	190															
12°00′	477	.052	100	140																		
13°00′	441	.054	100	140	D_{max} = 11°15′																	
14°00′	409	.055	100	150																		
16°00′	358	.058	100	160																		
18°00′	318	.059	110	160																		
20°00′	286	.060	110	160																		
21°00′	273	.060	110	160																		
		D_{max} = 21°00′																				

e_{max} = 0.06

D = degree of curve
R = radius of curve
V = assumed design speed
e = rate of superelevation

L = minimum length of runoff (does not include tangent runout)
NC = normal crown section
RC = remove adverse crown, superelevate at normal crown slope

NOTE: Lengths rounded in multiples of 25 or 50 ft permit simpler calculations

Table III-9. Values for design elements related to design speed and horizontal curvature.

D	R (ft)	V = 30 mph e	V = 30 mph L(ft) Two Lanes	V = 30 mph L(ft) Four Lanes	V = 40 mph e	V = 40 mph L(ft) Two Lanes	V = 40 mph L(ft) Four Lanes	V = 50 mph e	V = 50 mph L(ft) Two Lanes	V = 50 mph L(ft) Four Lanes	V = 55 mph e	V = 55 mph L(ft) Two Lanes	V = 55 mph L(ft) Four Lanes	V = 60 mph e	V = 60 mph L(ft) Two Lanes	V = 60 mph L(ft) Four Lanes	V = 65 mph e	V = 65 mph L(ft) Two Lanes	V = 65 mph L(ft) Four Lanes	V = 70 mph e	V = 70 mph L(ft) Two Lanes	V = 70 mph L(ft) Four Lanes
0°15′	22,918	NC	0	0	NC	0	0	NC	0	0	NC	0	0	NC	0	0	NC	0	0	NC	0	0
0°30′	11,459	NC	0	0	NC	0	0	NC	0	0	NC	0	0	RC	175	175	RC	190	190	RC	200	200
0°45′	7,639	NC	0	0	NC	0	0	RC	150	150	RC	160	160	.022	175	175	.025	190	190	.028	200	200
1°00′	5,730	NC	0	0	NC	0	0	.021	150	150	.025	160	160	.029	175	175	.032	190	190	.036	200	200
1°30′	3,820	NC	0	0	.021	125	125	.030	150	150	.035	160	160	.041	175	175	.046	190	200	.051	200	240
2°00′	2,865	RC	100	100	.027	125	125	.038	150	150	.045	160	170	.051	175	210	.058	190	250	.065	200	290
2°30′	2,292	.021	100	100	.033	125	125	.046	150	170	.053	160	200	.061	175	240	.068	190	300	.075	220	330
3°00′	1,910	.025	100	100	.038	125	125	.053	150	190	.060	160	230	.068	180	270	.075	210	320	.080	230	350
3°30′	1,637	.028	100	100	.043	125	140	.058	150	210	.067	170	260	.074	200	300	.079	220	350			
4°00′	1,432	.031	100	100	.047	125	150	.063	150	230	.071	180	270	.078	210	310						
5°00′	1,146	.038	100	100	.055	125	170	.071	170	260	.078	200	300									
6°00′	955	.043	100	120	.062	130	190	.077	180	280												
7°00′	819	.048	100	130	.067	140	210	.080	190	280												
8°00′	716	.053	100	140	.071	150	220															
9°00′	637	.056	100	150	.075	160	240															
10°00′	573	.060	110	160	.078	160	240															
11°00′	521	.063	110	170	.079	170	250															
12°00′	477	.065	120	180	.080	170	250															
13°00′	441	.068	120	180																		
14°00′	409	.070	130	190																		
16°00′	358	.074	130	200																		
18°00′	318	.077	140	210																		
20°00′	286	.079	140	210																		
22°00′	260	.080	140	230																		
		D_{max} = 22°45′			D_{max} = 12°15′			D_{max} = 7°30′			D_{max} = 6°00′			D_{max} = 4°45′			D_{max} = 3°45′			D_{max} = 3°00′		

e_{max} = 0.08

D = degree of curve
R = radius of curve
V = assumed design speed
e = rate of superelevation

L = minimum length of runoff (does not include tangent runout)
NC = normal crown section
RC = remove adverse crown, superelevate at normal crown slope

NOTE: Lengths rounded in multiples of 25 or 50 ft permit simpler calculations

Table III-10. Values for design elements related to design speed and horizontal curvature.

		V = 30 mph			V = 40 mph			V = 50 mph			V = 55 mph			V = 60 mph			V = 65 mph			V = 70 mph		
			L(ft)			L(ft)			L(ft)			L(ft)			L(ft)			L(ft)			L(ft)	
D	R (ft)	e	Two Lanes	Four Lanes	e	Two Lanes	Four Lanes	e	Two Lanes	Four Lanes	e	Two Lanes	Four Lanes	e	Two Lanes	Four Lanes	e	Two Lanes	Four Lanes	e	Two Lanes	Four Lanes
0°15′	22,918	NC	0	0	NC	0	0	NC	0	0	NC	0	0	NC	0	0	NC	0	0	NC	0	0
0°30′	11,459	NC	0	0	NC	0	0	NC	0	0	NC	0	0	RC	175	175	RC	190	190	RC	200	200
0°45′	7,639	NC	0	0	NC	0	0	RC	150	150	RC	160	160	.023	175	175	.025	190	190	.028	200	200
1°00′	5,730	NC	0	0	NC	0	0	.021	150	150	.025	160	160	.030	175	175	.033	190	190	.037	200	200
1°30′	3,820	NC	0	0	.021	125	125	.031	150	150	.037	160	160	.043	175	190	.048	190	220	.054	200	260
2°00′	2,865	RC	100	100	.028	125	125	.040	150	150	.048	160	180	.055	175	230	.062	190	290	.070	220	330
2°30′	2,292	.021	100	100	.034	125	125	.049	150	180	.056	160	210	.067	190	280	.075	220	330	.085	260	390
3°00′	1,910	.025	100	100	.040	125	125	.057	150	210	.067	170	250	.077	210	320	.087	250	380	.096	280	420
3°30′	1,637	.029	100	100	.046	125	140	.065	160	240	.075	190	290	.086	350	.095	.095	270	400	.100	300	450
4°00′	1,432	.033	100	100	.051	125	160	.072	180	260	.083	210	320	.093	250	380	.099	280	420			
5°00′	1,146	.040	100	110	.061	130	190	.083	200	300	.094	240	360	.098	270	400				D_{max} = 3°30′		
6°00′	955	.046	100	120	.070	150	220	.092	220	330	.099	250	380				D_{max} = 4°15′					
7°00′	819	.053	100	140	.078	160	240	.098	240	350				D_{max} = 5°15′								
8°00′	716	.058	110	160	.084	180	260	.100	240	360	D_{max} = 6°30′											
9°00′	637	.063	120	170	.089	190	280															
10°00′	573	.068	120	180	.094	200	290	D_{max} = 8°15′														
11°00′	521	.072	130	200	.097	200	310															
12°00′	477	.076	140	210	.099	210	310															
13°00′	441	.080	140	220	.100	210	320															
14°00′	409	.083	150	220																		
16°00′	358	.089	160	240	D_{max} = 13°15′																	
18°00′	318	.093	170	250																		
20°00′	302	.097	170	260																		
22°00′	260	.099	180	270																		
24°00′	239	.100	180	270																		
		D_{max} = 24°45′																				

e_{max} = 0.10

D = degree of curve
R = radius of curve
V = assumed design speed
e = rate of superelevation

L = minimum length of runoff (does not include tangent runout)
NC = normal crown section
RC = remove adverse crown, superelevate at normal crown slope

NOTE: Lengths rounded in multiples of 25 or 50 ft permit simpler calculations

Table III-11. Values for design elements related to design speed and horizontal curvature.

D	R (ft)	V = 30 mph e	V = 30 mph L(ft) Two Lanes	V = 30 mph L(ft) Four Lanes	V = 40 mph e	V = 40 mph L(ft) Two Lanes	V = 40 mph L(ft) Four Lanes	V = 50 mph e	V = 50 mph L(ft) Two Lanes	V = 50 mph L(ft) Four Lanes	V = 55 mph e	V = 55 mph L(ft) Two Lanes	V = 55 mph L(ft) Four Lanes	V = 60 mph e	V = 60 mph L(ft) Two Lanes	V = 60 mph L(ft) Four Lanes	V = 65 mph e	V = 65 mph L(ft) Two Lanes	V = 65 mph L(ft) Four Lanes	V = 70 mph e	V = 70 mph L(ft) Two Lanes	V = 70 mph L(ft) Four Lanes
0°15′	22,918	NC	0	0	NC	0	0	NC	0	0	NC	0	0	NC	0	0	NC	0	0	NC	0	0
0°30′	11,459	NC	0	0	NC	0	0	NC	0	0	NC	0	0	RC	175	175	RC	190	190	RC	200	200
0°45′	7,639	NC	0	0	NC	0	0	RC	150	150	.020	160	160	.023	175	175	.026	190	190	.029	200	200
1°00′	5,730	NC	0	0	NC	0	0	.022	150	150	.026	160	160	.030	175	175	.034	190	190	.038	200	200
1°30′	3,820	NC	0	0	.022	125	125	.032	150	150	.038	160	160	.044	175	175	.050	190	220	.056	200	250
2°00′	2,865	RC	100	100	.029	125	125	.042	150	150	.049	160	190	.058	175	230	.065	190	290	.073	220	330
2°30′	2,292	.022	100	100	.035	125	125	.051	150	180	.060	160	230	.070	190	280	.080	230	350	.090	270	410
3°00′	1,910	.026	100	100	.042	125	140	.060	150	220	.071	180	270	.082	220	330	.093	270	410	.106	320	480
3°30′	1,637	.030	100	100	.048	125	160	.069	170	250	.081	210	310	.094	250	380	.106	310	470	.118	350	530
4°00′	1,432	.034	100	100	.054	125	180	.077	190	280	.090	230	340	.104	280	420	.116	350	510	D_{max} = 3°45′		
5°00′	1,146	.041	100	110	.065	140	210	.092	220	330	.106	270	400	.117	310	470	D_{max} = 4°45′					
6°00′	955	.049	100	130	.075	170	250	.104	250	370	.116	290	440	D_{max} = 5°45′								
7°00′	819	.055	100	150	.085	180	280	.113	270	410	.120	300	460									
8°00′	716	.062	110	170	.094	200	300	.119	290	430	D_{max} = 7°00′											
9°00′	637	.068	120	180	.101	220	320	D_{max} = 9°00′														
10°00′	573	.074	130	200	.107	230	340															
11°00′	521	.079	140	210	.112	240	360															
12°00′	477	.084	150	230	.116	240	370															
13°00′	441	.089	160	240	.119	250	370															
14°00′	409	.093	170	250	.120	250	380															
16°00′	358	.101	180	270	D_{max} = 14°30′																	
18°00′	318	.108	190	290																		
20°00′	286	.113	200	310																		
22°00′	260	.116	210	310																		
24°00′	239	.119	210	320																		
26°00′	220	.120	220	320																		
		D_{max} = 26°45′																				

e_{max} = 0.12

D = degree of curve
R = radius of curve
V = assumed design speed
e = rate of superelevation

L = minimum length of runoff (does not include tangent runout)
NC = normal crown section
RC = remove adverse crown, superelevate at normal crown slope

NOTE: Lengths rounded in multiples of 25 or 50 ft permit simpler calculations

Table III-12. Values for design elements related to design speed and horizontal curvature.

Many agencies express this control as a single degree of curvature applicable to all design speeds. When using this method, care should be taken to ensure that the cross slope is sufficient to provide surface drainage and reduce the potential for vehicle hydroplaning and wet weather skidding, especially on flat, high-speed curves. Some agencies indicate a different curvature for each design speed. The latter method is more realistic and conforms to the previously discussed superelevation-speed-curve relations. The maximum curvature for normal crown section for each design speed should be determined by setting consistently low friction factor values, considering the effect of normal cross slope and both directions of travel. The result is a decreasing degree of curvature for successively higher design speed.

For an average rate of cross slope of 0.015 and the superelevation curves of Figure III-13 ($e_{max} = 0.10$), the corresponding maximum degree of curve for each design speed is shown in the third column of Table III-13. These are curvatures calling for superelevation to the amount of the normal cross slope, and therefore indicate the limit of normal crown sections. The friction factors developed because of adverse crown at both the design speed and the running speed are shown in the right columns. It is evident from their uniform and low value over the range in design speeds that these degrees of curve are logical limiting values for normal crown sections.

For a limited range of horizontal curves sharper than those in Table III-13, a superelevation adjustment can be obtained practically by retaining the shape of the crowned cross section but rotating it around the edge or centerline of pavement. This adjustment makes it unnecessary to change screeds on rigid pavements, and the construction procedure is the same on these curves as on tangent sections except that the side forms must be set to the proper difference in elevation. This method of eliminating crown results in a steeper slope at the low edge of pavement than otherwise obtained, which may be desirable for drainage. However, traffic operating on the high side of the pavement does not receive as much benefit as it does when changing from a normal crown section to a plane section for the full width.

On a curve sharp enough to require a superelevation rate in excess of about 0.020, a plane slope across the whole pavement should be used. A transition from the normal to a straight-line cross slope is required. For short lengths of highway requiring crown reversals, the difficulties and extra costs involved in constructing these transitions may supersede the desirable design refinement. This practical limit of 0.020 corresponds to degrees of curve ranging from 2°30′ for 30-mph design speeds to about 0°30′ for 70-mph design speeds. For curves between these values and those in Table III-13, the superelevation adjustment can be made by rotation of the normal crown section or, preferably, by change to a plane slope across the whole pavement.

Design Speed (mph)	Average Running Speed (mph)	Maximum Curve (degrees)	Minimum Curve Radius (ft)	Resulting Side Friction Factor, f, with Adverse Crown	
				At Design Speed	At Running Speed
20	20	3°23′	1,700	.031	.031
30	28	1°43′	3,340	.033	.031
40	36	1°02′	5,550	.034	.031
50	44	0°41′	8,320	.035	.031
55	48	0°35′	9,930	.035	.031
60	52	0°29′	11,690	.035	.030
65	55	0°26′	13,140	.036	.030
70	58	0°23′	14,690	.037	.030

Table III-13. Maximum curvature for normal crown section.

Effects of Grades

On long or fairly steep grades, drivers tend to travel somewhat faster in the downgrade than in the upgrade direction. In a refined design this tendency should be recognized, and some adjustment in superelevation rates would follow. In the case of a divided highway with each pavement independently superelevated, or on a one-way ramp, such an adjustment can be made readily. In the simplest practical form, values from Table III-7 to III-12 can be used directly by assuming a somewhat higher design speed for the downgrade and a somewhat lower design speed for the upgrade. The variation of design speed necessarily would depend on the particular conditions, especially the rate and length of grade and the relative value of the degree of curve as compared with other curves on the approach highway section.

It is questionable whether similar adjustments should be made on two-lane and multilane undivided roadways. In one respect the two directions of traffic tend to balance each other, and adjustment of superelevation is not needed. However, the downgrade speed is the most critical, and adjustment for it may be desirable in some cases. Although not common practice, it is possible to construct the lanes at different cross slopes in the same direction. More practical would be an adjustment for the whole pavement as determined by the downgrade speed, because the extra cross slope would not significantly affect upgrade travel, with the possible exception of heavy trucks on long grades. Also to be considered is the overall emphasis to avoid minor changes in design speed values. In general, it is advisable to follow the common practice of disregarding such superelevation adjustment on undivided roadways.

Transition (Spiral) Curves

General. Any motor vehicle follows a transition path as it enters or leaves a circular horizontal curve. The steering change and the consequent gain or loss of centrifugal force cannot be effected instantly. For most curves the average driver can effect a suitable transition path within the limits of normal lane width. However, with combinations of high speed and sharp curvature the resultant longer transition can result in crowding and sometimes actual occupation of adjoining lanes. In such instances transition curves would be appropriate because they make it easier for a driver to confine his vehicle to his own lane. The employment of transition curves between tangents and sharp circular curves and between circular curves of substantially different radii warrants consideration.

The principal advantages of transition curves in horizontal alinement are the following:

1. A properly designed transition curve provides a natural, easy-to-follow path for drivers, such that the centrifugal force increases and decreases gradually as a vehicle enters and leaves a circular curve. This transition curve minimizes encroachment on adjoining traffic lanes and tends to promote uniformity in speed.
2. The transition curve length provides a convenient desirable arrangement for superelevation runoff. The transition between the normal cross slope and the fully superelevated section on the curve can be effected along the length of the transition curve in a manner closely fitting the speed-radius relation for the vehicle traversing it. Where superelevation runoff is effected without a transition curve, usually partly on curve and partly on tangent, the driver approaching a curve may have to steer opposite to the direction of the curve ahead when on the superelevated tangent portion in order to keep his vehicle on tangent.
3. The spiral facilitates the transition in width where the pavement section is to be widened around a circular curve. Use of spirals provides flexibility in the widening of sharp curves.
4. The appearance of the highway or street is enhanced by the application of spirals. Their use avoids the noticeable breaks at the beginning and ending of circular curves as shown, for example, in Figure III-15, which are made more prominent by superelevation runoff.

Generally, the Euler spiral, which is also known as the clothoid, is used. The degree of curve varies from zero at the tangent end of the spiral to the degree of the circular arc at the circular curve end. By definition the degree of curve at any point of the spiral varies directly with the distance measured along the spiral. In the case of a combining spiral connecting two circular curves having different radii, there is an initial degree of curve rather than a zero value.

Length of spiral. The following equation, developed in 1909 by Shortt (**23**) for gradual attainment of centripetal acceleration on railroad track curves, is the basic expression used by some for computing minimum length of a spiral:

$$L = \frac{3.15V^3}{RC}$$

where:
- L = minimum length of spiral, ft;
- V = speed, mph;
- R = curve radius, ft; and
- C = rate of increase of centripetal acceleration, ft/sec^3.

The factor C is an empirical value indicating the comfort and safety involved. The value of C = 1 generally is accepted for railroad operation, but values ranging from 1 to 3 have been used for highways. Sometimes this formula is modified by taking into account the effect of superelevation, which results in much shorter lengths. Highways do not appear to need as much precision as is obtained from computing the length of spiral by this formula or its modification. A more practical control for the length of spiral is that in which it equals the length required for superelevation runoff.

Superelevation Runoff

Length required. Superelevation runoff is the general term denoting the length of highway needed to accomplish the change in cross slope from a section with adverse crown removed to a fully superelevated section, or vice versa. Tangent runout is the general term denoting the length of highway needed to accomplish the change in cross slope from a normal crown section to a section with the adverse crown removed, or vice versa. For added comfort and safety, the superelevation runoff should be effected uniformly over a length adequate for the likely operating speeds. To be pleasing in appearance the runoff pavement edges should not be distorted as the driver views them.

Some agencies employ the spiral and use its length in which to make the desired change in cross slope. One agency believes that the length of spiral should be based on a 4-sec minimum maneuver time at design speed. Other agencies do not employ the spiral but empirically designate proportional lengths of tangent and circular curve for the same purpose. In either case, as far as can be determined, the length of roadway to effect the superelevation runoff should be the same for the same rate of superelevation and degree of

Curve with and without spiral transition. The sharp "corners" at the juncture of curve and straight line in the top view are quite obvious from the driver's seat.

Manmade America, New Haven, Connecticut: Christopher Tunnard and Boris Pushkarev, Yale University Press, 1963. Used by permission.

Figure III-15. Transition spirals. (Source: Ref. 38).

curvature. The spiral simulates the natural turning path of a vehicle. On unspiraled curves the average vehicle tends to traverse a similar transitioned path within the limits of the pavement.

Review of current design practice indicates that the appearance aspect of superelevation runoff largely governs the length. Spiral lengths as determined otherwise often are shorter than that determined for general appearance, so that spiral formula values give way to long empirical runoff values. A number of agencies have established one or more control runoff lengths within a range of about 100 to 600 ft, but there is no universally accepted empirical basis, considering all likely pavement widths. In one widely used empirical expression the required length is indicated in terms of the slope of the outside edge of pavement relative to the centerline profile.

Current practice indicates that for appearance and comfort the length of superelevation runoff should not exceed a longitudinal slope (edge compared to centerline of a two-lane highway) of 1:200. In other words, when considering a two-lane pavement with plane sections, the difference in longitudinal gradient between the edge of pavement profile and its centerline profile should not exceed 0.5 percent.

In another source (**24**) the same 1:200 slope is used for a design speed of 50 mph and higher. Where design speeds are 30 and 40 mph, relative slopes of 1:150 and 1:175, respectively, are used. These slopes correspond to relative gradients of 0.67 and 0.58 percent. To reflect the importance of the higher design speed and to harmonize with the flatter curving elements, both horizontal and vertical, it appears logical to extrapolate the changing relative slope to the higher design speeds, as given in Table III-14.

The maximum relative gradients between profiles of the edges of two-lane pavements are double those given in Table III-14.

Length of runoff on this basis is directly proportional to the total superelevation, which is the product of the lane width and superelevation rate. However, there are certain minimum runoff lengths which should be provided for reasons of general appearance and to avoid undesirably abrupt edge-of-pavement profiles. These minimum values approximate the distance traveled in 2 seconds at the design speed. Table III-15 gives values for two-lane highways with 10- and 12-ft lanes. Values for the lower superelevation rates are on the basis of minimum required lengths while the values for the higher superelevation rates are based on the relative gradients for profiles between the edge of pavement and the centerline shown in Table III-14.

Although the values in Table III-15 often will be the actual design control, it is recognized that for high-type alinement, longer superelevation runoff lengths may be desirable. Further, the requirements of pavement drainage or smoothness in pavement edge profile may call for adjustment in values.

When a spiral is used, the superelevation runoff usually will be effected

Design Speed V_D(mph)	Maximum Relative Gradients (and Equivalent Maximum Relative Slopes) for Profiles Between the Edge of Two-Lane Pavement and the Centerline (percent)
20	0.75 (1:133)
30	.67 (1:150)
40	.58 (1:175)
50	.50 (1:200)
55	.47 (1:213)
60	.45 (1:222)
65	.41 (1:244)
70	.40 (1:250)

Table III-14. Relationship of design speed to maximum relative profile gradients.

over the whole of the spiral length. Depending on the formula and factors used, the length of spiral for a particular curve and design may be greater or less than the length of runoff given in Table III-15. For the most part the calculated values for length of spiral and length of runoff do not differ materially. In view of the empirical nature of both, an adjustment in one to avoid two sets of values is desirable for purposes of design control. The length of runoff is applicable to all superelevated curves, and it is concluded that this value also should be used for minimum lengths of spiral. This value is consistent with design practice, and experience indicates that the values in Table III-15 for lengths of runoff could also be satisfactory for use as lengths of spirals. Minimum design lengths are given in Table III-7 to III-12 in relation to degree of curve and superelevation rate. The values given are for 12-ft lanes. For other lane widths the lengths of runoff vary in proportion of the actual lane width to 12 ft. Shorter lengths could be applied for design with 10- and 11-ft lanes, but considerations of uniformity and practical use of the empirically derived values suggest that the values for 12-ft lanes should be used in all cases.

The length of runoff applicable to pavements wider than two lanes is subject to the same theoretical derivation as that for two-lane highways. On this basis the lengths for four-lane highways would be double the derived values for two-lane highways and those for six-lane highways would be triple. While lengths of this order may be considered desirable, especially with spirals where the changing superelevation rate is effected over its length, it is frequently not feasible to supply lengths on the basis of such direct ratios. On the other hand, most engineers agree that superelevation runoff lengths for wide pavements should be greater than those for a two-lane highway, but no generally accepted factor or criterion for length ratio has been established.

On a purely empirical basis it is concluded that minimum-design superelevation runoff lengths for pavements wider than two lanes should be as follows:

- Three-lane pavements, 1.2 times the corresponding length for two-lane highways;
- Four-lane undivided pavements, 1.5 times the corresponding length for two-lane highways; and

Superelevation Rate	L-Length of Runoff (ft) for Design Speed (mph) of:							
	20	30	40	50	55	60	65	70
				12-ft lanes				
.02	50	100	125	150	160	175	190	200
.04	60	100	125	150	160	175	190	200
.06	95	110	125	150	160	175	190	200
.08	125	145	170	190	205	215	230	240
.10	160	180	210	240	255	270	290	300
.12	195	215	250	290	305	320	350	360
				10-ft lanes				
.02	50	100	125	150	160	175	190	200
.04	50	100	125	150	160	175	190	200
.06	80	100	125	150	160	175	190	200
.08	105	120	140	160	170	180	190	200
.10	130	150	175	200	215	225	240	250
.12	160	180	210	240	255	270	290	300

Table III-15. Length required for superelevation runoff—two-lane pavements.

- Six-lane undivided pavements, 2.0 times the corresponding length for two-lane highways.

The four-lane lengths shown in Tables III-7 to III-12 are determined on this empirical basis. Proper design attention to obtain smooth edge profiles and to avoid distorted appearances may suggest lengths greater than these minimums. Runoff lengths for divided pavements are discussed in the section "Runoff With Medians."

The length of tangent runout is determined by the amount of adverse crown to be removed and the rate at which it is removed. This rate of removal should preferably be the same as the rate used to effect the superelevation runoff. However, where the superelevation is less than maximum the length of runout may be less as discussed in the following section on "Methods of Obtaining Superelevation."

Location with respect to end of curve. In alinement design with spirals the superelevation runoff is effected over the whole of the transition curve. The length of runoff is the spiral length with the tangent to spiral (TS) at the beginning and the spiral to curve (SC) at the end. The change in cross slope begins by removing the adverse cross slope from the lane or lanes on the outside of the curve on a length of tangent just ahead of TS (the tangent runout). (See Figure III-16.) Between the TS and SC (the superelevation runoff) the pavement is rotated to reach the full superelevation at the SC. This procedure is reversed on leaving the curve. By this design the whole of the circular curve has full superelevation.

In design of curves without spirals the superelevation runoff also is considered to be that length beyond the tangent runout. Empirical methods are employed to locate the superelevation runoff length with respect to the point of curvature (PC). No method for division between the tangent and the circular curve can be completely rationalized. With full superelevation attained at the PC, the runoff lies entirely on the approach tangent, where theoretically no superelevation is needed. At the other extreme, placement of the runoff entirely on the circular curve results in a portion of the curve having less than the desired amount of superelevation. Most agencies that do not use spirals design with part of the runoff length on the tangent and part on the curve. In this compromise design the tangent receives some but not the maximum unneeded superelevation, and the end portion of the circular curve receives somewhat less than the needed superelevation. Current design practice is to place approximately two-thirds of the runoff on the tangent approach and one-third on the curve.

In general, theoretical considerations favor the practice of placing a larger proportion of the runoff length on the approach tangent rather than on the circular curve. With the resultant superelevation on the tangent the driver may have to steer in a direction opposite to the direction of the curve ahead to stay in line, but the maximum side friction developed, which is equal to the rate of applied superelevation, is at all times below the rate of side friction considered comfortable and safe. A vehicle traveling at the design speed on the maximum degree of curve (with maximum rate of superelevation) develops the maximum side friction considered safe and comfortable. To apply rates of superelevation less than the maximum at any point on the curve means that vehicles traveling at the design speed develop side friction factors in excess of the allowable minimum. While the side friction factor developed on the tangent is undesirable, the development on curves of friction factors greatly in excess of the design basis results in a worse condition.

The resulting side friction factor depends on the actual vehicle path of travel. Regardless of the superelevation runoff, some form of transition path of travel can be expected. This actual transition path usually begins well back on the tangent and ends beyond the beginning of the circular curve, depending on the speed, sharpness of curvature, width available, and effect of other traffic. What might appear to be an undesirable cross slope on the tangent actually compensates for the curvilinear path of the vehicle. Also, what can be considered lack of superelevation at the beginning of the circular curve proper is compensated for by the vehicle's traveling a curvilinear path that is flatter than the roadway circular arc.

It is evident that in alinement design without spirals the placement of the length of runoff with respect to the PC cannot be determined exactly from available practice and information. In general, design with 50 to 100 percent of the length of superelevation runoff on the tangent can be considered as suitable. For a more precise design control it is concluded that from 60 to 80 percent of the length of runoff preferably should be located on the tangent at curves without spirals.

Methods of attaining superelevation. Change in cross slope should be effected with edge-of-pavement profiles that are rounded to smooth-flowing lines. The methods of changing cross slope are most conveniently discussed in terms of straight-line relations and controls, but it is emphasized that these straight-line profiles with angular breaks are to be rounded in the finished design as later discussed.

Four specific methods of profile design are practiced in attaining superelevation: (1) revolving a crowned pavement about the centerline profile, (2) revolving a crowned pavement about the inside-edge profile, (3) revolving a crowned pavement about the outside-edge profile, and (4) revolving a straight cross-slope pavement about the outside-edge profile. Figure III-16 illustrates these four methods diagrammatically. The centerline profile for Figures III-16A, III-16B or III-16C or the outside-edge profiles for Figure III-16D, drawn as a horizontal line, represents the calculated profile, which may be a tangent, a vertical curve, or a combination of the two. The small cross sections at the bottom of each diagram indicate the pavement cross slope condition at the lettered points.

Figure III-16A illustrates the method where a crowned pavement section is revolved about the centerline profile. This general method is the most widely used in design because the required change in elevation of the edge of the pavement is made with less distortion than with the other methods. The usual calculated centerline profile is the baseline, and one-half of the required elevation change is made at each edge.

Figure III-16B illustrates the method where the pavement section is revolved about the inside-edge profile. In this case the inside-edge profile is determined as a line parallel to the calculated centerline profile. One-half of the required change in cross slope is made by raising the centerline profile with respect to the inside pavement edge and the other half by raising the outside pavement edge an equal amount with respect to the centerlinc profile. The third method involves similar geometrics, where the pavement section is revolved about the outside-edge profile, as shown in Figure III-16C, except that the change is effected below the outside-edge profile instead of above the inside-edge profile.

The fourth method, as shown in Figure III-16D, revolves a pavement section having a typical straight cross-slope instead of the typical crowned cross section illustrated for the first three methods. The pavement is shown revolved about the outside-edge profile in Figure III-16D because this point is most often used for revolvement of two-lane one-way roadways, with profile along the median edge of pavement. The pavement could also be revolved about the lane division or about the inside-edge profile similar to Figures III-16A or III-16B. For the straight cross slope, the outside-edge profile is calculated, and the required change in cross slope is made by lowering the inside edge, as shown in Figure III-16D.

The design controls for attaining superelevation are nearly the same for all four of the methods. Cross section A at one end of the tangent runout is a normal crown or straight cross slope section. At cross section B, the other end of the tangent runout and the beginning of the spiral or length of runoff, the lane or lanes on the outside of the curve are made horizontal with the centerline profile for Figures III-16A, III-16B, and III-16C; there are no change in cross slope for Figure III-16D. On the basis of the relative slopes previously established, the tangent runout distance on two-lane roads varies from about 25 to 45 ft for the crown rate value of 0.015 and where there are 12 ft lanes. Where the crown rate is 0.020, about 35 to 60 ft are required where the lanes are 12 ft wide. These are the lengths for curves with maximum superelevation. Where there is less than maximum superelevation, tangent runout lengths will be longer if the same relative slope as for superelevation runoff are retained. It is desirable that these relative slopes be retained but where this is not possible the runout lengths should be at least equal to those required for a curve with maximum superelevation where the same relative slopes for tangent runoff and runout are retained. On curves where parabolic or circular crowns are not retained, the normal crown should be changed to straight cross slopes in the tangent runout length.

At cross section C the pavement is a plane, superelevated at the normal crown or straight cross slope rate. Between cross sections B and C for Figures III-16A, III-16B, and III-16C, the outside lane or lanes change from a level condition to one of superelevation at the crown rate, which rate is retained on the inner lanes. There is no change between cross sections B and C for Figure III-16D. Between cross sections C and E the pavement section is revolved to the full rate of superelevation. The rate of cross slope at any intermediate point, cross section D, is proportional to the distance from cross section C.

Considering the infinite number of profile arrangements and in recognition of such specific problems as drainage, avoidance of critical grades, esthetics, and fitting the pavement to the ground, the adoption of any specific axis of rotation, i.e., making a choice between methods A, B, C, or D in Figure III-16, cannot be recommended. To obtain the most pleasing and functional results, each runoff section should be considered an individual problem. In practice, any longitudinal profile line for the axis of revolution may be the most adaptable for the problem at hand. In any case a smooth-edge profile is desired as discussed in the section following.

In an overall sense the method of rotation about the centerline usually is the most adaptable. For example, in Figure III-16A, the change in longitudinal grade required for each profile with a relative slope of, say, 1:200 to the centerline, is 0.5 percent at point E. In Figure III-16B, no change is required in the direction of the inside edge of pavement profile, whereas double this amount, or 1.0 percent, is required in the outside edge of pavement profile at point E.

The method shown in Figure III-16B is preferable to the other three in cases where the lower edge profile is a major control, as for drainage. With uniform profile conditions its use results in the greatest distortion of the upper edge profile. Where the overall appearance is to be emphasized, the methods of Figures III-16C and III-16D are advantageous in that the upper-edge profile—the edge most noticeable to drivers—retains the smoothness of the control profile. The shape and direction of the centerline profile may determine the preferred method for attaining superelevation in the first three methods.

Design of smooth profiles for pavement edges. In the diagrammatic profiles of Figure III-16 the tangent profile control lines result in angular breaks at cross sections A, C, and E. For general appearance and safety, these breaks

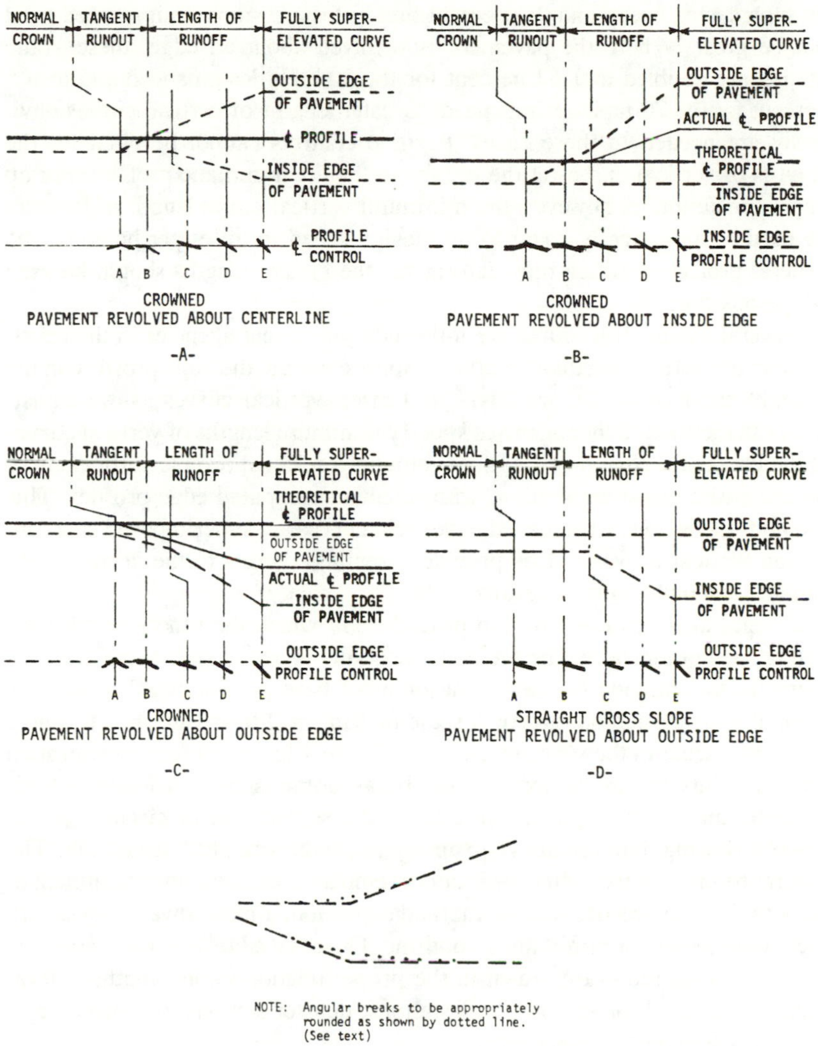

Figure III-16. Diagrammatic profiles showing methods of attaining superelevation for a curve to the right.

should be rounded in final design by insertion of vertical curves. With the method of Figure III-16A, usually short vertical curves are required. Even where the maximum relative slopes are used (minimum length of runoff), the length of vertical curve required to conform to the 0.67 percent break at the

30-mph design speed and 0.4 percent break at the 70-mph design speed need not be great. Where the pavement is revolved about an edge, these grade breaks are doubled to 1.33 percent for the 30-mph design speed and to 0.8 percent for the 70-mph design speed. Greater lengths of vertical curves obviously are needed in these cases. Positive controls cannot be cited for the lengths of vertical curves at the breaks in the diagrammatic profiles. For an approximate guide, however, the minimum vertical curve length in feet can be used as numerically equal to the design speed in miles per hour. As the general profile condition may determine, the greater lengths should be used where possible.

Several design procedures are followed by different agencies in the development of profiles for runoff sections. Some compute the edge profiles on the straight-line basis of Figure III-16 and insert vertical curves as eye adjustments in the field. Other agencies specify minimum lengths of vertical curves at the breaks for edge profiles. Some employ a selected reverse vertical curve for the entire transition section, with resultant computed edge profiles. This method is laborious when the edge vertical curves are superimposed on a centerline vertical curve. It does provide essential controls to the designer and should yield uniformity of results.

Several agencies use a design procedure in which the runoff profiles are determined graphically. The method essentially is one of spline-line development. In this method the centerline or other base profile, which usually is computed, is plotted on a vertical scale of 1 in. = 2 ft or 1 in. = 1 ft, and a horizontal scale in the range of 1 in. = 50 ft to 1 in. = 20 ft. Superelevation control points are in the form of the break points shown in Figure III-16. Then by means of a spline, curve template, ship curve, or circular curve, smooth-flowing lines are drawn to approximate the straight-line controls. The natural bending of the spline, or if curve templates are used, the attainment of smooth-flowing profiles without marked distortion, almost always will satisfy the requirements for minimum smoothing. Once the edge profiles—and lane profiles if required—are drawn in the proper relation to one another, elevations can be read for stations, half stations, quarter stations, or otherwise as necessary for construction control.

An important advantage of the graphical or spline-line method is the infinite study possibilities it affords the designer. Alternate profile solutions can be developed with a minimum expenditure of time. The net result is a design that is well suited to the particular control conditions. The engineering design labor required for this procedure is minimal. These several advantages make this method preferable over the methods of developing profile details for runoff sections.

Runoff with medians. In the design of divided highways, streets, and parkways, the inclusion of a median in the cross section alters somewhat the superelevation runoff treatment. Depending on the width of median and its cross section, there are three general cases for superelevation runoff design:

Case I — The whole of the traveled way, including the median, is superelevated as a plane section.

Case II — The median is held in a horizontal plane and the two pavements separately are rotated around the median edges or, where applicable, around the inside gutter lines.

Case III — The two pavements are separately treated for runoff with a resultant variable difference in elevation at the median edges.

Case I necessarily is limited to narrow medians and moderate superelevation rates to avoid substantial differences in elevation of the extreme pavement edges because of the median tilt. For the most part the method of rotation about the median centerline is used. Diagrammatic profile controls will be similar to those in Figure III-16A except for the two median edges, which will appear as profiles only slightly removed from the centerline.

Case II applies for any width median but has most application with medians of intermediate width to about 30 ft. By holding the median edges level the difference in elevation of the extreme pavement edges is limited to that of the pavement superelevation. Although this type of cross section may be used with wide medians, especially in flat terrain, its general use is limited to medians not wide enough to favor the case III treatment. Runoff design for case II usually has the median-edge profiles as the control. One pavement is rotated about its lower edge and the other about its higher edge. The diagrammatic profile controls are those shown in Figures III-16B, III-16C, and III-16D, the centerline grade control being the same for the two pavements.

The case III design is preferable on other-than-narrow medians in that the differences in elevation of the extreme pavement edges are minimized by a compensating slope across the median. This design may be used on a median of intermediate width by use of a sharply sloped section, but a fairly wide median is usually necessary to develop left shoulder areas and the desired gentle slopes between. With medians of about 40 ft or more in width, it is possible to design separately the profiles and superelevation runoff for the two pavements, a tie-in control being at the median edges. Accordingly, the pavement rotation can be made by the method otherwise considered appropriate, i.e., any of the methods in Figure III-16.

Superelevation runoff lengths for four- and six-lane undivided highways have been shown as 1.5 and 2 times, respectively, the lengths for two-lane highways. For case I designs of divided highways the length of runoff properly should be increased in the proportion to the total width, including the median. Because case I applies mainly to narrow medians, the added length usually will be insignificant. With medians of the order of 4 to 8 ft wide, any increase may well be ignored.

Under case II conditions with narrow medians in a horizontal plane, the runoff lengths should be the same as those for undivided highways, as shown in Tables III-7 to III-12 for four-lane highways. This length applies to highways with medians about 15 ft or less in width. However, with medians about 40 ft or more in width, the two-lane values should be used for the one-way roadways because the extreme pavement edges are at least 80 ft apart and independent of each other. Values for the one way roadways of six-lane highways when separated by a wide median should be 1.2 times the two-lane values of Tables III-7 to III-12. The one-way pavements of highways with medians between 15 and 40 ft might be designed on the basis of either the two-lane or multilane suggested lengths.

With case III cross sections, the median generally will be 40 ft or more in width, and the two-lane values for length of runoff are applicable for one-way pavements of four-lane divided highways. The values for the one-way pavements of six-lane divided highways should be somewhat greater. In situations where the median is less than about 40 ft, the runoff length will be determined in the same manner as for case II.

Divided highways warrant a greater refinement in design and greater attention to appearance than two-lane highways do because they serve much greater traffic volumes and because the cost of refinements is insignificant compared with the cost of construction. Accordingly, the indicated values for length of runoff should be considered minimums, and an effort should be made to use yet longer values. Likewise, there should be emphasis on the development of smooth-flowing pavement-edge profiles of the type obtained by spline-line design methods.

Design for Low-Speed Urban Streets

As previously discussed, the maximum allowable side friction factor for use in the design of horizontal curves is the point at which the centrifugal force causes the driver to experience a feeling of discomfort when driving a curve at a certain design speed. Figure III-5 summarized the test results of side friction factors developed on curves at these apparent limits of comfort. Use of the solid line in Figure III-7 and method 5 was recommended for distributing e and f in the design of all rural highways and high-speed urban streets. Method 2 is recommended for the design of horizontal curves on low-speed urban streets where, through conditioning, drivers have developed a higher threshold of discomfort. By this method, none of the centrifugal force is counteracted by superelevation so long as the side friction factor is less than the maximum allowable for the degree of curve and the design speed. For sharper curves f remains at the maximum and e is used in direct proportion to the continued increase in curvature until e reaches e_{max}. The recommended design values for f that are applicable to low-speed urban streets are shown in Figure III-17 as a solid line superimposed on the analysis curves from Figure III-5. They are based on a tolerable degree of discomfort and provide a reasonable margin of safety against skidding under normal driving conditions in the urban environment. These values vary with the design speed from 0.30 at 20 mph to about 0.17 at 40 mph, 40 mph being the upper limit for low speed established in the design speed discussion of Chapter II.

Although superelevation is advantageous for traffic operations, various factors often combine to make its use impractical in many built-up areas. Such factors include wide pavement areas, need to meet the grade of adjacent property, surface drainage considerations, and frequency of cross streets, alleys and driveways. Therefore, horizontal curves on low-speed streets in urban areas are frequently designed without superelevation, counteracting the centrifugal forcc solely with side friction. On these curves traffic entering a curve to the left has an adverse or negative superelevation in the slope of the normal crown, but with flat curves the resultant friction required to counteract both the centrifugal force and the negative superelevation is small. However, on successively sharper curves for the same design speed, the maximum degree of curvature or sharpest curve without superelevation is reached when the side friction factor developed to counteract centrifugal force and adverse crown reachcs the maximum allowable value based on safety and comfort considerations. For travel on sharper curves, superelevation is needed. The maximum superelevation rate of zero in Table III-16 establishes the minimum radius for each speed below which superelevation is not provided on local streets in residential and commercial areas but should be considered in industrial areas or other streets where operating speeds will be higher. A maximum superelevation rate of 0.040 or 0.060 is commonly used. The maximum curvature for a given design speed is defined for low-speed urban streets when both the maximum superelevation rate and the maximum allowable side friction factors are utilized.

Maximum Safe and Comfortable Speed on Horizontal Curves

Figures III-18 and III-19 and Table III-16, for low speed urban streets, are derived from the simplified curve formula:

$$e + f = V^2/15R$$

Figure III-18 has been prepared by using the recommended values of f for low-speed urban streets from Figure III-17 and by varying the rates of superelevation from -0.050 to $+0.060$. This figure may be used for the

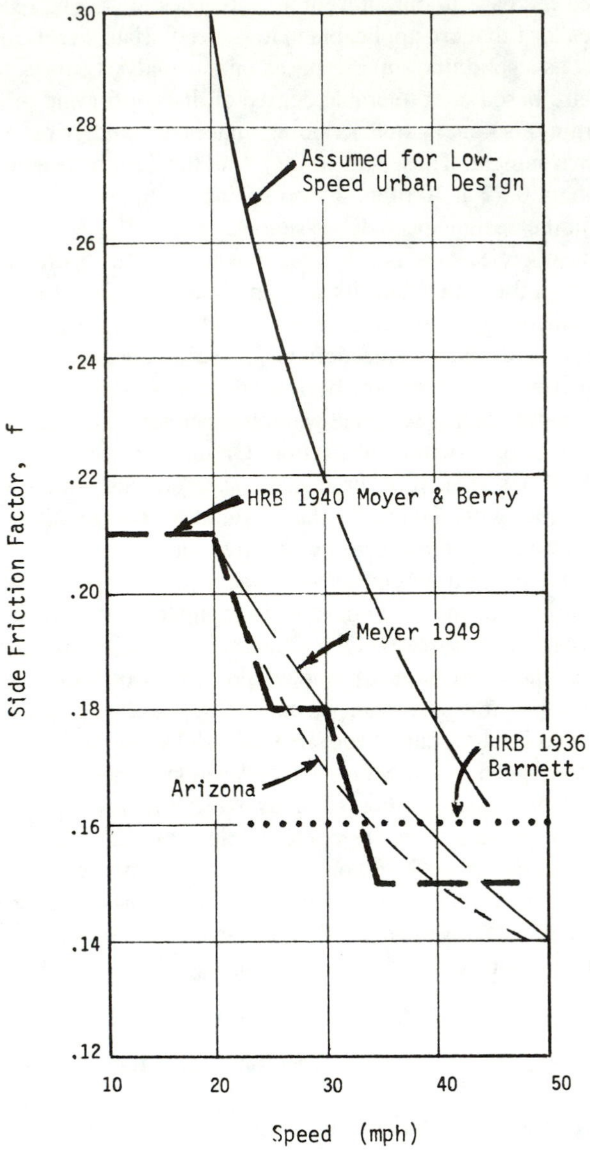

Figure III-17. Side friction factors for low-speed urban streets.

determination of the maximum safe and comfortable speeds for horizontal curves on low-speed urban streets.

Superelevation

Table III-16 shows minimum curve radii for maximum values of superelevation. There will be occasions, however, when low-speed urban streets will require some superelevation but not the maximum value for the rate selected. For example, say a 500 ft radius is proposed on a project with a design speed of 40 mph and a maximum superelevation rate of 0.060. From Table III-16 it is evident superelevation is required but not the maximum value. An accepted procedure for determining the appropriate superelevation is found in Figure III-19. By employing this procedure it will be found that the appropriate superelevation would be 0.038.

In the beginning of the low-speed urban street discussion it was noted that urban drivers are more conditioned to operating on curves without superelevation. However, once superelevation is introduced drivers expect consistency. In order to attain this consistency consideration should be given to balancing other curves with the minimum radii. The amount of superelevation for radii larger than the minimum is shown graphically on Figure III-19 for maximum superelevation rates of up to 0.06. For each design speed the value of e is distributed uniformly from a maximum of 0.040 or 0.060 to a minimum of 0.015 in direct proportion to the curve radius and hence to a degree of curvature. The maximum superelevation is not provided for all curves requiring superelevation. This figure should be used for balancing a series of superelevated curves in urban areas where speed is low but uniform from curve to curve.

Required Superelevation Runoff Length

The following formula for deriving the required superelevation runoff length is based on the maximum allowable side friction factor, where C is the rate of change of the side friction factor obtained from Table III-16.

$$L = \frac{47.2f\, V_D}{C}$$

Where L = length of superelevation runoff, ft;
f = side friction factor;
V_D = design speed, mph; and
C = rate of change of f, ft/sec^3.

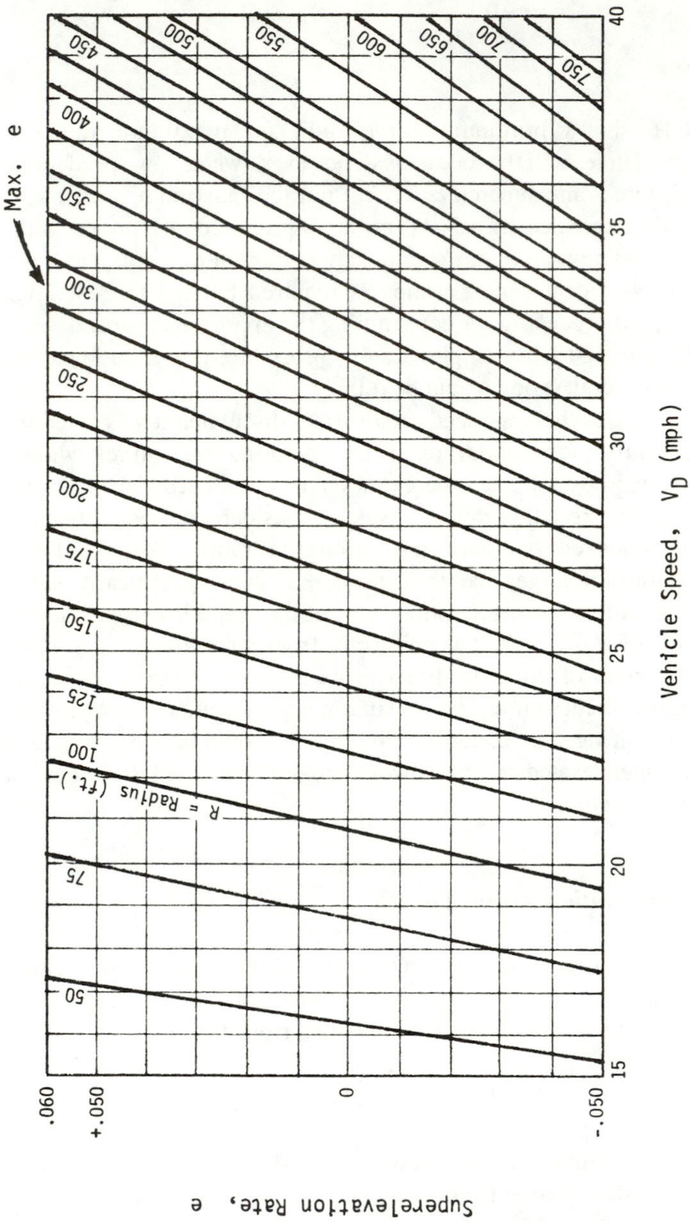

Figure III-18. Maximum safe and comfortable speed for horizontal curves on low-speed urban streets.

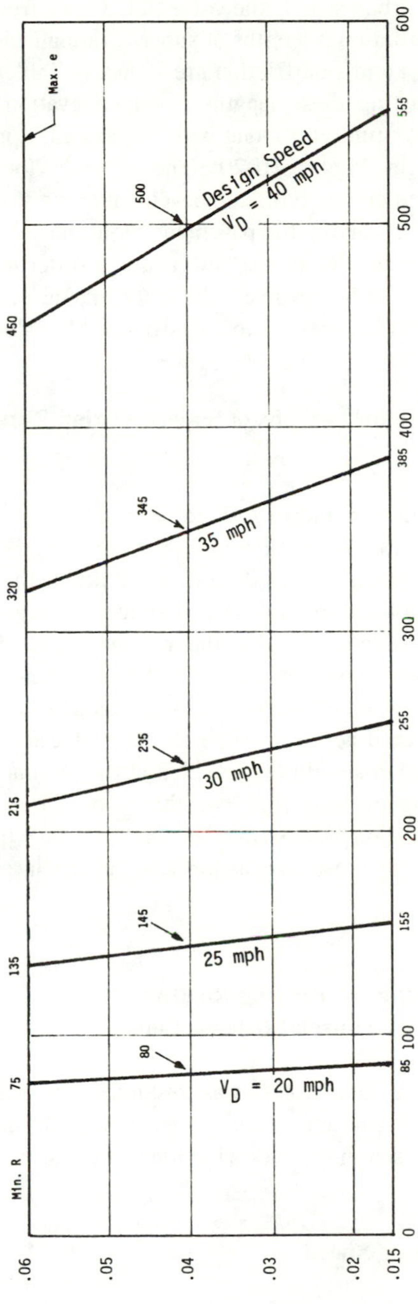

For Radii Larger Than Minimum e = $\frac{R\ Min.\ (e_{Max.} + f)}{R} - f$

Figure III-19. Superelevation for low-speed urban streets.

For design on low-speed urban streets the value of C varies from 4.0 at 20 mph to 3.0 at 40 mph. The minimum lengths of superelevation runoff for limiting values of superelevation and side friction are shown in Table III-16. An acceptable method for adjusting these lengths of superelevation runoff for radii that are larger than the minimum along with a graphical application of this method is contained in Figure III-20. The formula for length of superelevation runoff is based on revolving the pavement about the centerline of the street. On flat grades, revolving the pavement about the centerline may result in low spots on the inner edge of the pavement. In order to avoid this condition, the pavement should be revolved about the inside edge and the length of superelevation runoff shown on Table III-16 should be doubled.

Minimum Radii and Maximum Lengths of Superelevation Runoff for Limiting Values of e and f

Table III-16 gives the minimum radii for three rates of superelevation 0.0, 0.040 and 0.060. The 0.060 rate is considered to be the desirable maximum superelevation for low-speed urban street design. In addition, the table contains the minimum lengths of superelevation runoff for each superelevation rate. From this table it is possible to calculate the minimum desirable tangent between two reversing curves of minimum radii. The superelevation rate of zero is included because an intervening length of tangent is necessary between reversing curves even if neither is superelevated. Because two-thirds of the maximum superelevation should be provided at the PC and PT of the curves, the minimum tangent length is two-thirds of the sum of the superelevation runoff lengths. As has been noted Figure III-20 provides an acceptable method for adjusting these lengths for radii that are larger than the minimum.

Curvature of Turning Roadways and Curvature at Intersections

Curvature for through roads and streets was discussed previously. Curvature of turning roadways and curvature at high-speed at-grade intersections are special cases to be discussed in the following four sections.

Minimum Radius for Turning Speed

As further discussed in Chapter IX, vehicles turning at intersections designed for minimum-radius turns have to operate at low speed, perhaps less

Design Speed (mph)	Max e 0.06	Max f	Total (e+f)	Min R (ft)	Max e 0.04	Max f	Total (etf)	Min R (ft)	Max e 0.00	Max f	Total (etf)	Min R (ft)	C	Min L (ft)
20	0.06	0.300	0.360	75	0.04	0.300	0.340	80	0	0.300	0.300	90	4.00	75
25	0.06	0.252	0.312	135	0.04	0.252	0.242	145	0	0.252	0.252	165	3.75	80
30	0.06	0.221	0.281	215	0.04	0.221	0.261	230	0	0.221	0.221	275	3.50	90
35	0.06	0.197	0.257	320	0.04	0.197	0.237	345	0	0.197	0.197	415	3.25	100
40	0.06	0.178	0.238	450	0.04	0.178	0.218	490	0	0.178	0.178	600	3.00	115

Table III-16. **Minimum radii and maximum lengths of superelevation runoff for limiting values of e and f (low-speed urban streets).**

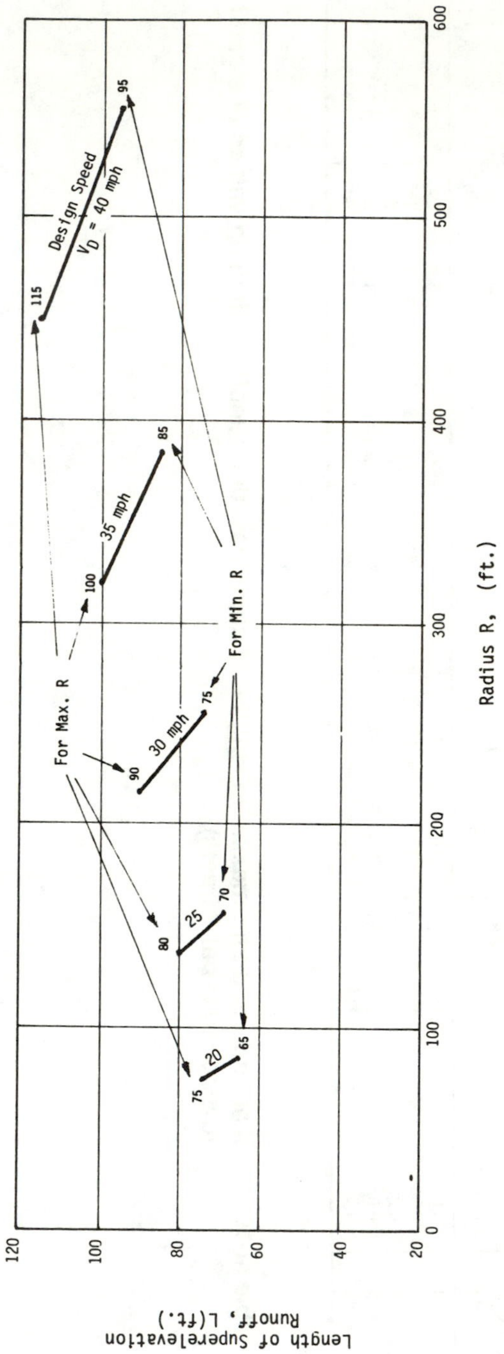

For Radii Larger Than Minimum $L = \frac{R_{Min.}}{R} \times L_{Max.}$

Figure III-20. Minimum length of superelevation runoff for low-speed urban streets.

than 10 mph. While it is desirable and often feasible to design for turning vehicles operating at higher speeds, it is often necessary for safety and economy to use lower turning speeds at most at-grade intersections. The speeds for which these intersection curves should be designed depend on vehicle speeds on the approach highways, type of intersection, and volumes of through and turning traffic. Generally, a desirable turning speed for design is the average running speed of traffic on the highway approaching the turn. Designs at such speeds offer little hindrance to smooth flow of traffic and may be justified for some interchange ramps or for at-grade intersections for certain movements involving little or no conflict with pedestrians or other vehicular traffic.

Curves at intersections need not be considered in the same category as curves on the open highways because the various warnings provided and the anticipation of more critical conditions at an intersection permit the use of less liberal design factors. Drivers generally operate at higher speeds in relation to the degree of curvature on intersection curves than on open highway curves. This increased speed is accomplished by the drivers' acceptance and use of higher side friction in operating around curves at intersections as opposed to side friction accepted and used on the through highway.

Several studies (**25, 26, 27**) have been conducted to determine lateral vehicle placement and distribution of speeds on intersection curves. Results of these studies pertinent to speed-curvature relationships are plotted in Figure III-21. In the analyses of these data the 95-percentile speed of traffic was assumed to be that closely representing the design speed, which generally corresponds to the speed adopted by the faster group of drivers. Side friction factors (taking superelevation into account) actually developed by drivers negotiating the curves at the 95-percentile speed are indicated for 34 locations in Figure III-21. The dashed lines at the upper left shows the side friction factors used for design of curves on rural highways and high-speed urban streets (Figure III-7). Use of this control limit for high speeds, and a friction factor of about 0.5 that could be developed at a low speed as the other limit, gives an average or representative curve through the plottings of individual observations—a relation between design (95-percentile) speed and side friction factor that is considered appropriate for rural and high-speed urban curve design for at-grade intersections.

With this relation established and with logical assumptions for the superelevation rate that can be developed on intersection curves, minimum radii for various design speeds are derived from the standard curve formula, (see the section "Theoretical Considerations.") Obviously, different rates of superelevation would produce somewhat different radii for a given design speed and side friction factor. For design of intersection curves it is desirable to establish a single minimum radius for each design speed. This is done by assuming a likely minimum rate of superelevation (a conservative value) that could nearly always be obtained for certain radii. If more superelevation than this minimum is actually provided, drivers will either be able to drive the curves a little faster or drive them more comfortably because of less friction.

In selecting a minimum rate of superelevation it is recognized that the sharper the curve the shorter its length and the less of an opportunity for developing a large rate of superelevation. This condition applies particularly to at-grade intersections where the turning roadway is often close to the intersection proper, where much of its area is adjacent to the through pavement, and where the complete turn is made through a total angle of about 90°. Assuming the more critical conditions and considering the lengths likely to be available for developing superelevation on curves of various radii, the minimum rate of superelevation for derivation purposes is taken as that varying from zero at 15 mph to 0.080 at 35 mph. By using these rates and the side

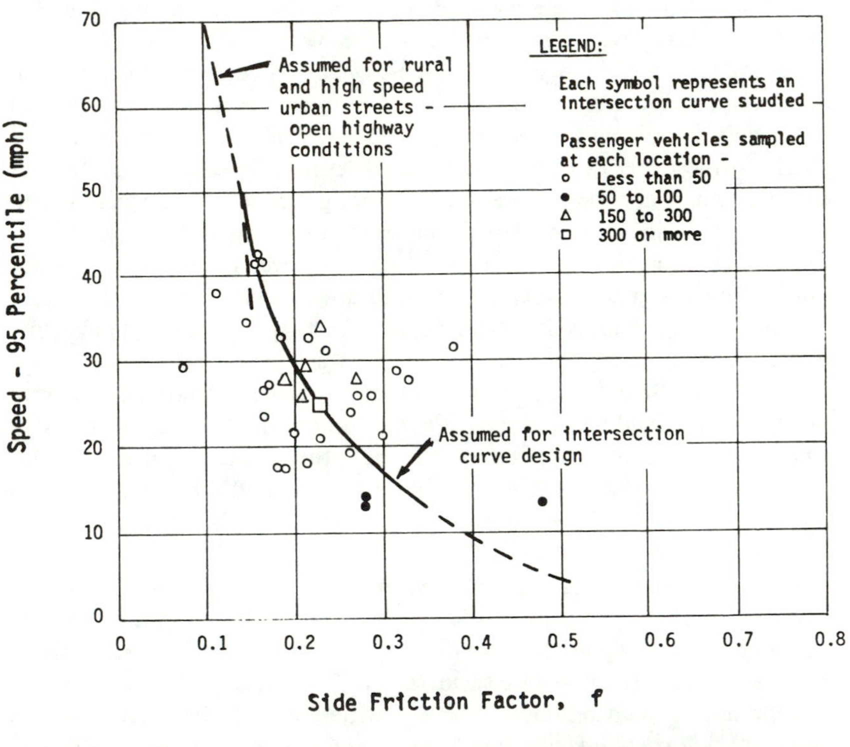

Figure III-21. Relation between speed and side friction factor on curves at intersections.

friction factors of Figure III-21 in the simplified curve formula, the minimum safe radii for intersection curves for operation at design speed are derived as shown in Table III-17.

Design (turning) speed, V (mph)	10	15	20	25	30	35	40
Side friction factor, f	0.38	0.32	0.27	0.23	0.20	0.18	0.16
Assumed minimum superelevation, e	0.00	0.00	0.02	0.04	0.06	0.08	0.09
Total e + f	0.38	0.32	0.29	0.27	0.26	0.26	0.25
Calculated minimum radius, R (ft)	18	47	92	154	231	314	426
Suggested curvature for—design:							
Radius—minimum (ft)	25	50	90	150	230	310	430
Degree of curve—maximum	—	—	64	38	25	18	13
Average running speed (mph)	10	14	18	22	26	30	34

NOTE: For design speeds of more than 40 mph, use values for open highway conditions.

Table III-17. Minimum radii for intersection curves.

The minimum radii of Table III-17 are represented by the heavy solid line at the left in Figure III-22. The thin solid line at the upper right shows the relation between design speed and minimum radius for open highway conditions, as derived with e values shown at the upper left. The joining of the heavy and thin lines indicates that open-highway conditions on intersection curves are approached when the curvature is sufficiently flat to permit operation between 40 and 50 mph. Thus, in design of intersection curves for design speeds of above 40 mph, open highway conditions should be assumed and the design based on Tables III-6 to III-12. The square points in Figure III-22 are the observed 95-percentile speeds from the same studies represented in Figure III-21 for the locations where more than 50 vehicles were sampled. The heavy line fits these points closely, indicating further that the assumptions made in the derivation of minimum radii in Table III-17 are appropriate and that a group of the higher speed drivers will use the design speed assumed.

In addition to the design speed, the average running speed is also used in consideration of certain elements of intersection design. The points indicated by crosses in Figure III-22 are actual average speeds observed on the same intersection curves referred to. The long dashed line through these points is assumed to represent the average running speed for intersection curves. At the right this curve crosses the short dashed line, which indicates the average speed for open highways. For a given design speed on the solid (upper) curve in Figure III-22, the assumed running speed lies vertically below on the dashed curve. These running speeds are shown in the last line of Table III-17.

The above established minimum radii should be used for design preferably on the inner edge of pavement rather than on the middle of the vehicle path or the centerline of the pavement. In all cases, as much superelevation as feasible up to a practical limit should be developed. For the suggested radii shown in Table III-17, a rate of at least 0.080 should be strived for at all locations, and a rate of 0.080 to 0.100 for locations where snow or ice is not a factor. On those intersection legs where all traffic comes to a stop, as at stop signs, a lesser amount of superelevation usually is in order. Also where large trucks will be using an intersection, use of superelevation may have to be limited because these larger trucks may have trouble negotiating intersection curves with superelevation. This is particularly true where they must crossover from a roadway or ramp sloping in one direction to one sloping the other way. This will require that where there is a significant number of large trucks for each design speed flatter curves and less superelevation will have to be provided. Superelevation for curves at intersections is further discussed under that heading in Chapter IX.

Transitions and Compound Curves

Drivers turning at at-grade intersections and at interchange ramp terminals naturally follow transitional travel paths just as they do at higher speeds on the open highway. If facilities are not provided for driving in this natural manner, many drivers deviate from the intended path and develop their own transition, sometimes to the extent of encroachment on other lanes or on the shoulder. Provision for natural travel paths is best effected by the use of transition or spiral curves that may be inserted between a tangent and a circular arc or between two circular arcs of different radii. Practical designs fitting transitional paths may also be developed by the use of compound circular curves. Transitioned roadways have the added advantage of providing practical means for changing from a crowned to a superelevated cross section.

Length of Spiral

Lengths of spirals for use at intersections are determined in the same manner as they are for open highways. On intersection curves, lengths of spirals may be shorter than they are on the open highway curves, because drivers accept a more rapid change in direction of travel under intersection conditions. That is, C (the rate of change of centripetal acceleration on intersection

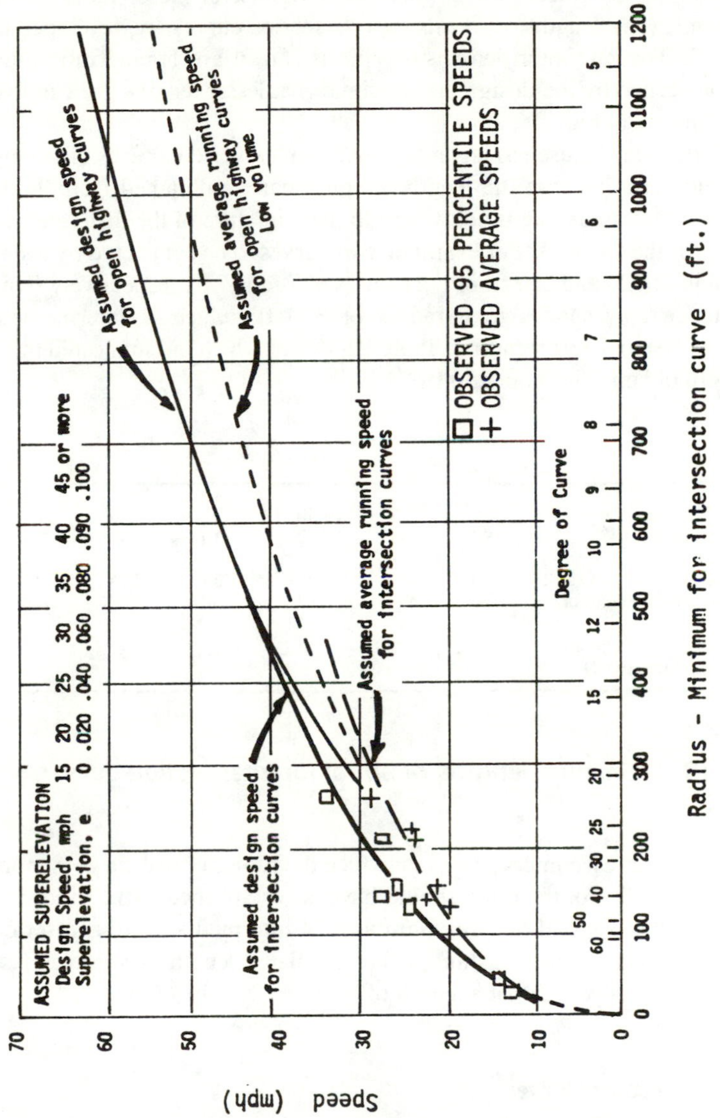

Figure III-22. Minimum radii for curves at intersections.

curves) may be higher on intersection curves than on open highway curves, where values of C ranging from 1 to 3 ft/sec^3 generally are accepted. Rates for curves at intersections are assumed to vary from 2.5 ft/sec^3 for a turnout speed of 50 mph to 4 ft/sec^3 for 20 mph. With the use of these values in the Shortt formula (**23**), lengths of spirals for intersection curves are developed in Table III-18. The minimum lengths of spirals shown are for the minimum radius as governed by the design speed; somewhat lesser lengths are suitable for above-minimum radii.

Spirals also may be used to advantage between two circular arcs of widely different radii. In this case the length of spiral can be obtained from Table III-18 by using a radius that is equivalent to the curvature of the difference in the degrees of the two arcs. For example, two curves to be connected by a spiral have radii of 820 and 260 ft, or curvatures of 7° and 22°, respectively. This 15° difference corresponds to a radius of 380 ft, which is intermediate between the 310-and 430-ft radii in Table III-18, for which the suggested minimum length of spiral is about 150 ft.

Design (turning) speed (mph)	20	25	30	35	40	45
Minimum radius (ft)	90	150	230	310	430	550
Assumed C	4.0	3.75	3.5	3.25	3.0	2.75
Calculated length of spiral (ft)	70	87	105	134	156	190
Suggested minimum length of spiral (ft)	70	90	110	130	160	200
Corresponding circular curve offset from tangent (ft)	2.2	2.2	2.2	2.3	2.5	3.0

Table III-18. Minimum lengths of spiral for intersection curves.

Compound curves at intersections on which the radius of one curve is more than twice the radius of the other should have, as established in the following paragraphs, either a spiral or a circular curve of intermediate radius inserted between the two. If in such instances the calculated length of spiral is less than 100 ft, it is suggested that a length of at least 100 ft be used.

Compound Circular Curves

Compound circular curves are advantageous in effecting desirable shapes of turning roadways at at-grade intersections and of ramps at interchanges. When circular arcs of widely different radii are joined, however, the alinement appears abrupt or forced, and the travel paths of vehicles require considerable steering effort.

On compound curves for open highways it generally is accepted that the ratio of the flatter radius to the sharper radius should not exceed 1.5:1. For compound curves at intersections where drivers accept more rapid changes in direction and speed, the radius of the flatter arc can be as high as 100 percent greater than the radius of the sharper arc, a ratio of 2:1. The ratio of 2:1 on the sharper curves used at intersections results in approximately the same difference (about 5 mph) in average running speeds for the two curves. These curves are compounded as for a ratio of 1.5:1 on the flatter curves used on the open highway. General observations on ramps having differences in radii with a ratio of 2:1 indicate that both operation and appearance normally are satisfactory.

Where feasible, a smaller difference in radii should be used; a desirable maximum ratio is 1.75:1. Where the ratio is greater than 2:1, a suitable length of spiral or a circular arc of intermediate radius should be inserted between the two curves. In the case of very sharp curves designed to accommodate minimum turning paths of vehicles, it is not practical to apply this ratio control. Here it is a matter of closely fitting compound curves to the path of the design vehicle to be accommodated, for which higher ratios may be necessary as shown in Chapter IX.

Curves that are compounded should not be too short or their effect in enabling change from tangent or flat-curve to sharp-curve operation is lost. In a series of curves of decreasing radii each curve should be long enough to enable the driver to decelerate at a reasonable rate, which at intersections is assumed to be not more than 3 mph/sec, although 2 mph/sec is desirable. Minimum lengths on this basis, using running speeds as shown in Figure III-22, are indicated in Table III-19. They are based on a deceleration of 3

Radius (ft)	100	150	200	250	300	400	500 or more
Length of circular arc (ft)							
Minimum	40	50	60	80	100	120	140
Desirable	60	70	90	120	140	180	200

Table III-19. Lengths of circular arc for a compound Intersection curve when followed by a curve of one-half radius or preceded by a curve of double radius.

mph/sec, and a desirable minimum deceleration of 2 mph/sec. The latter indicates very light braking, because deceleration in gear alone results generally in overall rates between 1 and 1.5 mph/sec.

These compound curve values are developed on the premise that travel is in the direction of sharper curvature. For the acceleration condition the 2:1 ratio is not as critical and it may be exceeded.

Widths for Turning Roadways at Intersections

The pavement and roadway widths of turning roadways at intersections are governed by the volumes of turning traffic and the types of vehicles to be accommodated and may be designed for one-way or two-way operation, depending on the geometric pattern of the intersection.

Pavement Widths

Pavement widths for turning roadways are classified for the following types of operations:

Case I — one-lane, one-way operation with no provision for passing a stalled vehicle;

Case II — one-lane, one-way operation with provision for passing a stalled vehicle; and

Case III — two-lane operation, either one-way or two-way.

Widths under case I usually are used for minor turning movements and for moderate turning volumes where the connecting roadway is relatively short. The chance of a vehicle breakdown is remote under these conditions, but one of the edges of pavement preferably should either have a mountable curb or be flush with the shoulder.

Under case II, widths are determined to allow operation at low speed and with restricted clearance past a stalled vehicle. These widths are applicable to all turning movements of moderate to heavy traffic volumes that do not exceed the capacity of a single-lane connection. In the event of a breakdown, traffic flow can be maintained at somewhat reduced speed. Many ramps and connections at channelized intersections are in this category. However, for case II the widths needed for the longer vehicles are very large as can be seen in Figure III-24. Case I values for these longer vehicles such as a WB-62, a WB-67, a WB-96, or a WB-114 may have to be used as the minimum values where they are in sufficient numbers to be considered the appropriate design vehicle.

Widths under case III are applicable where operation is two way or where operation is one way but two lanes are needed to handle the traffic volume.

In each category the required pavement width depends jointly on the size of the design vehicle and the curvature of the turning roadway. Selection of the vehicle for design is based on the size and frequency of vehicle types. The pavement width increases both with the size (for single unit vehicles or vehicles with the same number of trailers or semi-trailers) of the design vehicle and the sharpness of curvature. The width elements of turning vehicles that are used in determining the width of pavement, shown diagrammatically in Figure III-23, are explained separately in the following. They ignore the effects of insufficient superelevation and of surfaces with low frictional resistance that tend to cause the rears of vehicles traveling at other than low speed to swing outward, developing the necessary slip angles.

The track width of vehicle U is the width between the path of the outer front wheel and that of the inner rear wheel, out-to-out of tires. The normal track width of a vehicle moving on a tangent is the distance out-to-out of rear tires. When a vehicle travels on a curve, the rear wheels track inside the front wheels, resulting in a track width greater than that on the tangent. For P and SU vehicles the turning track width, measured radially, is the normal width plus the offtracking:

$$U = u + R_T - \sqrt{R_T^2 - L^2}$$

where: u = normal track width, 6 ft for the P design vehicle and 8.5 ft for the SU design vehicle;

R_T = turning radius of the outer front wheel, ft; and

L = wheelbase, 11 ft for the P design vehicle and 20 ft for the SU design vehicle.

Computed values of U for the P and SU design vehicles for a range of turning radii are shown by curves at the left in Figure III-24A.

Combination truck units involve two or more wheelbase lengths, each of which results in some offtracking on curves. The offtracking for semitrailer combinations cannot be obtained directly from this expression. Data for semitrailer combinations were obtained by using scale and computer models. Values of U for the WB-40, WB-50, WB-62, WB-67, WB-114, MH/B and P/T design vehicles on the basis of these data are shown by curves at the right in Figure III-24A. Values for an A-Bus closely approximate those of a WB-40; values for a WB-60 are not as wide as a WB-50 and are not shown; values for a WB-96 are not as wide as a WB-62 and are not shown.

The width of front overhang, F_A, is the radial distance between the outer edge of tire path of the outer front wheel and the path of the outer front edge of the body. These paths are computed by geometry from the dimensions of the assumed design vehicles and are shown by the curves in Figure III-24B.

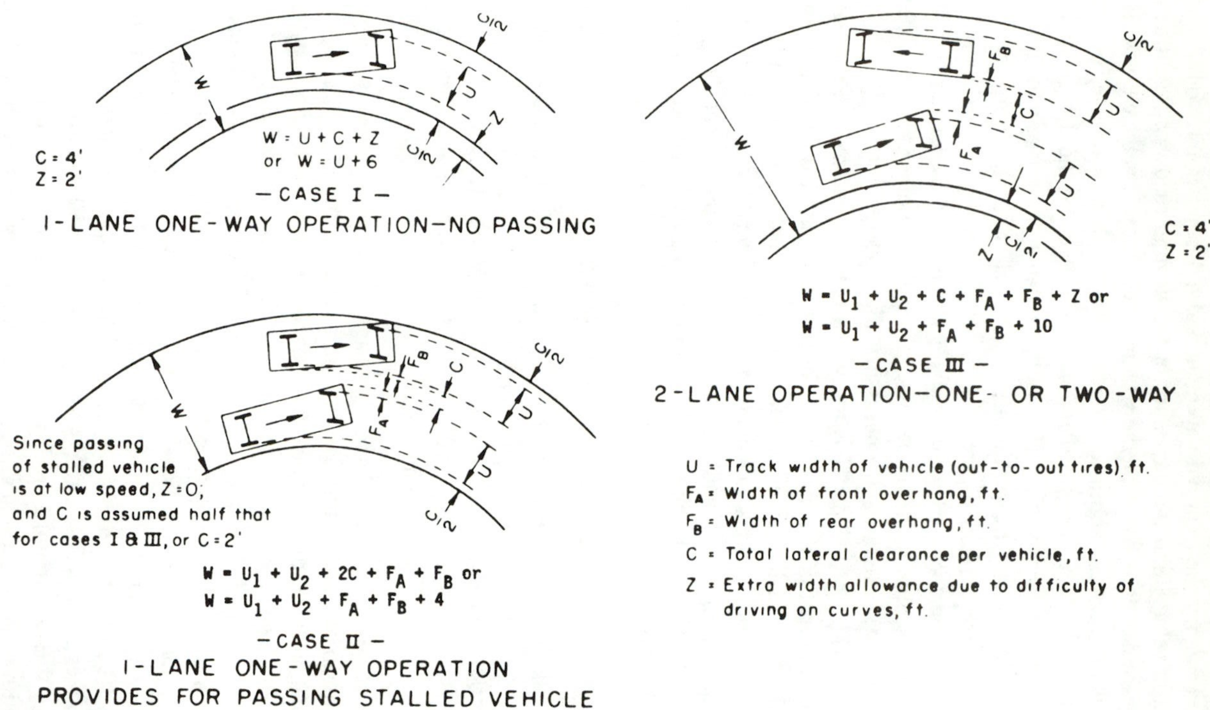

Figure III-23. Pavement widths on curves at intersections—basis of derivation.

The width of rear overhand, F_B, is the radial distance between the outer edge of tire path of the inner rear wheel and the inner side of the body. For the P design vehicle the width of body is 1.0 ft greater than the width out-to-out of rear wheels, making $F_B = 0.5$ ft. In the design truck vehicles, the width of body is the same as the width out-to-out of rear wheels, and $F_B = 0$.

The extra width allowance, Z, is additional radial width of pavement to allow for the difficulty of maneuvering on a curve and the variation in driver operation. This additional width is an empirical value that varies with the speed of traffic and the radius of the curve. For open-alinement roads the additional width has been expressed as the following:

$$Z = \frac{V}{\sqrt{R}}$$

where: Z = extra width allowance, ft;
V = design speed, mph; and
R = radius of curve, ft.

This expression, used primarily for pavement widening on open highways, is also applicable to intersection curves. For the normal range of curve radii at intersections, Z resolves into a nearly constant value of 2 ft by using the speed-curvature relations in Figure III-22 for radii in the range of 50 to 500 ft. This added width, as shown diagrammatically in Figure III-23, is located on the inner edge of pavement; actually it should be assumed to be evenly distributed over the pavement width to allow for the inaccuracy in steering on curved paths.

The lateral clearance allowance, C, provides for the clearance between edge of pavement and nearest wheel path and for the body clearance between vehicles passing or meeting, exclusive of Z. On modern highways, wheel clearances to the edge of pavement are generally 2 to 3 ft, and there are 4 ft or more between bodies of passing or meeting vehicles. This clearance allowance results in total lateral clearances per lane of traffic of 4 ft or more, i.e., about 2 ft on each side of the vehicles. For case II a lower value applies.

The total pavement width, W, for separate roadways at intersections is derived by summations of the proper width elements. Values of U and F are obtained from Figure III-24. The value of Z is constant, about 2 ft for all radii of 500 ft or less. A lateral clearance, C, of 4 ft per line of vehicles is considered appropriate for cases I and III. For case II, a total clearance of half this amount is assumed, i.e., 2 ft for the stopped vehicle and 2 ft for the passing vehicle.

The separate formulas and factor values for each case are shown in Figure III-23. In case I the body overhang need not be considered because no pass-

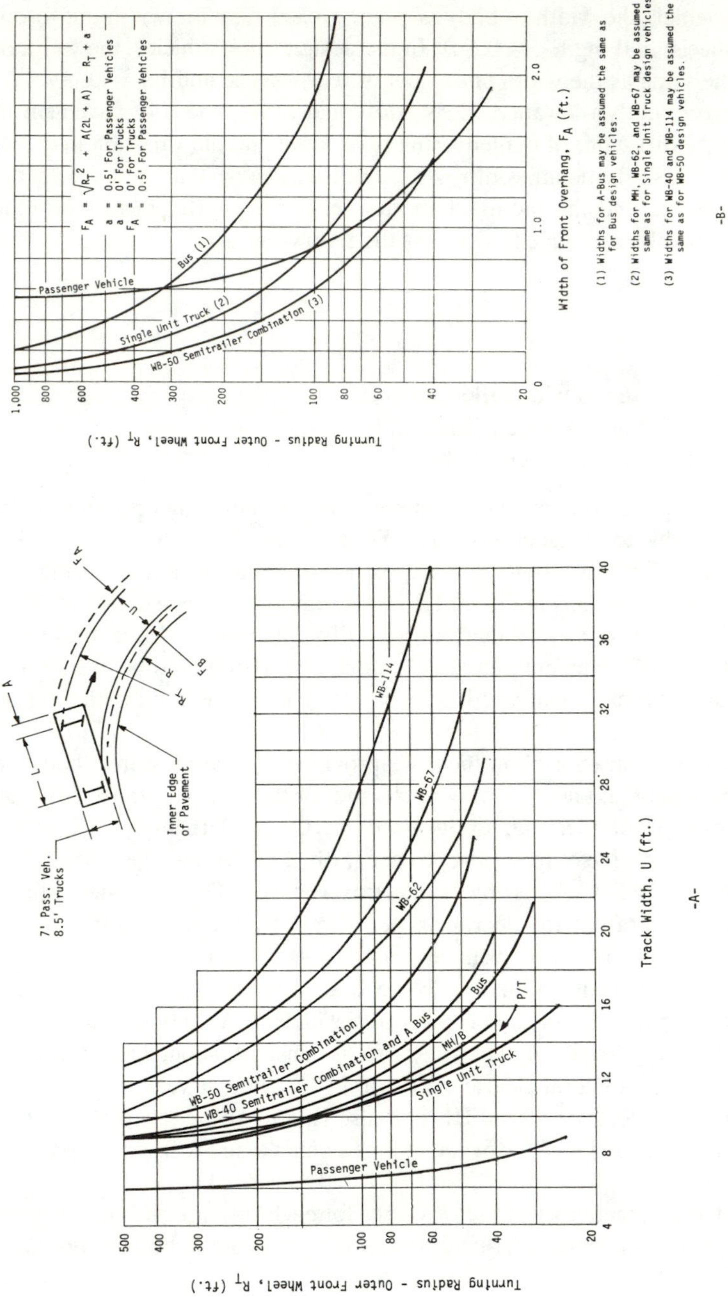

Figure III-24. Track width and overhang for design vehicles operating on curves.

ings are involved. For case II the pavement width involves U and C for the stopped vehicle and U and C for the passing vehicle. To this, is added extra width due to the front overhang of one vehicle and the rear overhang (if any) of the other vehicle. Because passing the stalled vehicle is accomplished at low speeds, the extra width allowance due to difficulty of driving on curves, Z, is omitted. For case III all width elements apply.

The derived pavement widths for various radii for each design vehicle are given in Table III-20. In the derivation, values of U and F were read from Figure III-24 for the turning radii R_T, which were closely approximated by adding the track width and proper clearances to the radius of the inner edge of pavement, R.

For general design use, the required pavement widths given in Table III-20 seldom apply directly, because the pavements usually must accommodate more than one type of vehicle. Even parkways designed primarily for P vehicles are used by buses and maintenance trucks. At the other extreme, few if any public highways are designed to accommodate the WB-50 or longer vehicles fully. Widths needed for some combination of separate design vehicles become the practical design guide for intersection pavements. Such design widths are given in Table III-21 for three logical conditions of mixed traffic. However, where the larger design vehicles such as a WB-62 or a WB-114 will be using a turning roadway or ramp the facility should accommodate as a minimum their turning paths for the case I condition. Therefore, case I widths for the appropriate design vehicle and radius shown in Table III-20 should be checked to determine whether they exceed widths shown in Table III-21. If they do, consideration should be given to using the widths for case I shown in Table III-20 as the minimum widths for the turning roadway or ramp.

Traffic Condition A. Predominantly P vehicles, but some consideration is also given to SU trucks; values in Table III-21 are somewhat above those for P vehicles in Table III-20.

Traffic Condition B. Sufficient SU vehicles to govern design, but some consideration is also given to semitrailer vehicles; values in Table III-21 for cases I and III are those for SU vehicles in Table III-20. For case II, values are reduced as explained later in this section.

Traffic Condition C. Sufficient semitrailer, WB-40 or WB-50, vehicles to govern design; values in Table III-21 for cases I and III are those for WB-40 in Table III-20. For case II, values are reduced.

Traffic conditions A, B, and C are described in broad terms because data regarding traffic volume, or percentage of the total, for each type of vehicle are not available to define these traffic conditions with precision in relation to

Radius on Inner Edge of Pavement R (ft)	Case I, One-Lane, One-Way Operation, No Provision for Passing a Stalled Vehicle										
	P	**PT**	**MH/B**	**SU**	**WB-40**	**WB-50**	**BUS**	**A-BUS**	**WB-62**	**WB-67**	**WB-114**
50	13	18	19	18	23	26	20	23	26	29	37
75	13	17	17	17	19	22	18	19	25	26	33
100	13	16	16	16	18	21	18	18	23	26	31
150	12	16	16	16	17	19	16	17	21	24	25
200	12	15	15	16	16	17	16	16	19	21	23
300	12	14	14	15	16	17	16	16	18	19	20
400	12	14	14	15	16	16	15	16	17	18	19
500	12	14	14	15	15	16	15	15	17	18	19
Tangent	12	14	14	15	15	15	15	15	15	15	15
	Case II, One-Lane, One-Way Operation with Provision for Passing a Stalled Vehicle by Another of the Same Type										
50	20	29	30	29	36	44	31	37	45	49	61
75	19	27	28	27	31	36	29	32	42	45	56
100	19	25	25	25	29	34	28	30	38	42	50
150	18	24	24	24	27	29	25	28	34	37	41
200	18	22	22	23	25	27	25	26	31	35	37
300	18	20	20	22	24	25	24	25	28	31	32
400	17	20	20	22	23	24	22	23	22	29	30
500	17	20	20	22	23	24	22	23	25	27	30
Tangent	17	20	20	21	21	21	21	21	21	21	21
	Case III, Two-Lane Operation, Either One or Two-Way (Same Type Vehicle in Both Lanes)										
50	26	35	36	35	42	50	37	43	51	55	67
75	25	33	34	33	37	42	35	38	48	51	62
100	25	31	31	31	35	40	34	36	44	48	56
150	24	30	30	30	33	35	31	34	40	43	47
200	24	28	28	29	31	33	31	32	37	41	43
300	24	26	26	28	30	31	30	31	34	37	38
400	23	26	26	28	29	30	28	29	33	35	36
500	23	26	26	28	29	30	28	29	31	33	36
Tangent	23	26	26	27	27	27	27	27	27	27	27

Table III-20. Derived pavement widths for turning roadways for different design vehicles.

Pavement Width (ft)

Radius on Inner Edge of Pavement R (ft)	Case I One-Lane, One Way Operation - No Provision for Passing a Stalled Vehicle			Case II One-Lane, One Way Operation - With Provision for Passing a Stalled Vehicle			Case III Two-Lane Operation - Either One-Way or Two-Way		
	Design Traffic and Condition								
	A	B	C	A	B	C	A	B	C
50	18	18	23	23	25	29	31	36	42
75	16	17	19	21	23	27	29	33	37
100	15	16	18	20	22	25	28	31	35
150	14	16	17	19	21	24	27	30	33
200	13	16	16	19	21	23	27	29	31
300	13	15	16	18	20	22	26	28	30
400	13	15	16	18	20	22	26	28	29
500	12	15	15	18	20	22	26	28	29
Tangent	12	15	15	17	19	21	25	27	27

Width Modification Regarding Edge of Pavement Treatment:

	Case I	Case II	Case III
No stabilized shoulder	None	None	None
Mountable curb	None	None	None
Barrier curb: one side two side	 Add 1 ft Add 2 ft	 None Add 1 ft	 Add 1 ft Add 2 ft
Stabilized shoulder, one or both sides	None	Deduct shoulder width; minimum pavement width as under case	Deduct 2 ft where shoulder is 4 ft or wider

Note: Traffic Condition A = predominately P vehicles, but some consideration for SU trucks.
Traffic Condition B = sufficient SU vehicles to govern design, but some consideration for semitrailer vehicles.
Traffic Condition C = sufficient bus and combination-types of vehicles to govern design.

Table III-21. Design widths of pavements for turning roadways.

pavement width. In general, traffic condition A may be assumed to have a small volume of trucks or only an occasional large truck; traffic condition B, a moderate volume of trucks, e.g., in the range of 5 to 10 percent of the total traffic; and traffic condition C, more and larger trucks.

In Table III-21, smaller vehicles in combination are assumed for deriving case II widths than for deriving case III widths, because passings in the former are apt to be very infrequent. Moreover, full offtracking need not be assumed for both the stalled and the passing vehicles. Often the stalled vehicles can be drifted adjacent to the inner edge of pavement, thereby providing additional clearance for the passing vehicle.

The design vehicles or combination of different design vehicles used in determination of values given in Table III-21 for the three traffic conditions, assuming full clearance for the design vehicles indicated, are:

	Design Traffic Condition		
Case	A	B	C
I	P	SU	WB-40
II	P – P	P – SU	SU – SU
III	P – SU	SU – SU	WB-40 – WB-40

The combination of letters, such as P – SU for case II, means that the design width in this example allows a P design vehicle to pass a stalled SU design truck or vice versa. In assuming full clearance, allowance was made for the values of C as discussed.

In negotiating pavements designed for smaller vehicles, larger vehicles will have less clearance and will require lower speed and more caution and skill by drivers; but there is a limit to the size of vehicles that can be so operated on these narrower pavements. The larger vehicles that can be operated on turning roadways of the widths shown in Table III-21, but with partial clearance varying from about one-half the total values of C, as discussed for the sharper curves, to nearly full values for the flatter curves, are:

	Design Traffic Condition		
Case	A	B	C
I	WB–40	WB–40	WB–50
II	P – SU	P – WB–40	SU – WB–40
III	SU – WB–40	WB–40 – WB–40	WB–50 – WB–50

The pavement widths in Table III-21 are subject to some modification with respect to the treatment at the edge of pavement, as shown at the bottom of the table. An occasional large vehicle can pass another on a pavement designed for small vehicles if there is space and stability outside the pavement and there is no barrier to prevent its occasional use. In such cases the pavement width can be a little narrower than the tabulated dimension. Barrier curbs along the edge of pavement give drivers a sense of restriction, and an occasional large vehicle has no additional space in which to maneuver; for this reason the pavement should be a little wider than the values in Table III-21.

When there is an adjacent stabilized shoulder, the pavement widths for cases II and III may be reduced. Case II values may be reduced by the additional width of stabilized shoulder but not below the widths for case I; similarly, case III values may be reduced by 2 ft. Case I values are minimums for the design vehicles and further reduction is not in order even with a usable shoulder. When barrier curbs are used on both sides, the tabulated pavement widths for cases I and III should be increased by 2 ft, and a 1-ft increase is needed for case II, because operation on passing a stalled vehicle is at low speed. Where such a curb is on one side only, the additional pavement width may be only 1 ft for cases I and III, and none need be considered for case II.

Design use of Table III-21 is illustrated by the following example. Assume that the geometric layout and traffic volume for a specific movement are such that one-lane, one-way operation with provision for passing a stalled vehicle is called for (case II), and traffic volume includes 10 to 12 percent trucks with an occasional large semitrailer combination for which traffic condition C is deemed applicable. Then, with a radius on the inner edge of pavement of 150 ft, the tabular width value is 24 ft. With a 4-ft stabilized shoulder, the width may be reduced to 20 ft (see lower part of Table III-21). With a barrier curb on each side, it should be not less than 25 ft.

Widths Outside Pavement Edges

The roadway width for a turning roadway, as distinct from pavement width, includes the shoulders or equivalent lateral clearance outside the edges of pavement. Over the whole range of intersections, the required shoulder width varies from none, or minimal, on curbed urban streets to that of an open highway cross section. The more general cases are discussed in the following paragraphs.

Within a channelized intersection, shoulders for turning roadways usually are unnecessary. The lanes may be outlined by curbs, pavement markings, or islands. The islands may be of curbed or other cross section, and the general dimensional controls for the islands provide the necessary lateral clearances outside the edges of the turning roadway pavement. In most instances the turning roadways are relatively short, and it is not necessary to provide shoulder sections along them for the temporary storage of vehicles. A discussion of island dimensions can be found in Chapter IX.

Where there is a separate roadway for right turns, the left edge of this pavement outlines one side of the triangular island. If the island is small or especially important in directing movements, it may be outlined by curbs or pavement markings. On the other hand, when the turning radius is large, the side of the island may be outlined by guideposts, by delineators, or simply by pavement markings and the edge of pavement of the turning roadway. In any case a developed left shoulder is normally unnecessary. However, there should be either an offset if curbs are used or a fairly level section of sufficient width on the left to avoid affecting the lateral placement of vehicles.

A shoulder usually is provided on the right side of a right-turning roadway in rural areas. In cross section and general treatment, the right shoulder is essentially the same as the shoulder of the open highway section, possibly somewhat scaled down in width because of the intersection conditions. Because turning vehicles have a tendency to encroach, consideration should be given to providing heavy-duty right shoulders to accommodate these more frequent wheel loads. Although a curb on the right side might be advantageous in reducing maintenance operations made necessary by vehicles hugging the inside of the curve and causing edge depressions or raveling, the introduction of curbing adjacent to high-speed highways is a safety hazard and, as such, should be discouraged. Because curbs are a hazard, they should not be located within the highway clear zone. For low-speed urban conditions, curbing of the right edge is normal practice. Curbs are discussed in greater detail in Chapter IV.

On large-scale channelized layouts and at interchanges, there may be turning roadways of sufficient curvature and length to be well removed from other roadways. These turning roadways should have a shoulder space on both sides. Curbs, when used, should be located at the outside edge of the shoulder and should be mountable.

Some turning roadways, particularly ramps, pass over drainage structures, pass over or under other roadways, or are adjacent to a wall or rock cut on one or both sides. For these locations, the minimum clearances for structures as established in later chapters and in the current edition of the AASHTO bridge specifications (**53**), apply directly. In addition, the design should be tested for adequate sight distance, as the sharp curve may require above-minimum lateral clearance.

Table III-22 is a summary of the range of design values for the described general turning roadway conditions. On roadways without curbs or those with mountable curbs, the adjacent shoulder should be of the same type and section as that on the approach highway. The widths shown are for usable shoulders. Where roadside barriers are provided, the width indicated should be measured to the face of the barrier, and the graded width should be about 2 ft greater. For other than low-volume conditions, it is desirable that right shoulders be surfaced, surface treated, or otherwise stabilized for a width of 4 ft or more.

Turning Roadway Condition	Shoulder Width or Lateral Clearance Outside of Pavement Edge (ft)	
	Left	Right
Short length, usually within channelized intersection	2 to 4	2 to 4
Intermediate to long length, or in cut or on fill	4 to 10	6 to 12

Table III-22. Range of shoulder widths or equivalent lateral clearances outside the pavement edges of turning roadways, not on structure. All dimensions should be increased, where necessary, for sight distance.

Pavement Widening on Curves

Pavements on curves sometimes are widened to make operating conditions on curves comparable to those on tangents. Pavement widening is needed on certain curves for one of the following reasons: (1) The vehicle or truck occupies a greater width because rear wheels generally track inside front wheels in rounding curves, or (2) the drivers experience difficulty in steering their vehicles in the center of the lane. On earlier highways with narrow pavements and sharp curves there was considerable need for pavement widening on curves, even though speeds generally were low. On modern highways and streets with 12-ft lanes and high-type alinement the need for widening has lessened considerably in spite of high speeds, but for some conditions of speed, curvature, and width it remains necessary to widen pavements.

When a vehicle traverses a curve without superelevation at low speed, the rear axle is radial and the rear wheels track inside the front wheels. The added width of pavement occupied as compared with travel on tangent can be computed by geometry for any combination of radius and wheelbase. When a vehicle traverses a superelevated curve, the rear wheels may track inside the front wheels more or less than the amount computed on the above basis. This is because of the slip angle assumed by the tires with respect to the direction of travel, which results from the side friction developed between pavement and rolling tires. The relative position of the wheel tracks depends on the speed and the amount of friction developed to counteract the centrifugal force not compensated by superelevation or, when traveling slowly, by the friction developed to counteract the effect of superelevation not compensated by centrifugal force. With excessive speeds the rear wheels may even track outside the front wheels, an extreme situation. The effect of variation in lateral placement of the rear wheels with respect to the front wheels and resultant difficulty of steering should be allowed for by widening on curves, but the amount cannot be determined as positively as that for simple offtracking.

Derivation of Design Values

Pavement widening on curves is the difference in pavement width required on a curve, W_c, and that used on a tangent, W_n. The required width W_c has several components related to operation on curves, namely the following: track width of each vehicle meeting or passing, U; lateral clearance per vehicle, C; width of front overhang of the vehicle occupying the inner lane or lanes, F_A; and width allowance for difficulty of driving on curves, Z.

Track width U is the sum of the track on tangent u (8.5 ft for a design truck) and the amount of offtracking (formula 3 of Figure III-25). The amount a vehicle offtracks depends on the radius of the turn, the number of articulation points, and the lengths of the wheelbases. Formula 3 of Figure III-25 can be used for any combination of radius and number and length of wheelbases. The radius is that which the midpoint of the front axle is desired to follow; however, for most design purposes for two-lane pavements the radius of the curve at the centerline may be used for simplicity of calculations. Lateral clearance C per vehicle is assumed to be 2, 2.5, and 3 ft for tangent lane widths W_n of 20, 22, and 24 ft, respectively. The width F_A of the front overhang to be accounted for on curves depends on the radius of the curve, the extent of the front overhang of the design vehicle, and the wheelbase of the unit itself. Formula 4 of Figure III-25 can be used for this computation. In the case of tractor-trailer combinations, only the wheelbase of the tractor unit is used. The extra width allowance Z is an additional radial width of pavement to allow for the difficulty of maneuvering on a curve and the variation in operation of drivers. This additional width is an empirical value that varies with the speed of traffic and the radius of the curve. For open-alinement roads the additional width has been expressed as shown by formula 5 in Figure III-25.

To determine width W_c it is necessary to choose an appropriate vehicle to represent traffic on the road and to serve as a basis for design. The vehicle should be a design truck because offtracking of the truck is much greater than that of a passenger car. The differences in track widths of the SU, WB-40, WB-50, WB-62, WB-67 and WB-114 design trucks are substantial for the sharp curves associated with intersections, but for open highways on which

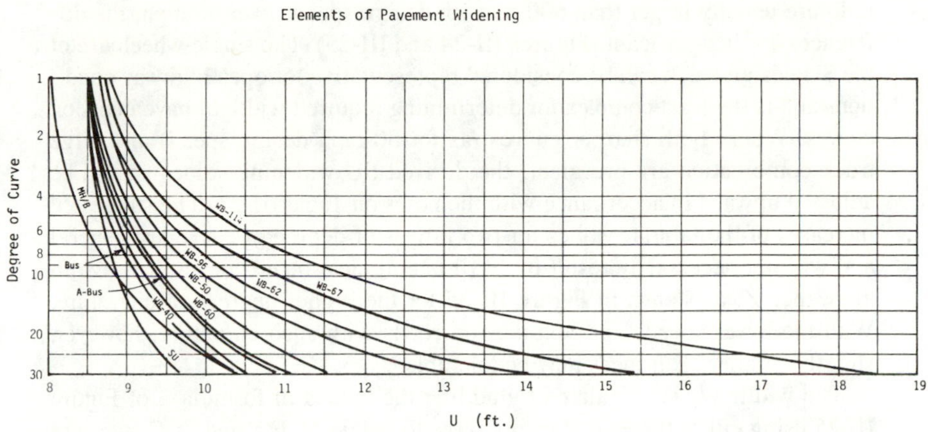

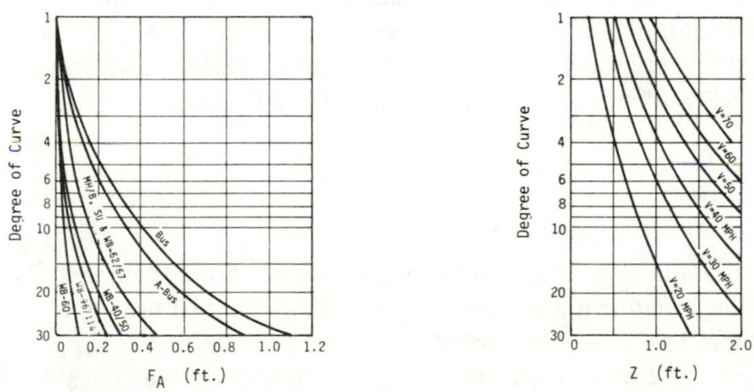

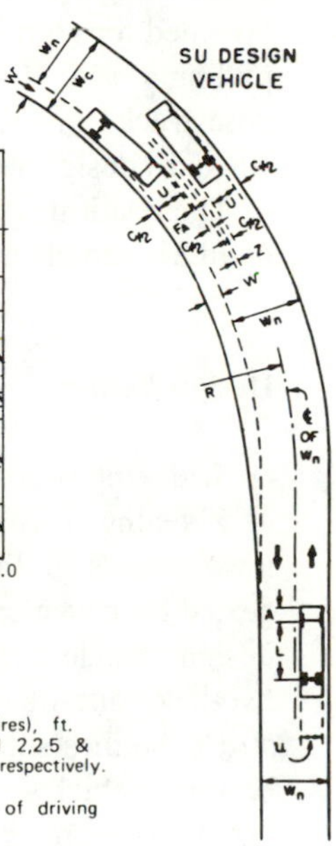

(1) $w = W_C - W_n$

(2) $W_C = N(U+C) + (N-1)F_A + Z$

N = Number of lanes
w = widening for pavement on curve, ft.
W_C = width of pavement on curve, ft.

W_n = width of pavement on tangent, ft.
U = track width of vehicle (out-to-out tires), ft.
C = lateral clearance per vehicle; assumed 2, 2.5 & 3 ft for W_n of 20, 22 & 24 ft, respectively.
F_A = width of front overhang, ft.
Z = extra width allowance for difficulty of driving on curves, ft.

(3) $U = u + R - \sqrt{R^2 - L^2}$

(4) $F_A = \sqrt{R^2 + A(2L + A)} - R$

(5) $Z = V/\sqrt{R}$

u = track width on tangent (out-to-out) 8.5 ft
R = radius on centerline of 2-lane pavement, ft
L = wheelbase
A = front overhang
V = design speed of highway, mph

Figure III-25. Pavement widening on curves—basis of derivation.

radii are usually larger than 600 ft, with design speeds over 40 mph, the differences are insignificant (Figures III-24 and III-25). The single wheelbase of the SU design truck can be considered representative for open highway conditions and is the least complex for determining required width of pavements on curves. Where both sharper curves (as for 30-mph design speed) and large truck combination are prevalent, the derived SU widening values should be adjusted upward in accordance with the notes on Table III-23. The suggested increases of the tabular values for two ranges of degree of curve are general and will not necessarily result in a full lateral clearance C or an extra width allowance Z, as shown in Figure III-25 for the higher degrees of curvature. With the lower speeds and volumes on roads with such curvature, however, slightly smaller clearances may be tolerable.

The width W_c is calculated by adding the values in formula 2 of Figure III-25 using either the formulas or curves to obtain U, F_A, and Z. Curves are provided for eight design trucks plus the bus design vehicle and motor home pulling a boat. If it is desired to consider other vehicles, such as special purpose trucks or recreational vehicles with characteristics different from the standard design vehicles, the formula must be used.

The resulting pavement widening values for the assumed design condition of an SU vehicle and two-lane pavement are given in Table III-23.

Design Values

Widening is costly and very little is actually gained from a small amount of widening. It is suggested that a minimum widening of 2.0 ft be used and lower values in Table III-23 be disregarded. Note that no widening is suggested for pavements 24 ft wide where curves are less than 9° and the largest design vehicle is a WB-50/60. Values in Table III-23 also are applicable to two-lane, one-way pavements, i.e., to each roadway of a divided highway or street. Studies show that on tangent alinement somewhat smaller clearances between vehicles are used in passings of vehicles traveling in the same direction as compared with meeting of opposing vehicles. There is no evidence that these smaller clearances are obtained on curved alinement of one-way pavements. Moreover, drivers are not in position to judge clearances as well when passing vehicles on a curved two-way pavement as when meeting opposing vehicles on a curved two-way pavement. For this reason and because all geometric elements on a divided highway are maintained to a relatively high standard, widening on a two-lane, one-way pavement of a divided highway should be the same as that on a two-lane, two-way pavement, as noted in Table III-23.

On four-lane undivided highways or streets the pavement widening should be double the design values indicated in Table III-23; also, the values of 1.0- and 1.5-ft widening should be doubled and used in design.

Degree of Curve	24 ft Design Speed (mph)					22 ft Design Speed (mph)					20 ft Design Speed (mph)			
	30	40	50	60	70	30	40	50	60	70	30	40	50	60
1	0.0	0.0	0.0	0.0	0.0	0.5	0.5	0.5	1.0	1.0	1.5	1.5	1.5	2.0
2	0.0	0.0	0.0	0.5	0.5	1.0	1.0	1.0	1.5	1.5	2.0	2.0	2.0	2.5
3	0.0	0.0	0.5	0.5	1.0	1.0	1.0	1.5	1.5	2.0	2.0	2.0	2.5	2.5
4	0.0	0.5	0.5	1.0	1.0	1.0	1.5	1.5	2.0	2.0	2.0	2.5	2.5	3.0
5	0.5	0.5	1.0	1.0		1.5	1.5	2.0	2.0		2.5	2.5	2.5	3.0
6	0.5	1.0	1.0	1.5	—	1.5	2.0	2.0	2.5		2.5	3.0	3.0	3.5
7	0.5	1.0	1.5	—		1.5	2.0	2.5			2.5	3.0	3.5	
8	1.0	1.0	1.5			2.0	2.0	2.5			3.0	3.0	3.5	
9	1.0	1.5	2.0			2.0	2.5	3.0			3.0	3.5	4.0	
10-11	1.0	1.5				2.0	2.5				3.0	3.5		
12-14.5	1.5	2.0				2.5	3.0				3.5	4.0		
15-18	2.0					3.0					4.0			
19-21	2.5					3.5					4.5			
22-25	3.0					4.0					5.0			
26-26.5	3.5					4.5					5.5			

NOTES: Values less than 2.0 may be disregarded.
3-lane pavements: multiply above values by 1.5.
4-lane pavements: multiply above values by 2.

Where semi-trailer volumes are significant up to WB-50/60 increase tabular values of widening by 0.5 for curves of 10° to 16° and by 1.0 for curves 17° and sharper. For WB-62's increase values of widening by 0.5 for curves 4° to 8°, by 1.0 for curves 9° to 11° 1.5 for curves 12° to 20°, 2.0 for curves 21° to 27°, 2.5 for curves 28° to 30°.

For WB-67's increase values of widening by 0.5 for curves 3° to 7° by 1.0 for curves 8° to 10° by 1.5 for curves 11° to 15°, by 2.0 for curves 16° to 20° by 2.5 for curves 21° to 25°, 3.0 for curves 25° to 30°.

For WB-114's increase values for widening by 0.5 for curves 1° to 2°, 1.0 for curves 3° to 5° 1.5 for curves 6° to 8°, 2.0 for curves 9° to 10°, 2.5 for curves 11° to 13°, 3.5 for curves 14° to 17°, 4.0 for curves 18° to 20°, 4.5 for curves 21° to 23°, 5.0 for curves 24° to 26°, 6.0 for curves 27° to 28° and 6.5 for curves 29° to 30°.

Table III-23. Calculated and design values for pavement widening on open highway curves (two-lane pavements one-way or two-way).

The above values are applicable for open highway curves. For intersection conditions, with generally smaller radii on turning roadways, the criteria for design pavement widths are somewhat different. These criteria are presented in the section "Widths for Turning Roadways at Intersections" in this chapter, and design values are given in Table III-21.

Attainment of Widening on Curves

Widening should be attained gradually on the approaches to the curve to ensure a reasonably smooth alinement of the edge of pavement and to fit the paths of vehicles entering or leaving the curve. The following are the principal points of concern in design. They apply to both ends of highway curves.

1. On simple (unspiraled) curves, widening should be applied on the inside edge of the pavement only. On curves designed with spirals, widening may be placed on the inside or divided equally between the inside and the outside of the curve. In the latter method, extension of the outer-edge tangent avoids a slight reverse curve on the outer edge of the pavement. In either case the final marked centerline, and desirably any central longitudinal joint, should be placed midway between the edges of the widened pavement.
2. Curve widening should be attained gradually over a length sufficient to make the whole of the pavement fully usable. Although a long transition is desirable for traffic operation, it may result in narrow pavement slivers that are difficult and expensive to construct. Preferably, widening should be attained over the superelevation runoff length, but shorter lengths are sometime used. Changes in width normally should be effected in a distance of 100 to 200 ft.
3. From the standpoints of usefulness and appearance, the pavement edge through the widening transition should be a smooth, graceful curve. A tangent transition edge should be avoided. On minor highways or in cases where plan details are not available, a curved transition staked by eye generally is satisfactory and better than one laid out on a tangent. In any event, the transition ends should be rounded to avoid an angular break at the pavement edge.
4. On highway alinement without spirals, smooth and fitting alinement results from attaining widening with one-half to two-thirds of the length along the tangent and the balance along the curve. This fits a common method for attaining superelevation. The inside edge may be designed as a modified spiral, with control points determined either by a width/length ratio of a triangular wedge on the tangential approach or by tabular calculated values based on a parabolic or cubic

curve; also the edge transition may be effected by a larger radius (compounded) curve. Otherwise it may be lined in the eye in the field. On highway alinement with spiral curves the increase in width usually is distributed along the length of the spiral, and most or all the widening is attained at the SC point.

5. Pavement-widening areas can be fully detailed on construction plans, or else general controls can be cited on construction or standard plans and final details left to the field engineer.

Sight Distance on Horizontal Curves

Another element of horizontal alinement is the sight distance across the inside of curves. Where there are sight obstructions (such as walls, cut slopes, buildings, and longitudinal barriers) on the inside of curves, a design to provide adequate sight distance may require adjustment in the normal highway cross section or change in alinement if the obstruction cannot be removed. Because of the many variables in alinement and cross sections and in number, type, and location of possible obstructions, specific study usually is necessary for each condition. Using design speed and a selected sight distance as a control, the designer should check the actual condition and make the necessary adjustments in the manner most fitting to provide adequate sight distance.

Stopping Sight Distance

For general use in design of a horizontal curve, the sight line is a chord of the curve, and the applicable stopping sight distance is measured along the centerline of the inside lane around the curve. Figures III-26A and III-26B are design charts showing the required middle ordinates for clear sight areas to satisfy the lower and upper values, respectively, of stopping sight distance required for curves of various degrees.

These design charts utilize the stopping sight distance values of Table III-1. The overlap of the ranges at the higher design speeds in this table precludes development of a single design chart. However, use of the two charts in combination provides the same result: Namely, a value that exceeds the minimum established in Figure III-26A, but not the value established in Figure III-26B, will provide acceptable stopping sight distance. As was the case with the stopping sight distances in Table III-1, a value at or approaching the upper limit should be used as a minimum wherever conditions permit because of the increased safety that is provided.

The value at or approaching the upper limit is an application of geometry for the several dimensions, as indicated in the diagrammatic sketch and formulas on the figures. These formulas apply only to circular curves longer than the sight distance for the pertinent design speed. For any design speed the relation of D to m is a straight line. Relations of D, m, and V in these chart forms can be quickly checked. For example, with a 50-mph design speed and a 5° curve, a clear sight distance with a middle ordinate between 17.5 ft (lower value) and 24.5 ft (upper value) is needed for stopping sight distance. As another example, for a sight obstruction condition with m = 20 ft on a 10° curve, the resulting sight distance is approximately at the midpoint of the range for a speed of 40 mph.

Horizontal sight restrictions may occur where there is a cut slope on the inside of the curve. For the height criteria used for stopping sight distance of 3.50-ft height of eye and 6-in. height of object, a height of 2.0 ft may be used as the midpoint of the sight line where the cut slope usually obstructs sight. This assumes that there is little or no vertical curvature. For a highway with 22-ft pavement, 6-ft shoulders, 2-ft ditch section, and 2:1 cut slopes, the sight obstruction is about 18 ft outside the centerline of the inner lane. This is sufficient for adequate sight distance at 30 mph when curves are about 21° or flatter and at 50 mph when the curves are about 5° or flatter. Curves sharper than these would require flatter slopes, benching, or other adjustments. At the other extreme, highways with normal lateral dimensions of more than 30 ft provide adequate stopping sight distances at curves over the entire range of design speeds and curves.

In some instances retaining walls, concrete median safety barriers, and other similar features constructed on the inside of curves may be obstructions and must be checked for stopping sight distance adequacy. As an example, an obstruction of this type offset 4 ft from the inside edge of pavement has a middle ordinate of about 10 ft. At 50 mph this provides adequate sight distance when curves are about 3° or flatter. If the offset width is increased to 11 ft, a 5° (or flatter) curve provides adequate sight distance at the same 50-mph speed. The same would be true for existing buildings or similar obstructions on the inside of curves.

When the needed stopping sight distance would not be available because the railing or a longitudinal barrier constitutes the obstruction, alternatives should be considered for both safety and economic reasons. The alternates are: increase the offset to the obstruction, decrease the degree of curve or reduce the design speed. However, any alternative selected should not require the width of the shoulder on the inside of the curve to exceed 12 ft because the potential exists that drivers will create a safety hazard by using shoulders in excess of that width as a passing or travel lane.

As can be seen from Figure III-26A and III-26B, the method presented is only exact when both the vehicle and the sight obstruction are located within the limits of the simple horizontal curve. When either the vehicle or the sight obstruction is situated beyond the limits of the simple curve, the values obtained are only approximate. The same is true if either the vehicle or the sight obstruction, or both, is situated within the limits of a spiral or a compound curve. In these instances, the value obtained would result in middle ordinate values slightly larger than those needed to satisfy the selected stopping sight distance. In many instances, the resulting additional clearance will not be significant. Whenever Figures III-26A and III-26B are not applicable, it is advisable to check the designs either by utilizing graphical procedures or by utilizing a computational method. Reference (**56**) provides a method for computing the needed values.

Passing Sight Distance

The minimum passing sight distance for a two-lane road or street is about four times as great as the minimum stopping sight distance at the same speed. To conform to those greater sight distances, clear sight areas on the inside of curves must have a width greatly in excess of those discussed. The formulas indicated on Figures III-26A and III-26B are directly applicable for passing sight distance but are of limited practical value except on long curves. A chart demonstration similar to Figures III-26A and III-26B would be of value primarily in reaching negative conclusions—that it would be difficult to maintain passing sight distance on other than very flat curves.

Passing sight distance is measured between a height of eye of 3.50 ft and height of object of 4.25 ft. The sight line near the center of the area inside a curve is about 1.75 ft higher than for stopping sight distance. The resultant lateral dimension for normal highway cross sections in cut between the centerline of the inside lane and the midpoint of the sight line is from 3.75 to 10.5 ft greater than that for stopping sight distance shown above. It is obvious that for many cut sections, practical design for passing sight distance must be confined to tangent and very flat alinement conditions. Even in flat terrain, these design criteria call for keeping the areas inside of curves clear for a width that in some instances may extend beyond the normal right-of-way line.

In general, the designer must use graphical methods to check sight distance on horizontal curves. This method is in Figure III-3 and described in the accompanying discussion.

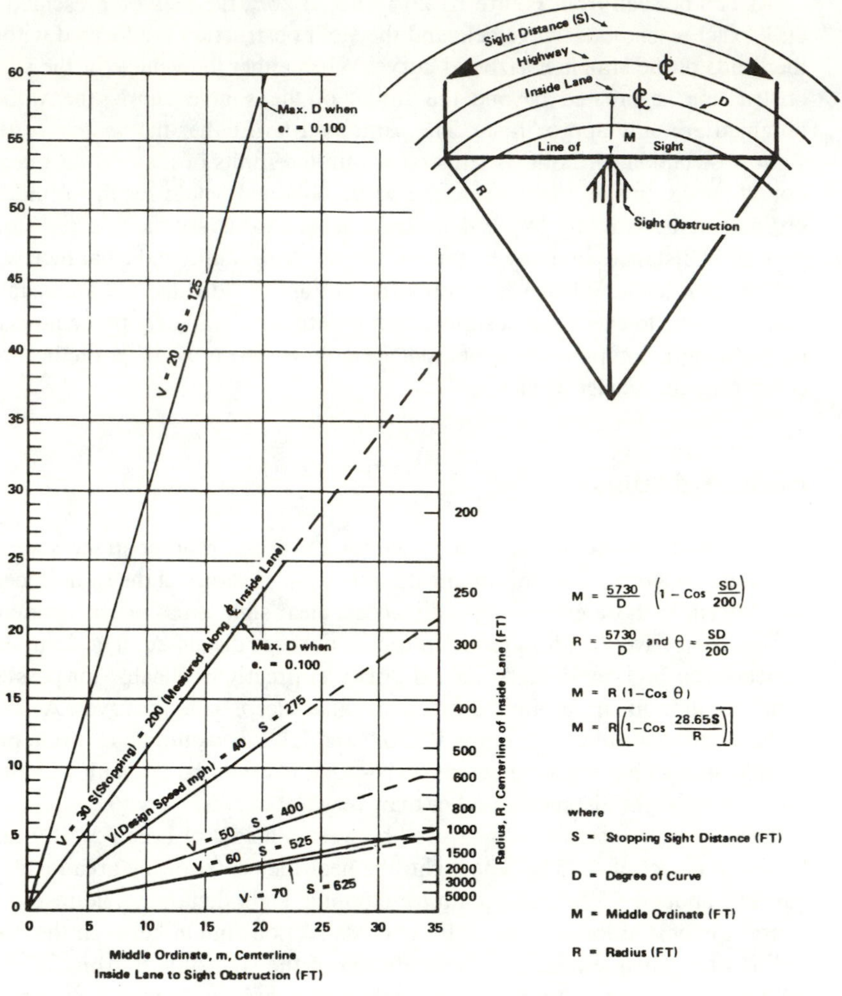

$$M = \frac{5730}{D}\left(1 - \cos\frac{SD}{200}\right)$$

$$R = \frac{5730}{D} \text{ and } \theta = \frac{SD}{200}$$

$$M = R\,(1 - \cos\theta)$$

$$M = R\left[\left(1 - \cos\frac{28.65S}{R}\right)\right]$$

where

S = Stopping Sight Distance (FT)

D = Degree of Curve

M = Middle Ordinate (FT)

R = Radius (FT)

Figure III-26A. Range of lower values—relation between degree of curve and value of middle ordinate necessary to provide stopping sight distance on horizontal curves under open road conditions.

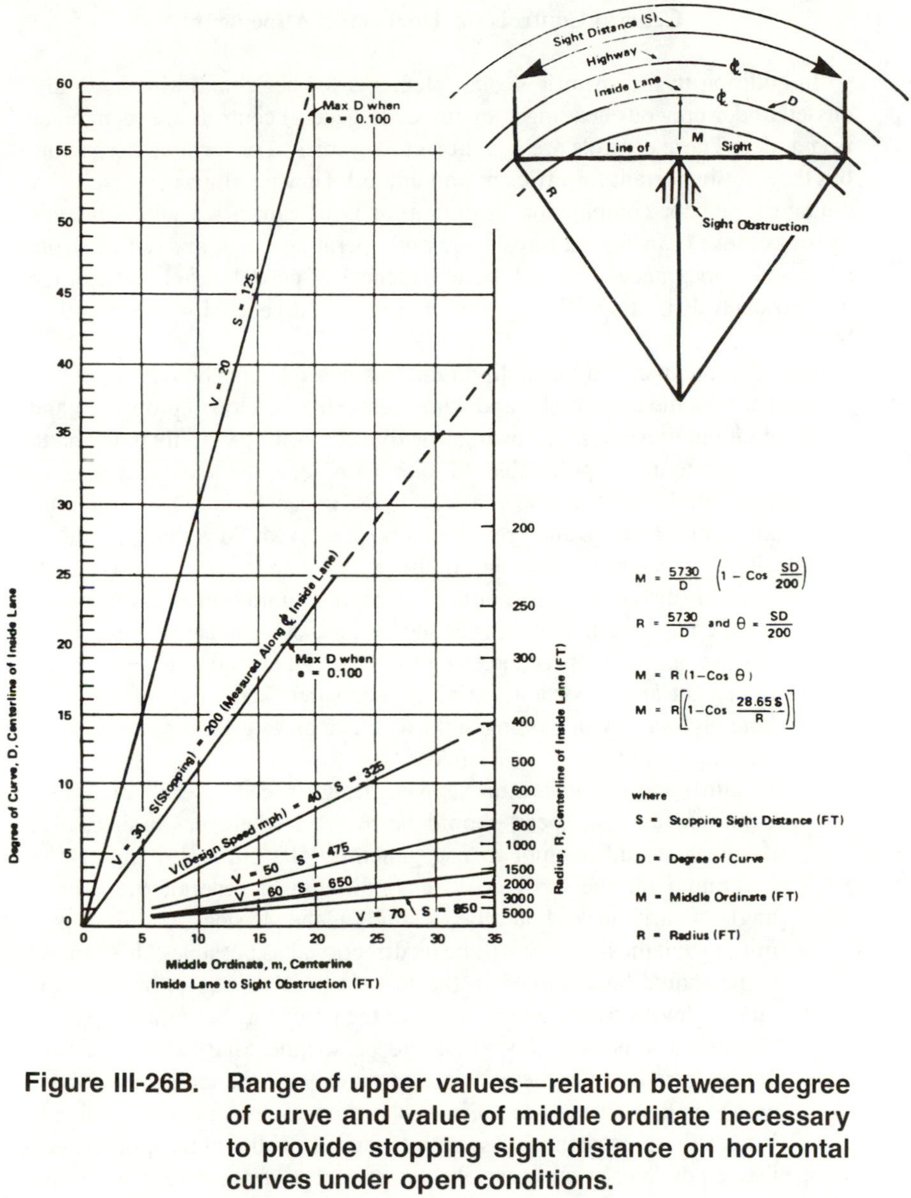

Figure III-26B. Range of upper values—relation between degree of curve and value of middle ordinate necessary to provide stopping sight distance on horizontal curves under open conditions.

General Controls for Horizontal Alinement

In addition to the specific design elements for horizontal alinement discussed under previous headings, a number of general controls are recognized in practice. These controls are not subject to empirical or formula derivation, but they are important for efficient and smooth-flowing highways. Excessive curvature or poor combination of curvature limit capacity, cause economic losses because of increased travel time and operating costs, and detract from a pleasing appearance. To avoid these evidences of poor design practices, the general controls in the following paragraphs should be used where practicable.

1. Alinement should be as directional as possible but should be consistent with the topography and with preserving developed properties and community values. A flowing line that conforms generally to the natural contours is preferable to one with long tangents that slashes through the terrain. The construction scars can be kept to a minimum and natural slopes and growth can be preserved. Such design is desirable from a construction and maintenance standpoint. In general, the number of short curves should be kept to a minimum. Winding alinement composed of short curves should be avoided because it usually is a cause of erratic operation. Although the esthetic qualities of curving alinement are important, passing necessitates long tangents on two-lane highways with passing sight distance on as great a percentage of the length of highway as feasible.
2. In alinement predicated on a given design speed, use of maximum curvature for that speed should be avoided wherever possible. The designer should attempt to use generally flat curves, retaining the maximum for the most critical conditions. In general, the central angle of each curve should be as small as the physical conditions permit, so that the highway will be as directional as possible. This central angle should be absorbed in the longest possible curve, but on two-lane highways the exception noted in the preceding paragraph applies.
3. Consistent alinement always should be sought. Sharp curves should not be introduced at the ends of long tangents. Sudden changes from areas of flat curvature to areas of sharp curvature should be avoided. Where sharp curvature must be introduced, it should be approached, where possible, by successively sharper curves from the generally flat curvature.
4. For small deflection angles, curves should be sufficiently long to avoid the appearance of a kink. Curves should be at least 500 ft long for a central angle of 5°, and the minimum length should be increased 100 ft for each 1° decrease in the central angle. The minimum length

of horizontal curve on main highways, L, should be about 15 times the design speed, or $L_{c\ min} = 15V$. On high speed controlled-access facilities that use flat curvature, a desirable minimum length of curve for esthetic reasons would be about double the minimum length, or $L_{c\ des} = 30V$.

5. Other than tangent or flat curvature should be avoided on high, long fills. In the absence of cut slopes, shrubs, and trees above the roadway, it is difficult for drivers to perceive the extent of curvature and adjust their operation to the conditions.
6. Caution should be exercised in the use of compound circular curves. While the use of compound curves affords flexibility in fitting the highway to the terrain and other ground controls, the simplicity with which such curves can be used often tempts the designer to use them without restraint. Preferably their use should be avoided where curves are sharp. Compound curves with large differences in curvature introduce the same problems that arise at a tangent approach to a circular curve. Where topography or right-of-way restrictions make their use necessary, the radius of the flatter circular arc, R_1, should not be more than 50 percent greater than the radius of the sharper circular arc, R_2, i.e., R_1 should not exceed 1.5 R_2. A several-step compound curve on this basis is suitable as a form of transition to sharp curves as previously discussed in the section "Compound Circular Curves." A spiral transition between flat curves and sharp curves is even more desirable. On one-way roads such as ramps, the difference in radii of compound curves is not so important if the second curve is flatter than the first. However, the use of compound curves on ramps, which results in a flat curve between two sharper curves, is not good practice.
7. Any abrupt reversal in alinement should be avoided. Such a change makes it difficult for a driver to keep within his own lane. It is also difficult to superelevate both curves adequately, and erratic operation may result. A reversal in alinement can be designed suitably by including a sufficient length of tangent between the two curves for superelevation runoff, or preferably an equivalent length with spiral curves. The distance between reverse curves should be the sum of the superelevation runoff lengths and the tangent runout lengths, as defined in the section "Superelevation Runoff". If sufficient distance (i.e., more than 300 ft) is not available to permit the tangent runout lengths to return to a normal crown section, there is a long length where the edges of pavement and centerline are at the same elevation and poor transverse drainage can be expected. In this case, the superelevation runoff lengths should be increased until they abut, thus providing one instantaneous level section. For pavements with straight

cross slopes, there is less difficulty in returning the edges of pavement to a normal section and 300 ft may be decreased.

8. The "broken-back" or "flat-back" arrangement of curves (having a short tangent between two curves in the same direction) should be avoided except where very unusual topographical or right-of-way conditions dictate otherwise. Except on circumferential highways, most drivers do not expect succeeding curves to be in the same direction, the preponderant condition of succeeding curves in opposite direction developing a subconscious habit in drivers for following them. Broken-back alinement is also not pleasing in appearance. Use of spiral transitions or a compound curve alinement, wherein there is some degree of continuous superelevation, is preferable for such conditions. The term broken-back usually is not applied when the connection tangent is of considerable length. Even in this case the alinement will be unpleasant in appearance when both curves are clearly visible for some distance ahead.
9. To avoid the appearance of inconsistent distortion the horizontal alinement should be coordinated carefully with the profile design. General controls for this coordination are discussed in the section "Combination of Horizontal and Vertical Alinement."

VERTICAL ALINEMENT

Terrain

The topography of the land traversed has an influence on the alinement of roads and streets. Topography does affect horizontal alinement, but it is more evident in the effect on vertical alinement. To characterize variations, engineers generally separate topography into three classifications according to terrain.

Level terrain is that condition where highway sight distances, as governed by both horizontal and vertical restrictions, are generally long or could be made to be so without construction difficulty or major expense.

Rolling terrain is that condition where the natural slopes consistently rise above and fall below the road or street grade and where occasional steep slopes offer some restriction to normal horizontal and vertical roadway alinement.

Mountainous terrain is that condition where longitudinal and transverse changes in the elevation of the ground with respect to the road or street are abrupt and where benching and side hill excavation are frequently required to obtain acceptable horizontal and vertical alinement.

Terrain classification pertain to the general character of a specific route corridor. Routes in valleys or passes or mountainous areas that have all the characteristics of roads or streets traversing level or rolling terrain should be classified as level or rolling. In general, rolling terrain generates steeper grades, causing trucks to reduce speeds below those of passenger cars, and mountainous terrain aggravates the situation, resulting in some trucks operating at crawl speeds.

Grades

Roads and streets should be designed to encourage uniform operation throughout. Use of a selected design speed as previously discussed is a means towards this end by correlation of various geometric features of the road or street. Design values have been determined and agreed upon for many highway features, but few conclusions have been reached on roadway grades in relation to design speed. Vehicle-operating characteristics on grades are discussed and relations of grades and their lengths to design speed are developed in this section.

Vehicle-Operating Characteristics on Grades

Passenger cars. The practices of passenger car operators with respect to grades vary greatly, but there is a general acceptance that nearly all passenger cars can readily negotiate grades as steep as 4 to 5 percent without appreciable loss in speed below that normally maintained on level highways, except for cars with high weight/horsepower ratios, including some compact and subcompact cars.

Studies show that operation on a 3-percent upgrade, compared with that on the level, has only a slight effect on passenger car speeds under uncongested conditions. On steeper grades, the speeds decrease progressively with an increase in the ascending grade. On downgrades, passenger car speeds generally are slightly higher than on level sections but local conditions govern.

Trucks. The effect of grades on truck speeds is much more pronounced than on speeds of passenger cars. Average speed of trucks on level sections of highway approximates the average speed of passenger cars. Trucks display up to about a 5 percent increase in speed on downgrades and about a 7 percent or more decrease in speed on upgrades as compared to operation on the level. On upgrades the maximum speed that can be maintained by a truck is dependent primarily on the length and steepness of the grade and the weight/horsepower ratio, which is the gross vehicle weight divided by the engine horsepower. Other vehicles that affect the average speed over the entire length of grade are the entering speed, wind resistance, and skill of the operator. The last two cause only minor variations in the average speed.

Extensive studies of truck performance have been conducted to determine the separate and combined effects of roadway grade, tractive effort, and gross vehicle weight (**30, 31, 32, 33, 34, 60**).

The effect of rate and length of grade on the speed of a typical heavy truck is shown in Figures III-27A and III-27B. From Figure III-27A it can be determined how far a truck, starting its climb from any speed up to 55 mph, travels up various grades or combinations of grades before a certain or uniform speed is reached. For instance, with an entering speed of 55 mph, the truck travels 1,650 ft up a 6-percent grade before its speed is reduced to 25 mph. If the entering speed is 30 mph, the speed at the end of a 1,000-ft-long climb is about 16 mph. This is determined by starting on the curve for a 6-percent grade corresponding to 30 mph, for which the distance is 1,400 ft, and proceeding along it to the point where the distance is 1,000 ft more, or 2,400 ft, for which the speed is about 16 mph. Figure III-27B shows performance when the truck approaches the grade at or below crawl speed. The truck is able to accelerate to a speed of 25 mph or more only on grades of less than 3.5 percent. These total data serve as a valuable guide for design in appraising the effect of trucks on traffic operation for a given set of profile conditions.

Travel time (and, therefore, speed) of trucks on grades is directly related to the weight/horsepower ratio. It has been assumed that vehicles of the same weight/horsepower ratio have similar operating characteristics. Hence, this rating is of considerable assistance in studying the performance of trucks. Normally, the weight/horsepower ratio is expressed in terms of gross weight and net horsepower. It has been found that trucks with a weight/horsepower ratio of about 300 have acceptable operating characteristics from the standpoint of the highway user. Such a weight/horsepower ratio assures a minimum speed of about 26.5 mph on a 3-percent upgrade. There is evidence that the automotive industry would find a weight/horsepower ratio of this magnitude acceptable as a minimum goal in the design of commercial vehicles. There is also evidence that carrier operators are voluntarily recognizing this ratio as the minimum performance control in the loads placed on trucks of different power, the overall result being that weight/horsepower ratio of trucks on highways has improved in recent years. This is illustrated by Figure III-28, which compares such ratios as developed from information obtained in conjunction with nationwide brake performance studies conducted in 1949, 1955, 1963, 1975, 1977 and 1985. The data show, for example, that for a gross vehicle

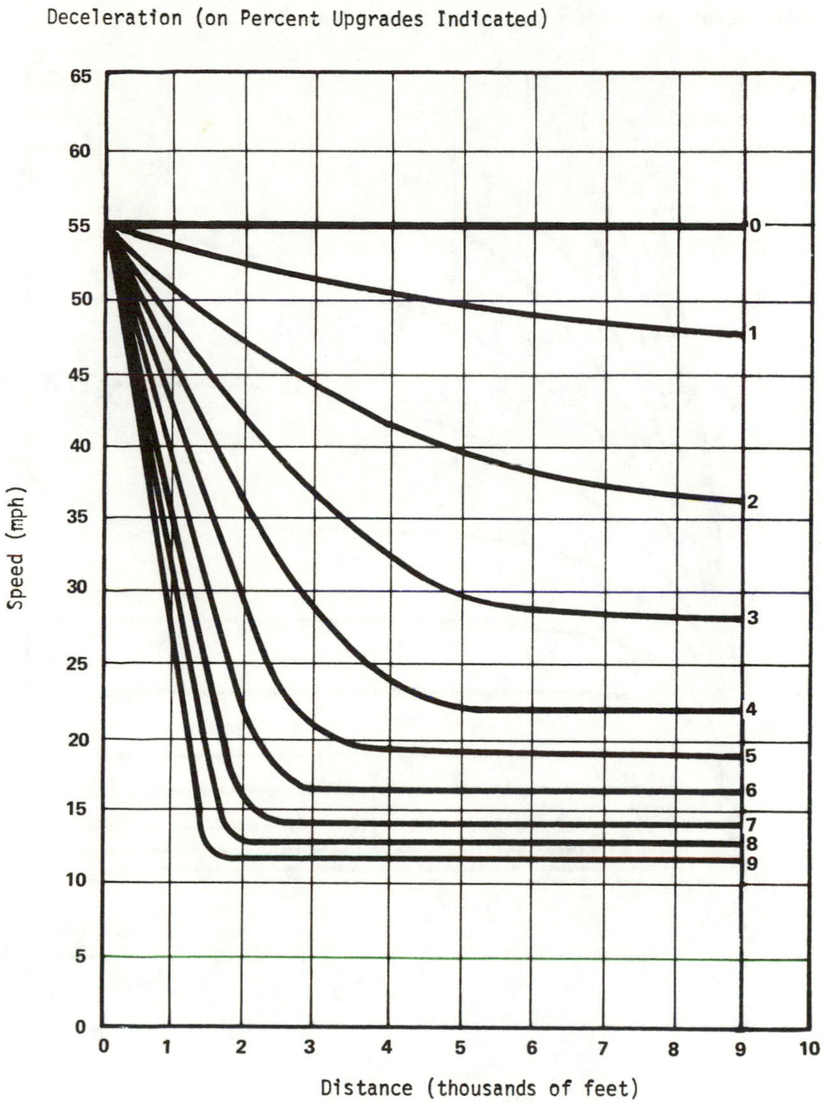

Figure III-27A. Speed-distance curves for a typical heavy truck of 300 lb/hp for deceleration (on percent upgrades).

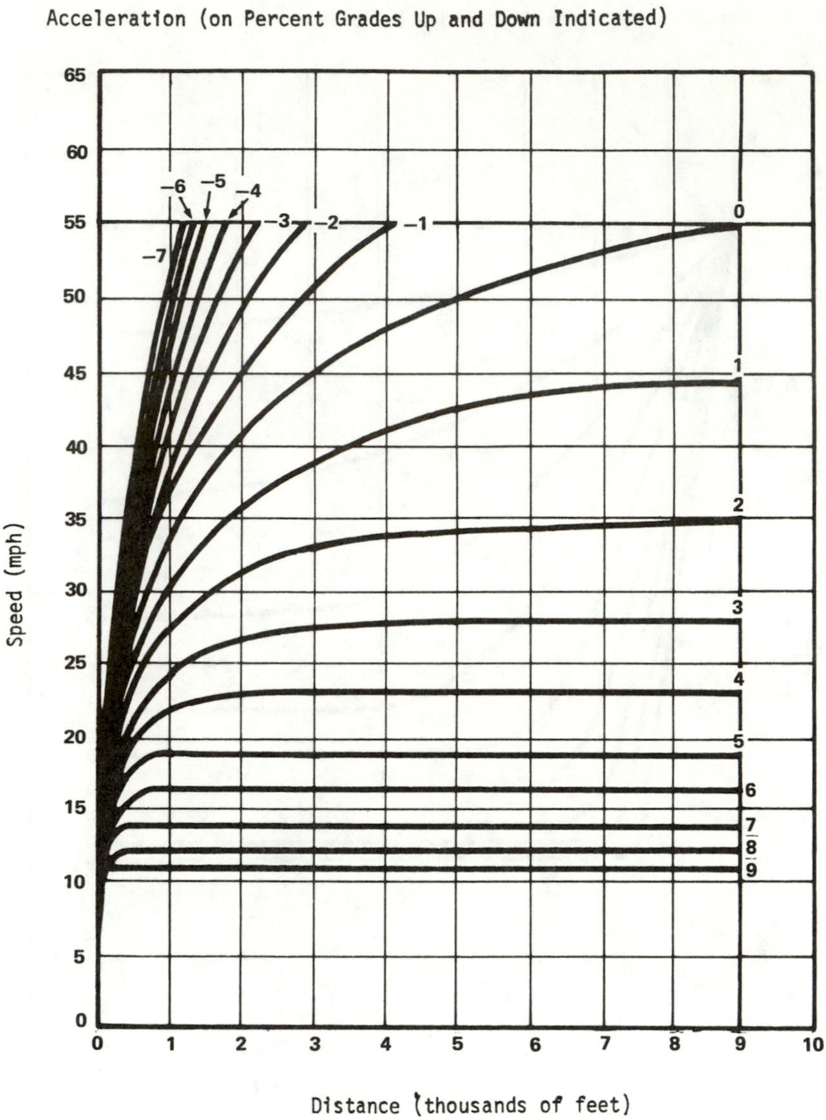

Figure III-27B. Speed-distance curves for a typical heavy truck of 300 lb/hp for acceleration (on percent grades up and down).

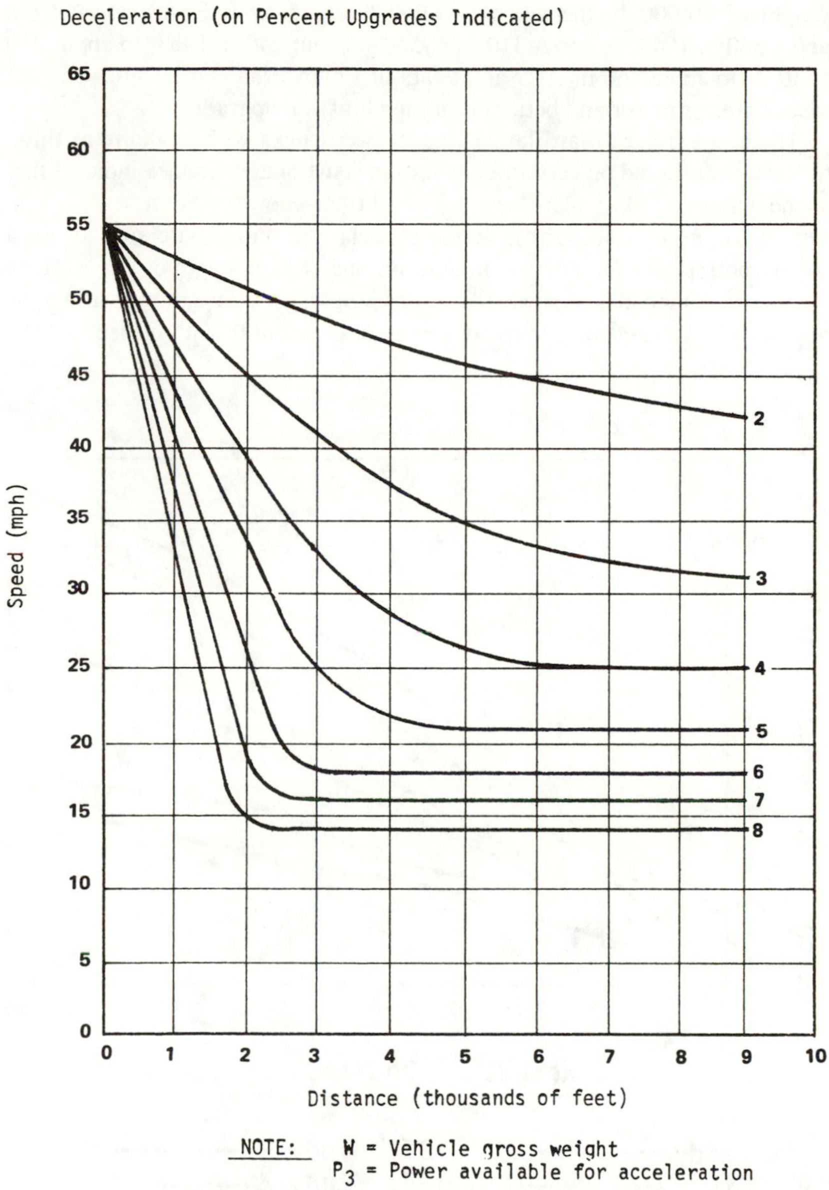

Figure III-27C. Speed loss for vehicles at W/P_3 values of 375 and 550, at 25 mph and 50 mph respectively—equals or exceeds 87.5% of all tractor trailers and straight trucks on all roads.

weight of 40,000 lb the average weight/horsepower ratio decreased from about 360 in 1949, to about 310 in 1955, to about 240 in 1963, to about 210 in 1975, to about 160 in 1977 and to about 130 in 1985. This decreased ratio means greater power and better climbing ability on upgrades.

There is a trend toward larger and heavier trucks with as many as three trailer units allowed on certain highways in some States. Studies indicate that as the number of axles increases, the weight/horsepower ratio increases. Taking all factors into account it seems prudent and on the safe side to use a weight/horsepower ratio of 300 in determining critical length of grade. However, there will be locations where the 300 weight/horsepower ratio will not be appropriate. Where this occurs, designers are encouraged to utilize either a

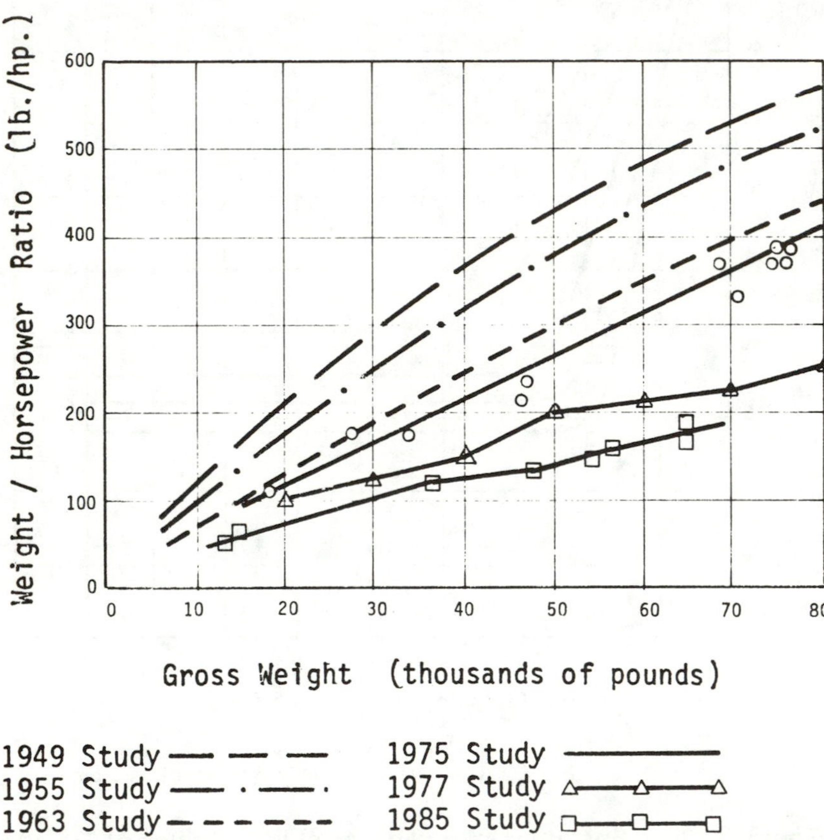

Figure III-28. Trend in weight-power ratios from 1949 to 1985 based on average data for all types of vehicles.

more representative ratio or an alternate method which more closely fits the conditions. A 1986 study (**60**) indicates measurable differences among certain truck classes. It recommends that the speed-distance relationship for a 87.5 percentile vehicle given in Figure III-27C provides a better indication of straight trucks, those smaller 2-axle trucks with gross vehicle weight (GVW) rating over 10,000 lbs, and to the tractor-trailer with GVW rating to the maximum allowable on the highways. Performance data levels reflected here indicate that longer values for critical length of grades are appropriate for these truck types. Figure III-27C should, therefore, be considered as an alternate to Figure III-27A, speed-distance curves on roads where truck traffic essentially consists of one of these classes.

Recreational vehicles. Consideration of recreational vehicles on grades is not as critical as that of trucks. However, on certain routes, such as designated recreational routes, where the low percentage of trucks may not otherwise warrant a truck climbing lane, sufficient recreational vehicle traffic may indicate a demand for an additional lane. This can be evaluated by using the design charts in Figure III-29 in the same manner as for trucks described in the preceding section of this chapter. Recreational vehicles are of many types including self-contained motor homes, pickup campers, and towed trailers of numerous sizes. Because the characteristics of recreational vehicles vary so much, it is difficult to establish a design vehicle. However, a recent study on the speed of vehicles on grades included recreational vehicles. The critical vehicle was considered to be a vehicle pulling a travel trailer and the charts in Figures III-29 and III-32 for a typical recreational vehicle are based on that assumption.

Control Grades for Design

Maximum grades. On the basis of the data in Figures III-27 through III-30, and according to the grade controls now in use in a large number of States, it is possible to arrive at reasonable guide values for maximum grades for design. Maximum grades of about 5 percent are considered appropriate for a design speed of 70 mph. For a design speed of 30 mph, maximum grades generally are in the range of 7 to 12 percent, depending on topography, with an average of about 8 percent. If only the more important highways are considered, it appears that a maximum grade of 7 or 8 percent would be representative for 30-mph design speed. Control grades for 40-, 50-, and 60-mph design speeds are intermediate between the above extremes.

For the maximum grade controls for each functional class of highway and street, see Chapters V through VIII.

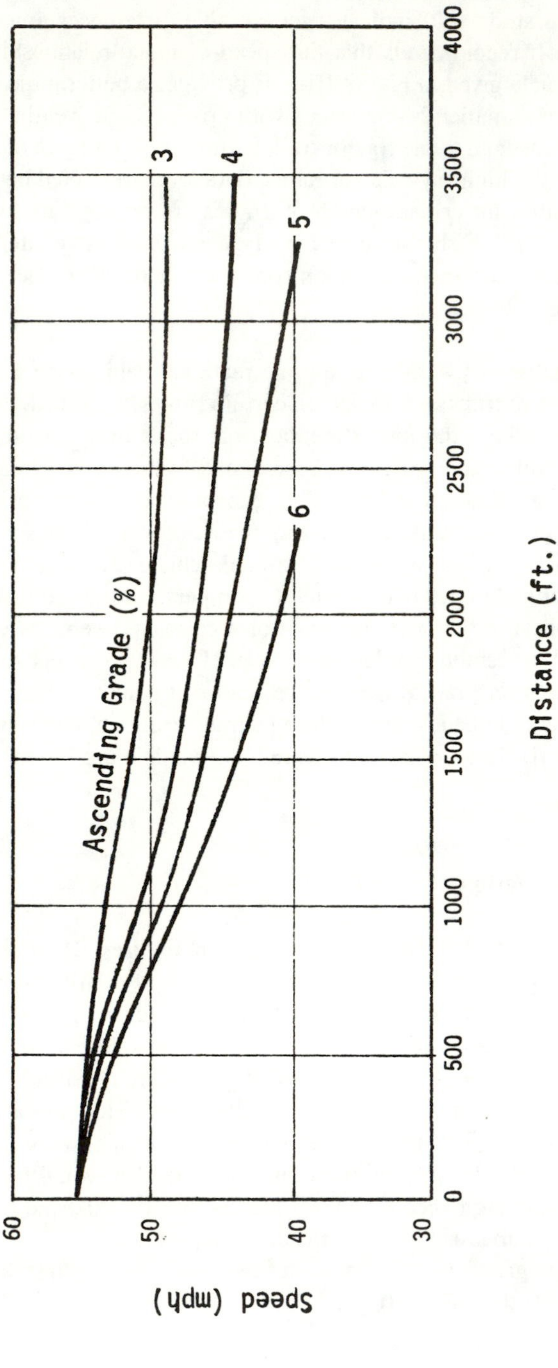

Figure III-29. Speed-distance curves for a typical recreational vehicle on the selected ascending grades. (Source: Ref. 28)

The maximum design grade should be used infrequently rather than as a value to be used in most cases. At the other extreme, for short grades less than 500 ft and for one-way downgrades, the maximum gradient may be about 1 percent steeper. For low-volume rural highways, grades may be 2 percent steeper.

Minimum grades. Flat and level grades on uncurbed pavements are virtually without objection when the pavement is adequately crowned to drain the surface laterally. With curbed pavements, longitudinal grades should be provided to facilitate surface drainage. A minimum grade for the usual case is 0.5 percent, but a grade of 0.30 percent may be used where there is a high-type pavement accurately crowned and supported on firm subgrade. Use of even flatter grades may be justified in special cases as discussed in subsequent chapters. Particular attention should be given to the design of storm water inlets and their spacing to keep the spread of water on the traveled way within tolerable limits. Roadside channels and median swales frequently require grades steeper than the roadway profile for adequate drainage. Drainage channels are discussed in Chapter IV.

Critical Lengths of Grade for Design

Maximum grade in itself is not a complete design control. It is necessary also to consider the length of a particular grade in relation to desirable vehicle operation. The term "critical length of grade" is used to indicate the maximum length of a designated upgrade on which a loaded truck can operate without an unreasonable reduction in speed. For a given grade, lengths less than critical ones result in acceptable operation in the desired range of speeds. If the desired freedom of operation is to be maintained on grades longer than critical ones, design adjustment such as change in location to reduce grades or addition of extra lanes should be made. The data for critical lengths of grade are used with other pertinent considerations (such as traffic volume in relation to capacity) to determine where added lanes are warranted.

To establish design values for critical lengths of grade for which gradeability of trucks is the determining factor, data or assumptions are needed for the following:

1. Size and power of a representative truck or truck combination to be used as a design vehicle along with the gradeability data for this vehicle.

   A loaded truck, powered so that the weight/horsepower ratio is about 300, is representative of the size and type of vehicle normally used for design control on main highways. Data in Figure

III-27A&B apply to such a vehicle. The data in Figure III-27C applies to vehicles of the straight truck and tractor-trailer classes with weight/horsepower ratios which are greater than 87.5% of the vehicles in these classes.

2. Speed at entrance to critical length of grade.

   The average running speed as related to design speed can be used to approximate the speed of vehicles beginning an uphill climb. It is, of course, subject to adjustment as approach conditions may determine. Where vehicles approach on nearly level grades, the running speed can be used directly. For a downhill approach it should be increased somewhat, and for an uphill approach it should be decreased.

3. Minimum speed on the grade below which interference to following vehicles is considered unreasonable.

   No specific data are available on which to base minimum tolerable speeds of trucks on upgrades. It is logical to assume that these minimum speeds are in direct relation to the design speed. Minimum truck speeds of about 25 to 40 mph for the majority of highways (on which design speeds are about 40 to 60 mph) probably are not unreasonably annoying to following drivers unable to pass on two-lane roads, if the time interval during which they are unable to pass is not too long. The time interval is not likely to be annoying on two-lane roads with volumes well below their capacities, whereas it is likely to be annoying on two-lane roads with volumes near capacity. Lower minimum truck speeds probably can be tolerated on multilane highways rather than on two-lane roads because there is more opportunity for and less difficulty in passing. Highways should be designed so that the speeds of trucks will not be reduced enough to cause intolerable conditions for the following drivers.

Studies show that regardless of the average speed on the highway, the greater a vehicle deviates from this average speed the greater its chances of becoming involved in an accident. One such study (**35**) used the speed distribution of vehicles traveling highways in one state, and related it to the accident involvement rate (the number of vehicles involved in accidents per 100,000,000 mi of travel) to obtain the rate for trucks of four or more axles operating on level grades. The accident involvement rates for truck speed reductions of 5, 10, 15, and 20 mph were developed assuming the reduction in the average speed of all vehicles on a grade was 30 percent of the truck speed reduction on the same grade. The results of this analysis are shown in Figure III-30.

The common basis for determining critical length of grade is a reduction in speed of trucks below the average running speed. The ideal would be for all traffic to operate at this speed. This, however, is not practical. In the past, the general practice has been to use a speed reduction of 15 mph in truck running speed below the average running speed of all traffic. As shown in Figure III-30, the accident involvement rate increases significantly when the truck speed reduction exceeds 10 mph with the involvement rate being 2.4 times greater for a 15-mph reduction than for a 10-mph reduction. On the basis of these relationships, it is recommended that a 10-mph reduction criterion be used as the general guide for determining critical lengths of grade.

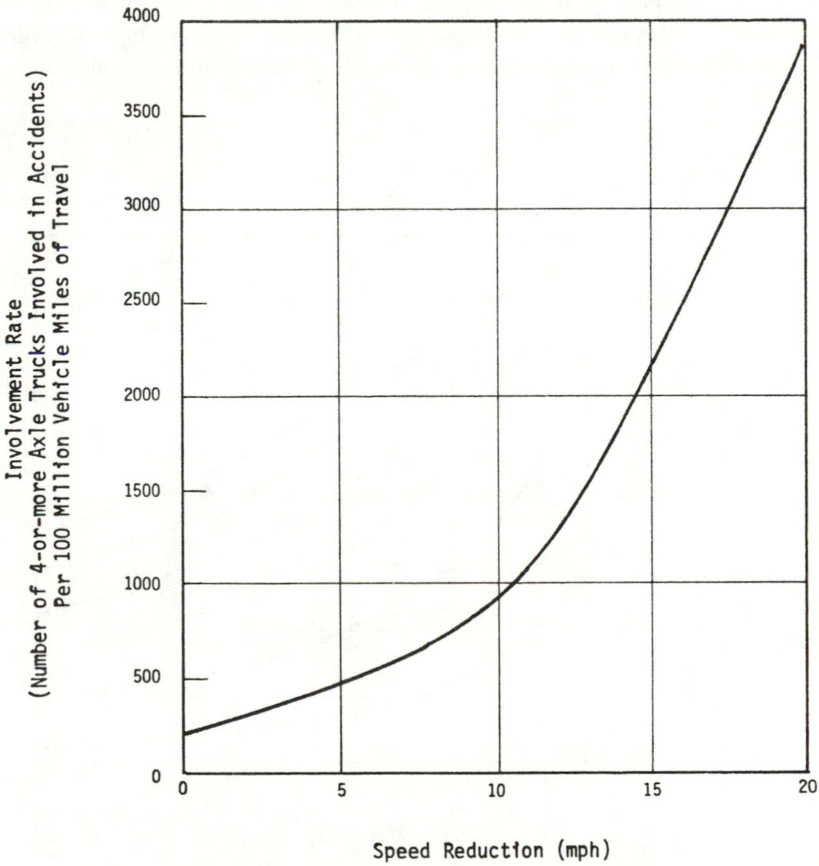

Figure III-30. Accident involvement rate of trucks for which running speeds are reduced below average running speed of all traffic. (Source: Ref. 35).

The length of any given grade that will cause the speed of a representative truck (300 lb/hp) entering the grade at 55 mph to be reduced by various amounts below the average running speed of all traffic is shown graphically in Figure III-31. The curve showing a 10-mph speed reduction (the general design guide for determining critical length of grade) is emphasized as the heavier one of the family of curves. Similar information for recreational vehicles may be found in Figure III-32.

Where the entering speed is less than 55 mph, as might be the case where the approach is on an upgrade, the speed reductions shown in Figures III-31 and III-32 are effected over shorter lengths of grade. Conversely, where the approach is on a downgrade, the probable approach speed is greater than 55 mph and the truck or recreational vehicle will ascend a greater length of grade than shown in the figures before the speed is reduced to the values shown.

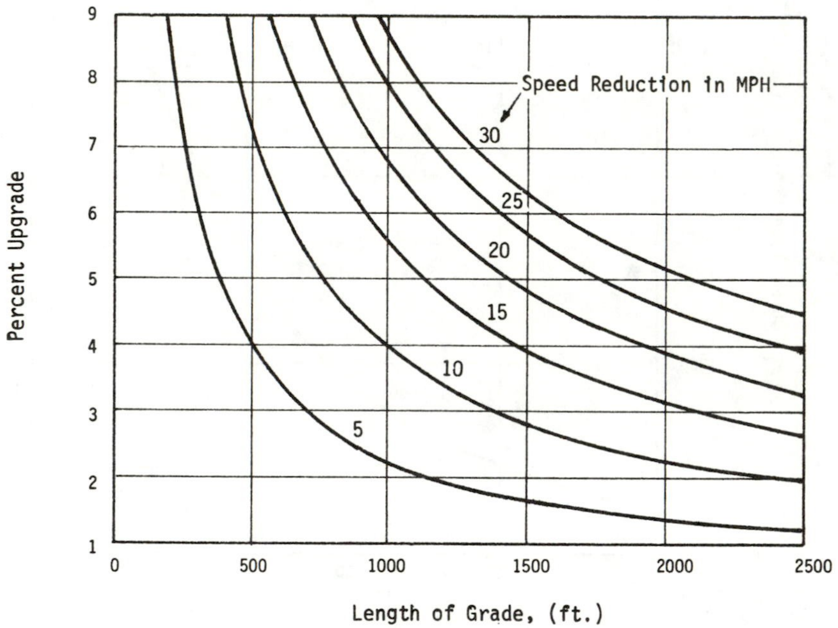

Figure III-31. Critical lengths of grade for design, assumed typical heavy truck of 300 lb/hp, entering speed = 55 mph.

The method of using Figure III-31 to determine critical lengths of grade is demonstrated in the following examples.

Assume that a highway is being designed for 60 mph and has a fairly level approach to a 4 percent upgrade; direct reading on the heavy line in Figure III-31 shows the critical length to be 1,000 ft. If instead, the design speed were 40 mph, the initial and minimum tolerable speeds on the grade would be different, but for the same permissible speed reduction the critical length would still be 1,000 ft.

In another distance, the critical length of a 5 percent upgrade approached by a 1,500 ft length of 2 percent upgrade is unknown. Figure III-31 shows that 1,500 ft of 2 percent upgrade results in a speed reduction of about 6 mph. The chart further shows that the remaining allowable speed reduction of 4 mph will be made on 275 ft of 5 percent upgrade.

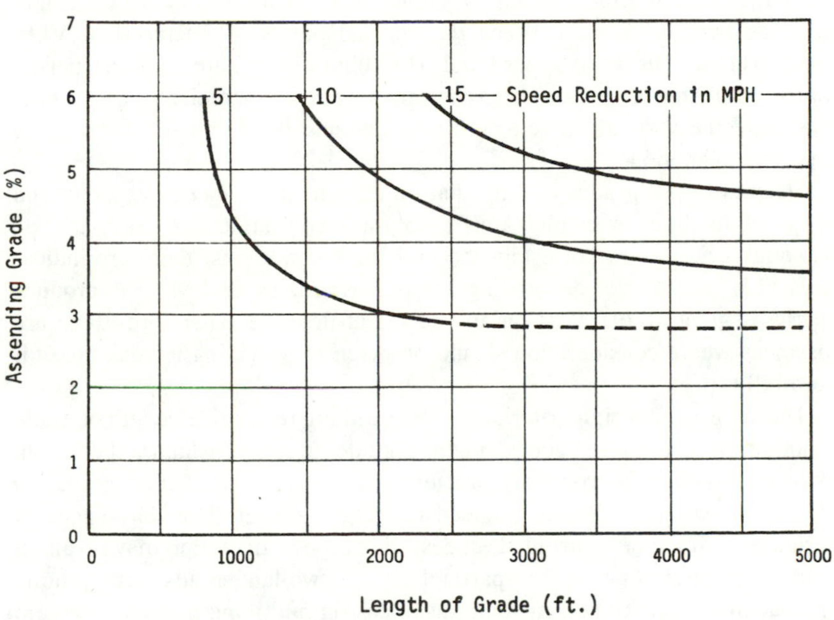

Figure III-32. Critical lengths of grade using an approach speed of 55 mph for typical recreational vehicle. (Source: Ref. 28).

Where an upgrade is approached on a momentum grade, heavy trucks often increase speed, sometimes to a considerable degree in order to make the climb in the upgrade at as high a speed as possible. This factor can be recognized in design by an increase in the permissible speed reduction. It remains for the designer to judge to what extent the speed of trucks would increase at the bottom of the momentum grade above that generally found on level approaches. It appears that a speed increase of about 5 mph can be considered for moderate downgrades and a speed increase of 10 mph for steeper grades of moderate length or longer. On this basis, the permissible speed reduction with momentum grades becomes 15 or 20 mph. For example, where there is a moderate length of 4 percent downgrade in advance of a 6 percent upgrade, a permissible speed reduction of 15 mph can be assumed. For this case the critical length of the 6 percent upgrade is about 925 ft.

The critical length of grade in Figure III-31 is derived as the length of tangent grade. Where a vertical curve is part of a critical length of grade, an approximate equivalent tangent grade length must be used. Where the condition involves vertical curves of Types II and IV in Figure III-40 and the algebraic difference in grades is not too great, the measurement of critical length of grade may be made between the vertical points of intersection (VPI). Where vertical curves of Types I and III in Figure III-40 are involved, particularly where the algebraic difference in grades is appreciable, about one-quarter of the vertical curve length may be considered as part of the grade under consideration.

Steep downhill grades can also have a detrimental effect on capacity and safety of facilities with high traffic volumes and numerous heavy trucks. Although criteria are not established for these conditions, there are indications (**13**) that trucks descending steep downgrades in low gear produce nearly as great an effect as on an equivalent upgrade. Therefore, there are instances where consideration should be given to providing a truck lane for downhill traffic.

The suggested design criterion for determining the critical length of grade is not intended as a strict control but as a guide. In some instances the terrain or other physical controls may preclude shortening or flattening grades to meet these controls. Where a speed reduction greater than the suggested design guide cannot be avoided, undesirable type of operation may result on roads with numerous trucks, particularly on two-lane roads with volume approaching capacity and in some instances on multilane highways. Where the length of critical grade is exceeded, consideration should be given to providing an added uphill lane for slow-moving vehicles, particularly where volume is at or near capacity and the truck volume is high. Data in Figure III-31 can be used along with other pertinent considerations, particularly volume data in relation to capacity and volume data for trucks, to determine where such added lanes are warranted.

Climbing Lanes

Climbing Lanes for Two-Lane Highways

General. Freedom and safety of operation on two-lane highways, besides being governed by the extent and frequency of passing sections, are adversely affected by heavily loaded vehicle traffic operating on grades of sufficient length to result in speeds that could impede following vehicles. In the past, provision of extra climbing lanes to improve safe operation has been rather limited because of extra construction cost. However, because of the increasing number of serious accidents occurring on grades, these lanes are more commonly included in original construction plans and additional lanes on existing highways built as safety improvement projects. The accident potential created by this condition has been dramatically illustrated in Figure III-30.

A highway section with a climbing lane is not considered a three-lane highway, but a two-lane highway with an extra lane for vehicles moving slowly uphill so that other vehicles using the normal lane to the right of the centerline are not retarded. These vehicles pass the slower vehicles moving upgrade, but not in the lane for opposing traffic, as on the customary two-lane road. A separate climbing lane exclusively for flow-moving vehicles is preferred to the addition of an extra lane carrying mixed traffic. Designs of two-lane highways with climbing lanes are demonstrated in Figure III-33 (A and B). Climbing lanes are designed for each direction independently of the other. Depending on the alinement and profile conditions, they may not overlap (Figure III-33A) or they may overlap as in Figure III-33B where there is a crest with a long grade on each side.

It is desirable to provide a climbing lane, as an extra lane on the upgrade side of a two-lane highway where the grade, traffic volume and heavy vehicle volume combine to degrade traffic operations from those on the approach to the grade. Where they have been provided there has been a high degree of compliance in their use by truck drivers. On highways with low volumes, only an occasional car is delayed, and climbing lanes, although desirable, may not be justified economically even where the critical length of grade is exceeded.

For these cases slow-moving vehicle turnouts should be considered where high volumes of slow-moving vehicles seriously impede normal flow of traffic. These are discussed under the section on "Operational Lane Improvements on Two-Lane Highways."

The following three conditions and criteria, reflecting economic considerations, should be satisfied to justify a climbing lane:

1. Upgrade traffic flow rate in excess of 200 vehicles per hour.
2. Upgrade truck flow rate in excess of 20 vehicles per hour.

3. One of the following conditions exists:
    - A 10 mph or greater speed reduction is expected for a typical heavy truck.
    - Level-of-service E or F exists on the grade.
    - A reduction of two or more levels-of-service is experienced when moving from the approach segment to the grade.

The upgrade flow rate is determined by multiplying the predicted or existing design hour volume by the directional distribution factor for the upgrade direction and dividing the result by the peak hour factor (the peak hour and directional distribution factors are discussed in Chapter II). The number of upgrade trucks is obtained by multiplying the upgrade flow rate by the percentage of trucks in the upgrade direction.

Trucks. As indicated in the preceding section, only one of the three conditions specified in item 3 must exist. The critical length of grade to effect a 10 mile per hour truck speed reduction is found using Figure III-27C or Figure III-31. This critical length is compared with the length of the particular grade being examined. If the critical length of grade is less than the length of the grade being studied item 3 is satisfied. This evaluation should be done first because the critical length of grade will be exceeded for most instances where the addition of a climbing lane is being considered, hence no further evaluations under item 3 will be required.

Justification for climbing lanes where the critical length of grade is exceeded may be considered from the standpoint of highway capacity. The procedures used are those from the HCM (**13**) for analysis of specific grades on two-lane highways. The remaining conditions in item 3 are evaluated using these HCM procedures. The effect that trucks have on capacity is primarily a function of the difference between the average speed of the trucks and the average running speed of the passenger cars on the highway. Physical dimensions of heavy trucks and their poorer accelerating characteristics also have a bearing on the space they occupy in the traffic stream; the average dual-tired truck controls about 130 to 210 percent (0 percent grade, Table III-24) the space of the average passenger car depending on the average speed of the car.

On individual grades the effect of trucks is more severe than their average effect over a longer section of highway. Thus, for a given volume of mixed traffic and a fixed roadway cross section, a higher degree of congestion is experienced on individual grades than for the average operation over longer sections that include downgrades as well as upgrades. Determination of design service volume on individual grades requires the use of truck factors derived from the geometrics of the grade and the level of service adopted by the highway agency for use as the basis for design of the highway under consideration. Passenger car equivalents so derived are given in Table III-24 for

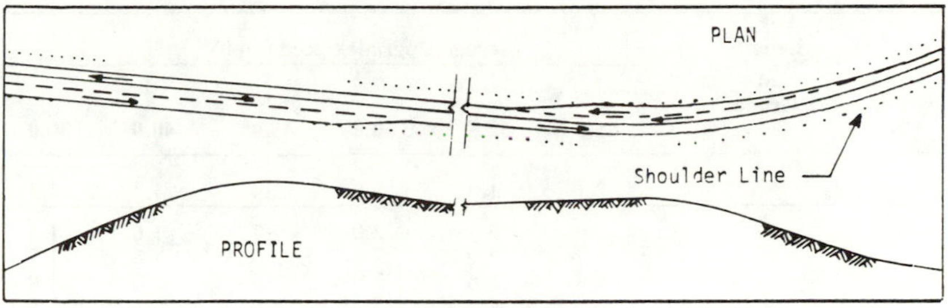

Climbing Lanes

-A-

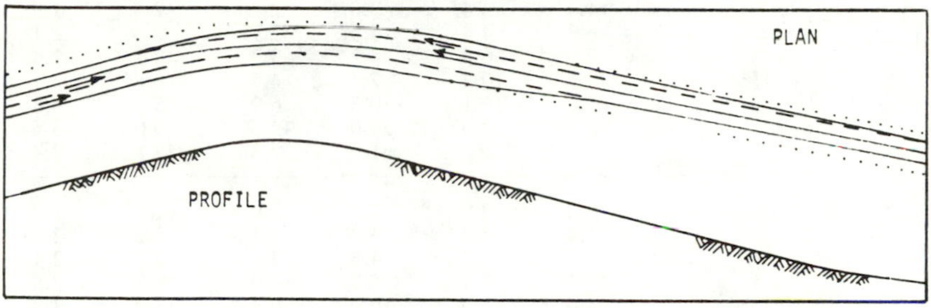

Climbing Lanes Overlapping on Crest

-B-

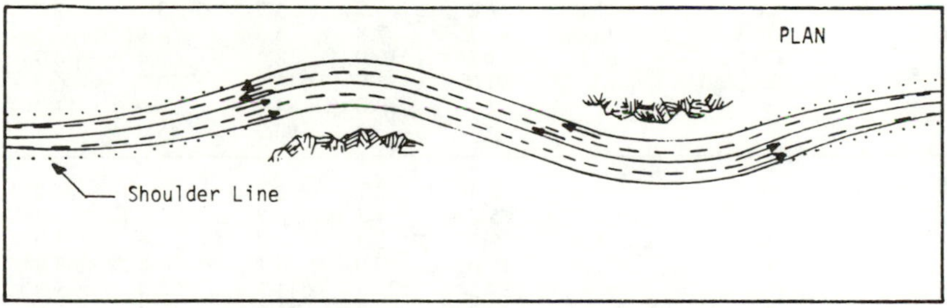

Passing Lane Sections on Winding Alinement

-C-

Figure III-33. Climbing lanes and passing lane sections on two-lane highways.

Grade (%)	Length of Grade (Mi)	Average Upgrade Speed (mph)					
		55.0	52.5	50.0	45.0	40.0	30.0
0	All	2.1	1.8	1.6	1.4	1.3	1.3
3	¼	2.9	2.3	2.0	1.7	1.6	1.5
	½	3.7	2.9	2.4	2.0	1.8	1.7
	¾	4.8	3.6	2.9	2.3	2.0	1.9
	1	6.5	4.6	3.5	2.6	2.3	2.1
	1½	11.2	6.6	5.1	3.4	2.9	2.5
	2	19.8	9.3	6.7	4.6	3.7	2.9
	3	71.0	21.0	10.8	7.3	5.6	3.8
	4	*	48.0	20.5	11.3	7.7	4.9
4	¼	3.2	2.5	2.2	1.8	1.7	1.6
	½	4.4	3.4	2.8	2.2	2.0	1.9
	¾	6.3	4.4	3.5	2.7	2.3	2.1
	1	9.6	6.3	4.5	3.2	2.7	2.4
	1½	19.5	10.3	7.4	4.7	3.8	3.1
	2	43.0	16.1	10.8	6.9	5.3	3.8
	3	*	48.0	30.0	12.5	9.0	5.5
	4	*	*	51.0	22.8	13.8	7.4
5	¼	3.6	2.8	2.3	2.0	1.8	1.7
	½	5.4	3.9	3.2	2.5	2.2	2.0
	¾	8.3	5.7	4.3	3.1	2.7	2.4
	1	14.1	8.4	5.9	4.0	3.3	2.8
	1½	34.0	16.0	10.0	6.3	3.9	3.8
	2	91.0	28.3	17.4	10.2	7.5	4.8
	3	*	*	37.0	22.0	14.6	7.8
	4	*	*	*	55.0	25.0	11.5
6	¼	4.0	3.1	2.5	2.1	1.9	1.8
	½	6.5	4.8	3.7	2.8	2.4	2.2
	¾	11.0	7.2	5.2	3.7	3.1	2.7
	1	20.4	11.7	7.8	4.9	4.0	3.3
	1½	60.0	25.5	16.0	8.5	6.4	4.7
	2	*	50.0	28.2	15.3	10.7	6.3
	3	*	*	70.0	34.0	23.9	11.3
	4	*	*	*	90.0	45.0	18.1
7	¼	4.5	3.4	2.7	2.2	2.0	1.9
	½	7.9	5.7	4.2	3.2	2.7	2.4
	¾	14.5	9.1	6.3	4.3	3.6	3.0
	1	31.4	16.0	10.0	6.1	4.8	3.8
	1½	*	39.5	23.5	11.5	8.4	5.8
	2	*	88.0	46.0	22.8	15.4	8.2
	3	*	*	*	66.0	38.5	16.1
	4	*	*	*	*	*	28.0

*Speed not attainable on grade specified.
NOTE: Round "percent grade" to next higher integer value.

Table III-24. Passenger car equivalents for specific grades on two-lane rural highways, E and E_o. (13)

various combinations of percent of grade, length of grade, and average upgrade speed. The relationship between the average upgrade speed and the level of service is shown in Table III-25.

Level of Service	Average Upgrade Speed (mph)
A	≥ 55
B	≥ 50
C	≥ 45
D	≥ 40
E	$\geq$ 25-40[a]
F	$<$25-40[a]

[a] The exact speed at which capacity occurs varies with the percentage and length of grade, traffic compositions, and volume; computational procedures are provided to find this value.

Table III-25. Level of service criteria for specific grades. (13)

If there is no 10 mile per hour reduction in speed—i.e., the critical length of grade is not exceeded—the level-of-service on the grade should be examined to determine if level of service E or F exists. This is done by calculating the limiting service flow rate for level-of-service D and comparing this rate to the actual flow rate on the grade. The actual flow rate is determined by dividing the volume of traffic by the peak hour factor. If the actual flow rate exceeds the service flow rate at level D, item 3 is satisfied. When the actual flow rate is less than the limiting value, a climbing lane is not warranted by this criteria.

The following formula is used to determine the limiting flow rate for level-of-service D:

$$SF_D = 2800 \text{ x } (V/C)_D \text{ x } f_d \text{ x } f_w \text{ x } f_g \text{ x } f_{hv}$$

Where:

SF_D = total service flow rate in both directions for prevailing roadway and traffic conditions, for level of service D in vph.

$(V/C)_D$ = ratio of flow rate to ideal capacity for level-of-service D.

f_d = adjustment factor for directional distribution.

f_w = adjustment factor for narrow lanes and restricted shoulder width.

f_g = adjustment factor for the operation of passenger cars on grades.

f_{hv} = adjustment factor for the presence of heavy vehicles in the traffic stream.

$(V/C)_D$ is determined from Table III-27 using an average upgrade speed of 40 mph and the percent of no passing zones for the grade being examined.

f_d is obtained from Table III-26 using the known or projected directional distribution on the upgrade.

f_w is obtained from Table III-28 using appropriate lane and shoulder widths.

f_g is obtained from the formula:

$$f_g = 1/[1 + (P_p I_p)]$$

Where:

P_p = proportion of passenger cars and;

$I_p = 0.02 (E - E_o)$

E is the base passenger car equivalent for a given percent of grade, length of grade, and speed, selected from Table III-24 and E_o is the base passenger car equivalent for a 0 percent grade and a speed also selected from Table III-24.

f_{hv} is determined from the formula:

$$f_{hv} = 1/\left[1 + P_{hv} (E_{hv} - 1)\right]$$

Where:

P_{hv} = proportion of heavy vehicles in the traffic stream or the proportion of trucks plus buses plus recreational vehicles.

$$E_{hv} = 1 + (0.25 + P_t/P_{hv}) (E - 1)$$

Where:

P_t = proportion of trucks in the heavy vehicle traffic stream and;

E is determined as described previously from Table III-24.

The HCM provides additional details and a worksheet to guide and organize the computations required by this procedure. This procedure is also available as software for desk top computers, reducing the hand calculations. A sample calculation is shown in Figure III-34.

The remaining condition to examine if neither of the other conditions of item 3 are satisfied is a two-level reduction in the level of service. To evaluate this criteria, the level-of-service for the grade and the approach segment must be determined. Since this criteria would be considered in only a very limited number of cases, it is not discussed in detail herein. The procedure uses the above equation for SF_D, but with appropriate values for the variables for level-of-service A, B, C. The values are selected from Table III-24 through Table III-28. The level-of-service on the grade is compared to the level-of-service on the approach to determine the reduction in operations.

Because there are so many variables involved that hardly any given set of conditions can be properly described as typical, a detailed analysis such as the one described is recommended wherever climbing lanes are being considered.

The point where an added lane should begin depends on the speeds at which trucks will approach the grade and on the extent of sight distance restrictions on this approach. Where there are no sight distance restrictions or other conditions causing low-speed approaches, the extra lane may be introduced on the grade some distance beyond the beginning of the upgrade because the speed of a truck will not be reduced to the point where it creates intolerable conditions for the following drivers until it has traveled a certain distance up the grade. The optimum for capacity would occur for a reduction in truck speed to 40 mph, but a 10-mph decrease in truck speed below the operating speed, as discussed in the preceding section "Critical Lengths of Grade for Design," would be the most practical reduction obtainable from standpoints of level of service and safety. This 10-mph reduction, which generally corresponds to the speed variation between adjacent levels of service, is the accepted basis for determining the location at which to begin climbing lanes. The distance from the bottom of the grade to the point where truck

Percent of Traffic on Upgrade	Adjustment Factor
100	0.58
90	0.64
80	0.70
70	0.78
60	0.87
50	1.00
40	1.20
30	1.50

Table III-26. Adjustment factor for directional distribution on specific grades f_d.

speeds fall 10 mph below the operating speed may be determined from Figure III-27C or Figure III-31 depending on the weight/horsepower ratio of the appropriate truck. For example, assuming an approach condition on which trucks with a 300-weight/horsepower ratio are traveling within a flow having an operating speed of 55 mph, the resulting 10-mph speed reduction occurs at distances of 600 to 1,000 ft for grades varying from 7 to 4 percent. With a downgrade approach these distances would be longer, and with an upgrade approach they would be shorter. Distances thus determined may be used to establish the point at which a climbing lane begins. Where restrictions, upgrade approaches, or other conditions indicate low speed of approaching trucks, the extra lane would be introduced near the foot of the grade. The beginning of the added lane should be preceded by a tapered section desirably with a ratio of 25:1, but at least 150 ft long.

The ideal design would be to extend the climbing lane to a point beyond the crest, where a typical truck could attain a speed that is within 10 mph of the speed of the other vehicles with the desirable speed being at least 40 mph, approximately at level of service D. Even this may not be practical in many instances because of the unduly long distance required for trucks to accelerate to the desired speed. For such a condition a practical point to end the added lane is where the truck can return to the normal lane without undue hazard—in particular, where the sight distance becomes sufficient to permit passing with safety when there is no oncoming traffic or, preferably, at least 200 ft beyond this point. In addition, a corresponding length of taper should be provided to permit the truck to return to the normal lane. For example, on a highway where the safe passing sight distance becomes available 100 ft beyond this point, the truck lane should extend 100 ft plus 200 ft or 300 ft beyond the crest, plus an additional length for taper, preferably at a ratio of 50:1 but with a length of at least 200 ft.

A climbing lane should desirably be as wide as the through lanes. It should be so constructed that it can immediately be recognized as an extra lane in one direction. The centerline of the normal two-lane pavement should be clearly marked in the required manner as for any two-lane road, including yellow barrier lines for no-passing zones. Adequate signs at the beginning of the upgrade such as "Slower Traffic Keep Right" or "Trucks Use Right Lane" may be used to direct slow-moving vehicles into the climbing lane. These and other appropriate signs and markings for truck climbing lanes may be found in sections 2B-23, 2B-24, 2C-19, and 3B-8 of the MUTCD (**8**).

The cross slope of a climbing lane is usually handled in the same manner as the addition of a lane to a multilane highway. Depending on agency practice, this design results in either a continuation of the cross slope or a lane with slightly more cross slope than the adjacent through lane. On a superele-

	Average Upgrade	Percent No Passing Zone					
Percent	Speed (mph)	0	20	40	60	80	100
3	55	0.27	0.23	0.19	0.17	0.14	0.12
	52.5	0.42	0.38	0.33	0.31	0.29	0.27
	50	0.64	0.59	0.55	0.52	0.49	0.47
	45	1.00	0.95	0.91	0.88	0.86	0.84
	42.5	1.00	0.98	0.97	0.96	0.95	0.94
	40	1.00	1.00	1.00	1.00	1.00	1.00
4	55	0.25	0.21	0.18	0.16	0.13	0.11
	52.5	0.40	0.36	0.31	0.29	0.27	0.25
	50	0.61	0.56	0.52	0.49	0.47	0.45
	45	0.97	0.92	0.88	0.85	0.83	0.81
	42.5	0.99	0.96	0.95	0.94	0.93	0.92
	40	1.00	1.00	1.00	1.00	1.00	1.00
5	55	0.21	0.17	0.14	0.12	0.10	0.08
	52.5	0.36	0.31	0.27	0.24	0.22	0.20
	50	0.57	0.49	0.45	0.41	0.39	0.37
	45	0.93	0.84	0.79	0.75	0.72	0.70
	42.5	0.97	0.90	0.87	0.85	0.83	0.82
	40	0.98	0.96	0.95	0.94	0.93	0.92
	35	1.00	1.00	1.00	1.00	1.00	1.00
6	55	0.12	0.10	0.08	0.06	0.05	0.04
	52.5	0.27	0.22	0.18	0.16	0.14	0.13
	50	0.48	0.40	0.35	0.31	0.28	0.26
	45	0.49	0.76	0.68	0.63	0.59	0.55
	42.5	0.93	0.84	0.78	0.74	0.70	0.67
	40	0.97	0.91	0.87	0.83	0.81	0.78
	35	1.00	0.96	0.95	0.93	0.91	0.90
	30	1.00	0.99	0.99	0.98	0.98	0.98
7	55	0.00	0.00	0.00	0.00	0.00	0.00
	52.5	0.13	0.10	0.08	0.07	0.05	0.04
	50	0.34	0.27	0.22	0.18	0.15	0.12
	45	0.77	0.65	0.55	0.46	0.40	0.35
	42.5	0.86	0.75	0.67	0.60	0.54	0.48
	40	0.93	0.82	0.75	0.69	0.64	0.59
	35	1.00	0.91	0.87	0.82	0.79	0.76
	30	1.00	0.95	0.92	0.90	0.88	0.86

Ratio of flow rate to ideal capacity of 2,000 pcph, assuming passenger-car operation is unaffected by grades.

NOTE: Interpolate for intermediate values of "Percent No Passing Zone"; round "Percent Grade" to the next higher integer value.

Table III-27. Values of v/c ratio[a] versus speed, percent grade and percent no passing zones for specific grades (13).

Usuable[a] Shoulder Width (ft)	12-ft Lanes		11-ft Lanes		10-ft Lanes		9-ft Lanes	
	Los A-D	Los[b] E	Los A-D	Los[b] E	Los A-D	Los[b] E	Los A-D	Los[b] E
≥6	1.00	1.00	0.93	0.94	0.84	0.87	0.70	0.76
4	0.92	0.97	0.85	0.92	0.77	0.85	0.65	0.74
2	0.81	0.93	0.75	0.88	0.68	0.81	0.57	0.70
0	0.70	0.88	0.65	0.82	0.58	0.75	0.49	0.66

[a] Where shoulder width is different on each side of the roadway, use the average shoulder width.
[b] Factor applies for all speeds less than 45 mph.

Table III-28. Adjustment factors for the combined effect of narrow lanes and restricted shoulder width, f_w (13).

vated section the cross slope is generally a continuation of the slope used on the through lane.

Desirably, the shoulder on the outer edge of a climbing lane should be as wide as the shoulder on the normal two-lane section. Conditions, however, may dictate otherwise, particularly when the climbing lane is added to an existing highway. A usable shoulder of 4 ft width or greater is acceptable. Although not wide enough for a stalled vehicle to completely clear the climbing lane, this 4-ft shoulder in combination with the climbing lane generally provides sufficient width for both the stalled vehicle and a slow-speed passing vehicle without need for the latter to encroach on the through lane.

In summary, climbing lanes offer a comparatively inexpensive means of overcoming losses in capacity and providing improved operation where congestion on grades is caused by slow trucks in combination with high traffic volumes. As discussed earlier in this section, climbing lanes also provide improved safety. On some existing two-lane highways the addition of climbing lanes could defer reconstruction for many years or indefinitely. On new designs climbing lanes could make a two-lane highway adequate whereas a much more costly multilane highway would be necessary without them. Climbing lanes are warranted where the critical length of grade is exceeded; level of service E or F exists on the grade or, the level of service on the grade is at least two levels lower than the level of service on the approach. An exception to these warrants would be where overriding safety considerations dictate the addition of the lane regardless of grade or traffic volume.

Climbing Lanes on Freeways and Multilane Highways

General. Climbing lanes, although they are becoming more prevalent, have not been used as extensively on freeways and multilane highways, perhaps for the reason that multilane facilities more frequently had sufficient capacity to handle their traffic load, including the normal percentage of slow-moving vehicles with high weight/horsepower ratios without being congested. Climbing lanes generally are not as easily justified on multilane facilities as on two-lane highways because on two-lane facilities vehicles following other slower moving vehicles on upgrades are frequently prevented by opposing traffic from using the adjacent traffic lane for passing, whereas such a barrier to its use is not present on multilane facilities. A slow-moving vehicle in the normal right lane does not impede the following vehicles that can readily move left to the abutting lane and proceed without difficulty, although there is evidence that safety is enhanced when vehicles in the traffic stream move at the same speed.

WORKSHEET FOR SPECIFIC GRADES Page 1

Site Identification:______ Date:______ Time:______

Name:______ Checked by:______

I. GEOMETRIC DATA

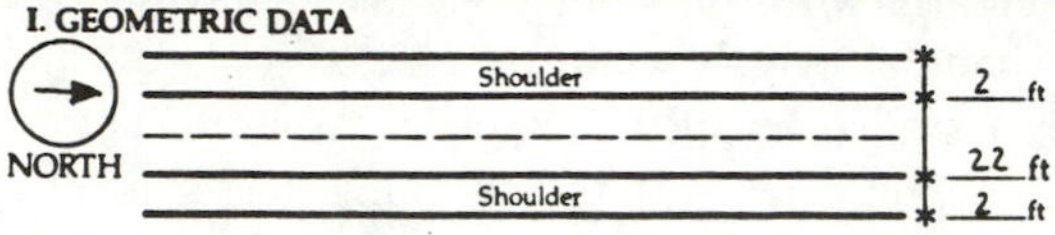

Design Speed: 60 mph
Grade: 6 %, 2 mi
% No Passing Zones: 80

II. TRAFFIC DATA

Total Volume, Both Dir. 500 vph
Flow Rate = Volume ÷ PHF
549 = 500 ÷ .91

Directional Distribution: 60/40
Traffic Composition: 16 %T, 5 %RV, 2 %B
PHF: .91

III. SOLVING FOR ADJUSTMENT FACTORS f_g AND f_{HV}

$f_g = 1 / [1 + P_p I_p]$
$I_p = 0.02 (E - E_o)$

$f_{HV} = 1 / [1 + P_{HV} (E_{HV} - 1)]$
$E_{HV} = 1 + (0.25 + P_{T/HV}) (E - 1)$

Speed (mph)	P_p	I_p	E	E_o	f_g
55					
52.5	.77	.964	50	1.8	.57
50	.77	.532	28.2	1.6	.71
45	.77	.278	15.3	1.4	.82
40	.77	.188	10.7	1.3	.87
30	.77	.100	6.3	1.3	.93

Speed (mph)	P_{HV}	E_{HV}	$P_{T/HV}$ (P_T/P_{HV})	E	f_{HV}
55					
52.5	.23	47.5	.70	50	.09
50	.23	26.3	.70	28.2	.15
45	.23	14.6	.70	15.3	.24
40	.23	10.2	.70	10.7	.32
30	.23	6.04	.70	6.3	.46

IV. SOLVING FOR SERVICE FLOW RATE

Speed (mph)	SF	2,800 ×	v/c ×	f_d ×	f_w ×	f_g ×	f_{HV}
55 (LOS A)	0	2,800	.05	.87	.75	—	—
52.5	13	2,800	.14	.87	.75	.57	.09
50 (LOS B)	54	2,800	.28	.87	.75	.71	.15
45 (LOS C)	212	2,800	.59	.87	.75	.82	.24
40 (LOS D)	483	2,800	.81	.87	.88	.87	.32
30	899	2,800	.98	.87	.88	.93	.46

Figure III-34. Sample capacity calculation, grades on two-lane highway.

WORKSHEET FOR SPECIFIC GRADES Page 2

V. PLOT SF vs Speed

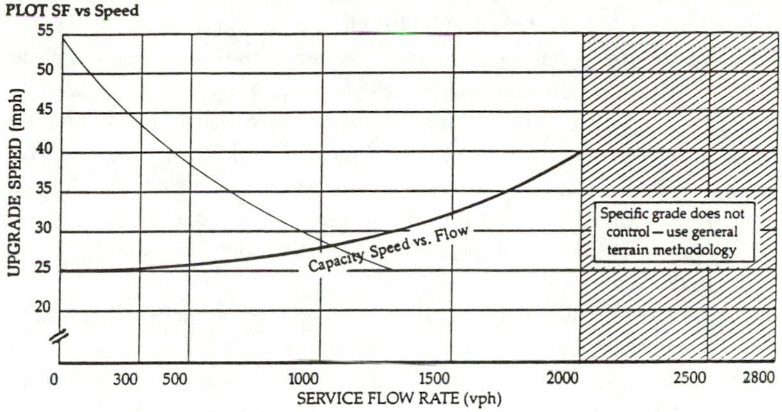

Intersection of Capacity Speed vs Flow curve with Service Flow Rate vs Speed curve defines Capacity, SF_E, and Speed at Capacity, S_c

VI. LEVEL OF SERVICE ANALYSIS

LOS	SF (from Worksheet)	Actual Flow Rate
A	0	549
B	54	
C	212	
D	483	
E	1050	Level of Service: E

Comments:

Since the actual flow rate exceeds the service flow rate for level-of-service D a climbing lane should be considered.

Figure III-34. Sample capacity calculation, grades on two-lane-highway. (Con't.)

Because highways are normally designed for 20 years or more in the future, there is less likelihood of climbing lanes being justified on multilane facilities than on two-lane roads for several years after construction even though they are deemed desirable for the peak hours of the design year. Where this the case, there is economic advantage in designing for, but deferring construction of, climbing lanes on multilane facilities. In this situation, grading for the future climbing lane should be provided initially. The necessary additional grading for a climbing lane is small when compared to that required by the overall cross section. If, however, even this additional grading creates a problem, it is acceptable, although not desirable, to use a narrower shoulder adjacent to the climbing lane rather than the full shoulder provided on a normal section.

Although primarily a rural condition, there are instances where climbing lanes will be necessary in urban areas. Climbing lanes are particularly important for freedom of operation on urban freeways where traffic volumes are high in relation to capacity. On older urban freeways and arterial streets with appreciable grades and no climbing lanes, it is a common occurrence for heavy traffic, which may otherwise move well, to queue up on grades.

Trucks. As in the case for two-lane highways discussed earlier in this chapter, the principal determinants of need for climbing lanes are critical lengths of grade, effects of trucks on grades in terms of equivalent passenger cars, service volumes for the desired level of service, and flow rates for the next poorer level of service.

Critical length of grade has been discussed previously in this chapter. It is the length of a particular upgrade which reduces the operating speed of high weight/horsepower trucks 10 mph below the operating speed of the remaining traffic. The critical length of grade to effect a 10 mph truck speed reduction is found using Figure III-27C or Figure III-31 which is compared to the length of the particular grade being examined. If the critical length of grade is less than the length of grade being studied a climbing lane is warranted.

In determining service volume the passenger car equivalent for trucks is a significant factor. It is generally agreed that trucks on multilane facilities have less effect in deterring following vehicles than on two-lane roads. Comparison of percent of grade, length of grade, percent of trucks, and level of service between Table III-29, III-30, or III-31, and Table III-24 clearly illustrates the difference in passenger car equivalents of trucks for two-lane and multilane facilities.

In the matter of justifying the cost of providing a climbing lane, a larger volume should be the criterion, as in the case of justifying climbing lanes for two-lane roads, because highway users accept and will tolerate a higher degree of congestion (poorer level of service) on individual grades than over long sections of highway. As a matter of practice, the service volume on an individual grade should not exceed that attained by using the next poorer level of service from that used for the basic design. The one exception is that the service volume derived from employing level-of-service D should not be exceeded.

Generally, climbing lanes should not be considered unless the upgrade directional traffic volume is equal to or greater than the level-of-service D service volume for two of the lanes of a four-lane highway with the normal percentage of trucks. In most cases when the service volume, including trucks, is greater than 1,700 vph and the length of the grade and the percentage of trucks are sufficient to consider climbing lanes, the volume in terms of equivalent passenger cars is likely to approach or even exceed the capacity. An increase in the number of lanes throughout the highway section would represent a better investment than the provision of climbing lanes.

A climbing lane is also generally not warranted on four-lane highways for a directional service volume below 1,000 regardless of the percentage of trucks. Although a truck driver will occasionally pass another truck under such conditions, the inconvenience and hazard with this low volume are not sufficient to justify the cost of a climbing lane in the absence of appropriate criteria.

The procedures outlined below and detailed in the Highway Capacity Manual are used to consider the traffic operational characteristics on the grade being examined. The service flow rate for the desired level of service is determined, together with the flow rate for the next poorer level of service. If the flow rate on the grade determined by dividing the upgrade DDHV by the peak hour factor (previously discussed in Chapter II) exceeds that of the next poorer level of service, a climbing lane is warranted.

The service flow rate on a freeway grade (upgrade) is determined from the equation:

$$SF = C_j \times (v/c) \times N \times f_w \times f_p \times f_{hv}$$

Where:

- SF = service flow rate for the segment to be analyzed, vph.
- C_j = capacity under ideal conditions. It is 2000 pcphpl for 60 mph and 70 mph freeway elements and 1900 for 50 mph freeway elements.
- V/C = volume-to-capacity ratio and is determined from Table III-34;
- N = number of lanes (each direction);
- f_w = factor to adjust for the effects of restricted lane widths and/or lateral clearances and is determined from Table III-35;

Grade (%)	Length (mi)	Passenger-Car Equivalent, E_T 4-Lane Freeways								6-8 Lane Freeways							
Percent Trucks		2	4	5	6	8	10	15	20	2	4	5	6	8	10	15	20
<1	All	2	2	2	2	2	2	2	2	2	2	2	2	2	2	2	2
1	0-½	2	2	2	2	2	2	2	2	2	2	2	2	2	2	2	2
	½-1	3	3	3	3	3	3	3	3	3	3	3	3	3	3	3	3
	>1	4	3	3	3	3	3	3	3	4	3	3	3	3	3	3	3
2	0-¼	4	4	4	3	3	3	3	3	4	4	4	3	3	3	3	3
	¼-½	5	4	4	3	3	3	3	3	5	4	4	3	3	3	3	3
	½-¾	6	5	5	4	4	4	4	4	6	5	5	4	4	4	4	4
	¾-1½	7	6	6	5	4	4	4	4	7	5	5	5	4	4	4	4
	> 1½	8	6	6	6	5	5	4	4	8	6	6	5	4	4	4	4
3	0-¼	6	5	5	5	4	4	4	3	6	5	5	5	4	4	4	3
	¼-½	8	6	6	6	5	5	5	4	7	6	6	6	5	5	5	4
	½-1	9	7	7	6	5	5	5	5	9	7	7	6	5	5	5	5
	1-1½	9	7	7	7	6	6	5	5	9	7	7	6	5	5	5	5
	>1½	10	7	7	7	6	6	5	5	10	7	7	6	5	5	5	5
4	0-¼	7	6	6	5	4	4	4	4	7	6	6	5	4	4	4	4
	¼-½	10	7	7	6	5	5	5	5	9	7	7	6	5	5	5	5
	½-1	12	8	8	7	6	6	6	6	10	8	7	6	5	5	5	5
	> 1	13	9	9	9	8	8	7	7	11	9	9	8	7	6	6	6
5	0-¼	8	6	6	6	5	5	5	5	8	6	6	6	5	5	5	5
	¼-½	10	8	8	7	6	6	6	6	8	7	7	6	5	5	5	5
	½-1	12	11	11	10	8	8	8	8	12	10	9	8	7	7	7	7
	> 1	14	11	11	10	8	8	8	8	12	10	9	8	7	7	7	7
6	0-¼	9	7	7	7	6	6	6	6	9	7	7	6	5	5	5	5
	¼-½	13	9	9	8	7	7	7	7	11	8	8	7	6	6	6	6
	½-¾	13	9	9	8	7	7	7	7	11	9	9	8	7	6	6	6
	> ¾	17	12	12	11	9	9	9	9	13	10	10	9	8	8	8	8

NOTE: If a length of grade falls on a boundary condition, the equivalent for the longer grade category is used. For any grade steeper than the percentage shown, use the next higher grade category.

Table III-29. Passenger-car equivalents for typical trucks (200 lb/hp) (13).

Grade (%)	Length (mi)	Passenger-Car Equivalent, E_T 4-Lane Freeways								6-8 Lane Freeways							
Percent Trucks		2	4	5	6	8	10	15	20	2	4	5	6	8	10	15	20
<2	ALL	2	2	2	2	2	2	2	2	2	2	2	2	2	2	2	2
3	0-¼	3	3	3	3	3	3	3	3	3	3	3	3	3	3	3	3
	¼-½	4	4	4	3	3	3	3	3	4	4	4	3	3	3	3	3
	½-¾	4	4	4	4	3	3	3	3	4	4	4	3	3	3	3	3
	¾-1	5	4	4	4	3	3	3	3	5	4	4	4	3	3	3	3
	>1	6	5	5	5	4	4	4	3	6	5	5	4	4	4	3	3
4	0-¼	4	4	4	3	3	3	3	3	5	4	4	4	3	3	3	3
	¼-½	5	5	5	4	4	4	4	4	5	4	4	4	4	4	4	4
	½-1	6	5	5	5	4	4	4	4	6	5	5	4	4	4	4	4
	> 1	7	6	6	5	4	4	4	4	7	5	5	5	4	4	4	4
5	0-¼	6	5	5	5	4	4	4	3	6	5	5	5	4	4	4	3
	¼-1	8	7	7	6	5	5	5	5	8	7	7	6	5	5	5	5
	> 1	9	7	7	6	5	5	5	5	8	7	7	6	5	5	5	5
6	0-¼	7	5	5	5	4	4	4	4	7	5	5	5	4	4	3	3
	¼-1	9	7	7	6	5	5	5	5	8	7	7	6	5	5	5	5
	> 1	9	7	7	7	6	6	5	5	9	7	7	6	5	5	5	5

NOTE: If a length of grade falls on a boundary condition, the equivalent from the longer grade category is used. For any grade steeper than the percentage shown, use the next higher grade category.

Table III-30. Passenger-car equivalents for light trucks (100 lb/hp) (13).

Grade (%)	Length (mi)	Passenger-Car Equivalent, E_T 4-Lane Freeways								6-8 Lane Freeways							
	Percent Trucks	2	4	5	6	8	10	15	20	2	4	5	6	8	10	15	20
<1	All	2	2	2	2	2	2	2	2	2	2	2	2	2	2	2	2
1	0-¼	2	2	2	2	2	2	2	2	2	2	2	2	2	2	2	2
	¼-½	3	3	3	3	3	3	3	3	3	3	3	3	3	3	3	3
	½-¾	4	4	4	4	3	3	3	3	4	4	4	3	3	3	3	3
	¾-1	5	4	4	4	3	3	3	3	5	4	4	4	3	3	3	3
	1-1½	6	5	5	5	4	4	4	3	6	5	5	4	4	4	3	3
	>1½	7	5	5	5	4	4	4	3	7	5	5	5	4	4	3	3
2	0-¼	4	4	4	3	3	3	3	3	4	4	4	3	3	3	3	3
	¼-½	7	6	6	5	4	4	4	4	7	5	5	5	4	4	4	4
	½-¾	8	6	6	5	5	4	4	4	8	6	6	6	5	5	4	4
	¾-1	8	6	6	6	5	5	5	5	8	6	6	6	5	5	5	5
	1-1½	9	7	7	7	6	6	5	5	9	7	7	6	5	5	5	5
	>1½	10	7	7	7	6	6	5	5	10	7	7	6	5	5	5	5
3	0-¼	6	5	5	5	4	4	4	3	6	5	5	5	4	4	4	3
	¼-½	9	7	7	6	5	5	5	5	8	7	7	6	5	5	5	5
	½-¾	12	8	8	7	6	6	6	6	10	8	7	6	5	5	5	5
	¾-1	13	9	9	8	7	7	7	7	11	8	8	7	6	6	6	6
	> 1	14	10	10	9	8	8	7	7	12	9	9	8	7	7	7	7

Table III-31. Passenger-car equivalents for heavy trucks (300 lb/hp).

4	0-¼	7	5	5	5	4	4	4	4	7	5	5	5	4	4	3	3
	¼-½	12	8	8	7	6	6	6	6	10	8	7	6	5	5	5	5
	½-¾	13	9	9	8	7	7	7	7	11	9	9	8	7	6	6	6
	¾-1	15	10	10	9	8	8	8	8	12	10	10	9	8	7	7	7
	> 1	17	12	12	10	9	9	9	9	13	10	10	9	8	8	8	8
5	0-¼	8	6	6	6	5	5	5	5	8	6	6	6	5	5	5	5
	¼-½	13	9	9	8	7	7	7	7	11	8	8	7	6	6	6	6
	½-¾	20	15	15	14	11	11	11	11	14	11	11	10	9	9	9	9
	>¾	22	17	17	16	13	13	13	13	17	14	14	13	12	11	11	11
6	0-¼	9	7	7	7	6	6	6	6	9	7	7	6	5	5	5	5
	¼-½	17	12	12	11	9	9	9	9	13	10	10	9	8	8	8	8
	>½	28	22	22	21	18	18	18	18	20	17	17	16	15	14	14	14

NOTE: If a length of grade falls on a boundary condition, the equivalent from the longer grade category is used. For any grade steeper than the percentage shown, use the next higher grade category.

Table III-31. Passenger-car equivalents for heavy trucks (300 lb/h p). (Con't.)

Grade (%)	Length (mi)	Passenger-Car Equivalent, E_R 4-Lane Freeways								6-8 Lane Freeways							
Percent RV'S		2	4	5	6	8	10	15	20	2	4	5	6	8	10	15	20
<2	ALL	2	2	2	2	2	2	2	2	2	2	2	2	2	2	2	2
3	0-½	3	2	2	2	2	2	2	2	2	2	2	2	2	2	2	2
	>½	4	3	3	3	3	3	3	3	4	3	3	3	3	3	3	3
4	0-¼	3	2	2	2	2	2	2	2	3	2	2	2	2	2	2	2
	¼-¾	4	3	3	3	3	3	3	3	4	3	3	3	3	3	3	3
	>¾	5	4	4	4	3	3	3	3	4	4	4	4	3	3	3	3
5	0-¼	4	3	3	3	3	3	3	3	4	3	3	3	2	2	2	2
	¼-¾	5	4	4	4	4	4	4	4	5	4	4	4	4	4	4	4
	>¾	6	5	4	4	4	4	4	4	5	5	4	4	4	4	4	4
6	0-¼	5	4	4	4	3	3	3	3	5	4	4	3	3	3	3	3
	¼-¾	6	5	5	4	4	4	4	4	6	4	4	4	4	4	4	4
	>¾	7	6	6	6	5	5	5	5	6	5	5	5	4	4	4	4

NOTE: If a length of grade falls on a boundary condition, the equivalent from the longer grade category is used. For any grade steeper than the percentage shown, use the next higher grade category.

Table III-32. Passenger-car equivalents for recreational vehicles (13).

Grade (%)	Passenger-Car Equivalent, E_B
0-3	1.6
4[a]	1.6
5[a]	3.0
6[a]	5.5

[a] Use generally restricted to grades more than ¼ mi long.

Table III-33. Passenger-car equivalents for buses (13).

f_p = factor to adjust for the effect of driver population and is determined from Table III-36;

f_{hv} = factor to adjust for the effect of heavy vehicles (trucks, buses, and recreational vehicles) in the traffic stream.

Where f_{hv} is computed from the formula:
$f_{hv} = 1/[1 + P_T (E_T - 1)] + P_R (E_R - 1) + P_B (E_B - 1)]$

Where:

E_T, E_R, E_B = the passenger car equivalents for trucks, recreational vehicles and buses respectively and may be obtained from Tables III-29 through III-33 as appropriate for the section for which a climbing lane is being considered.

P_T, P_R, P_B = the proportion of trucks, recreational vehicles and buses, respectively in the traffic stream.

When the analysis indicates that a climbing lane may improve operations or maintain the level of service on the section with a grade an additional check must be made to determine if the number of lanes required on the grade are sufficient even with a climbing lane. This is accomplished by estimating the capacity of the truck climbing lane from the formula:

$C_t = 2000/E_t$ – for 60 and 70 mph design speeds or
$C_t = 1900/E_t$ – for a 50 mph design speed

Where:

C_t = climbing lane truck capacity;
E_t = the truck passenger car equivalent and is obtained from Tables III-29, III-30, and III-31.

The number of trucks that are estimated will use the climbing lane is determined by multiplying the capacity of the truck climbing lane by the V/C ratio for the level of service desired. The level of service for the trucks in the climbing lane should be, desirably, the same as that provided for the mixed traffic in the other lanes on the grade. This V/C ratio is obtained from Table III-34 using the maximum permissible value for the level of service being considered. This number of trucks projected to use the climbing lane is subtracted from the total trucks in the traffic stream. The remaining trucks are used to calculate a new percentage of trucks in the mixed traffic stream. The level of service for the mixed stream is determined using the equation for service flow rate given above. An example of a freeway analysis is shown in Figure III-35.

When the grade being investigated is located on a multilane highway, one additional factor must be considered in determining the service flow rate. This factor accounts for the type of multilane highway—divided or undivided—and is selected from Table III-37. The equation become $SF = C_j$ x (v/c x f_w x f_p x f_{hv} x f_E where all the factors are as defined previously except for the f_E value which is obtained from Table III-37 and is selected appropriate for the type of highway.

Climbing lanes on multi-lane roads are usually placed on the outer or right-hand side of the roadway as shown in Figure III-36. The principles for cross slopes, for locating terminal points, and for designing terminal areas or tapers for climbing lanes are discussed earlier in this chapter in conjunction with two-lane highways. These principles are equally applicable to climbing lanes on multilane facilities, the prime factor being that of determining the location of the uphill terminus of the climbing lane at the point where a satisfactory speed is attained by trucks, preferably about 10 mph below the operating speed of the highway. Passing sight distance need not be considered.

Operational Lane Improvements on Two-Lane Highways

Passing Lane Sections on Two-Lane Roads

Passing lane sections, four lanes in width, are constructed on two-lane roads to provide the desired frequency of safe passing zones or to eliminate interference from low-speed heavy vehicles, or both. Where a sufficient num-

LOS	Density (PC/MI/LN)	70 mph Design Speed			60 mph Design Speed			50 mph Design Speed		
		Speed[b] (MPH)	v/c	MSF[a] (PCPHPL)	Speed[b] (MPH)	v/c	MSF[a] (PCPHPL)	Speed[b] (MPH)	v/c	MSF[a] (PCPHPL)
A	≤ 12	≥ 60	0.35	700	—	—	—	—	—	—
B	≤ 20	≥ 57	0.54	1,100	≥ 50	0.49	1,000	—	—	—
C	≤ 30	≥ 54	0.77	1,550	≥ 47	0.69	1,400	≥ 43	0.67	1,300
D	≤ 42	≥ 46	0.93	1,850	≥ 42	0.84	1,700	≥ 40	0.83	1,600
E	≤ 67	≥ 30	1.00	2,000	≥ 30	1.00	2,000	≥ 28	1.00	1,900
F	> 67	< 30	[c]	[c]	< 30	[c]	[c]	< 28	[c]	[c]

[a]Maximum service flow rate per lane under ideal conditions.
[b]Average travel speed.
[c]Highly variable, unstable.
NOTE: All values of MSF Rounded to the nearest 50 pcph.

Table III-34. Levels of service for basic freeway sections (13).

Distance Traveled Pavement[a] (ft)	Adjustment Factor, f_w							
	Obstructions on One Side of the Roadway				Obstructions on Both Sides of the Roadway			
	Lane Width (ft)							
	12	11	10	9	12	11	10	9
	4-Lane Freeway (2 Lanes Each Direction)							
≥6	1.00	0.97	0.91	0.81	1.00	0.97	0.91	0.81
5	0.99	0.96	0.90	0.80	0.99	0.96	0.90	0.80
4	0.99	0.96	0.90	0.80	0.98	0.95	0.89	0.79
3	0.98	0.95	0.89	0.79	0.96	0.93	0.87	0.77
2	0.97	0.94	0.88	0.79	0.94	0.91	0.86	0.76
1	0.93	0.90	0.85	0.76	0.87	0.85	0.80	0.71
0	0.90	0.87	0.82	0.73	0.81	0.79	0.74	0.66
	6-or 8-Lane Freeway (3 or 4 Lanes Each Direction)							
≥6	1.00	0.96	0.89	0.78	1.00	0.96	0.89	0.78
5	0.99	0.95	0.88	0.77	0.99	0.95	0.88	0.77
4	0.99	0.95	0.88	0.77	0.98	0.94	0.87	0.77
3	0.98	0.94	0.87	0.76	0.97	0.93	0.86	0.76
2	0.97	0.93	0.87	0.76	0.96	0.92	0.85	0.75
1	0.95	0.92	0.86	0.75	0.93	0.89	0.83	0.72
0	0.94	0.91	0.85	0.74	0.91	0.87	0.81	0.70

[a]Certain types of obstructions, high-type median barriers in particular, do not cause any deleterious effect on traffic flow. Judgement should be exercised in applying these factors.

Table III-35. Adjustment factor for restricted lane width and lateral clearance (13).

Traffic Stream Type	Factors, f_p
Weekday or Commuter	1.0
Other	0.75-0.90[a]

[a]Engineering judgement and/or local data must be used in selecting an exact value.

Table III-36. Adjustment factor for the character of the traffic stream (13).

Type	Divided	Undivided
Rural	1.00	0.95
Suburban	0.90	0.80

Table III-37. Adjustment factor for type of multilane highway and development environment, f_E (13).

ber and length of safe passing sections cannot be obtained in the design of horizontal and vertical alinement alone, an occasional section of four lanes may be introduced as shown in Figure III-33C to provide more sections and length safe for passing. Such sections are particularly advantageous in rolling terrain, especially where alinement is winding or the profile includes critical lengths of grade. In rolling terrain a highway on tangent alinement may have restricted passing conditions even though the grades are below critical length. Use of a four-lane section over some of the crests provides added passing sections in both directions where they are most needed. Four-lane sections should be sufficiently long to permit several vehicles in line behind a slow-moving truck to pass before reaching the normal section of two-lane highway.

Sections of four lanes introduced explicitly for passing purposes need not be divided because there is no separation of opposing traffic on the two-lane portions of the highway. The use of a median, however, is advantageous and should be considered on highways carrying 500 vph or more, particularly on highways to be ultimately converted to four lanes divided.

The introduction of a four-lane section on a two-lane highway does not necessarily involve much additional grading. A full shoulder width is not as

1. Geometry and Given Conditions: 4-12' lanes, no obstructions.

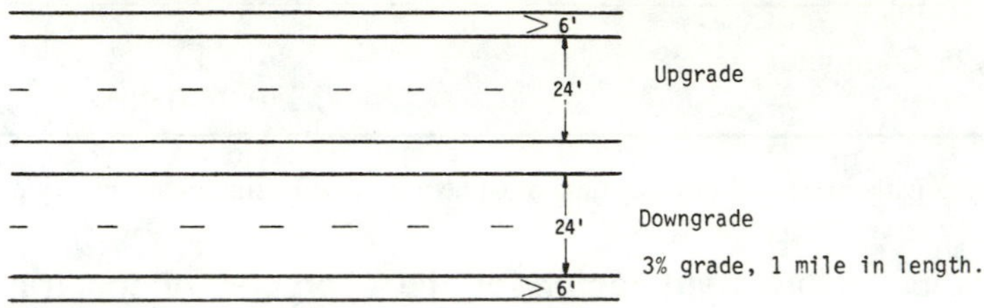

DDHV = 2200 vph; PHF = .95; Percent Trucks = 20%; Design Speed = 70 mph. SF = 2200/.95 = 2316; E_T = 5 (Table III-29; f_{HV} = 1/1 + .2(S-1) = .56; v/c = 0.77 (Table III-34).

2. Solution: LOS C is desired on this upgrade section of freeway

For LOS C, SF_c = 2000 x .77 x 1.0 x .56 x 1.0 x 2 = 1725 vph.

Next poorer LOS:

For LOS D, SF_D = 2000 x .93 x .56 x 1.0 x 2 = 2083 vph.

The LOS for the existing section is below LOS D (2314> 2083 vph) - warrant for climbing lane is established.

3. Capacity of the Truck Climbing Lane = C_T = 2000/5 = 400 tph

Number of trucks using the climbing lane = 400(0.77) = 308 tph (LOS C)

Total Trucks on Freeway = 2200(.20) = 440 tph

Remaining freeway lanes would serve = 2200 - 308 = 1892 vph; SF = 1897/.95 = 1992 vph

Percent trucks on freeway with LOS C = 440 - 308/1892 = 7% f_{HV} = 1/[1 + .07(5-1)] = .78

For LOS C = SF = 2000 x .77 x 1.0 x 2.0 x .78 = 2402 vph

The desired LOS C (1992<2402) for this freeway section is attained by the use of a climbing lane.

Figure III-35. Example problem—freeway truck climbing lane analysis—four-lane freeway.

necessary because likely volumes are much less than four-lane capacity, the vehicles likely to stop are few, and there is little difficulty in passing a vehicle with only two wheels on the shoulder. A shoulder width of 4 to 6 ft may be adequate. Thus, if the normal shoulder width on the two-lane highway is 10 ft, a 6- to 8-ft widening of the roadbed on each side is all that may be required. The additional lanes should be at least 10 ft and preferably 11 or 12 ft wide, and in no case should the traveled way be less than 40 ft wide.

The transitions between the two-lane and the four-lane pavements should be located where the change in width is in full view of the driver. Sections of four-lane highway, particularly divided sections, longer than about 2 mi may cause the driver to lose his sense of awareness that the highway is basically a two-lane facility. It is essential, therefore, that transitions from four lanes to two lanes be properly marked and identified with pavement markings and signs to alert the driver that he is approaching a section of two-lane highway on which he must remain in the right-hand lane (except to pass) and on which

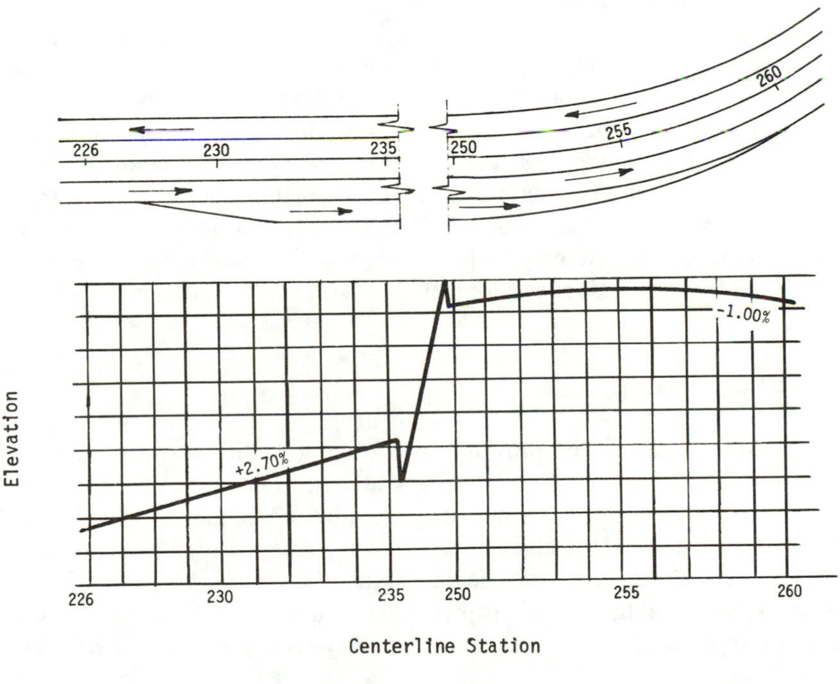

Figure III-36. Climbing lane placed on right-hand side of multilane highway.

passing must be performed with caution. An advance sign before the end of the four-lane pavement is particularly important to inform driver of the narrow pavement ahead. (See section 3B-8 of the MUTCD (**8**)).

The following is a summary of the design procedure to be followed in providing passing sections on two-lane highways:

1. Horizontal and vertical alinement should be designed to provide as much of the highway as feasible with sight distance safe for passing. (See Table III-5.)
2. Where design service volume approaches capacity, the effect of lack of passing sections in reducing the level of service should be recognized. (See Table III-27.)
3. Where design service volume approaches the service volume for the next poorer level of service because of trucks climbing up long grades, consideration should be given to constructing added uphill lanes on critical lengths of grade as shown in Figure III-33.
4. Where the extent and frequency of passing sections made available by application of items 1 and 3 are still too few, consideration should be given to the construction of four-lane sections.

Several highway agencies have pioneered successful methods for providing more passing opportunities along two-lane roads in addition to those discussed in preceding sections relating to climbing lanes and four-lane sections. Some of the more recognized of these methods, three-lane sections, turnouts, shoulder driving and shoulder sections are described in the Federal Highway Administration's informational guide "Low Cost Methods for Improving Traffic Operations on Two-Lane Roads" (**36**). A synopsis of portions of material found in this guide pertaining to these designs is presented in the succeeding sections. More detailed criteria for these methods will be found in the guide.

Three Lane Sections

An added lane can be provided in one or both directions of travel to improve traffic operations at a bottleneck at least to the same quality of service as adjacent road sections. Also they are provided to improve overall traffic operations on two-lane highways by reducing delays caused by inadequate passing opportunities over significant lengths of highways, typically 5 to 50 miles. When an additional lane is provided to improve traffic operations over a length of road, it is frequently constructed systematically at regular intervals.

The location of the added lane needs to appear logical to the driver. The value of it is more obvious at locations where passing sight distance is restricted than on long tangents which provide passing opportunities. On the other hand the selection of a site should recognize the need for adequate sight distance at both the lane addition and lane drop tapers. A minimum sight distance of 1,000 ft on the approach to each taper is recommended. The selection of an acceptable site also needs to consider the location of intersections and high volume driveways in order to minimize the volume of turning movements on a road section where passing is encouraged. Further, other physical constraints such as bridges and culverts should be avoided if they restrict provision of a continuous shoulder.

A minimum length of 1,000 ft, excluding tapers, is needed to assure that delayed vehicles have an opportunity to complete at least one pass in the added lane. Where such a lane is provided to reduce delays at a specific bottleneck the needed length is controlled by the extent of the bottleneck. A lane added to improve overall traffic operations should be long enough, over 0.25 mile, to provide a substantial reduction in traffic platooning. The optimal length is usually 0.5 to 1.0 miles. As the added lane is increased beyond a mile the reduction in platooning generally diminishes. Operational benefits typically result in reduced platooning for 3 to 8 miles downstream depending on volumes and passing opportunities. After that normal levels of platooning will occur until the next added lane is encountered.

The width of an added lane should normally be as wide as the lane widths of the two-lane highway. It is also desirable for the abutting shoulder to be a minimum of 4 ft wide and whenever possible the shoulder width in the added section should match that of the adjoining two-lane highway. The transition tapers at each end of the section should be designed to encourage safe and efficient operation. The lane drop taper length should be computed from the MUTCD (**8**) formula $L = WS$ (L = Length in feet, W = Width in feet, S = Speed in mph) while the recommended length for the lane addition taper is half to two-thirds of the lane drop length.

The signing and marking of an additional lane is partially addressed in the MUTCD (**8**) which indicates the appropriate centerline markings for these lanes as well as the signing and marking of lane drop transitions. However, the MUTCD (**8**) does not address signing in advance of and at the lane addition. A sign with the legend PASSING LANE 1/2 MILE should be placed in advance of each lane addition in order that drivers of both slow-moving vehicles and following vehicles can prepare to make effective use of the added lane. Additional signs 2 to 5 miles in advance are also desirable because they may reduce the frustration and impatience of drivers following a slow-moving vehicle by assuring them that they will soon have an opportunity to pass. In addition a sign should be installed at the beginning of the lane addition taper to assure that the slower-moving traffic keeps to the right.

Turnouts

A turnout is a widened, unobstructed shoulder which allows slow-moving vehicles to pull out of the through lane to give passing opportunities to following vehicles. The driver of the slow-moving vehicle, if there are following vehicles, is expected to pull out of the through lane and remain in the turnout only long enough for the following vehicles to pass before returning to the through lane. When there are only one or two following vehicles this maneuver can be accomplished without it being necessary for the driver of the vehicle in the turnout to stop. However, when this number is exceeded it will require the driver to stop in order for all the following vehicles to pass. Turnouts are most frequently used on lower volume roads where long platoons are rare and in difficult terrain with steep grades where construction of an additional lane may not be cost effective.

Such conditions are often found in mountains, coastal and scenic areas where more than 10 percent of the vehicle volumes are large trucks and recreational vehicles. The turnouts should only be provided where the following conditions exist or can be provided:

1. A minimum sight distance of 1,000 ft in each direction;
2. An available width from the edge of traveled way of 16 ft;
3. A firm and smooth surface and;
4. A minimum length of turnout including taper as shown in Table III-38 (tapers are generally 50 ft to 100 ft in length).

The minimum width of the turnout is 12 ft with widths of 16 ft considered desirable. Proper signing and marking are also necessary both to maximize usage and assure safe operations.

Approach Speed (MPH)	Minimum Length (ft)[a]
25	200
30	250
40	300
50	400
55	450
60	550

[a]Maximum length should be 600 feet to avoid use of the turnout as a passing lane.

Table III-38. Recommended lengths of turnouts including taper.

Shoulder Driving

In parts of the United States a long-standing custom has been established for slow-moving vehicles to move to the shoulder when another vehicle approaches from the rear and return to the travel lane after the following vehicle has passed. The practice generally occurs where adequate paved shoulders exist and in effect these shoulders function as continuous turnouts. This custom is regarded as a courtesy to other motorists requiring little or no sacrifice in speed by either vehicle operator. While highway agencies may want to permit such use as a means of improving passing opportunities without a major capital investment, they need to recognize that in all likelihood shoulder driving is currently prohibited by law. Thus, a highway agency considering shoulder driving as a passing aid probably will need to propose legislation to authorize such use as well as develop a public education campaign to familiarize vehicle operators with the new law.

Highway agencies should evaluate the mileage of two-lane highways with paved shoulders as well as their structural quality before deciding whether to allow their use as a passing aid. A consideration should be that where shoulder driving becomes common, it will not be limited to selected sites but rather will occur anywhere on the system where paved shoulders are provided. Another consideration is that shoulder widths of at least 10 ft and preferably 12 ft are required. Because the practice has grown up through local signing no special signing to promote such use has been created.

Shoulder Use Sections

Another approach is to permit slow-moving vehicles use of paved shoulders at selected sites designated by specific signing. This is a more limited application of shoulder use by slow-moving vehicles than shoulder driving described in the previous section. Typically, drivers move to the shoulder only long enough for following vehicles to pass and then return to the through lane. Thus, the section functions as an extended turnout. This approach enables a highway agency to promote shoulder use only where the shoulder is adequate to handle anticipated traffic loads and the need for more frequent passing opportunities has been established by the large amount of vehicle platooning.

Shoulder use sections generally range in length from 1,000 ft to 3 miles. Use should only be marked where shoulders are at least 10 ft and preferably 12 ft wide. Adequate structural strength to support the anticipated loads along with good surface conditions are required. Particular attention needs to be placed on the condition of the shoulder because drivers are unlikely to move over onto a shoulder if it is rough, broken or covered with debris. Signs need to be erected at both the beginning and end of the section where shoulder use is allowed. However, since signing of shoulder use sections is not addressed in the MUTCD (**8**) it will be necessary to create special signing developed by others.

Emergency Escape Ramps

General

Where long, descending grades exist or where topographic and location controls require such grades on new alinement, the design and construction of an emergency escape ramp at an appropriate location is desirable for the purpose of slowing and stopping an out-of-control vehicle away from the main traffic stream. An out-of-control vehicle generally is the result of an operator losing control of the vehicle because of loss of brakes either through overheating or mechanical failure, or failure to downshift at the appropriate time. Specific guidelines for the design of escape ramps are lacking at this time. However, considerable experience with ramps constructed on existing highways has led to the design and installation of effective ramps that are saving lives and reducing property damage. Reports and studies of the existing ramps indicate that their operational characteristics are providing acceptable deceleration rates and affording good driver control of the vehicle on the ramp.

There are several resistance forces that act on every vehicle to affect the vehicle's speed. These include engine, braking, and tractive resistance forces. Engine and braking resistance forces can be ignored in the design of escape ramps because they should be designed for the worst case, in which the vehicle is out of gear and the brake system has failed. Tractive resistance contains four subclasses: inertial, air, rolling, and gradient. Inertial and negative gradient resistance forces act to maintain motion of the vehicle, while rolling, positive gradient, and air resistance forces act to retard its motion. Figure III-37 illustrates the action of the various resistance forces on a vehicle.

Inertial resistance can be described as a force that will resist movement in a vehicle at rest or will keep a moving vehicle in motion, unless the vehicle is acted on by some external force. Inertial resistance must be overcome to either increase or decrease the speed of a vehicle. Rolling and positive gradient resistance forces are available to overcome the inertial resistance. Rolling resistance is a general term used to describe the resistance to motion at the area of contact between a vehicle's tires and the road surface and is only applicable when a vehicle is in motion. It is influenced by the type and displace-

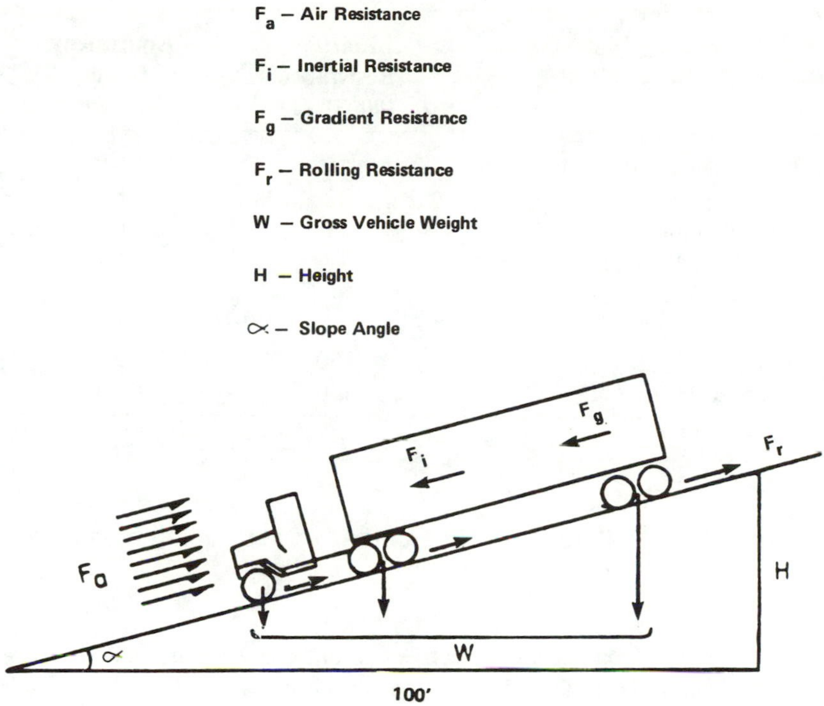

Figure III-37. Forces acting on vehicle in motion.

ment characteristics of the surfacing material of the roadway. Each surfacing material has a coefficient, expressed in lb/1,000 lb of gross vehicle weight (Table III-39), which determines the amount of rolling resistance of a vehicle. The values shown in Table III-39 for rolling resistance have been obtained from various sources throughout the country and are a best available estimate.

Gradient resistance is due to the effect of gravity and is expressed as the force required to move the vehicle through a given vertical distance. For gradient resistance to provide a beneficial force on an escape ramp, the vehicle must be moving upgrade, against gravity. In the case where the vehicle is descending a grade, gradient resistance is negative, thereby reducing the available forces to slow and stop the vehicle. The amount of gradient resistance is influenced by the total weight of the vehicle and the magnitude of the grade. For each percent of grade, the gradient resistance is 10 lb/1,000 lb whether the grade is positive or negative.

Surfacing Material	Rolling Resistance (lb/1000 lb GVW)	Equivalent Grade (percent)[a]
Portland cement concrete	10	1.0
Asphalt concrete	12	1.2
Gravel, compacted	15	1.5
Earth, sandy, loose	37	3.7
Crushed Aggregate, loose	50	5.0
Gravel, loose	100	10.0
Sand	150	15.0
Pea gravel	250	25.0

[a]Rolling Resistance expressed as equivalent gradient.

Table III-39. Rolling resistance of roadway surfacing materials.

The remaining component of tractive resistance is air resistance, the force resulting from the retarding drag on the various surfaces of the vehicle. Air causes a significant resistance at speeds above 50 mph and is negligible under 20 mph. The effect of air resistance has been neglected in determining the length of the arrester bed in the example below, thus introducing a small safety factor.

Types

There are four basic types of emergency escape ramps currently in use. These types are the sandpile, descending grade, horizontal grade, and ascending grade. The four types are illustrated in Figure III-38.

Sandpiles composed of loose, dry sand are usually no more than 400 ft in length. The influence of gravity is dependent on the grade of the sandpile. The increase in rolling resistance is supplied by the loose sand.

The horizontal-grade and descending-grade ramps can be rather lengthy because the gravitational effect is not acting to help reduce the speed of the vehicle. For the horizontal-grade ramp, the force of gravity is zero; for the descending-grade ramp, the force of gravity acts in the direction of vehicle movement. The increase in rolling resistance is supplied by an arresting bed composed of loose aggregate.

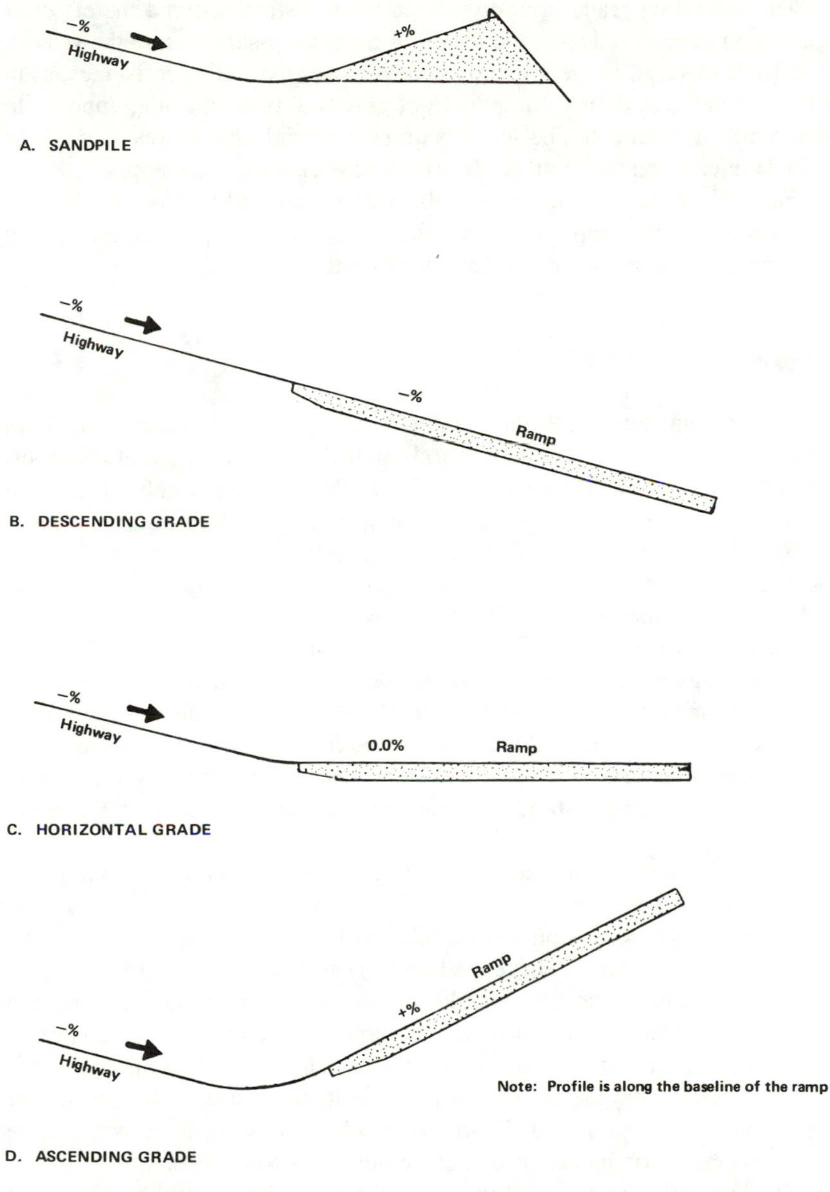

Figure III-38. Basic types of emergency escape ramps.

The ascending-grade ramp uses both the arresting bed and the effect of gravity, in general reducing the length of ramp necessary to stop the vehicle. The loose material in the arresting bed increases the rolling resistance, as in the other types of ramps, while force of gravity acts downgrade, opposite to the vehicle movement. The loose bedding material also serves to hold the vehicle in place on the ramp grade after it has come to a safe stop.

Each one of the ramp types is applicable to a particular situation where an emergency escape ramp is desirable and must be compatible with established location and topographic control at possible sites.

Design Considerations

The combination of the above external resistance and numerous internal resistance forces not discussed acts to limit the maximum speed of an out-of-control vehicle. Speeds in excess of 80 to 90 mph will rarely, if ever, be attained. Therefore, an escape ramp should be designed for a minimum entering speed of 80 mph, a 90-mph design being preferred. For the escape ramp to be effective it must stop the largest vehicle expected to use the ramp. This vehicle would generally be a WB-40 or a WB-50.

Selection of escape ramp locations on existing highways is usually based on accident experience. Analysis of accident data pertinent to a prospective site should include evaluation of the section of highway immediately uphill. An integral part of the evaluation would be the determination of the maximum speed which an out-of-control vehicle could attain at the proposed site. This highest obtainable speed can then be used as the minimum design speed for the ramp.

The design and construction of effective escape ramps involve a number of considerations as follows:

1. To safely stop an out-of-control vehicle, the length of the ramp must be sufficient to dissipate the kinetic energy of the moving vehicle.
2. The width of the ramp should be adequate to accommodate more than one vehicle because it is not uncommon for two or more vehicles to have need of the escape ramp within a short time. The minimum width of 26 ft may be all that is possible in some areas, though greater widths are preferred. Desirably, a width of 30 to 40 ft would more safely accommodate two or more out-of-control vehicles.
3. The surfacing material used in the arrester bed should be clean, not easily compacted, and have a high coefficient of rolling resistance. When aggregate is used, it should be rounded, predominantly single size, and as free from fines as possible. The use of large predominantly single-size aggregate will minimize the problems due to mois-

ture retention and freezing as well as minimizing required maintenance, which must be performed by scarifying when the material compacts. Such material will maximize the percentage of voids, thereby providing optimum drainage and minimizing interlocking and compaction. In addition, a positive means of draining the arrester bed should be provided to help protect the bed from freezing. Pea gravel is representative of the material used most frequently, although loose gravel and sand are also used. A gradation with a top size of 1.5 in. has been used with success in several States.

4. Arrester beds should be constructed with a minimum aggregate depth of 12 in. Successful ramps have used bed depths between 12 and 36 in. As the vehicle enters the arrester bed, the wheels of the vehicle displace the surface, thus increasing the rolling resistance and reducing the possibility of the vehicle rolling back down the grade. To assist in decelerating the vehicle smoothly, the depth of the bed should be tapered from 3 in. at the entry point to the full depth of aggregate in a minimum of 100 ft.
5. The entrance to the ramp must be designed so that a vehicle traveling at a high rate of speed can enter safely. The main roadway surface should be provided to a point at or beyond the exit gore so that both front wheels of the out-of-control vehicle will enter the arrester bed simultaneously, and the operator will have preparation time before actual deceleration begins.
6. Access to the ramp must be made obvious by exit signing, with sufficient sight distance in advance to allow the operator of an out-of-control vehicle time to react, so as to preclude the possibility of missing the ramp. Advance signing is required to inform a driver of the existence of an escape ramp and to prepare the operator well in advance of the decision point so that the operator will have enough time to decide whether or not to use the escape ramp. A regulatory sign near the entrance should be used to discourage other motorists from entering the ramp. (See the MUTCD (**8**) for sign designs.) Illumination of the approach and ramp is desirable.

A surfaced service road located adjacent to the arrester bed is needed so the wrecker and maintenance vehicles can use it without becoming trapped in the bedding material. The width of this lane should be at least 10 ft.

Wrecker anchors, usually located adjacent to the arrester bed at 300-ft intervals, are needed to secure the tow truck when removing a vehicle from the arrester bed. The local wrecker operators can be very helpful in properly locating the anchors.

Escape ramps should exit to the right of the main line. On divided multilane highways, where a left exit may appear to be the only feasible location, difficulties may be expected by the refusal of vehicles in the left lane to yield to out-of-control vehicles attempting to shift lanes. The alinement of the escape ramp should be tangent or of very flat curvature to relieve the driver of undue vehicle control problems. Escape ramps generally may be built at any feasible location where the main road alinement is tangent. They should be built in advance of main line curvature that cannot be negotiated safely by an out-of-control vehicle and in advance of populated areas.

Each grade has its own characteristics. Highway alinement, gradient, length, and descent speed contribute to the potential for out-of-control vehicles. An escape ramp should be provided as soon as a need is established. Unnecessary escape ramps should be avoided. For example, if an escape ramp is provided before a sharp horizontal curve, a second ramp would not be required just beyond the curve that created the need for the initial ramp. Another ramp above the initial ramp would not be necessary, except as a safety valve in cases where there have been unusual numbers of out-of-control vehicle incidents on the grade. Turnouts or pulloff areas at the summit can be used for equipment inspection to help reduce the number of out-of-control vehicle incidents on the grade.

The 80- to 90-mph entering speed, recommended for design, is intended to represent the extreme condition and therefore should not be used as the basis for selecting locations of escape ramps. Although the variables involved make it impossible to establish a maximum truck speed warrant for location of escape ramps, it is evident that anticipated speeds should be below the range used for design. Rather, the principal determinations as to the need should be the safety of the other traffic on the roadway, the operator of the out-of-control vehicle, and the residents along and at the bottom of the grade. An escape ramp, or ramps if the conditions dictate the need for more than one, should be located wherever grades are of a steepness and length to be potential hazards and topographic conditions will permit construction.

The most commonly used escape ramp is the ascending type with an arrester bed. Ramp installations of this type use gradient resistance to advantage, supplementing the effects of the aggregate on the arrester bed.

As a vehicle rolls upgrade, it loses momentum and will eventually stop because of the effect of gravity. To determine the distance required to bring the vehicle to a stop with consideration of the rolling resistance and gradient resistance, the following equation may be used (**30**):

$$L = \frac{\frac{W}{1.075\,(32.2 + k_n)}(V_i^2 - V_F^2)}{W(R \pm G) - TE}$$

Since the mass equivalent constant for the gear in which the vehicle is operating, k_n, final velocity, V_F, and tractive effort, TE, are zero because the vehicle is to be free rolling and is brought to a stop on the escape ramp, the equation simplifies to:

$$L = \frac{V^2}{30(R \pm G)}$$

where: L = distance to stop (i.e., the length of the arrester bed), ft;
V = entering velocity, mph;
G = percent grade divided by 100; and
R = rolling resistance expressed as equivalent percent gradient divided by 100 (Table III-39).

For example, assume that topographic conditions at a site selected for an emergency escape ramp limit the ramp to an upgrade of 10 percent (G = + 0.10). The arrester bed is to be constructed with loose gravel. By using Table III-39, R is determined to be 0.10. The ramp is to be designed for an entering speed of 90 mph. The length required is determined from $L = V^2/30(R + G)$. In this case the length of the arresting bed is 1,350 ft.

When conditions dictate a design where the arrester bed must be constructed using more than one grade along its length, such as shown in Figure III-39, the velocity of the vehicle at the end of the first grade is determined with the knowledge of entering velocity, length of grade, gradient, and resistance of the material in the bed by using the equation: $V^2_f = V^2_i - 30L(R \pm G)$. The velocity of the vehicle is determined at each change of grade on the ramp until sufficient length is provided to stop the out-of-control vehicle.

Figure III-39 shows a plan and profile of an emergency escape ramp with typical appurtenances.

After each use, aggregate arrester beds should be smoothed and the aggregate loosened as necessary. In addition, the bedding material should be cleaned of contaminants and loosened periodically to retain the retarding characteristics of the bedding material and maintain free drainage. Maintenance of the appurtenances should be accomplished as necessary.

When the only feasible location for an escape ramp will not provide sufficient length and grade to completely stop an out-of-control vehicle, it should be supplemented with an acceptable positive attenuation device. At least one escape ramp has been constructed with an array of crash cushions installed to prevent an out-of-control vehicle from leaving the end of the ramp.

Where a full-length ramp is to be provided with full deceleration capability for the design speed, a "last chance" device should be considered when the consequences of leaving the end of the ramp are serious. A mound of the bedding material has been used at the end of ramps in several instances as the "last chance" device. The mounds are between 2 and 5 ft high with 1.5:1 slopes.

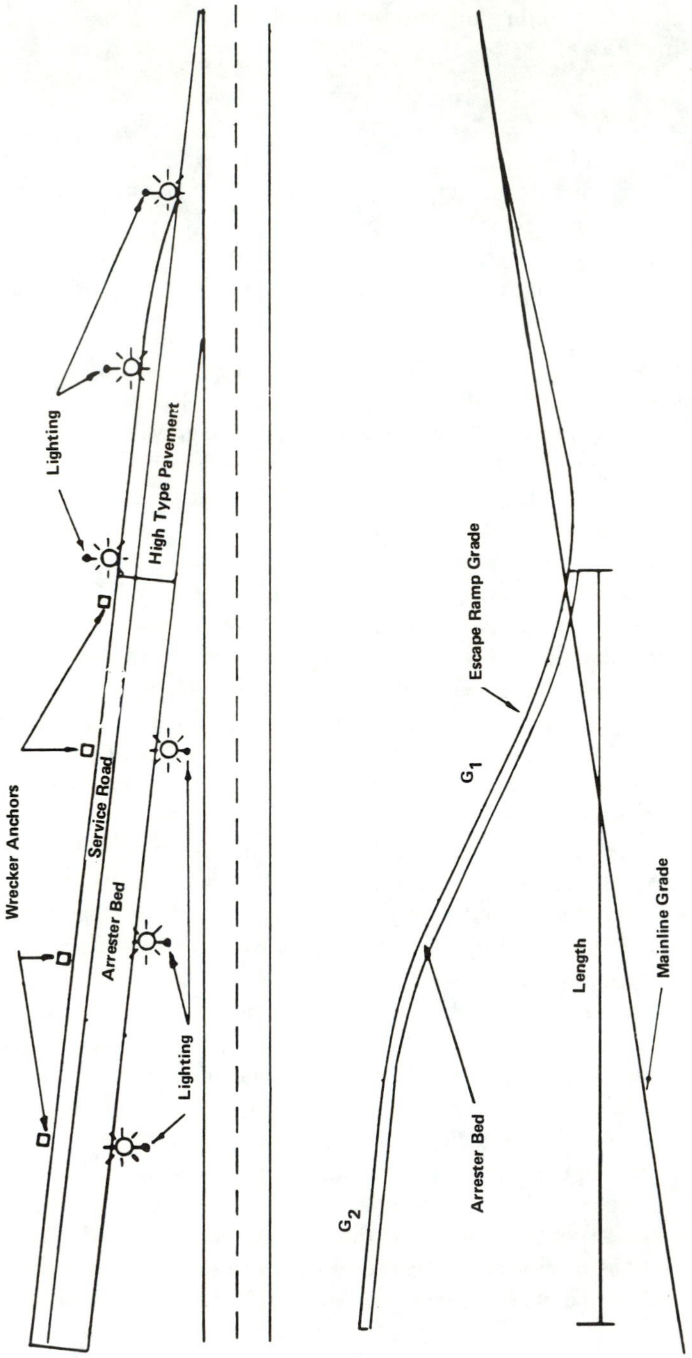

Figure III-39. Typical emergency escape ramp.

Vertical Curves

General Considerations

Vertical curves to effect gradual change between tangent grades may be any one of the crest or sag types depicted in Figure III-40. Vertical curves should be simple in application and should result in a design that is safe, comfortable in operation, pleasing in appearance, and adequate for drainage. The major control for safe operation on crest vertical curves is the provision of ample sight distances for the design speed. Minimum stopping sight distance should be provided in all cases. Wherever economically and physically feasible, more liberal stopping sight distances should be used. Further additional sight distance should be provided at decision points.

Consideration of motorists' comfort requires that the rate of change of grade be kept within tolerable limits. This consideration is most important in sag vertical curves where gravitational and vertical centrifugal forces act in the same direction. Appearance also should be considered. A long curve has a more pleasing appearance than a short one, which may give the appearance of a sudden break in the profile due to the effect of foreshortening.

Drainage of curbed pavements on sag vertical curves, Type III, Figure III-40, requires careful profile design to retain a gradeline of not less than 0.5 percent or, in some cases, 0.35 percent for the outer edges of the pavement. Although not desirable, flatter grades may be necessary in extenuating circumstances.

For simplicity the parabolic curve with an equivalent vertical axis centered on the vertical point of intersection (VPI) is usually used in roadway profile design. The vertical offsets from the tangent vary as the square of the horizontal distance from the curve end (point of tangency). The vertical offset from the tangent grade at any point along the curve is calculated as a proportion of the vertical offset at the VPI, which is AL/800, where the symbols are as shown in Figure III-40. The rate of change of grade at successive points on the curve is a constant amount for equal increments of horizontal distance, and equals the algebraic difference between intersection tangent grades divided by the length of curve in feet, or A/L in percent per foot. The reciprocal L/A is the horizontal distance in feet required to effect a 1 percent change in gradient and is, therefore, a measure of curvature. The quantity L/A, termed "K," is useful in determining the horizontal distance from the vertical point of curvature (VPC) to the apex of Type I curves or to the low point of Type III curves. This point where the slope is zero occurs at a distance from the VPC equal to K times the approach gradient. The K value is also useful in determining minimum lengths of vertical curves for various design speeds.

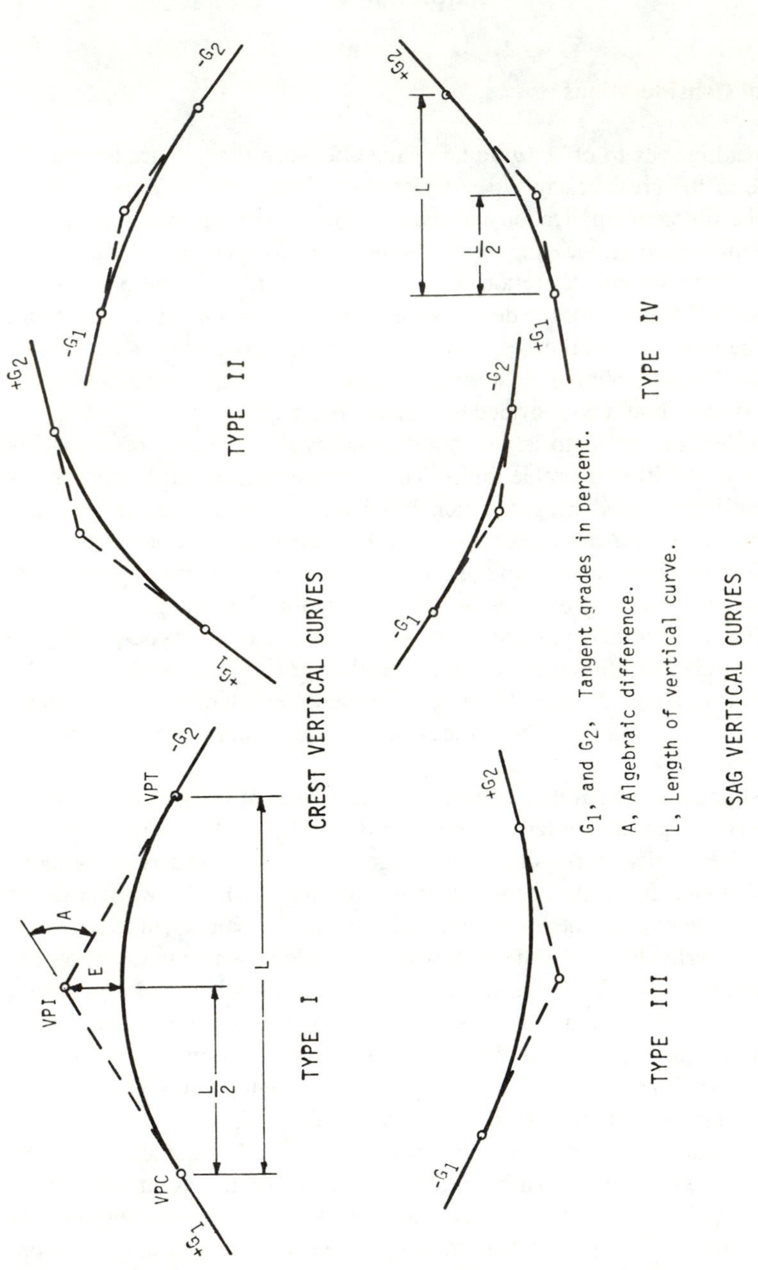

Figure III-40. Types of vertical curves.

Other details on parabolic vertical curves are found in textbooks on highway engineering.

On certain occasions, because of critical clearance or other controls, the use of unsymmetrical vertical curves may be required. Because the conditions dictating the need for these curves are infrequent, the derivation and use of the appropriate formulas have not been included herein. For use in such limited instances, refer to unsymmetrical curve data found in a number of highway engineering texts.

Crest Vertical Curves

Minimum lengths of crest vertical curves as determined by sight distance requirements generally are satisfactory from the standpoint of safety, comfort, and appearance. An exception may be at decision areas, such as sight distance to ramp exit gores, where longer lengths are necessary. Refer to the section in this chapter concerning decision sight distance.

The basic formulas for length of a parabolic vertical curve in terms of algebraic difference in grade and sight distance follow:

$$\text{When S is less than L, } L = \frac{As^2}{100\,(\sqrt{2h_1} + \sqrt{2h_2})^2} \qquad (1)$$

$$\text{When S is greater than L, } L = 2S - \frac{200\,(\sqrt{h_1} + \sqrt{h_2})^2}{A} \qquad (2)$$

where: L = length of vertical curve, ft;
S = sight distance, ft;
A = algebraic difference in grades, percent;
h_1 = height of object above roadway surface, ft.
h_2 = height of object above roadway surface, ft.

When the height of eye and the height of object are 3.50 ft and 6 in., respectively, as used for stopping sight distance.

$$\text{When S is less than L, } L = \frac{AS^2}{1{,}329} \qquad (3)$$

$$\text{When S is greater than L, } L = 2s - \frac{1{,}329}{A} \qquad (4)$$

Design controls—stopping sight distance. The required lengths of vertical curves from formulas 3 and 4 for different values of A to provide the upper value of the range of stopping sight distances for each design speed are shown in Figure III-41. The solid lines give the required lengths, on the basis of rounded values of K as determined from these equations. The dotted line for K = 110.2 gives unrounded values for 45 mph for comparison.

The short dashed curve at the lower left, crossing these lines, indicates where S = L. Note that to the right of the S = L line, the value of K, or length of vertical curve per percent change in A, is a simple and convenient expression of the design control. For each design speed this single value is a positive whole number that is indicative of the rate of vertical curvature. The design control in terms of K covers all combinations of A and L for any one design speed; thus, A and L need not be indicated separately in a design value tabulation. The selection of design curves is facilitated because the required length of curve in feet is equal to K times the algebraic difference in grades in percent. L = KA. Conversely, the checking of plans is simplified by comparing all curves with the design K value.

Table III-40 shows the computed K values for lengths of vertical curves as required for the range of values of stopping sight distances, Table III-1, for each design speed. For direct use in design, values of K are rounded as shown in the right column. The upper, rounded values of K are plotted as the solid lines in Figure III-41. Rounded values of K are higher than computed values, but the differences are not significant.

Where S is greater than L (lower left in Figure III-41), computed values plot as a curve (as shown by the dashed line for 45 mph) that bends to the left, and for small values of A the required lengths are zero because the sight line passes over the apex. This relation does not represent desirable design prac-

Design Speed (mph)	Assumed Speed for Condition (mph)	Coefficient of Friction f	Stopping Sight Distance Rounded for Design (ft)	Rate of Vertical Curvature, K (length (ft) per percent of A)	
				Computed[a]	Rounded for Design
20	20-20	0.40	125-125	8.6-8.6	10-10
25	24-25	0.38	150-150	14.4-16.1	20-20
30	28-30	0.35	200-200	23.7-28.8	30-30
35	32-35	0.34	225-250	35.7-46.4	40-50
40	36-40	0.32	275-325	53.6-73.9	60-80
45	40-45	0.31	325-400	76.4-110.2	80-120
50	44-50	0.30	400-475	106.6-160.0	110-160
55	48-55	0.30	450-550	140.4-217.6	150-220
60	52-60	0.29	525-650	189.2-302.2	190-310
65	55-65	0.29	550-725	227.1-394.3	230-400
70	58-70	0.28	625-850	282.8-530.9	290-540

[a]Using computed values of stopping sight distance.

Table III-40. Design controls for crest vertical curves based on stopping sight distance.

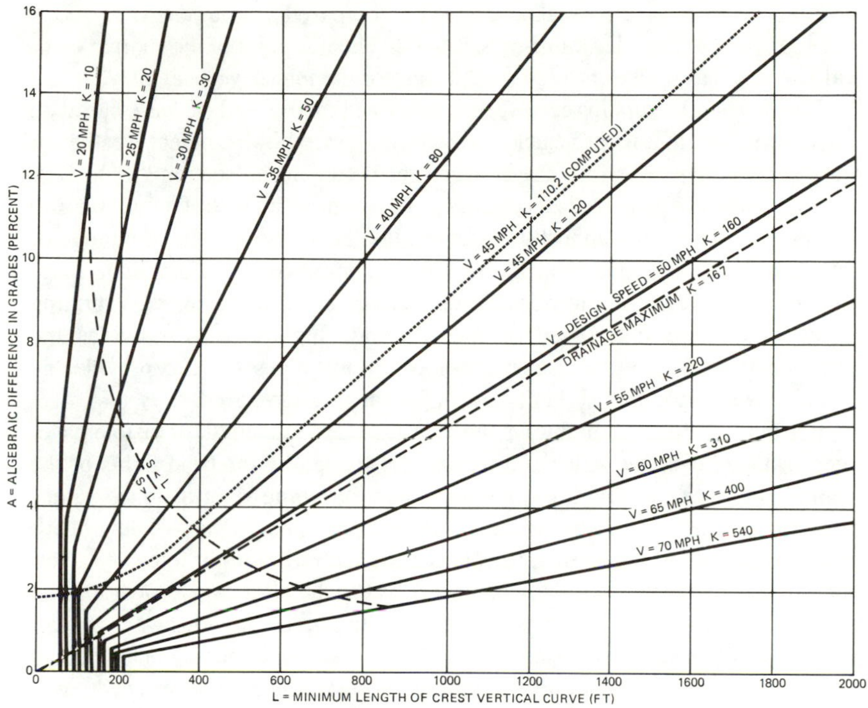

Figure III-41. Design controls for crest vertical curves, for stopping sight distance and open road conditions—upper range.

tice. Most of the States use a minimum length of vertical curve, expressed as either a single value, a range for different design speeds, or a function of A. Values now in use range from about 100 to 300 ft. To recognize the distinction in design speed and to approximate the range of current practice, minimum lengths of vertical curves are expressed as about 3 times the design speed, or $L_{min} = 3V$. These terminal adjustments show as the vertical lines at the lower left of Figure III-41.

The above values of K derived when S is less than L also can be used without significant error where S is greater than L. As shown in Figure III-41, extension of the diagonal lines to meet the vertical lines for minimum lengths of vertical curves results in appreciable differences from the theoretical only where A is small and little or no additional cost is involved in obtaining longer vertical curves.

For example, the computed length of vertical curve for a design speed of 45 mph when A = 2.5 percent is 268 ft, compared with a length of 300 ft where the upper rounded value of K is used. The midpoint of the shorter vertical curve is less than 2 in. higher than that for the longer vertical curve.

The lower K factors for each speed developed in Table III-40 were based on the assumption that most vehicles reduce their speed in inclement weather or on wet pavements. As discussed in greater detail in the section "Sight Distance" in this chapter, studies conducted in recent years of traffic on wet and dry pavements fail to support this stated hypothesis. Nevertheless, a range of values of stopping sight distance have been computed based on friction factors for wet pavements and on vehicular speeds equal to the average running speed and the design speed of the highway. Figure III-42 provides the required lengths of vertical curves for different values of A to provide the lower range of stopping sight distances for each design speed.

For night driving on highways without lighting, the length of visible pavement is that pavement which is directly illuminated by the head lights of the vehicle. For certain conditions, the minimum stopping sight distance values used for design exceed the length of visible pavement. First, vehicle headlights have limitations on the projection distance for the light intensity levels that are required for visibility purposes. When headlights are operated on low-beam, the reduced candlepower at the source plus the downward projection angle significantly restrict the length of visible pavement. Thus, particularly for high-speed conditions, stopping sight distance values exceed pavement visibility distances afforded by the low-beam headlights regardless of whether the roadway profile is level or vertically curving. Secondly, for crest vertical curves the pavement area forward of the headlight beam point of tangency with the pavement surface is shadowed and receives only indirect illumination. Since the headlight mounting height (typically about 2 ft) is lower than the driver eye height (3.5 ft for design), the sight distance to an illuminated object 6 in. in height is controlled by the height of the vehicle headlights rather than by the direct line of sight. Any object within the shadow zone must be high enough to extend into the headlight beam to be directly illuminated. On the basis of formula 1, the bottom of the headlight beam is about 16 in. above the pavement at a distance ahead of the vehicle equal to the low value of the range of stopping sight distance. Although the vehicle headlight system does limit pavement visibility length as mentioned above, there is some mitigating effect in that other vehicles, whose taillight height typically varies from 1.5 ft to 2.0 ft, and other sizable hazardous objects receive direct lighting from headlights at stopping sight distance values used for design. It also may be rationalized that drivers are aware that visibility at night is less than during the day regardless of road and street design features, and that vehicle operators are thus more attentive and alert.

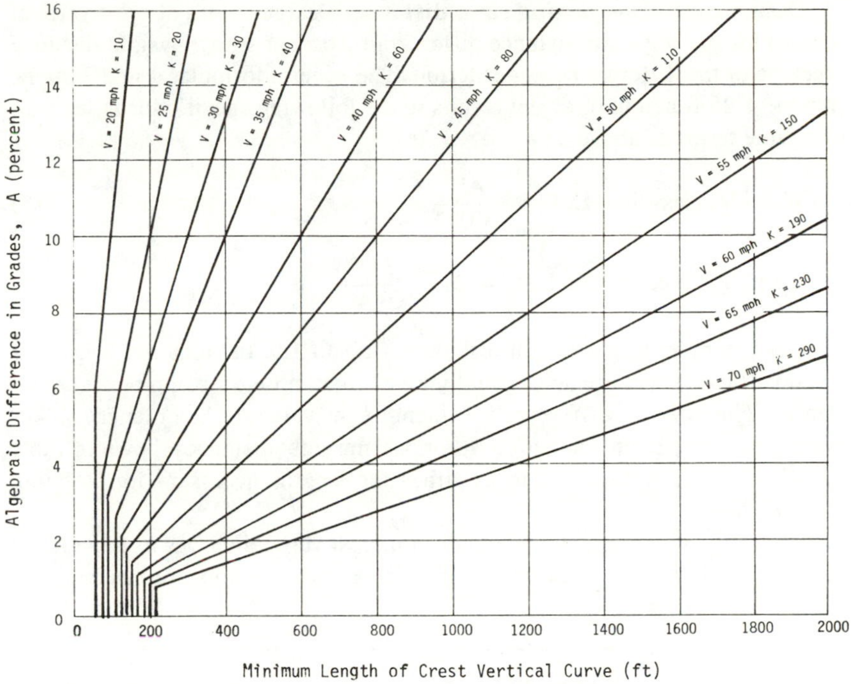

Figure III-42. Design controls for crest vertical curves for stopping sight distance and open road conditions—lower range.

There is a level point of minute length on a crest vertical curve of Type I (Figure III-40), but no difficulty with drainage on curbed pavements is experienced if the curve is sharp enough so that a minimum grade of 0.30 percent is reached at a point about 50 ft from the crest. This corresponds to an A value per station of 0.6 for which K = 100/0.6 = 167 ft per percent change in grade; this line is plotted in Figure III-41 as the drainage maximum. All combinations above or to the left of this line would satisfy this criterion for drainage. The combinations below and to the right of this line involve flatter vertical curves. Special attention is needed in these cases to ensure proper pavement drainage near the apex of crest vertical curves. It is not intended that a K value of 167 be considered a design maximum, but merely the value beyond which drainage must be more carefully designed.

Design controls—passing sight distance. Design values of crest vertical curves for passing sight distance differ from those for stopping sight distance because of the different height criterion. The general formulas 1 and 2 apply, but the 4.25-ft height of object results in the following specific formulas with the same terms as above:

When S is less than L, $L = \frac{AS^2}{3{,}093}$ (5)

When S is greater than L, $L = 2s - \frac{3{,}093}{A}$ (6)

For minimum passing sight distances, Table III-5, the required lengths of crest vertical curves are substantially longer that those for stopping sight distances. The extent of difference is evident by the values of K, or length of vertical curve per percent change in A, for passing sight distances shown in Table III-41. These lengths are 7 to 21 times the lengths necessary for stopping sight distances.

Generally, it is impracticable to design crest vertical curves to provide for passing sight distance because of high cost where crest cuts are involved and

Design Speed (mph)	Minimum Passing Sight Distance, Rounded for Design (ft)	Rate of Vertical Curvature, K,[a] Rounded for Design (length (ft) per percent of A)
20	800	210
25	950	300
30	1,100	400
35	1,300	550
40	1,500	730
45	1,650	890
50	1,800	1,050
55	1,950	1,230
60	2,100	1,430
65	2,300	1,720
70	2,500	2,030

[a]Computed from rounded values of minimum passing sight distance.

Table III-41. Design controls for crest vertical curves based on passing sight distance.

the difficulty of fitting the required long vertical curves to the terrain, particularly for high-speed roads. Passing sight distance on crest vertical curves may be feasible on roads with an unusual combination of low design speed and gentle grades or higher design speeds with very small algebraic differences in grades. Ordinarily, passing sight distance is provided only at places where combinations of alinement and profile do not require the use of crest vertical curves.

Sag Vertical Curves

At least four different criteria for establishing lengths of sag vertical curves are recognized to some extent. These are (1) headlight sight distance, (2) rider comfort, (3) drainage control, and (4) a rule-of-thumb for general appearance.

Headlight sight distance has been used directly by some authorities and for the most part is the basis for determining the length of sight distance used herein. When a vehicle traverses a sag vertical curve at night, the portion of highway lighted ahead is dependent on the position of the headlights and the direction of the light beam. General use is being given to a headlight height of 2.0 ft and a 1° upward divergence of the light beam from the longitudinal axis of the vehicle. The upward spread of the light beam provides some additional visible length but this is generally ignored. The following formulas show the S, L, and A relation, using S as the distance between the vehicle and point where the 1° angle of light ray intersects the surface of the roadway:

$$\text{When S is less than L, } L = \frac{AS^2}{400 + 3.5S} \qquad (7)$$

$$\text{When S is greater than L, } L = 2S - \frac{400 + 3.5S}{A} \qquad (8)$$

where: L = length of sag vertical curve, ft;
S = light beam distance, ft; and
A = algebraic difference in grades, percent.

For overall safety on highways, a sag vertical curve should be long enough so that the light beam distance is nearly the same as the stopping sight distance. Accordingly, it is pertinent to use stopping distances for different design speeds as the S value in the above formulas. The resulting lengths of vertical curves for the upper value of the range of stopping sight distances for each design speed are shown in Figure III-43 with solid lines using rounded K values as was done for crest vertical curves and the dotted line for K = 84.2 being an unrounded value for 45 mph for comparison. Figure III-44 provides the lengths of sag vertical curves for various algebraic differences in grades for the lower range of stopping sight distance.

The comfort effect of change in vertical direction is greater on sag than on crest vertical curves because gravitational and centrifugal forces are combining rather than opposing forces. Comfort due to change in vertical direction is not measured readily because it is affected appreciably by vehicle body suspension, tire flexibility, weight carried, and other factors. The limited attempts at such measurements have led to the broad conclusion that riding is comfortable on sag vertical curves when the centrifugal acceleration does not exceed 1 ft/sec^2. The general expression for such a criterion is:

$$L = \frac{AV^2}{46.5} \tag{9}$$

where L and A are the same as in previous formulas, and V is the design speed, mph.

The length of vertical curve required to satisfy this comfort factor at the various design speeds is only about 50 percent of that required to satisfy the headlight sight distance requirement for the normal range of design conditions.

Drainage affects design of vertical curves of Type III (Figure III-40) where curbed sections are used. An approximate criterion for sag vertical curves is the same as that expressed for the crest conditions, that is, providing a minimum grade of 0.30 percent within 50 ft of the level point. This criterion plots as the broken line shown in Figure III-43 for K = 167. The drainage requirement differs from other criteria in that the length for any other criterion is greater than the minimum length for other criteria up to 60 mph and is nearly equal for other criteria for 70 mph for minimum-length vertical curves.

For general appearance, some use formerly was made of a rule-of-thumb for length of sag vertical curves wherein the minimum value of L is 100A or, in Figure III-43, K = 100. This approximation is a generalized control for small or intermediate values of A. Compared with headlight sight distance, it corresponds to a design speed between 45 and 50 mph. On high-type highways longer curves are deemed appropriate to improve appearance.

From the preceding it is evident that design controls for sag vertical curves differ from those for crests, and separate design values are needed. The headlight sight distance basis appears to be the most logical for general use, and the values determined for stopping sight distances are within the limits recognized in current practice. It is concluded to use this criterion to establish design values for a range of lengths of sag vertical curves. As in the case of crest vertical curves, it is convenient to express the design control in terms of the K rate for all values of A. This entails some deviation from the computed

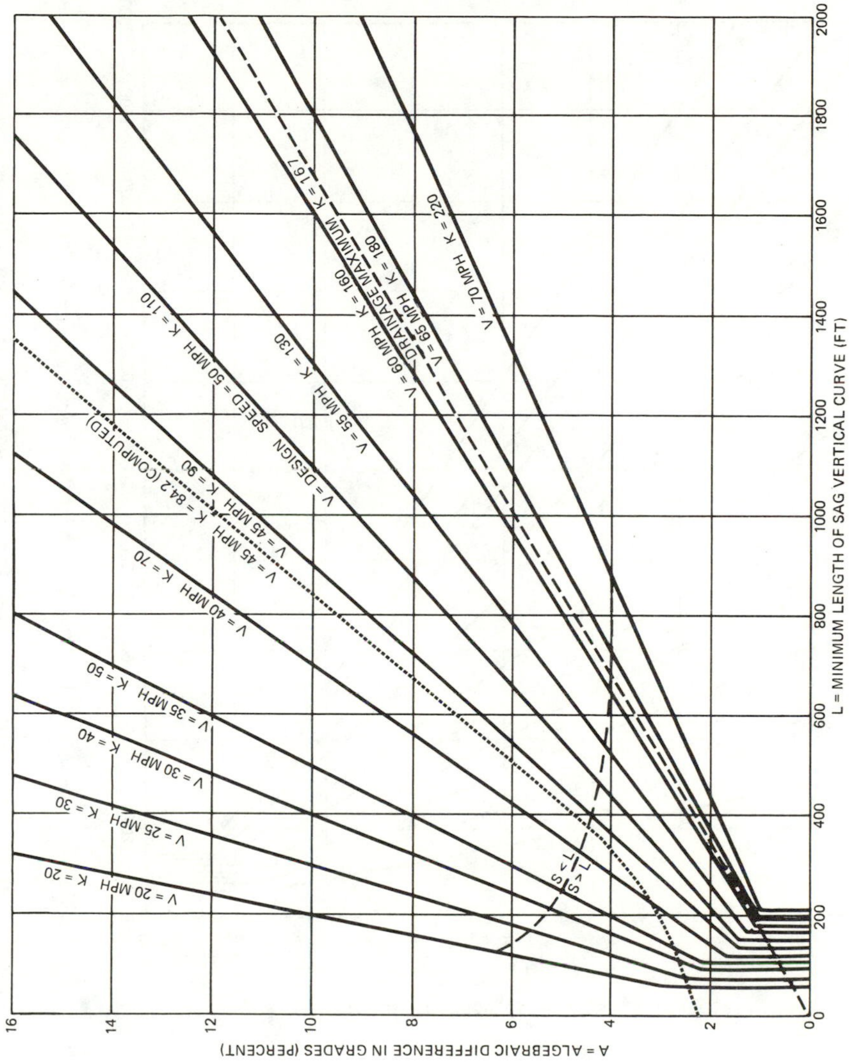

Figure III-43. Design controls for sag vertical curves, open road conditions—upper range.

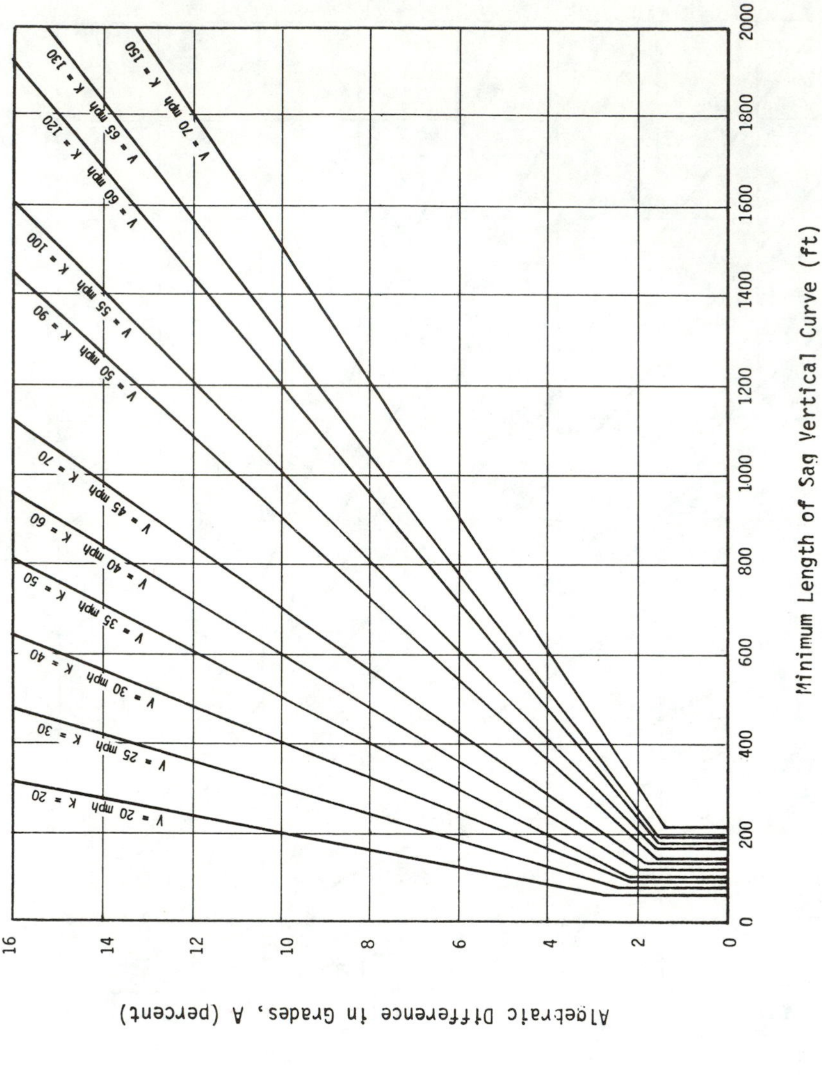

Figure III-44. Design controls for sag vertical curves—lower range.

values for small values of A, but the differences are not significant. Table III-42 shows the range of computed values and the rounded values of K selected as design controls. The lengths of sag vertical curves on the basis of the design speed values of K are shown by the solid lines in Figure III-43. It is to be emphasized that these lengths are minimum values based on design speed; longer curves are desired wherever feasible, but special attention to drainage must be exercised where a K value in excess of 167 is used.

Minimum lengths of vertical curves for flat gradients also are recognized for sag conditions. The values determined for crest conditions appear to be generally suitable for sags. Lengths of sag vertical curves, shown as vertical lines in Figure III-43, are equal to 3 times the design speed.

Sag vertical curves shorter than the lengths computed from Table III-42 may be justified for economic reasons in cases where an existing element, such as a structure not ready for replacement, controls the vertical profile. In certain cases ramps may also be designed with shorter sag vertical curves. Fixed source lighting is desirable in these cases. For street design, some engineers accept design of a sag or crest where A is about 1 percent or less without a length of calculated vertical curve. However, field modifications during construction usually result in constructing the equivalent to a vertical curve, even if short.

Design Speed (mph)	Assumed Speed for Condition (mph)	Coefficient of Friction f	Stopping Sight Distance Rounded for Design (ft)	Rate of Vertical Curvature, K (length (ft) per percent of A)	
				Computed[a]	Rounded for Design
20	20-20	0.40	125-125	14.7-14.7	20-20
25	24-25	0.38	150-150	21.7-23.5	30-30
30	28-30	0.35	200-200	30.8-35.3	40-40
35	32-35	0.34	225-250	40.8-48.6	50-50
40	36-40	0.32	275-325	53.4-65.6	60-70
45	40-45	0.31	325-400	67.0-84.2	70-90
50	44-50	0.30	400-475	82.5-105.6	90-110
55	48-55	0.30	450-550	97.6-126.7	100-130
60	52-60	0.29	525-650	116.7-153.4	120-160
65	55-65	0.29	550-725	129.9-178.6	130-180
70	58-70	0.28	625-850	147.7-211.3	150-220

[a]Using computed values of stopping sight distance.

Table III-42. Design controls for sag vertical curves based on stopping sight distance.

General Controls for Vertical Alinement

In addition to the above specific controls for vertical alinement, there are several general controls that should be considered in design.

1. A smooth gradeline with gradual changes, as consistent with the type of highways, roads, or streets and the character of terrain, should be strived for in preference to a line with numerous breaks and short lengths of grades. Detailed design values are the maximum grade and the critical length of grade, but the manner in which they are applied and fitted to the terrain on a continuous line determines the suitability and appearance of the finished product.
2. The "roller-coaster" or the "hidden-dip" type of profile should be avoided. Such profiles generally occur on relatively straight horizontal alinement where the roadway profile closely follows a rolling natural ground line. Examples of these undesirable profiles are evident on many older roads and streets. They are unpleasant esthetically and more difficult to drive. Hidden dips contribute to passing maneuver problems for drivers, the passing driver being deceived by the view of the road or street beyond the dip free of opposing vehicles. Even with shallow dips, this type of profile is disconcerting because the driver cannot be sure whether or not there is an oncoming vehicle hidden beyond the rise. This type of profile is avoided by use of horizontal curves or by more gradual grades.
3. Undulating gradelines, involving substantial lengths of momentum grades, should be appraised for their effect on traffic operation. Such profiles permit heavy trucks to operate at higher overall speeds than is possible when an upgrade is not preceded by a downgrade, but may encourage excessive speeds of trucks with attendant conflicts with other traffic.
4. A broken-back gradeline (two vertical curves in the same direction separated by short section of tangent grade) generally should be avoided, particularly in sags where the full view of both vertical curves is not pleasing. This effect is very noticeable on divided roadways with open median sections.
5. On long grades it may be preferable to place the steepest grades at the bottom and lighten the grades near the top of the ascent or to break the sustained grade by short intervals of lighter grade instead of a uniform sustained grade that might be only slightly below the allowable maximum. This is particularly applicable to low-design-speed roads and streets.
6. Where intersections-at-grade occur on roadway sections with moderate to steep grades, it is desirable to reduce the gradient through the

intersection. Such a profile change is beneficial for all vehicles making turns and serves to reduce the potential hazards.

7. Sag vertical curves should be avoided in cuts unless adequate drainage can be provided.

COMBINATION OF HORIZONTAL AND VERTICAL ALINEMENT

General Considerations

Horizontal and vertical alinements are permanent design elements for which thorough study is warranted. It is extremely difficult and costly to correct alinement deficiencies after the highway is constructed. On freeways there are numerous controls such as multilevel structures and costly right-of-way. On most arterial streets heavy development takes place along the property lines, which makes it impractical to change the alinement in the future. Thus, compromises in the alinement designs must be weighed carefully, because any initial savings may be more than offset by the economic loss to the public in the form of accidents and delays.

Horizontal and vertical alinements should not be designed independently. They complement each other, and poorly designed combinations can spoil the good points and aggravate the deficiencies of each. Horizontal alinement and profile are among the more important of the permanent design elements of the highway, for which thorough study is warranted. Excellence in their design and in the design of their combination increase usefulness and safety, encourage uniform speed, and improve appearance, almost always without additional cost.

General Design Controls

It is difficult to discuss the combination of horizontal alinement and profile without reference to the broader subject of location. The subjects are mutually interrelated and what may be said about one generally is applicable to the other. It is assumed here that the general location has been fixed and that the problem remaining is the specific design and harmonizing of the vertical and horizontal lines, such that the finished highway, road, or street will be an economical, pleasant, and safe facility on which to travel. The physical controls or influences that act singly or in combination to determine the type of alinement are the character of roadway justified by the traffic, topography, and subsurface conditions, existing cultural development, likely future developments, and location of the terminals. Design speed is thought of when determining the general location, but as design proceeds to more detailed alinement and profile it assumes greater importance, and the speed chosen for design acts to keep all elements of design in balance. Design speed determines limiting values for many elements such as curvature and sight distance and influences many other elements such as width, clearance, and maximum gradient; all are discussed in the preceding parts of this chapter.

Proper combination of horizontal alinement and profile is obtained by engineering study and consideration of the following general controls:

1. Curvature and grades should be in proper balance. Tangent alinement or flat curvature at the expense of steep or long grades and excessive curvature with flat grades are both poor design. A logical design that offers the most in safety, capacity, ease and uniformity of operation, and pleasing appearance within the practical limits of terrain and area traversed is a compromise between the two extremes.
2. Vertical curvature superimposed on horizontal curvature, or vice versa, generally results in a more pleasing facility, but it should be analyzed for effect on traffic. Successive changes in profile not in combination with horizontal curvature may result in a series of humps visible to the driver for some distance, an undesirable condition as previously discussed. The use of horizontal and vertical alinements in combination, however, may also result in certain undesireable arrangements, as discussed later in this section.
3. Sharp horizontal curvature should not be introduced at or near the top of a pronounced crest vertical curve. This condition is undesireable in that the driver cannot perceive the horizontal change in alinement, especially at night when the headlight beams go straight ahead into space. The difficulty of this arrangement is avoided if the horizontal curvature leads the vertical curvature, i.e., the horizontal curve is made longer than the vertical curve. Suitable design can also be made by using design values well above the minimums for the design speed.
4. Somewhat allied to the above, sharp horizontal curvature should not be introduced at or near the low point of a pronounced sag vertical curve. Because the road ahead is foreshortened, any but flat horizontal curvature assumes an undesirable distorted appearance. Further, vehicular speeds, particularly of trucks, often are high at the bottom of grades, and erratic operation may result, especially at night.
5. On two-lane roads and streets the need for safe passing sections at frequent intervals and for an appreciable percentage of the length of the roadway often supersedes the general desirability for combination of horizontal and vertical alinement. In these cases it is necessary to

work toward long tangent sections to secure sufficient passing sight distance in design.

6. Horizontal curvature and profile should be made as flat as feasible at intersections where sight distance along both roads or streets is important and vehicles may have to slow or stop.
7. On divided highways and streets, variation in width of median and the use of separate profiles and horizontal alinements should be considered to derive design and operational advantage of one-way roadways. Where traffic justifies provision of four lanes, a superior design without additional cost generally result from the concept and logical design basis of one-way roadways.
8. In residential areas the alinement should be designed to minimize nuisance factors to the neighborhood. Generally, a depressed facility makes a highway less visible and less noisy to adjacent residents. Minor horizontal adjustments can sometime be made to increase the buffer zone between the highway and clusters of homes.
9. The alinement should be designed to enhance attractive scenic views of the natural and manmade environment, such as rivers, rock formations, parks, outstanding buildings, and golf courses. The highway should head into rather than away from those views that are outstanding, it should fall toward those features of interest at a low elevation, and it should rise toward those features best seen from below or in silhouette against the sky.

Alinement Coordination in Design

Coordination of horizontal alinement and profile should not be left to chance but should begin with preliminary design, during which stage adjustments can readily be made. Although a specific order of study for all highways cannot be stated, a general procedure applicable to most facilities can be outlined.

The designer should use working drawings of a size, scale, and arrangement so that he can study long, continuous stretches of highway in both plan and profile and visualize the whole in three dimensions. Working drawings should be of sufficiently small scale, generally 1 in. = 100 ft or 1 in. = 200 ft, with the profile plotted jointly with the plan. A continuous roll of plan-profile paper usually is suitable for this purpose.

After study of the horizontal alinement and profile in preliminary form, adjustments in each, or both, can be made jointly to obtain the desired coordination. At this stage the designer should not be concerned with line calculations other than known major controls. The study should be made largely on the basis of a graphical analysis. The criteria and elements of design covered in this and the preceding chapter should be kept in mind. For the selected design speed the values for controlling curvature, gradient, sight distance, and superelevation runoff length should be available and checked graphically. Design speed may have to be adjusted during the process along some sections to conform to likely variations in speeds of operation. This need may occur where noticeable changes in alinement characteristics are necessitated by unusual terrain or right-of-way controls. In addition, the general design controls, as enumerated separately for horizontal alinement, vertical alinement, and their combination, should be considered. All aspects of terrain, traffic operation, and appearance can be considered and the horizontal and vertical lines adjusted and coordinated before the costly and time-consuming calculations and the preparation of construction plans to large scale are started.

The coordination of horizontal alinement and profile from the viewpoint of appearance usually can be accomplished visually on the preliminary working drawings. Generally, this visual method results in a satisfactory product when done by an experienced designer. This means of analysis may be supplemented by models or perspective sketches at locations where the effect of certain combinations of line and grade are questionable.

For highways with gutters the effects of superelevation transitions on gutter line profiles needs to be examined. This can be particularly significant when flat grades are involved and can result in local depressions. Slight shifts in profiles in relation to horizontal curves can sometimes eliminate the problem.

The above procedures obviously must be modified for the typical case of local road or street design as compared with a higher type highway. The alinement of any local road or street, whether for a new one or for an existing reconstruction, is governed by the existing or likely future developments along it. The crossroad or street intersection and land-service driveway are rather dominant controls. Although they must be fully considered, they should not override the broader desirable features described above. Even for street design, it is desirable to work out long, flowing alinement and profile sections rather than a connected series of block-by-block sections.

Some examples of poor and good practice are illustrated in Figure III-45.

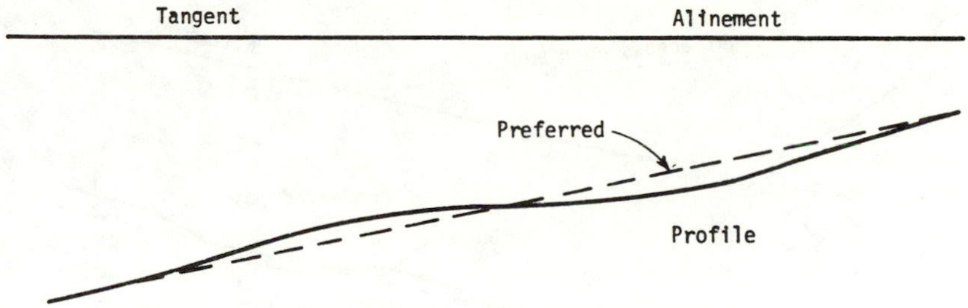

A. Profile with Tangent Alinement

Note: Avoid designing little local dips in an otherwise long, uniform grade. These usually result from zeal to balance cut and fill exactly and reduce overhaul.

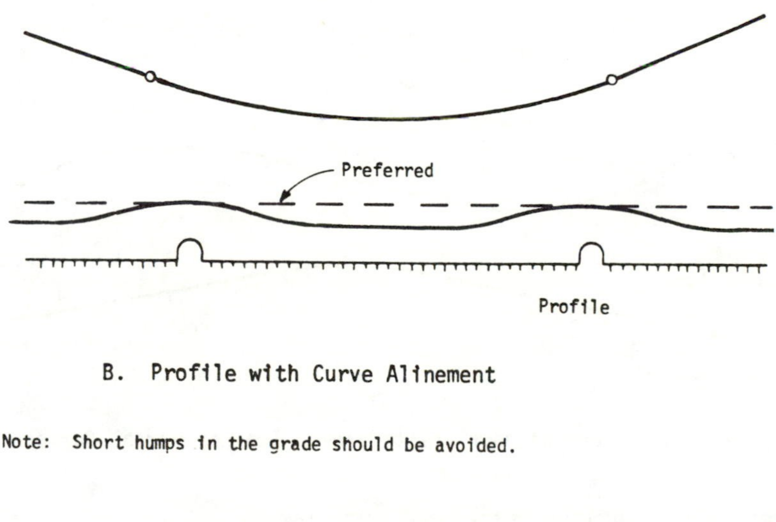

B. Profile with Curve Alinement

Note: Short humps in the grade should be avoided.

Figure III-45. Alinement and profile relationships in roadway design. *A and B.* **(Ref. 37, 38, & 39).**

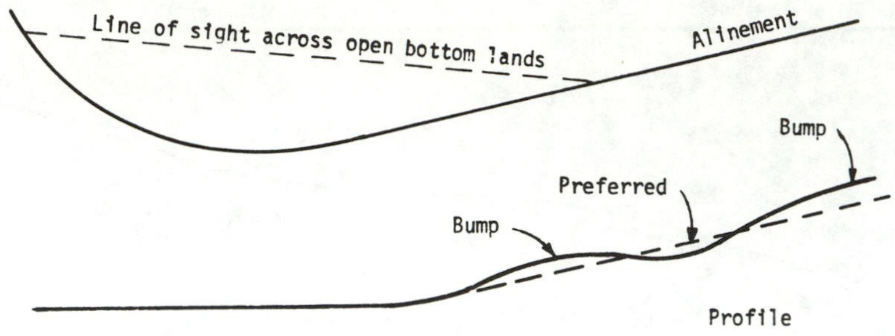

C. Distant View Showing Bumps in Profile Grade Line

Note: A distant side view of a long grade on tangent will reveal every bump on it.

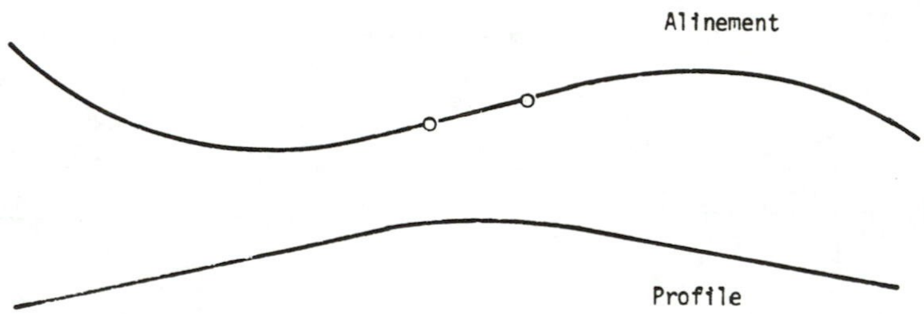

D. Short tangent on a Crest Between Two Horizontal Curves

Note: This combination is deficient for two reasons. The tangent between the curves is too short, and the reverse occurs on a crest.

Figure III-45. Alinement and profile relationships in roadway design. *C and D.* **(Ref. 37, 38, & 39).**

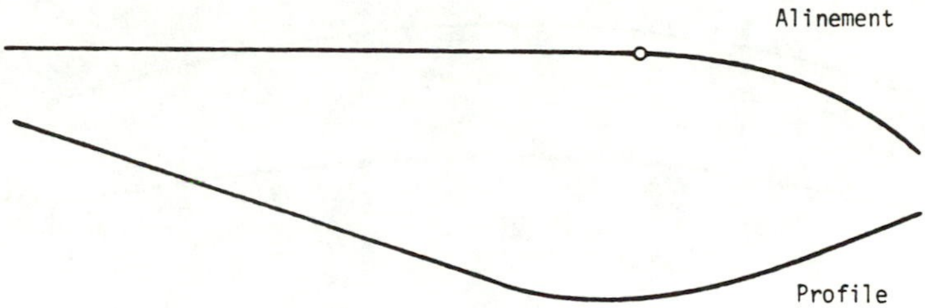

E. Sharp Angle Appearance

Note: This combination presents a poor appearance-
the horizontal curve looks like a sharp angle

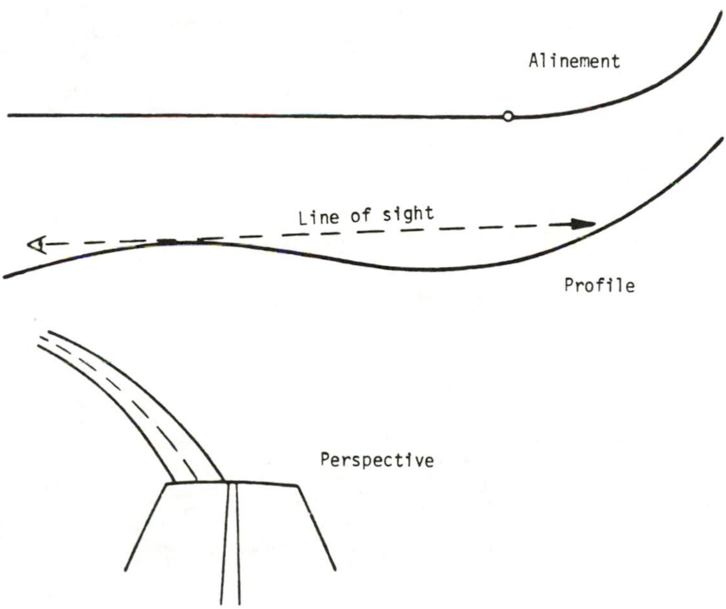

F. Disjointed Effect

Note: A disjointed effect occurs when the beginning of a horizontal curve is hidden from the driver by an intervening crest while the continuation of the curve is visible in the distance beyond the intervening crest.

Figure III-45. Alinement and profile relationships in roadway design. *E and F.* **(Ref. 37, 38, & 39).**

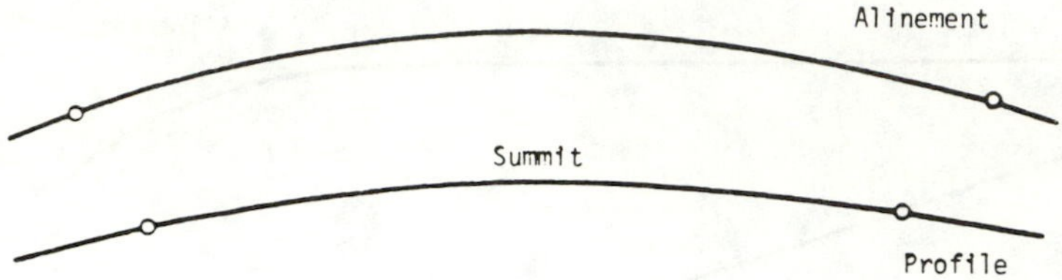

G. Coinciding Curves in Horizontal and Vertical Dimensions

Note: When horizontal and vertical curves coincide, a very satisfactory appearance results.

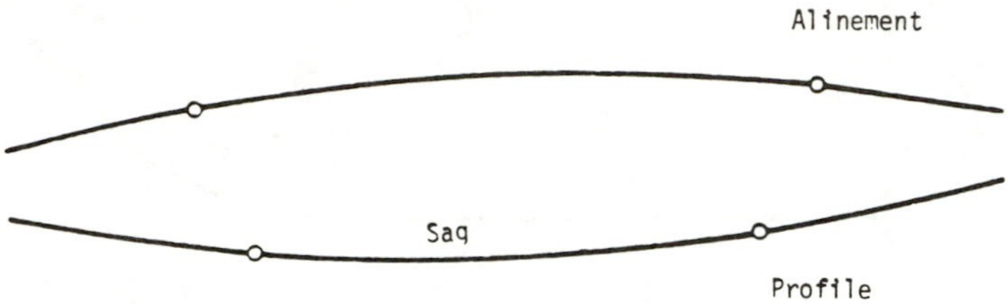

H. Opposing Curves in Horizontal and Vertical Dimensions

Note: When horizontal and vertical curves oppose, a very satisfactory appearance results.

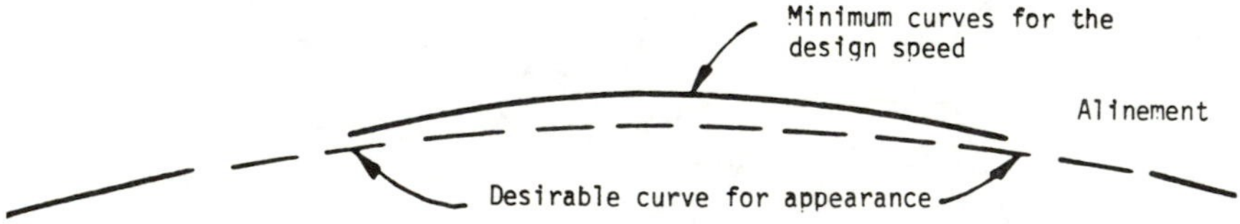

I. Flat Curves Appropriate for Horizontal with Small Central Angles Regardless of Profile.

Note: Very long flat curves, even where not required by the design speed, also have a pleasing appearance when the central angle is very small.

Figure III-45. Alinement and profile relationships in roadway design. *G, H and I.* **(Ref. 37, 38, & 39).**

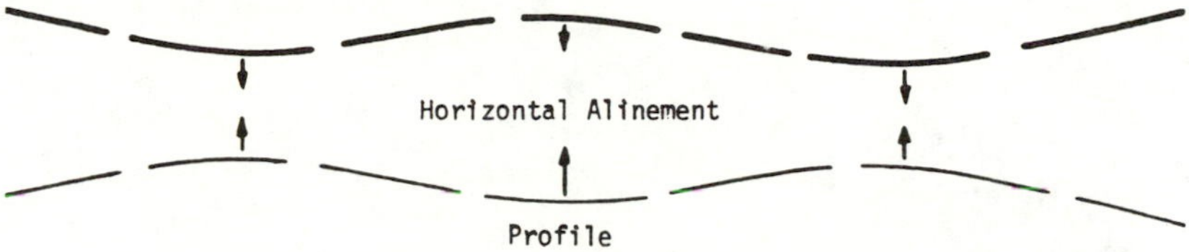

J. Coinciding Vertices in Horizontal and Vertical Dimensions

Note: The classic case of coordination between horizontal and vertical alinement in which the vertices of horizontal and vertical curves coincide, creating a rich effect of three-dimensional S-curves, composed of convex and concave helixes.

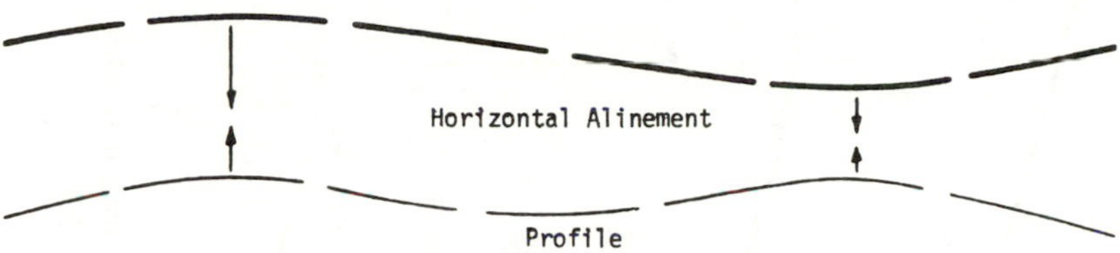

K. Coinciding Vertices with Single-Phase Skip

Note: A legitimate case of coordination: one phase is skipped in the horizontal plane, but vertices still coincide. The long tangent in plan is softened by vertical curvature.

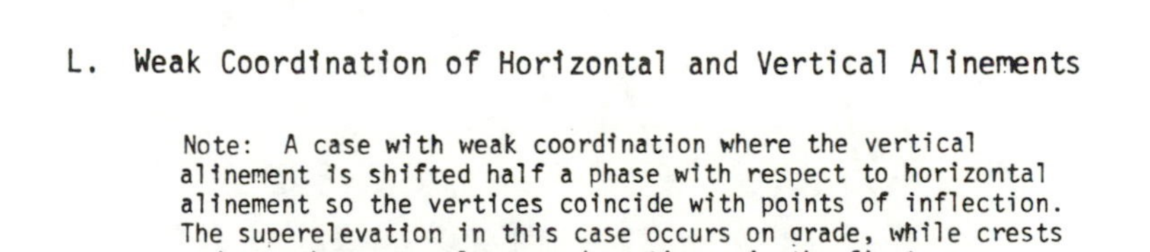

L. Weak Coordination of Horizontal and Vertical Alinements

Note: A case with weak coordination where the vertical alinement is shifted half a phase with respect to horizontal alinement so the vertices coincide with points of inflection. The superelevation in this case occurs on grade, while crests and sags have normal crowned sections; in the first case, superelevation occurs on crests and sags, while grades have normal crowned sections.

Figure III-45. Alinement and profile relationships in roadway design. *J, K and L.* **(Ref. 37, 38, & 39).**

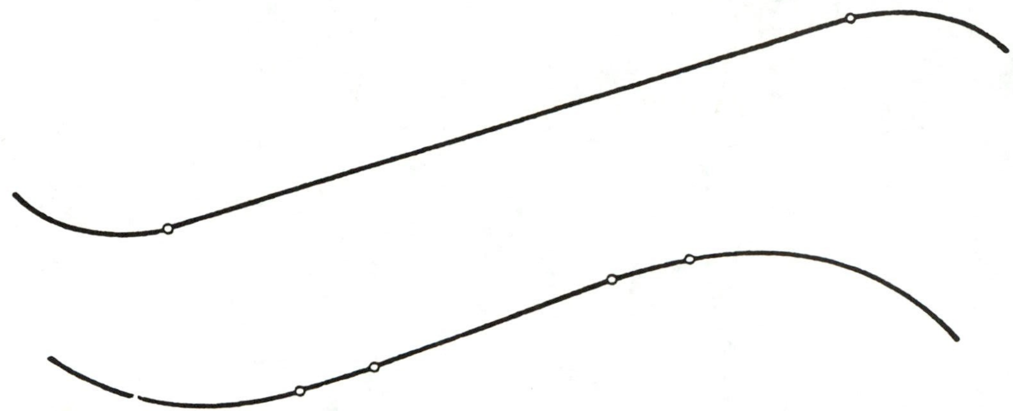

M. Horizontal Alinement Should Be Balanced

Note: The upper line is an example of poor design because the alinement consists of a long tangent with short curves, whereas the balance between the curves and tangents in the lower alinement is the preferred design.

Figure III-45. Alinement and profile relationships in roadway design. *M.* **(Ref. 37, 38, & 39).**

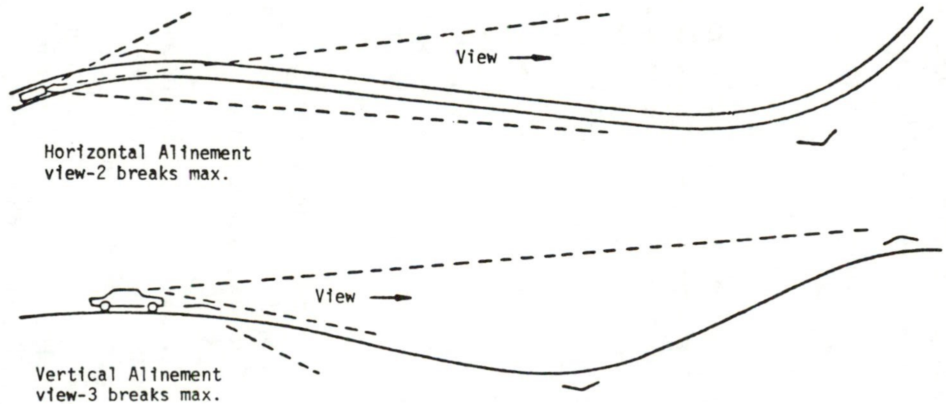

N. Good Coordination of Horizontal and Vertical Alinements

Note: Guideline to be used for coordination of horizontal and vertical alinement.

Figure III-45. Alinement and profile relationships in roadway design. *N.* **(Ref. 37, 38, & 39).**

OTHER ELEMENTS AFFECTING GEOMETRIC DESIGN

In addition to the design elements discussed previously, there are several elements that affect or are affected by the geometric design. Each of these elements is explained only to the extent necessary to show its relation to geometric design and how it, in turn, is affected thereby. Details for completed design are not covered herein.

Drainage

Highway drainage facilities provide for carrying water across the right-of-way and for removal of storm water from the road itself. These facilities include bridges, culverts, channels, curbs, gutters, and various types of drains. Hydraulic capacities and locations of such structures should be designed to take into consideration damage to upstream property and to secure as low a degree of risk of traffic interruption by flooding as is consistent with the importance of the road, the design traffic service requirements, and available funds. While drainage design considerations are an integral part of highway geometric design, specific drainage design criteria are not included herein. The AASHTO *Highway Drainage Guidelines* (**40**) should be referred to for a more complete discussion of highway drainage subjects. These are published as individual guidelines on major areas of highway hydraulic design.

Many State highway agencies have excellent highway drainage manuals that may be used for reference for hydraulic design procedures. In addition, other publications (**41**) are widely used and are available to highway agencies from FHWA.

Hydraulic requirements for stream crossings and flood plain encroachments frequently affect highway alinement and profile. The probable effects of a highway encroachment on flood hazards to other property and the risk of flood damage to the highway should be evaluated when a flood plain location is under consideration. Water surface elevations for floods of various return periods will influence decisions regarding the highway profile where an encroachment is necessary. Highway profiles at stream crossings will often be determined by hydraulic considerations. To the extent practicable stream crossings and other highway encroachments on flood plains should be located and aligned to preserve the natural flood flow distribution and direction. Stream stability and the stream environment are also important and complex considerations in highway location and design. *Highways in the River Environ-*

ment: Hydraulic and Environmental Design Consideration (**42**) is an excellent and comprehensive reference that may be used to analyze the effects of highway encroachment on streams and probable future activities of streams that adversely affect encroaching highways.

Surface channels are used to intercept and remove surface runoff from roadways, wherever practicable. They should have adequate capacity for the design runoff, and should be located and shaped to avoid creating a traffic hazard. Channels are usually lined with vegetation, and rock or paved channel linings are used where vegetation will not control erosion. Runoff from pavement surfaces normally drains down grassed slopes to roadside or median channels. Curbs or dikes, inlets, and chutes or flumes are used where runoff from the pavement would erode fill slopes. Where storm drains are necessary, pavement sections are usually curbed. Care should be exercised to ensure that this curb does not encroach on the clear zone of the highway; refer to the horizontal clearance to obstructions discussion in Chapter IV.

Drainage inlets should be so designed and located as to limit the spread of water on traffic lanes to tolerable widths. Because grates may become blocked by trash accumulation, curb openings or combination inlets with both grate and curb openings are advantageous for urban conditions. Grate inlets and depressions or curb-opening inlets should be located outside the through-traffic lanes to minimize the shifting of vehicles attempting to avoid riding over them. Inlet grates should also be designed to accommodate bicycle and pedestrian traffic where appropriate. Discontinuous sections of curbing, as at the gore of ramps, and variable curb offsets should not be used as expedients to handle pavement drainage where these features would detract from highway safety. Inlets should be designed and located to prevent silt and debris carried in suspension from being deposited on the traveled way where the longitudinal gradient is decreased. Extra inlets should be installed near low points of sag vertical curves to take any overflow from blocked inlets. Inlets should be located so that concentrated flow and heavy sheet flow will not cross traffic lanes. Where pavement surfaces are warped, as at cross streets or ramps, surface water should be intercepted just before the change in cross slope. Also, inlets should be located just upgrade of pedestrian crossings. Storm drains should have adequate capacity to avoid a hazardous condition on the roadway and bridges, especially in sag vertical curves where water can pond. The general effect of drainage on the geometrics of pavements, shoulder ditches, or gutters and side slopes is discussed further in Chapter IV.

Drainage is usually more difficult and costly for urban than for rural highways because of more rapid rates and larger volumes of runoff, costlier potential damage to adjacent property by flooding, higher overall costs because of more inlets and underground systems, greater restrictions because of urban development, lack of natural areas of water bodies to receive flood water, and higher volumes of traffic, including pedestrians. There is greater need to intercept concentrated storm water before it reaches the highway and to remove over-the-curb flow and surface water without interrupting traffic flow or causing a hazard to vehicle occupants or pedestrians. To accommodate such runoff, underground systems and numerous inlets, curbs, and gutters are usually required. Often new outfall drains of considerable length must be constructed because existing storm water systems usually lack capacity for highway surface drainage volumes. A joint use storm water system, shared by the highway agency with others, can have economic advantages to both parties, because it is normally more economical to build a common system rather than two independent systems.

Reduction of peak flows can be achieved by the storage of water that falls on the site in detention basins, storm drainage pipes, swales and channels, parking lots, and rooftops. Storm water is released to the downstream conveyance facility or stream at a reduced flow rate. The concept should be considered for use in highway drainage design where existing downstream conveyance facilities are inadequate to handle peak flow rates from highway storm drainage facilities, where the highway would contribute to increased peak flow rates and aggravate downstream flooding problems, and as a technique to reduce the construction costs of outfalls from highway storm drainage facilities.

The cost of drainage is neither incidental nor minor on most roads. Careful attention to requirements for adequate drainage and protection of the highway from floods in all phases of location and design will prove to be effective in reducing costs in both construction and maintenance.

Erosion Control and Landscape Development

Erosion prevention is one of the major factors in design, construction and maintenance of highways. It should be considered early in the location and design stages. Some degree of erosion control can be incorporated into the geometric design, particularly in the cross section elements. Of course, the most direct application of erosion control occurs in drainage design and in the writing of specifications for landscaping and slope planting.

Erosion and maintenance are minimized largely by the use of flat side slopes, rounded and blended with natural terrain; serrated cut slopes; drainage channels designed with due regard to width, depth, slopes, alinement, and protective treatment; inlets located and spaced with erosion control in mind; prevention of erosion at culvert outlets; proper facilities for ground water interception; dikes, berms, and other protective devices; sedimentation devices to trap sediment at strategic locations; and protective ground covers and planting.

Landscape development should be in keeping with the character of the highway and its environment. Programs include the following general areas of improvement: (1) preservation of existing vegetation, (2) transplanting of existing vegetation where feasible, (3) planting of new vegetation, (4) selective clearing and thinning, and (5) regeneration of natural plant species and material.

The objectives in planting or the retention and preservation of natural growth on roadsides are closely related. In essence, they are to provide (1) vegetation that will be an aid to esthetics and safety, (2) vegetation that will aid in lowering construction and maintenance costs, and (3) vegetation that creates interest, usefulness, and beauty for the pleasure and satisfaction of the traveling public.

Landscaping of urban highways and streets assumes additional importance in mitigating the many nuisances associated with urban traffic. Landscaping can reduce this contribution to urban blight and make the urban highways and streets better neighbors.

Further information concerning landscape development and erosion control is presented in AASHTO's *A Guide for Highway Landscape and Environmental Design* (**43**).

Safety Rest Areas, Information Centers, and Scenic Overlooks

Safety rest areas, information centers, and scenic overlooks are functional and desirable elements of the complete highway development and are provided for the safety and convenience of the highway user. A safety rest area is a roadside area, with parking facilities separated from the roadway, provided for the motorist to stop and rest for short periods. The area may provide drinking water, toilets, tables and benches, telephones, information facilities, and other facilities for travelers. A safety rest area is not intended to be used for social or civic gatherings or for such active forms of recreation as boating, swimming, or organized games. An information center is a manned or unmanned facility at a safety rest area for the purpose of providing informational services to the motorist. A scenic overlook is a roadside area provided for motorists to park their vehicles, beyond the shoulder, primarily for reviewing the scenery or for taking photographs in safety. Scenic overlooks need not provide comfort and convenience facilities.

Site selection for safety rest areas, information centers, and scenic overlooks should consider the scenic quality of the area, accessibility, and adaptability to development. Other essential considerations include an adequate source of water (**57**) and a means to treat and/or properly dispose of sewage (**58**). Site plans should be developed by the use of a comprehensive site planning process that should include the location of ramps, parking areas for cars and trucks, buildings, picnic areas, water supply, sewage treatment facilities, and maintenance areas. The objective is to give maximum weight to the appropriateness of the site rather than adherence to constant distance or driving time between sites.

Facilities must be designed to accommodate the needs of the handicapped and the elderly.

Further information concerning safety rest area design is presented in AASHTO's *A Guide on Safety Rest Areas for the National System of Interstate and Defense Highways* (**44**).

Lighting

Lighting may improve the safety of a highway or street and the ease and comfort of operation thereon. Statistics indicate that the nighttime accident rate is higher than that during daylight hours, which, to a large degree, may be attributed to impaired visibility. There is evidence that in urban and suburban areas, where there are concentrations of pedestrians and roadside intersectional interferences, fixed-source lighting tends to reduce accidents. Lighting of rural highways may be desirable but the need for it is much less than on streets and highways in urban areas. The general consensus is that lighting of rural highways seldom is justified except on certain critical portions, such as interchanges, intersections, railroad grade crossings, narrow or long bridges, tunnels, sharp curves, and areas where roadside interference is a factor. Most modern rural highways should be designed with an open cross section and horizontal and vertical alinement of a fairly high type. Accordingly, they offer an opportunity for near maximum use of vehicle headlights and resultant diminishing justification for fixed highway lighting throughout.

On freeways where there are no pedestrians, roadside entrances, or other intersections at grade, and where rights-of-way are relatively wide, the justification for lighting differs from that of noncontrolled streets and highways. AASHTO's *An Informational Guide for Roadway Lighting* (**45**) was prepared to aid in the selection of sections of freeways, streets, and highways other than controlled access facilities, on which fixed-source lighting may be warranted, and to present design guide values for their illumination. This guide also contains a section on the lighting of tunnels and underpasses.

Whether or not rural intersections at grade should be lighted depends on the layout and the traffic volumes involved. Intersections that normally do not require channelization are frequently left unlighted. On the other hand, intersections with substantial channelization, particularly multi-road layouts and those designed on a broad scale, often are lighted. It is especially desirable to illuminate large-scale channelized intersections. Because of the sharp curvatures, little of such intersections is within the lateral range of headlights, and the headlights of other vehicles are a hindrance rather than an aid because of the variety of directions and turning movements. There is need to obtain a reduction in the speed of vehicles approaching some intersections. The indication of this need should be definite and visible at a distance from the intersection that is beyond the range of headlights. Illumination of the intersection with fixed-source lighting accomplishes this.

At interchanges it also is desirable, and sometimes necessary, to provide fixed-source lighting. Drivers should be able to see not only the road ahead, but also the entire turning roadway area to discern properly the paths to be followed. They should also see all other vehicles that may influence their own behavior. Without lighting there may be noticeable decrease in the usefulness of the interchange at night, when there would be more cars slowing down and moving with uncertainty than during daylight hours. Consideration should be given to making visible at night by roadway lighting (or reflectorizing devices) the parts of grade separation structures that may be hazardous—such as curbs, piers, and abutments. The greater the volume of traffic, particularly turning traffic, the more important the fixed-source lighting at interchanges becomes. Illumination should also be considered on those sections of major highways where turning movements to and from roadside (ribbon) development are definite hazards.

Floodlighting or highway lighting may be desirable at railroad grade crossings when there are nighttime movements of trains. In some cases such treatment may apply also to crossings operated with flashing signals or gates, or both.

Tunnels nearly always are lighted and so are bridges of substantial length in urban and suburban areas. It is questionable whether the cost of lighting long bridges in rural areas is justified or desirable.

To minimize the effect of glare and to provide the most economical lighting installation luminaires are mounted at heights of at least 30 ft. Lighting uniformity is improved with higher mounting heights, and in most cases, mounting heights of 35 to 50 ft are usually preferable. High mast lighting, special luminaires on masts of 100 ft is used to light large highway areas such as interchanges and rest areas. This lighting furnishes a uniform light distribution over the whole area and may provide alinement guidance. However, it also has a disadvantage in that the visual impact from spurious light is increased to the surrounding community.

Lighting standards (poles) should be placed outside the clear zones whenever practical. The appropriate clear zone dimensions for the various functional classifications will be found in the horizontal clearance to obstructions discussion of Chapter IV. Where poles are located within the clear zone regardless of distances from the traveled way, they should be designed to have a suitable impact attenuation feature; normally a breakaway design is used. Breakaway poles should not be used on streets in densely developed areas, particularly with sidewalks. When struck, these poles could be hazardous to pedestrians and cause damage to adjacent buildings. Because of lower speeds and parked vehicles, there is much less chance of fatalities to vehicle occupants from striking fixed poles on a street as compared to a highway. Poles should not be erected along the outside of curves on ramps where they are more susceptible to being struck. Poles located behind longitudinal barriers (installed for other purposes) should be offset sufficiently to allow for deflection of the longitudinal barriers under impact.

On a divided highway or street, lighting standards may be located either in the median or on the right. Where lighting standards are located on the right, the light source is closer to the more heavily used traffic lanes. However, with median installation, the cost is generally lower and illumination is greater on the high-speed lanes. On median installations, dual-mast arms should be used, for which 40- or 50-ft mounting heights are favored. These should be protected with a suitable longitudinal barrier. On narrow medians, it is usually preferable to place the lighting standards so they are integral with the median barrier.

Where highway lighting is being considered for future installation, considerable savings can be effected through design and installation of necessary conduits under pavements and curbs as part of initial construction.

Highway lighting on freeways is intimately associated with the type and location of highway signs. For full effectiveness the two should be designed jointly.

Utilities

All highway and street improvements, whether upgraded within the existing right-of-way or entirely on new right-of-way, generally entail adjustment of utility facilities. Although utilities generally have little effect on the geometric design of the highway or street, full consideration should be given to measures, reflecting sound engineering principles and economic factors, necessary to preserve and protect the integrity and visual quality of the highway or street, its maintenance efficiency, and the safety of traffic. The costs of utility adjustments vary considerably because of the large number of companies, type and complexity of the facility, and the degree of involvement with the improvement. Depending on the location of a project, involved utilities could include (1) sanitary sewers; (2) water supply lines; (3) oil, gas, and petroleum products pipelines; (4) overhead and underground power and communications lines; (5) drainage and irrigation lines; (6) heating mains; and (7) special tunnels for building connections.

The following factors should be considered in the location and design of utility installations.

General

Utility lines should be located to minimize need for later adjustment, to accommodate future highway or street improvements, and to permit servicing such lines with minimum interference to traffic.

Longitudinal installation should be located on uniform alinement as near as practicable to the right-of-way line so as to provide a safe environment for traffic operation and preserve space for future highway or street improvements or other utility installations. To the extent practicable, utilities along freeways should be constructed so they can be serviced from outside the controlled access lines.

To the extent feasible and practicable, utility line crossings of the highway should cross on a line generally normal to the highway alinement. Those utility crossings that are more likely to require future servicing should be encased or installed in tunnels to permit servicing without disrupting the traffic flow.

The horizontal and vertical location of utility lines within the highway right-of-way limits should conform to the clear roadside policies applicable for the system, type of highway or street, and specific conditions for the particular section involved. Safety of the traveling public should be a prime consideration in the location and design of utility facilities on highway and street right-of-way.

Where it is feasible and reasonable to locate utility lines elsewhere, attachments to bridge structures should be avoided. Where there are no feasible alternate locations, such installations on bridge structures should be concealed from view.

On new installations or adjustments to existing utility lines, provision should be made for known or planned expansion of the utility facilities, particularly those located underground or attached to bridges.

All utility installations on, over, or under highway or street right-of-way and attached structures should be durable materials designed for long service-life expectancy, relatively free from routine servicing and maintenance, and meet or exceed the requirements of the applicable industry codes or specifications.

Utilities that are to cross or otherwise occupy the right-of-way of rural or urban freeways should conform to the requirements of AASHTO's *A Policy on the Accommodation of Utilities Within Freeway Right-of-Way* (**46**). Those on noncontrolled access highways and streets should conform to AASHTO's *A Guide for Accommodating Utilities Within Highway Right-of-Way* (**47**).

Urban

Because of lack of space in most metropolitan areas, special consideration should be given in the initial highway design to the potential for joint usage of the right-of-way that would be consistent with the primary function of the highway or street.

Appurtenances to underground installations, such as vents, drains, markers, manholes, and shutoffs, should be located so as not to interfere with the safety or maintenance of the highway or street, nor to be concealed by vegetation. Preferably they should be located near the right-of-way line.

Where there are curbed sections, utilities should be located in the border areas between the curb and sidewalk, at least 1.5 ft behind the face of the curb, and where feasible, behind the sidewalk. For non-curbed urban sections, refer to the clear zone discussions in Chapter IV.

Existing development and limited right-of-way widths may preclude location of some or all utility facilities outside the paved surface of the street or highway. Under some conditions, it may be necessary to reserve the area outside the paved surface exclusively for the use of overhead lines with all others located under the paved surface, and in some instances it may require location of all the facilities under the paved surface. Location under the paved surface is an exception to the stated policy and as such requires special consideration and treatment. Accommodation of these facilities under the paved surface should be accomplished in a manner that will ensure a minimum adverse effect on traffic as a result of future utility service and maintenance activities.

Publications, such as APWA-ASCE's *Accommodation of Utility Plant Within the Rights-of-Way of Urban Streets and Highways* (**48, 49**), furnish information on accommodating utilities within urban right-of-way.

Rural

On new construction no utility should be situated under any part of the pavement, except where it must cross the highway.

Normally, no poles should be located in the median of divided highways.

Utility poles, vent standpipes, and other above-ground utility appurtenances that would constitute hazards to errant vehicles should not be permitted within the highway clear zone. The only exceptions permitted would be where the appurtenance is breakaway or could be installed behind a traffic barrier erected to protect errant vehicles from some other hazard. The clear zone dimension to be maintained for a specific functional classification will be found in Chapter IV.

Signing and Marking

Signing and marking are directly related to the design of the highway or street and are features of traffic control and operation that the designer must consider in the geometric layout of such a facility. The signing and marking should be designed concurrently with the geometrics. The possibility of future operational problems can be reduced significantly if signing and marking are treated as an integral part of design. The extent of which signs and markings are used depends on the traffic volume, the type of facility, and the degree of traffic control required for safe and efficient operation. Arterial highways usually are numbered routes of fairly high type and have relatively high traffic volumes. On such highways, signs and markings are employed extensively. Local roads are low-type highways and usually have low volumes and speeds. On these facilities the use of complex traffic control devices is limited.

Although safety and efficiency of operation depend to a considerable degree on the geometric design of the facility, the physical layout must also be supplemented by effective signing as a means of informing, warning, and controlling drivers. Signing plans coordinated with horizontal and vertical alinement, sight distance obstructions, operational speeds and maneuvers, and other applicable items should be worked out before completion of design.

Highway signs are of three general types: regulatory signs, used to indicate the required method of traffic movement; warning signs used to indicate conditions that may be hazardous to highway users; and guide signs, used to direct traffic along a route or toward a destination. Uniformity in the use of signs and other traffic control devices is the main objective of the standards contained in the MUTCD (**8**).

Location, reflectorization, and lighting of signs are important considerations in signing. For details regarding design, location, and application of signs, reference should be made to the MUTCD (**8**).

Because supports for highway signs have the potential of being a hazard to motorists, signs should be placed on structures, outside the clear zone, or behind traffic barriers required to shield other hazards. If these measures are not feasible, the sign supports must be breakaway or, for overhead sign supports, shielded by appropriate traffic barriers. The AASHTO *Standard Specifications for Structural Supports for Highway Signs, Luminaires, and Traffic Signals* (**50**) provides the requirements for breakaway sign supports.

Markings and markers, like signs, have the function of controlling traffic to encourage safe and expeditious operation. Markings or markers either supplement the regulations or warnings of traffic signs or serve independently to indicate certain regulations or hazardous conditions. For highways and streets there are three general types of markings in use—pavement markings, object markings, and delineators.

Pavement markings or markers consist of centerline stripes, lane lines, no-passing barriers, and edge striping. These may be supplemented by other pavement markings, such as approach to obstructions, stop and crosswalk lines, and various word and symbol markings. Although essentially a design element, pavement contrast of adjoining lanes for certain conditions may be considered as another form of marking to guide and regulate drivers.

Physical obstructions in or near the roadway should be removed in order to provide the appropriate clear zone. Where removal is impractical, such objects should be adequately marked by painting or by use of other high-visibility material. Where the object is in the direct line of traffic, the obstruction and marking thereon preferably should be illuminated at night by floodlighting; where this is not feasible the object markings should be effectively reflectorized.

Post-mounted delineators are another type of marking device used to guide traffic, particularly at night. Reflector units are installed at certain heights and spacings to delineate the roadway where alinement changes may be confusing and not clearly defined.

Refer to the MUTCD (**8**) for criteria, methods, and standards of marking.

Noise Barriers

In recognition of the adverse effect noise can have on people living on, working on, or otherwise using land adjacent to highways, noise barriers are being used to an increasing extent. Such noise barriers have been constructed on both new and existing highways.

Careful consideration must be exercised to ensure that the construction of these noise barriers will not compromise the safety of the highway. Every effort should be made to locate noise barriers to provide the horizontal clearances to obstruction outside the edge of pavement established in Chapter IV. It is recognized, however, that such a setback may sometimes be impractical. When this clearance is not obtainable, the largest feasible width commensurate with cost effective consideration should be provided. Safe stopping sight distance is another important consideration. Therefore, horizontal clearances should be checked for adequate sight distances. Construction of a barrier should be avoided at the selected location if it would reduce the safe stopping sight distance. This situation could be particularly critical where the location of the noise barrier is along the inside of a curve. Some designs use a concrete safety shape either as an integral part of the noise barrier or as a separate roadside barrier between the edge of pavement and the noise barrier. On non-tangent alinements a separate concrete roadside barrier may obstruct sight distance even though the noise barrier does not. In such instances it may be necessary to install metal rather than concrete roadside barriers in order to retain adequate sight distance. Care must be exercised in the location of noise barriers near gore areas. Barriers at these locations should begin or terminate, as the case may be, at least 200 ft from the theoretical nose.

For further discussion on noise barriers, see the section on noise control in Chapter IV.

Maintenance of Traffic Through Construction Areas

Maintaining a safe flow of traffic during construction must be carefully planned and executed. Although it is often better to provide detours, frequently it will be necessary to maintain the flow of traffic through the construction area. Sometimes it will also be necessary to encroach on the through-traffic lanes or shift lanes entirely in order that the construction can be undertaken. When this is necessary, it is essential that in the development of construction plans, designs for traffic control should produce as minimal an effect on traffic operations as is possible by minimizing the frequency or time of impediment to normal traffic flow. The development of traffic control plans is an essential part of the overall project design and may affect the design of the facility itself. The traffic control plan depends on the nature and scope of the improvement, volumes of traffic, highway or street pattern, and capacities of available highways or streets. A well-thought-out and carefully developed plan for the movement of traffic through a work zone will contribute significantly to the safe and expeditious flow of traffic as well as the safety of the construction forces. It is desirable that these plans have some built-in flexibility to accommodate unforeseen changes in work schedule, delays, or traffic patterns.

The goal of any traffic control plan should be to safely route traffic at a controlled speed through or around construction areas with geometrics and traffic control devices as nearly as possible comparable to those for normal operating situations, while providing room for the contractor to work effectively. Too frequently this goal is not achieved, the result being that accidents occur because unsuspecting motorists are surprised and trapped by inadequate geometrics or traffic controls. The solution is largely one of providing the driver with adequate advance warning and sufficient followup information. Standards for the use and application of signs and other traffic control devices when highway construction occurs are set forth in Part VI of the MUTCD (**8**). It cannot be emphasized too strongly that the principles must be applied and a plan developed for the particular type of work performed. There is no standard application in the MUTCD (**8**) covering every kind of construction activity.

Adequate advance notice must be given to drivers to prepare them for the changed operating conditions in construction areas. The size of and the distance that such signing should be located in advance of construction operations vary with the speed on the affected facility. These construction operations frequently create the need for adjustments in traffic patterns including the shifting of lanes. The minimum taper length for lane transitions in construction areas can be computed by a formula found in the MUTCD (**8**). Because the values obtained are minimum there will be numerous instances, particularly on high-speed highways, where longer lengths should be used. Generally, on low-speed highways lengths in excess of the minimum are not needed and are not attainable on most urban streets. Various configurations for these and other conditions are illustrated in the MUTCD (**8**) and should be used in developing traffic control plans.

The stopping of traffic by a flagger or any other means should be avoided wherever possible. Stopping vehicles is usually difficult and hazardous; it presents a constant opportunity for rear-end collisions. Designs that provide for constant movement around an obstruction in the roadway, even if it is slow, are more acceptable and are less irritating to drivers than requiring them to stop.

When construction operations are scheduled to take place adjacent to passing traffic, a clear zone should be called for in the traffic control plans, wherever possible, between the work and the passing traffic. Under certain conditions, a positive barrier is justified.

Traffic operation controls for a detour are speed, capacity, travel distance, and safety. The speed for a detour may be less than that on the facility being improved but should be high enough so as not to affect the capacity. When an existing highway or street is used as a detour, higher volumes result and it may be necessary to increase the capacity of such a route in advance. The capacity is generally increased by eliminating troublesome turning movements, rerouting transit vehicles and trucks, adopting and enforcing a loading-unloading ban during peak hours, eliminating or adjusting certain transit stops, coordinating signal timing, and sometimes physically widening the traveled way. An effective means of obtaining the necessary capacity is by instituting a one-way detour system, coupled with parking restrictions. A detour plan is tested by comparing the traffic volumes expected to use the rearranged plan to the calculated capacity of the detour system.

The roadway near falsework openings should be well lighted and delineated. Channelization of traffic should be accomplished by the use of signing on yielding supports, pavement markings, and barricades.

Construction areas, detours, and temporary connections often include geometric features and roadway environments that may require more caution and alertness than can normally be expected of drivers. Care in the layout of these areas, in the use of delineation and warning devices, and in the establishment of areas for contractor operations is necessary to minimize the impact on the safety of both motorists and workers. Items that should be considered in developing traffic control plans include the following:

1. Detour alinement and surface that will allow traffic to pass smoothly around the work zones. The surface of the traveled way, whether located within the construction area or on a detour, should be maintained in a condition that will permit the safe movement of traffic at a reasonable speed.
2. Adequate tapers are needed for lane drops or at locations where traffic must be shifted laterally. Appropriate values for taper lengths can be found in Part VI of the MUTCD (**8**).
3. Adequate traffic control devices and pavement markings for complete day and night effectiveness must be provided, including specification of temporary marking materials that can be removed when traffic lane patterns change.
4. Roadway illumination and warning lights should be used where justified. Steady burning lights are used to delineate a continuous travel path through or around a work zone. The very short "on" time of flashing lights does not enable motorists to focus on the light and make a depth perception estimate. The use of flashers is limited to marking a single hazard, marking the start of a section using steady burn lights, and for use with traffic control signs.

5. The location of cones, delineators, drums, barriers, or barricades, as means of channelizing traffic, should be detailed in the traffic control plans when special conditions exist or if not shown in the standard plans.
6. If not provided for in the specifications, requirements regarding the removal of signs and markings from the job site, when they are no longer needed, should be noted on the traffic control plans.
7. The contract specifications or traffic control plans should require, except in extenuating circumstances, the removal of contractor equipment completely off the roadways, medians, and shoulders at night, on weekends, and whenever equipment is not in operation. In those instances where such removal is not feasible, appropriate signing, lighting, barricades, barriers, and similar devices to protect the motorist from collision with the equipment should be specified. The storage of hazardous materials, however, should never be permitted on roadways, medians, or shoulders near the flow of traffic.
8. A requirement for controlling or prohibiting the parking of employee's private vehicles in those areas on the project that would compromise the safety of workers and through traffic should be included in the plans or specifications.

Signals

Traffic control signals are devices that control vehicular and pedestrian traffic by assigning the right-of-way to various movements for certain pretimed or traffic-actuated intervals of time. They are one of the key elements in the function of many urban streets and of some rural intersections. For this reason the planned signal operation for each intersection of a facility should be integrated with the design so as to achieve optimum operational efficiency. Careful consideration should be given in plan development to intersection and access locations, horizontal and vertical curvature with respect to signal visibility, pedestrian requirements, and geometric schematics to ensure the best possible signal operation (individual signal phasing and traffic coordination between signals). In addition to initial installation, possible future signal needs should also be evaluated. The design of traffic signal devices and warrants for their use are covered in Part IV of the MUTCD (**8**).

Lane arrangement is the key to successful operation of signalized intersections. The crossing distances for both vehicles and pedestrians should be kept as short as possible to reduce the startup time lost at the beginning of the green phase and the clearance time lost at the end of the phase. The time saved from these two periods results in more green time for through traffic. Therefore, the first step in the development of intersection geometrics should be a complete analysis of current and future traffic demand. The need to provide right- and left-turn lanes to minimize the interference of turning traffic with the movement of through traffic should be evaluated concurrently with the potential for obtaining any necessary additional right-of-way. Along a highway or street with a number of signalized intersections, the locations where turns will, or will not, be accommodated should also be examined to ensure a good fit with two-way signal coordination. Because of the large volume of traffic turning into and out of large parking areas, parking area entrances and exits should be designed in a manner that will simplify the operation of the affected traffic signals.

There are some controls applicable to signal installations that may influence the design of intersections and their approaches. These are discussed in the MUTCD (**8**).

REFERENCES

1. Johansson, G. and Rumar, K. "Drivers' Brake Reaction Times." *Human Factors*, Vol. 13, No. 1, February 1971, pp. 23-27.
2. *Report of the Massachusetts Highway Accident Survey.* CWA and ERA project. Cambridge, Mass.: Massachusetts Institute of Technology, 1935, 117 pp.
3. Normann, O.K. "Braking Distances of Vehicles from High Speeds." *Proc. HRB*, Vol. 22. Washington, D.C.: Highway Research Board, 1953, pp. 421-436.
4. Shelburne, T.E. and Sheppe, R. L. "Skid Resistance Measurements of Virginia Pavements." *Research Report No. 5-B.* Washington, D.C.: Highway Research Board, April 1948, 27 pp.
5. Moyer, R. A. and Shupe, J. W. "Roughness and Skid Resistance Measurements of Pavements in California." *HRB Bulletin 37.* Washington, D.C.: Highway Research Board, August 1951, pp.1-37.
6. Schulze, K. H. and Beckmann, L. "Friction Properties of Pavements at Different Speeds." *ASTM Special Technical Publication No. 326.* Baltimore: American Society of Testing and Materials, December 1962, pp.42-49.
7. *Guidelines for Skid Resistant Pavement Design.* Washington, D.C.: AASHTO, 1976, 20 pp.

8. Federal Highway Administration. *Manual on Uniform Traffic Control Devices.* Washington, D.C.: U.S. Government Printing Office, 1988.
9. King, G. F. and Lunenfeld, H. "Development of Information Requirements and Transmission Techniques for Highway Users." *NCHRP Report 123*, 1971, 239 pp.
10. McGee, H. W., Moore, W., Knapp, B. G. and Sanders, J. H. (BioTechnology, Inc.). *Decision Sight Distance for Highway Design and Traffic Control Requirements.* U.S. Department of Transportation, Federal Highway Administration, Washington, D.C.: U.S. Government Printing Office, 1978, 70 pp.
11. Robinson, G.H., Erickson, D.J., Thurston, G.L. and Clark, R.L. "Visual Search by Automobile Drivers." *Human Factors*, Vol. 14, No. 4, August 1972, pp. 315-323.
12. Prisk, C.W. "Passing Practices on Rural Highways." *Proc. HRB*, Vol. 21. Washington, D.C.: Highway Research Board, 1941, pp. 366-378.
13. *Highway Capacity Manual, Special Report 209*. Washington, D.C.: Transportation Research Board, 1985.
14. Boyd, N.W., Littleton, A.C., Pilkington, G.B., II, and Boenau, R.E. *Determination of Motor Vehicle Eye Height for Highway Design*, U.S. Department of Transportation, Federal Highway Administration. Washington, D.C.: U.S. Government Printing Office, 1978, 37 pp.
15. Motor Vehicle Manufacturers Association. *Driver Eye Height Comparison for 1976 and 1978*. Motor Vehicle Manufacturers Association, Engineering Research Committee, Human Factors Engineering Subcommittee. Unpublished report for AASHTO, 1978, 14 pp.
16. Bryar, P.R. and Mojasiak, W. (Ontario Ministry of Transportation and Communications). *Highway Vertical Alignment*. Toronto, Ontario: Ontario Ministry of Transportation and Communications, February 1978, 70 pp.
17. Segar, E.E. and Brink, R.S. "Trends of Vehicle Dimensions and Performance Characteristics from 1960 through 1970." *Highway Research Record 420*. Washington, D.C.: Highway Research Board, 1972, pp. 1-15.
18. Moyer, R.A. "Skidding Characteristics of Automobile Tires on Roadway Surfaces and Their Relation to Highway Safety." *Bulletin No. 120*. Ames, Iowa: Iowa Engineering Experiment Station, 1934, 128 pp.
19. Stonex, K.A. and Noble, C.M. "Curve Design and Tests on the Pennsylvania Turnpike." *Proc. HRB*, Vol. 20. Washington, D.C.: Highway Research Board, 1940, pp. 429-451.
20. Moyer, R.A. and Berry, D.S. "Marking Highway Curves with Safe Speed Indications." *Proc. HRB*, Vol. 20. Washington, D.C.: Highway Research Board, 1940, pp. 399-428.

21. Barnett, J. "Safe Side Friction Factors and Superelevation Design." *Proc. HRB*, Vol. 16. Washington, D.C.: Highway Research Board, 1936, pp. 69-80.
22. Hajela, G.P. *Compiler, Resume of Tests on Passenger Cars on Winter Driving Surfaces, 1939-1966*. Chicago: National Safety Council, Committee on Winter Driving Hazards, 1968, 165 pp.
23. Shortt, W.H. "A Practical Method for Improvement of Existing Railroad Curves." *Proc. Institution of Civil Engineering*, Vol. 76. London: Institution of Civil Engineering, 1909, pp. 97-208.
24. Barnett, J. *Transition Curves for Highways.* Federal Works Agency, Public Roads Administration. Washington, D.C.: U.S. Government Printing Office, 1940, 211 pp.
25. Bureau of Public Roads (Federal Highway Administration). *Study of Speed Curvature Relations of Pentagon Road Network Ramps.* Unpublished data, Federal Works Agency, Public Roads Administration, 1954.
26. Cysewski, G.R. "Urban Intersectional Right Turning Movements." *Traffic Engineering*, Vol. 20, No. 1, October 1949, pp. 22-37.
27. George, L.E. "Characteristics of Left-Turning Passenger Vehicles." *Proc. HRB*, Vol. 31. Washington, D.C.: Highway Research Board, 1952, pp. 374-385.
28. Walton, C.M. and Lee, C.E. "Speed of Vehicles on Grades." *Research Report 20-1F.* Austin: Center for Highway Research, University of Texas at Austin, August 1975, 164 pp.
29. Werner, A. *Effect of Recreational Vehicles on Highway Capacity.* Calgary, Alberta: Department of Civil Engineering, University of Calgary, April 1974, 223 pp.
30. Taragin, A. "Effect of Length of Grade on Speed of Motor Vehicles." *Proc. HRB*, Vol. 25. Washington, D.C.: Highway Research Board, 1945, pp. 342-353.
31. Willey, W.E. "Survey of Uphill Speeds of Trucks on Mountain Grades." *Proc. HRB*, Vol. 29. Washington, D.C.: Highway Research Board, 1949, pp. 304-310.
32. Huff, T.S. and Scrivner, F.H. "Simplified Climbing-Lane Design Theory and Road-Test Results." *Bulletin 104*. Washington, D.C.: Highway Research Board, 1955, pp. 1-11.
33. Schwender, H.C., Normann, O.K., and Granum, J.O. "New Method of Capacity Determination for Rural Roads in Mountainous Terrain." *Bulletin 167*. Washington, D.C.: Highway Research Board, 1957, pp. 10-37.
34. Hayhoe, G.F. and Grundmann, J.G. *Review of Vehicle Weight/Horsepower Ratio as Related to Passing Lane Design Criteria*. Final Report pre-

pared by the Pennsylvania Transportation Institute for the Transportation Research Board, National Cooperative Highway Research Program. University Park: Pennsylvania State University, October 1978, 37 pp.

35. Glennon, J.C. "An Evaluation of Design Criteria for Operating Trucks Safely on Grades." *Highway Research Record 312*. Texas Transportation Institute, Texas A&M University. Washington, D.C.: Highway Research Board, 1970. pp. 93-112.
36. Harwood, D.W. and Hoban, C.J. "Low Cost Methods of Improving Traffic on Two-Lane Roads." *Report No. FHWA-1P-87-2*. McLean, VA.: Federal Highway Administration, 1987, 103 pp.
37. Cron, F. W. "The Art of Fitting the Highway to the Landscape." W.B. Snow, ed., *The Highway and the Landscape.* New Brunswick, N.J.: Rutgers University Press, 1959, pp. 78-109.
38. Tunnard, C. and Pushkarev, B. *Man Made America: Chaos or Control?* New Haven: Yale University Press, 1963, 479 pp.
39. Leisch, J.E. *Application of Human Factors in Highway Design.* Unpublished paper presented at AASHTO Region 2 meeting, June 1975.
40. *Highway Drainage Guidelines.* Vols. 1-8. Washington D.C.: AASHTO, 1988.
41. Federal Highway Administration Publications, U.S. Department of Transportation. Washington:
    a. Available from U.S. Government Printing Office.
       "Hydraulic Charts for the Selection of Highway Culverts," *Hydraulic Engineering Circular (HEC)* No. 5, 1965, 54 pp.; "Drainage of Highway Pavements," HEC No. 12, 1969, 90 pp.; "Hydraulic Design of Energy Dissipators for Culverts and Channels," HEC No. 14, 1975, 320 pp.; "Design of Stable Channels with Flexible Linings," HEC No. 15, 1975, 136 pp.
       "Hydraulics of Bridge Waterways," *Hydraulic Design Series (HDS) No. 1,* 2nd ed., 1978, 111 pp.
    b. Available from Federal Highway Administration.
       "Selected Bibliography of Hydraulic and Hydrologic Subjects," *Hydraulic Engineering Circular (HEC) No.1*, 1979, 36 pp.; "Hydrology of a Highway Stream Crossing," *HEC No. 3*, 1961, 10 pp.; "Debris-Control Structures," *HEC No.9*, 1971, 38 pp.; "Capacity Charts for the Hydraulic Design of Highway Culverts," *HEC No. 10*, 1972, 90 pp.; "Use of Riprap for Bank Protection," *HEC No. 11*, 1967, 43 pp.; "Hydraulic Design of Improved Inlets for Culverts," *HEC No. 13*, 1980, 172 pp.; "Addendum to Highways in the River Environment: Hydraulic and Environmental Considera-

tions," *HEC No. 16*, 1980, 42 pp.; "The Design of Encroachments of Flood Plains Using Risk Analysis," *HEC No. 17*, 1980, 134 pp.
"Design Charts for Open-Channel Flow," *Hydraulic Design Series (HDS) No. 3*, 1961, 105 pp.; "Design of Roadside Drainage Channels," *HDS No. 4*, 1973, 56 pp.
"Hydraulic Analysis for Pipe-Arch Culverts," *Electronic Computer Programs (HY) HY-2*, 1969, 48 pp.; "Hydraulics of Bridge Waterways," HY-4, 1969, 95 pp.; "Hydraulic Analysis of Culverts (Box and Circular)," *HY-6*, 1979, 160 pp.
"Hydraulic Design of Improved Inlets for Culverts Using Programmable Calculators (COMPUCORP-325)," *Calculator Design Series (CDS) No. 1*, 1980, 133 pp.; "Hydraulic Design of Improved Inlets for Culverts Using Programmable Calculators, (HP-65)," *CDS No. 2*, 1980, 147 pp.; "Hydraulic Design of Improved Inlets for Culverts Using Programmable Calculators, (TI-59)," *CDS No. 3*, 1980, 133 pp.

42. Richardson, E.V., et al. *Highways in the River Environment: Hydraulic and Environmental Design Considerations*. Prepared by the Civil Engineering Department, Engineering Research Center, Colorado State University for the U.S. Department of Transportation, Federal Highway Administration. Washington, D.C.: U.S. Government Printing Office, May 1975, 453 pp.
43. *A Guide for Highway Landscape and Environmental Design*. Washington, D.C.: AASHTO, (current edition).
44. *A Guide on Safety Rest Areas for the National System of Interstate and Defense Highways*. Washington, D.C.: AASHTO, 1968, 49 pp.
45. *An Informational Guide for Roadway Lighting*. Washington, D.C.: AASHTO, 1984.
46. *A Policy on the Accommodation of Utilities Within Freeway Right-of-Way*. Washington, D.C.: AASHTO, 1989, 7 pp.
47. *A Guide of Accommodating Utilities Within Highway Right-of-Way*. Washington, D.C.: AASHTO, (current edition).
48. Bert, K.E., et al. *Accommodation of Utility Plants Within the Rights-of-Way of Urban Streets and Highways, Manual of Improved Practice. ASCE Manual No. 14*. Chicago: American Public Works Association, and New York: American Society of Civil Engineers, July 1974, 102 pp.
49. Bert, K.E., et al. *Accommodation of Utility Plants Within the Rights-of-Way of Urban Streets and Highways, State-of-the-Art. APWA Special Report No.44*. Chicago: American Public Works Association, July 1974, 160 pp.
50. *Standard Specifications for Structural Supports for Highway Signs, Lumi-*

naries, and Traffic Signals. Washington, D.C.: AASHTO, 1986, 69 pp.

51. Alexander, G. J. and Lunenfeld, H. *Positive Guidance in Traffic Control*. U.S. Department of Transportation, Federal Highway Administration. Washington, D.C.: U.S. Government Printing Office, 1975, 57 pp.
52. Mullin, E. F. "The Part Visibility Could Play in Road Design." *Australian Road Research*, Vol. 2, No. 9, September 1966, pp. 15-43.
53. *Standard Specifications for Highway Bridges*. Washington, D.C.: AASHTO, 1983.
54. Weaver, G.D. and Glennon, J.C. "Passing Performance Measurements Related to Sight Distance Design," *Report 134-6*. College Station, Texas: Texas Transportation Institute, Texas A&M University, July 1971.
55. Weaver, G.D. and Woods, D.L. (Texas Transportation Institute). "Passing and No-Passing Signs, Markings, and Warrants." *Report No. FHWA-RD-79-5*, U.S. Department of Transportation, Federal Highway Administration. Washington, D.C.: U.S. Government Printing Office, September 1978, 239 pp.
56. Raymond, W.L., Jr., "Offsets to Sight Obstructions Near the Ends of Horizontal Curves." *Civil Engineering*, ASCE, Vol. 42, No. 1, January 1972, pp. 71-72.
57. Folks, N.E. (Ultrasystems, Inc.). "Manual for Safety Rest Area Water Supply Systems." *Report No. FHWA-RD-77-113*, U.S. Department of Transportation, Federal Highway Administration. Washington, D.C.: U.S. Government Printing Office, September 1977, 296 pp.
58. Hughes, G.W., Averett, D.E. and Francingues, N.R., Jr. (Environmental Effects Laboratory). "Wastewater Treatment Systems for Safety Rest Areas." *Report No. FHWA-RD-77-107*, U.S. Department of Transportation. Federal Highway Administration. Washington, D.C.: U.S. Government Printing Office, September 1977, 291 pp.
59. MacAdam, C.C., Fancher, P.S., Segal, L. (University of Michigan Transportation Research Institute) "Side Friction for Superelevation on Horizontal Curves" *Report No. FHWA-RD-86-024*. U.S. Department of Transportation, Federal Highway Administration. Springfield, Va.: National Technical Information Service, August 1985.
60. Gillespie, Thomas D. "Methods for Predicting Truck Speed Loss on Grades." *Final Report prepared by Transportation Research Institute, the University of Michigan for the Federal Highway Administration,* October 1986.

Chapter IV
CROSS SECTION ELEMENTS

PAVEMENT

Surface Type

The selection of pavement type is determined by the volume and composition of traffic, soil characteristics, weather, performance of pavements in the area, availability of materials, energy conservation, the initial cost and the overall annual maintenance and service life cost. The structural design of pavements is not included in this policy, but may be found in the AASHTO *Guide for Design of Pavement Structures* **(26)**.

Pavements may be considered as three general types—high, intermediate, and low. High-volume traffic justifies high-type pavements, which have smooth riding qualities and good skid-resistant properties in all weather. The surface should retain the qualities and cross section and adequately support the expected volume and weights of vehicles without fatigue failure, so that nonroutine maintenance and the resultant annoyance to traffic operations are kept to a minimum. Those pavements classified as intermediate-type surfaces range from surface treatments to pavements built to standards only slightly less strict than those for high-type pavements. Low-type surfaces range from surface-treated earth roads and stabilized materials to loose surfaces such as earth, shell, crushed stone, and bank run gravel.

Important pavement characteristics in relation to geometric design are the ability of a surface to retain its shape and dimensions, the ability to drain, the ability to retain adequate skid resistance and the effect on driver behavior. High-type pavements retain their shape and do not ravel at the edges if placed on a stable subgrade. Their smoothness and proper cross slope design enable drivers to steer easily and keep their vehicles in proper paths.

At the other extreme, low-type surfaces have a tendency toward raveling, which reduces the effective width, and require greater steering effort to maintain a correct path. Loose and other low-type surfaces are used where traffic volume is light.

High-type surfaces provide for higher safe operating speeds than do low-type surfaces. Although the choice of design speed is influenced to a degree by the type of surface, the choice is dependent on other more important factors. In most cases it is also probable that the surface will ultimately be improved or replaced with a higher type. A surface type should be provided commensurate with the design speed selected for the highway.

Cross Slope

Two-lane and wider undivided pavements on tangents or on flat curves have a crown or high point in the middle and slope downward toward both edges. The downward cross slope may be a plane or curved section or a combination of the two. With plane cross slopes there is a cross slope break at the crown line and a uniform slope on each side. Curved cross sections usually are parabolic, with a slightly rounded surface at the crown line and increasing cross slope toward the pavement edge. Because the rate of crown slope is variable, the parabolic section is described by the crown height, i.e., the vertical drop from the center crown line to the pavement edge. The advantage of the curved section lies in the fact that the cross slope steepens toward the pavement edge, thereby facilitating drainage. The disadvantages are that curved sections are more difficult to construct, the cross slope of the outer lanes may be excessive, and warping of pavement areas at intersections may be awkward or difficult to construct.

On divided highways each one-way pavement may be crowned separately, as on two-lane highways, or it may have a unidirectional slope across the entire width of pavement, which is almost always downward to the outer edge. Where freeze-thaw conditions are a problem, each pavement of a divided highway should be crowned separately.

A cross section with each roadway crowned separately, as shown in Figures IV-1A through IV-1C, has an advantage in rapidly draining the pavement during rainstorms. In addition, the difference between high and low points in the cross section is kept to a minimum. Disadvantages are that more inlets and underground drainage lines are required, and treatment of at-grade intersections is more difficult because of the several high and low points on the cross section. Use of such sections preferably should be limited to regions of high rainfall or where snow and ice are major factors. Sections having no curbs and a wide depressed median are particularly well-suited for these conditions.

Roadways that slope in only one direction, as shown in Figures IV-1D through IV-1G, are more comfortable to drivers because vehicles tend to be pulled in the same direction when changing lanes. Roadways having a unidirectional slope may drain away from or toward the median. Drainage away from the median may effect a savings in drainage structures and simplify treatment of intersecting streets. Drainage toward the median has an advantage over sections sloped to the outer edges in that outer lanes, which are used by most traffic, are freer of surface water and there is economy in the drain-

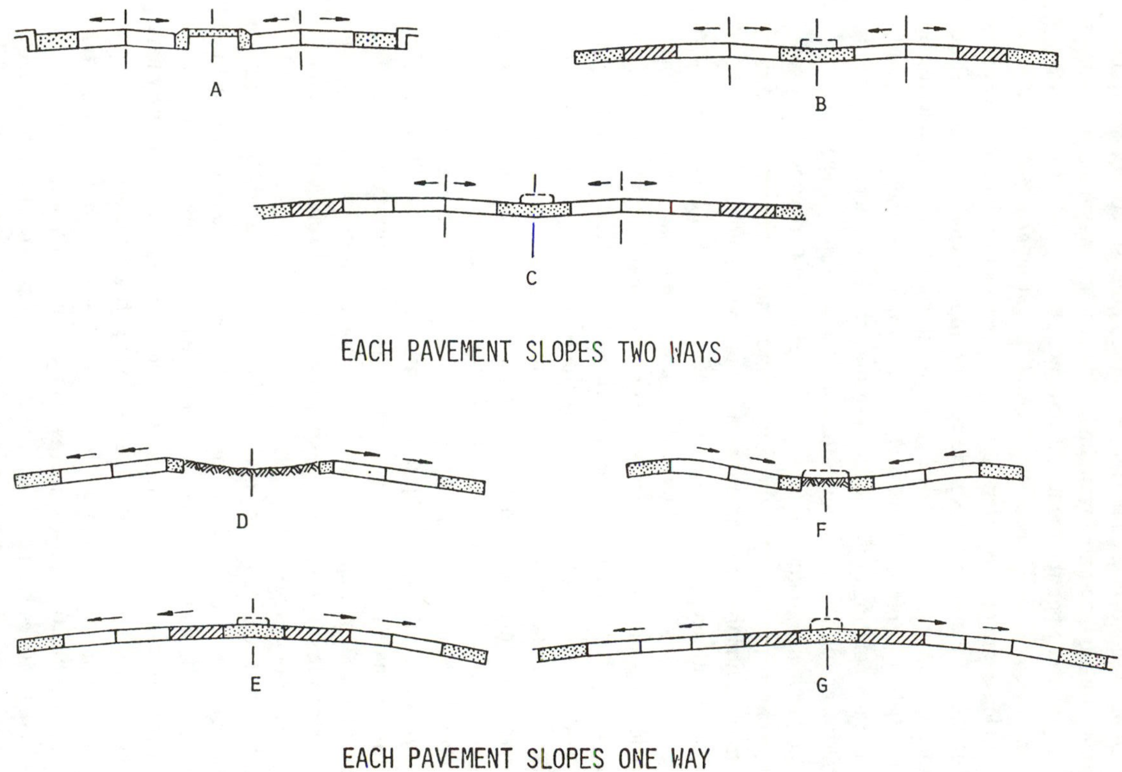

Figure IV-1. Roadway sections for divided highway (basic cross slope arrangements).

age system in that all surface runoff is collected into a single conduit under the median. A major disadvantage of this section is that all the pavement drainage must pass over the inner, higher speed lanes. Where curbed medians exist, drainage is concentrated next to or on the higher speed lanes. This concentration results in the annoying and hazardous splashing on the windshields of opposing traffic when the median is narrow.

The rate of cross slope on curves as well as on tangent alinement is an important element in cross section design. Pavement superelevation on curves is determined by the speed-curvature relationships given in chapter III, but cross slope or crown on tangents or on long-radius curves is complicated by two contradictory controls. A reasonably steep lateral slope is desirable to minimize water ponding on flat sections of uncurbed pavements as a result of pavement imperfections or unequal settlement; a steep cross slope is desirable on curbed pavements to confine the flow of water to a narrow width of pavement adjacent to the curb. On the other hand, pavements with steep cross slopes are objectionable in appearance and may be annoying and uncomfortable in operation. Steep cross slopes are undesirable on tangents because of the tendency of vehicles to drift toward the low edge of pavement. This drifting becomes a major concern in areas where snow and ice are common. Cross slopes up to 2 percent are barely perceptible as far as effect on vehicle steering is concerned, but cross slopes steeper than 2 percent are noticeable. The latter rate requires a conscious effort in steering and increases the susceptibility to lateral skidding when vehicles brake on icy or wet pavements and even on dry pavements when stops are made under emergency conditions.

The prevalence of high winds may significantly alter the effect of pavement crown on steering. In rolling terrain with alternate cut-and-fill sections or in areas alternately forested and cleared, any substantial cross wind produces an intermittent impact on a vehicle moving along the highway and affects its steering. In areas where such conditions are likely, it is desirable to avoid high rates of cross slope.

On high-type two-lane pavements, crowned at the center, the accepted rate of cross slope is from 1.5 to 2 percent. When three or more lanes are inclined in the same direction on multilane pavements, each successive pair of lanes or portion thereof outward from the first two lanes from the crown line may have an increased slope. The two lanes adjacent to the crown line should be pitched at the normal minimum slope, and on each successive pair of lanes or portion thereof outward, the rate may be increased by about 0.5 to 1 percent.

Use of cross slopes steeper than 2 percent on high-type, high-speed pavements with a central crown line is not desirable. In passing maneuvers, drivers must cross and recross the crown line and negotiate a total rollover or cross-slope change of over 4 percent. The reverse curve path of travel of the passing vehicle causes a reversal in the direction of centrifugal force, which is further exaggerated by the effect of the reversing cross slopes. Trucks with high body loads crossing the crown line are caused to sway from side to side when traveling at high speed, at which time control is difficult.

In areas of intense rainfall a somewhat steeper cross slope may be necessary to facilitate pavement drainage, even though undesirable from the operational point of view. In such areas the cross slope on high-type pavements may be increased to 2.5 percent, with a corresponding crown line crossover of 5 percent. Where three or more lanes are provided in each direction, the maximum pavement cross slope should be limited to 4 percent. Use of this increased cross slope should be limited to the condition described; for the reasons cited in the preceding discussion, 2 percent should be taken as the maximum cross slope on high-type pavements for all other conditions. In locations of maximum cross slope and intense rainfall, consideration should be given to the use of grooving or open graded mixes.

On pavements designed with rounded crown section, the rollover is not generally objectionable. In an operational sense, rounded crown sections are good designs for two-lane tangent pavements. In a design and construction sense, they cause difficulty in warping from a tangent section to a superelevated section. On pavements of more than two lanes the parabolic or circular crown should be used with caution. Any such curved section determined to have a desired slope across the center lane (or lanes) could result in an undesirably steep slope across the outer lanes. For example, an undivided four-lane highway has a parabolic or circular crown section designed so that the outer edge of the inner lane is 0.12 ft below the centerline. With equal-width lanes there is a fall of 0.36 ft across the outer lane. Lane widths of 12 ft result in an average cross slope of 3 percent on the outer lane. Another approach for using parabolic crowns lies in the determination of the maximum crossfall rate. This maximum rate is then used as the control, thus decreasing the slope near the center of the roadway.

In some cases on multilane nondivided highways, a rounded crown section is used for the central two lanes, with a plane section on the outer lanes having a cross slope rate of about 1.5 to 2 percent. Here the outer lane cross slope is made the same or slightly steeper than that at the end of the curved section.

The cross slope rates discussed above pertain largely to high-type surfaces. Without curbs, these types qualify for minimum cross slopes on tangents because of the excellent control over construction methods and smoothness of the finished surface. For intermediate and low types of pavement, a greater cross slope should be utilized. Table IV-I shows a range of values applicable to each type of surface.

On intermediate-type surfaces the running speed may approximate that on high-type pavements because of the generally lower volumes and fewer trucks and buses. A greater cross slope rate for intermediate-type surfaces is used

Surface Type	Range in Cross Slope Rate (percent)
High	1.5-2
Intermediate	1.5-3
Low	2-6

Table IV-1. Normal pavement cross slope.

because of the likelihood of lesser control in construction procedures and the proneness to settlement and warping of the cross section. On these surfaces a longitudinal grade will assist drainage in the event of a "rutting" type of settlement; otherwise a cross slope greater than normal should be used.

Because of the nature of the surfacing materials used and surface irregularities, low-type surfaces such as loose earth, gravel, or crushed stone require a still steeper cross slope on tangents to prevent the absorption of water into the surface. Cross slopes greater than 2 percent may be used on these types of surfaces.

Where pavements are designed with outer curbs, the lower values in the ranges of cross slopes in Table IV-1 are questionably low because of the likelihood of a sheet of water over a substantial part of the traveled way adjacent to the curb. For any rate of rainfall the width of traveled way inundated varies with the rate of cross slope, roughness of gutter, frequency of discharge points, and longitudinal grade. A cross slope greater than 1 percent is desirable, and in some cases, more than 1.5 percent is needed to limit inundation to about half of the outer traffic lane. A cross slope of 1.5 percent is suggested as a practical minimum for curbed high-type pavement and 2 percent for intermediate-type pavements. Curbs with adjacent steeper gutter sections may permit the use of lesser rates of pavement cross slope. In lieu of outer curbs a preferred cross section treatment is the use of a straight shoulder slope and the avoidance of curbs whenever possible.

Skid Resistance

Skidding accidents are a major concern in highway safety. It is not sufficient to attribute skidding accidents merely to "driver error" or "driving too fast for existing conditions." The roadway should provide a level of skid resistance for safely accommodating the braking and steering maneuvers which can reasonably be expected for the particular site. Recent research **(15)** has emphasized the influence of highway geometrics on skidding.

A skid-resistant pavement surface is one key element in the prevention of skidding accidents. Most dry pavements provide adequate skid resistance. Special consideration, however, must be given to the development of skid resistance of wet pavements. The four main causes of poor skid resistance on wet pavements are rutting, polishing, bleeding, and dirty pavements. **Rutting** causes water accumulation in the wheel tracks. **Polishing** reduces, and **bleeding** covers, the micro texture. In both cases, the harsh surface features needed for penetrating the thin water film are diminished. **Dirty pavements,** pavements contaminated by oil drippings or layers of dust or organic matter, will lose their skid resistance.

Skid resistance should be "built in" on new construction and major reconstruction. Vertical and horizontal alinement and pavement texture can be designed to provide high initial skid resistance. Measures taken to correct or improve skid resistance should result in the following characteristics: high initial skid resistance, durability to retain skid resistance with time and traffic, and minimal decrease in skid resistance with increasing speed.

On roadways of portland cement, tining has proved to be effective in reducing the potential for hydroplaning. The use of surface courses or overlays constructed with polish-resistant coarse aggregate is the most widespread method for improving the surface texture of bituminous pavements. Overlays of open-graded asphalt friction courses are quite effective because of their frictional and hydraulic properties. For further discussion, refer to the AASHTO *Guidelines for Skid Resistant Pavements* **(13)**.

LANE WIDTHS

No feature of a highway has a greater influence on the safety and comfort of driving than the width and condition of the surface. There is obvious need for a smooth, skid-resistant, all-weather surface on highways. Ten-to 13-ft lane widths are generally used, with a 12-ft lane predominant on most high-type highways. The extra cost of 12-ft over 10-ft lanes is offset to some extent by a reduction in cost of shoulder maintenance and a reduction in surface maintenance due to lessened wheel concentrations at the pavement edges.

Studies on two-lane two-way rural highways **(1)** show that undesirable conditions (inadequate vehicle clearances and edge-of-pavement clearances) exist on surfaces less than 22 ft wide carrying even moderate volumes of mixed traffic. A 24-ft surface is required to permit desired clearance between commercial vehicles. It is generally accepted that lane widths of 12 ft should be provided on main highways.

The capacity of a highway is affected by the lane width. Narrow lanes force drivers to operate their vehicles closer to each other laterally than they would normally desire. Restricted clearances have much the same effect. In a capacity sense the effective width of traveled way is reduced when adjacent obstructions such as retaining walls, bridge trusses or headwalls, and parked cars restrict the lateral clearance. The extent of this reduction is shown in Table IV-2. In addition to the capacity effect, the resultant erratic operation has an undesirable effect on driver comfort and accident rates.

An effective width of 20 ft is considered adequate only for low-volume roads where meeting and passing is infrequent and the proportion of trucks is low.

Where unequal-width lanes are used, it makes little difference on which side the wider lane is located. Locating the wider lane on the outside (right)

Usable shoulder width or clearance to obstruction (Ft.)	Capacity of Narrow Lanes with Restricted Lateral Clearance (Percent of Capacity of 12-Ft. Lane)[a]		
	12 Ft. Lanes	11 Ft. Lanes	10 Ft. Lanes
	Two Lane Roadway		
6	100	93	84
4	92	85	77
2	81	75	68
0	70	65	58
Four Lane Undivided—One Direction Travel-Obstruction One Side[b]			
6	100	95	89
4	98	94	88
2	95	92	86
0	88	85	80

[a]Uninterrupted flow Level of Service B, High Type Pavement

[b]Obstructions include retaining walls, bridge trusses or headwalls, abutments, poles, parked cars or other features which restrict lateral clearance.

[c]Source: Ref. 4.

Table IV-2. Combined effect of lane width and restricted lateral clearance on capacity.[c]

provides more space for large vehicles that usually occupy that lane and allows drivers to keep their vehicles at a greater distance from the right pavement edge. Where a curb is used adjacent to only one edge of the pavement, the wider lane is placed adjacent to that curb. The basic design determination is the total pavement width, and the placement of pavement stripes actually governs the lane width.

Although lane widths of 12 ft are desirable on both rural and urban facilities, there are circumstances that necessitate the use of lanes less than 12 ft wide. In urban areas where right-of-way and existing development become stringent controls, the use of 11 ft lanes is acceptable. Ten ft lanes are acceptable on low-speed facilities. Under unusual and severe constraints, 9 ft lanes can be used in residential areas where traffic volumes are low.

Auxiliary lanes at intersections and interchanges often help to facilitate traffic movements. Such added lanes should be as wide as the through traffic lanes but not less than 10 ft.

When continuous two-way left-turn lanes are provided, a lane width of 10 to 16 ft provides the optimum design.

SHOULDERS

General Characteristics

A shoulder is the portion of the roadway contiguous with the traveled way for accommodation of stopped vehicles, for emergency use and for lateral support of subbase, base, and surface courses. It varies in width from only 2 ft or so on minor rural roads, where there is no surfacing or the surfacing is applied over the entire roadbed, to about 12 ft on major roads, where the entire shoulder may be stabilized or paved.

The term "shoulder" is variously used with a modifying adjective to describe certain functional or physical characteristics.

The following meanings apply to the terms used herein:

1. The "graded" width of shoulder is that measured from the edge of the traveled way to the intersection of the shoulder slope and the foreslope planes as shown in Figure IV-2A.
2. The "usable" width of shoulder is the actual width that can be used when a driver makes an emergency or parking stop. Where the sideslope is 4:1 or flatter, the "usable" width is the same as the "graded" width since the usual rounding 4 to 6 feet wide at the shoulder break will not lessen its useful width appreciably. Figures IV-2B and IV-2C illustrate the usable shoulder width.

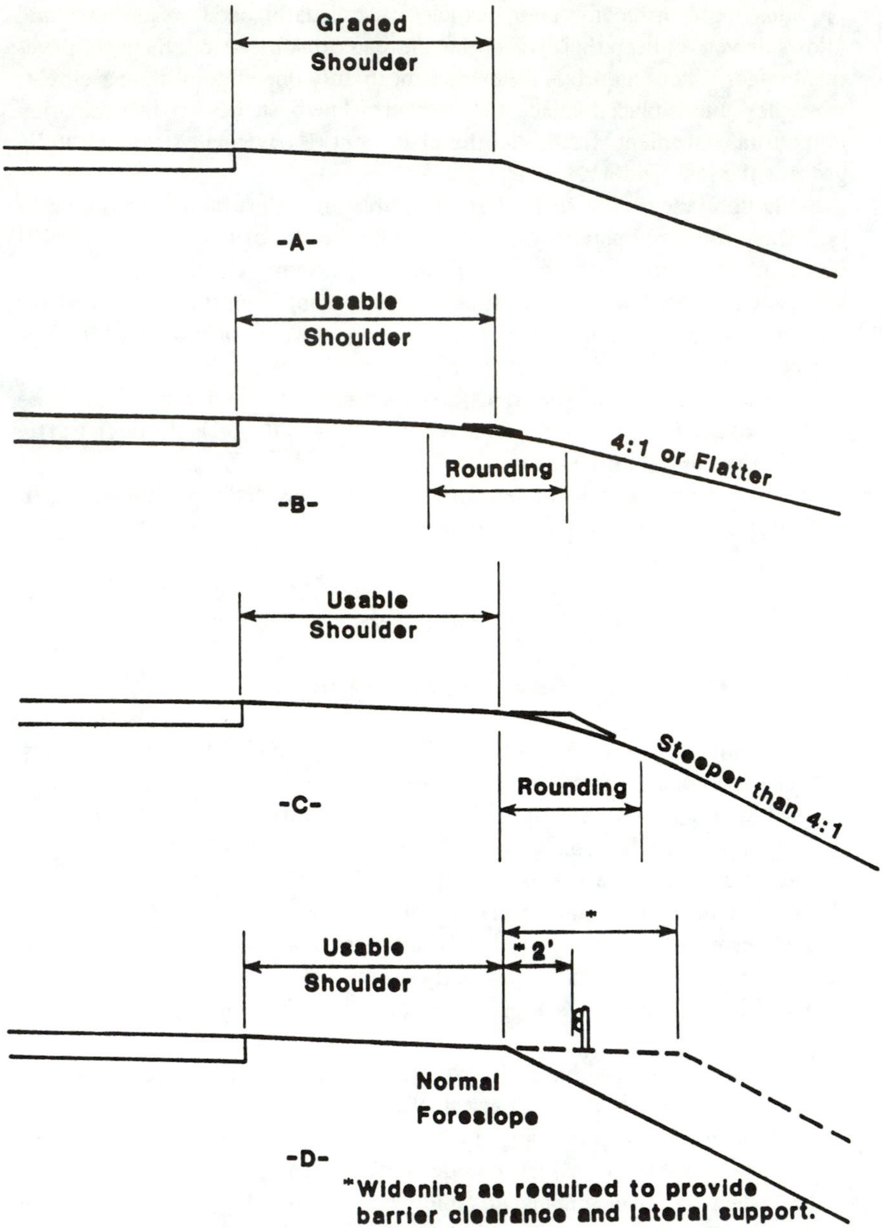

Figure IV-2. Graded and usable shoulders.

Shoulders may be surfaced either full-or partial width to provide a better all-weather load support than that afforded by the native soils. Materials used to surface shoulders include gravel, shell, crushed rock, mineral or chemical additives, bituminous surface treatments and various forms of asphaltic or concrete pavements.

The shoulder on minor rural roads with low traffic volume serves essentially as structural lateral support for the surfacing and as an additional width for the narrow traveled way. It permits drivers meeting or passing other vehicles to drive on the very edge of the roadway without leaving the surfacing, thus making use of the shoulder itself. Such operation is fitting only when meetings and passing occur infrequently. Where there is appreciable traffic volume, roads with narrow surfacing and narrow shoulders give poor service, have a relatively higher accident experience, and require frequent and costly maintenance.

Well-designed and properly maintained shoulders are necessary on rural highways with an appreciable volume of traffic, on freeways, and on some types of urban highways. Their more important advantages are as follows:

1. Space is provided for stopping free of the traffic lane because of mechanical difficulty, a flat tire, or other emergency.
2. Space is provided for the occasional motorist who desires to stop to consult road maps, to rest, or for other reasons.
3. Space is provided to escape potential accidents or reduce their severity.
4. The sense of openness created by shoulders of adequate width contributes much to driving ease and freedom from strain.
5. Sight distance is improved in cut sections, thereby improving safety.
6. Some types of shoulders enhance the esthetics of the highway.
7. Highway capacity is improved; uniform speed is encouraged.
8. Space is provided for maintenance operations such as snow removal and storage.
9. Lateral clearance is provided for signs and guardrails.
10. Storm water can be discharged farther from the pavement, and seepage adjacent to the pavement can be minimized. This may directly reduce pavement breakup.
11. Structural support is given to the pavement.
12. Space is provided for pedestrian and bicycle use.
13. Space is provided for bus stops.
14. Improved lateral placement of vehicles and space for occasional encroachment of vehicles is provided.

For further information on other uses of shoulders, refer to NCHRP 254 **(3)**.

Urban highways generally have curbs along the outer lanes. A stalled vehicle during peak hours disturbs traffic flow in all lanes in that direction when the outer lane serves through traffic. Where on-street parking is permitted, the parking lane provides some of the same services listed above for shoulders. Parking lanes are discussed further in the section "On-street Parking."

Width of Shoulders

Desirably, a vehicle stopped on the shoulder should clear the pavement edge by at least 1 ft, preferably by 2 ft. This preference has led to the adoption of 10 ft as the normal shoulder width that should be provided along high-type facilities. In difficult terrain and on low-volume highways, shoulders of this width may not be feasible. A minimum shoulder width of 2 ft should be considered for the lowest-type highway, and a 6 or 8 ft width would be preferable. Heavily traveled and high-speed highways and those carrying large numbers of trucks should have usable shoulders at least 10 ft and preferably 12 ft wide; however, widths greater than 10 ft may encourage unauthorized use as a travel lane. Shoulder widths for specific classes of highways are enumerated as parts of the total cross sections discussed in following chapters.

Where roadside barriers, walls, or other vertical elements are used, the graded shoulder should be wide enough that these vertical elements can be offset a minimum of 2 ft from the outer edge of the usable shoulder as shown in Figure IV-2D. It may be necessary to provide a graded shoulder wider than used elsewhere to provide lateral support for guardrail posts and/or clear space for lateral dynamic deflection required by the particular barrier in use. On low-volume roads, roadside barriers may be placed at the outer edge of the shoulder; however, a minimum of 4 feet should be provided from the traveled way to the barrier.

Although it is desirable that a shoulder be wide enough for a vehicle to be driven completely off the traveled way, narrower shoulders are better than none at all. When a vehicle making an emergency stop can drive onto the shoulder to occupy only 1 to 4 ft of a traveled way of adequate width, the remaining pavement width can be used by passing vehicles, although at considerable hazard. Partial shoulders are sometimes used where full shoulders are unduly costly, as on long bridges or in mountainous terrain.

Regardless of the width, a shoulder should be continuous. The full benefits of a shoulder are not available unless there is space where a driver can leave the pavement at any point. With a continuous shoulder almost all drivers making emergency stops will leave the pavement, but with intermittent sections of shoulder some drivers will find it necessary to stop on the pavement, creating a hazardous condition. Even though shoulders should be continuous, narrow shoulders and intermittent shoulders are superior to no shoulders. For the different conditions of major streets and ground-level parkways, parking bays will provide useful offlane stopping space where a continuous shoulder is not feasible. Intermittent shoulders, mentioned above, are discussed later.

Shoulders on structures should have the same width as the usable shoulders on the approach roadways, both right and left. This design is essential on freeways and is desirable on all arterials where shoulders are provided. As previously discussed, the narrowing or loss of shoulders, especially on structures, may cause serious operating and safety problems. Long, high-cost structures usually warrant detailed special studies to determine feasible dimensions. Wherever feasible, full shoulders should be included, but as has been indicated, for some rare cases it may be judged proper to use only partial-width shoulders.

Shoulder Cross Sections

Shoulders are important links in the lateral drainage systems. Shoulders should be flush with the roadway surface and abut the edge of the travel lane. On divided highways with a depressed median, all shoulders should be sloped to drain away from the traveled way. With a raised narrow median, the median shoulders may slope in the same direction as the traveled way, but in regions of heavy snowfall, median shoulders should be sloped to drain away from the traveled way. All shoulders should be sloped sufficiently to rapidly drain surface water but not to the extent that vehicular use would be hazardous. Because the type of shoulder construction has a bearing on the cross slope, the two should be determined jointly. Bituminous and concrete-surfaced shoulders should be sloped from 2 to 6 percent, gravel or crushed-rock shoulders from 4 to 6 percent, and turf shoulders about 8 percent. Where curbs are used on the outside of shoulders, the cross slope should be appropriately designed with the drainage system to prevent ponding on the traveled way.

Rigid adherence to the slope rates outlined presents a minor traffic operation problem when applied without regard to the cross section of the paved surface. On tangent or long-radius curved alinement with normal crown and turf shoulders, the maximum shoulder slope produces an algebraic difference in the pavement and shoulder grades of 6 to 7 percent at the edge of the paved surface. Although this is not desirable, it is tolerable due to the benefits gained in pavement stability by avoiding storm water detention at the pavement edge.

Shoulder slopes that drain away from the paved surface on the outside of well-superelevated sections should be designed to avoid too great a cross slope break. For example, use of a 4 percent shoulder cross slope in a section with a pavement superelevation of 8 percent results in a 12-percent algebraic difference in the pavement and shoulder grades at the high edge of pavement. Grade breaks of this order are not desirable and should not be permitted (Figure IV-3A). For desirable operation, all or part of the shoulder should be

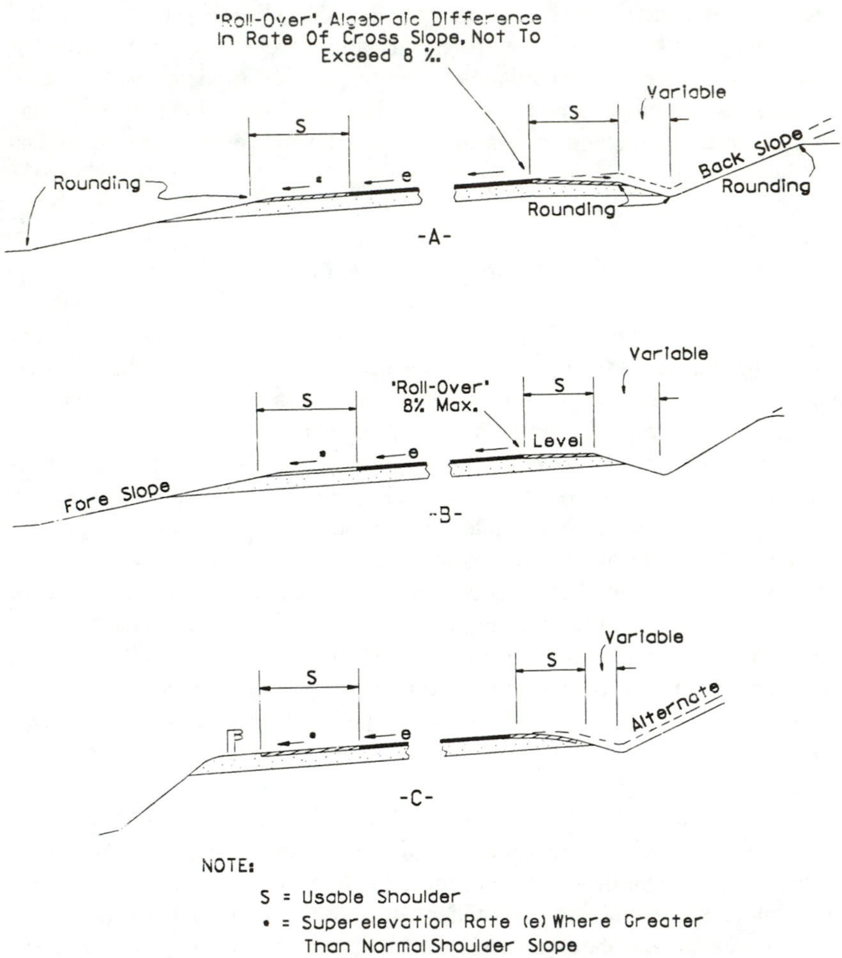

Figure IV-3. Typical cross section, superelevated.

sloped upward at about the same rate or at a lesser rate than the superelevated pavement (Figure IV-3A-Alternate). Where this is not desirable because of storm water and melting snow or ice draining over the paved surface, a compromise might be used in which the grade break at the edge of the paved surface is limited to approximately 8 percent by flattening the shoulder on the outside of the curve (Figure IV-3B). One means of avoiding too severe a grade break is the use of a continuously rounded shoulder cross section on the outside of superelevated pavement (Figure IV-3C). The shoulder in this case is a convex section continuing from the superelevation slope instead of a sharp grade break at the intersection of shoulder and pavement slope. In this method some surface water will drain upon the paved surface, but this disadvantage is balanced by the benefits to the vehicles that may accidentally or purposely be driven upon the shoulder. It should also be noted that convex shoulders present more construction difficulty than do planar sections. An alternate method to the convex shoulder consists of a planar shoulder section having multiple breaks in the cross slope. On the high side of a superelevated section, shoulder cross slopes substantially less than those previously given generally are not detrimental to shoulder stability; there is no discharge of storm water from the pavement to the shoulder and, therefore, little likelihood of shoulder erosion damage.

In some areas shoulders are designed with a curb or gutter at the outer edge to confine runoff to the surfaced shoulder area. Drainage for the entire roadway is handled by these curbs, with the runoff directed to selected outlets. The shoulder necessarily is surfaced, the outer portion serving as the longitudinal gutter. Cross slopes are as previously indicated, except that the slope may be increased somewhat on the outer portion. This type of shoulder is advantageous in that the curb on the outside of the shoulder does not deter motorists from driving off the traffic lanes, and the shoulder serves as a gutter in keeping the storm water off the traveled lanes. In fact, unless carefully engineered for color and texture, in contrast to the pavement proper, it may appear inviting as a through-traffic lane.

Shoulder Stability

If shoulders are to function effectively, they must be sufficiently stable to support occasional vehicle loads in all kinds of weather without rutting. Evidence of rutting, skidding, or vehicles mired down, even for a brief seasonal period only, is sufficient to discourage and to prevent the shoulder from being used as intended.

All types of shoulders should be constructed and maintained flush with the paved surface if they are to fulfill their intended function. Unstabilized shoulders generally undergo consolidation with time and the elevation of the shoulder at the pavement edge tends to become 0.5 inch to several inches lower than the pavement. Regular maintenance is necessary to provide a flush shoulder. The dropoff can adversely affect driver control when driving onto the shoulder at any appreciable speed. In addition, there is loss of the operational advantage of drivers staying close to the edge of the pavement when there is not visible assurance of a flush stable shoulder.

Paved or stabilized shoulders offer numerous advantages. They provide (1) area for vehicles to pull onto during emergency situations, (2) elimination of rutting and dropoff adjacent to the edge of travel lane, (3) adequate cross slope for drainage of roadway, (4) reduced maintenance, and (5) lateral support for roadway base and surface course.

Shoulders with turf growth may be appropriate, under favorable climatic and soil conditions, for local roads and some collectors. Where climate and soil conditions permit, turf growth may be developed in conjunction with compacted granular material that is suitably stable for occasional use, even in wet weather. Turf shoulders are subject to a buildup which may inhibit proper drainage of the pavement unless adequate cross slope is provided. When wet, the turf may be slippery unless closely mowed and on granular soil. Turf shoulders offer good pavement delineation and do not invite use as a traffic lane. Stabilized turf shoulders require little maintenance other than mowing.

Because of experience on unstabilized shoulders, the traveling public is wary of driving on them. Experience on high-volume highways, such as suburban expressways, has led to replacement of unstabilized shoulder with some form of a stabilized or surfaced shoulder.

In some areas, main rural highways are built with surfacing over the whole width, including shoulders. Depending upon the conditions, this surfacing may be from about 28 to 44 ft wide for two-lane roads. This type of treatment protects shoulders from erosion and also protects the subgrade from moisture penetration, thereby enhancing the strength and durability of the pavement. Where feasible, the actual shoulder width is made to contrast, but in some cases there is no indication of the edge of traveled way. Such design obviously provides a desired continuous shoulder, even if its separate width is not apparent. Pavement edge stripes are generally used to delineate the edge of the traveled way.

Experience on heavy-volume facilities shows that, on occasion, traffic will use smooth-surfaced shoulders as through traffic lanes. On moderate-to-steep grades trucks may pull to the right and encroach upon the shoulder. Obviously, the elimination of the surfaced shoulder is not the answer because it is the high volume and type of traffic that in part provide the warrant for a high-type shoulder.

Shoulder Contrast

It is desirable that the color and texture of shoulders be different from those of the through-traffic lanes. This contrast serves to clearly define the through-traffic lanes at all times, particularly at night and during inclement weather, while discouraging the use of shoulders as additional through lanes. Bituminous, crushed stone, gravel, and turf shoulders all offer excellent contrast with concrete pavements. Satisfactory contrast with bituminous pavements is more difficult to achieve. Various types of stone aggregates and turf offer good contrast. Several states have attempted to achieve contrast by seal-coating shoulders with lighter color stone chips. Unfortunately, the color distinction may diminish in a few years. The use of pavement edge lines as described in the MUTCD (**2**) reduces the need for shoulder contrast. Pavement edge lines should be applied where shoulder use by bicycles is expected.

Intermittent Shoulders or Turnouts

It will not always be economically feasible to provide desirably wide shoulders continuously along the highway through deep rock cuts or where other conditions limit the cross section width. In such cases, consideration should be given to use of intermittent sections of shoulder or turnouts along the highway. Such turnouts provide an area for emergency stops and also allow slower moving vehicles to pull out of the through lane to permit following vehicles to pass.

Proper design of turnouts should consider turnout length, including entry and exit tapers, turnout width and the location of the turnout with respect to horizontal and vertical curves which limit sight distance. Turnouts should be located so that approaching drivers will have a clear view of the entire turnout in order to determine whether the turnout is available for use (**29**).

HORIZONTAL CLEARANCE TO OBSTRUCTIONS

The term "clear zone" is used to designate the unobstructed, relatively flat area provided beyond the edge of the traveled way for the recovery of errant vehicles. The traveled way does not include shoulders or auxiliary lanes.

The width of the clear zone is influenced by the traffic volume, speed and embankment slopes. The AASHTO *Roadside Design Guide* (**12**) discusses clear zone widths as related to speed, volume and embankment slope. The *Guide* may be used as a reference for determination of clear zone for freeways, rural arterials and high-speed rural collectors. For low-speed rural collectors and rural local roads, a minimum clear zone width of 10 ft should be provided.

For urban arterials, collectors and local streets where curbs are utilized space for clear zones is generally restricted. A minimum distance of 1.5 feet should be provided beyond the face of the curb, with wider clear zones provided where possible. Where shoulders are provided rather than curbs, a clear zone commensurate with rural conditions should be provided.

CURBS

General Considerations

The type and location of curbs affects driver behavior and, in turn, the safety and utility of a highway. Curbs serve any or all of the following purposes: drainage control, pavement edge delineation, right-of-way reduction, esthetics, delineation of pedestrian walkways, reduction of maintenance operations, and assistance in orderly roadside development. To be considered a curb, some raised aspect or vertical element is required.

Curbs are used extensively on all types of urban highways. On rural highways, caution should be exercised in the use of curbs. In the interest of safety, curbs should be omitted on rural highways when the same objective can be attained by other acceptable means. For the intermediate design speed range, between 40 and 50 mph, the conditions of the roadway should control. If the location is predominantly urban or rapidly developing urban, a curb may be acceptable at the intermediate speed. On the other hand, if typically rural conditions predominate, the high speed design controls should prevail.

Types of Curbs

The two general classes of curbs are barrier curbs and mountable curbs and each has numerous types and detail designs. Each may be designed as a separate unit or integrally with the pavement. Barrier and mountable curb designs may include a gutter to form a combination curb and gutter section.

Barrier curbs are relatively high and steep-faced, designed to inhibit or at least discourage vehicles from leaving the roadway. As shown in Figure IV-4A, they range from 6 to 9 in. in height. A sloping face is desirable, but generally the face batter does not exceed about 1 in. per 3 in. of height. The upper corner may be rounded on a radius from 0.5 to 1 in.

Barrier curbs should not be used on freeways and are considered undesirable on other high-speed arterials. Generally, barrier curbs should not be used where design speeds exceed 40 mph, except as noted above in the intermediate design speed range. When impacted at high speeds, it is difficult for the operator to retain control of the vehicle. In addition, barrier curbs are not adequate to prevent a vehicle from leaving the roadway. Where positive action is required, such as along narrow medians or adjacent to bridge substructures, a suitable traffic barrier should be provided.

Generally, barrier curbs should not be used with traffic barriers. Curbs may contribute to vehicle vaulting of all types of traffic barriers and they are not compatible with yielding traffic barriers. However, traffic barriers may be appropriate on low-speed streets where curbs may be used in conjunction with sidewalks placed on the traffic side of railings on structures. It should be recognized in such instances that the barrier curb has limited capacity as a shield for pedestrians, but acts to discourage the mingling of vehicular and pedestrian traffic.

Barrier curbs and safety walks may be desirable along the faces of long walls and tunnels, particularly if full shoulders are not provided. The curb tends to prevent vehicles from scraping the wall, and the safety walk reduces the danger to persons walking from disabled vehicles.

Mountable curbs are designed so that vehicles can cross them readily when required. As shown in Figures IV-4B to IV-4G, they are low with flat sloping faces. The curbs shown in Figures IV-4B, IV-4C, and IV-4D are considered to be mountable under emergency conditions. Although mountable under emergency conditions, these curbs will drag the undersides of some vehicles. When the face slope is steeper than 1:1 and a mountable curb is desired, the height should be limited to 4 in. maximum, preferably less. When the face slope is intermediate between 1:1 and 2:1, the height should be limited to about 6 in. For ease in crossing, mountable curbs should be well rounded as in Figure IV-4B to IV-4G. Some highway agencies construct a vertical section on the lower face (Figures IV-4C, IV-4D, and IV-4F) as an allowance for future resurfacing. This vertical portion should not exceed about 2 in., and where the total height exceeds 6 in. it might not be considered a mountable curb.

Extruded curbs of either portland cement or asphaltic concrete are being built in many states. Extruded curbs usually have sloping faces because they provide better initial stability, are easier to construct, and are more economical than steep faces. Typical extruded designs are shown in Figures IV- 4C, IV-4E, and IV-4G.

Precast portland cement concrete curbs are installed by some highway agencies. Granite curbs are used where the local supply makes it competitive. Because of its durability, granite is preferred over portland cement concrete where deicing chemicals are used for snow and ice removal.

Mountable curbs can be used at median edges to outline channelizing islands in intersection areas or sometimes at the outer edge of the shoulder. Curbs should not be placed adjacent to high-speed lanes unless they serve a definite purpose. Any of the mountable sections in Figure IV-4 might be used for a median curb. Where drainage curbs are used in conjunction with metal median barriers, they should be installed directly under or behind the barrier. Drainage curbs should not be used with concrete median barriers. Improperly placed curbs may cause errant vehicles to vault a barrier or to strike it so that the vehicle overturns.

Shoulder curbs are mountable curbs placed at the outer edge of the shoulder to control drainage, improve delineation, and reduce erosion. They may be part of the longitudinal drainage system being combined with a gutter sec-

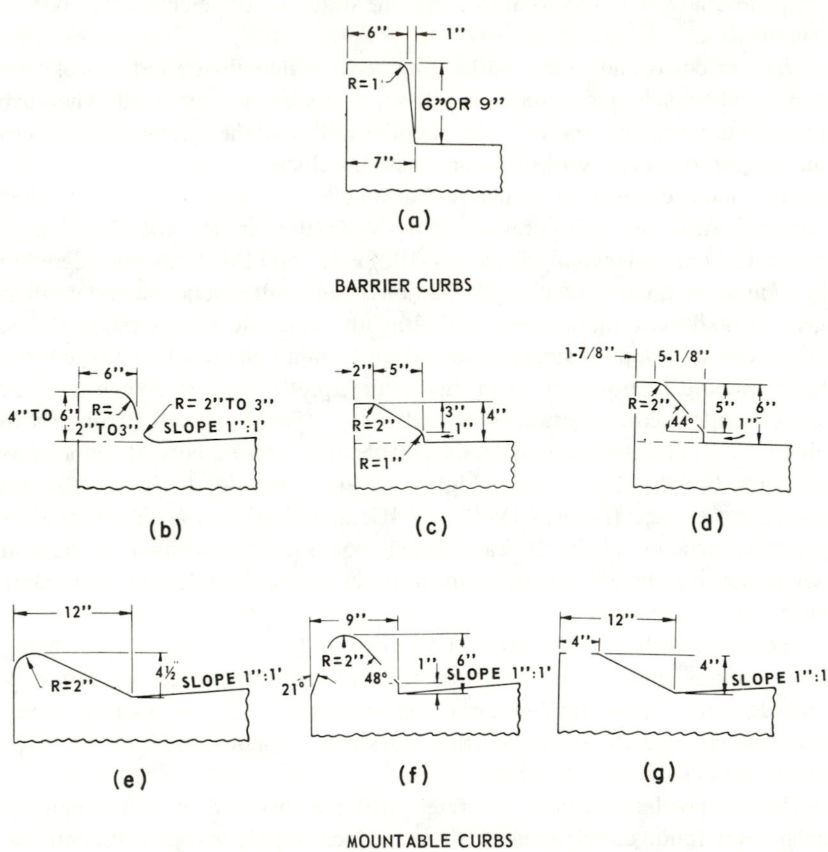

Figure IV-4. Typical highway curbs.

tion. All of the mountable sections are applicable for this use. If the surfaced shoulders are not wide enough for a vehicle to park, the shoulder curb must appear to be easily mountable to encourage motorists to park clear of the traveled way.

Gutter sections may be provided on the traveled-way side of a barrier or mountable curb to form the principal drainage system for the roadway. Inlets are provided in the gutter or the curb, or both. Gutters are generally 1 to 6 ft in width, with a cross slope of 5 to 8 percent to increase the hydraulic capacity of the gutter section. In general, the 5- to 8-percent slope should be confined to the 2 or 3 ft adjacent to the curb. Shallow dished gutters without a curb have small flow capacity and have limited value for drainage. Generally, it is not practical to design gutter sections to contain all of the runoff even from frequent rains, and some overflow onto the traveled surface can be expected. The spread of water on the traveled way is kept within tolerable limits by the proper size and spacing of inlets. Grate inlets and depressions for curb-opening inlets should not be placed in the traveled lane because of their adverse effect on drivers who veer away from them. Warping of the gutter for curb-opening inlets should be limited to the portion within 2 to 3 ft of the curb to minimize adverse driving effects.

The width of barrier and mountable curbs are considered as cross section elements entirely outside the traffic lane width. When a gutter has the same surface color and texture as the traffic lane, and it is not much steeper in cross slope than the adjoining pavement, it might be considered part of the traffic lane. This arrangement is used frequently in urban areas where undesirable right-of-way width would result from consideration of gutters as elements outside the traffic lane width. However, with any form of curb there is some lateral shying distance by drivers, particularly on their right, which reduces effective through-lane width. A gutter with an evident longitudinal joint and somewhat steeper cross slope than the adjacent lane is a greater deterrent to driving near the gutter than the situation in which the traffic lane and gutter are integral. A gutter of contrasting color and texture should never be considered part of the traffic lane width.

High-visibility curbs are advantageous at night, particularly in areas subject to fog or extended rains. Conventional concrete or bituminous curbs offer little contrast in visibility to normal pavements, particularly at night and when wet. Channelizing island curbs or continuous curbs along through-lane edges may be constructed with reflective surfaces. Curbs of special design with depressions and ribs to reflect headlight beams are visible during heavy rains when delineation is needed most and when plain curbs may be barely noticeable.

The higher installation cost and maintenance cleaning problems result in only limited use of such curbs.

In another form of high-visibility treatment, curbs are made conspicuous by painting and striping or by use of reflectorized paints or applied surfaces such as thermoplastic on conventional curb types. Reflector buttons have been installed on a limited basis on continuous curbs, particularly on narrow medians. These devices, however, are not in general use. For better visibility of continuous median curbs, some use is being made of curbs with faces or tops painted with reflectorized white or yellow paint. Painted curbs are not subject to traffic wear but they are subject to splash and dirt. To be kept fully effective, they need frequent cleaning or repainting, which entails a significant maintenance cost. Curb markings are to be in accordance with the MUTCD **(2)**.

Curb Placement

Mountable curbs at the edge of traffic lanes may have some effect on lateral placement of moving vehicles, depending on their shape and appearance. Those with low, sloping faces may encourage drivers to operate with wheels relatively close to them, but others should have some offset consideration. Curbs at the pavement edge, although considered mountable in emergencies, can be mounted satisfactorily only at reduced speed, thus discouraging many drivers from using the shoulder when stopping. To encourage proper use of the shoulders, any mountable curbs should preferably be placed at or beyond the outer shoulder edge. For low speed street conditions, mountable curbs may be placed at the edge of a through lane, although it is preferable that the curbs be offset 1 or 2 ft.

Data **(4)** on the lateral placement of vehicles with respect to high barrier curbs show that drivers shy away from curbs that are high enough to damage the body and fenders of vehicles. The exact relationship is not known precisely, but it is known that the lateral placement varies with the curb height and steepness and the position of other obstructions outside the curb. The average clearance to the curb is somewhat greater where the curb is first introduced than where the curb is continuous for some distance. The shying away at the beginning of the curb will be lessened if the curb is introduced with the end flared away from the pavement edge.

Neither barrier nor mountable curb should be used on freeways or other high-speed arterials, but if provided in special cases, the curb should not be closer than the outer edge of the shoulder, and a sloped end treatment must be provided. On intermediate speed roadways, the use of curbs should be based upon how nearly the other characteristics of the facility relate to urban or rural conditions.

Barrier curbs introduced intermittently along streets should be offset 2 ft from the edge of pavement; where continuous, as along a median, barrier curbs should be offset at least 1 ft, and preferably 2 ft.

WALKWAYS

Sidewalks are integral parts of city streets, but few are provided in rural areas. Yet, a need exists in many rural areas because the higher speed and general absence of lighting increase the accident potential to those walking on or adjacent to the traveled way. The limited data available suggests that sidewalks in rural areas do reduce pedestrian accidents.

Sidewalks in rural and suburban areas are most justified at points of community development such as at schools, local businesses, and industrial plants that result in pedestrian concentrations near or along the highways. When suburban residential areas are developed, initial facilities for vehicles are necessary if the development is to function at all, but the construction of sidewalks may be deferred. The cost of sidewalks should be justified by a study of the local conditions. Shoulders might obviate the need for sidewalks if they are of a type to encourage pedestrian use in all weather, but if sidewalks are justified, they should be separated from the shoulder. If the sidewalk is raised above the level of the shoulder, the cross section approaches that of an urban highway.

In suburban and urban locations a border area generally separates the roadway from the homes and businesses of the community. The main function of the border is to provide space for sidewalks. Other functions of the border are to provide space for street lights, fire hydrants, street hardware, and esthetic vegetation and to serve as a buffer strip. The width of border varies considerably, with 8 ft considered a minimum. Swale ditches may be located in these borders to provide an economical alternative to curb and gutter sections.

Sidewalks in residential areas may vary from 4 to 8 ft. The width of a planted strip between the sidewalk and traveled way, if provided, should be a minimum of 2 ft to allow maintenance activities. Sidewalks covering the full border width are generally justified and often required in commercial areas, through adjoining multiple-residential complexes, near schools and other pedestrian generators, and where border width is restricted.

Where sidewalks are placed adjacent to the curb, the walkway widths should be approximately 2 ft wider than those widths used when a planted strip separates the walk and curb. This space provides for street and highway hardware and allows for the proximity of moving traffic or the opening of doors of parked cars.

Justification for the construction of sidewalks depends upon the vehicle-pedestrian hazard, which is governed chiefly by the volumes of pedestrian and vehicular traffic, their relative timing, and the speed of vehicular traffic. Traffic volume-pedestrian warrants for sidewalks along highways are not established. In general, wherever the roadside and land development conditions are such that pedestrians regularly move along a main or high-speed highway, they should be furnished a sidewalk or path area, as suitable to the conditions.

As a general practice, sidewalks should be constructed along any street or highway not provided with shoulders, even though pedestrian traffic may be light. Where sidewalks are built along a rural highway, they should be well removed from the traveled way.

To insure their intended use, sidewalks should have all-weather surfaces. Without them, pedestrians often choose to use the traffic lanes.

Pedestrian crosswalks regularly are marked in urban areas but seldom so on rural highways. However, where there are pedestrian concentrations, appropriate traffic control devices should be used, together with necessary walkways constructed within the right-of-way.

If two urban communities are not far apart, consideration should be given to connecting the two communities with sidewalks, even though pedestrian traffic may be light. The additional psychological hazard of drivers meeting pedestrians on these sections of a through route thus may be avoided.

Pedestrian facilities such as walkways and sidewalks must be designed to accommodate the physically handicapped. See the section "Pedestrian Crossings" later in this chapter for further discussion on this point.

In general the guidelines set forth in this section for the accommodation of pedestrians are applicable on bridges. However, because of the high cost of bridges and some operational features that may be unique to bridge sites, pedestrian way details on a bridge will often differ from those on its approaches. For example, where a planted strip between a sidewalk and the traveled way approaches a bridge, continuation of the sidewalk offset effected by the planted strip will seldom be justified.

In rural areas or other sites where flush shoulders approach a bridge and light pedestrian traffic is anticipated on the shoulders, the shoulder width should be continued across the bridge and possibly even increased to account for the restriction to pedestrian escape imposed by the bridge rail. A flush roadway shoulder should never be interrupted by a raised walkway or a bridge. Where such installations already exist, and removal is not economically justified, the ends of the walkway should be shielded with a traffic barrier or ramped into the shoulder at a rate of approximately 20:1 with the grade of the shoulder.

On long bridges, if accommodation of pedestrian traffic is considered essential, a single walkway may be provided. However, care must be taken to ensure that approach walkways provide safe and relatively direct access to the bridge walkway. This may require the erection of fences to channelize the pedestrians and prevent or control pedestrian conflicts with vehicular traffic.

For a discussion of the problems sometimes associated with the introduction of a traffic barrier between a roadway and a walkway, see the section on "Bridge Rails," under "Traffic Barriers," in this chapter.

DRAINAGE CHANNELS AND SIDESLOPES

General Considerations

Modern highway drainage design should incorporate safety, good appearance, control of pollutants, and economy in maintenance (and frequently in construction). The above may be direct benefits of flat sideslopes, broad drainage channels, and liberal warping and rounding. These features avoid obsolescence, improve appearance, and invite favorable public reaction.

An important part of highway design is consistency, which prevents discontinuities in the highway environment and considers the interrelationship of all highway elements.

The interrelationship between the drainage channel and the sideslopes is important for safety because of their great influence on the sequence of events that can occur when a vehicle leaves the traveled way.

Drainage Channels

Drainage channels perform the vital function of collecting and conveying surface water from the highway right-of-way. Drainage channels, therefore, should have adequate capacity for the design runoff, should provide for unusual storm water with minimum damage to the highway, and should be located and shaped to avoid creating a hazard to traffic. Channels should be protected from erosion with the lowest cost protective lining that will withstand the flow velocities expected. Channels should be kept clean and free of material that would lower the capacity of the channel. Channel deterioration can reduce channel capacity and overflow may occur, often with erosion or deposition in the area adjacent to the channel.

Where the construction of a highway would have an adverse effect on drainage conditions downstream, drainage channels can be an effective means of flood storage within the highway right-of-way and thus mitigate adverse effects of flooding.

Drainage channels include (1) roadside channels in cut sections to remove water from the highway cross section, (2) toe-of-slope channels to convey the water from any cut section and from adjacent slopes to the natural watercourse, (3) intercepting channels placed back of the top of cut slopes to intercept surface water, and (4) chutes to carry collected water down steep cut or fill slopes.

The primary purpose for construction of roadside channels is to control surface drainage, and the most economical method usually entails formation of open-channel ditches by cutting into the natural roadside terrain to produce a drainage channel. From a standpoint of hydraulic efficiency, the most desirable channel contains steep sides. Limitations on slope stability usually require somewhat flatter slopes, and construction and maintenance factors have imposed still further restrictions on the degree of slope steepness that is practical alongside a highway. The offsetting factor of right-of-way costs must also be considered when selecting combinations of slopes to be used. The effect of slope combinations and safety during traversal by an errant vehicle is also an important consideration in designing the roadside.

In general, the severity of traversal of roadside channels less than about 4 to 8 ft wide is essentially the same for comparable slope combinations regardless of channel shape. Slope combinations forming these narrow channels can be selected to produce cross sections that can be safely traversed by an unrestrained occupant, the range of slopes being within the steepness usually found along highways.

The use of foreslopes steeper than 4:1 (horizontal to vertical) severely limits the range of backslopes. The flatter the foreslope the greater the choices of combinations of fore- and backslope that permit safe traversal. The flatter foreslope also provides greater recovery distance for an errant vehicle.

The depth of channel should be sufficient to remove the water without saturation of the pavement subgrade. The depth of water that can be tolerated, particularly on flat channel slopes, depends upon the soil characteristics. In open country of the northern regions, channel sideslopes of 5:1 or 6:1 are preferable to reduce snow drifts.

The broad, flat, rounded drainage channel also provides a sense of openness that relaxes driver tension. With a channel sideslope of 4:1 or flatter and a 10-ft shoulder, the whole roadside channel is visible to the driver. This lessens the feeling of restriction and adds measurably to the driver's willingness to use the shoulder in an emergency.

The desirable minimum grade for channels should be based upon obtaining velocities to avoid sedimentation. Flood depths should be limited to preclude saturation of the subgrade. The maximum desirable grade for unpaved channels should be based upon a tolerable velocity for vegetation and shear on soil types. AASHTO's *Highway Drainage Guidelines* **(5)** provides guidance in this area. The channel grade does not have to follow that of the roadbed, particularly if the roadbed is flat. Although desirable, it is unnecessary to standardize the design of roadside drainage channels for any length of highway. Not only can the depth and breadth of the channel be varied to meet different amounts of runoff, slopes of channel, types of lining, and distances between discharge points, but the lateral distance between the channel and the edge of the pavement can also be varied. Usually liberal offsets can be obtained where cuts are slight and where cuts end and fills begin. Care should be taken, however, to avoid abrupt major changes in the roadway section that would result in such a discontinuity of the highway environment as to violate the driver's expectancy. Care should also be taken to avoid major breaks in channel grade that would cause unnecessary scour or silt deposition.

Intercepting channels generally have a flat cross section preferably formed by a dike made with borrow material to avoid disturbing the natural ground surface. Intercepting channels should have ample capacity for the water they must carry and should be located with as little deviation from the contour as possible except when on top of a slope subject to sliding. In slide areas, storm water should be intercepted and removed as rapidly as practicable. Sections of channels that cross highly permeable soil might require lining with impermeable material.

Median drainage channels are generally shallow depressed areas, or swales, at or near the center of the median formed by the flat sideslopes of the divided roadways. The swale is sloped longitudinally for drainage, and at intervals the water is intercepted by inlets or transverse channels and discharged from the roadway in storm drains or culverts. Flat, traversable drainage dikes are sometimes used to increase the efficiency of the inlets. Refer to the section on medians for further discussion. Safety grates on median drains and cross drains, while enhancing safety for errant vehicles, can reduce the hydraulic efficiency of the drainage structures if not properly designed. The reduced inlet capacity is compounded by the accumulation of debris on the grates, occasionally resulting in pavement flooding. If significant hydraulic capacity or clogging problems could result from the use of grates, other methods, such as shielding the structure, should be considered.

Chutes or pipes are used to carry water collected by intercepting channels down cut slopes and to discharge the water collected by shoulder curbs. They can be either open channels or pipes. High velocities preclude sharp turns in open chutes and generally require some means of dissipating the energy of flow at the outlet of the chute. Closed chutes are preferred to avoid failure due to settlement and erosion. Generally in highly erodible soil, it is necessary to provide watertight joints to prevent failure of the facility. Caution must be exercised to avoid splash, which causes erosion.

Prevention of erosion in channels is accomplished by linings that withstand the velocity of storm runoff. The type of lining used in roadside channels depends upon the velocity of flow, type of soil, and the grade and geometry of the channel. In areas where grass will grow, it is usually the most economical channel lining except on steep slopes where the velocity of flow exceeds the permissible velocities for grass protection. Various materials including concrete, asphalt, stone, and nylon are used for linings where grass will not provide adequate protection. Smooth linings generate higher velocities than rough linings such as stone and grass. Provision must often be made to dissipate the energy of the high-velocity flow before it is released to avoid scour at the outlet and damage to the channel lining. If erosive velocities are developed because of smooth linings or steep grades, a special channel design or energy dissipater may be required.

Reference should be made to AASHTO's *Highway Drainage Guidelines* **(5)**, drainage design manuals, handbooks and publications of the Soil Conservation Service, Corps of Engineers, and Bureau of Reclamation for details as to design and protective treatments, including filter requirements. In addition, the publications made available to highway agencies from the FHWA, such as *Design of Stable Channels With Flexible Linings* **(17)**, provide excellent references. (See Chapter III.)

Sideslopes

Sideslopes should be designed to insure the stability of the roadway and to provide a reasonable opportunity for recovery for an out-of-control vehicle.

Three regions of the roadside are important when evaluating the safety aspects: the top of the slope (hinge point), the foreslope, and the toe of the slope (intersection of the foreslope with level ground or with a backslope, forming a ditch). Figure IV-5 illustrates these three regions.

The hinge point contributes to loss of steering control because the vehicle tends to become airborne in crossing this point. The foreslope region is important in the design of high slopes where a driver could attempt a recovery maneuver or reduce speed before impacting the ditch area. In many situations the toe of the slope is within the clear zone and the probability of reaching the ditch is high, in which case safe transition between fore- and backslopes should be provided.

Research **(6)** in these three regions of the roadside has found that rounding at the hinge point, though not necessary from a vehicle rollover standpoint, can significantly reduce its hazard potential. Rounded slopes reduce the chances of an errant vehicle becoming airborne, thereby reducing the hazard of encroachment and affording the driver more control over the vehicle. Foreslopes steeper than 4:1 are not desirable because their use severely limits the choice of backslopes. Slopes 3:1 or steeper are recommended only where site conditions do not permit use of flatter slopes. When slopes steeper than

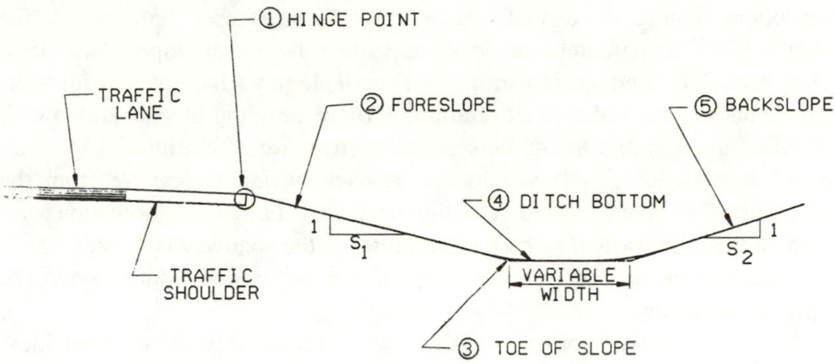

Figure IV-5. Designation of roadside regions.

3:1 must be used, consideration should be given to the use of a roadside barrier.

Another safety factor of significance for intersecting roadways is the angle of break between a sideslope and a transverse slope. Field observations indicate that more consideration should be given in roadway design to carrying the desirable flat sideslopes through intersections, driveway approaches, median crossovers, and cut sections. Providing a flatter slope between the shoulder edge and the ditch bottom, locating the ditch a little farther from the roadway, or even enclosing short sections of drainage facilities will enhance the safety of the roadside, often at a small increase in cost.

Earth cut and fill slopes should be flattened and liberally rounded as fitting with the topography and consistent with the overall type of highway. Effective erosion control, low-cost maintenance, and adequate drainage of the subgrade largely are dependent upon proper shaping of the sideslopes. Slope and soil data are used in combination to approximate the stability of the slopes and the erosion hazard potential. Overall economy depends not only on the element of first cost but also on the cost of maintenance, of which slope stability is a factor. In addition to these reasons for flat and rounded slopes on any highway, the proximity of any urban highway to the developments and people of the community call for additional attention to naturalized slopes and overall appearance factors.

Normally, backslopes should be 3:1 or flatter, to make it easier for motorized equipment to be used in maintenance. In developed areas, sufficient space may not be available to permit the use of desirable slopes. Backslopes steeper than 3:1 should be evaluated with regard to soil stability and traffic safety. Retaining walls of partial height should be considered where space restrictions require slopes that otherwise would be steeper than 2:1. On the other hand, soil characteristics may necessitate the use of slopes flatter than 2:1 or even 3:1. If adequate width is not available in such cases, partial walls may be necessary. The type of retaining structure should be compatible with the area traversed and the grade separation structures. To minimize the feeling of constriction, walls should be set back as far as feasible from the through-traffic lanes. Where retaining walls are used in combination with earth slopes, the walls may be located either at the roadway level adjacent to the shoulder or on the outer portion of the separation width above the depressed roadway.

On freeways and other arterials with reasonably wide roadsides, sideslopes on embankments and in cuts should be designed to provide a reasonable opportunity for recovery of an out-of-control vehicle. Where the roadside at the point of departure is reasonably flat, smooth, and clear of fixed objects, many potential crashes can be averted. A rate of slope of 6:1 or flatter on embankments can be negotiated by a vehicle with a good chance of recovery and should, therefore, be provided where feasible. For moderate heights with good roundings, steeper slopes up to about 3:1 can also be traversable. On intermediate-height fills, the cost of a continuous flat slope may be prohibitive, but it may be feasible to provide a recovery area with a reasonably flat and rounded area adjacent to the roadway. The recovery area should extend well beyond the edge of the shoulder as specific conditions may permit.

Consistent with traffic demand, roads and streets with wide borders should also be designed with a similar clear roadside recovery area approach. However, because of generally lower speeds and narrower side clearances along streets, the clear roadside area concept, at best, can only be partially used. This is also true for widening and other reconstruction within available limited right-of-way.

Desirably, slope combinations would be selected so that unrestrained occupants could be expected to sustain no injury and the vehicle would not incur major damage during traversal. However, site conditions such as restricted right-of-way or the cost effectiveness of such design may dictate the use of slope combinations steeper than desirable. If constraints other than safety dictate that the necessary recovery distance cannot be provided, the need for roadside barrier should be considered.

Where height and slope of roadway embankments are such that the severity of potential accidents will be reduced by the placing of a roadside barrier, the cross section should be designed to allow adequate slope rounding and to support the barrier.

Flat and well-rounded sideslopes simplify the establishment of turf and its subsequent maintenance. Grasses usually can be readily established on sideslopes as steep as 2:1 in favorable climates and 3:1 in semiarid climates. With slopes of 1.5:1 and steeper, it is difficult to establish turf even in areas of abundant rainfall. Because of the greater velocity of runoff, sufficient water for the maintenance of grass does not seep into the soil. Deep-rooted plants that do not depend upon surface water alone may be necessary where slopes are excessively steep. Slopes of the order of 3:1 and flatter can be machine mowed. Although steeper slopes reduce the mowing area considerably, the slow, time-consuming hand methods required add materially to the cost of maintenance.

With some types of soils, it is virtually mandatory that slopes be reasonably flat to gain the necessary stabilization. Soils predominantly clay or gumbo are particularly susceptible to erosion, and slopes of 3:1 or flatter should be used.

The intersections of slope planes in the highway cross section should complement the earth forms of the terrain being traversed. Some earth forms are well rounded and others steeply sloped; the designer should strive to create a natural look that is aesthetically pleasing. The rounded forms in nature are the result of erosion, the wearing away by water or wind; the use of well-rounded forms in the highway cross section will go far toward insuring stability.

Flat, well-rounded sideslopes are major elements in attaining the desired natural appearance; however, a uniform slope through a cut or fill section often results in a formal or stilted appearance. This appearance can be softened and made more natural by flattening the slopes on the ends where the cut or fill is light and by gradually steepening it toward the controlling maximum slope of the cut or fill. This design may be readily accomplished by liberal rounding of the hinge point in the transition area. On short cut or fill sections the result may be one of continuous longitudinal rounding, whereas on sections of substantial length the effect will be one of funneling. The transitioning of sideslopes is especially effective at the ends of cuts when combined with an increased lateral offset of the drainage channel and a widened shoulder.

The combination of flat slopes and rounding is frequently referred to as streamlined cross section. With this shape the cross winds sweep along the surface without forming eddies that contribute to the wind erosion and drifting of snow. The streamlined cross section usually results in a minimum of expenditure for snow removal because the winds blow the snow off the pavement instead of drifting it, as happens in cross sections with steep slopes without rounding. When combined with the design practice of making the level of the roadway above that of the surrounding land to also insure drainage of the subgrade, the streamlined cross section results in a road requiring a minimum of maintenance and operating costs and a maximum of safety.

In some cases an irregular slope stake line results from the strict adherence to specified cut or fill slopes. It may be more esthetically pleasing to vary the slope to yield a neat stake line.

Design slopes for rock vary widely, depending upon the materials. A commonly used standard slope for rock cuts is one-half to one. With modern construction methods such as presplitting, slopes ranging as steep as one-sixth to one to vertical may be used in good-quality rock. Deep cuts in rock often require the construction of benches in the slopes.

Slope stability as well as appearance may be enhanced in poor-quality rock by the establishment of vegetative cover. In some parts of the country, serrated cut slopes aid in the establishment of vegetative cover on decomposed rock or shale slopes. Serration **(7)** may be constructed in any material that can be ripped or that will hold a vertical face long enough to establish vegetation.

Desirably, the toe of the rock-cut slope should be located beyond the minimum lateral distance from the edge of the traveled way needed by the driver of an errant vehicle to either regain control and begin a return to the roadway or to slow the vehicle to a safe speed. Wide shelves at the bottom of rock cuts have advantages in that a safe landing area is provided for falling boulders and space is available for snow storage in colder climates. This width can also be shaped to provide a clear roadside recovery area.

Rock outcroppings have sometimes been left in place during construction of new highways for reasons of economy or esthetics. These are potentially hazardous if struck by an automobile and should be eliminated within the clear roadside recovery area where removal is feasible. If they cannot be removed, they should be shielded by the installation of a roadside barrier of appropriate design.

For additional guidance on sideslopes, refer to the AASHTO *Roadside Design Guide* **(12)**.

ILLUSTRATIVE OUTER CROSS SECTIONS

Figures IV-3 and IV-6 illustrate typical combinations of the marginal highway elements—shoulders, side-drainage channels, sidewalks, curbs, and sideslopes—for superelevated pavement sections and normal crowned pavement sections, respectively. Only a few of the desirable arrangements are illustrated, but other practical arrangements are discussed.

Normal Crown Sections

Figure IV-6A shows the most widely used cross section in modern highway practice. The combination of elements is simple and forms a streamlined

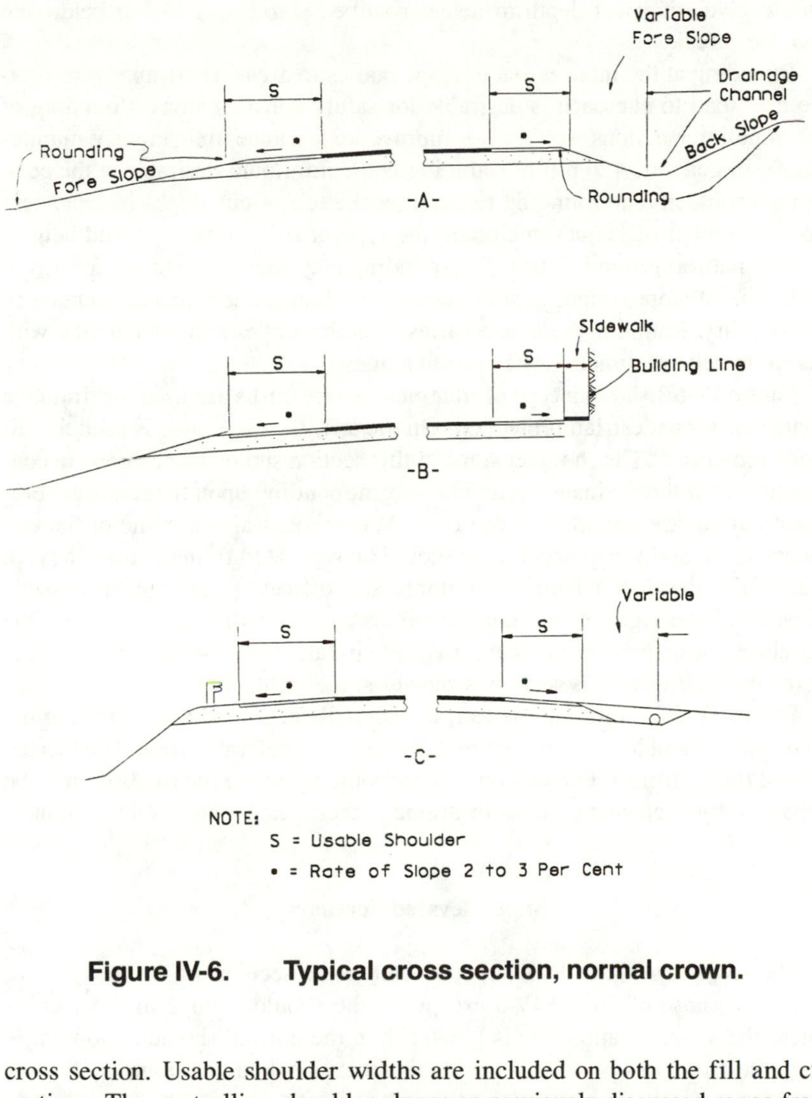

Figure IV-6. Typical cross section, normal crown.

cross section. Usable shoulder widths are included on both the fill and cut sections. The controlling shoulder slopes as previously discussed range from 2 percent for a paved or impervious surface to 8 percent as the maximum applicable to a turf surface.

The drainage channel at the right in Figure IV-6A is formed by the foreslope on the roadway side and the cut slope or backslope on the outer side. The foreslope and backslope combination should be selected to produce a cross section that can be safely traversed by an errant vehicle. The bottom width of the channel should be sufficient for the required drainage capacity and to give sufficient depth to insure roadbed stability, 1 to 4 ft below the shoulder break.

Rounding at the intersection of slope planes in areas where an errant vehicle may tend to encroach is desirable for safety considerations. Rounding of all slope intersections would also improve appearance and simplify maintenance. In general, 4 to 6 ft of rounding is the minimum desirable at the edge of the shoulder. The rounding required at the top of cut slopes is dependent upon a number of factors, including the type of soil, slope ratio and height, and the natural ground slopes. The rounding may vary from 4 ft to upward of 15 ft. Toe-of-slope rounding minimizes slope change and offers an increase in fill stability. Rounding here also varies with the slopes and fill heights, with the same general dimensions as on cut slopes.

Figure IV-6B shows a type of treatment where curbs are used for drainage control or for pedestrian protection. On the left, a curb or dike is used for fill slope protection. The shoulder slope of this section should be designed in conjunction with the drainage system to prevent ponding upon the roadway. Frequent outlets are required for drainage. Where sidewalks are one of the elements of the highway margin, they should be separated from the travelway. In areas fully developed with retail stores and offices it may not be possible because of the right-of-way considerations to offset the sidewalk from the travelway; in such cases curbs are used to separate the sidewalk from the edge of the traveled way. This section is shown on the right.

Figure IV-6C shows, on the left, a steep fill section with guardrail at the edge of the shoulder. Where required, a sidewalk preferably should be located behind the guardrail. For shallow fill sections, safety on the roadside may be enhanced by enclosing sections of drainage facilities as shown on the right.

Superelevated Sections

The low sides of the three superelevated cross sections of Figure IV-3 are similar to those of Figure IV-6 except for the shoulder slope in those cases where the superelevation rate is greater than the normal shoulder slope. It is desirable from an operational standpoint that the shoulder slope on the low side be the same as the pavement superelevation slope, except where the normal shoulder slope is greater.

In Figure IV-3A the direction of shoulder slope on the high side of the cross section is the same as that for normal crowned pavements, but its rate of slope should be limited. The algebraic difference in cross slopes at the pavement edge should not exceed about 8 percent in order to avoid an undesirable rollover effect. Accordingly, use of this section should be reserved for low rates of superelevation and low rates of shoulder slope. The shoulder slope on the alternate section of Figure IV-3A is a projection of the superelevated pavement.

In Figure IV-3B the level shoulder on the high edge of this cross section is a compromise that may be necessary to comply with the 8 percent rollover control and still keep the shoulder from draining to the pavement. The use of this cross section should be reserved for stable soils and, then, only where the percolation that is caused by the amount of water falling directly upon the shoulder is not very great. Where snowfall is prevalent, this cross section would tend to allow snow melt from a windrow on the shoulder to flow across the pavement, creating a potential icing situation.

Figure IV-3C shows the high-side shoulder rolled over in a well-rounded cross vertical curve so that the water falling upon the shoulder is divided between the pavement and the side channel or fill slope. On this rounded shoulder, any vehicle would stand nearly level as needed to facilitate tire changes and other repairs. The vertical curve should be not less than 4 ft long, and at least the inner 2 ft of the shoulder should be held on the superelevated slope. The shoulder slope on the alternate section of Figure IV-3C is a planar section having multiple breaks.

Superelevation is advantageous for street traffic operations, as well as for rural highways and urban freeways, but in built-up areas the combination of wide pavement areas, proximity of adjacent development, control of cross slope and profile for drainage, frequency of cross streets, and other urban features combine to make its use impractical or undesirable. Usually, superelevation is not provided on local streets in residential, commercial, or industrial areas.

For additional discussion on superelevation, refer to Chapter III.

TRAFFIC BARRIERS

General Considerations

Traffic barriers are used to minimize the severity of potential accidents involving vehicles leaving the traveled way: where the consequences of errant vehicles striking a barrier are less than leaving the roadway. Because barriers are a source of accident potential themselves, their use should be carefully considered. For more detailed information regarding traffic barriers, refer to the AASHTO *Roadside Design Guide* **(12)**.

Research continues to develop improved and more cost-effective barriers. The criteria discussed herein will undoubtedly be refined and amended in the future. The designer is, therefore, obligated to remain current on new concepts and criteria.

The term "traffic barriers" refers to longitudinal barriers and crash cushions. Longitudinal barriers function primarily by redirecting errant vehicles. Crash cushions function primarily by decelerating errant vehicles to a stop or by redirection.

Longitudinal barriers are located along the roadside and in medians. Bridge parapets or rails are covered in AASHTO design criteria and specifications for highway bridges.

Longitudinal barriers, are generally denoted as one of three types: flexible, semirigid, or rigid. The major difference between the types is the amount of barrier deflection that takes place when the barrier is struck.

Flexible systems undergo considerable dynamic deflection upon impact and generally impose lower impact forces on the vehicle. The resistance of this system is due in most part to tensile force developed in the longitudinal member. The cable or beams tear away from the support post upon impact, the post thus offering negligible resistance in the impact zone. However, the posts outside the impact zone provide resistance essential to control the deflection of the longitudinal member to an acceptable limit. This system is designed primarily to contain rather than redirect the vehicle and requires more lateral clearance from fixed objects due to the deflection during impact.

In the semirigid system, resistance is achieved through the combined flexure and tensile strength of the rail. The posts near the point of impact are designed to break or tear away, thereby distributing the impact force by beam action to adjacent posts. However, posts outside the impact zone provide resistance essential to control the deflection of the longitudinal member to an acceptable limit and redirect the errant vehicle along the path of traffic flow.

The rigid system does not deflect upon impact. During collisions, energy is dissipated by raising and lowering the vehicle and by deformation of the vehicle sheet metal. As the angle of impact increases, barrier deceleration forces increase because of the absence of barrier deflection. Therefore, installation of a rigid system is normally only appropriate where shallow impact angles are expected, such as along narrow medians or shoulders. The rigid system has proved to be very effective as a protective shield during maintenance work on a roadway where deflection cannot be tolerated. Because this system suffers little or no damage on impact, hence requiring little maintenance, it should be considered where heavy traffic volumes hamper replacement of damaged rail.

Performance level, or barrier capability, lateral deflection characteristics and the space available to accommodate barrier deflection are important factors in the selection of a longitudinal system. Consideration should also be given to the adaptability of the system to operational transitions, end treatments and to the initial and future maintenance cost.

Evaluation of the roadside environment entails three options: (1) removal or modifications of the structure or environment so that it no longer needs to be shielded, (2) installation of a barrier, or (3) no action. The third option would normally be cost effective only on low volume and/or low speed facilities or where engineering studies show the probability of accidents is low.

Site preparation is an important consideration in the use of traffic barriers. To insure maximum barrier effectiveness, site conditions should be tailored to the performance characteristics of the particular barrier.

Roadway cross section significantly affects traffic barrier performance. Curbs, dikes, sloped shoulders, and stepped medians can cause errant vehicles to vault or submarine a barrier or to strike a barrier so that the vehicle overturns. Optimum barrier system performance is provided by a relatively level surface in front of the barrier and, for semirigid and flexible barriers, beneath and behind the barrier. Where curbs and dikes are used to control drainage, they should be located close to, directly in line with or behind the face of the barrier—slight curb protrusions being associated with relatively rigid bridge railings or semirigid barriers. In general, on new construction, in the zone affecting barrier performance, all curbs and dikes that are not a tested, integral part of the barrier system should be avoided and drainage controlled by gentle swales or other means that will not adversely affect barrier performance. Where a barrier is to be installed in the vicinity of an existing curb and the cost of removing the curb cannot be justified, the designer should select a barrier and locate it so that the adverse effect of the curb on barrier performance is minimized.

Longitudinal Barriers

Roadside Barriers

A roadside barrier is a longitudinal system used to shield motorists from natural or man-made hazards located along either side of a roadway. It may occasionally be used to protect pedestrians, bystanders and cyclists from vehicular traffic.

Elements which may warrant shielding by a roadside barrier include (1) embankment obstacles, and (2) roadside obstacles.

Recent studies indicate that rounding at the shoulder and at the toe of an embankment slope can reduce its accident/casualty potential. Rounded slopes reduce the chances of an errant vehicle becoming airborne, thereby reducing the potential consequences of an encroachment and affording the driver more control over the vehicle.

Height and slope of the embankment are the basic factors in determining barrier need through a fill section. The designer should refer to current warrants and criteria for determination of barrier needs **(12)**.

A clear, unobstructed, flat roadside is desirable. When these conditions do not exist, criteria to establish barrier need are necessary. Roadside obstacles are classified as nontraversable areas and fixed objects. These obstacles account for over 30 percent of all highway fatalities each year and their removal should be the first alternative considered. If it is not feasible or practicable to remove, modify or relocate an obstacle, then a barrier may be necessary. However, a barrier should be installed only if it is clear that it offers the least accident/casualty potential.

The objective of barriers is to enhance safety. Short lengths of roadside barriers are discouraged. Where needed in two or more closely spaced locations, barriers should be continuous.

Barriers should be located beyond the edge of the shoulder to insure that use of the full shoulder width is possible. The fill supporting the barrier must be sufficiently wide to provide lateral support. Roadside barriers should be aligned with the bridge rail and properly secured to the bridge to prevent the possibility of a vehicle striking the barrier and snagging or colliding with a bridge rail or curb. Proper treatment of the exposed end of the barrier is also important. The untreated or square approach end of a barrier presents a formidable roadside obstacle. Ends may be buried, covered with a mound of earth, flared back, or protected with a crash cushion or a breakaway cable terminal. Buried barrier ends should be designed to minimize ramping of impacting vehicles. The AASHTO *Roadside Design Guide* **(12)** provides more information on crashworthy end treatments.

Barrier need for rock cuts and large boulders is a matter of judgment by the highway engineer depending on the potential severity of a crash and lateral clearance.

For additional material on Roadside Barriers, refer to the AASHTO *Roadside Design Guide* **(12)**.

Median Barriers

A median barrier is a longitudinal system used to minimize the possibility of an errant vehicle crossing into the path of traffic traveling in opposite directions.

When traffic volumes are low, the probability of a vehicle crossing a median and colliding with a vehicle in the opposing direction is relatively low. Likewise, for relatively wide medians the probability of a vehicle crossing the median and colliding with a vehicle in the opposing roadway is also relatively low. In these instances, median barriers would generally be warranted only when there has been a history of across-the-median accidents or, for new roadways, where an incidence of high accident rates of this type would be expected. Although the incidence of the serious cross median collision may be reduced by median barriers, accident frequency will generally increase because the space available for return-to-the-road maneuvers is decreased.

Special consideration should be given to barrier needs for medians separating traveled ways at different elevations. The ability of an errant driver leaving the higher elevated roadway to return to the road or to stop diminishes as the difference in elevations increases. Thus, the potential for crossover, head-on accidents increases.

For all divided highways, regardless of median width and traffic volume, the median roadside must also be examined for other factors, such as obstacles and lateral dropoff, as discussed earlier.

Careful consideration should be given to the installation of median barriers on multilane expressways or other highways with partial control of access. It is recognized that medians, although narrow, do permit inadvertent encroachments with a chance for motorist recovery and can also include geometric features to accommodate crossing or left-turn traffic. With the addition of a barrier, the required openings and resultant terminals present formidable obstacles. Crash cushions, although requiring maintenance and high initial expense, may be necessary to shield an errant motorist and provide some degree of safety. Consequently an evaluation of the number of crossovers, accident history, alinement, sight distance, design speed, traffic volume, and median width should be conducted prior to installation of median barriers on nonfreeway facilities.

Barriers also should be considered for use on outer separations of about 50 ft or less where the frontage roads carry two-way traffic.

An important safety consideration in the design of median barriers is shielding of a motorist from the exposed end. Exposed ends may be buried, covered with a mound of earth, flared back, or protected with a breakaway cable terminal end or a crash cushion. For more information on crashworthy end treatments, refer to the AASHTO *Roadside Design Guide* **(12)**.

Common types of median barrier include (1) double-faced steel W-beam (blocked-out) installed on strong posts, (2) box beam installed on weak posts, and (3) concrete barrier with a slope face. Some use is also made of a two- or three-cable barrier installed on light steel posts, a double-faced steel W-beam installed on weak posts, a double-faced steel thrie beam (blocked-out) installed on strong posts, and a cable-chain-link-fence combination. For additional data on median barrier types, reference is made to the AASHTO *Roadside Design Guide* **(12)**.

In selecting the type of median barrier, it is important to match the dynamic lateral deflection characteristics to the site. The maximum deflection should be less than one-half the median width. The median barrier should be designed to prevent penetration into the opposing lanes of traffic and should redirect the colliding vehicle in the same direction as the traffic flow. In addition, the design should be esthetically pleasing.

On heavily traveled facilities with narrow medians, a concrete barrier with a slope face has many advantages. This type will deflect a vehicle striking it at a slight impact angle. It has a pleasing appearance and requires little maintenance. The latter is an important consideration on highways with narrow medians since maintenance operations on the barrier encroach on the high-speed traffic lanes and may require closure of one of the traffic lanes during repair time. The designer should also bear in mind that even though the concrete barrier may not deflect, there may be significant intrusion into the air space above and beyond the barrier by high center of gravity vehicles when struct at high speeds or large angles. A bus or tractor-trailer may lean enough to strike objects mounted on top of the barrier or within a distance of up to 10 ft of the barrier face. While piers and abutments may be able to withstand such impacts, sign trusses, light standards, etc. may become involved in secondary accidents.

The selection of median barriers differs for stepped median sections, that is, where the median is between roadways of different elevations. Cable, W-beam on weak posts, and box-beam systems are limited to flat medians. The AASHTO *Roadside Design Guide* **(12)** provides additional data in this regard.

During the selection and design of a median barrier, consideration should be given to the possible effect of the barrier on sight distance on horizontal curves.

The design of median barriers and terminals continually improves, and reference should be made to the latest developments on this subject.

Precast concrete median barrier can be used for temporary protection of work areas and for guiding traffic during construction. It can also be incorporated permanently as part of the completed facility.

Bridge Railings

A bridge railing prevents vehicles, pedestrians, or cyclists from falling off the structure. The AASHTO's *Standard Specifications for Highway Bridges* **(18)** specifies geometric, design load, and maximum allowable material stress requirements for the design of traffic railings, pedestrian railings, bicycle railings, and railings that combine traffic railing features with those of pedestrian or bicycle railings. Bridge traffic railings and combination railings are longitudinal traffic barriers, differing from other traffic barriers primarily in their foundations, with bridge railings being a structural extension of a bridge and other traffic barriers usually being set in or on soil.

The need for a traffic barrier virtually never ends at a bridge end and the need must be filled by extending the bridge railing with a roadside barrier, which in turn must have a crash-worthy terminal.

At the juncture between a bridge railing and roadside barrier an incompatibility nearly always exists in the stiffnesses of the two barrier types. This stiffness must be transitioned over a length and with details that will prevent the barrier system from pocketing or snagging an impacting vehicle.

Where a curbed sidewalk stands in front of a bridge railing, design of the traffic portion of the railing may need to consider the potential of a vehicle vaulting due to impacting the curb. *National Cooperative Highway Research Program Report 150* (**27**) contains information on vehicle trajectories for various vehicle-curb-impact conditions. Curbed sidewalks should not be used adjacent to traffic where design speeds exceed 50 mph.

In instances where a roadside barrier is provided between the edge of traveled way and the bridge railing, special attention must be given to the barrier end treatment. Functional yet safe end treatments are difficult to design. The end treatments should safely accommodate vehicles, yet not impede pedestrian usage of the walkway.

The recommended lateral clearances between travel lanes and bridge railings usually exceed curb offset distances. This may create a problem in the introduction of a bridge railing where a curbed cross section is used on a bridge approach and a flush cross section is used on the bridge. This condition may result from the length of the bridge and its approaches being such that it takes on the character of a controlled access facility where traffic will operate at speeds in excess of 50 mph even though the approach speeds are lower than 50 mph—thus making curbs acceptable away from the bridge and not on the bridge. In such cases it may be reasonable to drop the curb at the first intersection away from the end of the bridge or, well in advance of the introduction of the traffic barrier, reduce the curb to a low, mountable curb with a gently sloped traffic face, which would be reasonably compatible with the traffic barrier even if continued into the high-speed region of the bridge.

Crash Cushions

Crash cushions (AASHTO *Roadside Design Guide* (**12**)) are protective systems that prevent errant vehicles from impacting roadside obstacles by decelerating the vehicle to a safe stop when hit head-on or redirecting it away from the obstacle.

A common application of a crash cushion is ramp gores where a bridge-rail end exists in the gore. Where site conditions permit, a crash cushion should also be considered as an alternative to a roadside barrier for shielding rigid objects such as bridge piers, overhead sign supports, abutments, and retaining-wall ends. Crash cushions may also be used to shield roadside and median barrier terminals.

Site preparation is important in using crash cushion design. Site conditions not compatible with the cushion design can compromise cushion effectiveness.

New highway design should consider alternatives to use of these devices where possible. Where a crash cushion is the best alternative adequate level space free from curbs or other physical features should be provided.

MEDIANS

A median is highly desirable on arterials carrying four or more lanes. A median is defined as the portion of a divided highway separating the traveled way for traffic in opposing directions. The median width is expressed as the dimension between the through-lane edges and includes the left shoulders, if any. The principal functions of a median are to separate opposing traffic, provide a recovery area for out-of-control vehicles, provide a stopping area in case of emergencies, allow space for speed changes and storage of left-turning and U-turning vehicles, minimize headlight glare, and provide width for future lanes. Another benefit of a median in an urban area is that it may offer an open green space. For maximum efficiency, a median should be highly visible both night and day and contrast with the through traffic lanes. Medians may be depressed, raised, or flush with the pavement surface. Medians should be as wide as feasible but of a dimension in balance with other components of the cross section.

In determining median width, consideration should be given to the possible need for median barrier. Where possible, median widths should be such that a median barrier is not warranted. In general, the median should be as wide as practical. The general range of median widths is from a minimum of 4 ft to 80 ft or more. As far as the safety and convenience of motor vehicle operation are concerned, the farther the pavements are apart, the better. However, economic factors often limit the width of median that can be provided. Cost of construction and maintenance increases in proportion to increases in the median width but the additional cost may not be appreciable compared with the cost of the highway as a whole and may be justified in view of the benefits derived. Medians should be wide enough to allow selected design vehicles to safely make a selected maneuver. A consideration in the use of wider medians on roadways other than freeways is the provision of adequate shelter for vehicles crossing at intersections and at crossovers serving commercial and private drives. These median crossings may need to be controlled as at-grade intersections (See Chapter IX). Wide medians may be a disadvantage when at-grade intersections are required. The increased time for vehicles to cross the median may lead to inefficient signal operation.

If right-of-way is restricted, a wide median may not be justified if provided at the expense of narrowed border areas. A reasonable border width is required to adequately serve as a buffer between the private development along the road and the traveled way, particularly where zoning is limited or nonexistent. Space must be provided on the borders for sidewalks, highway signs, utility lines, parking, drainage channels and structures, and for proper slopes and any retained native growth. Narrowing these areas may tend to develop obstacles and hindrances similar to those that the median is designed to avoid.

A depressed median is generally preferred on freeways for more efficient drainage and snow removal. Median side slopes should preferably be 6:1, but slopes of 4:1 may be adequate. All drainage inlets in the median should at least be designed with the top flush with the ground or culvert ends provided with traversable safety grates.

Raised medians have application on arterial streets where it is desirable to regulate left-turn movements. They are also frequently used where the median is to be planted, particularly where the width is relatively narrow. It must be pointed out, however, that planting in narrow medians creates hazardous conditions for maintenance operations.

Flush medians are used to some extent on all types of urban arterials. When used on freeways, a median barrier may be required. The median should be slightly crowned or depressed for drainage. In warmer climates the crowned type is frequently used because it eliminates the need for collecting drainage water in the median. However, the slightly depressed type is generally preferred with a cross slope of about 4 percent or with a minor steepening of the roadway cross slope.

The concept of converting flush medians into two-way left turn lanes on urban streets has become widely accepted. This concept offers several advantages when compared to no median. Among these advantages are reduced travel time, improved capacity, reduced accident frequency, particularly of the rear-end type, more flexibility because the median lane can be used as a travel lane during closure of a through lane and public preference, both from drivers and owners of abutting properties. Median widths 10 to 16 ft wide provide the optimum design for two-way left turn lanes. Refer to the MUTCD **(2)** for appropriate lane markings. Refer to Chapter II for additional discussion and details.

Where there is no fixed-source lighting, headlight glare across medians or outer separations can be a nuisance, particularly where the highway has relatively sharp curves. Under these conditions, some form of antiglare treatment should be considered as part of the median barrier installation, provided it does not act as a snow fence and create drifting problems.

Insofar as through traffic is concerned, a desired ease and freedom of operation, in the sense of physical and psychological separation from opposing traffic, is obtained when medians are about 40 ft or wider. With such widths the facility truly is divided. The noise and air pressure of opposing traffic is not noticeable, and at night the glare of headlights is greatly reduced. With widths of 60 ft or more the median can be pleasingly landscaped in a parklike manner. Plantings used to achieve this parklike appearance should not compromise the roadside recovery area. It must be pointed out, however, that there is demonstrated benefit in any separation, raised or flushed.

FRONTAGE ROADS

Frontage roads serve numerous functions, depending on the type of arterial they serve and the character of the surrounding area. They may be used to control access to the arterial, to function as a street facility serving adjoining property, and to maintain circulation of traffic on each side of the arterial. Frontage roads segregate local traffic from the higher speed through traffic and intercept driveways of residences and commercial establishments along the highway. Cross connections between the through-traffic lanes and frontage roads, usually provided in conjunction with crossroads, furnish the means of access between through roads and adjacent property. Thus, the through character of the highway is preserved and is unaffected by subsequent development of the roadsides.

Frontage roads are used on all types of highways, and they are further discussed in the chapters following for each type of street or highway. Their greatest use is on freeways where their primary function is to distribute and collect traffic between local streets and the freeway interchanges. In some circumstances, frontage roads are also desirable on arterial streets both in downtown and suburban areas. Frontage roads not only provide more favorable access for commercial and residential development than the faster moving arterial street, but also help to preserve the safety and capacity of the latter. In rural areas, development of expressways may require separated frontage roads that are somewhat removed from the right-of-way and serve as access connections between crossroads and adjacent farms or other development.

Despite the advantages cited, the use of continuous frontage roads on relatively high-speed arterial streets with intersections at grade may be undesirable. At the cross streets, the various through and turning movements at several closely spaced intersections greatly increase the accident potential. The multiple intersections are also vulnerable to wrong-way entrances. Traffic operations are improved if the frontage roads are located a considerable distance from the main line at the intersecting cross roads in order to lengthen the spacing between successive intersections along the crossroads. In urban areas a minimum spacing of about 150 ft between the arterial and the frontage roads is desirable, and 300 ft is desirable in rural areas. For further discussion on frontage roads at intersections, refer to Chapter IX.

Frontage roads generally are, but need not be, parallel to the roadway for through traffic, they may or may not be continuous, and they may be provided on one or both sides of the arterial. Where the highway diagonally crosses a grid street system on a diagonal course or where the street pattern is irregular, the frontage roads may be a variable distance from the through-traffic roadways. Arrangements and pattern of frontage roads are shown in Figures IV-7 and IV-8. Figure IV-7A is the most common arrangement, which is two frontage roads running parallel and approximately equidistant from a freeway. In urban areas continuous frontage roads that are more or less parallel to the freeway permit the use of the frontage roads as a backup system in case of freeway disruption. Figure IV-7B shows a freeway with one frontage road. On the side without the frontage road the local streets serve to collect and distribute the traffic. Figure IV-8 shows an irregular pattern of frontage roads.

From an operational and safety standpoint, one-way frontage roads are much preferred to two-way. One-way operation inconveniences local traffic to some degree, but the advantages in reduction in vehicular and pedestrian conflicts at intersecting streets often fully compensate for this inconvenience. In addition, there is some saving in pavement and right-of-way width. Two-way frontage roads at busy at-grade intersections complicate crossing and turning movements. Where off ramps join a two-way frontage road, the potential for wrong-way entry is increased. This problem is greatest where the ramp joins the frontage road at an acute angle, thus giving the appearance of an onramp to the wrong-way driver.

Two-way frontage roads may be considered for partially developed urban areas where the adjoining street system is so irregular or so disconnected that one-way operation would introduce considerable added travel distance and cause undue inconvenience. Two-way frontage roads may also be necessary for suburban or rural areas where points of access to the through facility are infrequent, where only one frontage road is provided, where roads or streets connecting with the frontage roads are widely spaced, or where there is no parallel street within reasonable distance of the frontage roads in urban areas that are developed or likely to be developed.

Connections between the arterial and the frontage road are an important element of design. On arterials with slow-moving traffic and one-way front-

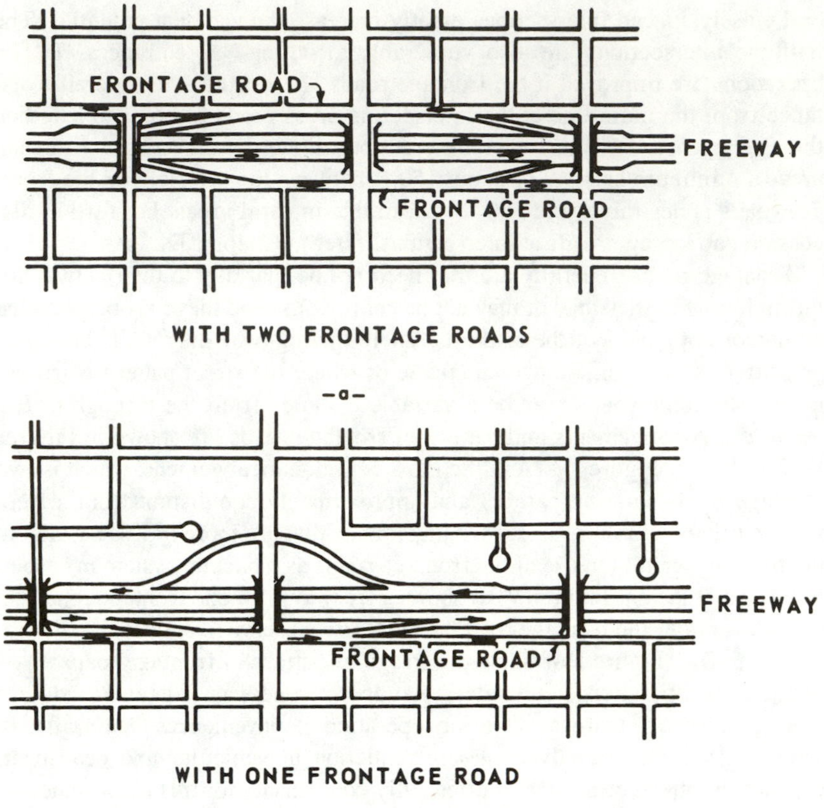

Figure IV-7. Typical frontage road arrangements.

age roads, slip ramps or simple openings in a narrow outer separation may work reasonably well.

Slip ramps from a freeway to two-way frontage roads are generally unsatisfactory because they tend to induce wrong-way entry to the freeway through lanes and cause accidents at the intersection of the ramp and frontage road. On freeways and other arterials with high operating speeds, the ramps and their terminals should be liberally designed to provide for speed changes and storage. Details of ramp design are covered in later chapters.

Figures IV-9 and IV-10 show various arrangements of frontage roads with entrance and exit ramps. These designs are applicable to freeways and other higher speed arterials.

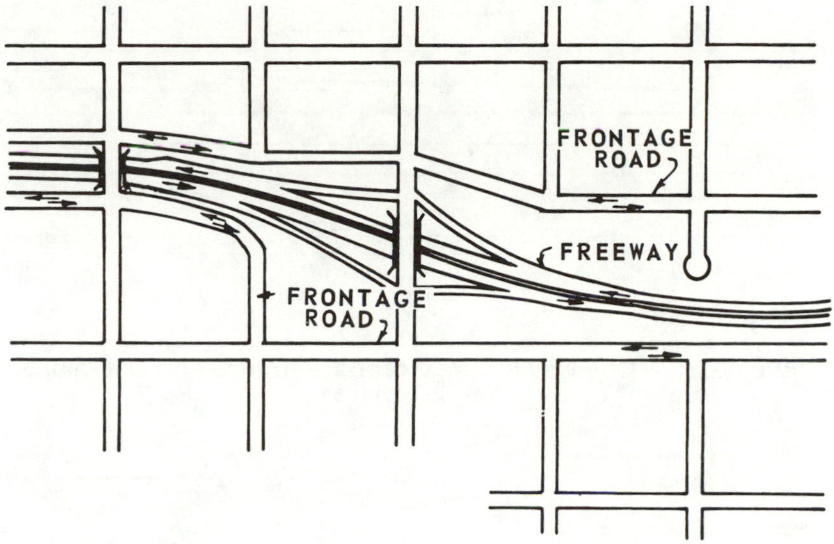

Figure IV-8. Frontage roads, irregular pattern.

Figure IV-9 also shows an arrangement with one-way frontage roads. All elements are designed to insure good operation on both the freeway and the frontage roads.

Figure IV-10 shows an arrangement of entrance and exit ramps at two-way frontage roads. This scheme requires a wide outer separation that is not always available in urban areas. The actual width would depend on design requirements of the ramps and their terminals. In most cases the width of outer separation would be greater than 200 ft in the area of the ramp terminals. The offramp is connected to the frontage road at a right angle to discourage wrong-way entry. Careful attention needs to be given to the placement of signs and the use of pavement markings to prohibit wrong-way movements. Because of the potential for wrong-way movements, the offramp should not intersect the frontage road opposite a two-way side street access.

The design of a frontage road is affected by the type of service it is intended to provide. Where a frontage road is continuous and passes through highly developed areas, its primary function is that of general service, and it assumes the character of an important street. Buses may be routed thereon and through-traffic overflow will use a frontage road when the through-traffic lanes are loaded. At the other extreme, where the frontage roads are only a few blocks long, follow an irregular pattern, border the rear and sides of

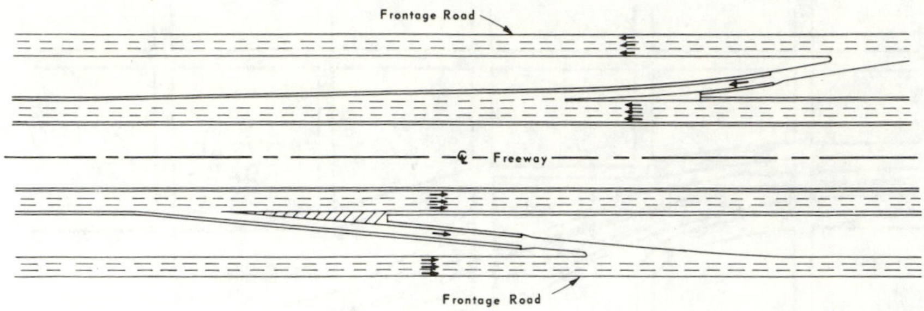

Figure IV-9. One-way frontage roads, entrance and exit ramps.

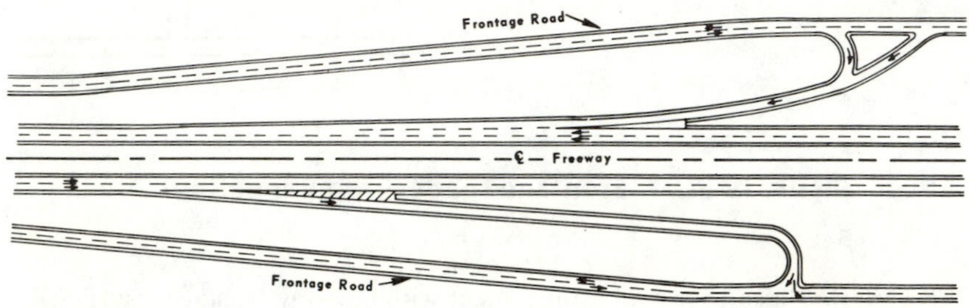

Figure IV-10. Two-way frontage roads, entrance and exit ramps.

buildings, or serve only scattered development, traffic will be light and operation will be local in character. For widths of one- or two-way, two-lane frontage roads for rural and urban collectors, see Chapter VI.

OUTER SEPARATIONS

The area between the traveled way of a roadway for through traffic and a frontage road or street is the outer separation. Such separations function as buffers between the through traffic on the arterial and the local traffic on the frontage road and provide space for a shoulder for the through pavement and ramp connections to or from the through facility.

The wider the outer separation, the less influence local traffic will have on through traffic. Wide separations lend themselves to landscape treatment and enhance the appearance of both the highway and the adjoining property. A substantial width of outer separation is particularly advantageous at intersections with cross streets. This outer separation minimizes vehicle and pedestrian conflicts.

Where ramp connections are provided between the through roadway and the frontage road, the outer separation must be substantially wider than normal. The required width will depend mostly upon design requirements at the ramp termini.

Where two-way frontage roads are provided, a driver on the through facility must contend with approaching traffic on the right (opposing frontage road traffic) as well as opposing arterial traffic on the left. Desirably, the outer separation should be sufficiently wide to minimize the effects of approaching traffic, particularly the confusing and distracting nuisance of headlight glare at night. With one-way frontage roads the outer separation need not be as wide as with two-way frontage roads.

Figure IV-11 illustrates a one-lane, one-way frontage road with parking serving businesses along a major undivided arterial street in a densely devel-

Figure IV-11. Frontage road in business area with narrow outer separation.

oped area of a large city. The raised and curbed outer separation separates through traffic from local traffic and provides a refuge for pedestrians.

The cross section and treatment of an outer separation depends largely upon its width, the type of arterial, and the type of frontage road. Preferably, the strip should drain away from the through pavement either to a curb and gutter at the frontage road or to a swale within the strip. Typical cross section of outer separations for various types of arterials are illustrated in Figure IV-12.

The design in Figure IV-12A would be applicable to low-speed arterial streets in densely developed areas.

Figure IV-12B shows a minimal outer separation that may be applicable to ground-level freeways and high-speed arterial streets. This outer separation consists simply of the shoulders of the through roadway and the frontage road and a physical barrier.

Figure IV-12C shows a depressed arterial with a cantilevered frontage road. In this example, the inside edge of the frontage road is directly over he outside edge of the through roadway.

Figure IV-12D illustrates a common type of outer separation along a section of depressed freeway.

Figure IV-12E shows a walled section at a depressed arterial with a ramp, and Figure IV-12F shows a typical freeway outer separation with a ramp.

NOISE CONTROL

General Considerations

Noise may be defined as unwanted sound. Motor vehicles generate traffic noise from the motor, aerodynamics, exhaust, and interaction of tires with the roadway. Efforts should be made to minimize the radiation of noise into noise sensitive areas along the highway. It is necessary that the designer evaluate probable noise levels and the effectiveness of reducing highway traffic noise through location and design features.

The physical measurement of human reaction to sound is difficult because there is no instrument that will measure this directly. A close correlation can be obtained by using the A-scale on a standard sound level meter. The meter yields a direct reading in effective decibels (dBA).

A few general relationships may be helpful in understanding some of the principles of sound generation and transmission. Because noise is measured on a logarithmic scale, a decrease of 10 dBA will appear to an observer to be a halving of the noise. For example, a noise of 70 dBA sounds only one-half as

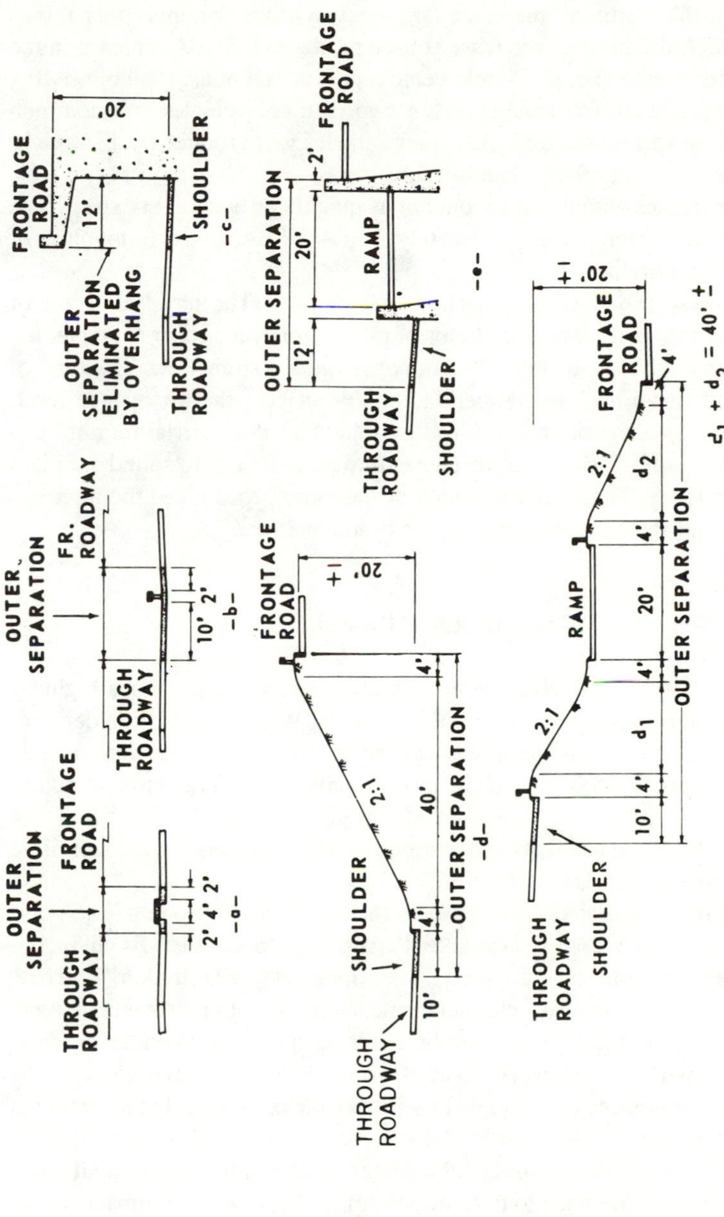

Figure IV-12. Typical outer separations.

loud as 80 dBA assuming the same frequency composition and other things being equal. A doubling of the noise source produces a 3 dBA increase in the noise level. For example, if a single vehicle produces a noise level of 60 dBA at a certain distance from the receiver, two of these vehicles at a common point of origin will produce 63 dBA, four vehicles will produce 66 dBA, eight vehicles will produce 69 dBA, and so forth.

Noise decreases with distance, but not as quickly as one might expect. The sound level will decrease approximately 3 to 4.5 dBA for each doubling of distance from a highway.

Traffic noise produces varying human reactions. The physical factor of noise is not, in itself, a good predictor of public annoyance. For example, the reaction is usually less if the noise source is hidden from view. The type of development in an area is another factor that affects the annoyance level. High traffic noise levels are usually more tolerable in industrial than in residential areas. Other factors that influence human reactions to sound are pitch and intermittency. The higher the pitch or the more pronounced the intermittency of the noise, the greater the degree of annoyance.

General Design Procedures

The first step in analyzing the effects of noise from a proposed highway facility is to define the criteria for noise impacts. With these criteria defined, the location of noise-sensitive areas can be indentified. These may include residential areas, schools, churches, motels, parks, hospitals, nursing homes libraries, etc.

The existing noise levels are determined by measurement for the identified noise-sensitive land uses or activities.

The highway-generated noise level is then predicted by one of the noise prediction methods presently available. Pertinent factors are traffic characteristics (speed, volume, and composition), topography (vegetation, barriers, and distance), and roadway characteristics (configuration, pavement type, grades, and type of facility). The prediction is normally based on the highway traffic which will yield the worst hourly traffic noise on a regular basis for the design year. References 20 through 24 will provide more detailed information on noise prediction.

Table IV-3 gives FHWA noise abatement criteria for various land uses. These sound levels are used to determine impact. Traffic noise impacts occur under two criteria: 1) when the predicted levels approach or exceed the noise abatement criteria and, 2) when predicted noise levels substantially exceed the existing noise level, even though the predicted levels may not exceed the noise abatement criteria. In order to adequately assess the traffic noise impact of a proposed project, both criteria must be analyzed.

Activity Category	Category Description	Design Noise Levels-dBA[2] $L_{eq}(h)^{(b)}$	$L_{10}(h)$
A	Tracts of land in which serenity and quiet are of extraordinary significance and serve an important public need and where the preservation of those qualities is essential if the area is to continue to serve its intended purpose. Such areas could include amphitheaters, particular parks or portions of parks, open spaces, or historic districts which are dedicated or recognized by appropriate local officials for activities requiring special qualities of serenity and quiet.	57	60 (Exterior)
B	Picnic areas, recreation areas, playgrounds, active sports areas, and parks not included in Category A and residences, motels, hotels, public meeting rooms, schools, churches, libraries, and hospitals.	67	70 (Exterior)
C	Developed lands, properties, or activities not included in Categories A or B above.	72	75 (Exterior)
D	Undeveloped lands which do not contain improvements or activities devoted to frequent human habitation or use and for which such improvements or activities are unplanned and not programmed.	—(c)	
E	Residences, motels, hotels, public meeting rooms, schools, churches, libraries, hospitals, and auditoriums.	52	55[d] (Interior)

[a]Source: *Federal Aid Highway Program Manual*, Vol. 7, Ch. 7, Sec. 3 Transmittal 348, August 9, 1982.

[b]Either $L_{10}(h)$ or $L_{eq}(h)$ (but not both) may be used on a project.

[c]Noise abatement criteria has not been established for these lands. They may be treated as developed lands if the probability for development is high. Provisions for noise abatement would be base on the need, expected benefits, and costs of such measures.

[d]Interior noise abatement criteria in this category apply to (1) indoor activities where no exterior noise-sensitive land use or activity is identified, and (2) exterior activities that are either remote from the highway or shielded so that they will not be significantly affected by the noise, but the interior activities will.

Table IV-3. Noise abatement criteria for various land uses.

Noise Reduction Designs

An effective method of reducing traffic noise from adjacent areas is to design the highway so that some form of solid material blocks the line of sight between the noise source and the receptors. Advantage should be taken of the terrain in forming a natural barrier so that the appearance remains esthetically pleasing. Early determination of potential noise problems should be attempted. Line, grade, earthwork balance, and right-of-way should all be worked out with noise in mind. Attenuation may be inexpensive and feasible if "built-in." Adding it after the design is tied down is expensive.

In terms of noise considerations, a depressed highway section is the most desirable. Depressing the roadway below ground level has the same general effect as erecting barriers, i.e., a shadow zone is created on the receiver plane wherein noise levels are reduced. (See Figure IV-13.)

Where a highway is constructed on an embankment in a noise sensitive area, additional width or the embankment beyond the shoulders will some-

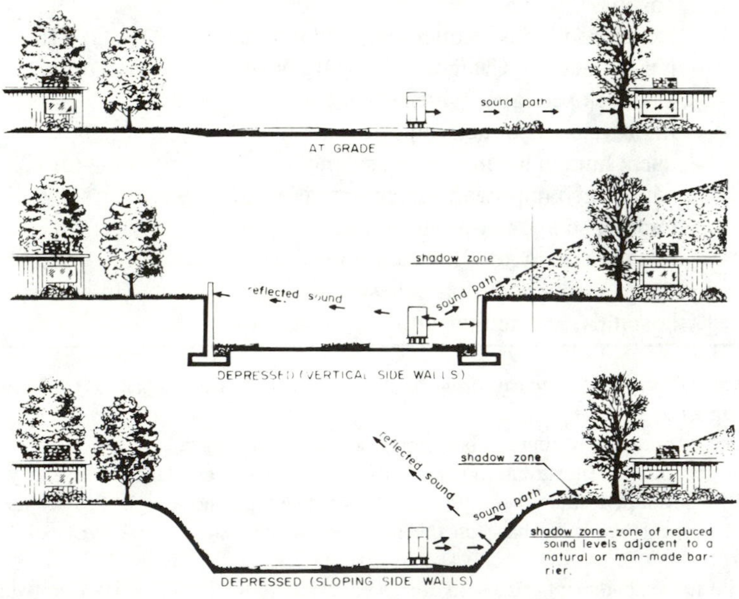

Figure IV-13. Effects of depressing the highway.

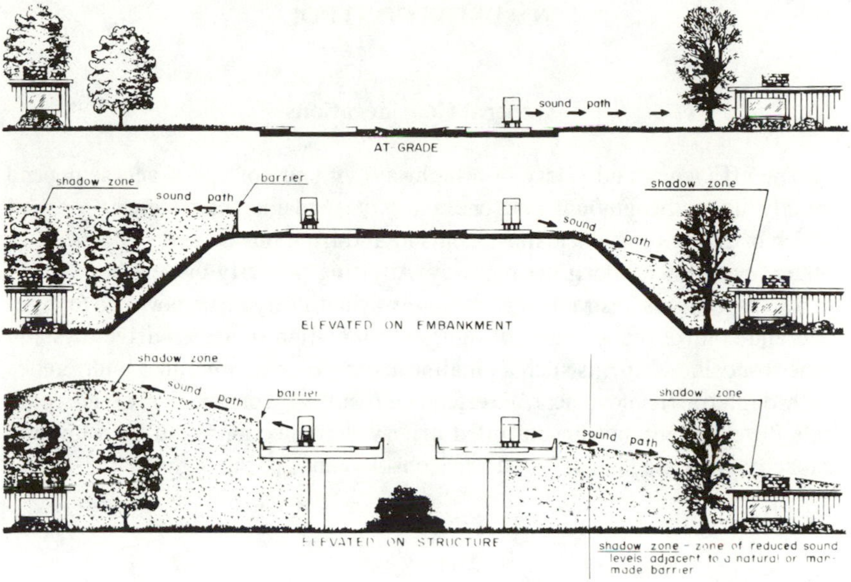

Figure IV-14. Effects of elevating the highway.

times block the line of sight to receptors near the highway, thus reducing the noise problem. (See Figure IV-14.)

Special sound barriers may be justified at certain locations, particularly along ground level or elevated highways through noise-sensitive areas. Concrete, wood, metal, or masonry walls are very effective. One of the more esthetically pleasing barriers is the earth berm that has been graded to achieve a natural form blending with the surrounding topography. The feasibility of berm construction should be considered as part of the overall grading plan for the highway. There will be instances where an effective earth berm can be constructed within normal right-of-way or with a minimal additional right-of-way purchase. If right-of-way is insufficient to accommodate a full-height earth berm, a lower earth berm can be constructed in combination with a wall or screen to achieve the desired height.

Shrubs, trees, or ground covers are not very efficient in shielding sound because of their permeability to the flow of air. However, almost all buffer plantings offer some noise reduction, and exceptionally wide and dense plantings may result in substantial reductions in noise levels. Even where the noise reduction is not considered significant, the esthetic effects of the plantings will produce a positive influence.

ROADSIDE CONTROL

General Considerations

The efficiency and safety of a highway without control of access depend greatly upon the amount and character of roadside interference, most of which originates in vehicle movements to and from businesses, residences, or other development along the highway. Abutting property owners have rights of access, but it is desirable that the highway authority be empowered to regulate and control the location, design, and operation of access driveways and other roadside elements such as mailboxes in order to minimize interference to through traffic movement. Interference resulting from indiscriminate roadside development and uncontrolled driveway connections results in lowered capacity, increased hazard, and early obsolescence of the highway.

Driveways

Driveway terminals are, in effect, low-volume intersections; thus their design and location merit special consideration.

Driveways are directly related to the functional classification of the particular roadway. Where driveways might adversely affect the operation of arterials, the driveways become important links on local streets, where the primary function of the street is to provide access to local establishments.

Driveways operated for right turns only are logical and desirable forms where the cross section includes a curbed median or a flush median and median barrier. Driveways operating with left turns offer considerably more interference to through traffic and are undesirable on arterial streets. However, there are instances on major streets with numerous drive-in businesses where the elimination of left turns at driveways would worsen traffic operations by forcing large volumes of traffic to make U-turns or travel around the block.

The regulation and design of driveways are intimately linked with the right-of-way and zoning control of the roadside. On new facilities the necessary right-of-way can be obtained to provide the desired degree of driveway regulation and control. In many cases on existing facilities additional right-of-way can be acquired or agreements made to improve existing conditions and prohibit undesirable access conditions. Often the desired degree of driveway control must be effected through the use of police powers to require permits for all new driveways and adjustment of those existing. The objective of driveway regulations is to preserve capacity and promote operational capacity by prescribing the desirable spacing and proper layout of driveways. Its attainment is dependent upon the type and extent of legislative authority granted the highway agency. Many States and local municipalities have developed standards for driveways and set up separate units to handle the details incident to checking requests and issuing permits for new, or for changes in existing, driveway connections to main highways. Major features are discussed and needed geometric controls are suggested in *An Informational Guide for Preparing Private Drivewav Regulations for Major Highways* **(8)**.

Driveway regulations, largely effected through issuance of permits, generally control right-of-way encroachment, driveway location, driveway design, sight distance, drainage, use of curbs, parking, setback, lighting, and signing. Some of the principles of intersection design apply directly to driveways. One of the important features is the elimination of large graded or paved areas adjacent to the traveled way upon which drivers can enter and leave the facility at will. Another feature is the provision of adequate driveway width and throat dimensions and layout to accommodate the types of vehicles patronizing the roadside establishment.

Sight distance is a significant design control. No advertising signs should be permitted in the right-of-way. Billboards or other elements outside the right-of-way that obstruct sight distance should be controlled by statutory authority or by purchase of easements.

A desirable ultimate plan for a main highway or street without control of access but with concentrated business frontage is the grouping of driveways to reduce the number of separate connections to the through traffic lanes by providing an outer parallel road (frontage road) to which the driveways connect. This type of control and design is particularly pertinent to a main highway or street on a new location for which sufficient right-of-way can be acquired for the ultimate development of needed frontage roads. In the first stage, intermittent sections of frontage roads are constructed to connect the few necessary driveways initially required. Then, in succeeding stages, extensions or additional sections of frontage roads are provided to intercept driveways resulting from further development of the roadsides. Thus, serious roadside interference is prevented at all stages, and the through character of the highway or street is preserved by gradual and judicious provision of frontage roads.

Mailboxes

Mailboxes and appurtenant newspaper tubes served by carriers in vehicles may very well constitute a safety hazard either directly or indirectly, depending upon the placement of the mailbox, the cross section dimensions of the highway or street, sight distance conditions in the vicinity of the mailbox, traffic volume, and impact resistance of the mailbox support. The safety of both the carrier and the motoring public is affected whenever the carrier slows for a stop and when the carrier resumes travel along the highway. The risk is greatly increased if the cross section of the highway and the lateral placement of mailboxes are such that the vehicle must occupy a portion of the traveled way while the mailbox is being serviced.

The mailbox and support, being fixed objects, may constitute a hazard to the passing motorist if the box is in an exposed location. The mounting height of the box places the box in a direct line with the windshield on many vehicles. This situation is more critical where multiple box installations are encountered. In many areas, the typical multiple mailbox installation consists of two or more posts supporting a horizontal member, usually a timber plank, which carries the group of mailboxes. This installation is hazardous because when struck by an automobile, the horizontal support element tends to penetrate the windshield and enter the passenger compartment. (Such installations are to be avoided where exposed to traffic.)

Mailboxes should be placed for maximum convenience to the patron, consistent with safety considerations for highway traffic, the carrier, and the patron. Consideration should be given to (1) minimum walking distance within the roadway for the patron (2) available stopping sight distance in advance of the mailbox site (especially on older roads), and (3) possible restriction to corner sight distance at driveway entrances. The placing of mailboxes along high-speed, high-volume highways should be avoided if other practical locations are available. New installations should, where feasible, be located on the far right side of an intersection with a public road or private driveway entrance. Boxes should be placed only on the right-hand side of the highway in the direction of travel of the carrier except on one-way streets where they may also be placed on the left-hand side.

Preferably, a mailbox should be placed so that it is not susceptible to being struck by an out-of-control vehicle. Where this placement is not feasible, the supports should be of a type that will yield or break away safely if struck. The mailbox should be firmly attached to the support to prevent it from breaking loose and flying through the windshield. The same safety criteria also apply to multiple box installations.

One of the primary considerations is the location of the mailbox in relation to the traveled way. Basically a vehicle stopped at a mailbox should be clear of the traveled way. The higher the traffic volume or the speed the greater the clearance should be. An exception to this may be reasonable on low volume, low-speed roads and streets.

Most vehicles stopped at a mailbox will be clear of the traveled way when the mailbox is placed outside an 8-ft wide usable shoulder or turnout. This position is recommended for most rural highways. For high-volume, high-speed highways, it is recommended that the width of shoulder in front of the mailbox or turnout be increased to 10 ft or even 12 ft for some conditions. However, it may not be reasonable to require even an 8-ft shoulder or turnout on low-volume, low-speed roads or streets. To provide space outside the all-weather surface for opening the mailbox door, it is recommended that the roadside face of a mailbox be set 8 to 12 in. outside the shoulder or turnout. Current postal regulations should be consulted for specific set-back requirements.

In areas of heavy or frequent snowfall, mailboxes may be placed at about the customary line of the plowed windrow, but no closer than about 10 ft to the edge of traveled way if the shoulder is wider than 10 ft. Cantilever mailbox supports may prove advantageous for snow plowing operations. Whenever possible mailboxes should be located behind existing guardrail.

In some urban and suburban areas mailboxcs are located along selected streets and highways where the local post office has established delivery routes. In these areas when the roadway has a curb and gutter section, which is the usual case, mailboxes should be located with the front of the box 6 to 12 in. back of the face of curb. On residential streets without curbs or shoulder and which carry low-traffic volumes operating at low speeds, the roadside face of a mailbox should be offset between 8 and 12 in. behind the edge of pavement.

For guidance on mailbox installations, refer to the latest edition of AASHTO's *A Guide for Erecting Mailboxes on Highways* (**9**).

TUNNELS

General Considerations

Development of streets or highways may require sections be constructed in tunnels either to carry the streets or highways under or through a natural obstacle or to minimize the impact of the freeway on the community. The general conditions under which tunnel construction may be warranted are the following:

1. Long, narrow terrain ridges where a cut section may be costly in money or in environmental consequences
2. Narrow rights-of-way where all of the surface area must be retained for street purposes

3. Large intersection areas or a series of adjoining intersections on an irregular or diagonal street pattern
4. Railroad yards, airport and runways, or similar facilities
5. Parks or other land uses, existing or planned
6. Right-of-way acquisition costs exceed cost of tunnel construction and operation.

Although the costs of operation and maintenance of tunnels are beyond the scope of this policy, these costs must nevertheless be considered.

Additional construction and design features of tunnel sections are discussed below. It is not intended that this section be considered a complete treatise on the subject of highway tunnels. The material that follows is for the purpose of providing highway planners and designers with general background information. To accomplish this basic objective, some simplification of subject matter is required. As with any highly specialized branch of engineering, such simplified information should be used with caution.

The ventilating, lighting, pumping, and other mechanical or electrical considerations in tunnel design are regarded as outside the scope of this policy.

Types of Tunnels

Tunnels can be classified into two major categories: (1) tunnels constructed by mining methods, and (2) tunnels constructed by cut-and-cover methods.

The first category refers to those tunnels that are constructed without removing the overlying rock or soil. Usually this category is subdivided into two very broad groups according to the required construction method. These methods are named to reflect the overall character of the material to be excavated, i.e., hard rock and soft ground.

Of particular interest to the highway designer are the structural requirements of these construction methods and their relative costs. As a general rule, hard-rock tunneling is less expensive than soft-ground tunneling. A tunnel constructed through solid, intact, and homogeneous rock will normally represent the lower end of the scale with respect to structural demands and construction costs. A tunnel located below water in material needing immediate and heavy support will require extremely expensive soft-ground tunneling techniques such as shield and compressed air methods.

The shape of the structural cross section of the tunnel varies with the type and magnitude of loadings. In those cases where the structure will be subjected to roof loads with little or no side pressures, a horseshoe-shaped cross section is used. As side pressures increase, curvature is introduced into the sidewalls and invert struts added. When the loadings approach a distribution similar to hydrostatic pressures, a full circular section is usually more efficient and economical. All cross sections are dimensioned to provide adequate space for ventilation ducts.

The second category of tunnel classification deals with the two types of tunnels where the major construction operations are accomplished from the surface, namely, cut-and-cover tunnels and trench tunnels. The latter are used exclusively for subaqueous work. In the trench method, prefabricated tunnel sections are constructed in shipyards or dry docks, floated to the site, sunk into a dredged trench, and joined together underwater; the trench is then backfilled. When conditions of subsurface soil, amount of river current, volume and character of river traffic, availability of construction facilities, and type of existing waterfront structures are favorable, the trench method may prove more economical than alternative methods.

By far the most common type of tunnel construction for shallow tunnels, which often occurs in urban areas, is the cut-and-cover method. As the name implies, the method consists of excavating an open cut, building the tunnel within the cut, and backfilling over the completed structure. Under ideal conditions, this method is the most economical for constructing tunnels located at a shallow depth; however, it must be noted that surface disruption and problems with utilities generally make this method very expensive and difficult.

General Design Considerations

The feeling of confinement in tunnels is unpleasant and traffic noises are magnified. Because tunnels are the most expensive highway structures they should be made as short as practicable. Keeping as much of the tunnel length as possible on tangent will not only minimize the length but also improve operating efficiency. Where tunnels are constructed with tight curves to the left, reduced stopping sight distance may result.

Grades in tunnels should be determined primarily on the basis of driver comfort while striving to reach a point of economic balance between the construction costs and the operating and maintenance expenses. Many factors have to be considered in tunnel lengths and grades and their effects on tunnel ventilation and lighting. Lighting expenses are highest near portals and depend heavily on availability of natural light and the need to make a good light transition. Ventilation costs depend on length, grades, natural and vehicle-induced ventilation, type of system, and air quality constraints.

The overall roadway design should avoid the necessity for guide signs within tunnels, because normal vertical and lateral clearances are usually insufficient for such and additional clearance can be provided only at very great expense. Exit ramps should be located a sufficient distance downstream from the tunnel portal to permit necessary guide signs between the tunnel and the point of exit. This distance should be a minimum of 1,000 ft. It is also highly undesirable that traffic be required to merge, diverge, or weave within a tunnel, as might be the case if the tunnel is located between two closely spaced interchanges. Forks and exit or entrance ramps should be avoided within tunnels.

Tunnel Sections

From the standpoint of service to traffic, tunnels should not differ materially from grade separation structures. The same standards for alinement and profile and for vertical and horizontal clearances apply generally except that minimum values generally are used because of high cost and restricted right-of-way.

Full left- and right-shoulder widths of the approach freeway desirably should be carried through the tunnel. Actually, the need for added lateral space is greater in tunnels than under separation structures because of the greater likelihood of vehicles becoming disabled in the longer lengths. If shoulders are not provided, intolerable delays may result when vehicles become disabled during periods of heavy traffic. However, the cost of providing shoulders in tunnels may be prohibitive, particularly on long tunnels that are constructed by the boring or shield-drive methods. Thus, the determination of the width of shoulders to be provided in a tunnel should be based on an in-depth analysis of all factors involved. Where it is not feasible to provide shoulders in a tunnel, arrangements should be made for around-the-clock emergency service vehicles that can promptly remove any stalled vehicles.

Figure IV-15 illustrates the minimum and desirable sections for two-lane tunnels. The minimum roadway width, between curbs, as shown in Figure IV-15A, should be at least 2 ft greater than the approach traveled way, but not less than 24 ft. The curb or sidewalk on either side should be a minimum of 1.5 ft. The total clearance between walls of a two-lane tunnel should be a minimum of 30 ft. The total width and curb or sidewalk width can be varied as needed or desired within the 30-ft minimum wall clearance; however, the widths should not be less than the minimum values stated above. The minimum vertical clearance is 16.5 ft for freeways and 14.5 ft for other highways.

However, the minimum clear height should not be less than the maximum height of load that is legal in a particular State, and it is desirable to provide an allowance for future repaving of the roadways.

Figure IV-15B illustrates the desirable section, a 44 ft-wide section with two 12 ft lanes, a 10-ft right shoulder, a 5-ft left shoulder, and a 2.5 ft curb or sidewalk on each side. The roadway width may be distributed to either side in a different manner if needed to better fit the dimensions of the tunnel

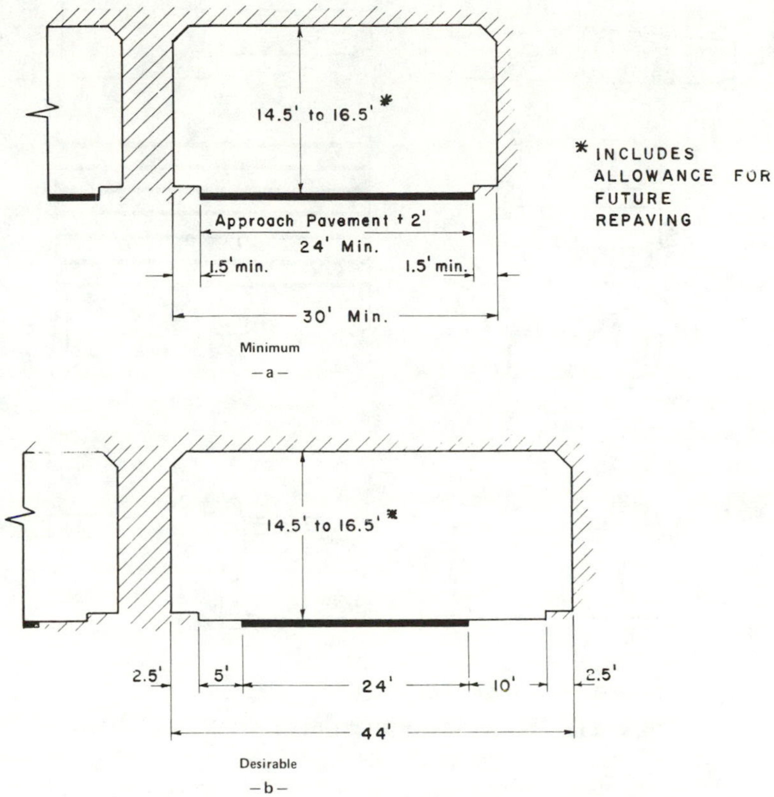

Figure IV-15. Typical two-lane tunnel sections.

approaches. The vertical clearance for the desirable section is 16.5 ft for freeways and 14.5 ft for other highways.

Normally, pedestrians are not permitted in freeway tunnels; however, space should be provided for emergency walking and for access by maintenance personnel. Raised sidewalks, 2.5 ft wide, are desirable beyond the shoulder areas to serve the dual purpose of a safety walk and an obstacle to prevent the overhang of the vehicles from damaging the wall finish or the tunnel lighting fixtures.

Figure IV-16 shows several diagrammatic tunnel sections and partially covered highways. Directional traffic should be separated for safety reasons and to relieve the dizzying effect of two-way traffic in the confined space. This separation can be achieved by providing a twin opening as shown in Figure IV-16A, by multilevel sections as shown in Figure IV-16B and IV-16C, or

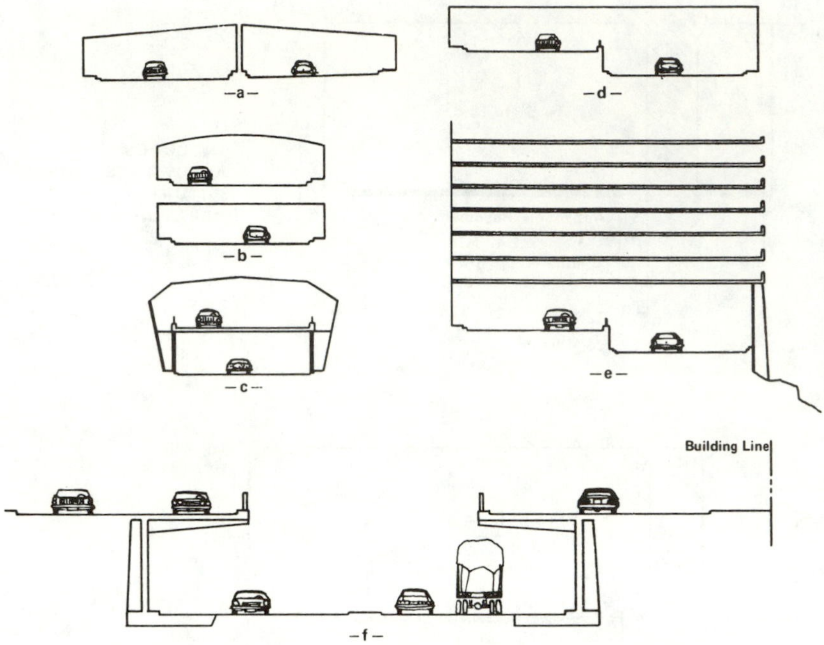

Figure IV-16. Diagrammatic tunnel sections.

by terraced structures as shown in Figure IV-16D. Figure IV-16E illustrates roadways that are tunneled under hillside buildings to mutual advantage. The terraced roadways are open on the outside for light, view, and ventilation. A partially covered section, as shown in Figure IV-16F, provides light and ventilation to the vehicle user while minimizing freeway intrusion on the community traversed. This type of cross section is covered in the section "Depressed Freeways" in Chapter VIII.

Examples of Tunnels

Figure IV-17 shows a freeway tunneling through a hillside. The portals are staggered and attractively designed. The interchange is located a sufficient distance from the tunnel to allow space for effective signing and the necessary traffic maneuvers.

Figure IV-18 shows the interior of a three-lane directional tunnel. Note the two rows of lighting fixtures on each wall in the foreground. The upper row

Figure IV-17. Entrance to freeway tunnel.

Figure IV-18. Interior of a 3-lane one-way tunnel.

of lights provides supplemental daytime lighting at the entrance portal to reduce the optical shock of traveling from natural to artificial lighting. The ceramic-tile finish on the walls and ceiling provides reflective surfaces that increase the brightness level and uniformity of lighting. A curb-to-curb width of 41 ft is provided with 2.5 ft wide safety walks along each wall.

PEDESTRIAN CROSSINGS

Street Overcrossings and Bridge Crossings

Provisions for pedestrians must often be considered on street overcrossings and on longer bridge crossings. On lower-speed streets, a barrier curb at the edge of the sidewalk is usually sufficient to protect pedestrians from vehicular traffic. Continuity of curb height should be maintained on the approaches to and over structures. For higher-speed roadways on structures, a barrier-type rail of adequate height may be used to separate the walkway and the vehicle lanes. A pedestrian-type rail or screen should be used at the outer edge of the walkway on structures.

Separations for Pedestrian Crossings

Pedestrian separations should be provided where pedestrian volume, traffic volume, intersection capacity, and other conditions favor their use. Local, State and Federal laws and codes should be consulted for possible additional requirements concerning need, as well as more stringent design requirements. Location and design of pedestrian undercrossings or overcrossings require individual study. They may be warranted where there are heavy peak pedestrian movements, such as at central business districts, factories, schools, or athletic fields, in combination with moderate to heavy vehicular traffic or where abnormal hazard or inconvenience to pedestrians would otherwise result. Pedestrian separations, usually overpasses, may be necessary at freeways or expressways where cross streets are terminated. On many freeways, highway overpasses for cross streets may be limited to three- to five-block intervals. As this situation imposes an extreme inconvenience on pedestrians desiring to cross the freeway at the terminated streets, pedestrian separations may be provided.

Where there are frontage roads adjacent to the arterial highway the pedestrian crossing may be designed to span only the through roadway or the entire roadway. Separations of both the through roadways and the frontage roads may not be justified if the frontage roads carry light and relatively slow-moving traffic; however, in some cases the separation should span the frontage roads as well. Fences may be required to prevent pedestrians crossing the arterial in spite of separations.

Pedestrian crossings or overcrossing structures at arterial streets are not likely to be used unless it is obvious to the pedestrian that it is easier to use such a facility than to cross the traveled way. Generally, pedestrians are more reluctant to use undercrossings than overcrossings. The usual pedestrian undercrossing nuisance hazard is minimized by locating the underpass on line with the approach sidewalk and ramping the sidewalk gently to permit continuous vision through the underpass from the sidewalk. Lighting is necessary on longer undercrossings.

Pedestrian ramps should be provided at all pedestrian separation structures. When warranted and possible, a stairway can be provided in addition to a ramp. Ramps should be deleted only when it would be infeasible for mobility-limited persons to reach the pedestrian separation structures due to unusual topographical or architectural obstacles or when alternate safe and reasonably convenient means are provided to cross the highway.

The ramp should have a maximum slope of 12:1 with a maximum rise of 30 in. between landings. Landings should have a minimum length of 5 ft and should be of sufficient width to allow wheelchairs to maneuver. (See Chapter II, "Handicapped Pedestrian Characteristics.")

Stairways, when provided, should have steps with a maximum rise of 7 in. The tread projection should be appropriate to normal stair formula to achieve the needed slope. Local, State, and Federal codes address this length. The nose of the stair tread should not be a square projection but should be rounded to allow reasonable usage by handicapped people. A landing should be provided for every 5- or 6-ft change in elevation. To limit the space required, stairs or ramps may be made more compact by reversing direction or by a winding alinement. (See Figures IV-19 and IV-20.) All ramps and stairways should have a handrail located between 30 and 34 in. above the ramp or above the tread nose.

The vertical clearance over roadways at pedestrian overpasses should be slightly higher than the minimum vertical clearance for vehicular structures. The lighter pedestrian structures are less resistant to impact if struck by an overheight vehicle. This additional height requirement results in a difference in elevation between a walkway at roadway level and a walkway of the pedestrian overpass of between 17 and 22 ft. It is generally desirable to minimize this difference in elevation by depressing the through roadway or raising the approach to the pedestrian overpass.

Pedestrian crossings should have a vertical clearance of 8 ft and.in some cases 10 ft. This clearance results in a minimum difference in elevation between the roadway and the walkway beneath of 10 to 12 ft or more. This difference in elevation can be diminished by raising the roadway or depressing the walkway approaching the pedestrian subway.

Walkways for pedestrian separations should have a minimum width of 8 ft. Greater widths may be required where there are exceptionally high volumes of pedestrian traffic, such as in the downtown areas of larger cities and around sports stadiums.

A serious problem associated with pedestrian and highway overpasses with sidewalks is the act of dropping objects into the path of the traffic moving under the structure. The seriousness of the consequences that can result from objects being thrown from bridges is brought out by scattered but continuous countrywide reports of fatalities and major injuries caused by this type of vandalism. There is no practical device or method yet devised that can be universally applied to prevent a determined individual from dropping an object from an overpass. Small objects can be dropped through mesh screens. Solid plastic enclosures are more effective deterrents but they are expensive and may be insufferably hot in the summer. Another undesirable feature may be that they obscure and darken the pedestrian traveled way, which is conducive to other forms of criminal activity. Any completely enclosed pedestrian overpass has an added problem of children walking or playing on top of the enclosure. In areas subject to snow and icing conditions, consideration should be given to the possibility of melting snow and ice dropping from the roof of a covered overpass and falling onto the roadway below.

At present it is not possible to establish absolute warrants as to when or where barriers should be installed to discourage the throwing of objects from structures. The general need for economy in design and desire to preserve the clear lines of a structure unencumbered by screens must be carefully balanced against the absolute requirement that the highway traveler and overpass pedestrian be provided maximum safety.

Locations where screens definitely should be considered at the time the overpass is constructed are as follows:

1. On an overpass near a school, a playground, or elsewhere where it would be expected that the overpass would be frequently used by children unaccompanied by adults.
2. On all overpasses in large urban areas used exclusively by pedestrians and not easily kept under surveillance by police.
3. On overpasses where experience on nearby structures indicates a need for screens.

Screens should also be installed on existing structures where there have been incidents of objects being dropped from the overpass and where it seems evident that increased surveillance, warning signs, or apprehension of a few individuals involved will not effectively alleviate the problem.

More complete information on the use of protective screens on pedestrian overpasses is available in *A Guide for Protective Screening of Overpass Structures* **(10)**.

Figures IV-19 and IV-20 illustrate pedestrian overcrossings of freeways. Figure IV-19 shows a partially enclosed pedestrian walkway over a six-lane divided freeway. The ramp leading up to the overcrossing is parallel to the freeway.

Figure IV-20 shows a spiral pedestrian ramp leading up to a pedestrian bridge over a divided freeway.

Figure IV-19. Pedestrian overpass.

Curb-Cut Ramps for the Physically Handicapped

When designing a project that requires or may require curbs and adjacent sidewalks to accommodate pedestrian traffic, proper attention should be given to the requirements of physically handicapped persons whose means of mobility is dependent upon wheelchairs and other devices. The street intersection with steep-faced curbs is one form of obstacle that can be alleviated while still providing adequate and reasonable mobility for the handicapped and without sacrificing the mobility of others. This can be done by proper design of curb cuts.

Design details of curb-cut and wheelchair ramps will vary in relation to the following factors:

Figure IV-20. Pedestrian separation structure.

1. Sidewalk width
2. Sidewalk location with respect to the back face of curb
3. Height and width of curb cross section
4. Design turning radius and length of curve along the curb face
5. Angle of street intersections
6. Planned or existing location of sign and signal control devices
7. Storm water inlets and public surface utilities
8. Possible sight obstructions
9. Street width
10. Border width
11. Roadway grade

These factors necessitate the establishment of basic ramp types and differing design applications predicated on acceptable design standards of each agency and the geometric characteristics of each intersection. Several agencies have design guides and standards that are reflected in the following discussions and design recommendations **(25)(28)**. The designer should consult current editions of guides and standards for state-of-the-art criteria.

Figure IV-21 illustrates curb ramps in varying conditions. The desirable slope should not exceed 12:1 in any location where pedestrians travel along or across the ramp. Figure IV-21A shows the condition where the entire grade differential is totally achieved outside the sidewalk. In this case a steep face curb can be used along the ramp. This condition is always preferable since it does not require anyone to walk across the ramped area.

In most areas where sidewalks are required, the ramp must be incorporated in the sidewalk as shown in Figures IV-21B, IV-21C and IV-21D. IV-21B shows the maximum slope along the sidewalk at 10:1; this is appropriate when at least 4 feet of level sidewalk is available beyond the ramp. Figures IV-21C and IV-21D show two possible designs where there is insufficient area for a 4 ft normal walk beyond the ramp. The maximum slope in this case should be 12:1.

In some designs a built-up ramp may be desirable. Figure IV-21E shows such a design. These should not project into the traffic lane. Also, drainage may be adversely affected if not properly considered.

Figure IV-21F is a cross-section and profile of a curb cut ramp. The surface of the ramp should be flush with the pavement at the face of the curb.

The location of the curb-cut opening and ramp must be carefully coordinated with respect to the pedestrian crosswalk lines. This planning must insure that the ramp opening (at the fully depressed curb) is situated within the parallel boundaries of the crosswalk markings.

Figure IV-22 shows a typical curb-cut ramp at the middle of the curb radius. In areas where pedestrian and/or vehicular traffic volumes are moderate to high, use of this configuration should be discouraged. Such placement forces the ramp users to enter diagonally into the intersection, perhaps misdirecting them and exposing them to hazards of traffic from two directions. This situation is of special concern to the blind. Also, for radii less than 10 ft, the requirement that a 3-ft "bottom discharge" platform be provided on the pavement side of the radius curb located within the marked crosswalk may cause encroachment of the inside edge-of-crosswalk line into the outer travel lane. This may expose crosswalk users to hazard from through traffic.

Figure IV-23 illustrates curb-cut ramps at the beginning and end points of the curb radius.

Figure IV-24 shows a ramp at the middle of the curb radius with an architectural texture.

Ramps for the handicapped are not limited to intersections and marked crosswalks, and ramps should also be provided at other appropriate or designated points of pedestrian concentration, such as loading islands, midblock pedestrian crossings, and locations where pedestrians could not otherwise recognize the proper place to cross the highway or street. Because nonintersection pedestrian crossings are generally unexpected by the motorist, warning signs should be installed and adequate visibility provided by prohibiting parking. For additional design guidance and recommendations with respect to pedestrian crosswalk markings, refer to the MUTCD (**2**). Figure IV-25 shows a midblock curb-cut ramp.

Ramps for the handicapped should have a textured nonskid surface. This texture not only provides a measure of safety for the user but also warns a blind person of the presence of the ramp.

As shown in Figure IV-26, when a major highway or secondary intersecting road involves pedestrian traffic and the roadway geometrics require convex islands or median dividers, the plan should include ramps for the handicapped. People in wheelchairs cannot safely take refuge in islands that are less than 4 ft wide because of the vulnerability to moving traffic. When they are propelled by attendants, a 6-ft island width is desirable.

Each intersection will differ with respect to the intersection angles, turning roadway widths, size of islands, drainage inlets, traffic control devices, and other variables previously described. An appropriate plan should be prepared (say 1:20 scale) that indicates all of the desired geometrics, including vertical profiles at the curb flow line. The plan could then be evaluated for the convenient and safe locations of the ramps for the handicapped. Drainage inlets should be located on the upstream side of all crosswalks and sidewalk ramps. This design operation will govern the pedestrian crosswalk patterns, stop bar locations, regulatory signs, and in the cases of new construction, establish the most desirable location of signal support standards.

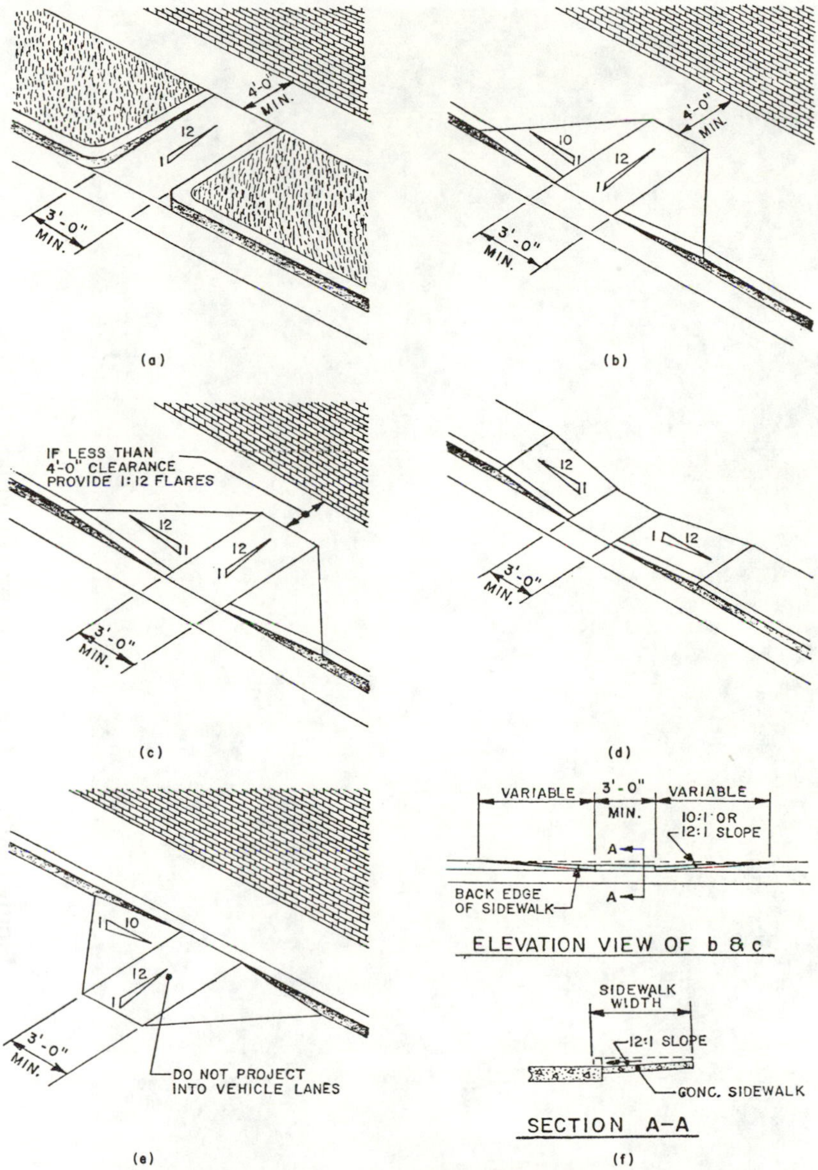

Figure IV-21. Curb-cut ramp details.

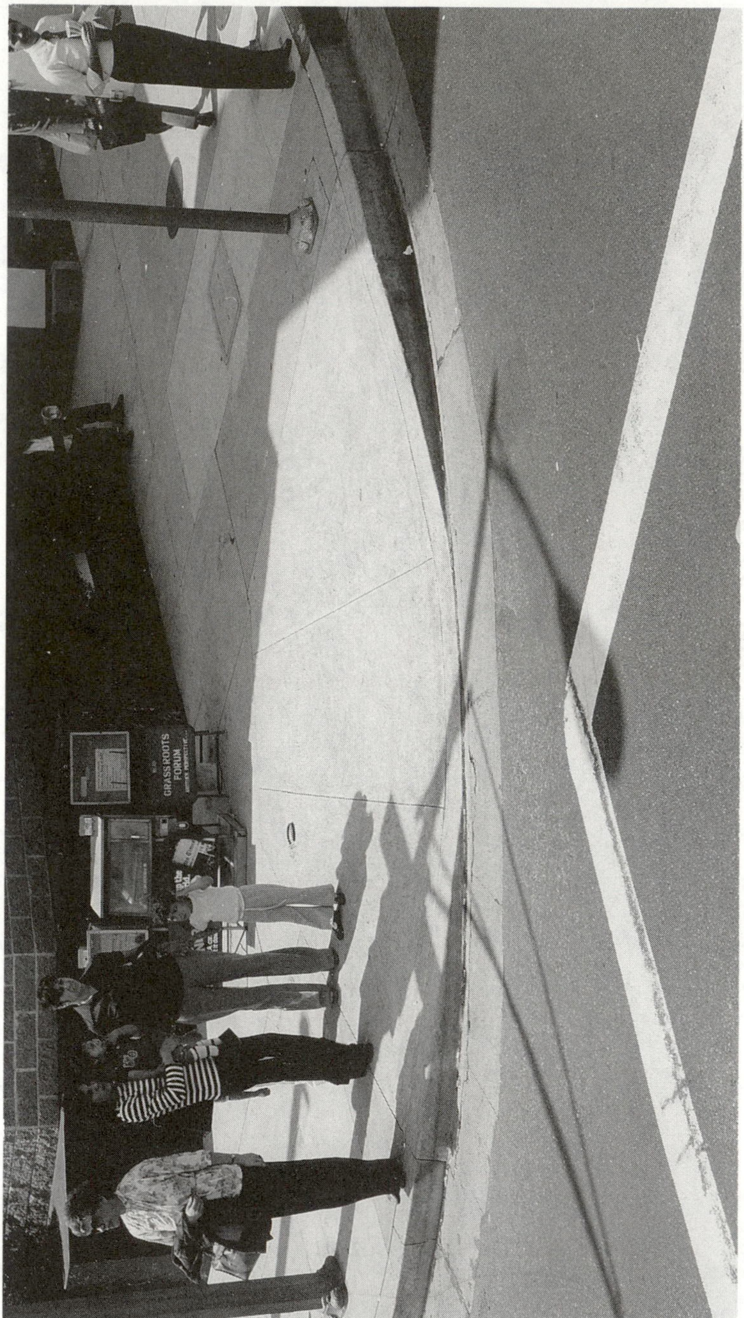

Figure IV-22. Curb-cut ramp at middle of radius.

Figure IV-23. Curb-cut ramps at ends of curb radius.

Figure IV-24. Curb-cut ramp with architectural texture.

Figure IV-25. Curb-cut ramp at mid-block.

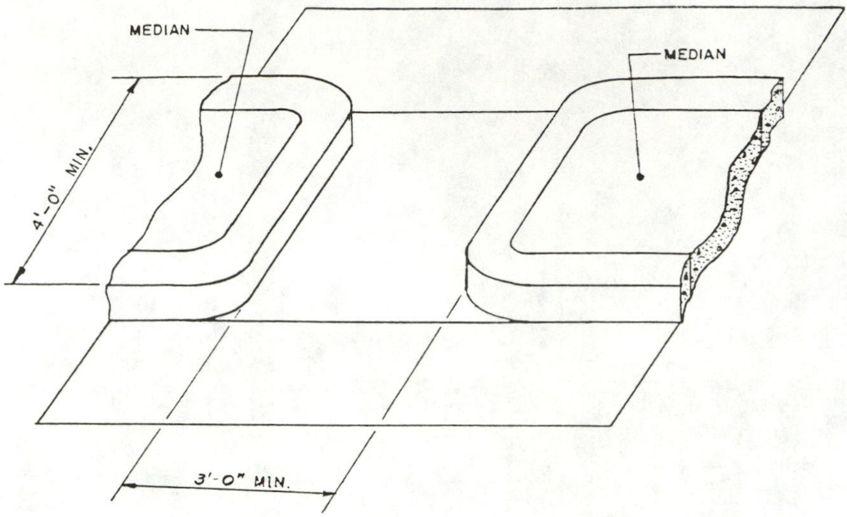

Figure IV-26. Median and island openings.

Ramps should be provided at all intersections where curb and sidewalk are provided even though the highway grade is greater than 12:1. This provision allows for wheelchairs when the occupant is assisted by another person.

For further information on curb-cut ramps for the physically handicapped, see the *Uniform Federal Accessibility Standards* **(28)**.

BICYCLE FACILITIES

It is practical for most of the mileage needed for bicycle travel to be comprised of the street and highway system generally as it presently exists. However, at certain locations, or in certain corridors, it is appropriate to supplement further the existing highway system by providing specifically designated bikeways.

Provisions for bicycle facilities should be in accordance with the AASHTO publication *Guide for Development of New Bicycle Facilities* **(11)**. In addition, appropriate consideration should be given to possible measures for enhancing probable bicycle travel on the highway, even if specific bicycle facilities are not provided.

Chapter II provides further discussion on the subject of Bicycle Facilities.

BUS TURNOUTS

Bus travel is an increasingly important mode of mass transportation. Bus turnouts serve to remove the bus from the traffic lanes. The location and design of turnouts should provide ready access in the safest and most efficient manner possible.

Freeways

The basic requirement for a bus turnout, when exclusive bus roadways are not provided, is that the deceleration, standing, and acceleration of buses be effected on pavement areas clear of and separated from the through-traffic lanes. Other elements in the design of bus turnouts include passenger platforms, ramps, stairs, railings, signs, and markings. Speed-change lanes should be long enough to enable the bus to leave and enter the through-traffic lanes at approximately the average running speed of the highway without undue inertial discomfort or sidesway to passengers. The length of acceleration lanes from bus turnouts should be well above the normal minimum values, as the buses start from a standing position and the loaded bus has a lower acceleration capability than passenger cars. Normal-length deceleration lanes are suitable. The bus standing area and speed-change lanes, which include the shoulders, should be 20 ft wide to permit passing a stalled bus. The pavement areas of turnouts should contrast in color and texture with the through-traffic lanes to discourage through traffic from encroaching on or entering the bus stop.

The dividing area between the outer edge of freeway shoulder and the edge of bus turnout lane should be as wide as feasible, preferably 20 ft or more. However, in extreme cases this width could be reduced to a minimum of 4 ft. A barrier is usually needed in the dividing area, and fencing is desirable to keep pedestrians from entering the freeway. Pedestrian loading platforms should not be less than 5 ft wide and preferably 6 to 10 ft wide. Platforms should be covered if required by the climate. Typical cross sections of turnouts are illustrated in Figure IV-27. This illustration shows (1) a normal section, (2) one through an underpass, and (3) a section on an elevated structure.

Arterials

The interference between buses and other traffic can be considerably reduced by providing turnouts clear of the lanes for through traffic. It is somewhat rare that sufficient right-of-way is available on the lower type arterial streets to permit turnouts in the border area, but advantage should be taken of every opportunity to do so.

To be fully effective, bus turnouts should incorporate (1) a deceleration lane or taper to permit easy entrance to the loading area, (2) a standing space sufficiently long to accommodate the maximum number of vehicles expected

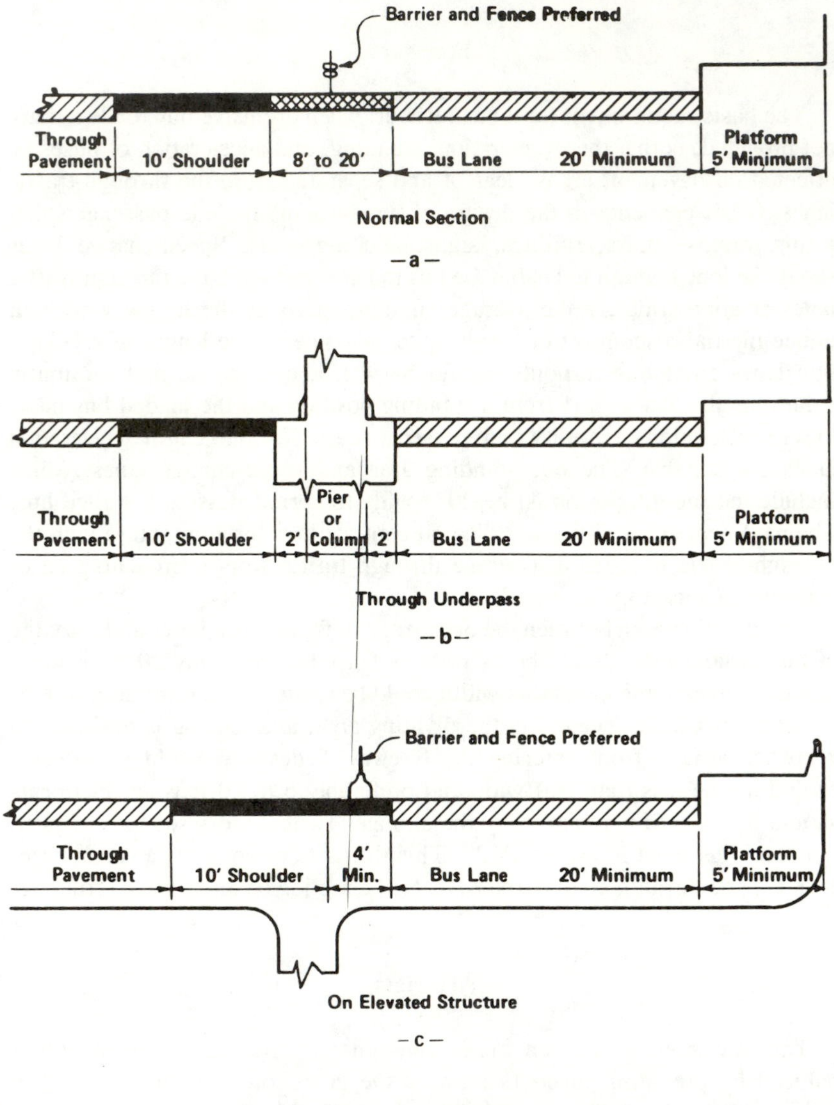

Figure IV-27. Bus turnouts.

to occupy the space at one time, and (3) a merging lane to enable easy reentry into the through-traffic lanes.

The deceleration lane should be tapered at an angle flat enough to encourage the bus operator to pull completely clear of the through lane. Usually it is not feasible to provide a length sufficient to permit deceleration from highway speeds clear of the traveled way. A taper of about 5:1, longitudinal to transverse, is a desirable minimum. When the bus stop is on the far side of an intersection, the intersection area may be used as the entry area to the stop.

The loading area should provide about 50 ft of length for each bus. The width should be at least 10 ft and preferably 12 ft.

The merging or reentry taper may be somewhat more abrupt than the deceleration taper but, preferably, should not be sharper than 3:1. Where the turnout is on the near side of an intersection, the width of cross street is usually great enough to provide the necessary merging space.

The total length of turnout for a two-bus loading area of minimum design should be about 180 ft for a midblock location, 150 ft for a near-side location, and 130 ft for a far-side location. These dimensions are based on a loading area width of 10 ft. They should be increased by 10 to 15 ft for a width of 12 ft. Greater lengths of bus turnouts expedite bus maneuvers, encourage full compliance on the part of bus drivers, and lessen interference with through traffic.

Figure IV-28 shows a bus turnout at a midblock location. The width of the turnout is 10 ft. Length of turnout, including tapers, is 210 ft. The deceleration and acceleration tapers are 4:1.

For more information on bus turnouts see the AASHTO *Guide for Design of High Occupancy Vehicle and Public Transfer Facilities* **(14)**.

Park-and-Ride Facilities

Location

Park-and-ride facilities should be located adjacent to the street or highway and be visible to the commuters whom they are intended to attract. Preferably, the parking areas should be located at points that precede the bottlenecks or points where there is significant traffic congestion; they should be located as close to the residential areas as possible in order to minimize travel by vehicles with only one occupant, and they should be located far enough out that land costs are not prohibitive.

Other considerations that affect parking lot location in addition to optimum distance are impacts on surrounding land uses, available capacity of the highway systems between the roadway and proposed sites, terrain, and the

Figure IV-28. Mid-block bus turnout.

costs to acquire the land.. The sites selected should be compatible with surrounding land uses.

Design

The size of the parking lot is dependent upon the design volume, the available land area, and the size and number of other parking lots in the area.

Kiss-and-ride (pickup and dropoff) facilities should be provided at each parking area. Twenty to 60 spaces represent a reasonable range. The area should involve dropoff close to the station entrance, plus a holding or short-term parking area for passenger pickup. This area should be clearly separated from the park-and-ride areas.

Vehicular priorities should be as follows: (1) bus loading and unloading, (2) taxi or kiss-and-ride, (3) bicycle parking, (4) handicapped parking, and (5) park-and-ride. Conflicts between pedestrians and vehicles should be minimized. Parking aisles should preferably be located normal to the bus roadway so that pedestrians are not required to cross the parking drives. All bus roadways should permit passing of standing buses. This requirement means that all bus roadways should have a minimum width of 20 ft. Facilities should be designed for self-parking. Parking spaces should be 9 by 20 ft for full-sized cars; where a special section is provided for subcompact cars, 8- by 15-ft spaces are sufficient. Where separate handicapped parking is provided, spaces should have a minimum width of 11 ft. Generally, aisles should be a minimum of 20 ft wide with right-angle parking and 18 ft for angle parking; however, bus designs and parking angle may require other widths.

Sidewalks should be a minimum of 5 ft wide and loading areas should be 12 ft wide. Principal loading areas should be provided with curb-cut ramps. Preferably, pedestrians should not have to walk more than 400 ft, although slightly longer distances may be permitted under some circumstances. Pedestrian paths from parking spaces to loading areas should be as direct as possible. Facilities for locking bicycles should be provided where needed.

Grades of parking areas can be relatively steep except for the roadways that accommodate buses. Grades for roadways that accommodate buses should be not greater than 7 percent, and acceleration grades should not be greater than 4 percent. Curvature, radius of planned vehicular paths within the parking area, and access roads should be sufficiently large to accommodate the vehicles that they are intended to serve.

Access to the lots should be at points where they will disrupt existing traffic as little as possible. The intersection of the access roads to the lots should be preferably at least 300 ft from other intersections, and there should be sufficient sight distance for vehicles to exit and enter the lot. This means that exits and entrances should not be located on crest vertical curves. There should be at least 300 ft corner sight distance.

There should be at least one exit and entrance for each 500 spaces in a lot, and preferably, exits and entrances are provided at separate locations. If possible, the exits and entrances should access different streets. It is also desirable to provide separate access for public transit vehicles.

Curb returns should be at least 30 ft in radius, although 15-ft radii are suitable for access points used exclusively by passenger vehicles.

Principal passenger-loading areas should be provided with shelters to protect waiting public transit patrons. Such shelters should as a minimum be sufficiently large to accommodate the volumes of offpeak waiting passengers and preferably be larger. Once the number of passengers that the shelter is to service has been determined, the size of the shelter can be arrived at by multiplying a factor of 3 to 5 sq. ft per person. Because the shelter can be expanded at a later date relatively easily, provided sufficient platform space is installed initially, the provision of a shelter at the time of original construction to accommodate all passengers is not critical. The shelter should be provided with lighting, benches, route information, trash receptacles, and often a telephone.

Figures IV-29 and IV-30 show typical park-and-ride facilities. The design of the bus loading area can be of a parallel or a sawtooth design; the best arrangement depends on the number of buses expected to use the facility. Where more than two buses are expected to be using a facility at one time the sawtooth arrangement is generally preferable, as it is easier for buses to bypass a waiting bus. A recommended design of a sawtooth arrangement is shown in Figure IV-34. The length of space that should be provided for a par-

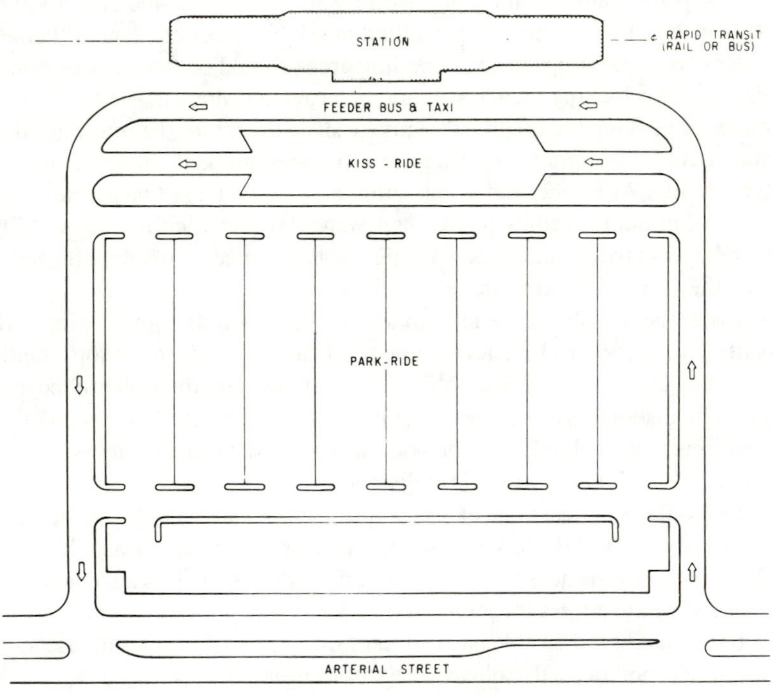

Figure IV-29. Typical park-and-ride lot.

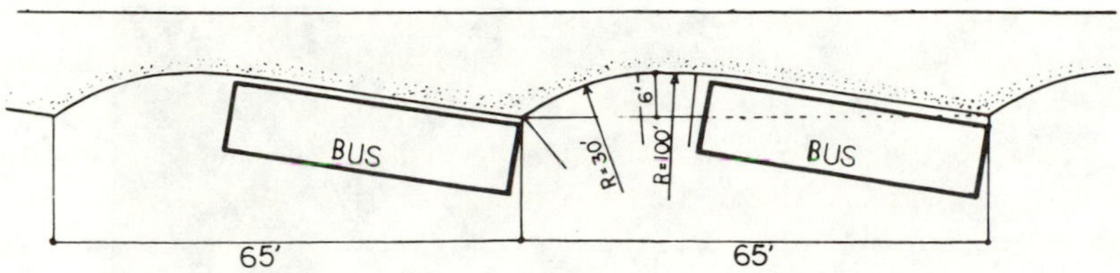

Figure IV-30. Sawtooth bus loading area.

allel design is 95 ft. This length will permit loading of two buses. For each additional space 45 ft should be allowed. The loading area should be at least 24 ft wide in order to permit passing a standing bus. The area delineating the passenger refuge area should be curbed in order to reduce the height between the ground and the first bus step and reduce encroachment by buses on the passenger areas. Parallel-type loading areas should not be located on curves as it makes it very difficult for drivers to park so that both the front and rear doors are close to the curb.

Special designs may be required to accommodate articulated buses, particularly where a sawtooth arrangement is used. A well-designed parking lot includes a buffer area around the lot with appropriate landscaping and often fence to separate land areas. The buffer should be at least 10 ft wide.

On all but the smaller lots, lighting should be provided. A level of 0.2 to 0.5-fc average maintained intensity will generally suffice.

Drainage systems should be designed so that parked cars will not be damaged by storm water. Under some circumstances, some ponding of water may be permitted or may even be desirable when the drainage is designed as part of a storm water management system. The intensity of storm that the drainage system should accommodate depends in some cases on the practice of the municipality. Permissible depths of ponding should generally not exceed 3 to 4 in. in areas where cars are parked, and there should be no ponding on pedestrian and bicycle routes or where persons wait for transit vehicles.

Figure IV-31 shows a typical park-and-ride facility. For additional information refer to the AASHTO *Guide for Design of High Occupancy Vehicle and Public Transfer Facilities* **(14)**.

ON-STREET PARKING

The design and development of a roadway network is to provide for the safe and efficient movement of the vehicles operating on the system.

Figure IV-31. Typical park-and-ride facility.

Although the movement of vehicles is the primary function of a roadway network, segments of the network may, as a result of land use, be required to provide for the parking vehicles.

In the design of freeways and control of access-type facilities, as well as on most rural arterials, collectors and locals, only emergency stopping or parking should be permitted or considered. Since the primary service of an arterial is the movement of vehicles it is desirable to prohibit parking on urban arterial streets and rural arterial highway sections. However, within urban areas and in rural communities located on arterial highway routes, existing and developing land uses necessitate the consideration of on-street parking. Usually, adequate off-street parking facilities are not available, so the designer must take cognizance of the fact that on-street parking must be a design consideration so that the proposed street or highway improvement will be compatible with the land use.

When on-street parking is to be an element of design, parallel parking should be considered. Under certain circumstances, angle parking is an allowable form of street parking. The type of on-street parking selected should depend on the specific function and width of the street, the adjacent land use, traffic volume, as well as existing and anticipated traffic operations. Angle parking presents special problems because of the varying length of vehicles and the sight distance problems associated with vans and recreational vehicles. The extra length of such vehicles may interfere with the traveled way.

It can generally be stated that on-street parking decreases through capacity, impedes traffic flow, and increases accident potential. For these reasons, it is desirable to prohibit parking on arterial streets. However, where parking is needed and adequate off-street parking facilities are not available or feasible, on-street parking may be necessary. A factor that compounds the urban parking problem is the distribution of off-street parking facilities within urban central business districts and the lack of off-street facilities in urban neighborhood commercial areas. As a consequence, there is a demand for onstreet parking to provide for the delivery and pick-up of goods. Frequently, alley and other off-street loading areas are not provided in many communities. Short-duration parking for business or shopping must, therefore, be accommodated.

Curb parking on urban arterial streets is acceptable when the available through-traffic lanes can accommodate the traffic volumes. As discussed in the section on arterials, provisions should be made on rural sections for emergency stopping only. When possible on projects of urban arterial street reconstruction or on projects where additional right-of-way is being acquired to upgrade an existing route to arterial status, parking should be eliminated to increase capacity and safety. The impacts on abutting land uses must, however, be carefully considered, as the loss of existing on-street parking can cause significant loss in the economic well being of the abutting property.

It has been found that most vehicles will park within 6 to 12 in. of the curb face when parking parallel and on the average occupy approximately 7 ft of actual street space. The desirable minimum width of a parking lane is 8 ft; however, to provide better clearance and the potential to use the parking lane during peak periods as a through-travel lane, a parking lane of 10 ft to 12 ft is desirable. This width will also accommodate delivery vehicles adequately. Where traffic volumes indicate that the parking lane on an arterial roadway will not be required for through use, the minimum parking lane width should be 8 ft.

On urban collector streets there is an equal demand for land access and mobility. Rural collector highways generally require provisions for emergency stopping only. On most urban collector roadways which must accommodate a wide variety of traffic operations and land uses, the desirable parking lane width is 8 ft. To provide better clearance and the potential to use the parking lane during peak periods as a through-travel lane, a parking lane of 10 ft to 12 ft is desirable. A 10-ft to 12-ft parking lane will also accommodate urban transit operations. On urban collector streets within residential neighborhoods where only passenger vehicles must be accommodated in the parking lane, 7-ft parking lanes have been successfully used. A common urban residential collector street width of 36 ft is frequently used. This section is designed to provide two (2) travel lanes of 11 ft and parking lanes of 7 ft on each side.

On-street parking is generally permitted on local streets. A 26-ft wide pavement is the typical section used in many urban residential areas. This width assures one freely moving lane even when parking occurs on both sides. Specific parking lanes are not usually designated on such local streets but rather the primary purpose of such local streets is to insure that there is one freely moving lane. User inconvenience to the motorist of the pavement section is occasioned by the lack of two moving lanes; however, the frequency of this situation has been found to be remarkably low. Random intermittent parking on both sides has been found to result in areas where two-way movement can be accommodated.

In the construction of new roadways, pavement construction procedures should be carefully considered so as to provide a longitudinal pavement joint at the boundary of the proposed parking lane. It has been found that such joints aid in insuring that the parked vehicle clears the parallel travel lane. On asphalt-surfaced streets, pavement markings are recommended to indentify the parking lane. The marking of parking spaces encourages more orderly and efficient use of parking spaces where parking turnover occurs and tends to prevent encroachment on fire hydrant zones, bus stops, loading zones, and approaches to corners.

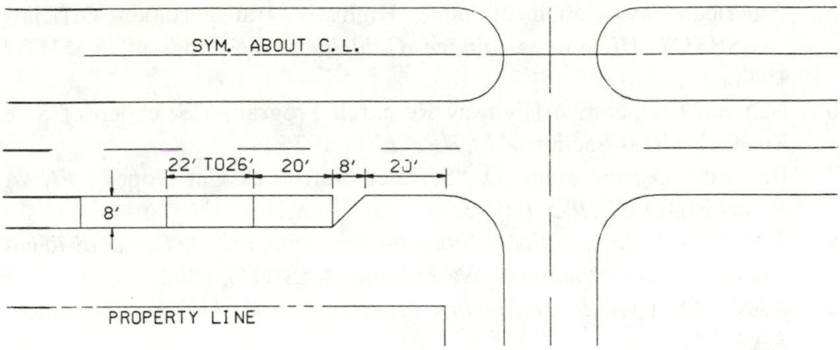

Figure IV-32. Parking lane transition at intersection.

In urban areas and, in particular, within central business districts and commercial areas where significant pedestrian crossings are likely to occur, the design of the parking lane/intersection relationship should be given consideration. When the parking lane is carried up to the intersection, motorists may develop on additional lane for right-turn movements. Such movements may cause operational problems and often result in turning vehicles mounting the curb and possibly striking such intersection elements as traffic signals, utility poles, or street-light standards. The transitioning out of the parking lane a minimum of 20 feet in advance of the intersection is one method of eliminating this problem. An example of such treatment is shown in Figure IV-32. A second method is to prohibit parking for such a distance as to create a short turn lane.

REFERENCES

1. Highway Research Board. "Relationship of Highway Geometry to Traffic Accidents." *HRR 312*. 1970.
2. National Joint Committee on Uniform Traffic Control Devices. *Manual on Uniform Traffic Control Devices*. Washington: U.S. Government
3. National Cooperative Highway Research Program. "Shoulder Geometrics and Use Guidelines." *NCHRP* 254, 1982.
4. Transportation Research Board. *Highway Capacity Manual. Special Report No. 209*. Washington: National Academy of Sciences, National Research Council, 1985.

5. American Association of State Highway Transportation Officials (AASHTO). *Highway Drainage Guidelines.* Washington: AASHTO 1988.
6. National Cooperative Highway Research Program. "Selection of Safe Roadside Cross Sections." *NCHRP 158* , 1975.
7. Richards, D., and Ham, D. "Serrated Soft-Rock Cut Slopes." *FHWA Report FHWA RDDP-5-1* 1973.
8. AASHTO. *An Informational Guide for Preparing Private Driveway Regulations for Major Highways.* Washington: AASHTO, 1960.
9. AASHTO. *A Guide for Erecting Mailboxes on Highways.* Washington: AASHTO, 1985.
10. AASHTO. *A Guide for Protective Screening of Overpass Structures.* Washington: AASHTO, 1969.
11. AASHTO. *Guide for Development of New Bicycle Facilities.* Washington: AASHTO, 1981.
12. AASHTO. *Roadside Design Guide.* Washington: AASHTO, 1989.
13. AASHTO. *Guidelines for Skid Resistant Pavement.* Washington: AASHTO, 1976.
14. AASHTO. *Guide for Design of High Occupancy, Vehicle and Public Transportation Facilities.* Washington: AASHTO, 1983.
15. National Cooperative Highway Research Program. "Influence of Combined Highway Grade and Horizontal Alignment On Skidding." *NCHRP 184,* 1978.
16. Federal Highway Administration, *Safety Aspects of Curb Parking.* FHWA-RD-79-76, 1978.
17. Federal Highway Administration, *Design of Stable Channels with Flexible Linings.* HEC-15, FHWA, Office of Engineering, Bridge Division, Washington, D.C., 1975.
18. AASHTO. *Standard Specifications for Highway Bridges.* Washington, AASHTO, 1977.
19. AASHTO. *Interim Specifications Bridges.* Washington, AASHTO, 1979 and 1980.
20. Simpson, M.A., *Noise Barrier Design Handbook,* FHWA-RD76-58, 1976, National Technical Information Service, Springfield, Virginia 22161.
21. Barry, T.M., Reagan, J.A., *FHWA Highway Traffic Noise Prediction Model,* FHWA-RD-77-108, 1978, National Technical Information Service, Springfield, Virginia 22161.
22. Bowlby, W., *"SNAP 1.1—Revised Program and User's Manual for the FHWA Level I Highway Traffic Noise Prediction Computer Program,"* FHWA Report FHWA-DP-45-4, 1980, National Technical Information Service, Springfield, Virginia 22161.

23. Bowlby, W., Higgins, J., and Reagan, J., *"Noise Barrier Cost Reduction Procedure STAMINA 2.0/OPTIMS: Manual."* FHWA Report FHWA-DP-58-1, 1982, Revised March 1983, National Technical Information Service, Springfield, Virginia, 22161.
24. Bowlby, W., "Sound Procedures for Measuring Highway Noise: Final Report," FHWA Report FHWA-DP-45-1R, 1981, National Technical Information Service, Springfield, Virginia 22161.
25. "Development for Priority Accessible Network, Provisions for the Elderly and Handicapped Pedestrians." FHWA Report FHWA-IP-80-8, 1980.
26. AASHTO, *Guide for the Design of Pavement Struc tures*, Washington, AASHTO, 1986.
27. Olson, R.M., Weaver, G.D., Ross, H.E. Jr., and Post, E.R. "Effect of Curb Geometry and Location on Vehicle Behavior." NCHRP Report 150, 1974, 88pp.
28. General Services Administration, *Uniform Federal Accessibility Standards,* "Federal Register," August 7, 1984.
29. Federal Highway Administration, *Low Cost Methods For Improving Traffic Operation On Two-Lane Roads*, FHWA IP-87-2, January 1987.

Chapter V
LOCAL ROADS AND STREETS

INTRODUCTION

The following guides apply to those roads functionally classified as local rural roads and local urban streets in accordance with the discussion in Chapter I. In a jurisdictional highway classification, these guides apply generally to village or city streets and township and county roads.

A local road or residential street primarily serves as access to the farm, residence, business, or other abutting property. Although local roads and streets may be planned, constructed, and operated with such service dominating, some of them properly include the "major street or highway" features of geometric design and traffic control to expedite the safe movement of through traffic. On these roads the through traffic is local in nature and extent rather than regional, intrastate, or interstate.

Local roads and streets constitute a high proportion of the roadway mileage in the United States. The traffic volumes generated by the abutting land uses are largely short trips or a relatively small part of longer trips where the local road connects with major streets or highways of higher classifications. Because of the relatively low traffic volumes and the extensive mileage, design standards for local roads and streets are of a comparatively low order as a matter of practicality. However, to provide the requisite traffic mobility and safety, together with the essential economy in construction, maintenance, and operation, they must be planned, located, and designed to be suitable for predictable traffic operations and must be consistent with the development and culture abutting the right-of-way.

In restricted or unusual conditions it may not be possible to meet these guide values. Every effort should be made to get the best possible alinement, grade, sight distance, and proper drainage that are consistent with the terrain, the development (present and anticipated), the safety, and the funds available.

Drainage on the pavement itself and from the sides and subsurface is very important. Drainage inadequacies can lead to high maintenance costs and adverse operational conditions affecting the safety of the motorist. In snow regions, it is necessary to design so that there is storage space for plowed snow and proper drainage for melting conditions.

Safety is an important factor in all roadway improvements. However, it may not be practical or possible to obtain obstacle-free roadsides. Every effort should be made to provide as much clear roadside as is practical. This becomes more important as speeds increase. Flatter slopes, guardrail, and warning signs help to achieve roadside safety.

LOCAL RURAL ROADS

General Design Considerations

Design Traffic Volumes

Roads should be designed for specific traffic volumes and level of service appropriate for functional classification as discussed in Chapters I and II. The average daily traffic (ADT), either current or projected to some future design year, should be the basis for design. Usually, the design year is about 20 years from the date of completion of construction but may range from the current year to 20 years depending on the nature of the improvement. The current ADT may be used for the design if the design hourly volume for the design year is less than 100; otherwise the DHV should be used.

Design Speed

Geometric design features should be consistent with a design speed selected as appropriate for environmental and terrain conditions. Low design speeds are generally applicable to roads with winding alinement in rolling or mountainous terrain or where environmental conditions dictate. High design speeds are generally applicable to roads in level terrain or where other environmental conditions are favorable. Intermediate design speeds would be appropriate where terrain and other environmental conditions are a combination of those described for low and high speed. Table V-1 lists values for minimum design speeds as appropriate for traffic needs and types of terrain. Terrain types are discussed further in Chapters II and III.

Sight Distance

Minimum stopping sight distance and passing sight distance should be as shown in Tables V-2 and V-3. Criteria for measuring sight distance, both vertical and horizontal, are as follows: For stopping sight distance the height of

Type of Terrain	Speeds (mph) for Design Volumes of						
	Current ADT Under 50	Current ADT 50-250	Current ADT 250-400	Current ADT 400 and Over	DHV 100-200	DHV 200-400	DHV 400 and Over
Level	30	30	40	50	50	50	50
Rolling	20	30	30	40	40	40	40
Mountainous	20	20	20	30	30	30	30

Table V-1. Minimum design speeds.

Design Speed (mph)	Assumed Speed for Condition (mph)	Stopping Sight Distance (Rounded for Design) (ft)	K Value[a] for Crest Vertical Curves (Rounded)	K Value[a] for Sag Vertical Curves (Rounded)
20	20-20	125-125	10-10	20-20
25	24-25	150-150	20-20	30-30
30	28-30	200-200	30-30	40-40
35	32-35	225-250	40-50	50-50
40	36-40	275-325	60-80	60-70
45	40-45	325-400	80-120	70-90
50	44-50	400-475	110-160	90-110
55	48-55	450-550	150-220	100-130
60	52-60	525-650	190-310	120-160

[a]K value is a coefficient by which the algebraic difference in grade may be multiplied to determine the length in feet of the vertical curve which will provide minimum sight distance. (See Chapter III for details.)

Table V-2. Minimum stopping sight distance (wet pavements).

Design Speed (mph)	Passing Sight Distance (Rounded for Design) (mph)	K Value[a] for Crest Vertical Curves (Rounded)
20	800	210
25	950	300
30	1,100	400
35	1,300	550
40	1,500	730
45	1,650	890
50	1,800	1,050
55	1,950	1,230
60	2,100	1,430

[a]K value is a coefficient by which the algebraic difference in grade may be multiplied to determine the length in feet of the vertical curve which will provide minimum sight distance. (See Chapter III for details.)

Table V-3. Minimum passing sight distance.

eye is 3.5 ft and the height of object is 0.5 ft; for passing sight distance the height of object is 4.25 ft. Chapter III provides a general discussion of sight distance.

Grades

Suggested maximum grades for local rural roads are shown in Table V-4.

Alinement

Alinement between control points should be designed to as high a standard as possible consistent with the environmental impacts, topography, terrain, design traffic volume, and the amount of reasonably obtainable right-of-way. Sudden changes between curves of widely different radii or between long tangents and sharp curves should be avoided. The design should include passing opportunities if at all feasible. Where crest vertical curves and horizontal

	Design Speed (mph)				
Type of Terrain	**20**	**30**	**40**	**50**	**60**
Level	-	7	7	6	5
Rolling	11	10	9	8	6
Mountainous	16	14	12	10	-

Table V-4. Maximum grades.

curves occur together, there should be greater than minimum sight distance to ensure that the horizontal curves are visible to approaching drivers.

Roadway Cross Slope

Pavement or surfacing crown should be adequate to provide proper drainage. Normally, cross slopes should be as shown in Table V-5.

Surface Type[a]	Range in Rate of Cross Slope (%)
High	1.5 to 2.0
Intermediate	1.5 to 3.0
Low	2.0 to 6.0

[a]See Chapter IV for pavement surface types description.

Table V-5. Normal cross slopes.

Superelevation

For rural roads of high- and intermediate-type surfaces, superelevation should be not more than 0.10 except where snow and ice conditions prevail, in

Design Speed (mph)	Maximum e	Maximum Degree of Curve Rounded	Minimum Radius (ft)
20	.04	45.0	127
30	.04	19.0	302
40	.04	10.0	573
50	.04	6.0	955
55	.04	4.75	1,186
60	.04	3.75	1,528
20	.06	49.25	116
30	.06	21.0	273
40	.06	11.25	509
50	.06	6.75	849
55	.06	5.5	1,061
60	.06	4.25	1,348
20	.08	53.5	107
30	.08	22.75	252
40	.08	12.25	468
50	.08	7.5	764
55	.08	6.0	960
60	.08	4.75	1,206
20	.10	58.0	99
30	.10	24.75	231
40	.10	13.25	432
50	.10	8.25	694
55	.10	6.5	871
60	.10	5.25	1,091
20	.12	62.0	92
30	.12	26.75	214
40	.12	14.5	395
50	.12	9.0	637
55	.12	7.0	807
60	.12	5.75	996

NOTE: With design speeds of 30 mph or less, conditions may warrant elimination of superelevation.

Table V-6. Maximum degree of curve and minimum radius for different values of maximum superelevation.

which case the superelevation should be not more than 0.08. For aggregate roads superelevation should be not more than 0.12.

Depending on the maximum superelevation value, the maximum degree of curvature to be used for different design speeds should be as given in Table V-6.

Superelevation runoff is the length of highway needed to accomplish the change in cross slope from a section with the adverse crown removed to a fully superelevated section. Minimum lengths of runoff are given in Table V-7. Adjustments in design runoff lengths may be necessary for smooth riding, surface drainage, and good appearance. General discussion on this topic appears in Chapter III.

Number of Lanes

Two travel lanes usually can accommodate the normal traffic volumes on rural local roads. If exceptional traffic volumes occur in specific areas, additional lanes may be provided in accordance with design criteria in Chapter II. Provisions for climbing and passing lanes are covered in Chapter III.

Superelevation Rate, e	Length of Runoff, L (ft) for Design Speed (mph)					
	20	30	40	50	55	60
Reverse Crown	50	100	125	150	160	175
.04	60	100	125	150	160	175
.06	95	110	125	150	160	175
.08	125	145	170	190	200	210
.10	160	180	210	240	250	270
.12	190	215	250	290	300	310

NOTE: Length of runoff on 10- and 11-ft lanes may be reduced proportionately but no shorter than the minimum length shown for reverse crown at the respective speed.

Table V-7. Minimum length for superelevation runoff for 2-12 ft lanes of pavement.

Width of Traveled Way, Shoulder, and Roadway

Graded shoulder width is measured from the edge of traveled way to the point of intersection of shoulder slope and foreslope. The minimum roadway width is the sum of the traveled way and graded shoulder width given in Table V-8. In mountainous terrain or sections with heavy earthwork the graded width of shoulder in cuts may be decreased 2 ft, but in no case should the roadway width be less than 18 ft.

	Width (ft) for Design Volume					
Design Speed (mph)	**Current ADT less than 250**	**Current ADT 250-400**	**Current ADT Over 400**	**DHV 100-200**	**DHV 200-400**	**DHV 400 and Over**
			Width of Traveled Way			
20	18	20	20	20	22	24
30	18	20	20	20	22	24
40	20	20	22	22	22	24
50	20	20	22	22	24	24
60	20	22	22	22	24	24
			Width of Graded Shoulder (Each Side)			
All Speeds	2	2	4	6	6	8

Table V-8. Minimum width of traveled way and graded shoulder.

Structures

The design of bridges, culverts, walls, tunnels, and other structures should be in accord with the current *Standard Specifications for Highway Bridges* **(1)**.

Except as otherwise indicated herein and in Chapter IV, the dimensional design of structures should also be in accord with reference **(1)**.

The minimum design loading for new bridges should be HS-20 on local roads.

For general discussion of structure widths see Chapter X.

The minimum clear roadway widths for new and reconstructed bridges should be as given in Table V-9.

Traffic	Min. Clear Roadway Width of Bridge	Design Loading Structural Capacity
[a]ADT-400 and under	Traveled way + 2 ft (each side)	HS 20
[a]ADT-over 400	Traveled way + 3 ft (each side)	HS 20
DHV-under 400	Traveled way + 3 ft (each side)	HS 20
DHV-400 and over	Approach roadway width	HS 20

[a]Current ADT

Table V-9. Minimum clear roadway widths and design loadings for new and reconstructed bridges.

Bridges to Remain in Place

Existing substandard structures should be improved, but because of their high replacement cost, reasonably adequate bridges and culverts that meet tolerable criteria may be retained. Some of the nontechnical factors that should be considered are the esthetic value and the historical significance attached to famous structures, covered bridges, and stone arches.

Where an existing road is to be reconstructed, an existing bridge that fits the proposed alinement and profile may remain in place when its structural capacity in terms of design loading and clear roadway width are at least equal to the values given in Table V-10 for the applicable traffic volume.

The values in Table V-10 do not apply to structures with total length greater than 100 ft. These structures should be analyzed individually, taking into consideration the clear width provided, traffic volumes, remaining life of the structure, pedestrian volumes, snow storage, design speed, accident record, and other pertinent factors.

Traffic	Design Loading Structural Capacity Minimum[e]	Roadway Clear Width (ft)[b] Minimum[c,e]
[a]ADT 0-50	H-10	20[d]
[a]ADT 51-250	H-15	20
[a]ADT 250 +	H-15	22
DHV 100-200	H-15	22
DHV 200-400	H-15	24
DHV over 400	H-15	28

[a]Current ADT.
[b]Clear width between curbs or rails, whichever is the lesser.
[c]Minimum clear widths that are 2 ft narrower may be used on roads with few trucks. In no case shall the minimum clear width be less than the approach traveled way width.
[d]For one-lane bridges use 18 ft.
[e]Does not apply to structures with total length greater than 100 ft.

Table V-10. Minimum structural capacities and minimum roadway widths for bridges to remain in place.

Vertical Clearance

Vertical clearance at underpasses should be at least 14 ft over the entire roadway width, to which up to 6 in. should be added to allow for future resurfacing.

Right-of-Way Width

The procurement of right-of-way to widths that accommodate the construction, adequate drainage, and proper maintenance of the highway is a very important part of the overall design. Wide rights-of-way permit the construction of gentle slopes, resulting in more safety for the motorist and allowing easier and more economical maintenance. The procurement of sufficient right-of-way at the time of the initial improvement permits the widening of the roadway and the widening and strengthening of the pavement at reasonable costs as traffic increases.

In developed areas, it may be necessary to limit the right-of-way width. However, the right-of-way width should not be less than that required for all the elements of the design cross sections, utility accommodation, and appropriate border areas.

Foreslopes

The maximum rate of foreslope depends on the stability of local soils as determined by soil investigation and local experience. Slopes should be as gentle as feasible, and other factors should be considered to determine the design slope. Gentle foreslopes increase safety by providing a maneuver area in emergencies, are more stable than steep slopes, aid in the establishment of plant growth, and simplify maintenance work. Vehicles that leave the traveled way can often be kept under control if slopes (Figure V-1) are gentle and drainage ditches are well-rounded. Such recovery areas should be provided where terrain and right-of-way controls permit.

Combinations of rate and height of slope should provide for vehicle recovery. Where controlling conditions (such as high fills, right-of-way restrictions, or the presence of rocks, watercourses, or other hazards) make this impractical, consideration should be given to the provision of guardrail, in which case the maximum rate of foreslope could be used.

Cut sections should be designed with adequate ditches. Preferably, the foreslope should not be steeper than 2:1. The ditch bottom and slopes should be well-rounded. The backslope should not exceed the maximum required for stability.

Horizontal Clearances to Obstructions

A clear zone of 10 ft or more from the edge of the traveled way, appropriately graded and having gentle slopes and rounded cross-sectional design, is desirable. An exception may be made where guardrail protection is being provided. The recovery area should be clear of all unyielding objects such as trees or unyielding sign supports, utility poles, light poles, and any other fixed objects that might severely damage an out-of-control vehicle.

To the extent feasible, where another highway or railroad passes over, the structure should be designed so that the pier or abutment supports have lateral clearance as great as the clear roadside area on the approach roadway.

Where it is not feasible to carry the full-width approach roadway across an overpass or other bridge, an appropriately transitioned roadside barrier should be provided.

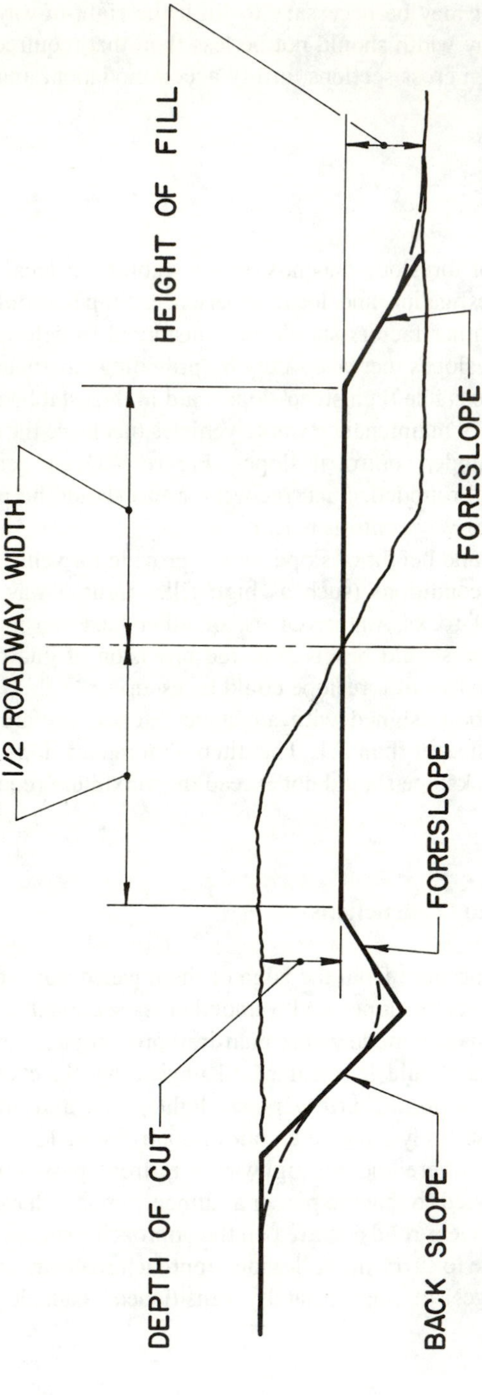

Figure V-1. Slope terms.

At selected locations, such as the outside of a sharp curve, a broader recovery area with greater horizontal clearances should be provided to any roadside obstruction.

Curbs

Use of curbs in conjunction with intermediate or high design speeds should be limited as discussed in Chapter IV.

Where curbs are to be used, refer to the discussion on curbs in the section "Local Urban Streets" in this chapter.

Intersection Design

Intersections should be carefully situated to avoid steep profile grades and to ensure adequate approach sight distance. An intersection should not be situated on a short-crest vertical curve, just beyond a short-crest vertical curve, or on a sharp horizontal curve. When there is no practical alternate to such a location, the approach sight distance on each leg should be checked carefully. Where necessary, backslopes should be flattened and horizontal or vertical curves lengthened to provide additional sight distance. Sight distance should be sufficient to permit a vehicle on the minor leg of the intersection to cross the traveled way without requiring the approaching through traffic to slow down. As a general rule, there should be a minimum of 7 sec available to the driver of a passenger vehicle crossing the through lanes. On this basis, the suggested corner sight distance for each design speed would be as given in Table V-11. For further details see the section on sight distance in Chapter IX.

Intersections should be designed with a corner radius of the pavement or surfacing that is adequate for the larger vehicles anticipated. For minimum edge radius, see Chapter IX. Where turning volumes are significant, consideration should be given to speed change lanes and channelization.

Intersection legs that operate under stop control preferably should be 90° if possible, but in no case less than 60°. For further details see Chapter IX.

Railroad—Local-Road Grade Crossings

Appropriate grade-crossing warning devices shall be installed at all railroad-local-road grade crossings. Details of the devices to be used are given in

Design Speed (mph)	**Corner Intersection Sight Distance (ft)[a]**
60	650[b]
50	515
40	415
30	310
20	210

[a]Corner sight distance measured from a point on the minor road at least 15 ft from the edge of the major road pavement and measured from a height of eye at 3.50 ft on the minor road to a height of object at 4.25 ft on the major road.
[b]At 60 mph stopping sight distance governs.

Table V-11. Corner sight distances at rural intersections.

the *Manual on Uniform Traffic Control Devices* (MUTCD) **(2)**. In some States the final approval of the devices to be used may be vested in the Public Utility Commission.

Sight distance is an important consideration at railroad grade crossings. There must be sufficient sight distance on the road for the driver to recognize the crossing, perceive the warning device as well as the trains, and stop if necessary. (See Chapter IX.)

The roadway width at all railroad crossings should be the same as the width of the approach roadway.

Traffic Control Devices

Signs, pavement and other markings, and, where pertinent, traffic signal controls are essential elements for all local roads and streets. Refer to the MUTCD **(2)** for details of the devices to be used and, for some conditions, warrants for their use.

Erosion Control

All slopes and drainage areas should be designed with proper regard for the desired natural ground cover and growth regeneration on areas opened during construction. Various acceptable methods of erosion control including seeding and mulching of slopes, sodding, or other protection of swales and other erodible areas should be included in the local road design. Consideration should also be given to maintenance requirements and overall economics.

In roadside design the preservation of natural ground covers and desirable growth of shrubs and trees should be considered, provided that such growth does not constitute a hazard in the recovery area.

LOCAL URBAN STREETS

General Design Considerations

A street is a public way for purposes of vehicular travel including public transit and refers to and includes the entire area within the right-of-way. The street also serves pedestrian and bicycle traffic and usually accommodates public utility facilities within the right-of-way. The improvement or development of streets should be based on a functional street classification that is part of a comprehensive community development plan. The design values should be those for the ultimately planned development.

Most urban functional classifications include three classes of streets: arterials, collectors, and local access routes, detailed discussion of which appears in Chapter I. Geometric design guides for arterial streets are given in Chapter VII, and those for collector streets are covered in Chapter VI. It is not practical to enumerate separate design guides for local streets for each item discussed herein. Where there is a significant difference, separate guide values are given.

The design features of local urban streets are governed by practical limitations to a greater extent than those of similar roads in rural areas. The two dominant design controls are (1) the type and extent of urban development with its limitations on rights-of-way, and (2) zoning or regulatory restrictions. Some streets primarily are land service streets in residential areas. In such cases the overriding consideration is to foster a safe and pleasant environment. The convenience of the motorist is a secondary consideration. Other streets are land service only in part, and features of traffic service may be predominant.

On streets serving industrial or commercial areas, the vehicle dimensions, traffic volumes, and vehicle loads differ greatly from those on residential streets, and different dimensional and structural design values are appropriate. Here, safety and service to traffic usually are the major design controls.

Where a particular design feature varies depending on the area served, viz., residential, commercial, or industrial, different guide values are enumerated for each condition. The designer should be apprised of local ordinances and resolutions that affect certain design features.

Design Traffic Volumes

Traffic volume is not usually a major criterion in determining the geometric values to be used in designing residential streets. Traditionally, these streets are designed to a standard two-lane cross section but may be a four-lane cross section in certain urban areas, as governed by traffic volume, administrative policy, or other community considerations.

For streets serving industrial or commercial areas, however, traffic volume is a major factor. The DHV projected to some future design year should be the basis of design. It usually is difficult and costly to modify the geometric design of an existing street unless provision is made at the time of initial construction. Design traffic in these areas should be that estimated for at least 10 years, and preferably 20 years, from the date of completion of construction.

Design Speed

Design speed is not a major factor for local streets. For consistency in design elements, design speeds ranging from 20 to 30 mph may be used, depending on available right-of-way, terrain, adjacent development, and other area controls.

In the typical street grid the closely spaced intersections usually limit vehicular speeds, making the effect of design speed of little significance. Design speeds exceeding 30 mph in residential areas may require longer sight distances and increased curve radii, which would be contrary to the basic function of a local street.

Sight Distance

Minimum stopping sight distance for local streets should range from 125 to 200 ft depending on the design speed (See Table III-1).

Design for passing sight distance seldom is applicable on local streets.

Grades

Grades for local residential streets should be as flat as is consistent with the surrounding terrain. The gradient for local streets should be less than 15 percent. Where grades of 4 percent or steeper are necessary, the drainage design may become critical. On such grades special care must be taken to prevent erosion on slopes and open drainage facilities.

For streets in commercial and industrial areas, gradient design should be less than 8 percent; desirably, it should be less than 5 percent, and flatter gradients should be emphasized.

To provide for proper drainage, the desirable minimum grade that should be used for streets with outer curbs is 0.30 percent, but a minimum grade of 0.20 percent may be used.

Alinement

Alinement in residential areas should fit closely with the existing topography to minimize the need for cuts or fills without sacrificing safety. There is an advantage in residential areas in purposely arranging the alinement to discourage through traffic. The alinement design should be such that the safety of the facility is not reduced.

Street alinement in commercial and industrial areas should be commensurate with the topography but should be as direct as possible.

Street curves should be designed with as large a radius curve as feasible, the minimum radius being 100 ft.

Where curves are superelevated, lower values may apply, but the radius should never be less than 75 ft for a 20-mph design speed.

Pavement Cross Slope

Pavement cross slope should be adequate to provide proper drainage. Cross slope normally should be as shown in Table V-5 where there are flush shoulders adjacent to the traveled way. Where there are outer curbs, cross slopes steeper by about 0.5 to 1 percent are desirable on the lane adjacent to the curb.

The center section of the pavement crown may be parabolic to permit smooth transition of cross slope. Further discussion of this element appears in Chapter IV.

Superelevation

Superelevation is advantageous for street traffic operation, but in built-up areas the combination of wide pavement areas, proximity of adjacent development, control of cross slope, profile for drainage, frequency of cross streets, and other urban features combine to make the use of superelevation impractical or undesirable. Usually, superelevation is not provided on local streets in residential and commercial areas; it should be considered on local streets in industrial areas to facilitate safe operation.

Where superelevation is used, street curves should be designed for a maximum superelevation rate of 0.04. If terrain dictates sharp curvature, a maximum superelevation of 0.06 is justified if the curve is long enough to provide an adequate superelevation transition. Minimum lengths of superelevation runoff are given in Table V-7. A detailed discussion of superelevation is found in Chapter III. (See Table III-16.)

Number of Lanes

On residential streets in areas where the primary function is to provide land service and foster a safe and pleasant environment, at least one unobstructed moving lane must be ensured even where parking occurs on both sides. The level of user inconvenience occasioned by the lack of two moving lanes is remarkably low in areas where single-family units prevail. Local residential street patterns are such that travel distances are less than 0.5 mi between trip origin and a collector street. In multifamily-unit residential areas a minimum of two moving traffic lanes to accommodate opposing traffic may be required. In many residential areas a 26-ft-wide roadway is typical. This curb-face-to-curb-face width provides for a 12-ft center travel lane and two 7-ft parking lanes. Opposing conflicting traffic will yield and pause on the parking lane area until there is sufficient width to pass.

In commercial areas where there will be several midblock left turns it may be advantageous to provide an additional continuous two-way left-turn lane in the center of the roadway.

Width of Roadway

Street lanes for moving traffic preferably should be at least 10 ft wide. Where feasible they should be 11 ft wide, and in industrial areas they should be 12 ft wide. Where available or attainable width of right-of-way imposes severe limitations, 9-ft lanes can be used in residential areas, as can 11-ft lanes in industrial areas. Added turning lanes where used at intersections should be at least 9 ft wide, and desirably 10 to 12 ft wide, depending on the percentages of trucks.

Where needed and where limitations exist in residential areas, a parallel parking lane at least 7 ft wide should be provided on one or both sides, as the conditions of lot size and intensity of development may require. In commercial and industrial areas, parking lanes should be at least 9 ft wide; usually they are provided on both sides.

Parking-lane-width determinations in commercial and industrial areas should include consideration for use of the parking lane for moving traffic during peak-hour traffic that may occur where industries have high employment concentrations.

Where curb and gutter sections are used, the gutter pan width may be included as a part of the parking lanc width.

Median

Medians provided on local urban streets primarily to enhance the environment and to act as buffer strips should be designed to minimize interference with the traffic needs of the land abutting the roadway. A discussion of the various median types appears in Chapter IV.

Openings should be situated only where there is adequate sight distance. The shape and length of the median openings depend on the width of median and the vehicle types to be accommodated. The minimum length of median openings should be that of the width of the projected roadway of the intersecting cross street or driveway. The desirable length of median openings, measured between the inner edge of the lane adjacent to the median and the centerline of the intersection roadway, should be great enough to provide for a 40-ft turning control radius for left-turning P vehicles.

Curbs

Streets normally are designed with curbs to allow greater use of available width and for control of drainage, protection of pedestrians, and delineation. The curb should be 4 to 9 in. high, depending on drainage considerations, traffic control, and safety.

On divided streets the type of median curbs should be compatible with the width of the median and the type of turning movement control to be effected.

Steep-faced curbs 6 in. or higher adjacent to moving traffic lanes should be offset at least 1 ft. Where there is combination curb and gutter construction, the gutter pan width, normally 2 ft, should be used as the offset distance.

Drainage

Drainage is an important consideration in an urban area because of high runoff and the flooding potential. Surface flow from adjacent tributary areas may be intercepted by the street system. There it is collected within the roadway by curbs and gutters, ditches, etc., and conveyed to appropriate outlets. Where drains are available under or near the roadway, the flow is transferred at frequent intervals from the street cross section by grating or curb-opening inlets to basins and from there by connectors to drainage channels or underground drains.

Economic considerations usually dictate that maximum practical use be made of the street sections for surface drainage. To avoid undesirable flowline conditions, the minimum gutter grade should be 0.30 percent. However, in very flat terrain and where drainage outlet is unavailable, a gutter grade as low as 0.20 percent may be necessary. Where a drainage outlet is available, the inlets should be spaced to provide a high level of drainage protection in areas of pedestrian concentrations or where adjacent property has an unusually important public or community purpose (e.g., schools and churches). For further details see the drainage section in Chapter III.

Cul-De-Sacs and Turnarounds

A local street open at one end only should have a special turning area at the closed end. This turning area desirably should be circular and have a radius appropriate to the types of vehicle expected. Minimum outside radii of 30 ft in residential areas and 45 ft in commercial and industrial areas are commonly used.

A dead-end street narrower than 40 ft usually should be widened to enable passenger vehicles, and preferably delivery trucks, to U-turn or at least turn around by backing once. The design commonly used is a circular pavement symmetrical about the centerline of the street sometimes with a central island, as shown in Figure V-2C, which also shows minimum dimensions for the design vehicles. Although this type of cul-de-sac operates satisfactorily and looks well, better operation is obtained if the design is offset so that the entrance-half of the pavement is in line with the approach-half of the street, as shown in Figure V-2D. One steering reversal is avoided on this design. Where

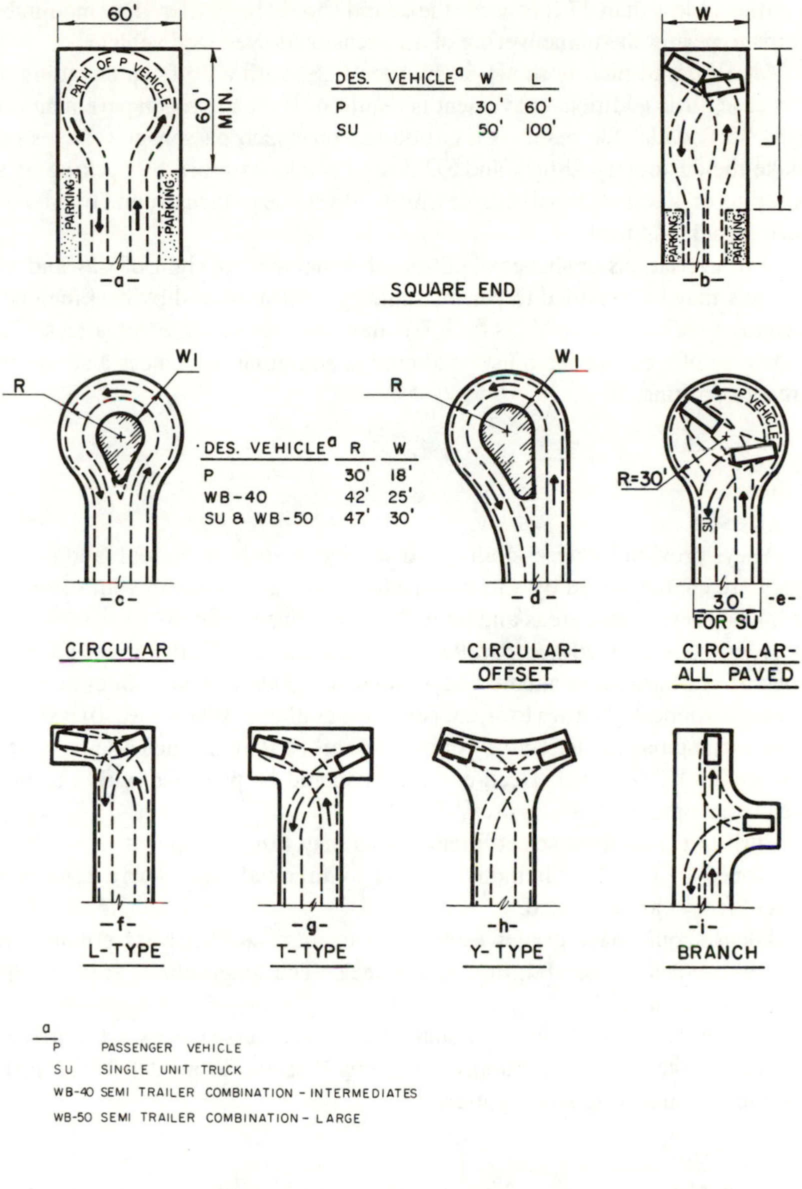

Figure V-2. Types of cul-de-sacs and dead-end streets.

a radius of less than 47 ft is used, the island should be bordered by mountable curbs to permit the maneuvering of an occasional oversized vehicle.

An all-paved plan is shown in Figure V-2E, with a 30-ft outer radius. In this case, little additional pavement is required. If the approach pavement is at least 30 ft wide, the result is a cul-de-sac on which passenger vehicles can make the customary U-turn and SU design trucks can turn by backing once. A radius of about 40 ft will enable a WB-50 vehicle to turn around by maneuvering back and forth.

Other variations or shapes of cul-de-sacs that include right-of-way and site controls may be provided to permit vehicles to turn around by backing once. Several types (Figures V-2F to V2I) may also be suitable for alleys. The geometry of a cul-de-sac must be altered if adjoining residences also use the area for parking.

Alleys

Alleys provide for accessibility and service to each individual land parcel. They are characterized by a narrow right-of-way and range in width from 16 to 20 ft in residential areas and up to 30 ft in industrial areas.

Alleys should be alined parallel to or concentric with the street property lines. It is desirable to situate alleys in such a manner that both ends of the alley are connected either to streets or to other alleys. Where two alleys intersect, a triangular corner cutoff of not less than 10 ft along each alley property line should be provided. Dead-end alleys should be provided with a turning area in accordance with Figure V-3.

Curb return radii at street intersections may range from 5 ft in residentially zoned areas to 10 ft in industrial and commercial areas where large numbers of trucks are expected.

Alleys should have grades established to meet as closely as possible the existing grades of the abutting land parcels. The longitudinal grade should not be less than 0.20.

Alley cross sections may be V-shaped with transverse slopes of 2.5 percent toward a center V gutter. Runoff is thereby directed to a catch basin in the alley or to connecting street gutters.

Sidewalks

Sidewalks used for pedestrian access to schools, parks, shopping areas, and transit stops and placed along all streets in commercial areas should be

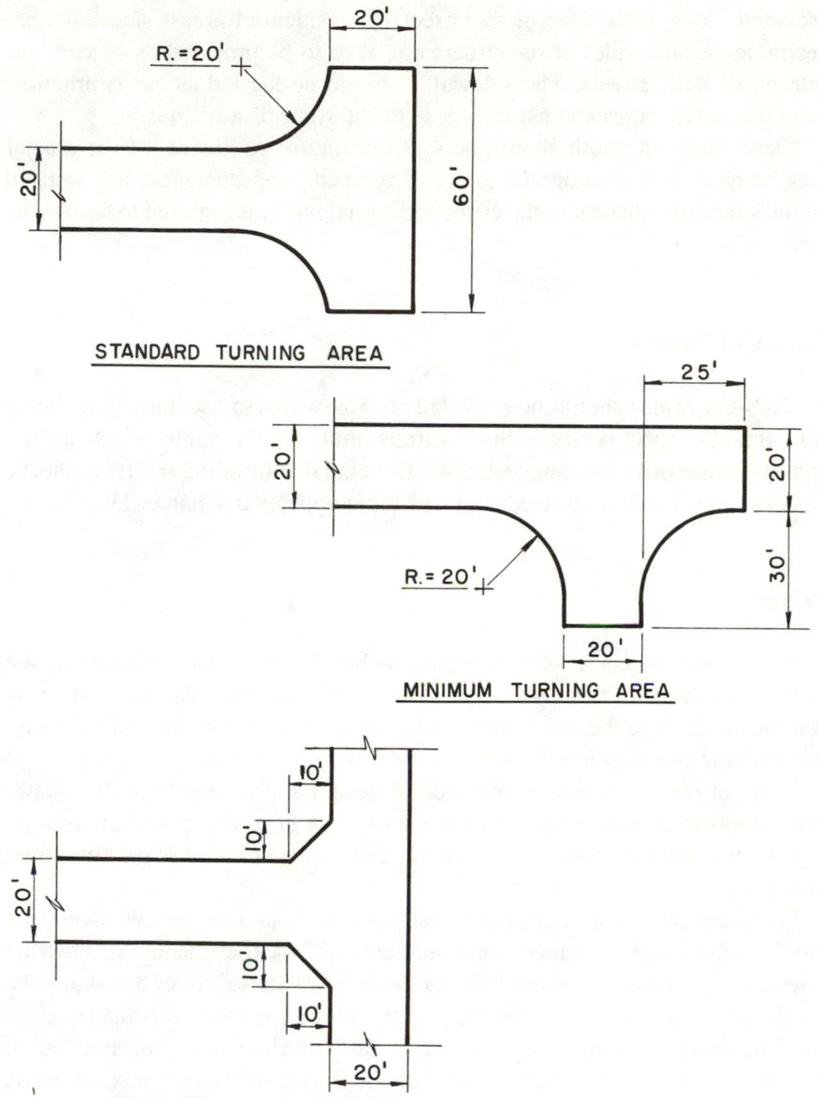

Figure V-3. Alleys.

provided along both sides of the street. In residential areas, sidewalks are desirable on both sides of the streets but need to be provided on at least one side of all local streets. The sidewalks should be located as far as practical from the traffic lanes and usually close to the right-of-way lines.

Clear sidewalk width should be 4 ft minimum; widths of 8 ft or greater may be needed in commercial areas. If roadside appurtenances are situated on the sidewalk adjacent to the curb, additional width is required to secure the clear width.

Curb-Cut Ramps

Curb-cut ramps should be provided at crosswalks to accommodate physically handicapped persons. Such ramps may be the same width as the approach sidewalks, the suggested width being 4 ft (minimum 3 ft) exclusive of sideslopes. Further discussion of this topic appears in Chapter IV.

Driveways

A driveway is an access constructed within the public way, connecting the public roadway with adjacent property and intended to be used in such a way that the access into the adjacent property will be complete and will not cause the blocking of any sidewalk border area or street roadway.

Some of the principles of intersection design apply directly to driveways. One important feature is the elimination of large graded or paved areas adjacent to the traveled way, which allow drivers to enter or leave the street randomly.

Sight distance is a significant design control, and driveways should be avoided where sight distance is not sufficient. Vertical elements that obstruct essential sight distances should be controlled by statute. Driveways should be regulated as to width of entrance, placement with respect to property lines and intersecting streets, angle of entry, vertical alinement, and number of entrances to a single property to provide for traffic safety and maximum use of curb space for parking where permitted. Driveways should be situated as far away from intersections as practicable, particularly if the local street intersects an arterial street.

Driveway returns should not be less than 3 ft in radius. Flared driveways are preferred because they are distinct from intersection delineations, can properly handle turning movements, and can minimize the problems for handicapped pedestrians. Flare grades should not exceed 12.5 percent.

Roadway Widths for Bridges

The minimum clear width for all new bridges on streets with curbed approaches should be the same as the curb-to-curb width of the approaches. For streets with shoulders and no curbs, the clear roadway width preferably should be the same as the approach roadway width and in no case less than the width shown in Table V-9. Sidewalks on the approaches should be carried across all new structures. There should be at least one sidewalk on all street bridges.

Horizontal Clearance to Obstructions

On all streets a minimum clearance of 1.5 ft should be provided between the curb face or shoulder edge and obstructions such as utility poles, lighting poles, and fire hydrants. In areas of dense pedestrian traffic the construction of barrier-type curbing (curbing with nearly vertical facing, 6 to 9 in. high) aids in protecting the high-volume pedestrian traffic.

Trees may be minimal hazards along local streets where speeds are low (40 mph or below) and barrier curbs exist. Generally, prudence should be exercised and careful analysis should weigh the benefits against the possibly adverse effects that tree removal could have on the roadside environment.

Guardrail is not used extensively on local streets except where there is a hazard to motorists and pedestrians, such as along sections with steep foreslopes and at the approaches to overcrossing structures.

Vertical Clearance

Vertical clearance at underpasses should be at least 14 ft over the entire roadway width, to which up to 6 in. should be added to allow for future resurfacing.

Border Area

A border area should be provided along streets for the safety of motorists and pedestrians as well as for esthetic reasons. Selection of the street alinement should be based on minimizing of slopes. However, the preservation and enhancement of the environment is of major importance in the design and construction of local streets.

The border area between the roadway and the right-of-way line should be wide enough to serve several purposes, including provision of a buffer space between pedestrians and vehicular traffic, sidewalk space, snow storage, an area for placement of underground and aboveground utilities, and an area for maintainable esthetic features such as grass or other landscaping. The border width may be a minimum of 5 ft, but desirably should be 10 ft or wider.

Where the available right-of-way is limited and in areas of high right-of-way costs, as in some industrial and commercial areas, a buffer width of 2 ft may be tolerated.

Wherever practical, an additional obstacle-free buffer width of 12 ft or more should be provided between the curb and the sidewalk for safety and environmental enhancement. In residential areas wider building setback controls can be used to attain these features.

Right-of-Way Width

The right-of-way width should be sufficient to accommodate the ultimate planned roadway including median (if used), shoulder (if used), landscaping strip, sidewalks, utility strips in the border areas, and necessary outer slopes unless provided for by easement. On local residential streets that use the 26-ft roadway section, a 50-ft right-of-way is commonly used to accommodate roadway, sidewalks, utilities, and landscaping strips. For commercial and industrial roadways where on-street parking is provided, the common right-of-way width is 66 ft. These dimensions should be increased when medians are provided. Smaller widths may also be justified in hilly areas or where access to adjacent properties is not required.

Provision for Utilities

In addition to the primary purpose of serving vehicular traffic and in accordance with State law or municipal ordinance, streets also often accommodate public utility facilities within the street right-of-way. Use of the right-of-way by utilities should be planned to cause the least interference with traffic using the street. References 3 and 4 give general principles for location and construction of utilities to minimize conflict between the use of the street right-of-way for vehicular movement and for its secondary purpose of providing space for location of utilities.

Intersection Design

Intersections, including median openings, should be designed with adequate corner sight distance as suggested in Table V-11, and the intersection area should be kept free of obstacles. Where stop control is not used, the corner sight distance for local streets should be a minimum of 200 ft and desirably should be 300 ft or more. To maintain the minimum sight distance, restrictions on height of embankment, locations of buildings, and screening fences may be necessary. Any landscaping in the sight distance triangle should be low-growing, and should not be higher than 3 ft above the level of the intersecting street pavements. Tree overhang should be trimmed to a line at least 8 ft above the level of the intersections.

Intersecting streets should meet at approximately a 90° angle. The alinement design should be adjusted to avoid an angle of intersection of less than 60°. Closely spaced offset intersections are undesirable.

The intersection area and approach areas where vehicles store while waiting to enter the intersection should be designed with a relatively flat grade; the maximum grade on the approach leg should not exceed 5 percent where practical. Where ice and snow create hazardous conditions the desirable grade on the approach leg should be 0.5 percent with no more than 2 percent wherever practical.

At street intersections in residential areas and areas where there are heavy pedestrian movements, the minimum radius of curb return where curbs are used or the outside edge of pavement where curbs are not used should be 15 ft. A minimum radius of 25 ft is desirable.

In industrial areas the radius of the curb return should not be less than 30 ft, and desirably, use should be made of a three-centered curve of sufficiently large radii to accommodate the largest vehicles expected.

Further information pertaining to intersection design appears in Chapter IX.

Railroad-Street Grade Crossings

Appropriate grade-crossing warning devices shall be installed at all railroad-street grade crossings. Details of the devices to be used are given in the MUTCD (**2**). In some States the final approval of the devices to be used may be vested in the Public Utility Commission.

Sight distance is an important consideration at railroad grade crossings. There must be sufficient sight distance on the street for the driver to recognize the crossing, perceive the warning device as well as the trains, and stop if necessary. (See Chapter IX.) Signalized intersections adjacent to signalized RR grade crossing should be designed with RR preemption.

The roadway width at all railroad crossings should be the same as the width of the approach roadway.

Sidewalks should be provided at railroad grade crossings to connect existing or future walkways that approach these crossings.

Street and Roadway Lighting

Good visibility under day or night conditions is one of the fundamental requirements enabling motorists to move on roadways in a safe and coordinated manner. Properly designed and maintained street lighting will produce comfortable and accurate visibility at night, which will facilitate and encourage both vehicular and pedestrian traffic. Thus, where adequate illumination is provided, efficient night use can be made of the large investment in roadway construction and motor vehicles. Determinations of need for lighting should be coordinated with crime prevention programs and other community needs.

Warrants for the justification of street lighting involve more than just identifying a roadway classification as being local, rural, urban, collector, arterial, or major. Pedestrian and vehicular volume, night-to-day accident ratios, roadway geometry, merging lanes, curves, and intersections all require careful consideration in establishing illumination levels. When illumination levels provided for the various roadway, walkway, and area classifications are not influenced by these considerations, the suggested minimum levels of illumination (expressed in average maintained horizontal foot-candles) are contained in Table V-12.

Illumination levels at intersections should be the sum of illumination levels on intersecting streets at the intersection.

The uniformity of the lighting is an indication of the quality of illumination and should be considered along with illumination levels. Uniformity of illumination can be represented by a uniformity ratio of the average-to-minimum foot-candle values on the roadway or walkway surface. Recommended uniformity ratios are as follows: residential roadways, 6:1; commercial roadways, 3:1; residential walkways, 10:1; and commercial walkways, 4:1.

Because glare is also an indication of the quality of lighting, the type of fixtures and the height at which the light sources are mounted are also factors in design of street lighting systems. The objectives of the designer should be to minimize visual discomfort and impairment of driver and pedestrian vision due to glare. For the condition of lighting intersections only, a gradual lighting transition from dark to light to dark should be provided so that drivers may have time to adapt their vision.

Classification	Industrial Commercial (ft-c)	Residential (ft-c)
Local	0.9	0.4
Alleys	0.6	0.2
Sidewalks	0.9	0.2

Table V-12. Minimum illumination levels.

More detailed discussion of this topic is contained in AASHTO's *An Informational Guide for Roadway Lighting* (**5**).

Traffic Control Devices

Consistent and uniform application of traffic control devices is important. Details of the standard devices and warrants for many conditions are found in the MUTCD (**2**).

Geometric design of streets should include full consideration of the types of traffic control to be used, especially at intersections where multiphase or actuated traffic signals are likely to be needed.

Erosion Control

Design of streets should include considerations for preservation of natural ground cover and desirable growth of shrubs and trees within the right-of-way as would be consistent with Federal and local statutes. Seeding, mulching, and sodding of slopes, swales, and other erodible areas should be included in the design plans. Other erosion-prevention measures may be acceptable, depending on local climatic conditions, natural environment, and resources.

Landscaping

Landscaping should be provided for esthetic and erosion control purposes in keeping with the character of the street and its environment. Landscaping should be arranged to permit sufficiently wide, clear, and safe pedestrian walkways. Combinations of turf, shrubs, and trees are desirable in border areas along the roadway. However, care should be exercised to ensure that guidelines for sight distances and clearance to obstructions are observed, especially at intersections.

Bicycle Facilities

The local roadway is generally sufficient to accommodate bicycle traffic; however, when special facilities are desired they should be in accordance with AASHTO's *Guide for Development of New Bicycle Facilities* **(6)**.

SPECIAL PURPOSE ROADS

Introduction

For the purpose of design, highways have been classified in this book by function with specific design values for each functional class. Subsequent chapters discuss the design of collectors, arterials, and freeways. The first two sections of this chapter discuss the design of local roads and streets. Another type of highway, however, is different because of its purpose and does not fit into any of the noted classifications. This type of highway is referred to as a special purpose road and because of its uniqueness, separate design criteria are provided. Special purpose roads include recreational roads, resource development roads, and local service roads. Roads in the special purpose category are generally lightly traveled and of low speed and for these reasons deserve special consideration relative to their appropriate design features.

Recreational Roads

General Considerations

Roads serving recreational sites and areas are unique in that they are also part of the recreational experience. Design criteria described herein meet the unusual requirements of roads for access to, through, and within recreational sites, areas, and facilities for the complete enjoyment of the recreationist. The criteria are intended to protect and enhance the existing esthetic, ecological, environmental, and cultural amenities that form the basis for distinguishing each particular recreational site or area.

Visitors to a recreational site need access to the general area, usually by a statewide or principal arterial highway. Secondly they need access to the site. This is the most important link from the statewide road system. For continuity beyond this point, design criteria requires that the visitor be made aware of the nature of the area. The design should be accomplished by a multidisciplinary team of varied backgrounds and experience, to ultimately provide a road system that is an integral part of the recreational site. Depending on the conditions, internal tributaries will have a variety of lower design features.

The criteria discussed in this chapter are applicable for public roads within all types of recreational sites and areas. Design criteria for recreational roads are discussed for primary access roads, circulation roads, and area roads. Primary access roads are defined as roads that allow through movement into and between access areas. Circulation roads allow movement between activity sites within an access area. Area roads allow direct access to individual activity areas such as campgrounds, park areas, boat launching ramps, picnic groves, and scenic and historic sites.

Figure V-4 depicts a potential road system serving a recreational area. Road links are labeled in accordance with the classification system noted.

Design Speed

The effect of design speed on various roadway features is considered in its selection; however, the speed is selected primarily on the basis of the character of the terrain and the functional classification of the road. The design speeds should be approximately 40 mph for primary access roads, 30 mph for circulation roads, and 20 mph for area roads. There may be instances where design speeds less than these may be appropriate because of severe terrain conditions or major environmental concerns. Design speeds on one-lane roads would usually be less than 30 mph. If a design speed of greater than 40 mph is used, the first section of this chapter should be consulted.

Design speed is the principal factor that must be correlated with the physical features of design to achieve a roadway that will accommodate the traffic safely for the planned use. Once a design speed is selected, all geometric features should be related to it to obtain a balanced design. Changes in terrain and other physical controls may dictate a change in design speed in certain sections. A decrease in design speed along the road should not be introduced abruptly, but be extended over a sufficient distance to allow the driver to adjust and make the transition to the slower speed.

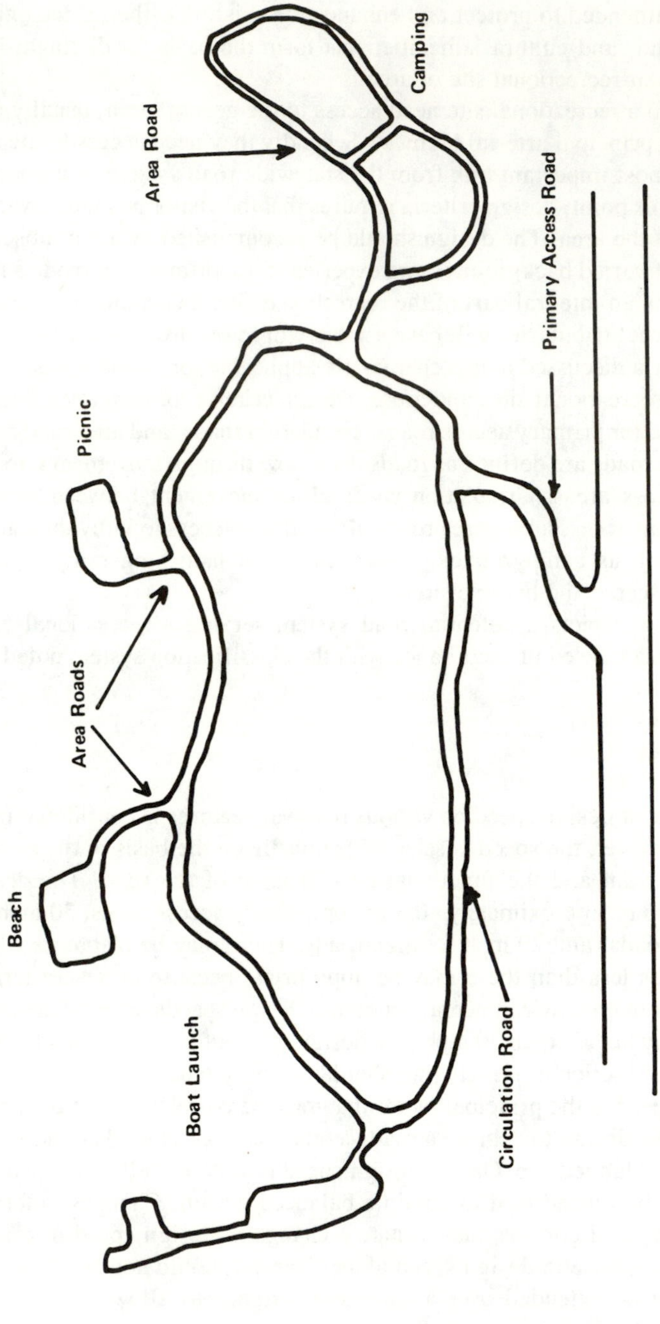

Figure V-4. Potential road network.

Design Vehicle

The physical dimensions and operating characteristics of vehicles and the percentage of variously sized vehicles using recreational roads are primary geometric design controls. Existing and anticipated vehicle types must be examined to establish representative vehicles for each functional roadway class. Each design vehicle considered should represent an ample percentage of the vehicles expected to use the facility during its design life.

Three categories of vehicles are common to recreational areas—motor homes, vehicles with trailers, and standard passenger vehicles. Critical physical dimensions for geometric design are the overall length, width, and height of the units. Minimum turning paths of the design vehicles are influenced by the vehicle steering mechanism, track width, and wheelbase arrangement. Figures V-5, V-6, and V-7, show minimum turn paths for motor homes (MH), passenger cars with 30-ft travel trailers (PT), and passenger cars with 20-ft boats (PB). Figure V-8 shows a minimum turning path for motor homes with 20-ft. boats (MHB). Turning path dimensions for other vehicle types such as buses and passenger cars are covered in Chapter II.

Sight Distance

Minimum stopping sight distance and passing sight distance are a direct function of the design speed. The subject of sight distance for two-lane roads is covered in Chapter III; however, values are not included for cases of very low design speeds and two-directional one-lane roads. On one-lane roads enough sight distance must be available for one vehicle to reach a turnout or for both to stop before colliding. Criteria for measuring stopping sight distance for this type of road assumes a height of eye of 3.50 ft and a height of opposing vehicle of 4.25 ft. The stopping sight distance for two-directional one-lane road is approximately twice the stopping sight distance for a two-lane road. Suggested stopping sight distances for two-directional one-lane roads are given in Table V-13.

Passing Sight Distance

Because of low operating speeds and the nature of travel on recreational roads, frequent passing maneuvers are not anticipated. Nevertheless, minimum passing sight distance should be provided as frequently as possible, particularly on primary access roads where users travel considerable distances to reach activity sites. Passing sight distance is not a factor on one-lane roads. It

THIS TURNING TEMPLATE SHOWS THE TURNING PATHS OF THE AASHTO DESIGN VEHICLES. THE PATHS SHOWN ARE FOR THE LEFT FRONT OVERHANG AND THE OUTSIDE REAR WHEEL. THE LEFT FRONT WHEEL FOLLOWS THE CIRCULAR CURVE, HOWEVER, ITS PATH IS NOT SHOWN.

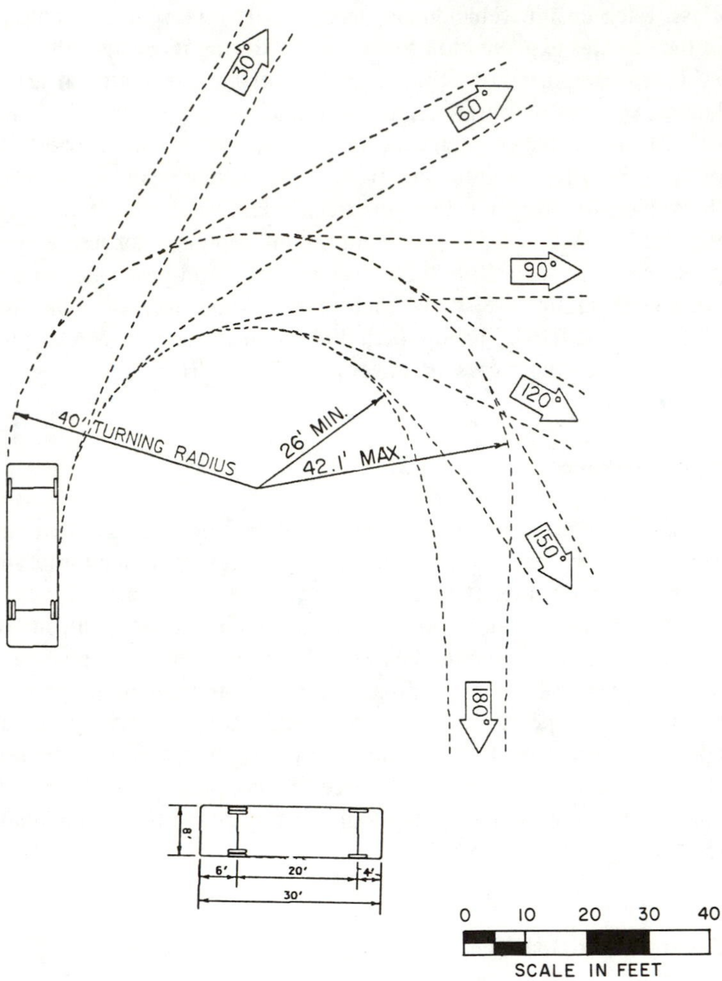

Source: Texas State Department of Highways and Public Transportation

Figure V-5. Minimum turning path for MH design vehicle.

THIS TURNING TEMPLATE SHOWS THE TURNING PATHS OF THE AASHTO DESIGN VEHICLES. THE PATHS SHOWN ARE FOR THE LEFT FRONT OVERHANG AND THE OUTSIDE REAR WHEEL. THE LEFT FRONT WHEEL FOLLOWS THE CIRCULAR CURVE, HOWEVER, ITS PATH IS NOT SHOWN.

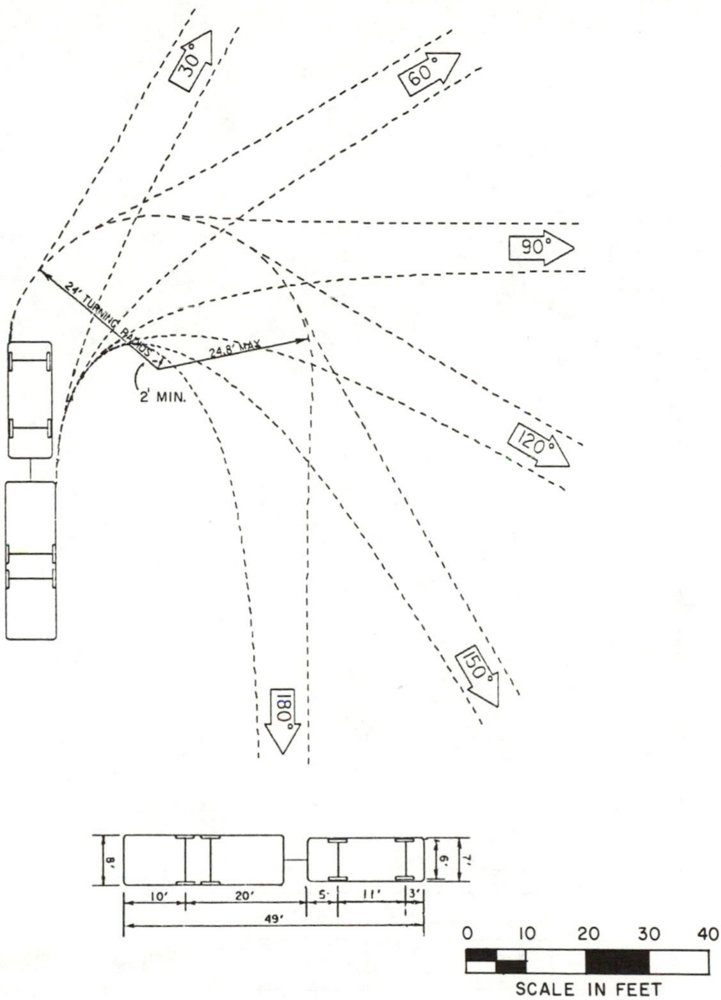

Source: Texas State Department of Highways and Public Transportation

Figure V-6. Minimum turning path for PT design vehicle.

THIS TURNING TEMPLATE SHOWS THE TURNING PATHS OF THE AASHTO DESIGN VEHICLES. THE PATHS SHOWN ARE FOR THE LEFT FRONT OVERHANG AND THE OUTSIDE REAR WHEEL. THE LEFT FRONT WHEEL FOLLOWS THE CIRCULAR CURVE, HOWEVER, ITS PATH IS NOT SHOWN.

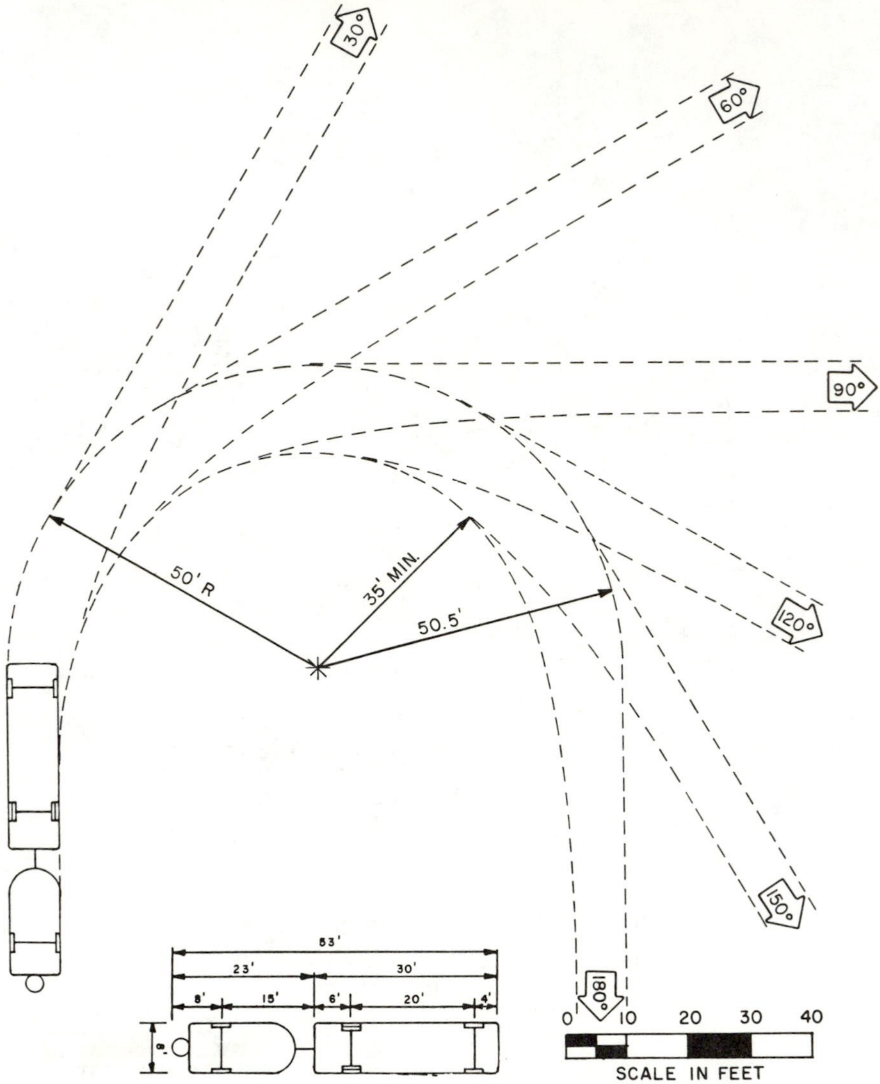

Source: CalTrans

Figure V-7. Minimum turning path for PB design vehicle.

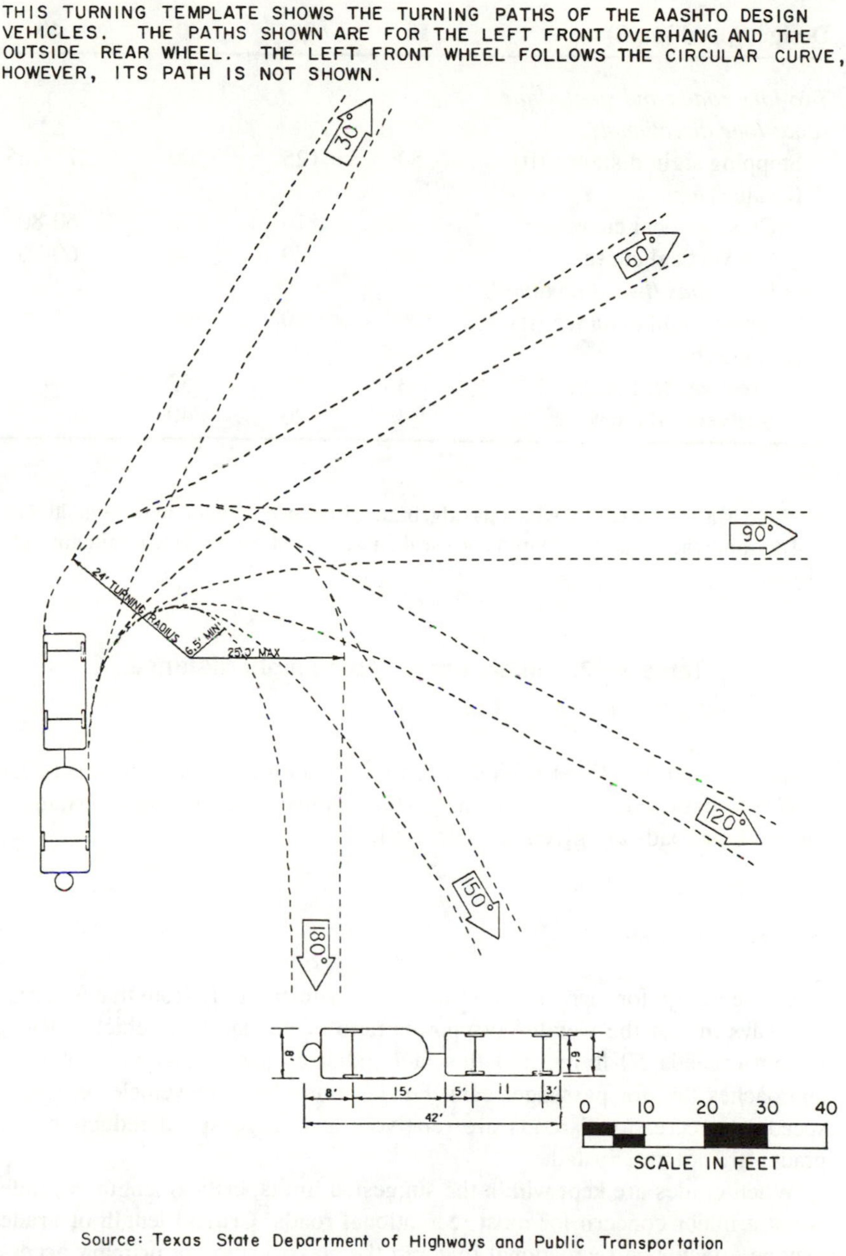

Figure V-8. Minimum turning path for MH-B design vehicle.

Design Speed (mph)	10	20	30	40
Two-lane roads and single-lane roads (one directional)				
Stopping sight distance (ft)	50	125	200	275-325
K value[a] for:				
Crest vertical curve	2	10	30	60-80
Sag vertical curve	4	20	40	60-70
One-lane roads (two-directional)				
Stopping sight distance (ft)	100	250	400	-
K value[a] for:				
Crest vertical curve	3	20	52	-
Sag vertical curve	4	20	40	-

[a]K value is a coefficient by which the algebraic difference in grade may be multiplied to determine the length in feet of the vertical curve, which will provide minimum sight distance.

Table V-13. Minimum stopping sight distance.

is assumed that the slower vehicle will pull into a turnout and allow the faster vehicle to pass when necessary. Suggested minimum passing sight distances for two-lane roads are given in Table V-14.

Grades

Grade design for recreational roads differs substantially from that for rural highways in that the weight/horsepower ratio of recreational vehicles (RV's) seldom exceeds 50 lb/hp, and this fact indicates that gradeability of RV's approaches that for passenger cars. Furthermore, because vehicle operating speeds on recreational roads are relatively low, large speed reductions on grades are not anticipated.

When grades are kept within the suggested limits, critical length of grade is not a major concern for most recreational roads. Critical length of grade may be a factor on recreational roads in the classification of primary access roads, and appropriate consideration should be given to this element for the design for these roads.

Design Speed (mph)	Minimum Passing Sight Distance (ft)[a]	K Value[b] Crest Vertical Curve
20	800	210
30	1,100	400
40	1,500	730

[a]Minimum passing sight distance measured from driver's eye height of 3.50 ft to top of object 4.25 ft above pavement.

[b]K value is a coefficient by which the algebraic difference in grade may be multiplied to determine the length in feet of the vertical curve, which will provide minimum sight distance.

Table V-14. Minimum passing sight distance for two-lane roads.

Table V-15 identifies suggested maximum grades for given terrain and design speed. Chapter III contains a more detailed discussion. The grades noted in Table V-15 relate primarily to operational performance of vehicles. A major item to be considered in selection of a maximum grade is the capability of the soil for erosion resistance. In many instances grades considerably less than those shown in Table V-15 should be chosen to satisfy this concern. In addition, type of surface should also be a factor in grade selection. Steep grades with dirt or gravel surfaces may cause driving problems in the absence of continued maintenance, whereas a bituminous surface will give better vehicle performance in general.

	Design Speed (mph)			
Type of Terrain	10	20	30	40
Level	8	8	7	7
Rolling	12	11	10	9
Mountainous	18	16	14	12

Table V-15. Maximum grades (%) for recreational roads.

Vertical Alinement

Vertical curves should be safe, comfortable in operation, pleasing in apperance, and adequate for drainage.; Minimum or greater stopping sight distance should be provided in all cases. The designer must exercise considerable judgment in designing vertical curves, because lengths in excess of the minimum may be needed at driver decision points, where drainage or esthetic problems exist, or simply to provide an additional margin of safety.

Vertical curve design for two-lane roads is discussed in Chapter III, which also contains specific design values. Table V-13 also includes additional information for very low design speeds not tabulated elsewhere. For two-directional one-lane roads, crest vertical curves are significantly longer than those for two-lane roads. As previously discussed under the section of sight distance, the stopping sight distance for a two-directional one-lane road is approximately twice the stopping sight distance for a two-lane road. Table V-13 includes K values for one-lane roads, from which vertical curve lengths can be determined.

Horizontal Alinement

Because the use of straight sections of roadway would be physically impractical and (for recreational roads) esthetically undesirable, horizontal curves are necessary elements. The proper relationship between design speed and horizontal curvature and the relationship of both with superelevation are discussed in detail in Chapter III. The guidance provided in Chapter III is generally applicable to paved recreational-type facilities; however, in certain instances variations are appropriate. At locations where there is a tendency to drive slowly, as with local and some circulation roads, a maximum superelevation rate of 0.06 is suggested. On roads with design speeds of 20 mph or less, superelevation may not be warranted.

The design values for maximum curvature and superelevation discussed in Chapter III are based on friction data for paved surfaces. Some lower volume recreational facilities may not be paved, and because friction values for gravel surfaces are less than those for paved surfaces, fiction values should be considered in curvature selection. Figure V-9 shows the relationship between minimum radius and superelevation for gravel-surfaced roads. This figure was developed by using f values from 0.12 at 10 mph to 0.10 at 30 mph.

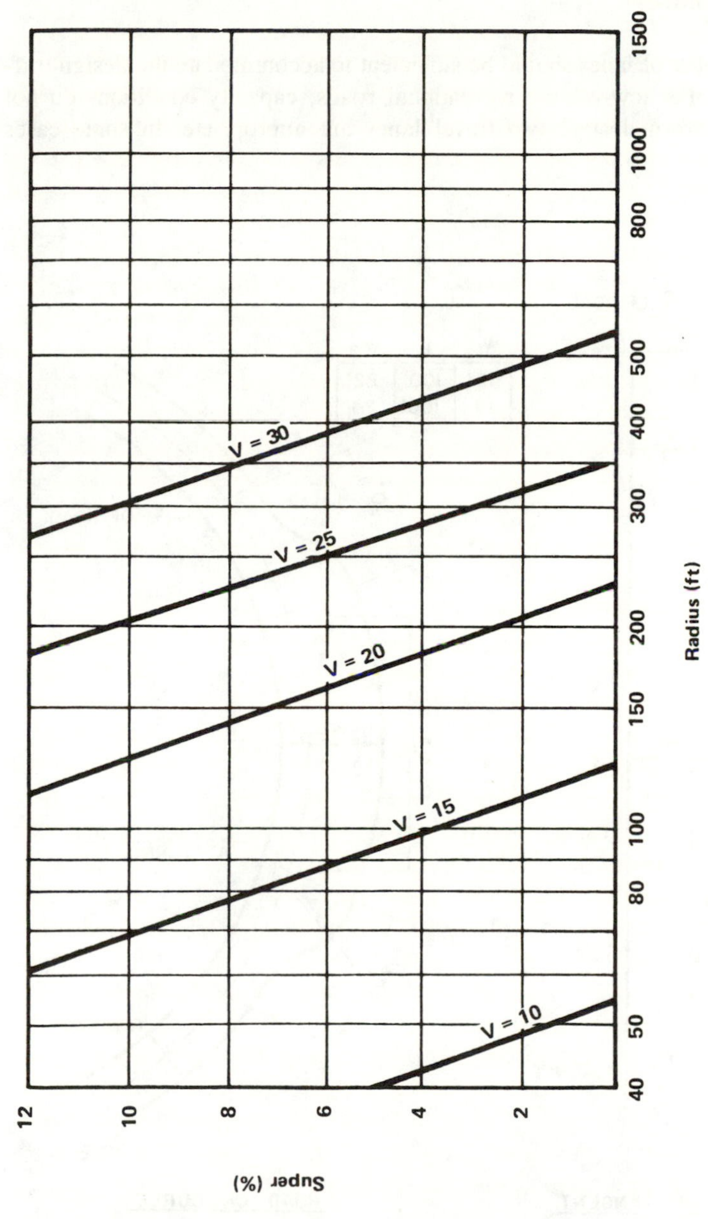

Figure V-9. Minimum-radius horizontal curve for gravel surface.

Number of Lanes

The number of lanes should be sufficient to accommodate the design traffic volume. For low-volume recreational roads, capacity conditions do not normally govern design; two travel lanes are appropriate. In some cases

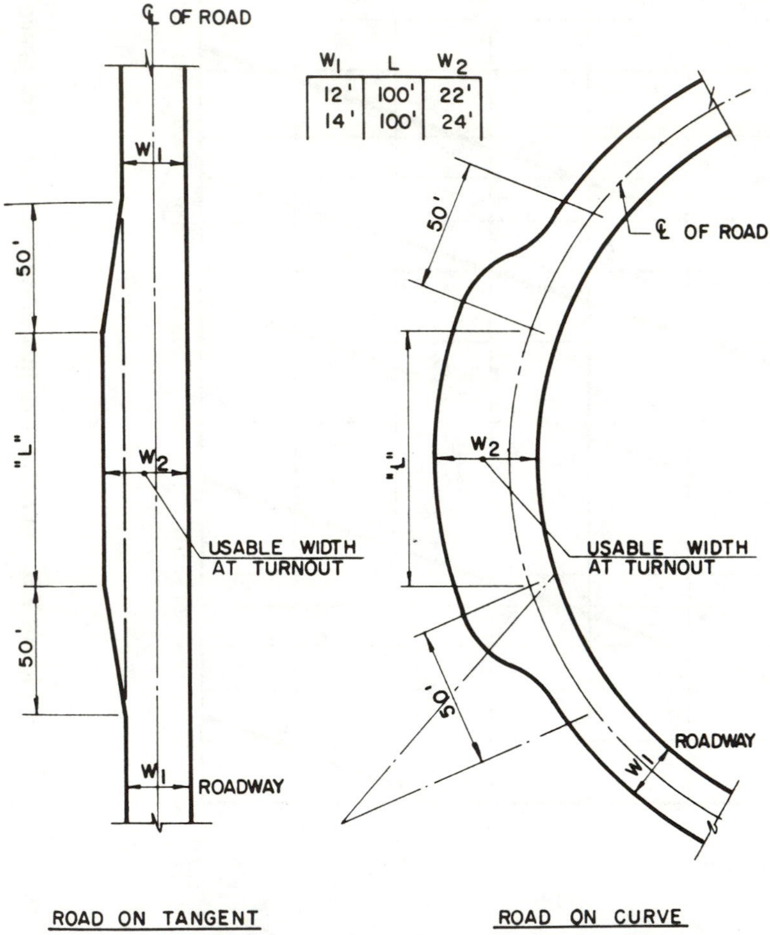

Figure V-10. Turnout design.

where traffic volumes are less than 100 vehicles per day, it may be feasible to use a two-directional one-lane roadway. This type of road is often desirable from an economic and environmental standpoint. When one-lane roadways with two-directional traffic are used, turnouts for passing should be provided. Traffic convenience requires that such turnouts be intervisible, provided on all blind curves, and supplemented as necessary so that the maximum distance between turnouts is no more than 1,000 ft. The turnouts should be a minimum of 10 ft wide for a length of 50 ft and should have a 25-ft taper on each end. For overwide and extra-long vehicles the values should be adjusted to accommodate the usage. Figure V-10 shows a typical design that may be used for turnouts on tangent and curve sections for two-directional one-lane roads.

Widths of Traveled Way, Shoulder, and Roadway

A roadway is defined as that portion of the highway including shoulders for vehicular use. Proper roadway width is selected on the basis of numerous factors including existing and anticipated vehicular and bicycle traffic, safety, terrain, and design speed. Table V-16 gives recommended traveled way widths and shoulder widths for the various types of roadways. The sum of the traveled way and shoulder widths given in Table V-16 constitutes the roadway width.

Type	Traveled Way Width (ft)[a]	Shoulder Width (ft)
Primary Access Roads (two lanes)	22-24	2-4
Circulation Roads (two lanes)	20-22	2-4
Area Roads (two lanes)	18-20	0-2
Area Roads (one lane)[b]	12	0-1

[a]Widening on the inside of sharp curves should be provided. Additional width equal to 400 divided by the curve radius in feet is recommended.

[b]Roadway widths greater than 14 ft should not be used because of the tendency for drivers to use the facility as a two-lane road.

Table V-16. Widths of traveled way and shoulder.

The low operating speeds and relatively low traffic volumes on recreational roads do not warrant wide shoulders. In addition, wide shoulders may be esthetically objectionable. These facts and concerns are reflected in the shoulder width values given in Table V-16. Under adverse terrain conditions, intermittent shoulder sections or turnouts may be suitable alternatives to continuous shoulders, particularly on lower functional roadway classes. Where guardrail is used, the graded width of the shoulder should be increased about 2 ft.

Surface Cross Slope

Surface cross slope must be provided to ensure adequate drainage. However, excessive surface sloping can cause steering difficulties. Cross slope rates given in the first section of this chapter are generally applicable to recreational roads.

On one-lane roads with low-type surfaces, a crown would not usually be provided. Roads of this type would be inslope graded (toward the cut ditch) or outslope graded (toward the embankment fill), depending on the resistance of the soil to erosion. Where the soil is unstable and subject to major erosion, inslope grading should be used. Foreslope or backslope grading should exceed the horizontal gradient of the road to a maximum of 5 percent.

Clear Recovery Area

Providing a clear zone adjacent to a road involves a trade-off between safety and esthetics. A driver who leaves the road should be provided a reasonable chance to regain control and avoid serious injury. On the other hand, the philosophy of recreational roads dictates that natural roadside features be preserved where possible. Because of the character of the traffic and the relatively low operating speeds on recreational roads, wide clear zones are not as important as on high-speed, high-volume facilities. For these reasons, dimensions smaller than those used on these higher order roads are appropriate. Desirably, 10 or more ft of recovery area, measured from the edge of the traffic lane, should be provided on the higher order recreational roads, i.e., the primary access roads. These values are recommended for the general case; however, where economic and environmental concerns are great, even smaller values are appropriate. Clear zone widths on the lower order recreational roads, i.e., circulation roads and areas, are even less critical than on primary access roads. In areas where the accident potential is greater than normal, such as on the outside of sharp horizontal curves at the end of long, steep downgrades, liberal clear zone widths should be provided.

Roadside Slopes

Where terrain conditions permit, backslopes, foreslopes, and roadside drainage channels should have gentle well-rounded transition. Foreslopes of 4:1 or flatter are safer, stable, and permit establishment and maintenance of turf. The maximum rate of foreslope depends on terrain conditions and the stability of local soils as determined by local experience. Cut sections should be designed with adequate ditches.

The ditch should be deep enough to accommodate the design flow and provide for satisfactory drainage of the pavement base and subbase. While foreslopes of 4:1 or flatter are preferable, there are other important considerations in ditch design for recreational roads. Surrounding terrain and physical feature preservation may dictate narrow width ditches. The lower speeds prevailing on recreational roads reduce the chance of personal injury for passengers in vehicles which drive into shallow-sided ditches.

Roadside Barrier

Roadside barriers should be installed at points of unusual danger, particularly those points that are unusual compared with the overall characteristics of the road. The criteria used in freeway design do not fit the low-volume recreational road situation. The AASHTO *Roadside Design Guide* **(8)** provides some insight into the application of roadside barriers on low-speed, low-volume facilities.

Signing and Marking

Although safety and efficiency of operation depend to a major extent on the geometric design of a road, they should be supplemented by standard signing and marking to provide information and warning to drivers. The extent to which signs and markings are used depends on the traffic volume, the type of highway, and the frequency and use by drivers unfamiliar with the area. The MUTCD **(2)** contains details regarding design, location, and application of highway signs and markings.

Structures

The design of bridges, culverts, walls, tunnels, and other structures should be in accordance with the AASHTO *Standard Specifications for Highway Bridges* **(1)**. The minimum design loading for new bridges should be H-15.

Higher design loadings are appropriate for highways carrying other than just recreational traffic. The vertical clearance at underpasses should be at least 14 ft over the entire roadway width. The clear roadway widths for new and reconstructed bridges should be a minimum of the surface width plus 4 ft. Where the approach roadway is surfaced for the full crown width, that surfaced width should be carried across structures.

Resource Development Roads

Resource development roads include mining and logging roads. Design criteria appropriate for this type of road in many areas are not significantly different from those for recreational roads. For this reason the criteria developed for recreational roads should be followed to the extent they are applicable. Several items are unique to this category of road and deserve special attention.

Traffic on this type of road is primarily composed of large, slow-moving, heavily loaded vehicles. For this reason, particular attention should be paid to superelevation of horizontal curves. The center of gravity of trucks is much higher than that of passenger cars, and this fact increases the tendency of trucks to overturn. When semitrailers are used, only part of the payload is on the drive axles. This situation increases the tendency of the drive wheels to spin and sideslip on slippery surfaces. For these reasons the maximum superelevation should be limited to 6 percent. On long sustained grades averse to the direction of haul, the superelevation should be reduced to accommodate the slow truck.

Gradients on this type of facility have an effect on the road maintenance costs and costs to users. An economic analysis is usually necessary to determine the most economical grade for the specific conditions encountered. Such an analysis must consider the increase in culvert installations to prevent ditch erosion on steeper grades and the more frequent surface replacement needs. Adverse grades are a special problem on roads planned for heavy hauling. Sections of adverse grades should not be so long that they slow a loaded truck to crawl speed. Except for short sections that can be overcome largely by momentum, adverse grades merit special analysis. In many instances, failure to use flatter grades may result in additional expenses for transportation during the life of the road far in excess of any saving in construction costs.

Geometric design features for resource development roads are similar to those for recreational roads in that they must be consistent with the design speed selected. Low design speeds 40 mph or below are generally applicable to roads with winding alinement in rolling mountainous terrain. Table V-17

Type of Terrain	Single Lane 100 VPD Maximum (mph)	Two Lane (mph)
Level	30	40
Rolling	20	30
Mountainous	10	20

Table V-17. Minimum design speeds for resource development and local service roads.

lists those minimum design speeds for both one- and two-lane roads for varying terrain conditions.

Because of the mechanical limitations of many of the vehicles using these roads, special attention should be given to the need for warning signs and markings. On long descending grades, consideration should be given to providing escape lanes to decelerate heavy vehicles that lose their brakes and run out of control. Deceleration may be artificially induced by the use of loose material or providing combinations of sufficient length and upgrade for freewheeling deceleration.

Many design considerations for resource development roads are based on the economics of the equipment operating on the facility. The effects of grades and curvature on operational cost are discussed in considerable detail in the *Logging Road Handbook* **(9)**.

In many instances, resource development roads are ultimately used for other (e.g., recreational) purposes. In instances such as these, the original design should take into account all the possible ultimate usages.

Local Service Roads

Local service roads are those serving isolated areas that have little or no potential for further development (or that require a higher type facility if further developed) and those serving a minimal number of parcels of land. Most of these roads will not be through roads (connected to public roads on both ends) but will dead end at the service to the last parcel on the road.

Traffic on this type of road is of very low volume (fewer than 100 vehicles per day) and is of a repeat type. The design criteria, therefore, can be basically the same as those developed for recreational roads. Those criteria should be followed where applicable.

REFERENCES

1. *Standard Specifications for Highway Bridges,* Washington, D.C.: AASHTO, 1973.
2. U.S. Department of Transportation, Federal Highway Administration. *Manual on Uniform Traffic Control Devices.* Washington, D.C.: U.S. Government Printing Office, 1988.
3. Bert, K. E., et al. *Accommodation of Utility Plants Within the Rights-of-Way of Urban Streets and Highways, Manual of Improved Practices. ASCE Manual No. 14*. Chicago: American Public Works Association, and New York: American Society of Civil Engineers, July 1974, 102 pp.
4. *Guide for Accommodating Utilities on Highway Rights-of-Way.* Washington, D.C.: AASHTO.
5. American Association of State Highway and Transportation Officials, Joint Task Force for Highway Lighting. *An Informational Guide for Roadway Lighting*. Washington, D.C.: AASHTO, 1976.
6. *Guide For Development of New Bicycle Facilities.* Washington, D.C.: AASHTO, 1981.
7. Construction Engineering Laboratory, Corps of Engineers. *Design Guidelines for Recreational Roads,* 1975.
8. *Roadside Design Guide*, Washington, D.C.: AASHTO, 1989.
9. U.S. Forest Service. *Logging Road Handbook—The Effect of Road Design on Hauling Costs, Agriculture Handbook No. 183,* 1960.
10. Glennon, John C. "Design and Traffic Control Guidelines for Low-Volume Rural Roads." *NCHRP Report 214,* 1979, 41 pp.

Chapter VI
COLLECTOR ROADS AND STREETS

INTRODUCTION

For purposes of design, highways are classified by function with enumeration of design values for each functional type. The following guides applicable to collector roads and streets are presented on a functional basis. This chapter will be subdivided into rural collectors and urban collectors.

A definition of the collector can be developed by referring to its upper and lower limits—the arterial and the local road or street. The collector is the road or street between these two limits.

Collectors also serve dual functions: collecting traffic between local roads and arterial streets and providing access to abutting properties.

The collector streets link neighborhoods or areas of homogeneous land use with arterial streets. These streets serve traffic movements between arterial and local streets and serve through traffic within local areas. Collector streets should be planned so as not to disrupt the activities within the areas they serve.

In many cases there may be no discernible difference between the collector and the local access street within a neighborhood. The local access street, from a design classification standpoint, is a local street that provides direct access to lower density residential lots and is intended to serve the collection function for a group of access roads and ideally *not* the immediate access needs of individual residences. However, the collector street does serve the access function for higher density residential development and for some neighborhood facilities.

The collector street is a public highway, usually serving moderate traffic volumes. To establish its own individuality, the design should reflect its specific status and should not be conceived or developed simply as a continuous access street. The collector road or street should allow access to abutting properties consistent with the level of service desired.

The use of more liberal values of design dimensions than the minimums described herein is recommended where economically feasible. In all cases every effort should be made to get the best possible alinement, grade, sight distance, and proper drainage that are consistent with the terrain, present and proposed development, and available funds.

Drainage, both on the pavement and on the sides and subsurface, is an important consideration. Unless drainage is properly accommodated, maintenance costs will be unduly high. In areas of significant snowfall, provision should be made for storing plowed snow with proper drainage as it melts.

Safety is an important factor in all roadway improvements. On low-volume roads or streets or in urban areas it may not be possible to obtain the desirable obstacle-free roadsides. As much clear roadside as practical should be provided. The judicious use of flatter slopes, roadside barriers, and warning signs helps to achieve roadside safety. Proper placement of utility features also assists in achieving safer roadsides. Noise abatement may need to be considered; see Chapter IV, section "Noise Control."

The *Highway Capacity Manual* (HCM) **(1)** provides the designer with the choice of level of service for the highway facility under consideration. Collector streets should generally be designed for level-of-service C to D. In heavily developed sections of metropolitan areas, conditions may necessitate the use of level-of-service D. In rural areas a level-of-service C is desirable for collector roads. However, level-of-service D is practical where unusually high traffic volumes exist or where terrain is rolling or mountainous. Collector roads and streets cannot be designed entirely on the basis of functional classification. Many other facets of design must also be considered. See Chapter II, section "Level of Service."

RURAL COLLECTORS

General Design Considerations

A major part of the rural highway system consists of two-lane collector highways. These roadways should be designed to accommodate the highest practical standards compatible with traffic and topography. Basic information necessary for design of collectors includes safety, traffic volumes, terrain controls, and alinement.

Design Traffic Volumes

Rural collector highways should be designed for specific traffic volumes and levels of service appropriate to accepted procedures. Usually, the design year is 20 years from the date of construction completion but may be anytime within a range of the present (for restoration-type projects) to 20 years in the future (for new construction-type projects).

The current average daily traffic (ADT) may be used for design if the design hourly volume (DHV) is less than 100; otherwise, the DHV should be used. See functional classifications as discussed in Chapter I.

Design Speed

Geometric design features should be consistent with a design speed appropriate for the conditions. Low design speeds (40 mph and below) are generally applicable to highways with curvilinear alinement in rolling or mountainous terrain or where environmental conditions dictate. High design speeds (50 mph and above) are generally applicable to highways in level terrain or where other environmental conditions are favorable. Intermediate design speeds would be appropriate where terrain and other environmental conditions are a combination of those described for low and high design speeds. Table VI-1 identifies minimum design speeds for rural collector roads as a function of the type of terrain and ADT and DHV traffic volumes. The designer should strive for higher values than the minimum where conditions of safety dictate and costs can be supported.

	Minimum Design Speeds (mph) for Design Volumes of:				
Type of Terrain	**Current ADT 0-400**	**Current ADT Over 400**	**DHV 100-200**	**DHV 200-400**	**DHV Over 400**
Level	40	50	50	60	60
Rolling	30	40	40	50	50
Mountainous	20	30	30	40	40

Table VI-1. Minimum design speeds (rural conditions).

Sight Distance

Minimum stopping sight distance and passing sight distance are a direct function of the design speed. A height of eye of 3.50 ft and a height of object of 0.50 ft is used to determine stopping sight distance. A height of eye of 3.50 ft and a height of object of 4.25 ft is used to determine passing sight distance. See Tables VI-2A and VI-2B and Chapter III, section "Sight Distance."

Design Speed (mph)	Assumed Speed for Condition (mph)	Stopping Sight Distance (Rounded for Design) (ft)	K Value[a] for Crest Vertical Curves (Rounded)	K Value[a] For Sag Vertical Curves (Rounded)
20	20-20	125-125	10-10	20-20
25	24-25	150-150	20-20	30-30
30	28-30	200-200	30-30	40-40
35	32-35	225-250	40-50	50-50
40	36-40	275-325	60-80	60-70
45	40-45	325-400	80-120	70-90
50	44-50	400-475	110-160	90-110
55	48-55	450-550	150-220	100-130
60	52-60	525-650	190-310	120-160
65	55-65	550-725	230-400	130-180
70	58-70	625-850	290-540	150-220

[a]K value is a coefficient by which the algebraic difference in grade may be multiplied to determine the length in feet of the vertical curve which will provide minimum sight distance. See Chapter III, section "Sight Distance."

Table VI-2A. Minimum stopping sight distance (wet pavements).

Grades

Vertical alinement must be designed in conjunction with the horizontal alinement. Both alinements must complement each other to provide the safety, capacity, and appearance for the type of improvement proposed. Table VI-3 identifies suggested maximum grades for the terrain and design conditions. See Chapter III, section "Vertical Alinement."

Alinement

The designer should strive for as high a standard as practical for rural collector alinements. As noted, horizontal and vertical alinements must complement each other and be considered in combination. Topography, traffic volume and composition, and right-of-way conditions are controlling features. Sudden changes in horizontal alinement should be avoided. Vertical curves

Design Speed (mph)	Passing Sight Distance (Rounded for Design) (ft)	K Value[a] for Crest Vertical Curves (Rounded)
20	800	210
25	950	300
30	1,100	400
35	1,300	550
40	1,500	730
45	1,650	890
50	1,800	1,050
55	1,950	1,230
60	2,100	1,430
65	2,300	1,720
70	2,500	2,030

[a]K value is a coefficient by which the algebraic difference in grade may be multiplied to determine the length in feet of the vertical curve which will provide minimum sight distance. See Chapter III, section "Sight Distance."

Table VI-2B. Minimum passing sight distance.

should meet the sight distance criteria for the design speed. Frequent opportunities for passing should be employed, where practical. See Chapter III, section "Horizontal Alinement."

Pavement Cross-slope

Pavement cross-slope should be adequate to provide proper drainage. Normally, cross-slopes range from 0.015 to 0.020 ft/ft for high-type pavements and from 0.015 to 0.030 ft/ft for intermediate-type pavements.

High-type pavements are those that retain smooth riding qualities and good non-skid properties in all weather under heavy traffic volumes and loadings with little maintenance.

Intermediate-type pavements are those designed to retain smooth riding qualities and good nonskid properties in all weather, but under lighter loads and lesser traffic volumes.

Low-type pavements are those with treated earth surfaces and those with

Rural Collectors	Design Speed (mph)					
Type of Terrain	**20**	**30**	**40**	**50**	**60**	**70**
			Grades (Percent)			
Level	7	7	7	6	5	4
Rolling	10	9	8	7	6	5
Mountainous	12	10	10	9	8	6

Urban Collectors	Design Speed (mph)					
Type of Terrain	**20**	**30**	**40**	**50**	**60**	**70**
			Grades (Percent)			
Level	9	9	9	7	6	5
Rolling	12	11	10	8	7	6
Mountainous	14	12	12	10	9	7

[a]Maximum grades shown for rural and urban conditions of short lengths, (less than 500 ft), on one-way down grades and on low-volume rural collectors may be 2% steeper.

Table VI-3. Maximum grades[a].

loose aggregate surfaces. A 0.030 ft/ft cross-slope is desirable for low-type pavements. See Chapter IV, section "Normal Crown Sections."

Superelevation

Many rural collector highways have curvilinear alinement. A superelevation rate compatible with the design speed must be used. For rural collectors, superelevation should not exceed 0.12 ft/ft. Where snow and ice conditions may be a factor, the superelevation should not exceed 0.08 ft/ft. Superelevation runout denotes the length of highway needed to accomplish the change in cross-slope from a section with the adverse crown removed to a fully superelevated section and vice versa. Adjustments in design runout lengths may be necessary for smooth riding, surface drainage, and good appearance. Chapter III, section "Horizontal Alinement," provides a detailed analysis on superelevation for appropriate design speeds.

Number of Lanes

The number of lanes should be sufficient to accommodate the design volumes for the selected level of service. Normally, capacity conditions do not govern rural collector roads, and two lanes are appropriate. See Chapter II, section "Highway Capacity."

Width of Traveled Way, Shoulder, and Roadway

Graded shoulder width is measured from the edge of the traveled way to the point of intersection of shoulder slope and foreslope. For intermediate and high-type surfaces the minimum roadway width is the sum of the traveled way and graded shoulder widths given in Table VI-4. Where roadside barriers are proposed, it is desirable to increase the shoulder width by 2 ft wherever practical. The minimum offset of a roadside barrier is 4 ft from the traveled way. The designer should address the foreslope and ditch to accommodate a clear zone as needed.

Foreslopes

The maximum rate of foreslope will depend on the stability of local soils as determined by soil investigation and local experience. Slopes should be as flat as feasible, and other factors should be considered to determine the design slope. Flat foreslopes increase safety by providing maneuvering area in emergencies, are more stable than steep slopes, aid in the establishment of plant growth, and simplify maintenance work. Roadside barriers may be used where topography and right-of-way are restrictive and a need is proven.

Drivers who inadvertently leave the traveled way can often recover control of their vehicles if foreslopes are 3:1 or flatter and shoulder and ditches are well rounded. Such recovery areas should be provided where terrain and right-of-way controls permit. Reference should be made to current AASHTO *Roadside Design Guide* (**3**). See Chapter IV, section "Traffic Barriers."

Where this construction is not practical, the combinations of rate and height of slope provided should be such that occupants of an out-of-control vehicle have a good chance for survival. Where high fills, right-of-way

	Width (ft) for Design Traffic Volumes of:				
Design Speed (mph)	**Current ADT Under 400**	**Current ADT 400 and Over**	**DHV 100-200**	**DHV 200-400**	**DHV Over 400**
20	20	20	20	22	24
30	20	20	20	22	24
40	20	22	22	22	24
50	20	22	22	24	24
60	22	22	22	24	24
70	22	22	22	24	24
	Width of Grade Shoulder - Each Side of Pavement				
All Speeds	2[a]	4	6	8	8

[a]Minimum width is 4 ft. if roadside barrier is utilized.

Table VI-4. Minimum width of traveled way and graded shoulder.[a]

restrictions, watercourses, or other problems make this provision impractical, roadside barriers should be considered, in which case maximum rate of fill slope could be used.

Cut sections should be designed with adequate ditches. Preferably, the foreslope should not be steeper than 3:1. The ditch bottom and slopes should be well rounded. The backslope should not exceed the maximum required for stability, preferably not steeper than 2:1.

Structures

The design of bridges, culverts, walls, tunnels, and other structures should be in accordance with the current AASHTO *Standard Specifications for Highway Bridges* **(2)**.

Except as otherwise indicated herein, the dimensional design of structures should also be in accordance with these specifications.

The minimum design loading for bridges should be HS-20 on collector roads.

The minimum roadway widths for new and reconstructed bridges should be as given in Table VI-5.

Current Traffic ADT	Traffic DHV	Minimum Roadway Width of Bridge	Design Loading Structural Capacity
Under 400	—	Width of traveled way plus 2 ft. each side	HS 20
400 and Over	—	Width of traveled way plus 3 ft. each side	HS 20
—	100-200	Width of traveled way plus 3 ft. each side	HS 20
—	200-400	Width of traveled way[b] plus 4 ft. each side	HS 20
—	Over 400	Approach roadway width[b]	HS 20

[a]Where the approach roadway width (traveled way plus shoulder) is surfaced, that surfaced width shall be carried across all structures.
[b]For bridges in excess of 100 ft. in length the minimum width of traveled way plus 3 ft. on each side will be acceptable.

Table VI-5. Minimum roadway widths for new and reconstructed bridges[a].

Bridges to Remain in Place

Because highway geometric and roadway improvements encourage higher speeds and attract larger vehicles, existing structures also should be improved correspondingly. Because of their high cost, reasonably adequate bridges and culverts that meet tolerable criteria should be retained.

Where an existing highway is to be reconstructed, an existing bridge that fits the proposed alinement and profile may remain in place when its structural capacity in terms of design loading and roadway width are at least equal to the value shown for the applicable traffic volume in Table VI-6.

The values in Table VI-6 do not apply to structures with a total length greater than 100 ft. These structures should be analyzed individually by tak-

Current Traffic ADT	Traffic DHV	Design Loading Structural Capacity	Roadway Clear Width (ft)[a]
Under 400	—	HS-15	22
400 and Over	—	HS-15	22
—	100-200	HS-15	22
—	200-400	HS-15	24
—	Over 400	HS-15	28

[a]Clear width between curbs or rails, whichever is the lesser, is considered to be at least the same as the approach traveled way width.

Table VI-6. Minimum structural capacities and minimum roadway widths for bridges to remain in place.

ing into consideration the clear width provided, safety, traffic volumes, remaining life of the structure, design speed, and other pertinent factors.

Vertical Clearance

Vertical clearance at underpasses should be at least 14.0 ft over the entire roadway width, plus an allowance for future resurfacing.

Horizontal Clearances to Obstructions

On a rural collector road with a design speed of 40 mph and below, a minimum clear zone of 10 ft from the edge of the through-traffic lane should be provided. A 3:1 or flatter foreslope with a rounded shoulder and ditch is preferred. The recovery area should be clear of all unyielding objects such as trees, sign supports, utility poles, light poles, and any other fixed objects. The benefits of removing these obstructions must be weighed again the safety aspects and the environmental effects.

On a rural collector road with a design speed of 50 mph and above, the AASHTO *Roadside Design Guide* **(3)** should be referred to for guidance in making a selection of clear zone. For the intermediate range between high and low design speeds, it is desirable to provide the greatest clear zone associated with the high design speeds. If, however, the conditions more nearly approach those associated with the low design speeds, it would be permissible to employ the lesser clear zone width. See Chapter IV, section "Horizontal Clearances to Obstructions."

To the extent feasible, where another highway or railroad passes over, the structure should be designed so that the pier or abutment supports have lateral clearance as great as the clear zone on the approach roadway.

The approach roadway (traveled way plus shoulders) should be carried across an overpass or other bridge when feasible. Approach roadside barriers, anchored to the bridge rail or parapet, should be provided. Sidewalks should be carried across a bridge if the approach roadway has sidewalks or sidewalk areas. Structures over the project roadway should have piers set back beyond the clear zone. When this is not feasible, roadside barrier protection should be provided at the piers.

Right-of-Way Width

The procurement of right-of-way to widths that accommodate the construction, adequate drainage, and proper maintenance of a highway is a most important part of the overall design. Wide right-of-way permits the construction of gentle slopes, resulting in greater safety for the motorist, and providing easier and more economical maintenance. The procurement of sufficient right-of-way at the time of the initial improvement permits the widening of the roadway and the widening and strengthening of the pavement at reasonable cost as traffic increases.

In developed areas it may be more feasible to limit the right-of-way width to a practical minimum. In every case, however, the right-of-way should be not less than that required for all elements of the design cross section and appropriate border areas.

Intersection Design

Intersections should be situated carefully to avoid steep profile grades and to ensure adequate approach sight distance. An intersection should not be situated on a sharp crest vertical curve, just beyond a sharp crest vertical curve, or on a sharp horizontal curve. Where there is no practical alternate to such a location, the approach sight distance on each leg should be checked carefully. Where necessary, cut slopes should be flattened, and horizontal or vertical curves should be lengthened to provide additional sight distance. There should be sufficient sight distance to permit a vehicle on the minor leg of the intersection to cross the traveled way without requiring the through approaching traffic to slow down. As a general rule, there should be a minimum of 7 sec. available to the driver crossing the through lanes. Chapter IX, section "Sight Distance," contains a detailed discussion of intersection design.

Intersections should be designed with a corner radius for pavement or surfacing adequate for the larger vehicles anticipated; usually, a minimum edge radius of 50 ft is applicable in rural areas. Where turning volumes are significant, speed-change lanes and channelization should be considered.

Intersection legs that will operate under stop sign control preferably should be at right angles. The designer should strive to provide angles no greater than 30° from normal, where practical. A landing platform as level as possible should be provided for approach vehicles that are required to stop.

Railroad—Road Grade Crossing

Appropriate grade crossing warning devices should be installed at all railroad-road grade crossings. Details of the devices to be used are given in the *Manual on Uniform Traffic Control Devices* **(4)**.

Sight distance is an important consideration at railroad-road grade crossings. There must be sufficient sight distance on the collector rural road for the driver to recognize the railroad crossing, perceive the warning device, and stop if necessary.

The roadway width at all railroad crossings should be the same width as the traveled way plus shoulders. See Chapter IX, section "Railroad Grade Crossing."

Traffic Control Devices

Signs, pavement markings, and traffic signal controls where pertinent are essential elements for all highways. The *Manual on Traffic Control Devices* **(4)** contains details of the devices to be used and warrants for their use.

A system of route numbering or lettering for rural collector roads is highly desirable for travelers unfamiliar with the area and simplifies essential signing in accordance with the *Manual on Uniform Traffic Control Devices* **(4)**.

Erosion Control

All slopes and drainage areas should be designed with proper regard for the desired natural ground cover and growth regeneration on areas opened during construction. Seeding and mulching of slopes and the treatment of swales and other erodible areas should be a part of the collector design.

In roadside design the preservation of natural ground covers and desirable growth of shrubs and trees should be considered if such growth does not constitute a problem in the recovery area. Placement of trees should not constitute a problem to the driver nor should it interfere with his sight distance in conducting necessary highway operations.

URBAN COLLECTORS

General Design Considerations

A collector street is a public way for purposes of vehicular travel and includes the entire area within the right-of-way. The urban collector street also serves pedestrian traffic and often accommodates public utility facilities within the right-of-way. The improvement or development of streets should be based on a functional street classification which should be part of a comprehensive community development plan. The design values should be those for the ultimate planned development. In general, design values for collector streets should be greater than the minimums cited.

Geometric design guides for arterial streets are covered in Chapter VII; local access streets are covered in Chapter V.

The function of urban collectors is equally divided between mobility and access. Few cities have effective access control restrictions along collector streets; almost all streets permit access to abutting properties, except where access rights have been acquired. Many new collectors are planned and constructed with little or no access restriction. Uncontrolled access facilities are subject to the deteriorating influences that cause obsolescence of the collector facilities. Management of driveway access to the urban collectors is a desirable undertaking.

When the objective is to expedite traffic mobility, there are many additional criteria for which guidelines are necessary. These criteria include the minimizing of conflict points, adequate storage for all turning movements, minimum conflict with pedestrians, coordination of driveways serving adjoining and opposite properties, locational criteria to meet signal progression needs, and efficient circulation for adequate ingress and egress capacity. By using these design criteria, an optimum system of access can be developed.

Access control on urban collector streets should be exercised primarily to ensure that access points conform to the adopted standards for safety, location, design, construction, and maintenance.

Design Traffic Volumes

Traffic volumes are a major criterion in determining the geometric values to be used in designing urban collector streets. Specifically, the DHV projected to the (future) design year should be the basis of design. It usually is difficult and costly to modify the geometric design of an existing collector street unless provision is made at the time of initial construction. The design traffic should be estimated for at least 10 years and preferably 20 years from the date of construction completion.

Design Speed

Design speed is a factor in the design of collector streets. For consistency in design elements a minimum design speed of 30 mph or higher should be used for urban collector streets, depending on available right-of-way, terrain, adjacent development, and other area controls. See Table VI-1 and Chapter II, section "Design Speed," for additional information.

In the typical street grid the closely spaced intersections usually limit vehicular speeds and thus make the effect of design speed of lesser significance. Nevertheless, the longer sight distances and curve radii commensurate with design speeds higher than the value indicated result in safer highways and should be used to the extent practicable and feasible.

Sight Distance

Stopping sight distance for urban collector streets should be in the range from 200 to 475 ft. Design for passing sight distance seldom is applicable on urban collector streets. See Tables VI-2A and VI-2B. See Chapter III, section "Sight Distance."

Grades

Grades for urban collector streets should be as level as consistent with the surrounding terrain. A 0.30 percent grade is acceptable to facilitate drainage.

However, it is recommended to use 0.50 percent grade or more, when possible, for drainage purposes. Urban street grades are generally depressed below the surrounding terrain to accommodate adjacent property drainage to the curb area and accumulation in the storm drain system. Applicable gradients, radii, and other pertinent features are discussed in Chapter III, section "Vertical Alinement." See Table VI-3 for maximum urban street grades.

Alinement

Alinement in residential areas should fit closely the existing topography to minimize the need for cuts or fills without sacrificing safety. Caution that the safety of the facility is not reduced should be taken in the design of both the vertical and horizontal alinement.

Pavement Cross-slope

Pavement cross-slope should be adequate to provide proper drainage. Cross-slope normally should be from 0.015 ft/ft to 0.030 ft/ft where there are flush shoulders adjacent to the traveled way or where there are outer curbs.

The center section of the pavement crown may be parabolic to permit smooth transition of cross-slope. Normally, parabolic sections are used on two-lane streets, and plane sections are used for divided streets and four-lane streets.

Superelevation

Superelevation, in specific locations, may be advantageous for urban collector street traffic operation, but in built-up areas the combination of wide pavement areas, proximity of adjacent development, control of cross-slope, profile for drainage, frequency of cross streets, and other urban features combine to make its use impractical or undesirable. When used, superelevation on urban collector streets should be 0.06 ft/ft or less. The absence of superelevation on urban collectors for low speeds (40 mph and below) is not detrimental to the user. Generally, some warping or partial removal of the tangent pavement crown facilitates the user's operation. See Chapter III, section "Horizontal Alinement."

Number of Lanes

Two moving traffic lanes plus additional width for shoulders and parking are sufficient for most urban collector streets. Where the street is developed by stages, initially a rural cross section with shoulders may be used. This street should be planned for later stage conversion of the shoulder width to a parking lane or a through lane, usually with outer curbs. Where the initial development utilizes a rural cross section, a clear zone consistent with rural conditions and commensurate with the design speed is to be provided. As the conversion takes place, the clear zone can be modified to that acceptable for urban conditions. If practical and economically feasible, the initial construction should be four lanes with curbs, allowing parking on the outer two lanes until later development necessitates use of all four lanes for moving traffic.

In some cases in commercial areas where there will be several mid-block left turns, it may be advantageous to provide an additional continuous two-way, left-turn lane in the center.

The number of lanes to be provided on urban collector streets with high traffic volumes should be determined by a highway capacity analysis. The capacity should be based on intersection analysis and mid-block analysis, when appropriate, for level of service. Desirably such analyses should be made for future design traffic utilizing the *Highway Capacity Manual* **(1)**. See Chapter II, section "Highway Capacity."

Width of Roadway

The width of an urban collector street should be planned as the sum of the ultimate lanes for moving traffic and parking, including median width where appropriate.

Street lanes for moving traffic should range in width from 10 to 12 ft. In industrial areas, lanes should be 12 ft wide. Where available or attainable, width of right-of-way imposes severe limitations, lanes 10 ft wide can be used in residential areas and lanes 11 ft wide in industrial areas. Added turning lanes at intersections, where used, should range in width from 10 to 12 ft depending on the percentage of trucks. When rural shoulders are used, reference should be made to Table VI-4.

Parking Lanes

Although on-street parking constitutes a safety problem and impedes traffic flow, parallel parking lanes currently are conventional on many collector streets. Parallel parking will normally be acceptable on urban collectors where sufficient street width is available to provide parking lanes. Where needed in residential areas, a parallel parking lane from 7 to 10 ft in width should be provided on one or both sides as the conditions of lot size and intensity of development may require. In commercial and industrial areas, parking lane widths should range from 8 to 10 ft; usually they are provided on both sides. The principal problem of diagonal or angle parking when compared to parallel parking is the lack of adequate visibility for the driver during the back-out maneuver. Design with diagonal or angle parking should only be in special cases. See Chapter IV, section "On-Street Parking."

Parking lane width determinations should include consideration for likely ultimate use as a lane for moving traffic either during peak hours or continuously.

Where curb and gutter sections are used, the gutter pan width may be included as a part of the minimum width of parking lane, but desirably the lane widths should be in addition to that of the gutter pan.

Median

Urban collector streets designed for four or more lanes should include width for appropriate median treatment whenever necessary and practical. There are substantial advantages for each increment in median width, and medians should be as wide as practical for the specific conditions. For the general types of street median treatment the following widths should be provided: (1) paint-striped separation, 2 to 4 ft wide; (2) narrow raised or curbed sections, 2 to 6 ft wide; (3) paint-striped or curbed sections, 10 to 16 ft wide providing space for separate left-turn lanes; and (4) curbed sections, 16 to 40 ft wide or more providing space for shielding protection of a vehicle crossing at an intersection and space for parkway landscape treatment.

On urban collector streets with raised median, openings should be provided only at other street intersections and at reasonably spaced driveways serving major traffic generators such as industrial plants and shopping centers. Whenever practical, median openings should be designed to include left-turn lanes.

Openings should be situated only where there is adequate sight distance. The shape and length of the median openings depend on the width of the median and the vehicle types that must be accommodated. The minimum length of median openings should be that of the projected roadway width of intersecting cross street or driveway. Desirably the length of median openings should be great enough to provide a 50 ft turning control radius or the design vehicle for left-turning vehicles between the inner edge of the lane adjacent to the median and the centerline of the intersection roadway.

On many urban collectors it may be impractical to place a raised median. A continuous left-turn lane, flush with the adjoining traffic lanes and painted, is an acceptable approach. Chapter IV, section "Medians," contains further discussion on medians.

Curbs

Streets normally are designed with curbs for high utilization of available width, control of drainage, protection of pedestrians, and delineation. The curb on the right of traffic should be 6 in. high, usually with appropriate batter. On lightly traveled residential streets with grades less than 2 percent, a lower curb that does not require modification at driveway entrances may be used. The minimum curb slope should be 6:1.

On divided streets the type of median curbs should be determined in conjunction with the median width and the type of turning movement control to be effected. Where mid-block left-turning movements are permitted and the median width is under 10 ft, a well-delineated flush or rounded raised median separator 2 to 4 in. high is effective in channelizing traffic and in serving to avoid excessive travel distances and concentrations of turns at intersections. Where traversable wider medians are appropriate, they may be either flush or bordered with low curb 1 to 2 in. high. On narrow and intermediate-width medians, and on some wide medians, where cross-median movements are undesirable or create problems, a barrier median curb should be used, usually 6 to 9 in. high and with an appropriate curb batter. A high barrier curb or a median barrier should be used where positive separation of opposing traffic is essential, where there is no need for pedestrian crossing, and where local regulations permit. See Chapter IV, section "Curbs."

Barrier curbs 6 in. high or more adjacent to moving traffic lanes should be offset from 1 to 2 ft. Where there is combination curb and gutter construction, the gutter pan width (normally 1 to 2 ft) may be used as the offset distance.

Often a good application of painted or striped median on collector streets can preclude the need for median curbs.

Drainage

Surface runoff is gathered by a system of gutters, inlets, catch basins, and sewers. The gutter grade should be 0.30 percent. However, a gutter grade of

0.50 percent or more should be used, when possible, for drainage purposes. Inlets or catch basins with an open grate should be located in the gutter line limits. Grates should be designed to accommodate bicycle and pedestrian traffic. See drainage portions of Chapters III and IV for additional details.

Sidewalks

Sidewalks should be provided along both sides of urban collector streets that are used for pedestrian access to schools, parks, shopping areas, and transit stops and along all collectors in commercial areas. In residential and be so spaced that ponding on the pavement would not exceed tolerable areas, sidewalks are desirable on both sides of the street, but should be provided on at least one side of all collector streets. The sidewalk should be situated as far as practical from the traffic lanes, usually close to the right-of-way line. See Chapter IV, section "Walkways."

Clear sidewalk width should be at least 4 ft in residential areas and range from 4 to 8 ft in commercial areas.

Driveways

Driveways should be regulated as to width of entrance, placement with respect to property lines and intersecting streets, angle of entry, vertical alinement, and number of entrances to a single property.

Curb-Cut Ramps

Curb-cut ramps should be provided at crosswalks to accommodate the physically handicapped person. Chapter IV discusses various design applications on ramps for the handicapped. See Chapter IV, section "Pedestrian Crossing."

Roadway Widths for Bridges

The minimum clear width for all new bridges on urban collector streets with curbed approaches should be the same as the curb-to-curb width of the approaches. The bridge rail should be placed immediately beyond the curb if no sidewalk is present to avoid vaulting of vehicles. For urban collector streets with shoulders and no curbs, the full width of approach roadways preferably should be extended across bridges. Sidewalks on the approaches should be extended across all new structures. Desirably there should be at least one sidewalk on all street bridges. Further discussion on this subject and the application of the concrete barrier rail is given in Chapter IV, section "Traffic Barriers." Table VI-5 applies for urban collector bridge widths.

Vertical Clearance

Vertical clearance at underpasses should be at least 14.0 ft. over the entire roadway width, plus an allowance for future resurfacing.

Horizontal Clearance to Obstructions

On all urban collector streets, which are curbed with no shoulders, a minimum clearance of 1.5 ft. or wider where possible beyond the face of the curb should be provided. Where a continuous parking lane is provided, no clearance is required but a minimum 2 ft. setback to obstructions is desirable to avoid interference with opening car doors. Preferably these obstructions are located at or near the right-of-way line and outside of the sidewalks. In areas of dense pedestrian traffic the construction of barrier type curbing between the traveled way and adjacent street fixtures would aid in protecting the high volume of pedestrian traffic. All urban collector streets with shoulders and without curbs should have clear zones the same as on rural collectors. See discussion in this chapter.

Other off-roadway obstacles such as trees that might seriously damage out-of-control vehicles should be removed from the roadside wherever feasible. However, the potential benefits from the removal of trees should be weighed against the adverse effects that their removal may have on the roadside environment; and they should be removed only when necessary for reasons of safety. Depending on the conditions, only those fixed objects in very vulnerable locations may be subject to removal. See Chapter IV, section "Horizontal Clearance to Obstructions."

A wide and level border area should be provided along collector streets for the safety of the motorist and pedestrian as well as for esthetic reasons. However, the preservation and enhancement of the environment are of major importance in the design and construction of collector streets. The street alinement should be selected on the basis of minimizing cut and fill slopes.

Roadside barriers are not used extensively on urban collector streets except where there are safety concerns or environmental considerations such as along sections with steep foreslopes and at approaches to overcrossing structures.

Right-of-Way Width

The right-of-way width should be sufficient to accommodate the ultimate planned roadway, including median, shoulder, grass strip, sidewalks, public utility facilities, and width for necessary outer slopes, except where they are on an obtained easement. The width of right-of-way for a two-lane urban collector street should range from 40 to 60 ft, depending on the conditions listed above.

Provision for Utilities

In addition to the primary purpose of serving vehicular traffic, urban collector streets may accommodate public utility facilities within the street right-of-way in accordance with state law or municipal ordinance. Use of the right-of-way by utilities should be planned to cause the least interference with traffic using the street. The AASHTO *Guide for Accommodating Utilities Within Highway Right-of-Way* **(5)** gives general principles for utility location and construction to minimize conflicts between the use of the street right-of-way for vehicular movements and a secondary purpose of providing the space for location of utilities.

Border Area

A border area between the roadway and the right-of-way line should be wide enough to serve several purposes including provision of a buffer space between pedestrians and vehicular traffic, sidewalk space, and an area for both underground and above-ground utilities such as traffic signals, parking meters, and fire hydrants. A portion of the border area would serve snow storage and esthetic features such as grass or other landscaping features. The border width should range from 8 to 11 ft, including the sidewalk width. Traffic signals, utility poles, fire hydrants, and other utilities should be placed as far

Intersection Design

The pattern of traffic movements at the intersections and the volume of traffic on each approach, including pedestrian traffic, during one or more peak periods of the day are indicative of the type of traffic control devices back of the curb as practical for safety reasons. Breakaway features should be built when feasible and as an aid to safety considerations.
necessary, the widths of pavements required including auxiliary lanes, and where applicable, the degree of channelization needed to expedite the movement of all traffic. The arrangement of islands and the shape and length of auxiliary pavements may differ depending on whether or not signal control is necessary. The composition and character of traffic is a design control; movements involving large trucks necessitate larger intersection areas and flatter approach grades than those required at intersections where traffic predominantly consists of passenger cars. Bus stops located near an intersection may further modify the arrangement. Approach speeds of traffic also have a bearing on the geometric design as well as on control devices and pavement markings. Traffic signals are a major element in the design of major urban collector street intersections. See Chapter III, section, "Signals."

The number and locations of approach roadways and their angles of intersection are major controls for the intersection geometric pattern, the location of islands, and the types of control devices. Intersections at-grade preferably should be limited to no more than four approach legs. When two crossroads intersect, the collector highway in close proximity should be combined into a single public road connection.

The more important design considerations for intersections at-grade fall into two major categories: First, the geometric design including a capacity analysis, and the second, the location and type of traffic control devices. For the most part, these considerations are applicable to both new and existing intersections; although on existing intersections in built-up areas, heavy development may make extensive design changes impractical.

Chapter IX contains the discussion on intersection design.

Railroad—Street Grade Crossings

Appropriate grade crossing warning devices should be installed at all railroad-street grade crossings. Details of the devices to be used are given in the *Manual on Uniform Traffic Control Devices* **(4).**

Sight distance is an important consideration at railroad-street grade crossings. There must be sufficient sight distance on the street for the driver to recognize the railroad crossing, perceive the warning device, and stop if necessary.

The roadway width at all crossings should be the same as the curb-to-curb distance of the approaches. Sidewalks should be provided at all railroad crossings where approach sidewalks now exist or are planned within the near future.

In specific instances the street section may be a non-curbed unit. The crossing should be consistent with the approach street width and shoulder width. See Chapter IX, section "Railroad Grade Crossing."

Provisions for future sidewalks should be incorporated into design, if they can be anticipated to avoid future crossing work on the railroad facility.

Street and Roadway Lighting

Good visibility under day or night conditions is a fundamental requirement that enables the motorist to move on roadways in a safe and coordinated manner. Street lighting, properly designed and maintained, should produce comfortable and accurate night visibility, which should facilitate vehicular and pedestrian traffic.

Determinations on lighting should be coordinated with safe operation, crime protection, and other community concerns. See Chapter III, section "Lighting."

The AASHTO publication *An Information Guide for Roadway Lighting* **(6)** provides discussion on street and roadway lighting.

Traffic Control Devices

Traffic control devices should be applied consistently and uniformly. Details of the standard devices and warrants for many conditions are found in the *Manual on Uniform Traffic Control Devices* **(4)**.

Geometric design of streets should include full consideration of the types of traffic control to be used, especially at intersections where multiple phases of actuated traffic signals are likely to be needed. Traffic signals are a major element in the design of major urban collector street intersections. Consideration for multiphasing, signal progression, pedestrian walk cycles, and traffic flow are necessary in major signalized intersection design. See Chapter III, section "Signals."

Erosion Control

Design of streets should include consideration for preservation of natural ground cover and desirable growth of shrubs and trees within the right-of-way. Seeding, mulching, sodding, or other acceptable measures of slopes, swales, and other erodible areas should be a part of the urban collector street design. See Chapter III, sections "Erosion Control" and "Landscape Development."

Landscaping

Landscaping should be provided in keeping with the character of the street and its environment for esthetic and erosion control purposes. Landscape designs should be arranged to permit a sufficiently wide, clear, and safe pedestrian walkway. Combinations of turf, shrubs, and trees should be considered in continuous border areas along the roadway. However, care should be exercised to ensure that sight distances and clearance to obstruction guidelines are observed, especially at intersections. The roadside should be developed to serve both the community and the traveling motorist. Landscaping should also consider maintenance problems and costs, future sidewalks, utilities, additional lanes, and possible bicycle facilities. See Chapter IV, section "Bicycle Facilities."

REFERENCES

1. *Highway Capacity Manual*, Washington D. C.: Highway Research Board, 1985.
2. *Standard Specifications for Highway Bridges,* Washington, D.C.: AASHTO, 1989.
3. *Roadside Design Guide:* Washington, D.C.: AASHTO, 1989.
4. U.S. Department of Transportation Federal Highway Administration, National Advisory Committee on Uniform Traffic Control Devices. *Manual on Uniform Traffic Control Devices.* U.S. Government Printing Office, 1989.
5. *Guide for Accommodating Utilities Within Highway Right-of-Way,* Washington D.C.: AASHTO, 1988.

ADDITIONAL REFERENCES

An Information Guide for Roadway Lighting .Washington, D.C.: AASHTO, 1984.

Drainage of Highway Pavements. Washington, D.C.: Federal Highway Administration, Office of Engineering, Bridge Division, 1984.

Schoppert, D.W. and Hoyt, D.W "Factors Influencing Safety at Highway-Rail Grade Crossings. *NCHRP Report 50*, 1968, 113 p.

Residential Streets. Published by American Society of Civil Engineers, National Association of Home Builders, and Urban Land Institute, 1974.

Design of Urban Streets. Prepared for the U.S. Department of Transportation, FHWA, by J.H.K. and Associates, 1980.

Chapter VII
RURAL AND URBAN ARTERIALS

INTRODUCTION

The principal and minor arterial road systems provide the high speed, high volume network for travel between major points in both rural and urban areas. Chapter I has an extensive discussion on the functional purposes of this class of facility, both rural and urban. The material in this chapter provides the general information required to establish the basis of design for these roadways.

The design of these arterials covers a broad range of roadways, from two-lane to multilane, and is the most difficult class of roadway design because of the need to provide a high standard of operation under sometimes unusual or limited conditions. The designer must be thoroughly familiar with the material in all chapters in this text in order to skillfully blend the various types of arterials into the functional network. Although freeways are included in the functional description of an arterial, the design is unique; therefore, that class of roadway is treated separately in Chapter VIII.

The material in this chapter considers the rural and urban arterials separately because of distinctive features of each. The designer must, however, be prepared to use both to provide adequate transitions as the arterial flows between the rural and urban settings.

RURAL ARTERIALS

General Characteristics

A major part of the rural highway system consists of the rural arterials, which range from two-lane roadways to multilane, divided controlled-access arterials. This part of the chapter relates to the design of new rural arterials and major reconstruction of existing ones. These roadways are designed on the basis of traffic volume needs and should be constructed to the highest standards possible.

Principal rural arterials include the Interstate System and most rural freeways. They usually warrant multilane roadways and connect most urban centers. Minor rural arterials link the urban centers to larger towns and are spaced to provide a relatively high level of service to developed areas of a State.

With available information regarding traffic volumes and composition, the type of terrain, and the general character of alinement, the design geometrics of the arterial needed may be readily determined from the selected design speed and design traffic volumes. Operational characteristics, design features, cross section, and right-of-way are discussed in this chapter.

Two-lane arterials constitute the bulk of the arterial system. These roadways are adequate where traffic volumes are light and long sight distances are generally available. These roadways are generally surfaced with all-weather-type surfaces and are marked and signed in accordance with the *Manual on Uniform Traffic Control Devices* (MUTCD) **(1)**.

General Design Considerations

Design Speed

Rural arterials, excepting freeways, are normally designed for speeds of 40 to 70 mph depending on terrain, driver expectancy and whether the design is construction on new location or reconstruction of an existing facility. Design speeds in the higher range, 60 to 70 mph, are normally used in level terrain, design speeds of 50 to 60 mph are normally used in rolling terrain and design speeds of 40 to 50 mph are used in mountainous terrain. When a lower design speed is required, refer to Chapters II, III and IV for design features.

Design Traffic Volume

When a rural arterial is to be improved, traffic counts are taken, or if on a new location, estimates are prepared. These values are then projected to the design year, normally 20 years, and are expressed in terms of average daily traffic (ADT). These ADT values can then be translated to the design hourly volume (DHV), usually 15 percent of ADT, and the estimated 30th highest hourly volume, 30 HV. Low-volume rural arterials are normally designed with the use of only the ADT values because neither capacity nor intersection operations govern the overall operation. These roadways are of minimum service design and normally provide free flow under all conditions. High-volume roadways are often not capable of providing free flow; therefore, the DHV is used to determine the design features required.

Levels of Service

Techniques and procedures for adjusting operational and highway factors to compensate for conditions other than ideal are found in the *Highway Capacity Manual* (HCM) **(8)**, which also presents a thorough discussion of the level-of-service concept. Although the choice of the design level of service is left to the user of the HCM **(8)**, designers should strive to provide the highest level of service feasible and consistent with anticipated conditions. Refer to Table II-5 for level-of-service characteristics. For acceptable degrees of congestion, rural arterials and their auxiliary facilities, i.e., turning lanes, passing sections, weaving sections, intersections, and interchanges, should generally be designed for level-of-service B except in mountainous areas where level-of-service C is acceptable.

Sight Distance

Sight Distance is a direct function of the design speed which greatly influences the level of service on rural arterials. Minimum stopping sight distance must always be provided as a safety requirement. Passing and decision sight distances influence the operational capability which provides the selected level of service. Provision for adequate sight distance on rural arterials where both high speeds and high volumes can be factors is complex. The designer is referred to Chapter III for a comprehensive discussion on the subject of sight distance and for tables of values suitable for the design of arterials.

Alinement

A smooth flowing alinement is desirable on a rural arterial. Changes in alinement, both horizontal and vertical, should be so gradual that the driver will not be surprised by any of the changes. Special consideration should be given to providing adequate decision sight distance at locations such as high-volume intersections and at transitions of roadway widths or numbers of lanes. Adequate alinement is usually more safe and effective than control by signing or pavement marking and should be achieved whenever feasible.

Grades

The length and steepness of grades directly affect the operational characteristics of an arterial. Table VII-1 gives recommended maximum grades for rural arterials. When vertical curves for stopping sight distance are considered, there are seldom advantages to using the maximum grade values except when grades are long. Grades below the maximum are always desirable, the minimum grades being considered primarily to provide natural roadside drainage.

	Design Speed (mph)			
	40	50	60	70
Type of Terrain	Grades (percent)			
Level	5	4	3	3
Rolling	6	5	4	4
Mountainous	8	7	6	5

Table VII-1. Relation of maximum grades to design speed for rural arterials.

Number of Lanes

The number of lanes required is determined by volume, level of service, and capacity conditions. A divided arterial, as discussed in this chapter, refers to four or more lanes.

Superelevation

When the use of curves is required on a rural arterial alinement, a superelevation rate compatible with the design speed must be used. Superelevation rates should not exceed 0.12; however, where ice and snow conditions are a factor, the maximum superelevation rate should not exceed 0.08. Superelevation runoff denotes the length of roadway needed to accomplish the change in cross slope from a section with adverse crown removed to a fully superelevated section and vice versa. Adjustments in design runoff lengths may be necessary for smooth riding, drainage, and appearance. Chapter III provides a detailed analysis and tables for superelevation for various design speeds.

Pavement Crown

Pavement crown is constructed to provide cross-slope drainage for a pavement. Two-lane rural pavements are normally designed with a centerline crown and cross slopes ranging from 1.5 to 2 percent with the higher values being most prevalent. Multilane pavements are crowned at the pavement centerline or sloped one way. When drainage is carried across adjacent lanes, the cross slope may be increased from one lane to another.

Vertical Clearances

New or reconstructed structures should provide 16-ft clearance over the entire roadway width. Existing structures that provide 14-ft, if allowed by local statute may be retained. In highly urbanized areas, a minimum clearance of 14-ft may be provided if there is one route with 16-ft clearance. Structures should provide additional clearance for future resurfacing of the underpassing road.

Structures

The full width for the approach roadways should normally be provided across all new bridges. Long bridges, defined as bridges having an overall length in excess of 200 feet may have a lesser width. On long bridges, offsets to parapet, rail or barrier shall be at least 4 feet measured from the edge of the nearest travel lane on both the left and right. See Chapter X for more information on bridge widths.

Bridges to remain in place should have adequate strength and at least the width of the traveled way plus 2-ft clearance on each side, but should be considered for ultimate widening or replacement if they do not provide at least 3-ft clearance on each side or are not capable of HS-20 loadings. As an interim measure, narrow bridges should be considered for special narrow bridge treatments such as signing and pavement marking.

Traffic Control Devices

Signs, pavement delineation, and pavement marking play an important role in the optimum operation of a rural arterial. Placement of these items should be considered early in the design stage while adjustments to the alinement and intersection design can easily be considered. Refer to the current MUTCD (**1**) for guidance in signing and marking.

Erosion Control

Consideration of erosion control features is important to the proper design of a rural arterial. By controlling erosion, the safety shape of the roadside is maintained and the environment downstream is protected from siltation and other possible harmful effects. Providing adequate ground treatment and cover has the additional benefit of assuring a pleasing roadside appearance at little or no additional cost. Topsoiling and seeding are considered minimum treatments for erosion control in rural areas.

Widths

An ideal two-lane arterial would consist of two 12-ft traffic lanes and have usable shoulders 8 ft wide. Under restrictive or special conditions, 11-ft lanes may be acceptable. The logical approach on shoulder widths is to provide a width related to the traffic demands.

Table VII-2 provides values for the usable width of shoulder that should be considered for the volumes indicated. Shoulders should be usable at all times regardless of weather conditions. On high-volume highways they should be paved, but because of economic limits, this type of surfacing may not always be feasible. The shoulder should be constructed to a uniform width for relatively long stretches of roadway. For additional information regarding shoulders, refer to Chapter IV.

Horizontal Clearance to Obstructions

A clear unobstructed roadside is highly desirable. When fixed objects or nontraversable slopes fall within the clear roadside zones discussed under horizontal clearance to obstructions in Chapter IV, refer to AASHTO *Roadside Design Guide* (**6**) for guidance in selecting the appropriate treatment. Utilities

Projected Design Traffic Volume	Current ADT Under 400	Current ADT 400 & Over	DHV 100-200	DHV Over 200
	Width of Traveled Way (ft)[a]			
Design Speed (mph)				
40	22	22	22	24
50	22	24	24	24
60	24	24	24	24
70	24	24	24	24
	Width of Usable Shoulder (ft)[b] Each Side of Pavement			
All Speeds	4	6	6	8

[a]Width of traveled way may remain at 22 ft on reconstructed highways where alinement and safety records are satisfactory.
[b]Usable shoulders on arterials should be paved.

Table VII-2. Minimum width of traveled way and usable shoulder for rural arterials.

and landscaping should be located near the right-of-way line and should be outside the selected clear zone.

Cross Section and Right-of-Way

The type of surfacing and shoulder treatment should fit the volume and composition of traffic. Two-lane rural arterials normally are crowned to drain away from the centerline except where superelevation is required. The treatment of cross slopes, pavements, shoulders, ditches, and sideslopes is discussed in Chapter IV. Right-of-way should be wide enough to include all of the cross-sectional elements throughout the project. A uniform width of right-of-way may be convenient, but there are special cases where additional right-of-way may be very desirable. These cases would be locations where the sideslopes extend beyond the normal right-of-way, where greater sight distance is desirable, at intersections and junctions with highways, at at-grade railroad crossings, and for environmental considerations.

Figure VII-1 indicates the approximate right-of-way for a two-lane arterial. The designer must also consider local conditions such as drainage and snow storage when determining right-of-way widths. Where additional lanes may be required in the future, the initial right-of-way should be adequate to provide the wider roadway section. It may be desirable to construct the initial two lanes off-center so that the future construction will not interfere with the traffic or waste the investment in the initial grading and surfacing stage.

Provision for Passing

In the design of two-lane arterials, accommodation of alinement and profile should provide sections for safe passing at frequent intervals and on a major portion of the highway length. Table VII-3 provides the relation of sight distance to the design speed.

Design Speed (mph)	30	40	50	60	65	70
Minimum stopping sight distance (ft)	200-200	275-325	400-475	525-650	550-725	625-850
Minimum passing sight distance (ft) (two-lane, two-way)	1,100	1,500	1,800	2,100	2,300	2,500

Table VII-3. Relation of sight distance to design speed.

There may be restrictive cases where passing sight distance is economically difficult to justify. In those instances, passing sections should at least provide the desired level of service. Where achievement of adequate passing sight distance is not practical auxiliary lanes; such as truck climbing lanes or perhaps passing lanes should be considered. Although climbing lanes are normally provided to prevent unreasonable reductions in operating speeds, they also provide opportunities for passing in areas that normally would not permit passing sections. Climbing lanes are usually provided to the right of the normal traffic lane and should be 12 ft wide, with a somewhat reduced shoulder width. A usable shoulder of 4 ft width or greater is acceptable as indicated in Chapter III.

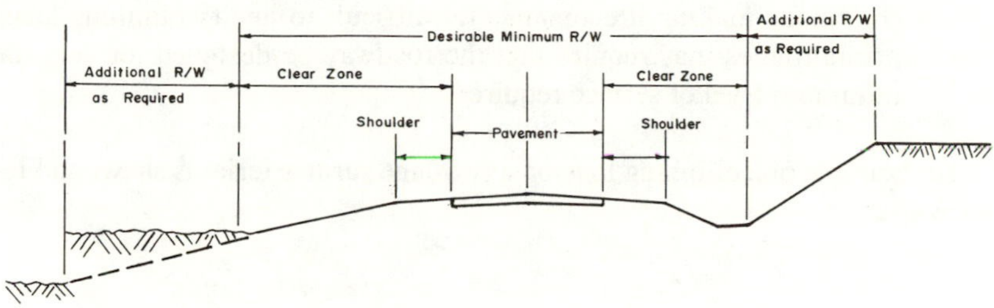

Figure VII-1. Desirable minimum R/W two-lane arterial.

Passing lanes should normally only be considered in areas where the roadway can be expanded to four lanes. The introduction of a short four-lane section requires that consideration be given to adequate transitions and particularly to marking and signing. Climbing lanes are preferred over passing lanes because they provide a normal type of treatment without the risk of the motorist thinking that he is entering a four-lane roadway of any length. Adequately designed and well-marked climbing lanes will usually be used by the slow-moving vehicles, both trucks and automobiles, and will allow passing by drivers who prefer to move at normal speeds.

Intersections and railroad crossings should be separated or provided with adequate sight distance. Decision sight distance should always be provided when economically attainable. Ideally, intersections should be placed in sag locations to allow maximum visibility of the roadway and pavement markings.

The design elements and requirements for the use of climbing lanes are discussed in Chapter III.

In conclusion, the following is a summary of the design procedures to be followed in providing passing sections on two-lane highways:

1. Horizontal and vertical alinement should be designed to provide as much of the highway length as feasible with sight distance for safe passing.
2. For design volumes approaching design capacity, the effect of insufficient passing sections in reducing capacity should be considered.
3. Where the DHV is approached because of trucks climbing up long grades, construction of climbing lanes on critical lengths of grade should be considered.
4. Where the extent and frequency of passing sections made available by application of items 1 and 3 are still too few, consideration should be given to the construction of a four-lane highway. However, restrictive

conditions making it economically difficult to justify climbing lanes or multilanes may require that the roadway be designed for only the minimum level of service required.

An example of a climbing lane on a two-lane rural arterial is shown in Figure VII-2.

Ultimate Development of Four-Lane Divided Arterials

Although many two-lane arterials will adequately serve the traffic demands in the future, there are numerous instances, particularly near urban areas, where two-lane arterials will require ultimate development to a higher type to handle the expected traffic.

Where it is anticipated that a DHV within about 10 to 20 years will be in excess of the design capacity of the two-lane arterial, the initial improvement should be patterned to the ultimate development of a four-lane divided arterial and provision made for acquisition of the necessary right-of-way. No two-lane rural arterial should be built with the thought that it will later be widened to a four-lane undivided highway. Even where right-of-way is restricted, some form of separator should be used in the ultimate facility, with a median at least 4 ft wide and preferably much wider. (Four-lane undivided arterials are discussed in a following section of this chapter.)

In the ultimate development of a four-lane divided arterial, the initial two-lane surfacing should be constructed to form one of the two-lane one-way surfaces. The advantages of this procedure over that of the initial two lanes in the middle of the right-of-way are the following:

1. There is no loss of investment in existing surfacing and in highway and railroad overcrossings when progressing to the higher type. Where undercrossings are constructed for the ultimate improvement, there exists a continuing cost without immediate benefit in the form of interest on the additional cost, but this cost may be less in total than the later cost of rebuilding the undercrossing.
2. There may be no loss of interest on investment in initial extra grading. If the entire ultimate development is graded when the two-lane arterial is constructed, there exists a continuing cost without immediate benefit in the form of interest on the investment in that portion of the grading not required for the two-lane arterial. If the initial two lanes are offset, the grading may be omitted on one side of the road until the higher type facility is constructed.

Figure VII-2. Climbing lane on two-lane rural arterial.

3. There may be no loss of interest on investment in initial extra right-of-way. If the entire right-of-way required for the ultimate development is purchased when the two-lane arterial is constructed, theoretically there exists a continuing cost of interest on investment in that portion of the right-of-way not required for the two-lane arterial. However, the increase in land values, particularly after the construction or improvement of the arterial, may more than offset the interest on investment in additional right-of-way. It is desirable to acquire initially sufficient right-of-way for the ultimate development, including that required for future intersection improvements and grade separations.

4. Traffic is subjected to little restriction or delay when the additional two-lane surfacing is constructed. The two-lane surfacing existing at that time continues in use as a two-lane highway, no detours are necessary, and contact with construction operations is restricted to intersections and turnouts on one side.
5. Later adjustment of minor road structures and plant growth are reduced to a minimum. When the entire grade for the ultimate four-lane divided arterial is constructed initially, all structures such as drains and culverts usually are completed and remain undisturbed when the final two lanes are added. If grading for one two-lane roadway is economically advisable, road structures may be completed on one side, and temporary headwalls and open drains may be provided on the side where additional lanes will be placed later.

Care must be exercised, however, that an appropriate clear zone is provided in the initial stage. A similar procedure may be adopted for topsoiling, seeding, planting, and any other work that is done to prevent soil erosion, the value of which increases with time.

Two-lane arterials planned for ultimate conversion to a divided arterial have sufficient volume initially to warrant pavements 24 ft wide and usable shoulders 8 ft wide, as shown in Figure VII-3A. These pavement and shoulder dimensions are commensurate with those recommended for four-lane divided arterials as discussed in the last part of this chapter.

When an arterial will ultimately be developed to a four-lane divided arterial with a wide median and the initial pavement is offset to one side of the right-of-way centerline, the pavement generally is crowned to drain both ways. Ultimately, the wide median is depressed to be self-draining and to receive surface runoff from one-half of each pavement (Figure VII-3B). Grading for the future development generally is deferred when the median is wide.

Where the right-of-way for the future four-lane arterial is restricted, a narrow median, which should be not less than 4 ft wide, may have to be used. As for a wide median, the initial work should be offset so that the ultimate development is centered on the right-of-way. To economize on the cost of drainage structures and to simplify construction, the initial and the future two-lane pavements may be positioned to drain to the outside (Figure VII-3C). Future grading may or may not be deferred, depending on local conditions and on the probable length of time to the full development.

On most two-lane arterials constructed many years ago, no provision was made for future improvement to a higher type. In such instances, if feasible, a new two-lane one-way roadway should be provided approximately parallel to the first, which then is converted to one-way operation to form a divided arterial. Where the frontage is developed, it may be feasible to construct another

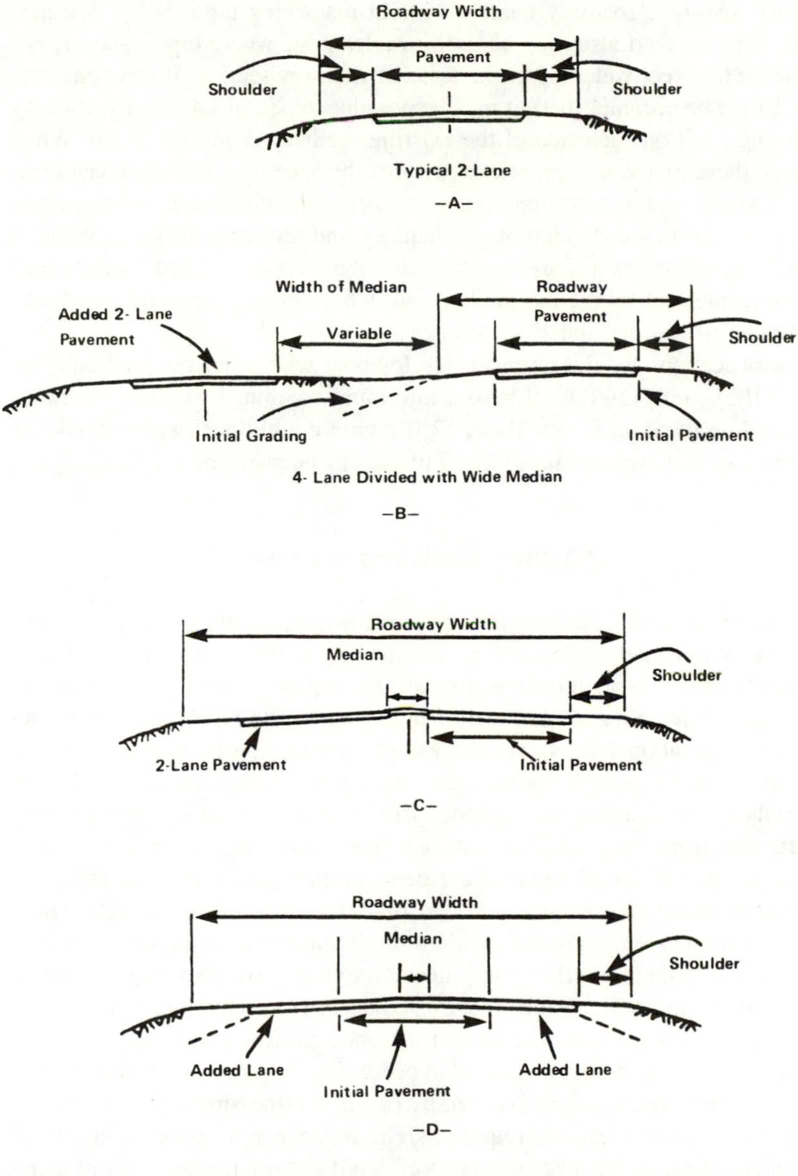

Figure VII-3. Two-lane arterial cross section with ultimate development to a four-lane arterial.

one-way two-lane roadway nearby without disturbing the existing development. This method also may be used to advantage where topography is not favorable to direct widening of the existing roadway section. If this construction cannot be accomplished, it may be possible to obtain a divided section by widening 14 ft on each side of the existing roadway (Figure VII-3D). When none of these methods is practicable, it may be necessary to find a new location. The old road then becomes a local facility and may also serve as an alternate route. From the standpoint of adequacy and service provided to through traffic, the last method is preferred because the arterial on a new location will not be influenced by the old facility and can be built to modern standards, preferably with some control of access.

Cross sections for the ultimate development with a narrow median (Figures VII-3C and VII-3D) all have a minimum combined width of roadways and median of 68 to 72 ft. About 12 ft or more additional width should be obtained so that median lanes for left turns may be provided at intersections.

Multilane Undivided Arterials

A multilane undivided arterial is the narrowest arterial on which each traffic lane is intended to be used by traffic in one direction, and passing is accomplished on lanes not subject to use by opposing traffic. The ability to pass without traveling in the lane of opposing traffic results in a freer and smoother operation and a large increase in arterial capacity over that of two-lane arterials. Because of the generally higher volumes, drivers on multilane arterials are confronted with added traffic frictions not only from opposing traffic and from the roadsides, but also from traffic in the same direction. The accident rate for all reported accidents on multilane undivided arterials, as an average for all volumes, is higher than that on two-lane arterials. There are individual cases where the accident rate is substantially higher. However, as a rule, multilane arterials carry heavier volumes and have more frequent intersections and more development of adjacent land. Frequency of at-grade crossings has appreciable bearing on accident experience and capacity.

The elements of design discussed in preceding chapters are applicable generally to multilane undivided arterials, except that passing sight distance is not essential. The minimum required sight distance at all points is the stopping sight distance, because passing is effected without the necessity of using a lane for opposing traffic. Longer than minimum required stopping sight distance is desirable, as it is on any type of arterial.

Adequate shoulders, which encourage all drivers to use them in emergencies, are essential on multilane undivided arterials. Vehicles stopping on through-traffic lanes are very hazardous, probably more so than on a divided arterial, because following vehicles that maneuver to pass encourage and sometimes force vehicles behind them and in the inner lane to edge over the centerline.

If traffic volumes require the construction of multilane arterials in rural areas where speeds are apt to be high, the consensus is that opposing traffic should be separated. All arterials on new locations requiring four or more lanes should be divided. Improvement of an existing two-lane arterial to a multilane facility preferably should include a median.

Undivided arterials with four or more lanes are most applicable in urban and suburban areas where there is concentrated development of adjacent land.

Divided Arterials

General Features

A divided arterial is one with separated lanes for traffic in opposite directions. It may be situated on a single roadbed, or two roadways may be widely separated. The width of median may vary and is governed largely by the type of area, character of terrain, intersection treatment, and economics. An arterial is not normally considered to be divided unless two full lanes are provided in each direction of travel and the median is 4 ft or more wide and constructed or marked in a manner to preclude its use by moving vehicles except in emergencies or for left turns.

A four-lane rural facility should have adequate median width to provide for protected left turns. Vehicles making a left turn should not be required to stop in the passing lane of a roadway designed for high volumes and speeds.

The principal advantages of dividing the multilane arterial are increased safety, comfort, and ease of operation. Of significance is the reduction in head-on collisions and virtual elimination of such accidents on sections with wide medians. These accidents usually are serious. Where median lanes for left turns are provided, rear-end collisions and other inconveniences to through traffic resulting from left-turn movements are greatly reduced. Pedestrians crossing the divided arterial are required to watch traffic in only one direction at a time and are given a welcome respite at the median. Where the median is wide enough, crossing and left-turning vehicles can slow down or stop between the one-way pavements to take advantage of breaks in traffic and cross when it is safe to do so. Divided multilane arterials make for more relaxed and pleasant operation, particularly in inclement weather and at night when headlight glare is bothersome. Headlight glare is reduced somewhat by narrow medians but can almost be eliminated by wide medians.

Lane Widths

Pavements on divided arterials should be designed with lanes 12 ft wide. The high speeds and large volumes associated with divided arterials will justify the construction of 12-ft lanes. On reconstructed arterials, it may be acceptable to retain 11-ft lanes if the alinement and safety records are satisfactory.

Cross Slope

Each pavement of a divided arterial may be sloped to drain to both edges, or each pavement may be sloped to drain to the outer edge, the arrangement depending on climatic conditions and the width of median. Pavements on divided arterials should be of high type and have a normal cross slope of 1.5 to 2 percent. Pavements with unidirectional slope may have the outer lane on a steeper slope than the inner lane. On an auxiliary lane, normally the cross slope should not exceed 3 percent on tangent alinement. In no case shall the cross slope of an outer and/or auxiliary lane be less than the adjacent lane.

Shoulders

Arterials having sufficient traffic volume to justify the construction of four lanes also justify the provision of full width shoulders. The width of usable outside shoulders preferably should be at least 8 ft.

The normal roadway section, including usable shoulders, should be carried over all structures. Shoulders on divided arterials should be usable during all seasons of the year. Paving of the usable width of shoulder is preferred.

Shoulder space on the left side of four-lane divided arterials within the median is not intended to serve the same purpose as that of the right shoulder. The shoulder on the right, through customary use on undivided arterials, is accepted by all drivers as the refuge space for stops. Where the median is flush with the pavement or is mountable, vehicles may encroach or drive on it momentarily when forced to do so to avoid an accident. Only on rare occasions would there be need for drivers to use the median for deliberate stops.

On divided arterials having two lanes in each direction, a paved shoulder strip 3 ft wide may satisfy the needs for a shoulder within the median. Such a shoulder strip will preclude rutting at the edge of pavement and will alleviate possible loss of driver control of vehicles that inadvertently encroach on the median.

On divided arterials having three or more lanes in each direction a driver in distress in the lane nearest the median might have difficulty maneuvering to the right-hand shoulder. Consequently, a full-width shoulder, within the median is desirable on divided arterials having six or more lanes.

Where the pavements of a divided arterial are at different levels, the graded shoulder space on the left assumes greater importance than when the pavements are at equal elevation because drivers may have a sense of insecurity where narrow shoulders are coupled with a downward sideslope and because there is need for lateral space for inadvertent encroachment by vehicles. The graded left shoulder width in such case preferably should be nearly the same as that on the right, 8 ft. Where lateral controls are rigid, the usable shoulder width of the left shoulder may be reduced somewhat but to a dimension of not less than 4 ft.

Median Barrier Clearance

In cases where a wall or median barrier is required on the median, the AASHTO *Roadside Design Guide* **(6)** may be referred to for guidance in selecting an appropriate lateral clearance from the normal edge of the pavement to the base of the wall or barrier.

Medians

On highways without at grade intersections, the median may be as narrow as 4 to 6 feet under very restricted conditions but a width of 60 ft or more should be provided wherever feasible. A wide median allows the use of independent profiles. Reduced frequency of crossover accidents and reduction of headlight glare are safety features associated with a wide median.

When intersections are to be provided, special concern must be given to the width of the median. While medians as narrow as 4 to 6 ft may be required under very restricted conditions, medians 12 to 30 ft wide provide protection for left turning vehicles at intersections.

Median widths from 30 to 50 ft should be carefully considered from an operational standpoint at intersections. These widths do not provide median storage space for larger vehicles crossing the median. Also, these widths may encourage the driver to attempt the crossings independently leaving a portion of the vehicle unprotected from through traffic. These widths, even with these problems, normally operate quite well and apparently are within the realm of normal operational expectations of the driver. Widths in the range of 50 to 80 ft have developed accident problems in some areas as the drivers apparently tend to become confused about the intended operational characteristics of the multiple intersections encountered. Medians wide enough to assure the driver that the intersection with the two sets of lanes operate separately have worked quite well. Research may prove that wider medians are not desirable for some facilities with at-grade intersections. Refer to Chapter IV for more information on median design.

Alinement and Profile

A divided arterial generally is designed for high-volume and high-speed operation, for which a flowing type of alinement should be provided. Because a divided arterial consists of two separated roadways, there may be instances where median widths and roadway elevations can be varied for special consideration. Special topographic or intersection requirements may necessitate such treatments for economic or operational reasons. Precaution should be taken that such variations do not adversely affect operations. Such problems as sharp reverse curves, headlight glare, and grades of intersection crossings must always be considered.

Profile design is less difficult for multilane highways than for two-lane highways. With two or more lanes for travel in each direction, stopping sight distance governs. For volumes well below capacity, grades may be steeper and longer on multilane highways than on two-lane arterials, because there is a continuous lane for passing of heavy, slow vehicles on upgrades.

Although vertical design controls may be lower on divided arterials than on two-lane arterials because only stopping sight distance is required, proper and adequate profile design for divided arterials requires design judgement and study.

Even though the profile may satisfy all of the design controls, the finished product can appear forced and angular. A flowing spline line on a broad scale should be used. Such design is of prime importance where a median of constant width is used in rolling terrain. The lack of a passing sight distance requirement tempts designers to use a roller coaster profile, which is more displeasing on a divided arterial than on a two-lane arterial because a driver can see the roadway for opposing traffic. With a wide divided arterial of fixed cross section, the profile vertical curves and combinations with tangent grades are more apparent in the driver's foreshortened view of the arterial. The relation of horizontal and vertical alinement should be studied to obtain a suitable combination. To avoid undesirable appearance, profile designs should be checked in long continuous plottings, wherein the foreshortened aspect can be simulated.

Climbing Lanes on Multilane Arterials

Multilane rural roads usually have sufficient capacity to handle their traffic load, including the normal percentage of heavy trucks, without becoming severely congested. Climbing lanes generally are not as easily justified on multilane arterials as on two-lane arterials, because on two-lane arterials vehicles following slow-moving trucks on upgrades are prevented psychologically and physically from using an adjacent traffic lane for passing; whereas such barriers to its use are not present on multilane arterials. Furthermore, a climbing lane on a two-lane two-way road has some use at all times, even during nonpeak hours; whereas on a multilane arterial a climbing lane has only limited use other than during periods of heavy flow. At all other times a slow-moving truck in the normal right lane does not impede following vehicles that can readily move left to the next lane and proceed without difficulty or hazard, although there is evidence that slow vehicles on through-traffic lanes may cause accidents.

Because arterials are designed for 20 years or more in the future there is little likelihood of climbing lanes being justified on multilane arterials for several years after construction, even though they are deemed desirable for the peak hours of the design year. There may be an economic advantage in designing for, but deferring construction of, climbing lanes on multilane arterials. In this situation, grading for the future climbing lane should be provided initially. Very little additional grading is required because a full shoulder is likely to be provided where there is no climbing lane, but only a narrow shoulder is required outside of a climbing lane, the climbing lane itself serving as an emergency area when needed.

A full discussion on the need for climbing lanes and their derivation is found in Chapter III.

Superelevated Cross Sections

A divided arterial on a curve should be superelevated to ensure safe traffic operation, pleasing appearance, and economy. Of specific concern is the means of effecting progressive changes in cross section to fit site conditions and to meet controls of intersection design.

General methods of attaining superelevated cross sections for divided arterials are discussed in Chapter III. The three general cases are the following: (1) The whole of the traveled way, including the median, is superelevated as a plane section; (2) the median is held in a horizontal plane, and the two pavements are rotated about the median edges; and (3) the two pavements are separately treated with a resultant variable difference in elevation at the median edges. Figure VII-4 demonstrates treatment of cross section for narrow and wide medians with superelevated pavement in relation to the width of median for the three cases noted. In cross sections A and D both roadways lie in the same plane. The pavements are rotated about a profile control on the centerline of the median. The same effect can be obtained by rotation about a pavement edge or any other convenient control line.

The projection of superelevation across wide medians, as in cross sections A and D of Figure VII-4, may be fitting in some instances, but its general use in conjunction with large rates of superelevation is not satisfactory in appearance and generally is uneconomical. It may fit at highway intersections where the profile of the intersecting road approximates the superelevated slope. Occasionally it will fit the natural slope of the terrain. Unless these conditions prevail, the large difference in elevation between the outer shoulder edges is likely to be objectionable. For example, the difference in elevation between the outer shoulder edges of a four-lane divided arterial with a median of 40 ft and a superelevation rate of 8 percent is about 8 ft. In level terrain and in terrain where the natural slope of the lane is adverse to the cross sectional slope, substantial improvement in appearance and economy in earthwork results if the wide median is made level as in cross section B, or sloped opposite to the superelevation plane as in cross section C of Figure VII-4.

Where cross section A of Figure VII-4 is used, the median should be graded in such a manner that surface water from the higher pavement does not drain across the lower pavement. On tangent alinement a shallow drainage swale can be provided in a median about 15 ft wide, and a well-rounded drainage channel in a width of about 60 ft (Figure VII-5). On a superelevated section rotated about the median centerline, as in section A of Figure VII-4, approximately 30 ft of median width is necessary for a rounded drainage channel and adequate left shoulders. In a median somewhat less than 30 ft wide, a channel with flat sideslopes can be provided if the superelevation rate is small, or a paved channel can be used in conjunction with higher rates of superelevation.

In cross sections B and E of Figure VII-4 the edges of pavement on the median sides are at the same elevation. Designs on this basis are pleasing in appearance and generally are desirable in an operational sense. With a wide separation between the one-way pavements, cross section B has considerable advantage over cross section A in the reduction in difference in elevation across the entire roadbed. On pavements having a superelevation rate near 0.10, the treatment in cross section B requires a minimum median width of about 30 ft to provide fully effective shoulder areas and a well-rounded swale.

In cross sections C and F of Figure VII-4 the two one-way pavements have a common centerline grade. The difference in elevation of the outer extremities of the superelevated pavements is a minimum, being the product of the

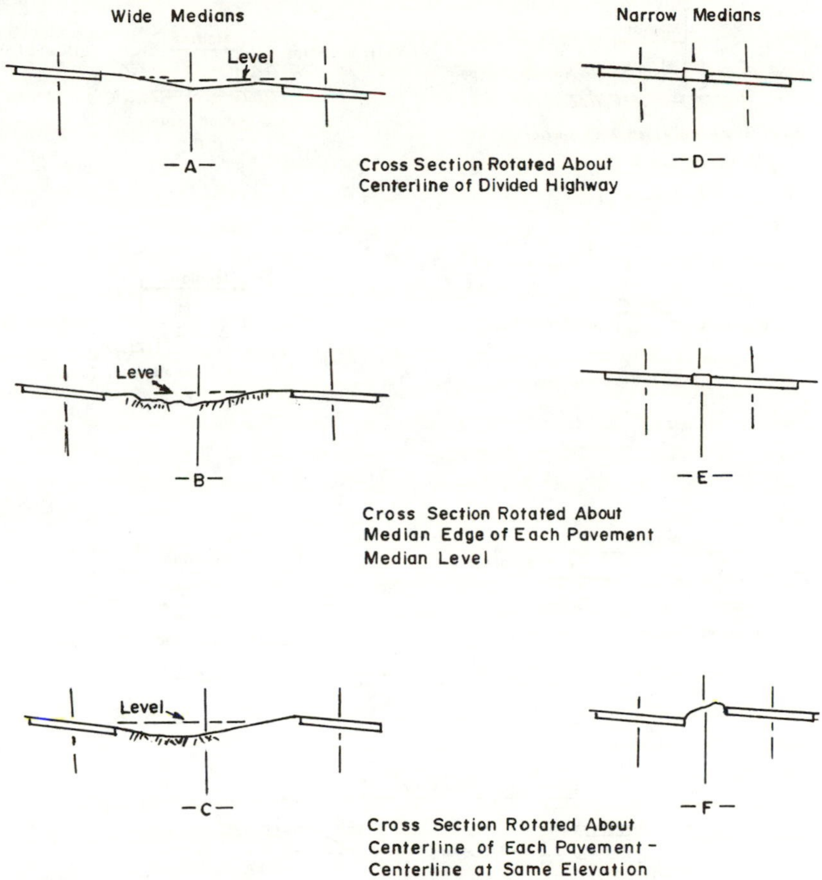

Figure VII-4. Methods of attaining superelevation on divided arterials.

superelevation rate and the width of one of the one-way pavements. The method of attaining superelevation runoff is directly applicable to each one-way pavement.

With a wide median the treatment of cross section C of Figure VII-4 allows the desired appearance to be maintained and permits economy in the wide-graded cross section. The roadway as a whole will appear fairly level to the motorist, who will not readily perceive the difference in elevation of the inside edges of pavement. This cross section generally is not suitable for important at-grade intersections unless the median is very wide. The median

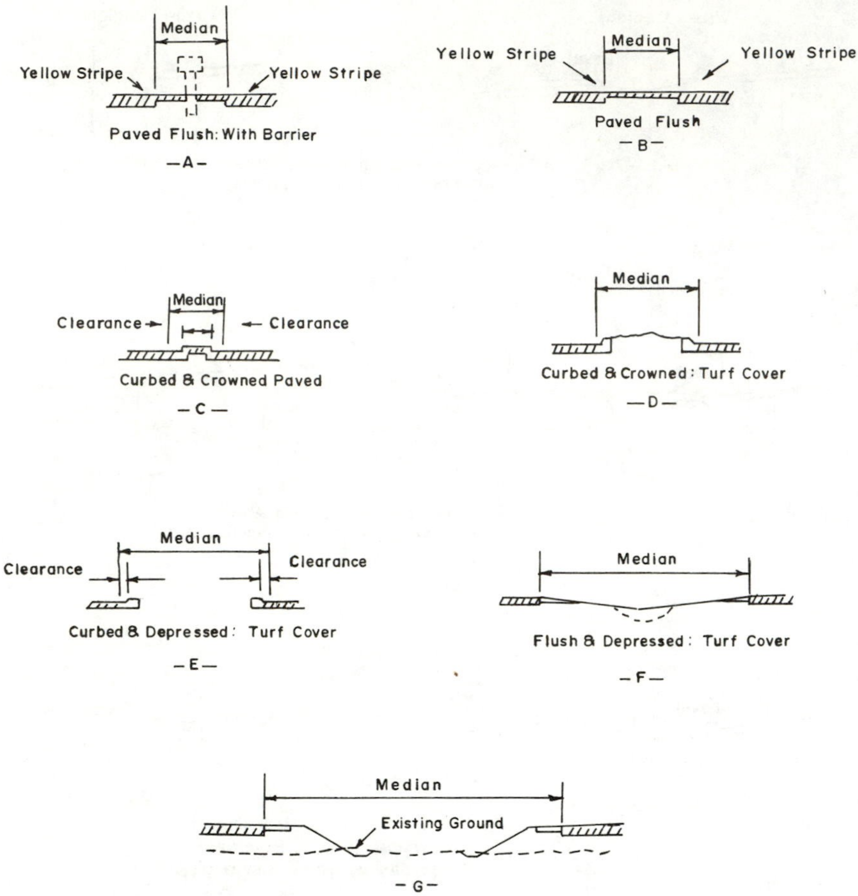

Figure VII-5. Typical medians on divided arterials.

should be sufficiently wide in relation to superelevation to afford a smooth S-shaped profile across its width. The width for this shape is somewhat more than that required for the previous sections; about 40 ft is necessary, with a superelevation rate of 10 percent and adequate shoulder areas. The width can be reduced to about 30 ft when a paved channel is provided.

On a divided arterial with variable width of median and difference in elevation between the two roadways, each roadway is designed with a separate profile. With a reasonable wide median, each roadway can be superelevated in any manner suitable for a single roadway with little effect on the median slope. With a narrow median, appreciable difference in elevation might require a retaining wall in the median. The manner of superelevating the roadways has some effect on the height of wall, but the amount is small and should have little bearing on design. Figure VII-4 shows various median configurations which may be used on rural arterials excepting Figures C, D and E. Curbs should never be used on rural high-speed arterials except where protected by guardrail.

Cross Section and Right-of-Way Widths

Cross sectional elements of divided arterials—the widths and details of pavements, shoulders, medians, sideslopes, and drainage channels—have been discussed separately in this and in other chapters. The required widths of right-of-way to include all elements in a composite arterial cross section are demonstrated in Figure VII-5. In an ideal situation, the topography, other physical controls, and economic feasibility permit the design of a well-balanced cross section of desirable dimensions, for which an adequate width of right-of-way is established and procured. On the other hand, the controls may be so rigid that if a divided arterial is to be provided at all, it must be done within a limited width of right-of-way, using minimum or near-minimum dimensions for each element of the arterial cross section. In the first instance the right-of-way is based on the higher design standards of the cross section; in the latter case the cross section is determined on the basis of the available width of right-of-way.

The widths of cross sectional elements should be proportioned to provide a well-balanced arterial section. Traffic lanes should be 11 or 12 feet wide as required by design speed and traffic volumes and outside shoulders should be at least 8 feet wide. The border width is affected directly by the depth of cut or fill. If the right-of-way is restricted, the border area or median width, rather than the lane or shoulder width, should be reduced. In most cases the border width should be reduced before that of the median, foreslopes permitting.

Figure VII-5C shows a desirable divided arterial cross section warranted for a high type facility where liberal width of right-of-way is attainable. Where these wider widths cannot be obtained, attempts should be made to provide a right-of-way width which permits the use of a median 30 feet or more and sufficient borders to provide for the needed clear zone. For additional information on clear zone refer to AASHTO *Roadside Design Guide* (**6**).

Sometimes the right-of-way may be so restricted that minimum or near minimum widths of cross sectional elements must be used. If at all feasible, the right-of-way should be wide enough to permit the use of median and borders of not less that 15 ft (Figure VII-6A). A 15-ft median is a near minimum within which a median lane can be provided at at-grade crossings. Figures VII-4 and VII-5 show some sections with curbs. Curbs are not recommended in rural roadways, except if required in restricted areas or if special treatment is needed in locations such as intersections. Adequate side clearances are required when curbs are installed.

The cross sections and right-of-way widths shown in Figure VII-6 pertain to four-lane facilities. Where provision is to be made for ultimate conversion to a six-lane or eight-lane facility, the right-of-way widths should be increased by the width of lanes added, with this width preferably included in the initial median.

The right-of-way widths and cross sectional arrangements shown in Figure VII-6 indicate generally balanced sections for what are termed desirable, minimum, and restricted rights-of-way. Some variation in these arrangements may be appropriate in individual cases. The right-of-way width need not be uniform and may be varied along the course of the arterial as necessary to fit grading requirements and other conditions. Where controls become rigid, the two roadways may have to be brought closer together. Where physical conditions are favorable and land is readily available, the roadways may be spread farther apart. Where future grade separations and ramps are envisioned, provision for the initial acquisition of additional corner rights-of-way should be considered.

The cross sections depicted in Figure VII-6 represent normally divided facilities in rural areas. Sometimes in rural areas, and particularly in and near urban districts, it is necessary to separate through traffic from local traffic. Where such is the case, frontage roads may be provided along the outer limits of the highway cross section (Figure VII-7). Frontage roads serve to collect and distribute local traffic to and from adjacent development, provide parking and service thereto removed from the main traveled way, and thus free through traffic from the disturbance and hazards introduced by local operation. The component parts of a typical cross section with frontage roads in generally flat terrain are shown in Figure VII-7A. The frontage roads are shown within the right-of-way limits, an arrangement that most often is the case. Frontage roads sometimes are provided outside the right-of-way limits, in which case the right-of-way can be narrower than shown. Where the profile of the through pavement passes over or cuts through the natural ground, the frontage roads are generally held at the level of the existing development, and the difference in elevation between the main traveled ways and the frontage roads is effected within the outer separations by earth slopes or retaining walls.

Where crossroads are grade separated from the through pavements and local service is provided on frontage or other roads, a high-type expressway

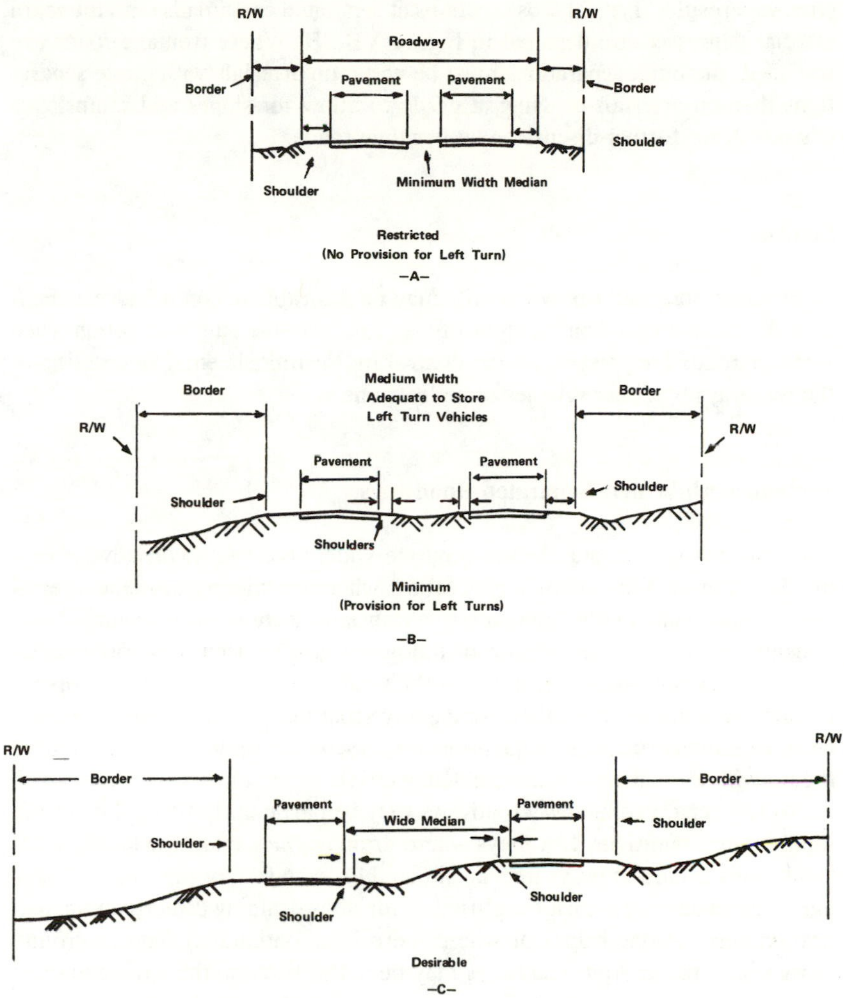

Figure VII-6. Cross section and right-of-way on divided arterials.

(freeway) results. Typical cross sections at separated crossroads with the main arterial depressed are depicted in Figure VII-7B. Where frontage roads are provided, the outer separations must be wider on arterials with grade separations than on arterials crossing at grade to allow for slopes and ramps. See Chapter X for further discussion on frontage roads.

Fencing

Fencing along right-of-way limits may be desirable to control access. Fencing should be located on the right-of-way line or in the outer separation when frontage roads are present. Fences control indiscriminate entry or crossing of the roadways by either vehicles or pedestrians.

Sections with Widely Separated Roadways

Occasionally, it is practical to separate widely the one-way roadways of a divided arterial. This situation may follow where an existing two-lane arterial proves inadequate and is improved to a four-lane section, but on which direct widening is not feasible because of topography or adjacent development. In such case the old roadway is not disturbed but is converted to one-way operation on construction of another, completely separate, one-way roadway. Sometimes this action results in acquisition of two separate rights-of-way to contain the individual roadways of the divided arterial.

Widely separated one-way roadways may be particularly fitting for certain topographic conditions. In valleys where drainage may make the location difficult, individual roadways may be situated on each side of the valley. Drainage of the roadways is then simplified, both sides draining directly to the natural channel. Along ridges or where there is a continual change in ground cross slopes the separate roadways may be better fitted to the terrain than an arterial on a single roadbed. Such arrangements simplify location problems because only half of the total roadway width is considered at a time. With reduced roadway prisms, construction scars are kept to a minimum and more of the natural growth is retained, particularly that between the separate roadways. In areas where right-of-way is not a problem, designs of this character often result in lower construction costs.

Crossing at grade with the one-way pavements are greatly simplified in design and operation. Accident potential is generally reduced and the capacity of intersections is increased. Moreover, operation on widely separated roadways provides the maximum in driver comfort. Strain is lessened by

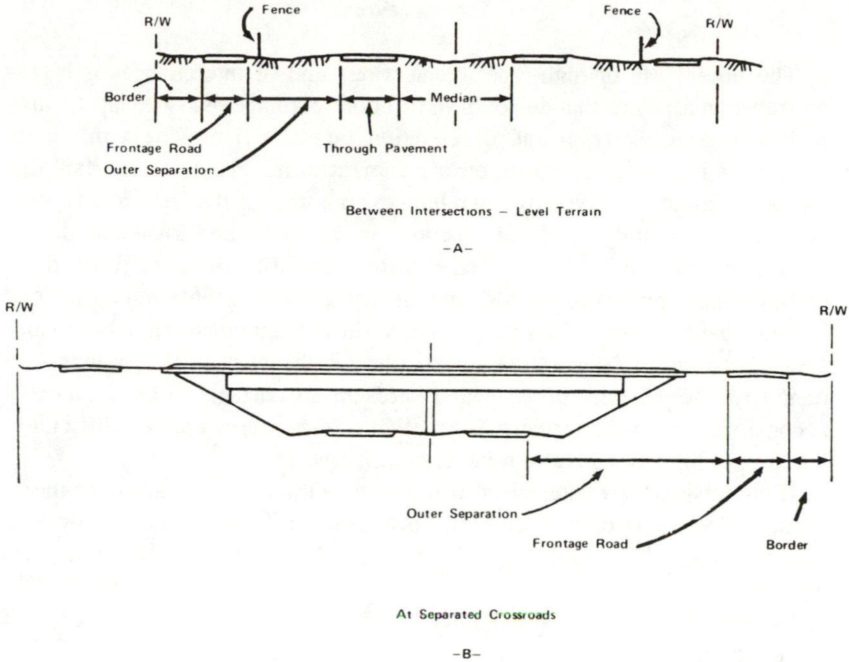

Figure VII-7. Cross section and right-of-way on divided arterials.

largely eliminating the view and influence of opposing traffic. Substantial reduction or elimination of headlight glare at night is especially helpful in easing driver tension and in minimizing hazards resulting therefrom.

Operational problems of intersections on widely spaced roadways must be considered. Desirably, a wide median is adequate to store the longest legal vehicles. All movements and their volumes should be considered to determine the number of intersection lanes required. Additional discussion on wide medians is located in this chapter.

If arterials of appreciable length have roadways separated so widely that each roadway cannot be seen from the other, drivers may develop a sense of being on a two-way instead of a one-way roadway and hesitate to pass slow-moving vehicles. This situation is not hazardous but decreases somewhat the usefulness of the divided arterial. This decrease can be alleviated by an occasional open view between the two roadways.

Intersections

The liberal use of high-type intersections and of interchanges is highly desirable on arterials that do not have full control of access. Adequate turning widths with acceleration and deceleration tapers will provide a minimum design for minor intersections on a minor arterial. Principal arterials that intersect should be served by interchanges, possibly of the free-flow type. A comprehensive study of all intersections is desirable, and a suitable design consistent with the level of service required should be selected. Rural intersection control by traffic signals is not desirable, nor are intersections divided by high barrier curbs. Drivers generally do not anticipate signals in rural areas that have high operating speeds, especially when traffic volumes are relatively low. Barrier curbed islands present an obstacle to the driver and become snow traps in northern States. Wet or dirty curbs are also difficult to see in poor light and adverse weather conditions.

If interchanges are intermixed with at-grade intersections, adequate merging distances must be provided to allow ramp traffic to operate freely. The merging driver should not have to be concerned with cross traffic while making the merging maneuver.

Design of intersections and interchanges should be in accordance with Chapters IX and X.

Control of Access

Rural arterials are designed and built with the idea that they will provide a higher type of service than that required or available on local roads and streets. Although an arterial may not have more traffic lanes, its ability to carry greater volumes is usually related to the amount of crossroad interference or side friction to which it is subjected. One of the most important considerations in the development of an arterial is the degree of access control that can be acquired. The ability to control access will relate directly to the safety and capacity of the project.

Access control is usually not too difficult to obtain in a rural area where development is light. Adequate access can normally be provided without great interference to land-use operations. Rural areas do pose distinct problems. Movement of large, slow-moving pieces of machinery is not uncommon; numerous field drives are also requested for the landowner's convenience. Because of these unique problems, access points must be situated in such a way as to minimize their harmful effects. Access points situated directly opposite will allow machinery to cross the arterial in the shortest possible time. Dual access points situated at property lines or land-use change points will reduce the number of access locations required. Adequate and uniform minimum spacing will also eliminate many conditions where a large vehicle on an intersection will hide another vehicle on a nearby approach. The location of the access point with relation to sight distance or important intersections must be considered. High-volume drives are especially bothersome if not properly situated. Short sections of rural frontage roads often may be needed to provide optimum intersection location conditions.

The degree of access control required depends on the demands placed on the arterial. Because the rural arterial has greater importance than the local roads and collectors that usually serve all access needs and cannot normally provide features associated with freeways, the arterial is most influenced by the use of access control. Provision of access control is vital to the concept of an arterial if it is to provide the service life for which it is designed.

Bikeways and Pedestrians

Rural arterials often are the only direct connection between areas of population and locations to which the public wishes to travel. Schools, parks, and rural housing developments are usually located to be readily accessible by automobile. However, pedestrians and bicycle riders may also wish to travel to the same destination points. When such a situation exists, the designer should consider the effects on the safety and operation of the arterial. A special effort should be made to provide the greatest degree of safety within the economic constraints that must always be considered.

Where use of the travel lane or wide paved shoulders is deemed unacceptable by some pedestrians or bicyclists, these may be accommodated on additional special bicycle facilities designed in accordance with AASHTO's *Guide for Development of New Bicycle Facilities* **(2)**.

Bus Turnouts

When bus routes are located on a rural arterial, provision must be made for loading the passengers. Because of the size of the vehicle, a bus cannot easily leave the roadway unless special provisions are made. A well-marked, widened shoulder or independent turnout is highly desirable and should be included, if possible. Although it may be impossible or impractical to provide, for example, school bus turnouts for every dwelling, they should be provided in locations where there are known concentrations of users. Adequate provisions for buses will provide greater capacity and safety to a rural arterial.

Railroad Crossings

Desirably, all railroad crossings should be separated on the rural arterial system. Practical aspects of the problem, however, require that many crossings will be at grade. Crossings can be treated in various ways, including adequate signing, signals, signals with gates, and finally separations. Judgement must be used in the selection process, which will involve the amount and speed of traffic on both the roadway and railroad and the available sight distance, and safety benefits. The tendency must always be to attempt to provide the best protection possible. Refer to the MUTCD **(1)** for guidance for traffic control systems for railroad-highway grade crossings. See Chapter IX for design criteria.

Safety Rest Areas

The provision of safety rest areas on the rural arterial system is a desirable feature particularly on principal arterials. Rest areas provide the high-speed, long-distance traveler with the necessary opportunity for short periods of relaxation, which relieves driver hypnosis. Public recognition of the hypnosis problem and the use of the rest area indicate that they provide a desirable safety feature.

The location of rest areas should be considered early in development of an arterial. Sites of special interest or visual quality provide additional reasons for the motorist to stop and usually extend the length of the stay. The spacing of rest areas is dependent on many considerations. Construction and operating costs are significant, but the driver value must also be considered. Additional information on the subject of rest areas may be found in Chapter III.

URBAN ARTERIALS

General Characteristics

Urban arterials carry large traffic volumes within and through urban areas. Their design varies from the freeway with fully controlled access to the two-lane street. The type selected is closely related to the type of service required. The principal characteristic of the urban arterial should be mobility with limited or restricted service to local development. If restriction to local access is not practical, special designs that consider these problems are desirable. These designs can vary from roadways providing separate turn lanes to one-way streets.

Urban arterials are capable of providing some access to abutting property. Such service should, however, be only incidental to the primary functional responsibility of travel service to major traffic movements.

Before designing an urban arterial, it is important to establish the extent and need for such a system. If the need is established, steps must be taken to protect the design capacity from such things as strip development or the unplanned location of a major traffic generator. Development should be anticipated regardless of the size of the city. With proper planning and design, the development along an arterial need not seriously affect its major purpose of long and safe through trips. Rather, it can complement such development and provide the service required.

Urban arterials are functionally divided into two classes, principal and minor. These are discussed in detail in Chapter I.

The urban arterial system, which includes arterial streets and freeways, serves the major centers of activity of a metropolitan area, the highest traffic volume corridors, and the longest trip desires.

The portion of the arterial system, either planned or existing, on which access is not fully controlled constitutes the arterial street system for the urban area. From the standpoint of design characteristics, all urban arterials, other than freeways, are treated as a single class in this chapter.

General Design Considerations

In the development of a transportation improvement program, the routes selected for improvement as arterials may comprise portions of an existing street system, or they may be projected locations on new alinements through relatively undeveloped areas. Usually, it will be the former because historically the need for improving existing streets has surpassed the availability of resources. As a consequence, street improvements tend to lag, rather than lead, land-use development.

Major improvement of existing arterials can be extremely costly, particularly where additional rights-of-way must be acquired through intensively developed areas. Accordingly, it is frequently necessary to use dimensional values that are somewhat less than desirable and are below those used where wide right-of-way is available or can be acquired economically. When restricted conditions are encountered, consideration should be given to providing above-desirable values for such items that tend to offset those created by the restriction. An example would be to eliminate left turns through an area having below normal lane widths.

Design Speed

Design speed for urban arterials generally range from 40 to 60 mph, and occasionally may be as low as 30 mph. The lower (40 mph and below) speeds apply in the central business district and intermediate areas. The higher speeds are more applicable to the outlying business and developing areas.

Design Traffic Volumes

The design of urban arterials should be based on traffic data developed for the design year, normally 20 years. The design hourly volume (DHV) is the most reliable method to determine design requirements. Sometimes, capacity is used as a design tool. The limitations and restrictions that are often encountered are recognized in capacity design, and a preselected level of service for those conditions is provided. Refer to Chapter II for information on traffic and capacity.

Levels of Service

For acceptable degrees of congestion, rural and suburban arterials and their auxiliary facilities, i.e., turning lanes, weaving sections, intersections, interchanges, and traffic control systems (traffic signals, etc.), should generally be designed for level-of-service C. Heavily developed sections of metropolitan areas may necessitate the use of level-of-service D. When level-of-service D is selected, it may be desirable to consider the use of one-way streets or alternative bypass routes to improve the level of service.

Sight Distance

The provision of adequate sight distance is important in urban arterial design. Sight distance affects normal operational characteristics, particularly where roadways carry high traffic volumes. The sight distance values given in Table VII-3 are also applicable to urban arterial design.

Grades

The grades selected for an urban arterial may have a significant effect on its operational characteristics. Steep grades affect truck speeds and overall capacity. On arterials having large numbers of trucks and operating near capacity flat grades should be considered to avoid undesirable reductions in speeds. Steep grades also result in operational problems at intersections, particularly during adverse weather conditions. For these reasons it is desirable to provide the flattest grades practicable while providing minimum gradients as required to ensure adequate longitudinal drainage in curbed sections. (See Table VII-4.)

	Design Speed (mph)			
	30	**40**	**50**	**60**
Type of Terrain	**Maximum Grade (percent)**			
Level	8	7	6	5
Rolling	9	8	7	6
Mountainous	11	10	9	8

Table VII-4. Maximum grades for urban arterials.

Alinement

Alinement of the urban arterial is ideally developed strictly in accordance with the design speed selected particularly when a principal arterial is located on a new location and is not restricted by normal right-of-way requirements. There are many cases, however, where this is not possible and deflections must be made in intersections. It is desirable to use the highest alinement design possible because urban arterials are often not superelevated in the low-speed range.

Cross Slope

Adequate cross slope for proper drainage is important on urban arterials. The normal problems related to splashing and hydroplaning are compounded in heavy traffic volume operations at intermediate to high speeds. Cross slopes should range from 1.5 to 3 percent, the lower values being in the center lanes with the cross slopes increasing about 1 percent for each additional lane over which water must drain until the maximum 3 percent is reached. Even higher rates are used for parking lanes. Adequate cross slopes offset problems associated with pavement rutting. The overall cross section should provide a smoothly rounded appearance without sharp breaks. Because urban arterials are often curbed, it is necessary to allow for longitudinal as well as cross-slope damage. The use of higher cross-slope rates reduces ponding and flow on the pavement, a feature which is desirable on arterial pavements. The discussion on cross slopes in Chapter IV provides additional discussion on the subject.

Superelevation

Superelevation is generally not used on low-speed curbed arterial streets. Problems such as drainage, ice formation, driveways, pedestrian crossings, and the effect on developed property must be recognized when superelevation is considered. The section on horizontal alinement in Chapter III provides discussion on superelevation in greater detail.

Vertical Clearances

New or reconstructed structures should provide 16-ft clearance over the entire roadway width. Existing structures that provide 14-ft clearance, if allowed by local statute may be retained. In highly urbanized areas, a minimum clearance of 14-ft may be provided if there is one route with 16-ft clearance. Structures should provide additional clearance for future resurfacing of the underpassing road.

Lane Widths

Lane widths may vary from 10 ft to 12 ft. The 10-ft widths are used in highly restricted areas having little or no truck traffic. The 11-ft lanes are used quite extensively for urban arterial street designs. The 12-ft lane widths are most desirable and are generally used on all higher speed, free-flowing, principal arterials. Under interrupted-flow operating conditions at low speeds up through 40 mph narrower lane widths are normally adequate and have some advantages. Reduced lane widths allow greater numbers of lanes in restricted right-of-way and allow better pedestrian cross movements because of reduced distance. They are also more economical to construct. An 11-ft lane width is adequate for through lanes, continuous two-way left-turn lanes, and a lane adjacent to a painted median. A 10-ft left-turn lane and a combination lane used for parking with traffic during peak hours are acceptable. Lane widths for such instances as bike lanes within the roadway require special evaluation.

These lane widths must be used with judgement. If heavy truck traffic is anticipated, an additional 1 ft in width is desirable, and situations having restrictions on trucks may allow the use of lanes 1 ft less in width. All the lane-width requirements and intersection design controls should be evaluated when making the lane width selection. For instance, a wider right-hand lane that provides right turns without adjacent lane encroachment may require a narrower left-turn lane. Local practice and experience regarding lane widths should also be evaluated. Special lane widths for such instances as bike lanes within the roadway require special attention.

If curbs are used, their placement should be in accordance with the discussion on curb placement in Chapter IV.

Number of Lanes

The number of lanes will vary, depending on traffic demand and availability of the right-of-way, but normally it will be within the range of four to eight lanes. A capacity analysis should be used to determine the proper number of lanes. Pavements are sometimes widened through intersections by the addition of one or two lanes to accommodate turning vehicles. See Chapter II for additional information on capacity analysis.

Width of Roadway

The street roadway width should be adequate to accommodate the traffic lanes, medians, curbs, and the required clearances from barrier faces. Parking on an arterial street should only be considered when provision is required because of existing conditions. When parking lanes are provided, the width should be adequate to allow ultimate operation as a traffic lane. In many instances the parking lane is initially used to provide a turn lane at intersections.

Medians

Medians are a desirable feature of arterial streets and should be provided where space permits. Medians and median barriers are discussed in Chapter IV. Where right-of-way is limited, it is frequently necessary to decide how best to allocate the available space between border areas, traffic lanes, and medians. On the less important arterials the decision is often resolved in favor of no median at all. A median only 4 ft wide is better than none; however, each additional foot provides an added increment of safety and improved operation. At intersections where left turns are made, a left-turn lane is always desirable from a capacity and safety standpoint. The median width to accommodate left-turning movements desirably should be at least 12 ft. Desirably, the median should be at least 18 ft wide for a 12-ft median lane and a 6-ft medial separator. At restricted locations, a 10-ft lane with a 2-ft medial separator may be used.

Figure VII-5 shows various configurations for medians that may be used on urban arterials. The type of treatment used is usually dependent on local practice and available right-of-way widths. The type selected should always be compatible with drainage and street hardware requirements.

Median openings on roadways provided with depressed or raised medians should be carefully considered when a divided arterial is planned. Openings should only be provided for street intersections or for major developed areas. Spacing between median openings must be adequate to allow for introduction of left-turn lanes and signal detection loops to operate without false calls.

Where intersections are widely spaced, e.g., 0.5 mi or more, the width of median may be varied by using a narrow width between intersections where necessary for economy and gradually widening on the approach to the intersection to accommodate the left-turn lane. This solution is hardly practical, however, and should never be used where intersections are closely spaced because the curved alinement of the lane lines results in excessive lane changes. It is far more desirable that the median be of uniform width.

A street with an odd number of lanes, perhaps three or five, may be used to advantage in providing a storage lane for left-turning vehicles. This lane is one form of utilizing a paved, flush median. Left-turn bays are marked in advance of the intersections. The length of lane between left-turn bays may be used for storage of vehicles making midblock left turns if this usage does not adversely affect traffic flow and safety on the arterial; otherwise, midblock left turns may be prohibited. Under some conditions it is better to permit midblock turns than to require that vehicles make U-turns at intersections or travel around a block to reach a destination. In some cases the center lane is designated for "Left-Turn Only" throughout, without specially marked bays at intersections. This type of operation works well where the speed on the arterial highway is relatively low (25 to 45 mph) and there are no heavy concentrations of left-turn traffic. Figure VII-8 is an example of a continuous left-turn lane. Refer to Chapter IV, "Medians" and Chapter IX, "Continuous Left Turn Lanes" for further discussions of this type of median.

Where an arterial must pass through a developed area having numerous street and driveway intersections, and where it is impractical to limit left

Figure VII-8. Continuous median left-turn lane.

turns, the continuous left-turn median lane is often the only practical solution. Because all left turns are thus protected, the interference to through-traffic lanes is minimized. Successful operation of a continuous left-turn lane requires adequate lane marking.

A raised median with barrier-type curbing may be used on low and intermediate speed arterial streets. This type is desirable where it is necessary to prevent midblock turns. On streets serving low-speed traffic this type will prevent most cross-median accidents. Raised medians also provide a refuge for pedestrians and a good location for signs, signals, and other appurtenances. In snow-belt areas the curbed median provides positive delineation, whereas a flush median becomes indiscernible under the lightest of snowfall conditions. However, raised medians present disadvantages that should be considered in deciding the type to be employed.

On streets serving high-speed traffic the raised median does not prevent cross-median accidents unless a median barrier is provided. If accidentally struck, raised medians may cause the driver to lose control of the vehicle. They are difficult to see at night except when used in conjunction with a good system of fixed-source lighting or with proper delineation. They cast a shadow from oncoming headlights that not only makes the curb difficult to see, but also places most of the adjacent lane in shadow. Raised medians are of little use as a place of refuge for disabled vehicles unless they are very low with flat sloping curbs. In some cases the prevention of midblock turns causes operational problems because of heavy concentrations of left-turning traffic at the intersections.

The foregoing disadvantages of raised medians are largely eliminated when flush or low-profile mountable medians are used. However, flush-paved medians are also difficult to see at night when the pavement is wet. Visibility can be improved by use of a contrasting texture such as a chip seal coat and by proper delineation. Raised bars or blocks have proved to be ineffective as a median treatment and should not be used.

In suburban areas and elsewhere where a median width of 24 ft or more can be provided, a flush or depressed landscaped median offers most of the advantages of a raised median with few of the unfavorable attributes.

Special consideration needs to be given to the width of medians when at-grade intersections are to be provided. Median widths from 30 to 50 feet should be carefully considered from an operational viewpoint. These widths do not always provide median storage space for longer vehicles crossing the median. Also, these widths may encourage the driver to attempt the crossings independently and leave the vehicle not fully protected from the through traffic. These widths, even with these problems, operate quite well and apparently are within the realm of normal operational expectations of the driver.

Widths in a range of 50 to 80 feet have developed accident problems in some areas as the drivers may tend to become confused about the intended operational characteristics of the multiple intersections encountered.

Experience indicates that drivers prefer medians that are obviously narrow or those that provide an adequate refuge area to allow independent roadway crossing operation.

Uncurbed narrow medians often present problems with turning movements at intersections in that vehicles tend to run off the pavement edges.

A median barrier may be desirable on some arterial streets with relatively fast-moving traffic. It permits a positive separation of traffic and discourages indiscriminate pedestrian crossing. Where the median barrier is terminated at cross streets and other median openings, it should be turned down or flared to reduce the hazard. Reference is made to the *Roadside Design Guide* **(6)** for further discussion on treatment of the end of barriers. Additional information on median barriers and median treatments at intersection areas is contained in Chapters IV and IX, respectively. The information on medians and median barriers in Chapter IV is especially pertinent to urban arterials since they require the most varied application of these features.

Curbs and Shoulders

Shoulders are desirable on any highway, and urban arterials are no exception. They contribute to safety by affording maneuver room and providing space for immobilized vehicles. They offer a measure of safety to the occasional pedestrian in sparsely developed areas where sidewalks are not appropriate. They serve as speed-change lanes for vehicles turning into driveways, and they provide storage space for plowed snow.

Despite the many advantages of shoulders on arterial streets, their use is generally limited by restricted right-of-way and the necessity of using the available right-of-way for traffic lanes. Where the abutting property is used for commercial purposes or consists of high-density residential development, a shoulder, if provided, is subject to such heavy use in serving local traffic that the pavement strength must be about the same as that for the travel lanes. In addition, a raised curb at the outer edge of the shoulder is usually necessary in heavily developed areas as a means of controlling access and preventing deterioration of the shoulder. These requirements usually result in a cross section having a uniform pavement design with barrier-type curbs.

When shoulders are curbed with barrier curbs, the shoulder width should be adequate to accommodate a disabled vehicle and be at least 6-ft wide. If permanent on-street parking is required, the effect on adjacent traffic lanes must be considered.

In newly developing areas, where rights-of-way can be acquired at reasonable costs, and in the extension or improvement of arterials through suburban areas, it is often feasible to include shoulders as a part of the ultimate cross section or possibly for interim use until the space is needed for additional traffic lanes. In these situations, where it is feasible to provide shoulders, curbs should preferably be omitted unless necessary to control drainage.

Wherever conditions require the construction of curbs or shoulders, the discussion in Chapter IV would apply to their design.

Drainage

An adequate drainage system to accommodate design runoff should be included in the design of every arterial street. Inlets that are bicycle safe should be located adjacent to and upstream of the intersections and at intermediate locations where necessary. Where a parking lane is provided, the full width of this lane may be utilized to conduct surface water to the drainage inlets. Where no parking lane is provided, one-half of the outside travel lane may be utilized to conduct surface drainage provided two or move travel lanes exist in each direction. Where three or more travel lanes are provided in each direction, the full outside travel lane may be utilized for drainage in a major design storm. Ponding of water on traffic lanes on arterial streets at low points should be kept within tolerable limits. The width of water spread on the pavement should not be substantially greater than the width of spread encountered on continuous grades. Additional inlets should be provided in sag locations to avoid ponding where the grade flattens to 0 percent. Chapters III and IV have further discussions concerning drainage.

Borders and Sidewalks

The border is the area between the roadway and the right-of-way line that separates the traffic from the homes and businesses of the community. For a minimum section in a residential area, a border should include a sidewalk and a buffer strip between walk and curb. In blocks fully developed with retail stores and offices, the entire border area usually is devoted to sidewalk.

Some factors to be considered in determining minimum border widths are pedestrian requirements, snow storage, storm drainage, traffic control devices, roadside appurtenances, and utilities. In all cases the minimum border should be 8 ft wide and preferably 12 ft or more. Every effort should be made to provide wide borders not only to serve functional needs but also as a matter of esthetics, safety, and reducing the nuisance of traffic to adjacent development. Where sidewalks are not included as a part of the initial construction, the border should be sufficiently wide for their future installation. See the section "Pedestrian Facilities" later in this chapter.

Urban arterials are seldom conducive to bicycle travel because of high traffic volumes and the high speeds. When bicycle traffic is anticipated or is to be served on arterial streets, provisions should be in accordance with AASHTO's *Guide for Development of New Bicycle Facilities* (**2**).

Figure VII-9 illustrates a divided arterial street in a residential area. A turfed buffer strip is provided between the sidewalk and the gutter. Barrier-curb and gutter sections are employed on the outside. Parking lanes, which may also serve as shoulders, are provided.

Railroad Crossings

Railroad crossings on an urban arterial can often be the most disruptive feature affecting the operation of an urban arterial. Crossings that are fre-

Figure VII-9. Divided arterial street with parking lanes.

quently occupied or those that are occupied during high-volume traffic periods can only be effectively treated by providing a separation. Crossings that are only occupied infrequently during off-peak traffic periods may be provided with high type at-grade treatments and are normally provided with gate-equipped automatic flashing signals.

Parking Lanes

Parallel parking may be considered where adequate capacity is available in the through lanes, and when parking is needed and adequate off-street parking facilities are not available or feasible. The parking lane width should be designed to provide a clearance of 3 to 5 ft from the edge of the through-traffic lane to the parked vehicles. Passenger vehicles parked 6 to 12 in. from the curb will occupy, on the average, approximately 7 ft of actual street space, and the total parking lane width for passenger cars should, therefore, be 10 to 12 ft. This width is also adequate for an occasional parked commercial vehicle. Where there is not, in the foreseeable future, a demand that the lane be used as a traffic lane, a parking lane as narrow as 8 ft may be acceptable.

A parking lane which is 10 to 12 ft in width can be converted to an additional traffic lane if future traffic volumes warrant additional capacity. It can also be used as an additional traffic lane during peak traffic hours by prohibiting parking during these hours. A parking lane 10 to 12 ft in width can also be used as a storage lane for turning vehicles at signalized intersections by prohibiting parking for some distance from the intersection. Widths less than 10 ft are considered substandard for arterial traffic lanes. Therefore, a parking lane of less than 10 ft in width is considered undesirable if usage of full curb-to-curb width for through traffic is anticipated. A parking lane of 9 ft may be acceptable as a storage lane for turning vehicles in areas where the average running speed on the arterial is 40 mph or less.

The marking of parking-space limits on arterial streets encourages more orderly and efficient use of parking spaces where parking turnover is substantial and tends to prevent encroachment on fire hydrant zones, bus stops, loading zones, approaches to corners, clearance space for islands, and other zones where parking is prohibited. Typical parking-space markings are shown in the MUTCD (**1**).

In downtown districts and in areas with large office or industrial buildings, it may be possible to provide parking turnouts as shown in Figure VII-10. The paved area of the turnout is 24 ft wide. A curbed island with entrance and exit tapers separates the through lanes from the turnout.

Roadway Width for Bridges

The minimum clear width for all new bridges on arterial streets should be the same as the curb-to-curb width of the street. Long bridges, defined as bridges having an overall length in excess of 200 feet may have a lesser width. Such bridges should be analyzed individually. On long bridges, offsets to parapet, rail or barrier shall be at least 4 feet measured from the edge of the nearest travel lane on both the left and right. For further relevant discussion, see the sections on "Curbs," "Walkways," "Traffic Barriers," and "Bridge Railings" in Chapter IV.

Horizontal Clearance to Obstructions

Clear roadside design is recommended for urban arterials whenever feasible. On curbed street sections, this design is often impracticable, particularly in restricted areas. In those areas a clearance from curb face to object of 1.5 ft or wider where possible should be the minimum. A 3-ft clearance will desirably be provided particularly near turning radii at intersections and driveways.

Figure VII-10. Taxi loading and unloading turnout.

This desirable offset provides the clearance required for overhang of trucks from striking the object. When pedestrians are not a factor, obstructions should be well set back, protected, or provided with breakaway features. Protective curbs or guardrail may be considered in special cases.

Right-of-Way and Cross Section

The width of right-of-way for the complete development of an arterial street is influenced by traffic requirements, topography, land use, cost, intersection design, and extent of ultimate expansion. The required width of right-of-way is the summation of the various cross-sectional elements through pavements, median, auxiliary lanes, shoulders, borders, and, where required, frontage roads, outer separations, sideslopes, and retaining walls. The width of right-of-way should be based on the preferable dimensions of each element to the extent feasible. In built-up districts, however, it often is necessary to consider less desirable dimensions. In such a case the designer is confronted with the problem of providing an overall cross section that will give maximum service within a limited width of right-of-way. Right-of-way widths in urban areas are governed primarily by economic considerations, physical obstructions, or environmental reasons. Along any arterial route, conditions of development and terrain vary, and accordingly, the availability of right-of-way varies. For this reason the right-of-way on a given facility should not be a fixed width predetermined on the basis of the most critical point in its location. Instead, every opportunity should be taken to provide a desirable width along most if not all of the facility.

The likely range of right-of-way widths for four- to eight-lane arterial streets is shown in Figure VII-11. These cross sections, both without and with frontage roads, are separated into three groups representing desirable, intermediate, and restricted widths. Dimensions of cross sectional elements fitting to each group are shown, although these can and should be varied to meet specific conditions. Right-of-way widths vary from a restricted width of about 64 ft for a four-lane undivided arterial to a desirable width of about 300 ft for an eight-lane arterial with frontage roads.

Figure VII-11 (A, B, and C) shows cross sections of arterial streets without frontage roads. In Figures VII-11A and VII-11B, 8-ft shoulders are provided, with borders of about 30 and 20 ft, respectively. Where the arterial must pass between buildings, walls, or other confining vertical elements, and where it is impractical to provide shoulders, a cross section as shown in Figure VII-11C may be used. Such arrangements within the right-of-way widths of about 64 and 100 ft for four- and six-lane facilities, respectively, may be applicable where an alternate location is not feasible and costs for additional rights-of-way are excessive in relation to the added service and safety that would result from more generous dimensions. Any added increment of width that can be made available should be incorporated in the median as a first priority.

Figure VII-11 (D, E, and F) shows cross sections for divided arterial streets with frontage roads. Desirable sections, as in Figure VII-11D, are usually practical in outlying areas. Intermediate widths are shown in Figure VII-11E. These widths, including the 20-ft outer separations with full shoulders and a 14-ft median, are suitable for most cases, both between and at intersections where one-way frontage roads are used. With two-way frontage roads, the width of right-of-way should be increased by 40 ft or more at intersecting streets. If a frontage road joins an intersecting street that has high volumes or is signalized, serious operational problems should be anticipated unless the intersection with the frontage road is set back adequately to allow the two intersections to operate independently.

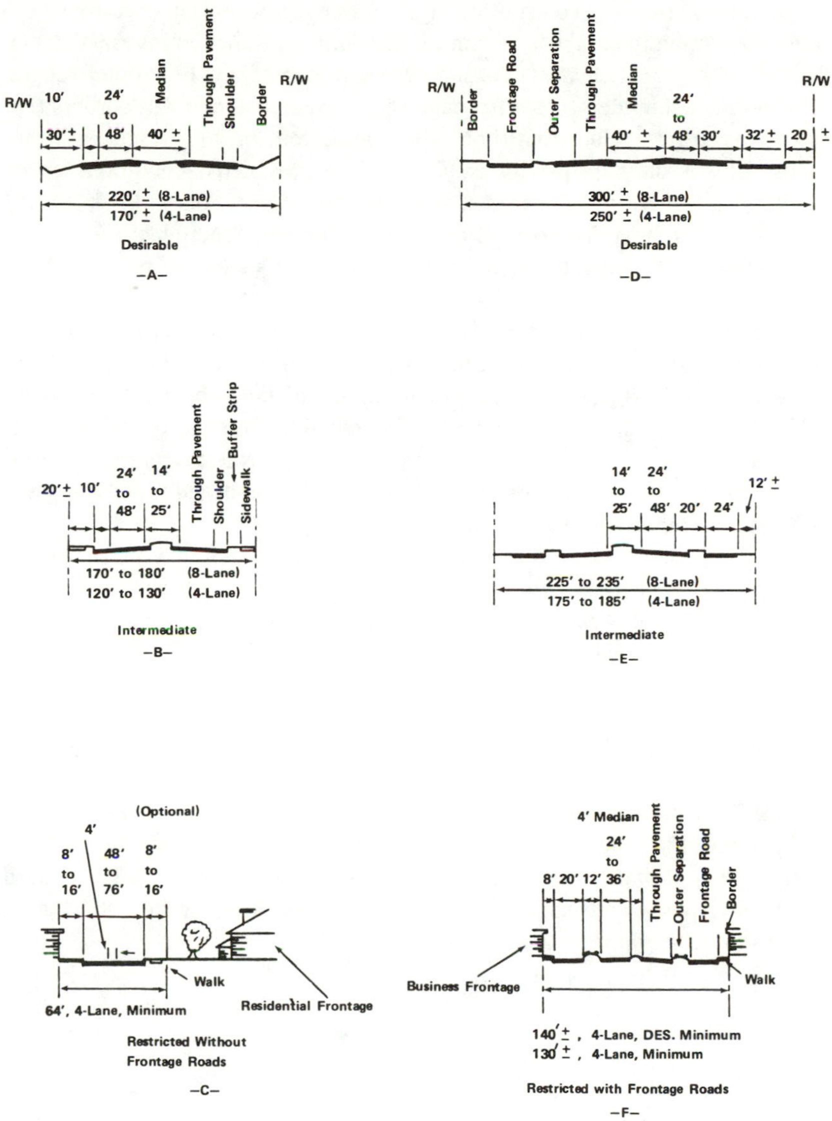

Greater Widths of Outer Separations Needed at Intersections

Figure VII-11. Cross sections of arterial streets with and without frontage roads.

Where right-of-way controls are rigid, it may be necessary to constrict the highway section to an absolute minimum width, as shown in Figure VII-11F. The cross-sectional elements consisting of a 4-ft median, 12-ft outer separations without shoulders, 20-ft frontage road pavements, and 8-ft borders represent the lowest values that should be considered for an arterial with four lanes. This section primarily is adaptable to restricted conditions either between intersections or at minor at-grade cross streets on which left turns from the through pavements are limited or prohibited. Where facilities for left turns must be provided, the median and total widths should be increased by at least 10 ft.

Arterials with frontage roads require right-of-way widths equivalent to a half- or a full-block width. Where a half-block width is to be taken, it will be preferable, both for the avoidance of remnants of land and from the standpoint of cost, to take the whole depth of a row of lots on one side of an existing street rather than half portions of the lots of opposite frontages. An arterial of this type has only limited application, being confined largely to stage construction for an ultimate freeway.

Widths greater than those indicated in Figure VII-11 may be necessary where fill or cut slopes or other highway development must be included within the right-of-way. An occasional grade separation with ramps generally requires additional width.

Traffic Barriers

Traffic barriers are sometimes used on urban arterials in restricted areas and at separations, and in medians. The barrier should be compatible with the desired visual quality, and shall be installed in accordance with standard practice. Exposed ends should be treated with breakaway design, attenuation devices, or other appropriate means.

Access Control

General Features

Partial control of access is highly desirable on an arterial facility. Such a provision will not only enhance its initial service capability but may also often preserve the original level of service. While service may be required to abutting property, it should be carefully regulated to limit the number of points and their locations. Access control is especially needed in areas approaching intersections where auxiliary and storage lanes may be required.

Access control may be exercised by statute, through application of zoning ordinances, driveway controls, turning and parking regulations, and geometric highway design acceptable to the community and property owners.

Access Control by Statute

Where a high degree of access control is desired, it is usually accomplished by statute. When statutory control is applied to an arterial street, access is usually limited to the cross streets or to other major traffic generators.

Figure VII-12 illustrates an arterial street with access control by statute. Access is limited to the major cross streets. In the upper middle of the picture, cul-de-sacs have been constructed on the two local streets that were terminated. Sidewalks are provided on sections of this facility. This example also features a painted, flush median of variable width and exclusive lanes for both left and right turns at the intersections.

Access Control by Zoning

Zoning can effectively control the property development along the arterial and thereby influence the type and volume of traffic generated. Property uses can be limited to those that attract very few people, excluding those uses that generate significant volumes of traffic during the hours of peak movements. Heavy volumes of traffic would have a serious adverse effect on the area. In certain cases it might be desirable to exclude land uses that would depend on heavy volumes of commercial traffic if, for various reasons, this class of vehicle cannot be accommodated readily or there are limitations in the design of the highway.

Zoning regulations should require ample off-street parking as a condition of approval of a building permit. The internal arrangement of the land-use development should be such that the parking spaces most distant from the arterial street are the most attractive to the user to minimize congestion in the vicinity of the entrance at the street. Vehicles exiting from the parking facility to the arterial should offer no impedance to traffic entering the parking facility from the arterial.

Subdivision or zoning ordinances should require that the developer of a major traffic generator provide a suitable connection to the arterial street

Figure VII-12. Arterial streets with access controlled by statute.

comparable to that for a well-designed street intersection serving a similar volume of traffic. It should be understood that the intersection is subject to the same traffic control measures, including restrictions to turning movements, as are applicable elsewhere on the arterial. The developer may be required to provide a frontage road between the property and the arterial street if needed to maintain a high level of operation and safety on the arterial.

Access Control through Driveway Regulations

Driveway controls can be effective in preserving the functional character of arterial streets. In heavily built-up areas and areas having potential for intensive development, the control of permits for curb cuts can be adopted to minimize marginal interference with the free flow of traffic. In more sparsely developed areas holding little promise for dense development, driveway controls are a useful device for ensuring that driveways are placed so that there will be minimum hazard to the highway user and least interference with the free movement of traffic.

Access Control through Geometric Design

Frontage roads and grade separations afford the ultimate in access control. Fully developed frontage roads effectively control access to the through lanes on an arterial street, provide access to adjoining property, separate local from through traffic, and permit circulation of traffic on each side of the arterial. They may be used in conjunction with grade separation structures at major cross streets, in which case the arterial takes on many of the operating characteristics of a freeway.

Continuous frontage roads in combination with at-grade intersections create hazardous traffic conflicts and are conducive to wrong-way entry into the through lanes. For this reason, continuous frontage roads should be limited in application to stage construction of an ultimate freeway design. Short sections of frontage road are not so hazardous and have wider application.

Frontage road connections to signalized cross streets are also a problem. They should not be used unless the intersection can be situated to avoid influencing the operation of the main signalized intersection.

General features of frontage roads and their design are discussed in Chapter IV. The effect of frontage roads on the design of intersections is covered in Chapter IX under the heading, "Intersection Design Elements with Frontage Roads."

Figures VII-13 through VII-16 are examples of access control by frontage roads. Figure VII-13 shows a divided arterial with one-way frontage roads in a residential area. The residences are well removed and shielded from the through traffic on the arterial. Minor roads are intercepted by the frontage roads. Only the more important streets are permitted to cross the arterial.

Figure VII-14 shows an undivided arterial with frontage roads. Other features are the grassed outer separations, traffic markings, and traffic signals.

Figure VII-15 shows a short section of a one-way frontage road along an undivided arterial in an intermediate area of a large city. The frontage road serves a row of small commercial establishments. This arrangement reduces congestion on the through roadway, but may cause problems mentioned previously.

Figure VII-16 shows an undivided major street with one-way frontage roads in a downtown district of a large city. The frontage roads are minimal, but greatly improve traffic operations by relieving the through lanes from most of the marginal friction. The raised outer separation provides a safety refuge for pedestrians.

Pedestrian Facilities

Arterial streets may accommodate both vehicles and pedestrians, and the design should include sidewalks, crosswalks, and sometimes grade separations for pedestrians. Pedestrian facilities and control measures will vary,

Figure VII-13. Divided arterial street with one-way frontage roads.

Figure VII-14. Undivided arterial street with frontage road.

Figure VII-15. Short segment of frontage road in intermediate area of large city.

Figure VII-16. One-way frontage road in downtown district of a large city.

depending largely on the volume of pedestrian traffic, the volume of vehicular traffic to be crossed, the number of lanes to be crossed, and the number of vehicles turning at intersections.

On many sections of arterial streets that traverse relatively undeveloped areas, the number of pedestrians will be so small that sidewalks are not needed immediately. Because these areas will usually be developed in the future, the design should allow for the ultimate installation of sidewalks. However, as a general practice, sidewalks should be constructed initially along all arterial streets that are not provided with shoulders, even though pedestrian traffic may be light.

The major pedestrian-vehicular conflict usually occurs at intersections. On the lower classes of arterials, especially at intersections with minor cross streets where turning movements are light, pedestrian facilities are usually limited to cross walk markings. Features that help the pedestrian include fixed-source lighting, refuge islands, barriers, and signals. Such features are discussed in Chapter IV.

On the more important arterials, i.e., six or eight lanes wide with heavy traffic volumes, the interference between pedestrians and vehicles at at-grade intersections sometimes presents serious problems. The problem is especially acute where the arterial traverses a business district and there are intersections with important cross streets. In extreme cases, grade separations for pedestrians provide the only satisfactory solution. (See Chapter IV for further details.) Although separations for pedestrians are justified in some instances, at-grade crosswalks will remain the predominant form of crossing. Minimum conflict and hazard result if the crosswalks are properly placed, designed, maintained, and operated.

The number of pedestrian crossings on heavily traveled arterials should be kept to a minimum, but in and near business districts it is usually necessary to provide crosswalks at every intersecting street. Enforcement of a ban on pedestrian crossings at an intersection is very difficult. A crossing should not be closed to pedestrians unless the benefits in improved safety and traffic operation are sufficient to offset the inconvenience to pedestrians.

Indiscriminate closing of pedestrian crossings will lead to illegal crossings that are more hazardous. Therefore, proper and reasonable design for pedestrians is important.

The pedestrian walk signal is especially desirable on wide arterials having frontage roads because of the great distances to be covered. On exceptionally wide arterial streets, pedestrian signals may be mounted in the median as well as on the far side of the intersection and, if necessary, in the outer separations. Refer to the current MUTCD **(1)** for additional information regarding installation of signals.

Curb-Cut Ramps

The use of crosswalks at typical curbed-street intersections is difficult for people in wheelchairs. Curb-cut ramps of sufficient width must be provided in curbed areas that have sidewalks. The subject of curb-cut ramps is covered in Chapter IV.

Provision for Utilities

The urban arterial system must also often serve as a utility corridor. Utilities desirably are located in the outer edge of the borders. Poles should be located as near the right-of-way lines as possible. Service manholes should never be located in traffic lanes and desirably outside the entire roadway; however, locations in the medians or parking lanes may be acceptable under special conditions. Utilities should seldom be added to an arterial by the open cut method. Additional installations are desirably bored or jacked to avoid interference with normal traffic movements.

Intersection Design

The design and operation of intersections have a significant effect on the operational quality of an arterial. Sight distance, capacity, grades, and provision for turning movements all affect intersection operation. Turning movements on principal arterials should not encroach on adjacent lanes, or if necessary, their effects should be considered. Chapter IX discusses intersection development in detail. It is recommended that each individual intersection be carefully evaluated in the early design phases.

Operational Control and Regulations

The efficiency of an arterial street system can be measured in large part by the adequacy of traffic control devices and the degree of enforcement of traffic regulations. If the demand for arterial route service exists within the traffic corridor, these measures are of such prominence that it may be possible to convert a street with only moderately good design standards into a major traffic artery.

The fact that operational controls can have a strong bearing on street use can be both good and bad from the standpoint of the life of the city and its environmental values.

If used unwisely, traffic operational controls can lead to blighted areas with depressed property values and major upheavals in land-use patterns. This undesirable result might be brought about, for example, by upgrading traffic control measures and regulations on a residential street to accommodate overflow traffic from a nearby parallel arterial serving the same traffic corridor. The potential economic loss resulting from such injudicious use of such expedients may more than offset the cost of improving the arterial to a standard to serve the major movements within the corridor.

The potential of traffic control measures to improve the capacity and traffic service should be exploited to the maximum degree on properly designated arterial streets. Control measures may be divided into the following categories: (1) traffic control devices, (2) regulatory measures, and (3) directional lane usage.

Traffic Control Devices

When traffic signals are required or anticipated during the roadway construction phase, the design should integrate the ultimate signal operation. The design should consider the reduction of signal phases by providing for concurrent opposing left turns and constructing the left-turn lanes in a manner which will allow free operation. Channelization, which provides for single or double left turns, and free right turns often results in better signal control and may make pedestrian crossings easier. The spacing of signals to allow free-flow timing in both directions is highly desirable and may be achieved by control of access during early development stages. If this cannot be achieved, suitable time-space diagrams based on traffic forecasts may be used to determine signal and access spacing. Such efforts will provide optimum signal progression to provide maximum vehicle capacity and minimum vehicle delay time. The location of driveways also affects signal operation significantly. Driveway locations positioned so as to allow driveway vehicles to interfere with major through movements should be avoided. Locations which will allow entering or exiting vehicles to generate false calls on signals are undesirable.

The physical location of signal standards can also often be improved for safety and visibility reasons when the location is considered in the roadway design stage.

The ultimate goal of any intersection design should be to safely carry the required traffic volumes at an acceptable level of service consistent with the overall arterial design. This goal requires that all intersection elements, including traffic signals be integrated into the entire design process.

Traffic control devices such as signs, markings, signals, and islands are placed on or adjacent to a street or highway to regulate, warn, or guide traffic. Each device is designed to fulfill a specific need with regard to traffic operation, control, or safety. The need for traffic control devices should be determined by an engineering study made in conjunction with the geometric design of the street or highway. To assure standard design and installation application of the various traffic control devices, refer to the current MUTCD **(1)**.

The importance of a signal system that is responsive to traffic needs cannot be overemphasized. Traffic needs fluctuate from hour to hour and from day to day. The signal system should likewise be flexible. The system should include detection and data processing components as well as a control component. The installation of a sophisticated signal system is generally the most economical method to improve traffic service in a densely developed area.

Among other considerations, the signal system should allocate the time in accordance with demand. Signals should be coordinated for progressive movement in the direction of predominant flow of traffic on an arterial street, although the optimum traffic service of the system as a whole must be considered. Where multiphasing is necessary to control various movements at intersections, the detecting and programming equipment should provide for skipping phases when there is no traffic demand. Idle and lost green time due to lack of demand, or too many phases must be kept to a minimum. Equipment to accomplish this goal is available. This method of improving traffic service, although expensive, is usually less costly than additional right-of-way and construction, which may be the only other alternative. The installation of sophisticated traffic signal systems requires considerable effort to maintain and operate; therefore, the operating agency must be prepared to assign adequate, capable manpower to the system if optimum benefits are to be realized.

Progression of through movements on an arterial is one of the most effective means of improving operations on both the arterial and the cross streets. By moving queues of traffic through the corridor, the corridor remains open, allowing more time for cross-street movements. The selection and operation of the optimum traffic control equipment is related to the design of the arterial.

Successful operation of an urban arterial depends largely on proper pavement marking, especially on arterials having multiple lanes and particularly when special provision is made for left turns. Recent developments in products for pavement marking show considerable promise in providing adequate long-life marking, even for areas where snow removal often obliterates ordinary markings in very short times. Overhead lane signing is usually very helpful. The sign allows the driver to make maneuvers well in advance and also is helpful under adverse weather conditions, such as rain or snow. Adequate crosswalk marking is required for the safety of pedestrians on the urban arterial.

Regulatory Measures

Regulatory measures take many forms as discussed in the "Uniform Vehicle Code" **(3)**, and the "Model Traffic Ordinance" **(4)**. The two measures that contribute most to the operational efficiency of arterial streets are restrictions of turning movements and prohibition of parking, stopping, or standing at curbside.

Restrictions against turning movements, both midblock and at intersections, promote good operation if applied wisely; however, such restrictions should not be applied indiscriminately. Some means must be provided for highway users to reach destinations, and this can be accomplished only by turning from the arterial, either right or left.

Vehicles turning right may interfere with through traffic where there are heavy pedestrian movements. Where corner curb radii are undesirably short, vehicles turning right may also interfere with through traffic to an extent that is far out of proportion to their numbers. In such cases it may be desirable to prohibit turns during certain hours if there are alternate routes to the destinations involved. As a general rule, however, interference from vehicles turning right is not a problem.

Vehicles turning left into cross streets or at midblock locations frequently create considerable interference, diminish the effectiveness of the arterial, and contribute to the incidence of accidents. Few of these movements can be eliminated, notwithstanding the popular notion that a "No Left Turn" sign at an intersection will cause the problem to disappear. To the contrary, origins and destinations on the left of traffic must be served and the prohibition of left turns at one or more locations may compound the problem at others.

The solution lies in discovering or anticipating the dimension of the problem and in providing for the movements through a combination of regulatory measures (that is, selective prohibition of turns, geometric design, and traffic control). It is difficult to discuss these factors independently, and no firm rules are applicable to all situations. Several principles and methods that, if properly considered and applied, will lead to a logical solution are outlined as follows:

1. The desires of traffic must be served. Left turns should not be prohibited unless alternative routings are available.
2. As a general rule, the fewer the number of left turns at any location, the less the interference with other traffic. Thus, for a given total number of left turns within a given length of highway, it may be better to encourage a few left turns at each of several locations than to concentrate the turns at a single location.
3. Separate signal phases for left-turn movements reduce the amount of green time available for other movements at the intersection. Multiphase signals are therefore advantageous only if traffic operation and safety are improved sufficiently to offset the loss in green time. This determination requires careful analyses in every case.
4. Where selective prohibition of left turns is necessary, there are operational advantages in concentrating left turns at intersections where the volume of cross traffic is low so that a large fraction of the signal time is available for green on the arterial. Where two arterial streets intersect, there may be advantage in requiring left-turning vehicles to bypass the main intersection. For instance, one manner in which the left-turn maneuver can be accomplished is to turn left from the first arterial a block in advance of the main intersection, proceed one block, turn right, proceed another block, and turn left. Clear guide signing is essential.
5. The routing of traffic around a block to the right, in lieu of left turns, is sometimes advantageous. There are disadvantages, also. The turn-

ing traffic destined to the left is required to pass through the intersection twice. The three right turns at the other intersection may be undesirable, and the miles of travel are increased. The applicability of this form of traffic regulation is usually limited to those intersections where only a few vehicles desire to turn left and where a separate turning lane is not feasible.

6. The desirability of exclusive left-turn lanes cannot be overstressed. These lanes may consist of left-turn slots in the median or a continuous center lane exclusively for left turns from both directions. Multiphase signal control is very inefficient if turning traffic and through traffic both use the same lane. Where turning traffic is light, a left-turn lane may eliminate the need for left-turn phases because the storage of left-turning vehicles will not affect through traffic. Traffic safety is greatly enhanced if turning vehicles can be stored clear of the lanes used by through vehicles.
7. With a separate indication for left turns, two left-turn lanes will accommodate up to about 180 percent of the volume that can be served by a single lane with the same available green time, depending on the width of the cross street and the radius of turn. Thus, where right-of-way, space for a long radius turn, and a wide cross street are available, the installation of two left-turn lanes may be a practical solution to a problem involving a heavy left-turn movement. Figure VII-17 shows an example of two left-turn lanes on an urban arterial.
8. Grade separations for left-turn movements are sometimes warranted, or other special treatments as outlined in Chapter X may be required.

In summary, left turns should be accommodated as near as practicable to the point of desire of the highway user. Shifting the left turns away from the point of desire sets up secondary problems. Nevertheless, if the point of desire for left turns is highly objectionable from the standpoint of design, traffic control, or safety, regulatory measures may be employed to shift the point where left turns may be made to a location that is more suitable. Only in exceptional cases should the point be shifted more than two blocks from the point of desire. Where left turns are to be permitted from an arterial street, the intersection design should incorporate left-turn storage lanes unless it is totally impractical to provide them.

Figure VII-18 shows an arterial with excellent provisions of left turns. This facility passes through a predominantly residential area of a large city, although most of the abutting property is commercial. The arterial has a wide median with crossroads limited to every two or three blocks. The left-turn lanes are offset where they join the crossroads so that opposing left-turn traffic does not conflict. Left turns are regulated. The intersection at the bottom of the photo has two lanes for left turns. Access is controlled through curb-cut

Figure VII-17. Urban arterial with dual left-turn lanes.

Figure VII-18. Arterial with provisions for left-turn lanes.

regulations, and in most blocks access is limited to one or two points well removed from the intersections.

During periods of heavy traffic movement curb parking is incompatible with arterial street service. Even when the traffic demand is well below the capacity of the street, vehicles parked at the curbs are accident hazards and impede the free flow of traffic. Desirably, curb parking on arterial streets should not be permitted. The unit cost per parking space is exorbitant in relation to that on off-street parking lots. Nevertheless, during off-peak periods, parking can be tolerated on some arterial streets.

The greatest adverse effect of parked vehicles results from their interference with free flow of traffic and the attendant loss in capacity. With arterials that are four to six lanes wide curb to curb, parked vehicles may be expected to limit capacity to only 55 to 65 percent of the capacity with parking prohibited. This loss is critical during the hours of peak traffic movement.

Where parking provisions are included in the design, cross-sectional dimensions should be such that the entire width can be used by moving traffic when parking is removed. At intersections, there should be a liberal distance from a corner to the first parking stall. This distance should be at least 20 ft from a crosswalk. It provides extra maneuvering space for turning traffic, reduces the conflict with through traffic, and increases sight distance.

No other single operational control can have as dramatic an effect on traffic flow on arterial streets as the proper regulation of parking, including the enforcement of "No Parking" regulations in loading zones and at bus stops.

Directional Lane Usage

Typically, the conventional arterial street is a multilane two-way facility with an equal number of lanes for traffic in each direction. Often, however, one-way operation is employed where conditions are suitable. Somewhat less frequently, reversible lane operation is used to improve the efficiency of operation. The conditions under which each form of operation is most suitable depend largely on traffic flow characteristics, the street pattern, and the geometric features of the particular street. Where a street system is undergoing expansion or improvement, the ultimate form of directional usage should be anticipated, and the design should be prepared accordingly. Once an arterial street is completed to a desirable standard, conversion from one form of directional usage to another may entail considerable expense and disruption to traffic. For existing streets of conventional design, this conversion may be a feasible alternative for increasing the traffic capacity. For information regarding signing for directional lane usage, refer to the current MUTCD (**1**).

One-Way Operation

An arterial facility consisting of one or more pairs of one-way streets is generally adaptable to the following combinations of conditions: (1) where a single two-way street does not have adequate capacity and does not lend itself readily to improvement to accommodate anticipated traffic demand, particularly where left-turning movements at numerous intersections are difficult to handle, (2) where there are two parallel arterial streets a block or two apart, and (3) where cross streets of sufficient number and appropriate spacing are available to permit circulation of traffic.

One-way streets have the following advantages:

1. Traffic capacity in most cases is increased as a result of reduced medial and intersectional conflicts and more efficient operation of traffic control devices.
2. Travel speed is increased as a result of reduction of medial conflicts and delays caused by left-turning vehicles. The increase in the number of lanes in one direction permits ready passing of slow-moving vehicles. One-way operation permits good progressive timing of traffic signals.
3. The number and severity of accidents are reduced by eliminating the hazardous head-on accidents and reducing some types of intersection conflicts.
4. An odd number of traffic lanes is used fully on streets that do not lend themselves to division into an even number of lanes of suitable width.
5. The available street width is used fully through the elimination of need for a median.
6. On-street parking that would otherwise have to be eliminated or curtailed may be retained.

There are also disadvantages to one-way operation. Among these are the following:

1. Travel distances are increased because certain destinations can be reached only by around-the-block maneuvers. This disadvantage is more acute if the street grid is composed of one-way streets.
2. One-way streets may be confusing to strangers.
3. Emergency vehicles may be blocked by cars in all lanes at intersections waiting for signals to change.

In summary, there are several advantages and disadvantages to one-way operation. The choice of one- or two-way operation depends largely on which type can be developed to meet the traffic requirements most economically and with greatest benefit to the adjacent property. Both types of operation should be considered. In many cases, the proper choice is immediately obvious. In other instances, a thorough study involving all of the many facets of the matter may be necessary.

Reverse-Flow Operation

The familiar imbalance in directional distribution of traffic during peak hours on principal radial streets in large and medium-sized cities often results in congestion in the direction of heavier flow and excess capacity for opposing traffic. Capacity during peak hours can be increased by using more than half the lanes for one direction.

Reverse-flow operation on undivided streets generally is justified where 65 percent or more of the traffic moves in one direction during peak periods, where the remaining lanes for the lighter flow are adequate for that traffic, where there is continuity in the route and width of street, where there is no median, and where left turns and parking can be restricted.

The conventional undivided street need not be changed appreciably for conversion to reverse-flow operation, and the cost of additional control measures is not great. On a five-lane street, three lanes would be operated in the direction of heavier flow. On a six-lane street with directional distribution of approximately 65 to 35 percent, four lanes can be operated inbound and two lanes outbound during the morning peak. Operation is reversed on the middle lanes so that two lanes are inbound and four lanes are outbound during the evening peak. During offpeak periods, traffic is accommodated on three lanes in each direction or on two lanes in each direction with curb parking.

Streets with three or four lanes can also be operated with a reverse flow; however, with only one lane in the direction of lighter flow, a slow vehicle or one picking up or discharging a passenger retards the entire movement, and a vehicle breakdown blocks traffic in the one direction completely. Occasionally, circumstances may be such that these streets can be adapted to complete reverse flow, i.e., one-way inbound in the morning and one-way outbound in the evening. At other times the street may be operated two ways, with or without parking.

Direct left turns by traffic in the minor direction should be carefully controlled. Left turns from the predominant flow are subject to the same considerations and regulations as they are for conventional operation with two-way traffic. This system is in contrast to a one-way reversible street where left turns at all intersections can be made readily.

Reverse flow requires special signing and additional control devices. More policing and manpower are necessary to operate the control devices. Stanchions or traffic cones are usually desirable to separate the opposing traffic. "No Left Turn" and "Keep Right" signs on pedestals or stanchions are also necessary.

Assigning traffic to proper lanes can be accomplished by placing overhead signs indicating lane usage for given hourly periods. However, with this arrangement there is a tendency for some motorists to pull out for left turns into a lane that is signed for traffic in the opposite direction. It is better to place separate lane-use control signals at intervals over each lane. This method is particularly adaptable to long bridges and sections of streets without side connections. Further efficiency is gained in the predominant direction by progressive timing of signals. With a three-dial interconnected system, signals are set for proper progression of the major movement in the peak periods; a third setting is used to fit the traffic flow during off-peak periods. Sometimes the signals for the center lane or lanes are set red in both directions during off-peak hours, thus converting the unused pavement into a median area for safety.

The principle of reverse-flow operation is also applicable to divided arterials, but the arrangement is more difficult with a median on a divided highway than on an undivided highway. The difficulty of handling cross and turning traffic, the confusion for pedestrians, and the questionable suitability of an at-grade facility for the high volumes may make other highway types preferable. For example, the capacity of an at-grade expressway with a reverse-flow section of three-two-three or three-three-three lanes (equivalent to a 10-lane or 12-lane conventional section) may be comparable to the capacity of a six-lane depressed or elevated freeway. For these widths, likely volumes would be 3,500 to 4,000 vph in one direction, or two-way ADT volumes of 50,000 to 60,000 vehicles, for which a freeway is warranted. Furthermore, traffic in some cases may not remain materially unbalanced in future years. Reverse-flow operation for at-grade facilities is applicable chiefly as a means of increasing capacity on existing highways.

Frontage Roads and Outer Separations

Frontage roads are sometimes used on arterial streets to control access, as discussed in the section "Access Control Through Geometric Design" in this chapter. Other important functions of frontage roads are to facilitate operations on the through-traffic lanes and to provide access to abutting properties. For data on widths and other features of frontage roads and outer separations, refer to Chapters IV, IX, and X.

Figure VII-19 is an example of a two-way frontage road along a divided arterial in a residential area.

Grade Separations and Interchanges

Grade separations and interchanges are covered in other chapters and many of the principles discussed are applicable to arterial streets. Although

Figure VII-19. Divided arterial street with two-way frontage road.

grade separations and interchanges are not often used on arterial streets, they may be the only means available for providing sufficient capacity at some critical intersections.

Grade separations can be constructed in some cases within the existing right-of-way. Locations where grade separations should be considered are:

1. At very high-volume intersections between principal arterials.
2. At high-volume intersections having more than four approach legs.
3. At high-type arterial street intersections where all other principal intersections are grade separated.
4. At all railroad crossings.
5. At sites where terrain conditions favor separation of grades.

Normally, where a grade separation is provided, it is provided as part of an interchange, usually of the diamond type where there are four legs. Other types of interchanges have application where more than four legs are involved. These types are discussed in Chapter X.

It is desirable to carry the entire roadway width of the approach, including parking lanes or shoulders, across or under the grade separation. However, in some cases with restricted right-of-way it will be necessary to reduce the width. This reduction is not as objectionable on arterial streets as on freeways because of lower speeds. As a minimum, the full approach width of the traffic lanes plus 2 ft on each side should be carried across or under the structure. The reduction in parking-lane or shoulder width should be accomplished with a taper. See Chapter X for taper design elements.

Interchange elements for arterial streets may have lower dimensional values as compared with freeways. Desirably, loops should have radii no smaller than 150 ft. Diamond-type ramps may have lengths as short as the minimum distance necessary to overcome the difference in elevation between the two roadways at suitable gradients. Speed-change lanes must be consistent with the design speed. Chapter X gives criteria for design of interchanges and grade separations.

Figure VII-20 shows grade separation structures built to eliminate bottlenecks in congested areas. Connections are provided between the major arterial street and the local street system.

Erosion Control

When an urban arterial is designed with an open ditch cross section, rural erosion control measures should be applied. Curbed cross sections usually require more intensive treatment to prevent damage to adjacent property and siltation in sewers and drainage systems. Seeding, mulching, and sodding are usually employed to protect disturbed areas from erosion. Landscaping features, such as ground cover plantings, bushes, and trees, are also employed to control erosion, enhance the beauty, and provide sound control of the roadside.

Lighting

Adequate lighting is very important to safe operation of an urban arterial. The higher volumes and speeds require the driver to make correct decisions with adequate time to make the proper maneuvers without creating undue conflict in the traffic lanes. Sudden braking and swerving are minimized where lighting is adequate. The visibility of signing and pavement marking also helps to smooth traffic flow. A safely designed, adequate lighting system is more important to optimum operation of an urban arterial than any other city street. The lighting should be continuous and an energy-saving type. Lighting in an urban area is also often a matter of civic pride and is a deterrent to crime. In the event that it is impossible to provide continuous lighting, consideration should be given to providing intermittent lighting at such locations as intersections and ramp termini. The AASHTO *Information Guide for Roadway Lighting* (**5**) is recommended as a source of lighting information.

Bikeways

Appropriate consideration should be given to possible measures for enhancing bicycle travel. If special bicycle facilities are desired, in addition to adequate roadway width, their design should be in accordance with the AASHTO *Guide for Development of New Bicycle Facilities* (**2**). Chapters II and IV also should be consulted.

Figure VII-20. Grade separation structure on arterial street.

Public Transit Facilities

Wherever there is a demand for highways to serve automobile traffic, there is likewise a demand for public transportation. With almost total elimination of fixed-rail transit vehicles from surface streets and the increased use of free-wheeled buses, public transit is becoming increasingly compatible with other highway traffic. With proper recognition of bus requirements and provisions for them in the design and operation of highways they can become even more compatible in the future.

Other high-volume passenger vehicles such as the minibus, taxicab, and limousine may merit serious consideration in the overall planning of a high-volume arterial. The transit vehicle is more efficient than the private automobile with regard to street space occupied versus passenger-carrying capacity.

However, the vehicle-carrying capacity of the through lanes is decreased when the transit vehicle and the other traffic use the same lanes. A bus stopping for passenger loading, for example, not only blocks traffic in that lane but affects operation in all lanes. It is desirable that this and other forms of interference be minimized through careful planning, design, and traffic control measures.

The requirements for public transit must be considered in the development of an urban highway improvement program. Information gathered during the planning process on the routings of transit vehicles (including turns and transfer points) and the volumes of buses (i.e., average or minimum headways) must be considered in the design of the highway. Design and operational features of the highway that will be affected by the foregoing information include the following: (1) locations of bus stops (general with respect to spacing, and specific as to near side, far side, and midblock), (2) design of bus stops and turnouts, (3) reservation of bus lanes, and (4) special traffic control measures. Some of the design and control measures that are beneficial to bus operation will have an adverse effect on other traffic and vice versa, thus calling for a compromise that is most favorable to all users.

Location of Bus Stops

The general location of bus stops, as opposed to the specific location, will be largely dictated by patronage and by intersecting bus routes or transfer points. These are functions of land-use patterns, and the highway should be designated to conform with the need for service. Convenience is a prime consideration.

The specific location is influenced not only by convenience to patrons but also by the design characteristics and operational considerations of the highway. Except where cross streets are widely spaced, bus stops are usually located in the immediate vicinity of intersections. This location facilitates crossing of streets by patrons without the necessity of midblock crosswalks. Where blocks are exceptionally long or where bus patrons are concentrated at places of employment or residence that are well removed from intersections, midblock locations for bus stops may be a more suitable alternate.

Bus stops at intersections may be located on the near (approach) side or the far (exit) side. There are advantages and disadvantages to each situation, and each case should be examined separately and a determination made as to the most suitable location, with consideration being given to such things as service to patrons, efficiency of transit operations, and traffic operation in general. Far-side stops are advantageous at intersections where (1) other buses may turn in either direction, (2) turning movements from the arterial in the direction being considered are heavy, particularly right turns, (3) cross traffic is heavy, and the curb lane, as well as other lanes, is needed for storage during periods when the signal is red, and (4) several streets meet at the intersection. Far-side stops have also proven to be an effective measure to reduce pedestrian accidents.

Sight distance conditions generally favor far-side bus stops, especially at unsignalized intersections. A driver approaching a cross street on the through lanes can see any vehicles approaching from the right. With near-side stops the view to the right may be blocked by a stopped bus. Where the intersection is signalized, the bus may block the view of one of the signal heads.

Another disadvantage of near-side bus stops is the difficulty and hazard of vehicles making turns when a bus is loading. Drivers frequently swing around the bus to turn right, which interferes first with other traffic and then the bus as it leaves the stop. This disadvantage is eliminated if the cross street is one way from right to left. Where the street pattern consists of a one-way grid, there is some advantage in having stops at alternate cross streets in advance of the streets crossing from right to left.

Where buses turn left, the bus stop should be located at least one block before the turn. Even with this arrangement the bus must cross all traffic lanes to reach the left lane for the turn. Midblock stops are used occasionally under these circumstances, but the bus maneuver is much more difficult.

On highly developed arterials with ample rights-of-way, bus turnouts, and speed-change lanes, there is a definite traffic advantage to the far-side stop, as shown in Figure VII-21A. Such stops can be combined with speed-change lanes for turning vehicles entering the arterial. Where the stop is located on the near side, vehicles turning right from the through lanes of the arterial cannot use the deceleration lane when occupied by a transit vehicle, and instead will maneuver around it on the through pavement. Where the bus stop is located on the far side, traffic turning right from the arterial does so freely.

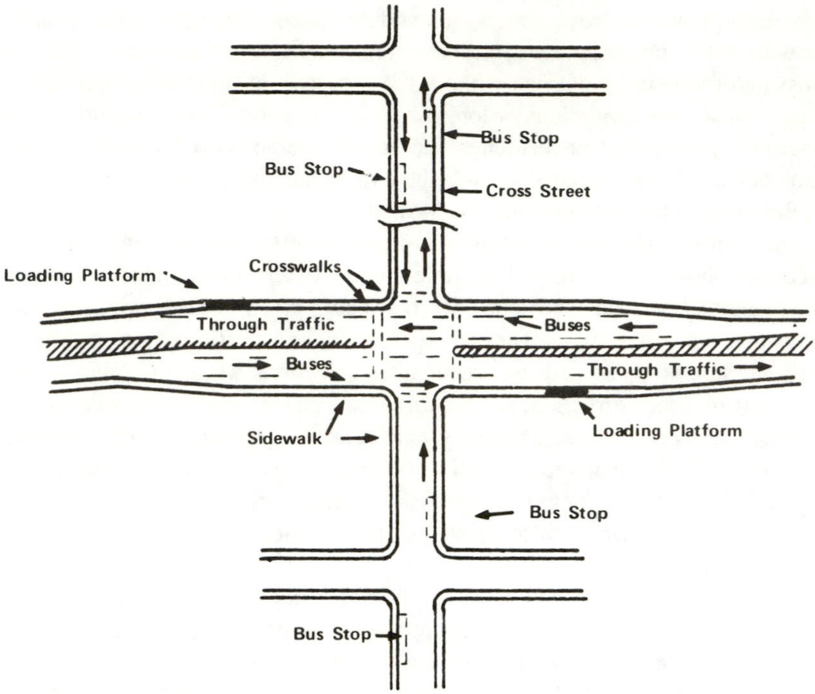

On Acceleration Lane

–A–

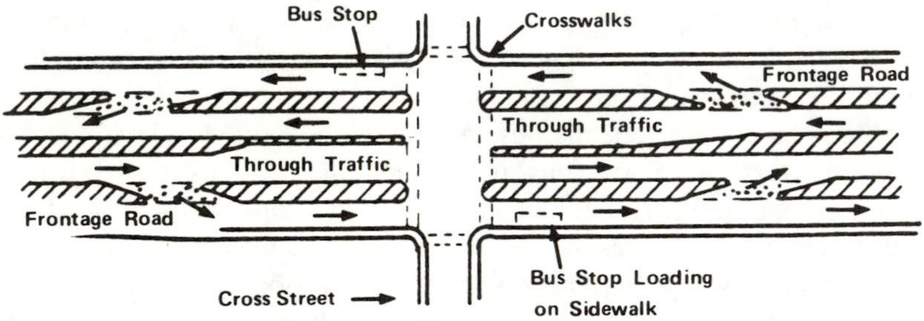

On Frontage Road

–B–

Figure VII-21. Bus stops at special locations adjacent to certain arterials.

The stop should be situated a short distance beyond the intersection so that vehicles making right turns onto the arterial can use a part of the acceleration lane as a maneuvering area, and so that succeeding buses are permitted to stop without blocking the cross street.

Bus stops on cross streets should be located and arranged so that transferring riders are not required to cross the arterial, regardless of the direction they wish to travel. Figure VII-21A shows such an arrangement.

In another form of bus stop for an arterial with frontage roads (Figure VII-21B), buses leave and return to the arterial by special openings in the outer separation in advance of and beyond the intersection. This arrangement has the advantage of buses stopping in a position well removed from the through pavements. Right-turning traffic to and from the arterial street may also use these special openings, thereby reducing conflicts at the intersection proper. In alternate arrangements, the slot in advance of the intersection is not provided, and buses cross to the frontage road at the intersection proper. Both slots may be eliminated where the frontage road is continuous between successive cross streets because buses can leave the through pavement at one intersection and use the frontage road to reenter at the next intersecting street. This type of operation is fitting where bus stops are widely spaced.

Midblock stops, like far-side stops, have an advantage over near-side stops in that the full pavement width at the intersection is made available for storage and turning maneuvers and maintaining as high a capacity as practicable. However, midblock stops are generally not suitable on streets where parking is permitted, as might be the case on some arterials during off-peak hours. Usually, a crosswalk is necessary to provide access to bus stops from either side of the arterial and to serve as an intermediate crosswalk for other pedestrian traffic. In this case signal control may be necessary to safeguard pedestrians. In conjunction with a major transit stop and heavy pedestrian movements, a pedestrian grade separation may be warranted.

Bus Turnouts

The interference between buses and other traffic can be considerably reduced by providing stops clear of the lanes for through traffic. The foregoing discussion and Figure VII-21 provide methods for reducing such interference on high-type arterials. It is somewhat rare that sufficient right-of-way is available on the lower-type arterial streets to permit turnouts in the border area, but advantage should be taken of every opportunity to do so. For geometric details, see "Bus Turnouts" in Chapter IV.

Reserved Bus Lanes

Some improvement in transit service can be realized by excluding other traffic from the curb lane of arterial streets. The success of such a regulatory measure is rather limited in most instances; however, for the very practical reason that vehicles making right turns must occupy this lane, it is not feasible to exclude them, even for a distance up to a block or two in advance of the turn. Vehicles preparing to turn right cannot be distinguished from through traffic, and compliance with the regulation is largely on a voluntary basis. Nevertheless, there are certain combinations of conditions under which a modest improvement, at least, can be achieved. These conditions are not always apparent or definable, and the only way to determine conclusively that there will be overall benefit is to test the regulation in practice wherever preliminary investigation indicates likelihood of success.

There is one arrangement wherein exclusive use of the reserved lane by buses can be ensured with very little policing. This usage is accomplished by offsetting the division line on a multilane street so that there is only one lane in the one direction of travel, which is reserved for this exclusive use of buses,

Figure VII-22. Exclusive bus lane.

with the remaining lanes being available for the normal traffic to flow in the opposite direction. Refer to the current MUTCD **(1)** for directions for marking the division line separating traffic in the two directions of travel. Permanently installed flexible stanchions may be used to accentuate further the division between the bus lane and the lanes for ordinary traffic. Figures VII-22 and VII-23 are photographs of a street used in this manner. The lane on the left is reserved for buses traveling toward the viewer. The foregoing arrangement is sometimes referred to as a "contra-flow" system. This system has shown advantages in facilitating bus movements under high-density traffic conditions.

Reserved bus lanes of the type just described may be justified for relatively few buses. Because the carrying capacity of a bus is many times that of a passenger car, a small number of buses can serve the same number of people as a large number of passenger cars.

The detailed discussion of bus facilities is not intended to indicate that other types of mass transit facilities should not be considered. The more sophisticated means, including fixed rail, present problems as varied as the types. Therefore, this discussion has concentrated on the means that most affect roadway design. Other types require specific studies for the type employed.

Figure VII-23. Exclusive bus lane.

Traffic Control Measures

Traffic control devices on arterial streets are usually installed with the intent of favoring automobile traffic, with only secondary consideration to transit vehicles. For express-bus operation, the control measures that are most favorable for one mode will generally be equally well suited for the other. Where local service is provided by buses, however, with frequent stops to pick up and discharge passengers, a signal system that provides for good progressive movement of privately operated vehicles may actually result in reverse progression for buses. The resulting slow travel speed for buses tends to discourage patronage, further adding to the already heavy volume of automobile traffic.

Limited effort has been devoted to development of traffic control systems that would be more favorable to bus service without serious adverse effects on other traffic. This approach holds some promise of improving travel speeds for buses and making public transit more attractive. This result can be achieved only at the expense of reduced travel speed for passenger cars, but in some instances this solution may be the most beneficial from the overall point of view. Development of a suitable signal system requires careful investigation by those possessing the proper technical skills and should be a part of the arterial improvement program, involving the joint efforts of traffic specialists, the transit industry, and the design team.

Although the major emphasis in the application of traffic control measures is in minimizing delay, the control measures can be used to facilitate bus operation in other respects, particularly where buses turn from the arterial onto a cross street.

Buses making right turns may create a problem where the cross street is narrow and the adjoining property is developed so intensively that it is not practicable to provide a curve of sufficiently long radius. Buses turning right from the curb lane may encroach beyond the centerline of the cross street. At signalized intersections, the space beyond the centerline would normally be occupied by vehicles stopped by the red light. Under such conditions the stop line on the cross street should be relocated to provide sufficient space for the turn, and if necessary, an auxiliary signal head should be placed at the relocated stop line to obtain compliance.

Examples of Arterial Streets

Typical examples of various types of arterial streets are shown in Figures VII-24 through VII-42. Other examples are included throughout this chapter.

Figure VII-24 shows a series of arterial and cross streets in the downtown

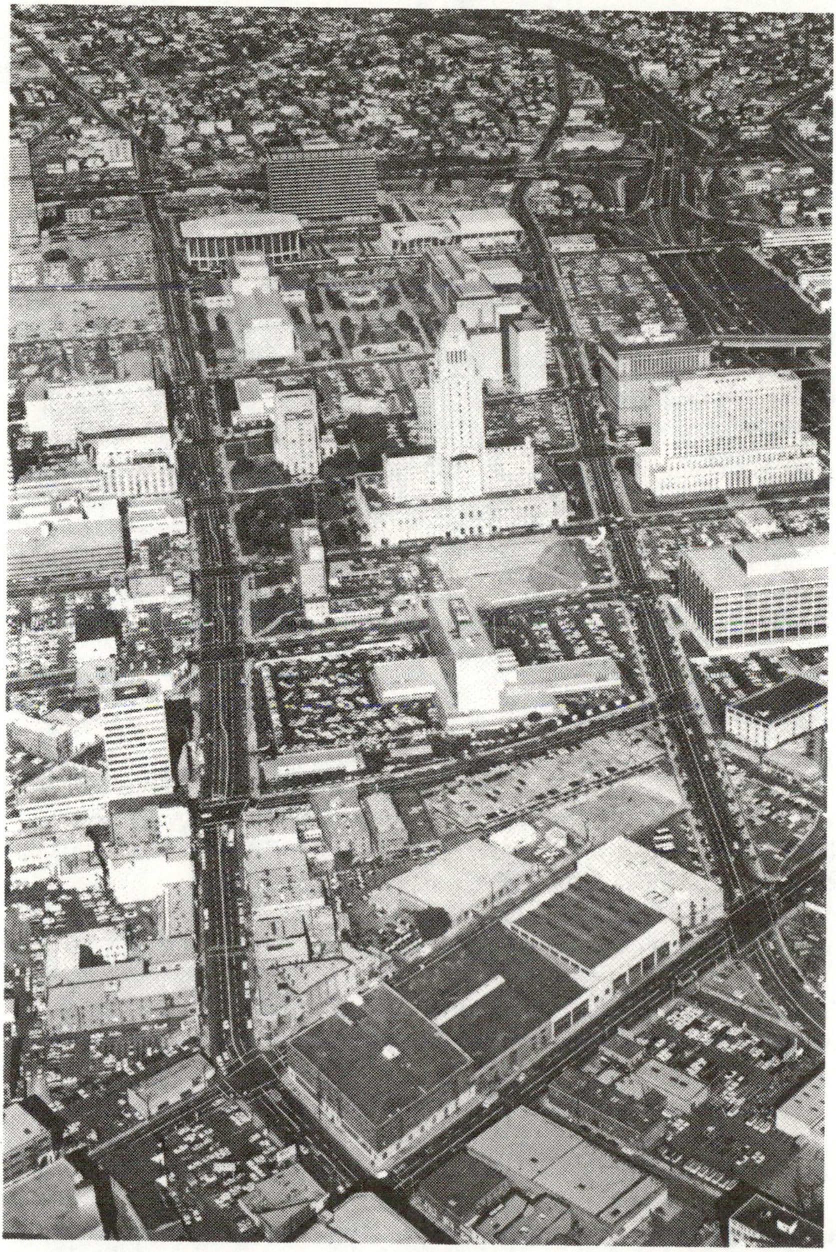

Figure VII-24. Network of arterial streets in downtown district.

district of a large city. Corner radii are minimal because of right-of-way restrictions and pedestrian requirements. The lane and crosswalk striping is excellent.

Figure VII-25 shows an arterial street in a densely developed and partly commercial suburban area. This facility has a painted flush median. Each intersection has left-turn lanes.

Figure VII-26 is an arterial street in a newer residential area with intermittent frontage roads. There is a flush median with left-turn lanes. The gently curving alinement blends in well with the surroundings.

Figure VII-27 is an arterial street in a residential area. It features barrier curbs, a parking lane, curb cuts for driveways, and sidewalks.

Figure VII-28 shows a high-type arterial street sometimes classed as an expressway or at-grade expressway. It has a high degree of access control, and intersections are either grade separated with interchanges or provided with high-type channelization and speed-change lanes. A square loop interchange is present at the crossing of another expressway (upper center of the picture).

Figure VII-29 is an arterial street in a suburban area.

Figure VII-30 shows a divided arterial street in a lightly developed suburban area. The median treatment is novel and quite attractive. The wide borders would permit the installation of sidewalks or parking lanes with a buffer area if needed in the future.

Figure VII-31 shows a divided arterial in a smaller city. Parking lanes are provided. The wide borders contain a sidewalk and a buffer strip. Liberal widths of design elements help make this arterial pleasing.

Figure VII-32 shows an older, divided arterial in a large city. A one-way frontage road is provided to serve the dense apartment development.

Figure VII-33 shows a divided arterial in an outlying area with full shoulders and a grass median of varying width.

Figure VII-34 shows a divided arterial with a full shoulder in a suburban area. The utility poles are placed well back from the full shoulders.

Figure VII-35 shows a divided arterial in an older, residential area. The heavy landscaping effectively screens the homes from the arterial. Although the raised median is quite narrow, it provides a positive separation for traffic in the opposite directions. No provision is made for left-turn storage. This situation can result in significantly reduced capacity. Unless U-turns can be adequately controlled, hazardous conditions can also result from inadequate turning radii.

Figure VII-36 shows an arterial in a large city with frontage roads and a wide parklike outer separation. Other features are excellent pavement striping, crosswalks, lighting, and painted left-turn channelization at the intersection.

Figure VII-25. Arterial street in a densely developed suburban area.

Figure VII-26. Arterial street with intermittent frontage roads.

Figure VII-27. Arterial street in a residential area.

Figure VII-28. Expressway at-grade.

Figure VII-29. Arterial street in a suburban area.

Figure VII-30. Divided arterial street in outlying area of a large city.

Figure VII-31. Divided arterial street with a depressed median in a small city.

Figure VII-32. Divided arterial street with a one-way frontage road.

Figure VII-37 shows a minor arterial constructed through an established neighborhood. Note the attractive landscaped border areas with pedestrian walks and provision for wheelchairs. The arterial was constructed to operate as a 4-lane street, but parking is being allowed until traffic requires the removal of the parking.

Figure VII-38 shows the use of tower lighting at the intersection of a freeway and a high-volume urban arterial.

Figure VII-39 shows a fully enclosed pedestrian walkway structure crossing a freeway.

Figure VII-33. Divided arterial street with a variable width median.

Figure VII-34. Divided arterial street with a full shoulder.

Figure VII-35. Divided arterial street with a narrow median through a residential area.

Figure VII-36. Arterial street with frontage roads and a wide park-like outer separation.

Figure VII-37. Minor arterial through established neighborhood.

Figure VII-38. Tower lighting.

Figure VII-39. Enclosed pedestrian mall.

REFERENCES

1. Federal Highway Administration, National Advisory Committee on Uniform Traffic Control Devices. *Manual on Uniform Traffic Control Devices.* Washington, D.C.: U.S. Government Printing Office, 1988.
2. American Association of State Highway and Transportation Officials (AASHTO). *Guide for Development of New Bicycle Facilities.* Washington, D.C.: AASHTO, 1981.
3. National Committee on Uniform Traffic Laws and Ordinances. "Uniform Vehicle Code."
4. National Committee on Uniform Traffic Laws and Ordinances. "Model Traffic Ordinance."
5. *Informational Guide for Roadway Lighting*. Washington, D.C.: AASHTO, 1976.
6. American Association of State Highway and Transportation Officials (AASHTO). *Roadside Design Guide.* Washington, D.C.: AASHTO, 1989.
7. Jens, S.N. "Design of Urban Highway Drainage, The State of The Art." *FHWA-TS-79-225*. Federal Highway Administration, Office of Research and Development, Implementation Division, Washington, D.C.: 1979.
8. *Highway Capacity Manual, Special Report No. 209*. Washington, D.C.: Highway Research Board, 1985.

Chapter VIII FREEWAYS

INTRODUCTION

The highest type of arterial highway is the freeway, which is defined as an expressway with full control of access.

Control of access is the condition where the right of owners or occupants of abutting land to access a highway is fully or partially controlled by public authority. Full control of access means that the authority to control access is exercised to give preference to through traffic by providing access connections with selected public roads only and by prohibiting crossings at grade or direct private driveway connections.

The principal advantages of control of access are the preservation of the as-built capacity of the highway, higher speed, and improved safety to highway users. Highways with fully controlled access have grade separations at all railroads and grade separations or interchanges at selected public crossroads. The remaining crossroads are interconnected or terminated.

Essential freeway elements include medians, grade separations at cross streets, ramp connections for entrance to and exit from the through pavements, and (in some cases) frontage roads.

Freeways are selected for principal arterial corridors that are intended to provide for high levels of safety and efficiency in the movement of high volumes of traffic at high speeds.

Chapters II, III, and IV cover in detail those design elements, controls, criteria, and cross-sectional elements applicable to all highway classes. This chapter identifies the various types of freeways, emphasizes selected features, and discusses other design details unique to these freeway types.

GENERAL DESIGN CONSIDERATIONS

Design Speed

As a general consideration, the design speed of urban freeways should reflect the desired safe operating speed during nonpeak hours, but should not be so high that it exceeds the limits of prudent construction, right-of-way, and socioeconomic costs because a large proportion of vehicles are accommodated during periods of peak flows when lower speeds are necessary. The design speed should not be less than 50 mph. Wherever this minimum design speed is used, the important safety considerations are to have a properly posted speed limit and to enforce the speed limit during off-peak hours. On many urban freeways, particularly in developing areas, a design speed of 60 mph or higher can be provided with little additional cost. The corridor of the main line may also be relatively straight, and the character and location of interchanges may permit high-speed design. Under these conditions, a design speed of 70 mph is desirable because the higher design speeds are closely related to the overall quality and safety of the facility. A design speed of 70 mph should be used for rural freeways. Where terrain is mountainous, a design speed of 60 mph or 50 mph, which is consistent with driver expectancy, may be used.

Design Traffic Volumes

General guidance toward the selection of appropriate highway design periods and design traffic is provided in Chapter II. Both urban and rural freeways, especially on new location, are normally designed to accommodate traffic projections for a 20-year period. Some elements of freeway reconstruction may be based on a lesser design period. Specific capacity requirements should be determined from directional design hourly volumes (DDHV) for the appropriate design period. In large metropolitan areas, the selection of appropriate design traffic and design periods may be influenced by system planning. Segments of freeways may be constructed or reconstructed to be commensurate with either intermediate traffic demands or traffic based on the completed systems, whichever may be most appropriate.

Levels of Service

Techniques and procedures for adjusting operational and highway factors to compensate for conditions other than ideal are found in the *Highway Capacity Manual* (HCM) **(1)**, which also presents a thorough discussion of the level-of-service concept. Although choice of the design level of service is left to the user of the HCM **(1)**, designers should strive to provide the highest level of service feasible and consistent with anticipated conditions. Refer to Table II-5 for level-of-service characteristics. For acceptable degrees of congestion, freeways and their auxiliary facilities, i.e., ramps, main lane weaving sections, and C-D roads in urban and developing areas, should generally be designed for level-of-service C. In heavily developed sections of metropolitan areas, conditions may necessitate the use of level-of-service D. In rural areas, level-of-service B is desirable for through and auxiliary lanes, although level-of-service C may be acceptable on auxiliary facilities carrying unusually high volumes.

Pavement and Shoulders

Freeways should have a minimum of two through-traffic lanes for each direction of travel. Through-traffic lanes should be 12 ft. wide. Pavements should have a high-type surface with adequate skid resistance and provide a high degree of structural adequacy. Cross slopes should range between 1.5 and 2 percent on tangent sections consisting of two lanes in each direction with a crown at the centerline of the pavement. The higher value is recommended for areas of moderate rainfall. For areas of heavy rainfall a cross slope of 2.5 percent may be necessary to provide adequate pavement drainage. Appropriate rates of cross slope are discussed in Chapter IV. For elevated freeways on viaducts, two-lane pavements usually are sloped to drain the full width of the roadway. On wider facilities, particularly in areas of heavy rainfall, surface drainage may be two ways on each traveled way, the crown being located on the lane line at one-third or one-half the total width from one edge. In snow areas, surface drainage should be two ways on each traveled way so that snow stored in the median will not melt and drain across the traveled way, or the median should be designed to prevent this occurrence.

There should be continuous paved shoulders on both the right and the left sides of all freeway facilities. The usable paved width of the right shoulder should be at least 10 ft and where truck traffic exceeds 250 DDHV it should preferably be 12 ft. On four-lane freeways, the median shoulder or left shoulder is normally 4 to 8 ft wide. At least 4 ft should be paved, and the remainder should be surfaced to some extent. On freeways of six or more lanes, the usable paved width of the median shoulder should also be 10 ft and preferably 12 ft where the truck traffic exceeds 250 DDHV. Ramp shoulder widths are usually constructed adjacent to acceleration and deceleration lanes with transitions to the freeway shoulder width at the taper ends. Shoulder cross slope should range between 2 and 6 percent and should be at least 1 percent more than the pavement cross slope on tangent sections to facilitate drainage. It is desirable that the color or texture of the shoulders be different from that on the through lanes. On viaducts, differentiation between traveled way and shoulders is sometimes effected by striping and pavement marking or corrugated depressions.

Curbs

Barrier curbs should not be used on freeways, but if provided in special cases, the curb should not be closer than the outer edge of shoulder. Mountable curbs, if used, should also be placed at the outer edge of shoulder. Refer to the discussion on curb placement in Chapter IV.

Superelevation

The full superelevation rates used on depressed, ground level, or embankment freeways generally are not applicable to elevated freeways on viaducts. Appearance and adjacent development somewhat limit the difference in elevation between the edges of multilane pavements. Superelevation rates of 0.06 to 0.08 are generally the maximum on viaducts. The lower value should be used where freezing and thawing are likely, because bridge decks generally freeze more rapidly than other roadway sections. Combinations of design speed and curvature that require superelevation rates greater than these values should be avoided. Where freeways are intermittently elevated on viaducts, the lower superelevation rates should be used throughout to promote consistently safe operation. In developing areas, urban and rural design criteria may be considered for freeways. Maximum superelevation rates of 0.08 to 0.12 are applicable for these facilities if snow and ice are not factors.

Grades

Grades on urban freeways should be comparable to those on rural freeways of the same design speed. Steeper grades may be tolerated in urban areas, but the closer spacing of interchange facilities and the need for frequent changes in speed make it desirable to use flat grades wherever possible. Maximum grades as a function of design speed and type of terrain are given in Table VIII-1. Where sustained upgrades are required, the need for climbing lanes, as discussed in Chapter III, should be investigated.

Structures

The design of bridges, culverts, walls, tunnels, and other structures should be in accordance with the principles of the current *Standard Specifications for*

	Design Speed (mph)		
	50	60	70
Type of Terrain	Grades (percent)		
Level	4	3	3
Rolling	5	4	4
Mountainous	6	6	5

[a] Grades 1 percent steeper than the value shown may be used for extreme cases in urban areas where development precludes the use of flatter grades and for one-way downgrades except in mountainous terrain.

Table VIII-1. Maximum grades for urban and rural freeways.[a]

Highway Bridges (**2**). Structures carrying freeway traffic should provide an HS-20 design loading.

The clear width on bridges carrying freeway traffic should be as wide as the approach pavement and paved shoulders as discussed in Chapter X. On long bridges, some economy in substructure costs may be gained by building a single structure rather than twin parallel structures. In such cases, the same shoulder widths are provided and a median barrier is carried across the bridge.

Structures carrying ramps should provide a clear width equal to the ramp width and paved shoulders. Clear widths for structures carrying auxiliary lanes are discussed in Chapter X.

The structure width and lateral clearance of highways and streets overpassing or underpassing the freeway are dependent on the functional classification of the highway or street as discussed in Chapters V, VI, and VII.

Vertical Clearance

The vertical clearance to structures passing over freeways should be at least 16 ft over the entire roadway width, including auxiliary lanes and the usable width of shoulders. In highly developed urban areas, where attainment of the 16-ft clearance would be unreasonably costly, a minimum clearance of 14 ft may be used if there is an alternate freeway facility with the minimum 16-ft clearance (with allowance for any future resurfacing).

Because of their lesser resistance to impacts, the vertical clearance to sign trusses and pedestrian overpasses should be 17 ft. The vertical clearance from the deck to the cross bracing of through-truss structures should also be a minimum of 17 ft.

An allowance of 6 in. should be added to all vertical clearances to accommodate future resurfacing.

Horizontal Clearance to Obstructions

Urban freeways at ground level and rural freeways should have clear zone widths consistent with their operating speed and side slopes as discussed under horizontal clearance to obstructions in Chapter IV. Nontraversable hazards within the clear zone should be shielded by appropriate barriers when the barrier presents a lesser hazard. Fixed objects which cannot be relocated beyond the clear zone should be made "breakaway" or protected by barriers or attenuators. Bridge piers and abutments should be located as near to the clear zone edge as practical. Where the width of the median is not twice that of the clear zone, shielding of median piers may be required depending on median width and traffic volumes.

Depressed freeways in urban areas have more restrictive right-of-way and may require the placement of retaining walls or bridge piers within the clear zone. Such piers and walls should not be located closer than the outer edge of shoulder and preferably 2 ft beyond. Retaining walls and pier crash walls should incorporate an integral concrete barrier shape or they should be offset from the shoulder to permit shielding with a barrier as discussed under lateral clearances in Chapter X. Where walls are located beyond the clear zone or are not required, back slopes should be traversable and fixed objects within the clear zone should be of a "breakaway design" or shielded.

Elevated freeways on embankments will generally require roadside barriers when slopes are steeper than 3:1, or the area beyond the toe of slope remaining within the clear zone is not traversable. The tops of retaining walls used in conjunction with embankment sections should be located no closer than the outer edge of shoulder and the walls should incorporate the concrete barrier shape or be appropriately shielded.

Ramps and Terminals

The design of ramps and connections for all types of freeways is covered in Chapter X.

Outer Separations, Borders, and Frontage Roads

The outer separation is the area between the traveled ways of a roadway for through traffic and a frontage road or street. The border is the area between the adjacent traveled way and the private development along the road. Where there are no frontage roads or local streets functioning as frontage roads, the area between the traveled way of the main lanes and the right-of-way should be referred to as the border. Because of the dense development along urban freeways, frontage roads are often necessary to maintain local service and to collect and distribute ramp traffic entering and leaving the freeway. Where the freeway occupies a full block, the adjacent parallel streets are usually retained as frontage roads, which are discussed in Chapter IV.

The outer separation or border provides for shoulders, sideslopes, drainage, access-control fencing, and in urban areas retaining walls and ramps. The outer separation or border may also provide for noise abatement measures in sensitive areas. Usually, the outer separation is the most flexible element of an urban freeway section. Adjustment in width of right-of-way, as may be necessary through developed areas, ordinarily is made by varying the width of the outer separation.

The outer separation or border should be as wide as can be attained economically to provide a buffer zone between the freeway and the adjacent area. The border should extend beyond the construction limits where possible to facilitate maintenance operations and safety. Wide outer separations also permit a high standard of ramp design between the freeway and the frontage road. The typical range in widths of outer separations is 80 to 150 ft, but much narrower widths may be used in urban areas if retaining walls are employed. In rural cut sections, drainage is carried in a side ditch. The foreslope, backslope, and ditch configuration should provide an acceptable recovery area in case drivers lose control and run off the traveled way. Where the right-of-way on a depressed freeway is insufficient to provide for a swale, the drainage is usually carried in a gutter section along the outside edge of the shoulder. Details of gutter sections are covered under curbs in Chapter IV.

URBAN FREEWAYS

General Characteristics

Urban freeways are capable of carrying high traffic volumes. They may have from 4 to 16 through-traffic lanes, although only in unusual cases are there more than four through lanes in one direction.

An urban freeway may be either depressed, elevated, ground level, or combinations of these, as is appropriate for conditions.

Medians

A wide separation is safer and more comfortable for the motorists, and therefore the median on urban freeways should be as wide and flat as feasible. Extra median width also can be used for mass transit or to provide additional lanes if more capacity is needed in the future. However, in densely developed areas with expensive right-of-way, the width available for a median is usually restricted. The minimum median width for a four-lane urban freeway is 10 ft, which provides for two 4-ft shoulders and a 2-ft median barrier. For freeways with six or more lanes, the minimum width is 22 ft, and preferably 26 ft when truck traffic exceeds 250 DDHV to accommodate the wider median shoulder. For these minimum median widths, a median barrier is always required. Additional clearance may be required to provide minimum stopping sight distance along the median lane on relatively short-radius curves. The narrow medians should generally be paved for the full width with a material which is contrasting in color or texture.

Median crossovers for emergency or maintenance purposes are generally not warranted on urban freeways due to the close spacing of interchange facilities and the extensive development of the abutting street network.

Depressed Freeways

A depressed freeway may occupy a full-block width and be parallel to the grid street system for most of its length. The freeway through roadways are approximately 16 ft plus the clearance for structural depth below the surface of the adjacent streets. They are often flanked on one or both sides by frontage roads at the street level. All important streets pass over the through roadways. Other streets are intercepted by the frontage roads or terminated at the right-of-way line. Interchange with the surface streets is usually accomplished by ramps that connect directly with the frontage roads or by diamond interchanges where there are no frontage roads. Higher types of interchanges are usually provided at the intersections of certain major arterials. Interchanges are discussed in Chapter X.

Depressed freeways are desirable in that they reduce the impacts on adjacent areas. They are less conspicuous than ground level or elevated freeways, permit surface streets to cross at their normal grade, and reduce freeway noise.

Structures passing over the depressed freeway and retaining walls, located in close proximity to traffic lanes, should be fenced to prevent the dropping or throwing of objects onto passing vehicles.

Slopes and Walls

Sideslopes of a depressed freeway are designed in the same manner as those for any cut slopes, except that restriction in width is more likely to control. Foreslopes, if used beyond the shoulder, should be safely traversable.

Normally foreslopes will not be used beyond the shoulder on depressed freeways and in such cases backslopes should not be steeper than 3:1. In developed areas space may not be available for desirable slopes, particularly where ramps are present, and full or partial height retaining walls may bc necessary. Various forms of retaining walls are appropriate for depressed freeways. Walls may be constructed of solid masonry, concrete, stone, precast panels, or metal. Types of walls include cantilevered, crib or bin walls, reinforced earth, or sheet piling. Where retaining walls are used in combination with earth slopes, the walls may be located either at the roadway level adjacent to the shoulder or on the outer portion of the separation above the depressed roadway. Retaining walls above the roadway are desirable from thc driver's viewpoint, because they provide a more open feeling at the roadway level. This arrangement also provides space for the storage of snow plowed from the freeway travel lanes and shoulders. However, it may be more advantageous for the overall plan in its relation to the surrounding neighborhood if the wall is located at the roadway level and a slope is located on the upper portion of the cross section. This arrangement permits planting to be used more effectively for screening the surrounding properties from the vehicular traffic. It also permits maintenance of the slopes to be performed more safely, and the abatement of noise may be more effective. Both designs should be evaluated to determine which is better for the particular situation.

Full- or partial-height walls should not be located closer than the outer edge of shoulder and preferably 2 ft beyond. Where the wall is located at or near the shoulder edge, bridge columns, light fixtures, and sign supports should not protrude from the lower portion of the wall and the contour of the lower portion of the wall should have the "safety shape" used on concrete median barriers as shown in Figure X-5B. Where such an arrangement is used, the base of the wall should not be placed farther than approximately 15 ft from the edge of the traffic lane to minimize the angle of impact.

Where the top of the retaining wall is at the level of a frontage road, the face of the parapet or rail should be at least 4 ft and preferably 6 ft from the edge of the traveled way. Where a retaining wall is adjacent to an auxiliary lane or ramp, it should be placed to provide normal ramp shoulder widths.

Sight distance should be checked when designing slopes and retaining walls. On curved alinement, the slopes, walls, and other side obstructions should be sufficiently removed from the pavement edge to provide the required stopping sight distance for a vehicle in the traffic lane nearest the obstruction.

Typical Cross Section and Right-of-Way

Cross sections of depressed freeways vary considerably through urban and suburban areas. They are influenced primarily by the need to accommodate the required number of traffic lanes, but an important factor is availability of right-of-way, which depends on such factors as the type and value of urban development, topography, soil and drainage conditions, and the frequency and type of interchanges required. The design of the cross section should be liberal, but sometimes it is necessary to compromise on certain elements because of physical and economic limitations to fit the cross section within a relatively narrow right-of-way. Figures VIII-1 through VIII-3 illustrate depressed cross sections for various conditions and the approximate range of widths of the cross-sectional elements.

In urban areas it is often not practical, nor is it necessary, to maintain a uniform right-of-way width on depressed freeways. Changes in street pattern, costly buildings, and curvature in the freeway alinement are apt to cause variations in right-of-way widths. In some instances, changes in right-of-way widths necessitate, or make desirable, adjustment in the highway cross section. In such cases a balance in width should be maintained among the various cross-sectional elements. In the construction of new freeway facilities, lane and shoulder widths are not subject to adjustment. Adjustment may be feasible in the widths of median and borders, but most adjustments must be accomplished within the width of the outer separation.

Where the freeway is bridged by closely spaced cross streets, a continuous full-depth section results. In outlying areas where separated crossroads are widely spaced, it usually is possible and economical to adjust the profile to decrease the depth of cut between structures, an approach resulting in a combination of depressed and ground-level-type freeway. By so doing, the design of ramps is simplified, excavation quantities are reduced, sideslopes can be flatter, and wider marginal areas at street level may be provided within the right-of-way. Generally, the result is a more pleasing freeway.

Figure VIII-1 shows a typical plan and cross section for depressed freeways. The section in Figure VIII-1B provides for a 10- to 22-ft median, 12-ft traffic lanes, 80- to 150-ft outer separations, and 50 ft for each frontage road

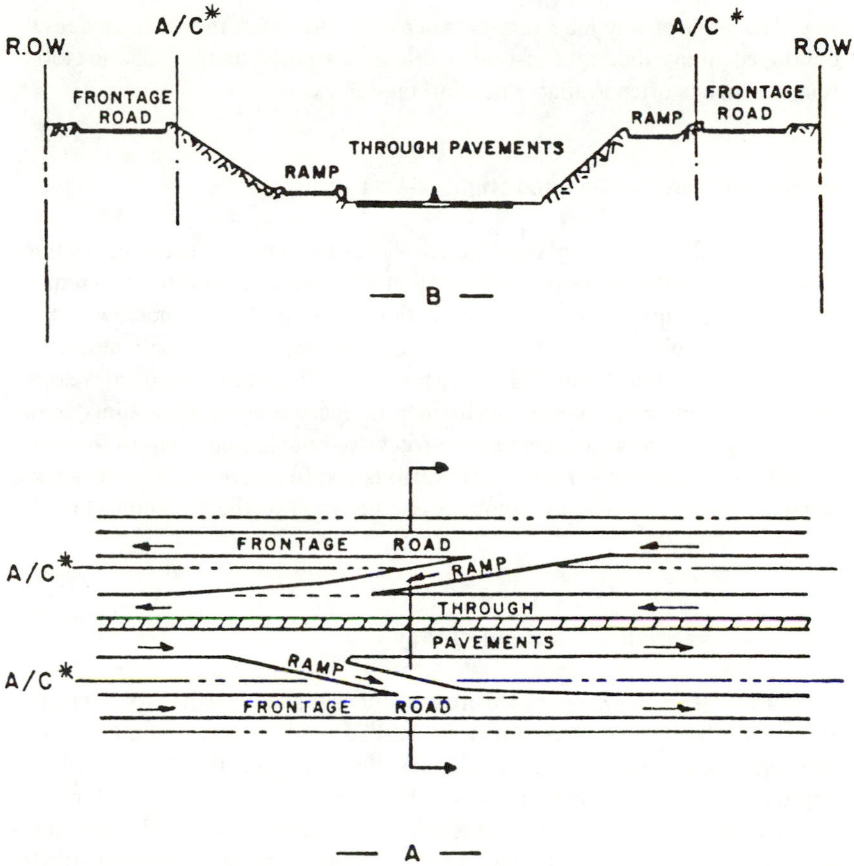

*ACCESS CONTROL LINE

Figure VIII-1. Typical cross sections and right-of-way for depressed freeways.

plus border. The minimum median width of 10 to 22 ft is based on the assumption that for depressed freeways the ultimate section is constructed initially. However, where additional width is provided in the median for stage construction, the median should be widened in multiples of 12 ft (width of a traffic lane). Where ramps are not required, the uniform-width section should be graded to provide slopes as flat as possible within the available right-of-way. The right-of-way may vary between 325 and 525 ft in width. In densely developed areas the right-of-way width is normally in the 325- to 450-ft range, which is often available in a full city block.

Restricted Cross Section and Right-of-Way

Figure VIII-2 shows cross-sectional widths that enable depressed sections to be constructed with earth slopes at locations without ramps but that require retaining walls at ramps. The cross section in Figure VIII-2 includes a 40-ft frontage road plus border, a 60-ft outer separation, 12-ft traffic lanes, and a 10- to 22-ft median. Figure VIII-2C shows a section at the ramps, developed within the same right-of-way. Walls may be located at various points in the cross section, such as adjacent to the freeway shoulder, adjacent to the ramp shoulder or at the tops of slopes, or various combinations of these locations. Some variations in wall arrangements can be noted on the left and right sides of Figure VIII-2C.

Walled Cross Section and Right-of-Way

Figure VIII-3 shows restricted cross sections that are feasible for depressed freeways. The freeway is continuously walled, and ramps are omitted. With no overhang, as shown in Figure VIII-3A, the width of right-of-way needed is approximately 160 ft for four lanes, 195 ft for six lanes, or 220 ft for eight lanes. These widths are based on the following dimensions: the width of frontage road plus border and parapet of 40 ft, outer separation of 12 ft, traffic lanes of 12 ft each, and the median of 10 to 22 ft.

In special cases where even less right-of-way is obtainable, this design can be further consolidated by an overhanging section with part of the frontage road cantilevered over the freeway shoulder width, as shown in Figure VIII-3B. By using a 10- to 12-ft overhang and the same dimensions as in Figure VIII-3A, the facility is contained within right-of-way widths of about 140 ft for four lanes, 175 ft for six lanes, and 200 ft for eight lanes. By increasing the cantilevered section to 12 to 14 ft and decreasing the width of the frontage road from 30 ft to 24 ft, the rights-of-way may be further reduced to a width of about 125 ft (four lanes), 165 (six lanes), and 190 ft (eight lanes), as illusated in Figure VIII-3C. This type of design may be applicable in special instances where large buildings or other obstructions cannot be avoided. A special feature of this design is its effectiveness in containing highway noise within the roadway and shielding abutting areas from such noise.

Although the restricted cross sections shown are acceptable, they should

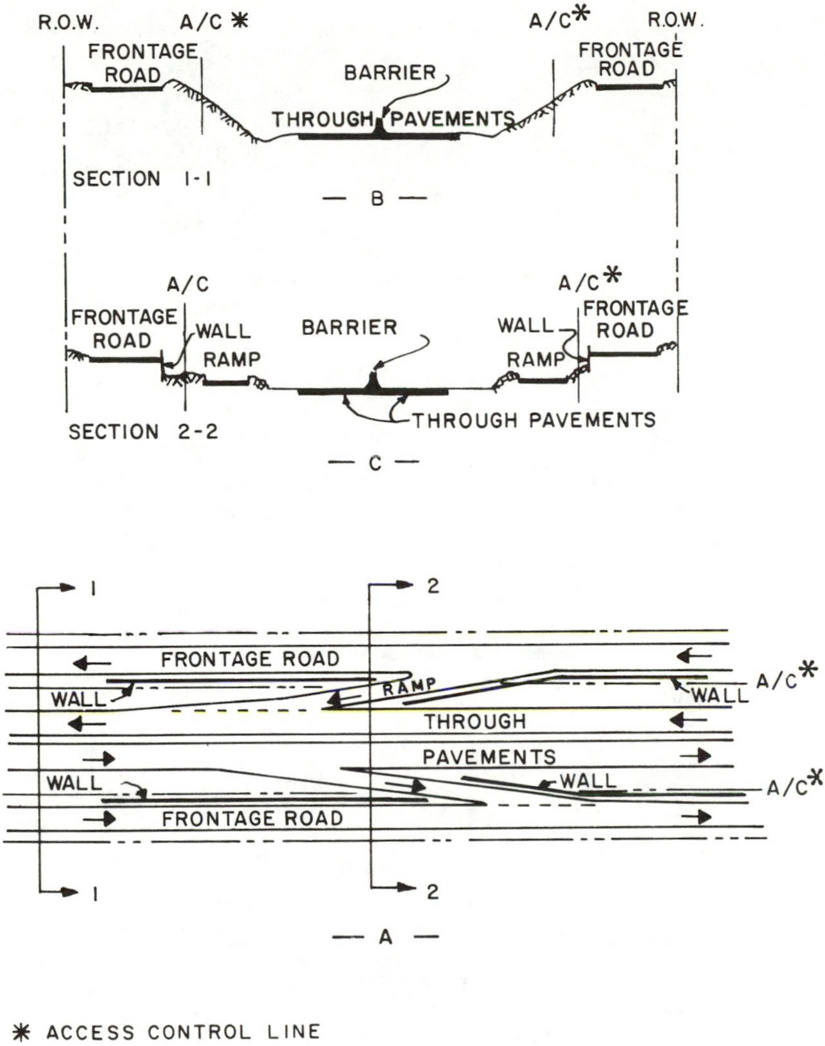

Figure VIII-2. Restricted cross sections and right-of-way for depressed freeways.

be used only where the purchase of additional right-of-way would be extremely costly or where this type of cross section is needed to preserve the surrounding environment.

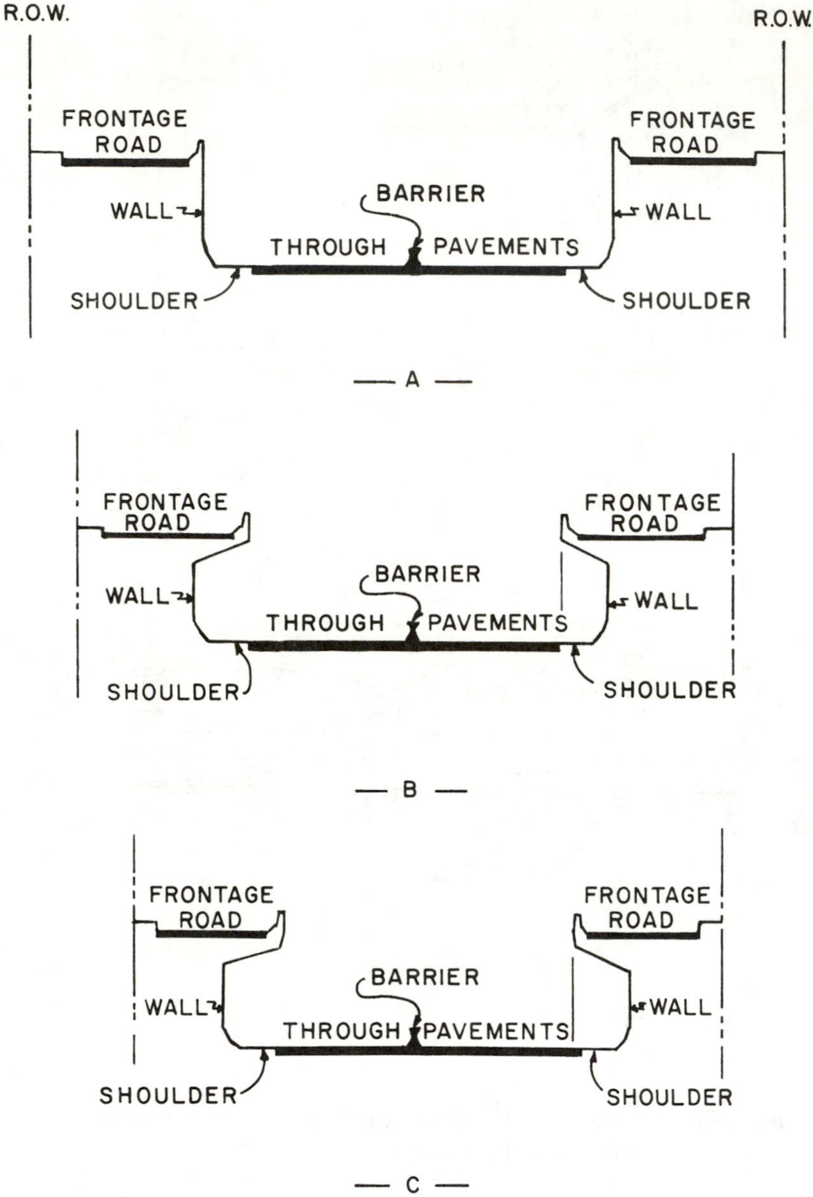

Figure VIII-3. Cross sections and right-of-way with retaining walls on depressed freeways without ramps.

Examples of Depressed Freeways

Figure VIII-4 is an aerial view of a six-lane depressed freeway in the downtown district of a large city. The freeway fits in well with the existing street system. There is a four-level directional interchange in the upper center of the photograph. In the more densely developed portion (lower right) the freeway has full-height retaining walls and frontage roads. In the center of the photograph sufficient right-of-way is available to use slopes in lieu of retaining walls.

Figure VIII-5 is an eight-lane depressed freeway in a fully developed residential area of a large city. Right-of-way is sufficient to use a sloped cross section throughout. The diagonal ramps are two lanes wide except at the freeway terminals. This configuration provides storage for the off ramp, as may be noted at the bottom of the picture. Highway overpasses are spaced at four- or five-block intervals. The two narrow structures near the top of the photograph are pedestrian bridges.

Figure VIII-6 is a four-lane depressed freeway in the downtown district of a large city. The shoulders and ramps are a different color from the main lanes. The different texture and color of the shoulder and ramps in conjunction with the edge striping are assets for the motorist. The earth-filled concrete median barrier contains plantings which can improve the esthetic appearance of the freeway and reduce the problem of headlight glare. Median planting should be restricted to locations where climatic conditions and the absence of salt spray enhance plant survival with a minimum of maintenance. Unhealthy plant material is not attractive and periodic maintenance activities which require vehicles to be parked along the barrier can result in potentially hazardous conditions. The bridge piers are enclosed inside the concrete median barrier. The attractive bridges, textured retaining walls, and landscape planting on the earth slopes all contribute to an esthetically pleasing facility. No luminaires are situated next to the roadway. The luminaire on the right is located on top of the retaining wall and cantilevered over the main roadway. The location of the luminaires improves the safety and appearance of the freeway.

Figure VIII-7 shows a depressed freeway flanked by major surface streets on the upper level. The auxiliary lanes of the surface street on the right side are partially cantilevered over the shoulder of the freeway. The cross streets over pass the freeway with level grades thus facilitating operations on the structures and at the adjacent intersections.

Figure VIII-4. Depressed freeway.

Figure VIII-5. Depressed freeway.

Figure VIII-6. Depressed freeway.

Figure VIII-7. Depressed freeway.

Elevated Freeways

General Characteristics

An elevated freeway may be on either a viaduct or an embankment. Continuous elevation of the freeway may be appropriate in level terrain where restricted right-of-way, high-water table, extensive underground utilities, close pattern of streets that must be retained, or other circumstances make depressed construction undesirable and perhaps uneconomical.

Several types of structures are used for viaducts carrying elevated freeways. The design of these viaducts is influenced by traffic requirements, right-of-way, topography, foundation conditions, character of urban development, requirements for interchange, availability of materials, and economic considerations. Viaducts are perhaps the most difficult of all freeway types to fit harmoniously into the environment.

The supporting columns for viaducts are located to provide reasonable clearance on each side and to leave much of the ground-level area free for other use. This design has advantages in that (1) practically all cross streets can be left open with little or no added expense, (2) existing utilities that cross the freeway right-of-way are hardly disturbed, and (3) surface traffic on cross streets usually can be maintained during construction with few, if any, detours. The space under the structure can be used for surface-street traffic, for parking, or for a transit line. If not needed for these purposes, the area under the viaduct has a high potential value to the community for some form of joint development use. This use may be any of a wide variety of types, ranging from playgrounds to major buildings. Conversely, the disadvantages of the design are the high costs of maintaining the structure and its closed drainage system, susceptibility to icing, difficulty in obtaining a pleasant appearance, and possible policing of undeveloped space beneath the structure.

An elevated highway on an earth embankment must be of sufficient height to permit intersecting surface roads to pass under it. Freeways on embankments are feasible in suburban areas where crossing streets are widely spaced and where wide right-of-way and fill material are available. Usually, an embankment section occurs on a combination-type freeway in rolling terrain where excavation material from depressed portions is used for embankment. Where necessary, the fill may be confined by partial- or full-height walls on one or both sides. The sloped areas are available for planting to improve the appearance of the freeway.

Medians

Where the freeway is on continuous viaduct the median width will generally be the minimum necessary to accommodate the median shoulders and a barrier. Where the freeway is on embankment and the median width is approximately 30 ft or less, to prevent vehicles from entering the median and dropping down to a stream or roadway below, consideration should be given to decking the median at intermittent structures. Where a median barrier is used, decking also permits continuity of the barrer.

Ramps and Terminals

The design of ramps and connections for all types of freeways is covered in Chapter X, but some details and controls peculiar to elevated sections are discussed herein.

Freeways on viaducts are generally located in densely developed areas where property values are high and space is limited. However, the various forms of ramp connections, such as loops, diagonal ramps, and semidirect connections, are as adaptable to elevated freeways as to depressed or other freeway types.

Despite the high cost of elevated freeways, the lengths of speed-change lanes should not be reduced. The length of acceleration and deceleration lanes should conform to the requirements discussed in Chapter X. Long acceleration lanes are especially needed because a ramp leading to an elevated structure is usually on a relatively steep upgrade. Trucks require a considerable distance to gain highway speed.

Gore areas at exits from an elevated structure have a high accident potential. There appears to be no practical way to design the parapet and rail at these gores that are not hazardous if struck by an out-of-control vehicle. Therefore, the design should provide as much space in the gore area as practical, not only for recovery but also to permit the installation of an impact attenuating device.

Frontage Roads

New frontage roads adjacent to viaduct freeways generally are not necessary because the local street network is usually not disturbed. The existing parallel and cross streets are usually adequate to provide local circulation and access; however, frontage roads may be needed for use with embankment freeways to provide adequate local circulation and access. Frontage roads are discussed in Chapter IV where their general features are given and their design values are developed.

Clearance to Building Line

The minimum lateral clearance between a freeway viaduct and adjacent buildings may be a significant cross-sectional element. Major factors where buildings are close to the roadway are (1) working space for maintenance and repairs of structure or buildings, (2) space to prevent salt and water spray damage, (3) protective space against possible fire damage, and (4) space for ladders and other fire-fighting equipment to reach upper floors of buildings from the street. Building offsets should be sufficient to ensure adequate sight distance to signs where the alinement is curvilinear.

All of these space requirements are reasonably well satisfied by a clearance of 15 to 20 ft. Such clearance provides a reasonable but near-minimum space for maintenance, repair, construction, or reconstruction of buildings and viaduct; precludes damage to the structure from a building fire or to buildings from a vehicle fire; and permits ladders and some fire-fighting equipment to be raised and used. Without such clearance the use of some fire-fighting equipment, such as that with mechanically raised ladders, would be hampered. Some of these units might be operable from the elevated freeway.

Basic cross sections and the approximate right-of-way required for elevated freeways on the basis of the minimum structure width and, where fitting, on the minimum clearance between structure and building (right-of-way) line are shown in Figures VIII-8 and VIII-9. Roadways directly under the structure usually are needed for accommodation of surface traffic, but the elements on their cross section are not considered as controls where existing right-of-way determines the structure section.

Typical Cross Section and Right-of-Way

Widths of elevated freeway sections vary considerably, as do widths of right-of-way in which they may be developed. For elevated freeways on embankments, the total widths required are about the same as those for depressed freeways. Elevated freeways on structures may be cantilevered over side roadways or sidewalks, but side roadways or sidewalks also may be cantilevered over depressed freeways. One advantage of elevated freeways on viaduct is that the space under the structure may be used for streets, parking, or other purposes. Space over a depressed freeway may also be devoted to these purposes, but only at considerable additional expense.

The difference in elevation between the local street and the elevated freeway, except in Figure VIII-8B, is approximately 20 ft. The vertical clearance between the local street and the freeway bridge varies with the legal requirements of the various States but usually ranges from 14 to 16.5 ft. Where a railroad is overpassed usually a 22-ft vertical clearance is required which makes it necessary to have the gradeline of the elevated freeway about 28 ft above the tracks. Vertical height, other than that determined at street or railway separations, is determined by ground elevation and freeway profile. At ramps it is advantageous to have the viaduct as low as feasible. Initial construction cost and the cost and ease of operation of vehicles using the ramps

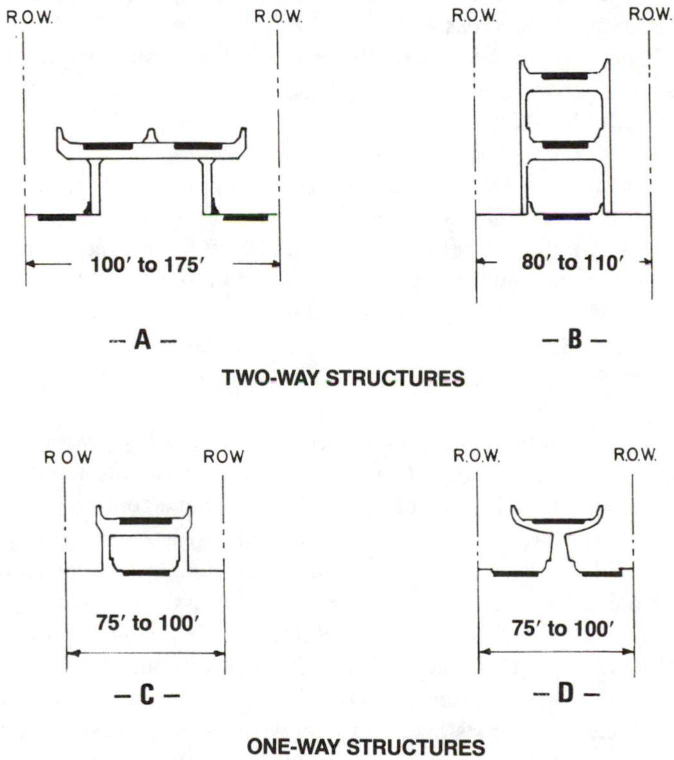

Figure VIII-8. Typical cross-sections and approximate right-of-way for elevated freeways on structures without ramps.

are favorably affected by low viaducts. These combined factors may justify a rolling freeway profile where it can be developed gracefully; however, a roller coaster effect should be avoided. Where a viaduct would profile a clearance of less than about 10 ft from the underside of the structure to the ground, it would usually be better to alter the design to provide retaining walls and fill unless the underneath space could be used for other purposes such as off-street parking.

Viaduct Freeways Without Ramps

Figure VIII-8 shows typical cross sections of elevated freeways without ramps on structures and approximate right-of-way widths. The following assumptions apply to all cross sections:

1. All space under the structure is available for surface streets or other use by the community.
2. Lane width is 12 ft.
3. Parapet width is 2 ft.
4. Shoulder width for four lanes is 10 ft right and 4 ft left; for six and eight lanes, shoulder width is 10 ft right and left.
5. Median width is 10 ft for four lanes and 22 ft for six and eight lanes.
6. Minimum clearance between structure and building line is 15 ft. For Figure VIII-8B, the minimum clearance should be 20 ft.

In Figure VIII-8A, the overhang enables surface roads to be provided outside the lines of columns, and the area between the columns can be used for vehicular traffic, public transit, or parking.

Where it is impractical to obtain the necessary right-of-way widths, it may be feasible to convert the normal two-way, one-level structure to one of two levels. The double-deck design in Figure VIII-8B is not a common type but is adaptable to narrow rights-of-way, particularly where few ramps are required. Double-deck structures have the disadvantage of long structural ramps to make the necessary change in elevation from the top roadway to the local city streets. Double-deck structures also may be adaptable where it is not feasible to continue the freeway as a single-deck structure because of large buildings or for other reasons. Conversion to double-deck construction through such confined areas may be the only solution. As shown in Figure VIII-8B, narrower rights-of-way are required.

Sometimes an elevated freeway is constructed on two one-way structures, as shown in Figures VIII-8C and VIII-8D. The structures may be separated by one or more city blocks. Right-of-way widths as narrow as 75 ft to 100 ft may be used. The structure may be either a two-column section, as in Figure VIII-8C, or a single-column, cantilevered section, as in Figure VIII-8D, which depends on the arrangements of understructure streets and other controls.

Two-Way Viaduct Freeways With Ramps

Elevated freeways generally are developed as single-level, two-way structures, of which the basic structure section is shown in Figure VIII-8A. Cross-sectional arrangements and the resultant right-of-way widths for elevated freeways on structure with ramps and frontage roads are illustrated in Figure VIII-9. The following dimensions are used for general illustration:

Median width	10 to 22 ft
Lane width	12 ft
Right shoulder width:	
Four-lane	10 ft
Six- and eight-lane	10 ft
Left shoulder width:	
Four-lane	4 ft
Six- and eight-lane	10 ft
Parapet width	2 ft
Outer separation	50 to 150 ft
Frontage road plus border	50 ft
Clearance between structure and building line	15 ft

The right-of-way width can vary considerably. The most flexible element is the outer separation. Generally, a width of about 50 ft is required to provide space for a parallel ramp. Depending on the number of lanes, the width of outer separation, and the width of other variable elements such as the median, the overall width of right-of-way for an elevated freeway with ramps and frontage roads is 300 ft or more, as shown in Figure VIII-9B.

An elevated section on structure has great flexibility in right-of-way arrangements. In tight locations where ramps are not provided, the frontage roads can be curved in under a cantilevered section of the structure, as shown in Figure VIII-9C. At these locations the minimum building-line clearance may provide sufficient space for the frontage roads.

Where it is not necessary to narrow the right-of-way width before and beyond ramps, it is possible to obtain liberal clearance between the structure and building line. As a result, there may be space for a green belt strip, parking space off the frontage road, or a wider border and the frontage road.

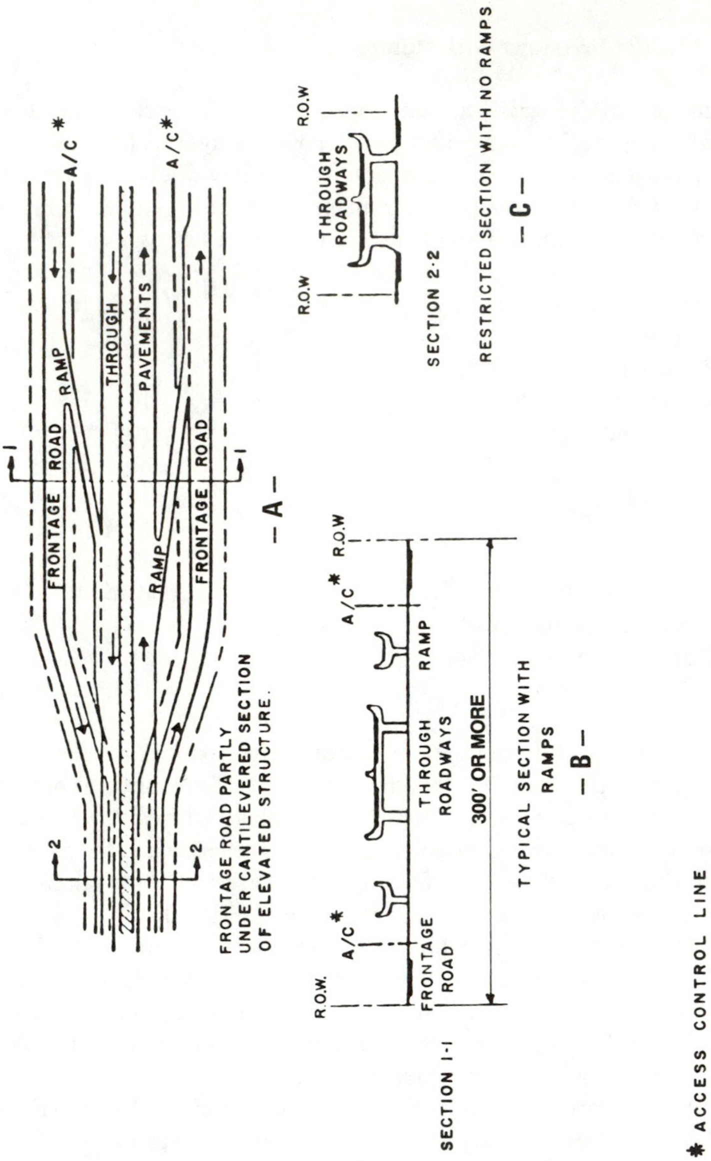

Figure VIII-9. Cross sections and right-of-way for elevated freeways on structure with frontage roads.

Freeways on Earth Embankment

Elevated freeways on embankments of sufficient height to permit surface cross streets to pass under them are fitting where the terrain is rolling and the right-of-way is sufficiently wide for slopes that can be pleasingly landscaped.

Figure VIII-10 shows typical and restricted cross sections and the approximate widths of right-of-way. The right halves of the sections illustrate outer separations without ramps within the same right-of-way width. The difference in elevation between the frontage road and the through pavement is approximately 20 ft. The following dimensions are applicable to the embankment sections in Figure VIII-10:

Median width	10 to 22 ft
Lane width	12 ft
Right shoulder width	10 ft
Outer separation	
Typical	80 or more ft
Restricted	60 ft
Border plus frontage road	
Typical	50 ft
Restricted	40 ft

The typical right-of-way width for an elevated freeway on an embankment with frontage roads illustrated in Figure VIII-10A is approximately 350 ft to 400 ft. Figure VIII-10B illustrates restricted cross-sectional widths with the use of retaining walls. The outer separation may permit the use of earth slopes at locations without ramps, but it requires retaining walls at ramps. Embankment slopes steeper than 3:1 will generally require a roadside barrier. The section requires right-of-way widths of approximately 250 ft to 325 ft. By omission of frontage roads and use of walled sections, widths may be reduced to values approaching those for elevated structures on viaducts. Special wall treatment or planting of trees and shrubs may be necessary to make the retaining walls esthetically pleasing.

Examples of Elevated Freeways

Figure VIII-11 shows a four-lane viaduct freeway on curved alinement located adjacent to the business district of a city. The freeway is situated to minimize the amount of land taking. Existing cross streets have not been disturbed. In the upper part of the photograph, the freeway section changes from viaduct to embankment.

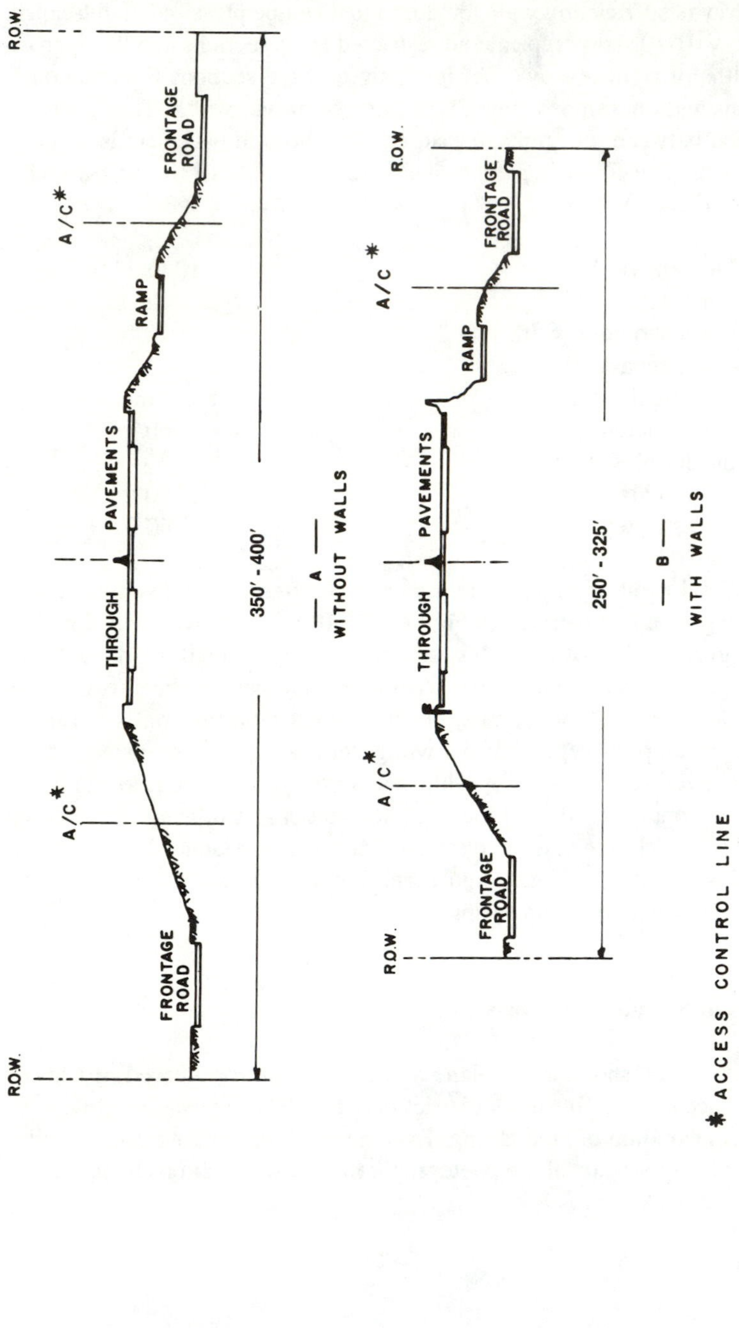

Figure VIII-10. Cross sections and right-of-way for elevated freeways on embankment.

Figure VIII-11. Viaduct freeway.

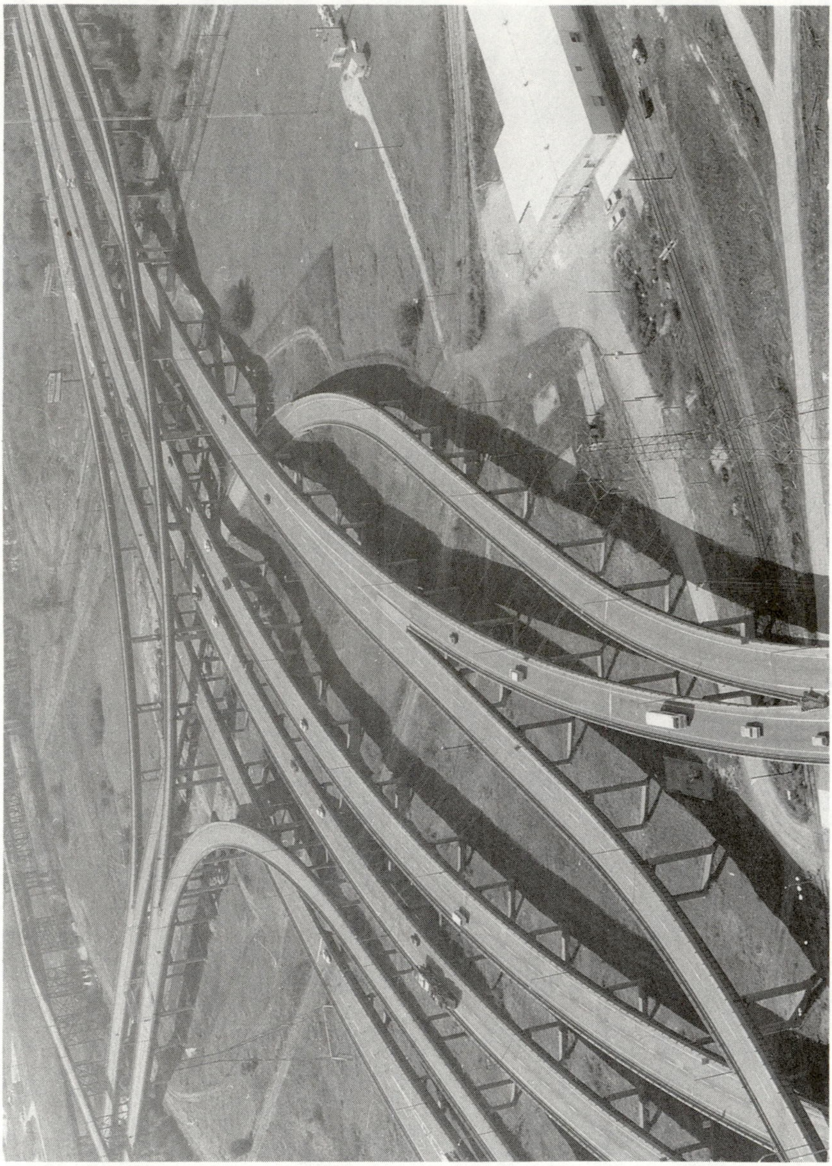

Figure VIII-12. Viaduct freeway.

Figure VIII-12 shows the interchange of two viaduct freeways overpassing a light industrial area located along a river front. Both freeways have a two-structure design with an open median. The reinforced concrete columns supporting the two-girder decks complement the openness of the ground area.

Figure VIII-13 shows an eight-lane, two-level viaduct freeway in a developed area of a large city. Frontage roads are provided on either side of the freeway. The parallel entrance ramp is designed to pass over the continuous frontage road. Except for the one-loop ramp, the freeway is constructed on a minimum amount of right-of-way.

Figure VIII-14 shows an eight-lane embankment freeway passing through

Figure VIII-13. Two-level viaduct freeway.

Figure VIII-14. Freeway on embankment.

a residential area of a large city. The shoulders, medians, and ramps contrast in color from the main lanes. The interchange in the left center of the photograph consists of ramps offset from the cross street, sometimes called lateral ramps. The use of offset lateral ramps saves the right-of-way at the major cross street and moves the ramp connections away from the major directional interchange partly shown at the upper left of the photograph. Offset ramps, however, can be subject to wrong-way movements on the exit ramps when connected to two-way frontage roads or streets.

Figure VIII-15 shows an eight-lane embankment freeway constructed with minimum earth slopes in a predominantly residential area of a large city. The ramps in the center of the photograph illustrate the use of a two-lane ramp design that narrows to one lane at the exit-and-entrance terminal of the freeway. The photograph also illustrates the use of two freeway exits (near center) in advance of an entrance connection. The second exit ramp is extended for a considerable distance to provide proper connection with the local street system.

Figure VIII-16 is a view of the underside of an eight-lane, two-structure viaduct freeway. The smooth lines of the enclosed box-girder design are quite attractive. The tapered piers and planting below the structures are esthetically pleasing. Parking is permitted under the structure. The lighting standards for the freeway are located on top of the bridge parapet.

Figure VIII-17 is a side view of an entrance ramp to an elevated freeway. The textured walls, selected planting, tapered piers, and smooth lines of the structure blend in well with the adjacent residential development.

Ground-Level Freeways

General Characteristics

Many freeways have substantial lengths that are constructed essentially at ground level. This design is often used in flat terrain and along railroads and water courses. Ground-level freeways are also suitable in suburban areas where cross streets are widely spaced. A major feature is that of profile change on each crossroad so that it passes over or under the freeway. These freeways have many of the characteristics of rural freeways. Substantial lengths of ground-level freeways are generally not feasible in heavily developed areas because the profiles of crossroads cannot be altered without severe impact on the community. The profile changes of cross streets are further discussed in the section "Combination-Type Freeways" in this chapter.

Where the ground-level freeway follows the grid of the city, it is usually

Figure VIII-15. Freeway on embankment.

Figure VIII-16. Underside view of viaduct freeway.

Figure VIII-17. Side view of freeway entrance ramp.

desirable to provide continuous one-way frontage roads that serve as a means of connection for streets that are not carried across. However, there will be situations where two-way frontage roads may be the only means to maintain local service, even though they are less desirable than one-way frontage roads. Ground-level freeways usually are employed in the outlying sections of metropolitan areas where right-of-way is not as expensive as it is in the downtown areas. As a result, the variable width elements of medians, outer separations, and borders are widened to increase the safety and appearance of the freeway.

Typical Cross Section and Right-of-Way

Figures VIII-18 and VIII-19 show typical sections for ground-level freeways with and without frontage roads. Where only four or six lanes are provided initially, it may be desirable to provide the same right-of-way width as proposed for six- and eight-lane construction. In these situations the median should be widened by multiples of 12 ft in anticipation of a need for the additional lanes. This step simplifies the installation of additional lanes. The cost would be nominal, and there would be a minimum of disruption to traffic.

Figure VIII-18 shows typical cross sections and approximate right-of-way widths for ground-level freeways. The section in Figure VIII-18A includes a 56-ft median, 12-ft lane widths, 50-ft outer separations, 30-ft frontage roads, and 15-ft borders. The right-of-way width requirements are about 300 ft to 350 ft.

Where additional right-of-way is available without severe impact on the community, the outer separations and borders should be widened to provide esthetically pleasing green belts and to insulate the freeway from the surrounding area. In areas where ramp connections are made to the frontage roads, the width of outer separations should be increased to at least 80 ft and preferably more, to allow space for liberal design of the ramp and ramp terminals.

Where fill material is available and the width of cross section is sufficient to construct safely traversable slopes, an earth berm may be desirable in the median, outer separation, or border. The earth berm shields the freeway from view, lessens highway noises in the adjacent areas, and minimizes headlight glare. Adequate provisions must be made for drainage to ensure that water is not ponded on the shoulder area.

Figure VIII-18B shows a typical cross section without frontage roads. The clearance from the edge of the pavement to the right-of-way line is a minimum of 50 ft. The right-of-way width requirement is about 225 ft.

Restricted Cross Section and Right-of-Way

Figure VIII-19 illustrates restricted cross sections and approximate right-of-way widths for ground-level freeways. The section in Figure VIII-19A includes a 30-ft, two-way frontage road with a 10-ft border, a 24-ft outer separation, 12-ft lane widths, and a 10-ft median with four lanes or a 22-ft median with six or eight lanes, and a 20-ft clearance from the edge of freeway pavement to the right-of-way line on the side opposite the frontage road. Right-of-way width requirements are about 150 ft to 200 ft.

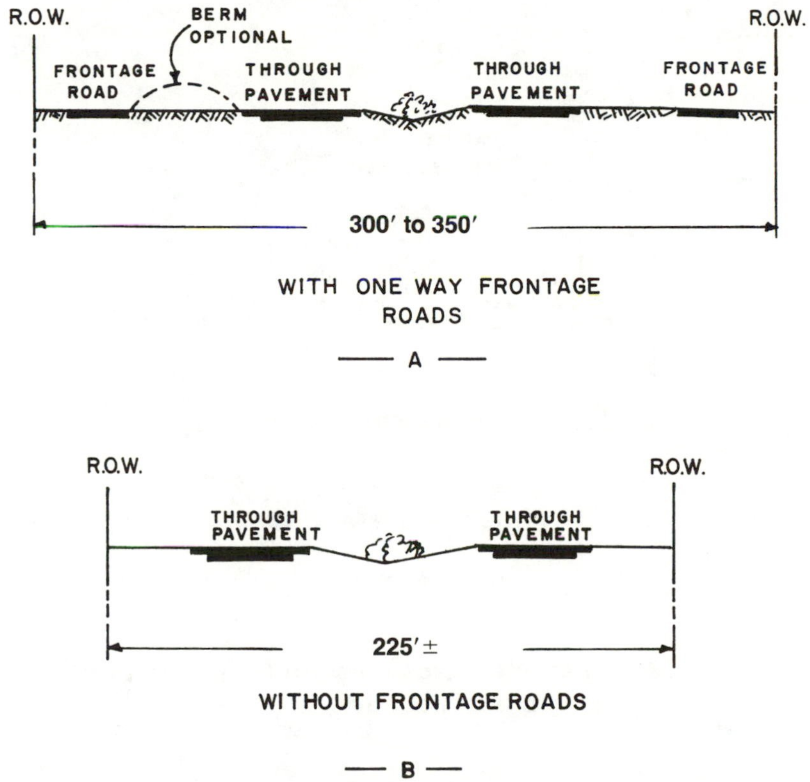

Figure VIII-18. Typical cross sections and right-of-way for ground-level freeways.

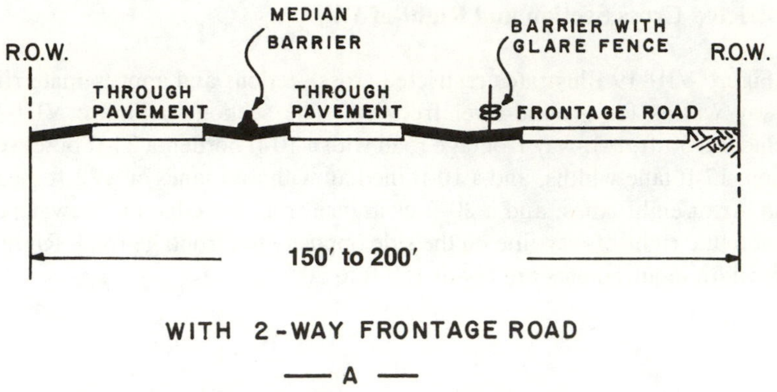

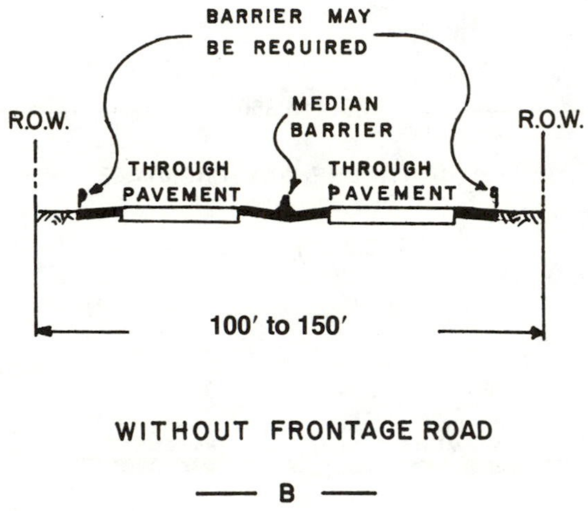

Figure VIII-19. Restricted cross sections and right-of-way, ground-level freeways.

With restricted cross sections, both the median and outer separation should be paved. On these narrow medians a median barrier is required. With two-way frontage roads, it is also desirable to provide a barrier in the outer separation in place of an access control fence. Preferably, the barrier should be located close to the frontage road to allow extra recovery space outside the freeway shoulders. Where there is no fixed-source lighting, a glare screen is desirable in the outer separation.

Figure VIII-19B shows a restricted cross section without frontage road. The section includes a 20-ft clearance from the outer edge of pavement to the right-of-way line, 12-ft lane widths, and a 10- to 22-ft median. Right-of-way width requirements are about 100 ft to 150 ft.

Figure VIII-20 is an example of a six-lane ground-level freeway without frontage roads on restricted right-of-way. The narrow median consists of two 10-ft paved shoulders and a concrete barrier. The barrier has been widened to accommodate double-masted light fixtures which illuminate both roadways.

Combination-Type Freeways

General Characteristics

In many cases, urban freeways are in part depressed, elevated, and at ground level. The variations and combinations of these types are treated in this section. Combination-type freeways result from variations in profile or cross section, and the following discussion is grouped on the basis of these two controlling conditions.

Profile Control

Rolling terrain. The typical plan and profile of a combination-type freeway in rolling terrain are shown in Figure VIII-21. The profile is developed to the best advantage by underpassing some cross streets and overpassing others. The facility is neither generally depressed nor elevated, although for short lengths it embodies the design principles for fully depressed or fully elevated highways. For instance, at A and C the facility is depressed, at B it is elevated on an earth embankment, and at each end of the illustration it approaches a ground-level section.

Between A and C the roadway is of fixed cross section with a narrow median, necessitated by lateral restrictions and cost of earthwork. Near each end of the illustration the profile and cross section are varied to fit cross slop-

Figure VIII-20. Ground-level freeway, restricted right-of-way without frontage roads.

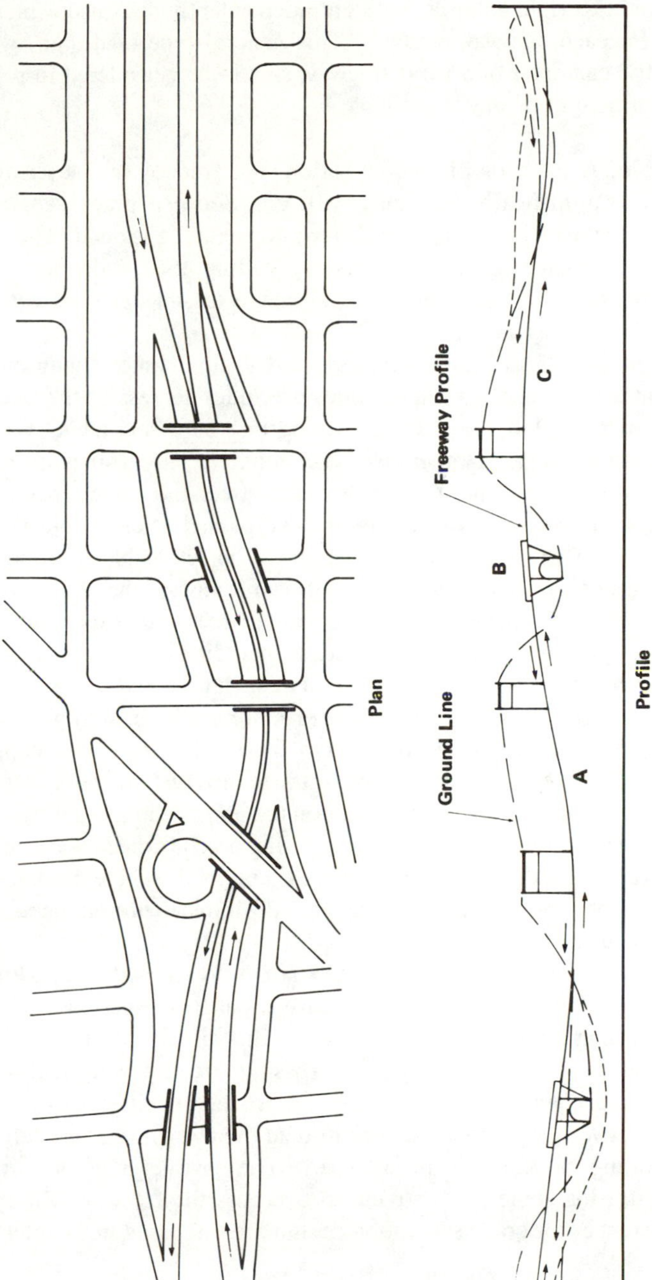

Figure VIII-21. Profile control—rolling terrain combination-type freeway.

ing terrain and less rigid controls, with an independently designed centerline and profile for each one-way roadway. This general type of design, which approaches the character of a rural freeway, should be considered in rolling terrain wherever right-of-way is available.

Flat terrain. A variation of a combination-type freeway in flat terrain is illustrated diagrammatically in Figure VIII-22. Between grade separation structures the highway profile closely follows the existing ground. The freeway overpasses important cross streets by rolling the gradeline to the required height above the surface streets, or where feasible, cross streets are carried over the freeway as at A.

This design fits flat terrain where soil and ground water conditions or heavy concentration of underground utilities precludes depressing the freeway to any great extent below the existing ground or where continuous viaduct construction is too costly or otherwise objectionable. The arrangement permits parallel or diagonal ramps to be provided between the grade separations. Where several successive cross streets are overpassed at close intervals, a roller coaster type of profile results, which is objectionable, particularly where the highway is on a tangent. On curvilinear highways the objection of a roller coaster profile is alleviated to some extent because drivers cannot see more than one or two grade separations ahead. In addition to poor appearance resulting from visual distortion, a succession of such high and low points can be disturbing to drivers when they see two or more grade separations ahead. A moving vehicle ahead can disappear into these blind dips and reappear again. The profile should be designed to eliminate dips that are more than 4 ft or so under the line of sight from any point on the approaching highway. Caution should also be exercised in designing a rolling profile grade to ensure that adequate vertical sight distance to exit ramps is obtained. Where truck traffic is heavy, grades of approximately 2 percent are desirable to prevent queuing at the base of grade changes.

The freeway may be carried over a cross street on an earth embankment with a conventional grade separation structure (as at B of Figure VIII-22) or on a relatively long structure (as at C). Availability of fill material and soil conditions would control. Depressing the crossing streets several feet below the ground surface and raising the freeway grade several feet above the ground level between separations will minimize the overall rise and fall and make the rolling profile less pronounced. The profile may be further improved by raising some cross streets to overpass the freeway. Minimum governing distances for grade separation design are discussed in Chapter X.

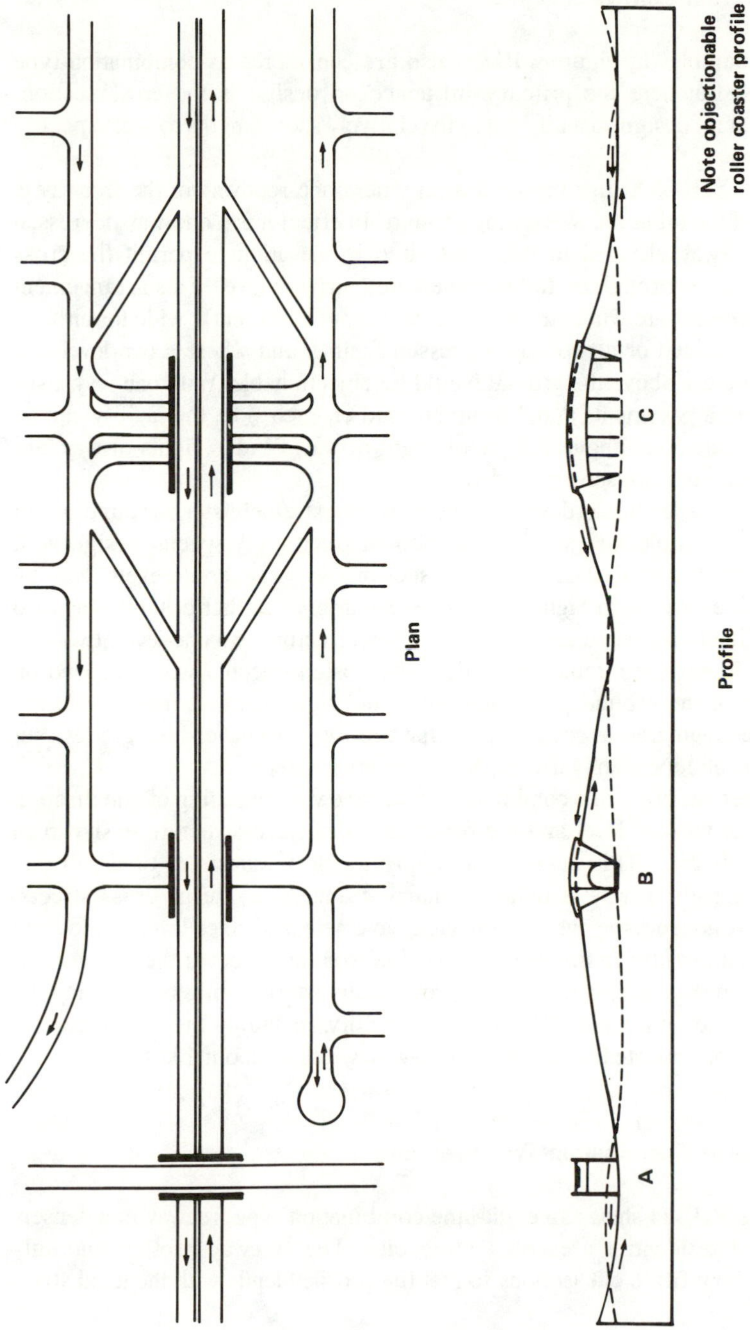

Figure VIII-22. Profile control—flat terrain combination-type freeway.

Cross-Section Control

The examples in Figure VIII-23 also are considered as combination-type highways, but here the primary influence on design is the cross section. These special designs usually apply to relatively short lengths to meet specific conditions.

Figure VIII-23A illustrates a design where one roadway of the freeway is above and the other below existing ground. In effect it is a one-way depressed and a one-way elevated facility, separated in elevation to permit the cross streets to pass through on the intermediate (surface) level. This arrangement may be appropriate where the right-of-way is not sufficiently wide for either a two-way elevated or a two-way depressed facility and where a two-level elevated structure above the ground would be objectionable. With one-way, two-lane through pavements and a frontage road on each side, the facility can be accommodated on about 125 ft of right-of-way. Wider rights-of-way are required where ramps are provided.

A city area left underdeveloped because of extremely steep ground might serve as a feasible location for a section of freeway. A special design with partly elevated and partly walled sections at staggered levels can be employed, as shown in Figure VIII-23B. A variety of other designs may also be used, including a one-level, two-way deck structure or a one-level, two-way cut-and-fill section retained by walls. The design selected would depend on the slope of the ground, soil conditions, and right-of-way width. Difficulty may be encountered at cross streets that are likely to have steep grades, but areas of such topography usually have few cross streets.

Another variation of a combination-type freeway consisting of one through roadway at surface level and the other on an elevated structure is shown in Figure VIII-23C. This design may apply along a waterfront or a railroad where the right-of-way is relatively narrow and there are no cross streets. Access to and from the one-way through pavement at surface level on the right is provided directly at streets that cross the frontage street at the left. Access to and from the elevated roadway is provided by lateral ramps overpassing the frontage road on the left. The complete facility, including the frontage road, can be accommodated within a right-of-way width of about 125 ft.

Examples of Combination-Type Freeways

Figure VIII-24 shows an eight-lane combination-type freeway in a densely developed residential area of a large city. The freeway profile smoothly changes from fill to cut sections so that the profile blends with the local street system.

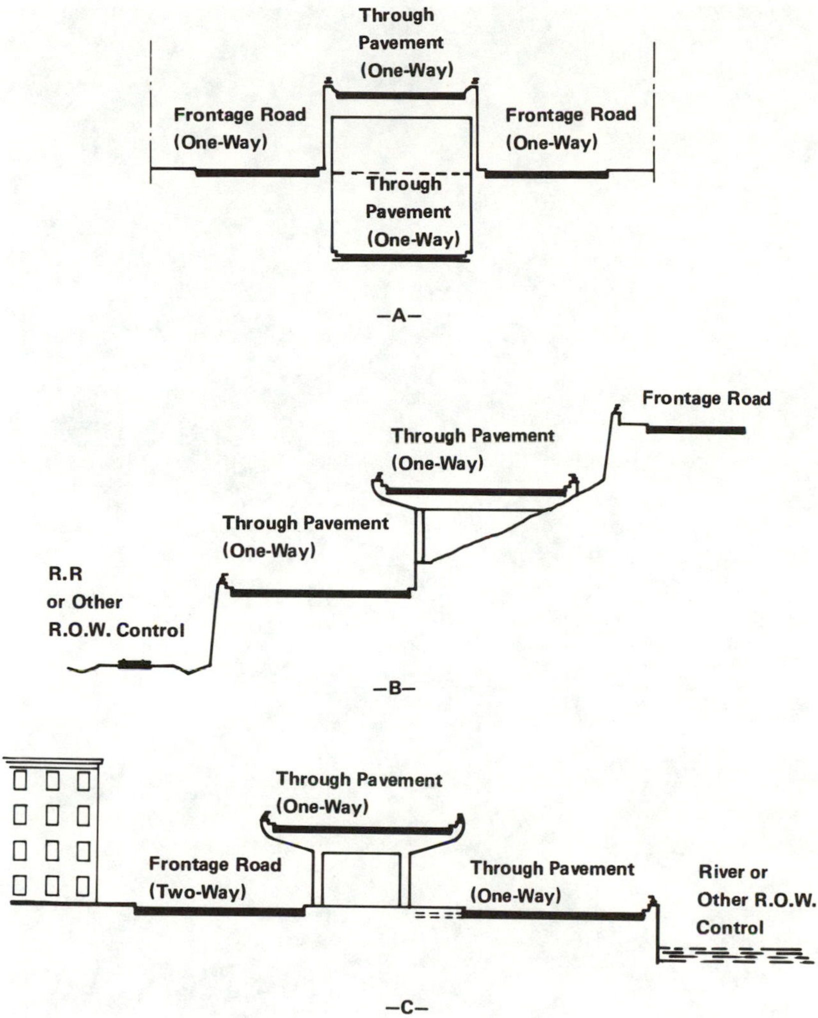

Figure VIII-23. Cross-section control combination-type freeway.

Figure VIII-25 shows a six-lane combination-type freeway through a developing suburban area (freeway not open to traffic). Vertical and horizontal alinements are of high standards. At the interchange in the lower center of the photograph, the freeway profile rolls over the cross street at ground level.

Figure VIII-24. Combination-type freeway.

The ramps are liberally designed. The entrance and exit ramps are two lanes wide at the cross street to provide for additional capacity and to improve operating conditions.

Figure VIII-26 shows a six-lane freeway within a very narrow right-of-way in a densely developed area of a large city. The four-level cantilever structure provided is similar to Figure VIII-23B. The second and third levels carry the directional roadways of the freeway. The top level is a promenade, which enhances the value of the adjacent apartment buildings and shields traffic noise. The ground level serves local traffic.

Figure VIII-25. Combination-type freeway.

Figure VIII-26. Four-level cantilevered freeway.

Special Freeway Designs

Reverse-Flow Roadways

Special conditions may suggest the use of a reverse-flow roadway as a part of the freeway design, which is usually accomplished by placing a separate reversible roadway within the normal median area, as shown in Figure VIII-27A. There are advantages and disadvantages of reverse-flow roadways. Reverse-flow roadways provide an opportunity for smoother operation for motorists, but they may have unused capacity much of the time because of the limited numbers of points of access. The costs of construction, maintenance, and operation for a freeway with a reverse-flow roadway also may differ considerably from those of a conventional freeway. A separate reverse-flow roadway should be considered when (1) the directional distribution during peak hours is substantially unbalanced (e.g., a 65:35 percent split) and traffic volumes require a conventional facility more than eight lanes wide, (2) design controls and right-of-way limitations are such that providing two or more par-

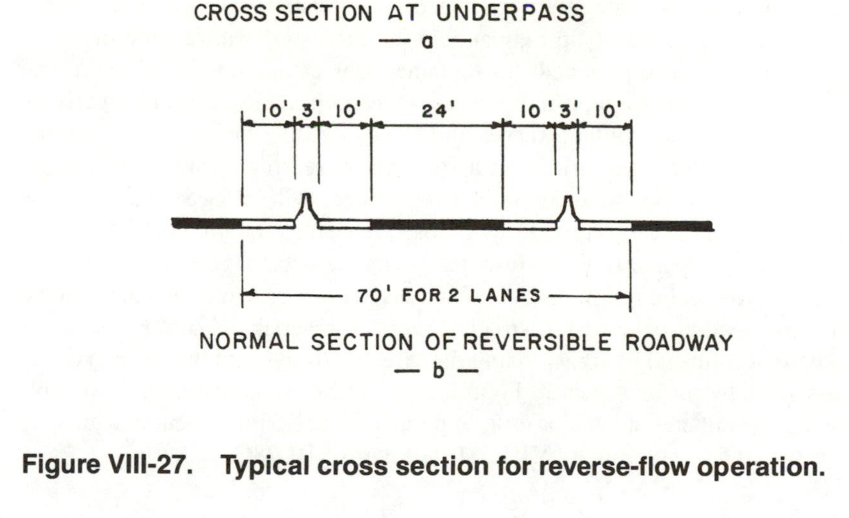

Figure VIII-27. Typical cross section for reverse-flow operation.

allel facilities on separate rights-of-way is not feasible, and (3) a sizable portion of traffic in the predominant direction during peak hours is destined for an area between the central portion of the city or another area of concentrated development and the outlying area; i.e., a large percentage of peak-hour traffic travels a long distance between principal points of origin and destination with little or no need for intermediate interchange.

In some large metropolitan areas, demand may be sufficiently great to justify the use of a reversible roadway exclusively for buses or other high-occupancy vehicles.

The right-of-way width required for a reverse-flow-type freeway is not greatly different from that required for a conventional-type freeway serving an equivalent traffic volume. In fact, with the dimensions shown in Figure VIII-27B, the right-of-way required for a three-two-three reversible freeway is the same as that required for a conventional 10-lane freeway with a 24-ft median. The cross section in Figure VIII-27B uses full right and left shoulders on the reverse-flow roadway because it carries one-way traffic in different directions at different times.

In the central business district it may be desirable to provide a separate collection and distribution system for the reversible roadway on radial freeways. The normal directional roadways would serve through traffic and freeway-freeway interchanges. Only the reversible roadway would connect directly to the downtown streets. This arrangement would enhance the usefulness of the reversible roadway for express bus operation.

Entrance and exit ramps on the reverse-flow roadway should be well spaced, and the entering traffic volume should be balanced with roadway capacity. Where there are major connectors, the design should provide for ramps going to and from the reverse-flow roadway and separated in grade from the outer freeway roadways. In cases to date, very few intermediate crossover connections (slip ramps) between the inner roadway and the outer roadway have been provided. Such connections entail substantial width and length for proper design, usually in areas where needed space is not available. Furthermore, the resultant weaves and left-side exits or entrances are unsafe and operationally undesirable on a freeway warranting a reverse-flow roadway. In reverse-flow operations, normally there will be two intervals daily during which the central roadway is closed to change the direction of flow.

Adequate reverse-flow roadway terminals are necessary to transfer traffic between the section of highway with reverse-flow lanes and the conventional freeway section or the local street system. A reversible roadway section is usually terminated by transitioning the three roadways into two normal directional roadways, as shown in Figure VIII-28A. In some cases the reversible section terminates at a major fork, and the terminal arrangement may be similar to that shown in Figure VIII-28B or Figure VIII-28C.

As illustrated in Figure VIII-28A, the end of the reversible roadway is Y-shaped and has the entrance and exit connections on the median side of the normal roadways. The entrance connection leading into the reversing roadway is relatively easy to provide, and there are usually no operational problems at this point. However, the exit connection from the reversing roadway

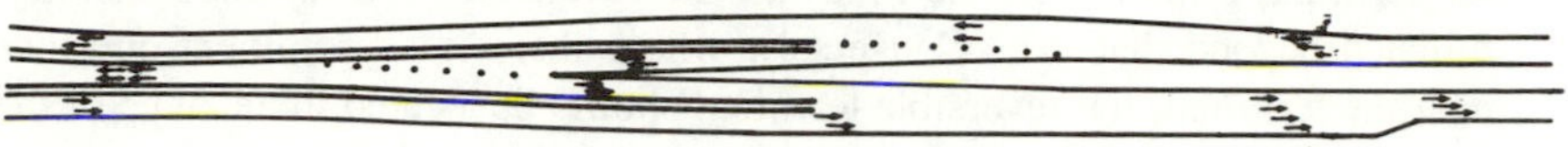

Common Terminal

-A-

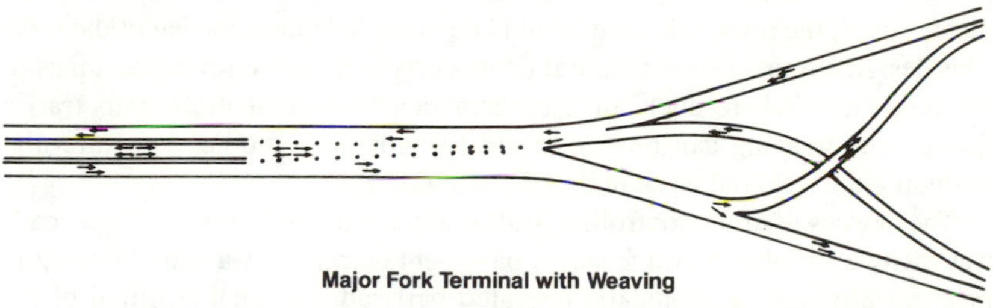

Major Fork Terminal with Weaving

-B-

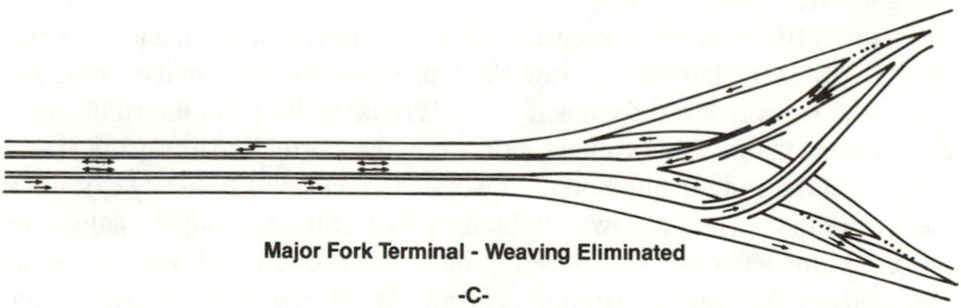

Major Fork Terminal - Weaving Eliminated

-C-

Note: Dotted lines represent removable barriers

Figure VIII-28. Diagrammatic reverse roadway terminals.

calls for careful consideration of many factors to prevent undesirable merging situations and backups during peak flows. As a minimum, the connections should be designed as major forks 1,200 to 1,800 ft long. Preferably, additional lanes should be provided on the normal roadway beyond the point of junction to the next exit or exits or for a distance of 2,500 to 3,000 ft. Such lanes will provide for adequate merging.

Where there is a prominent exit from the through lanes in the vicinity of the terminal of the reversible roadway, the reversible roadway should be terminated beyond that exit. Conversely, where there is a prominent entrance near the terminal, the reversible terminal should be located in advance of the entrance. This arrangement minimizes congestion and eliminates weaving conflicts.

Where the reversible lanes terminate at a major fork on the freeway, the single-structure arrangement shown in Figure VIII-28B would require weaving for traffic entering or exiting from the reversible roadway. This requirement is not desirable if weaving movements are heavy. Such designs may give rise to considerable operational problems. When the single-structure arrangement is used, the reversible lanes should be extended along one leg of the fork. This design eliminates weaving but denies access from the reversible lanes to the other leg. Where such an arrangement is not compatible with traffic desires, the weaving can be eliminated by providing another structure and designing the terminal as shown in Figure VIII-28C.

The devices used for controlling traffic at terminals of a reverse-flow roadway include variable message signs, pavement markings, warning lights, and mechanically and electronically operated barricades at each terminal of the reverse-flow roadway and at intermediate ramp terminals. Such devices have been under development for years, and several types of installations now in operation are being evaluated.

Figure VIII-29 shows a freeway having four traffic lanes in each direction and two reversible lanes in the middle. During the morning and evening rush hours, the freeway has the capacity of a 12-lane facility. On the right side of the freeway there is a two-lane exit from the normal roadway located in advance of the exit terminal on the reversible roadway. The heavy exit movement from the normal roadway lightens traffic at the reversible roadway terminal. On the left side of the freeway there is a two-lane entrance to the normal roadway downstream from the entrance to the reversible roadway. Thus, the reversible roadway removes considerable through traffic from the normal roadway, and this decrease in traffic results in an easier merge at the two-lane entrance. The reversible terminal in the middle of the photograph is an example of the common Y-shaped terminal, similar to that shown diagrammatically in Figure VIII-28A.

Figure VIII-29. A 4-2-4 reverse flow freeway.

Figure VIII-30 shows a three-two-three lane reverse-flow section in a suburban area. The reverse-flow roadway is 24 ft wide and has 10-ft shoulders on both sides. The normal directional roadway is 36 ft wide and has a 10-ft shoulder on the right and a 6-ft shoulder on the left. Each separator between the reverse-flow roadway and the normal roadway is 4 ft wide. Within each separator is a barrier 2 ft wide.

Dual-Divided Freeways

Where more than eight through lanes are required and the directional distribution is sufficiently balanced so that a reversible roadway is not applicable, a dual-divided freeway made up of two one-way roadways in each direction may be the optimum facility. All four roadways are within the

Figure VIII-30. A 3-2-3 reverse flow freeway.

control-of-access lines. This type of cross section is sometimes referred to as dual-dual. The outer freeway roadways usually serve all the interchange traffic, but they may also serve a substantial portion of the through traffic. For example, all trucks might be required to use the outer roadways only. Various arrangements are possible, depending on the character of traffic and crossroad conditions.

Dual-divided freeways usually function smoothly. They carry extremely high volumes of traffic efficiently. Motorists using the inner roadways are removed from the weaving movements at frequently spaced interchanges, and disabled vehicles in either the inner or outer roadway can quickly be steered to a shoulder by traversing a minimum number of lanes.

Dual-divided construction may be the most practical solution to widening an existing freeway where the present traffic volumes are so great that the disruption in traffic during complete reconstruction could not be tolerated. Where the future need can be anticipated and the necessary right-of-way can be reserved, it is possible to develop a dual-divided facility in two stages.

Dual-divided facilities have great flexibility in their operation and maintenance. For example, during maintenance or reconstruction operations, one of the directional roadways may be temporarily closed during off-peak hours. The accident potential is greatly reduced by eliminating traffic conflicts with the construction or maintenance work. Another advantage of dual-divided design is that the affected roadway can also be closed in case of an accident or other emergency, thus facilitating cleanup operations.

Dual-divided facilities also present disadvantages. The wide expanse of pavement and heavy traffic volumes may have a disruptive influence on an established community and tend to cut off the continuity of the area. The dual-roadway system reduces the flexibility of traffic to distribute, and as a result, uneven traffic distribution among lanes may be experienced. Because of the requirements for other medians, median barriers, and additional paved shoulders on a dual-divided facility, the costs for right-of-way, construction, and maintenance may be greater than those on a normal divided facility with an equal number of lanes.

Roadway arrangement for a dual-divided freeway are four-four-four-four, three-three-three-three, three-two-two-three, two-three-three-two, or other suitable combinations of lanes. Typical cross sections would be comparable to those previously described for various types of freeways with frontage roads, except that there are four, rather than two, main roadways. Each of the outer medians would have a median barrier and full shoulders on each side.

Figure VIII-31 shows a diagrammatic layout for a dual-divided freeway. All interchange connections are made to the outer roadways. Intermediate transfer connections are shown between the inner and outer roadways so that traffic on the inner roadways can use the interchanges. The number of such transfer connections should be kept to a minimum, one set serving several successive interchanges. There should be a spacing of 2,500 ft or more between the terminal of a transfer connection and an exit ramp. The adequacy of all weaving lengths must be checked.

Figure VIII-32 is an example of a dual-divided freeway with four-three-three-four roadway arrangement plus frontage roads. Rail rapid transit is carried in the median. Pedestrians have access to the transit station from the cross street. All ramp connections are made to the outer roadways.

Collector-Distributor Roads

An arrangement having cross-sectional elements similar to the dual-divided freeway is the collector-distributor (C-D) system. The purpose of a collector-distributor road is to eliminate weaving and reduce the number of ingress and egress points on the through roadways while satisfying the demand for access to and from the freeway. Collector-distributor roads may be provided within a single interchange (as discussed in Chapter X), through two adjacent interchanges, or continuously through several interchanges of a freeway segment. Continuous collector-distributor roads are similar to continuous frontage roads except that access to abutting property is not permitted.

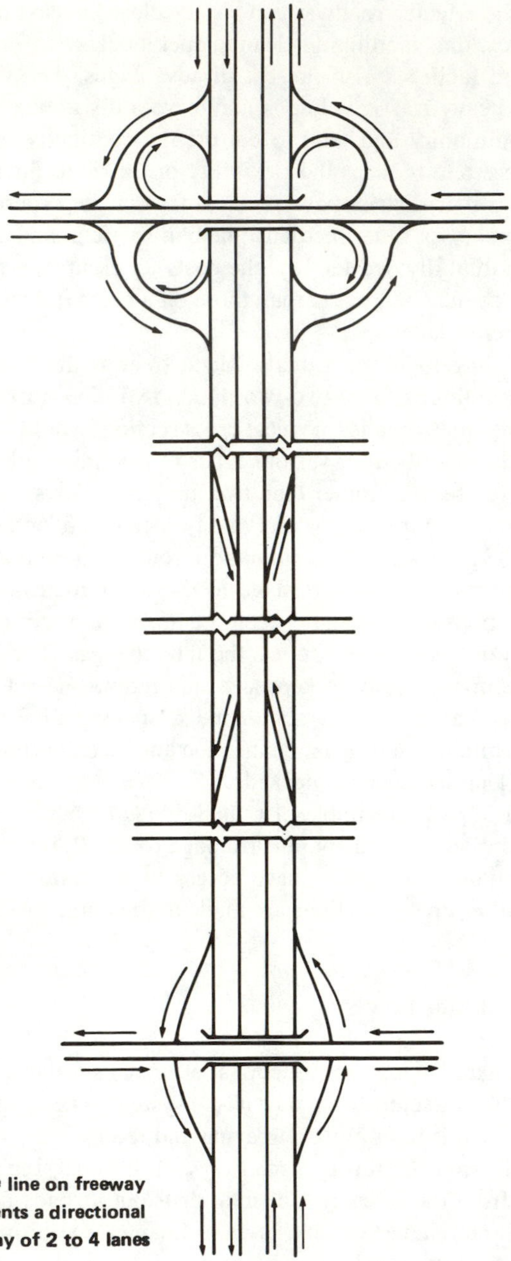

Note: Single line on freeway represents a directional roadway of 2 to 4 lanes

Figure VIII-31. Diagrammatic dual-divided freeway.

Figure VIII-32. Dual-divided freeway with a 4-3-3-4 roadway arrangement.

The inside high-speed through roadways are identified as core roads, and the outside slower speed roadways are identified as collector-distributor roads. Usually, the traffic volumes on the collector-distributor system are less than those encountered on the dual-divided freeway, and there are fewer lanes. The minimum lane combination on a collector-distributor system is two C-D, two core, two core, two C-D although other combinations may be developed as capacity requirements warrant. Continuous collector-distributor roads should be integrated into a basic lane design to develop an overall system. Capacity analysis and basic lane determination should be performed for the overall system rather than for the separate roadways.

Connections between the core roadways and collector-distributor roads are called transfer roads. Transfer roads may be either one lane or two lanes in width, and the principle of lane balance applies to the design of transfer roads on both the core roadways and the collector-distributor roadways. Both transfer roads and collector-distributor roads should have shoulders equal in width to those of the core roadways. The outer separation should be as wide as practical with an appropriate barrier. Terminals of transfer roads should be designed in accordance with the discussion of ramp terminals in Chapter X.

The design speed of collector-distributor roadways is usually less than that of the core roadways because most of the turbulence caused by weaving occurs on the collector-distributor roadways. A 10-mph reduction in design speed is preferably the maximum for continuous collector-distributor systems.

Accommodating Transit and High-Occupancy Vehicle Facilities

General Considerations

Combining mass transit or high-occupancy vehicle (HOV) facilities with the freeway may be a means for optimum transportation services in larger cities. This improvement can be accomplished by the joint use of right-of-way to include rail transit or separate roadway facilities for buses and other high occupancy vehicles such as car pools and van pools. The total right-of-way cost not only will be less than those for two separate land strips but the combination also will preserve taxable property, reduce the displacement of businesses and persons, and lessen impact on neighborhoods. In some cases mass transit can be incorporated into existing freeway systems. Reverse-flow roadways in the median and reserved lanes will work well for exclusive bus and HOV use during rush hours.

When transit, either bus or rail, occurs within the freeway median, access to the transit vehicles is generally obtained from the crossroad at interchange locations. Such an arrangement does not lend itself to intermodal transfer. Transfer to and from bus or private auto adds congestion to the interchange area, and off-street parking is usually so remote from interchange areas that it discourages some transit ridership. Bus roadways within the median, as shown in Figure VIII-33A, essentially restrict operations to the line-haul or express type, because ramps that would permit collection and distribution from the median area are expensive or operationally undesirable. When freeways undergo major repair or reconstruction, it is frequently necessary to construct crossovers and temporarily shift all traffic onto one roadway. Where transit is located within the median, such staging is not possible without a complete disruption of transit operations.

When the transit facility is parallel to the freeway but located to one side rather than in the median area, these objections are overcome. Figure VIII-33B shows a bus roadway located between the freeway and a parallel frontage road. Access to the bus roadway is obtained from the frontage road. The station is removed from the congestion of the interchange area, adequate space is available for auto or bus turnouts, and space for off-street parking may be more readily available. All factors combine to enhance intermodal transfer. Slip ramps from the bus roadway to the frontage road permit collec-

Figure VIII-33A. Exclusive bus roadway.

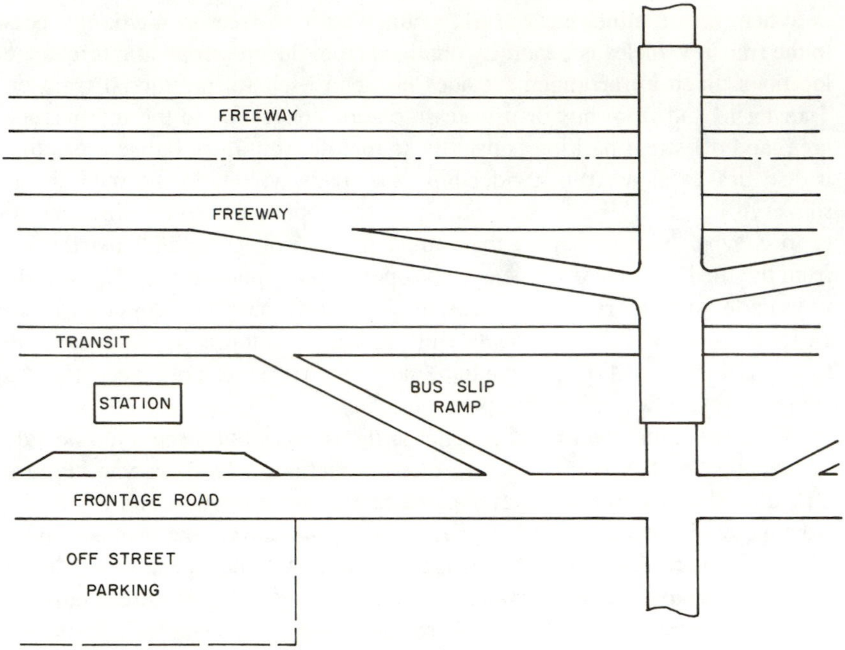

Figure VIII-33B. Freeway-transit outer parallel arrangement.

tion and distribution, in addition to line-haul or express operation, without disruption of freeway operations. A similar arrangement would serve fixed-rail transit except that the slip ramp would be omitted.

Buses

A truly rapid transit service by bus has had only limited application because normal bus service usually combines collection and distribution with suburb-to-city transportation and most street or highway facilities for such bus routes are not adaptable to high-speed operation. Many metropolitan areas have nonstop freeway express buses that operate on the conventional freeway roadways from suburban pickup points near the freeway to locations within the central business district or to other heavy traffic generators. The number of buses operating during peak hours, the spacing of bus stops, and the design of bus turnouts determine the efficiency of bus operation and its effect on highway operations. Buses operating with short headway and frequent pickup and discharge points are likely to accumulate at the stops and interfere with through traffic. Express bus operation with few, if any, stops along the freeway, on the other hand, gives superior transit service for the outer urban area and affects freeway operation the least.

Exclusive HOV lanes. In addition to express service, other operational means should be considered to reduce the travel time of the public transportation user when the demand warrants. An exclusive HOV roadway is an entire highway facility reserved at all times solely for the use of buses or buses and other high-occupancy vehicles. This facility offers buses and HOVs a high level of service and decreases the travel time for the users. A discussion of HOV lanes and roadways is contained in the AASHTO *Guide for the Design of High-Occupancy Vehicle and Public Transfer Facilities* **(3)**.

Bus stops. The spacing of bus stops largely determines the overall speed of buses. Bus stops on freeways should be spaced to permit buses to operate at or near the prevailing speed of traffic on the highway. This allowance normally requires spacing of 2 mi or more.

Bus stops along freeways are usually located at intersecting streets where passengers transfer to or from other lines or automobiles. Bus turnouts should be located where site conditions are favorable and, if possible, where gradients on the acceleration lane are flat or downward. Bus stops may be provided at the freeway level for which stairs, ramps, or escalators are necessary, or they may be located at the street level that buses reach by way of interchange ramps. The design of bus turnouts is discussed in Chapter IV.

Stairs, ramps, and escalators. With bus stops at the freeway level, stairs, ramps, escalators, or combinations of these are necessary for passenger access between the freeway and local street levels. Stairways and ramps at transit stops should be easy to climb and present an inviting appearance. This effect is partially accomplished by providing railings and ample lighting both day and night and by providing landings at every 6- to 8-ft change in elevation. A covering on the stairways, ramps, and platforms may also be desirable. Stairways should be located where the climb will be a minimum, preferably not more than 18 or 20 ft. Where space is available, the bus roadway under the structure might be raised 2 to 4 ft by reducing vertical clearance to about 12.5 ft, because only buses are to be served. Most intracity buses are less than 10 ft high. When the stairs are located a little distance from the point of loading and unloading, the connecting walkway may be inclined at about a 4 percent grade, and another 1 or 2 ft may be gained in elevation. Thus, it may be possible in some instances to reduce the height of stairways to 15 ft or less.

Stairs are likely to be installed at bus stops in built-up districts. Pedestrian ramps are well adapted to bus stops in suburban or parklike areas. Railings are desirable and usually necessary. Combinations of ramps and stairs may be fitting at some locations. If the busline serves a large percentage of the elderly population, is extremely busy, or the climb is extra long, the use of escalators should be considered. Where provisions for the handicapped should be included, the use of ramps, elevators, widening of passageways and doors, and the elimination of other barriers, are effective. Reference should be made to Chapter IV for guidance on the design of facilities for the handicapped.

Bus stop arrangements. With bus stops at the freeway level, buses consume little additional time other than that for stopping, loading, and starting; however, turnouts, stairways, and sometimes extra spans at separations must be provided. Passengers must use stairs or ramps. With bus stops at street level, less special construction is necessary and passengers do not have to use stairs or ramps. Buses, however, have to mix with traffic on the ramps and frontage roads and generally must cross the intersecting street at grade. Where traffic on the surface streets is light, these disadvantages are not serious, but where the streets are operating at or near capacity, buses crossing them will experience some delay. Generally, street-level stops are appropriate in and near downtown districts, and either street- or freeway-level stops are appropriate in suburban and outlying areas. Combinations of the two types may be used on any one freeway.

Stops at freeway level. Bus stops logically are located at street crossings where passengers can use the grade-separation structure for access from either side of the freeway. Figure VIII-34A shows an arrangement at an overcrossing street without an interchange. The turnouts and loading platforms are under the structure, requiring greater spans or additional openings. Each stairway should be located on the side of the cross street used by most passengers. Two additional stairways can eliminate any crossings of surface streets by transferring riders.

Figure VIII-34B shows an arrangement at an undercrossing street without an interchange. As indicated at the top left of Figure VIII-34B, platform exits and entrances may be connected directly to adjoining developments such as public buildings and department stores.

Sometimes transit stops are needed at locations other than at overcrossing streets, as in outlying areas or in built-up districts where it is neither feasible nor desirable to provide stops at cross-street structures. Such stops preferably should be located opposite cross streets intercepted by the frontage roads or major passenger walkways. A pedestrian overpass is necessary to make the bus stops usable from either side of the freeway. Figure VIII-34C illustrates

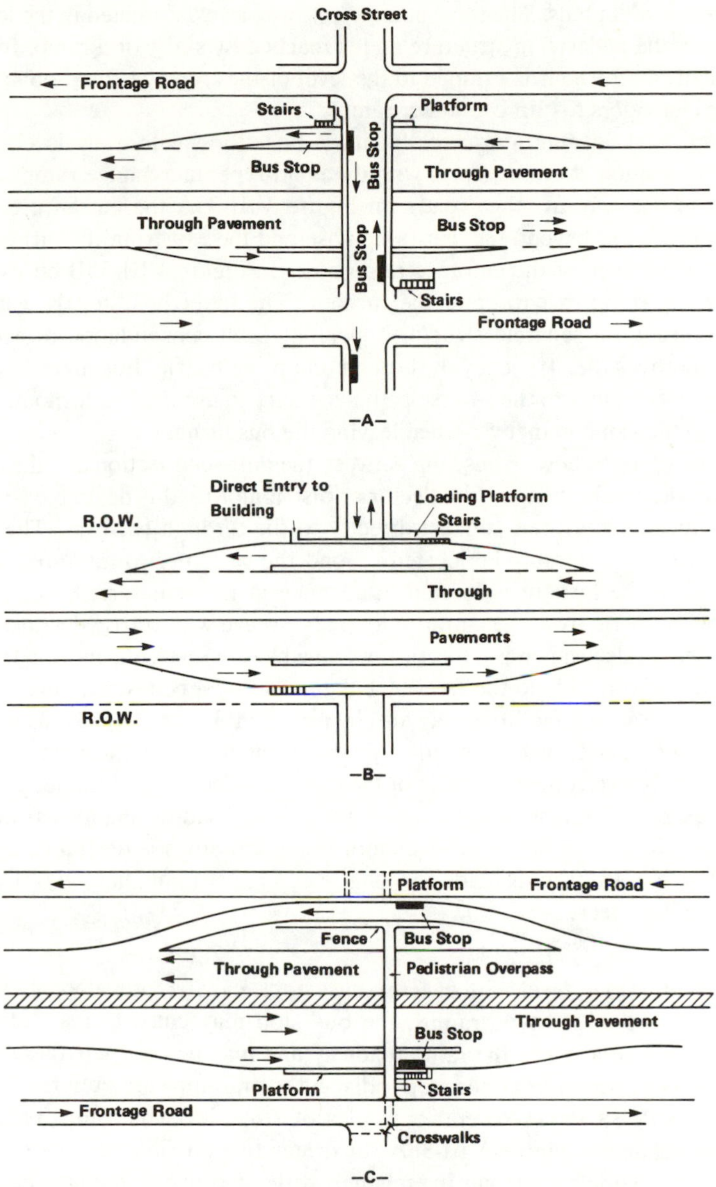

Figure VIII-34. Bus stops at freeway level.

two likely layout plans. The turnout at expressway level is situated in the lower half under the pedestrian structure and is reached by stairs or a ramp. In the upper half, the turnout is ramped to the level of the frontage road, obviating the need for riders to climb stairs or ramps.

Figure VIII-35 diagrammatically shows bus stops at freeway level on a depressed section at cross streets with diamond-type interchange ramps connecting to one-way frontage roads. In Figure VIII-35A the entrance to the turnout is located beyond the exit ramp nose and the exit from the turnout is located in advance of the entrance ramp nose. In Figure VIII-35B buses use the freeway ramp exit to enter the turnout. The latter bus stop is usually located through a separate structure opening. Such consolidation of access points improves the efficiency of through and ramp traffic. Bus drivers readily adapt themselves to the reverse path necessary to use the bus turnout. The buses use the same maneuver when leaving the bus turnout.

Figure VIII-36 shows a bus stop between the outer connection and the loop of a cloverleaf interchange. A collector or distributor road is desirable so that the bus turnouts will not connect directly to the through roadway. The bus turnout should preferably be located beyond the structure to minimize conflicts. When the turnout is located in advance of the structure, buses must merge with traffic from the entrance loop and weave with traffic destined for the exit loop. Also shown is a bus turnout on a nearby service drive of a frontage road that connects to the freeway bus turnout by a pedestrian underpass or overpass. Parking facilities may also be provided. Figure VIII-37 shows an existing interchange with a similar arrangement but with turnouts on the cross street between the terminals of the outer connection and the loop. The circuitous pedestrian walkways (white lines from loading platforms) avoid grade crossings of ramps. Elimination of grade crossings is desirable, but if the pedestrian walkway is made too long, e.g., greater than about 700 ft, bus patronage may decrease.

Stops at street level. Street-level bus stops can be provided at interchanges. On diamond-type ramps, the bus stop may consist of a widened shoulder area adjacent to the ramp roadway and may be on a separate roadway. Generally, street-level bus stops adjacent to on-ramps are preferred.

Figure VIII-38 shows several examples of street-level bus stops on diamond interchanges. Figure VIII-38A illustrates two possible locations for a bus stop at a simple diamond interchange without frontage roads. The bus stop can be provided on either the on-ramp or off-ramp by widening the ramp. The decision of which is better depends on an analysis of the turning conflicts.

Figure VIII-38B illustrates a street-level bus stop on a one-way frontage road at a diamond interchange. Buses use the off-ramp to reach the surface

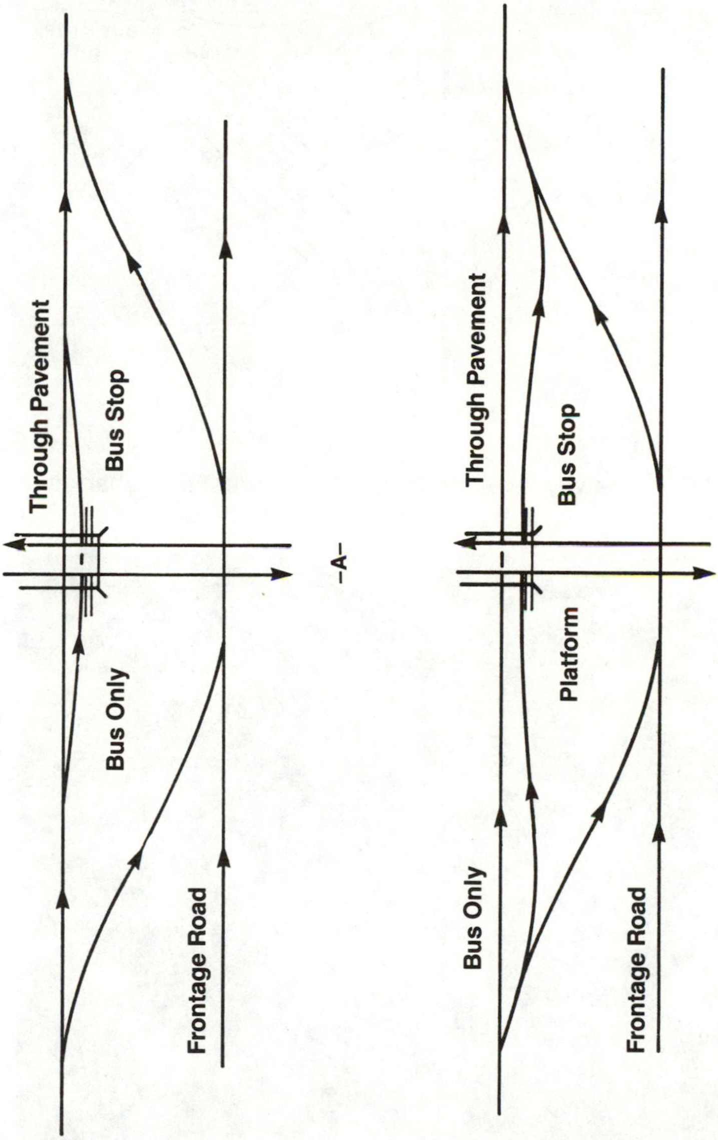

Figure VIII-35. Bus stops at freeway-level diamond interchange.

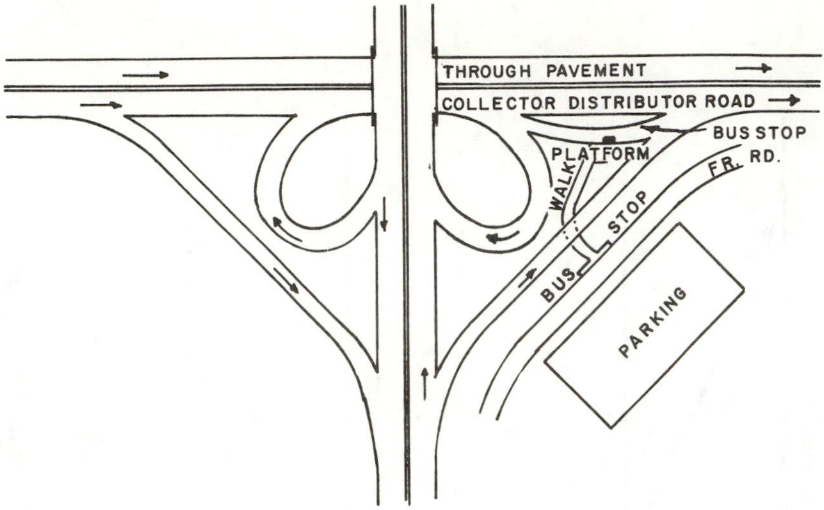

Figure VIII-36. Freeway-level bus stop at cloverleaf interchange.

Figure VIII-37. Freeway-level bus stop at cloverleaf interchange.

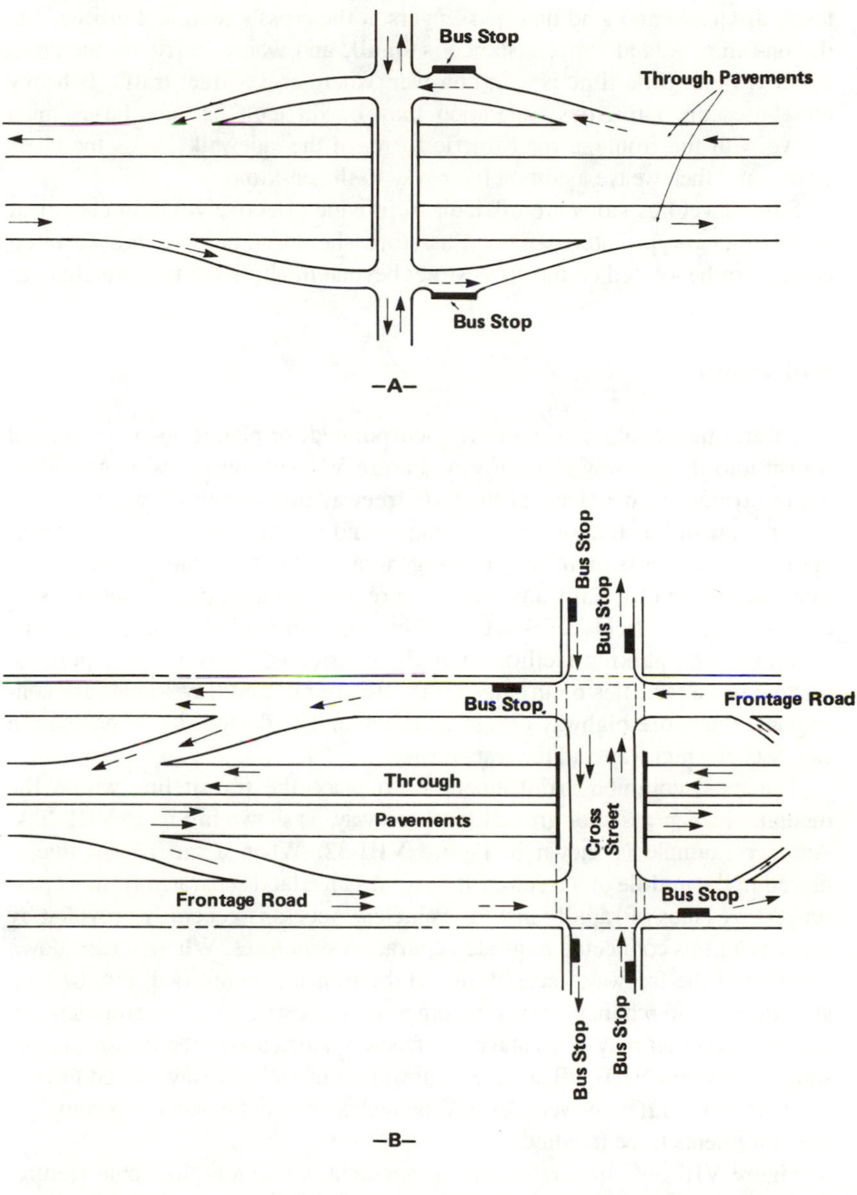

Figure VIII-38. Bus stops at street level on diamond interchange.

level, discharge and load their passengers at the cross street, and proceed via the on-ramp. Added travel distance is small, and where traffic on the cross street is light, little time is lost; however, where cross-street traffic is heavy and buses are numerous, operation may be difficult because buses must weave with the frontage road traffic to reach the sidewalk, cross the cross street, and then weave again on their way to the on-ramp.

Street-level bus stops are difficult to provide effectively within cloverleaf or directional-type interchanges. Bus stops should be omitted at such interchanges or be located on the cross street beyond the limits of the interchange.

Rail Transit

Several metropolitan areas have incorporated, or plan to incorporate, rail transit into the freeway right-of-way. Figure VIII-39 illustrates diagrammatically various arrangements of the joint freeway-transit right-of-way.

Rail transit installations are so unique and their design is so highly specialized that discussion of only a few general items is in order herein. Location and design of a rail transit facility are joint undertakings involving several specialized fields of interest. The location and design of stations, terminals, and parking facilities must be considered from the standpoint of serving these facilities by urban streets. Where rail and travel lanes are contiguous, the entire highway design is affected. The design should ensure the safety of the highway and the transit user.

The most common arrangement is to place the transit line within the median of a depressed or ground-level freeway, as shown in Figure VIII-39A. Another example is shown in Figure VIII-32. When a rail transit line is placed in the middle of a freeway, it becomes an island separated from its passengers by lanes of rapidly moving vehicles. Access is generally provided by stairs or ramps connected to grade-separation structures. Where it runs down one side of the freeway, accessibility to the transit is simplified, but the construction of interchange ramps becomes more costly. In some situations an alternate solution may be to stack the freeway lanes above the transit line, as shown in Figure VIII-39B, to use a minimum of right-of-way. An additional level for cross traffic of vehicles and pedestrians may be needed to complete the movements to be handled.

Figure VIII-39C illustrates an arrangement where a topographic feature, such as the river on the right, presents a natural deterrent to development on one side. The transit line is placed on the inside for easier access to the community. Where the area has scenic value, this arrangement presents motorists with an open view. Figure VIII-39D illustrates an arrangement whereby the alinement and grade of the transit line and freeway can be adjusted to serve better a community on one side of the transit facility.

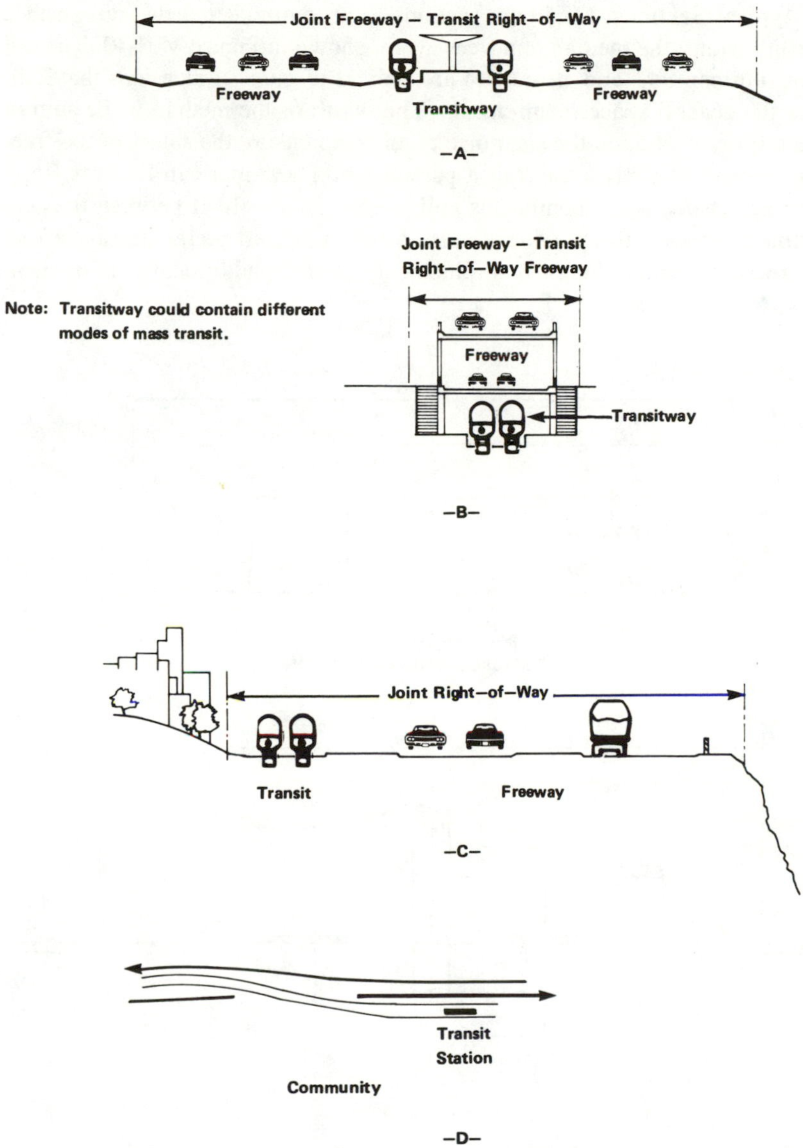

Figure VIII-39. Joint freeway-transit right-of-way.

Typical Sections. Minimum sections to accommodate a two-rail surface facility within the median of a freeway are shown in Figure VIII-40. The rail transit dimensions and clearances are typical of guide dimensions that indicate the general space requirements. The width of the median is the sum of the railway width and the clearance required to ensure the safety of the freeway drivers as well as the transit passengers. The American Railway Engineering Association recommends a distance of 13 to 14 ft between the centerline of adjacent tracks and requires an 8-ft minimum lateral clearance from the track centerline, for tangent alinements, with an additional 1 in. of clear-

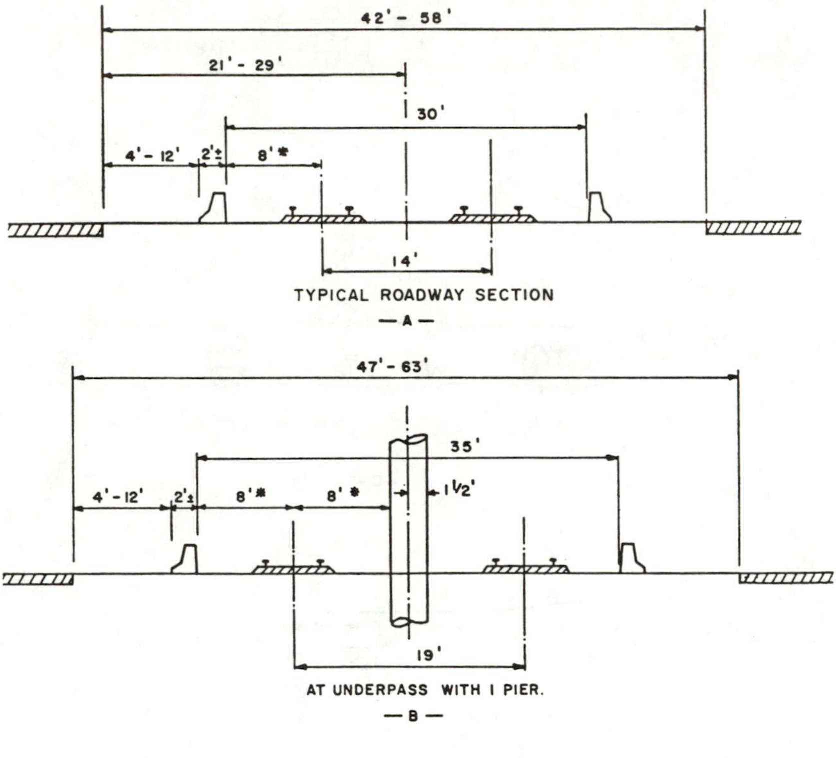

NOTE: RAIL TRANSIT DIMENSIONS AND CLEARANCES ARE NOT TO BE CONSIDERED AS STANDARD.

* ADD 1" FOR EACH DEGREE OF HORIZONTAL CURVATURE.

Figure VIII-40. Typical sections with rail transit in freeway median.

ance for each degree of curvature up to a maximum of an additional 18 in. of clearance. Figure VIII-40A illustrates a minimum section without piers. Figure VIII-40B illustrates a minimum section with one median pier. The piers should be provided with crash walls to deflect transit cars in the event of a derailment. A fence should be mounted on or adjacent to the barrier to prevent pedestrians from entering the rail area. A screen or fence may also be needed to screen the motorists from the headlights of the mass transit vehicle. If a semirigid barrier is utilized at the shoulder edge, the dynamic deflection of the barrier should be added to the dimensions shown.

Where the rail line is placed along one side of the freeway, the dimensions shown in Figure VIII-40 would be appropriate between the outer lane and the rail line, and ample clearance should be allowed between the rail line and the frontage or local road. A minimum of 10 ft should always be maintained between the freeway through lane and the barrier.

Stations. The transit station location and spacing should be in keeping with the environment and passenger flow. Frequent stations may be required within the central business district and other heavy traffic generators, but few stations would be required in the outlying or suburban areas. The downtown stations should be within easy walking distance of the business and working centers or a feeder bus system. The outlying stations should provide ample parking and storage for waiting cars; access to local buses and taxis also should be available. Two general layouts for a rail transit station at a local cross street or pedestrian overcrossing are shown in Figure VIII-41 with typical control dimensions. The median required for either type is approximately 80 ft. The station shown in Figure VIII-41B makes better use of the required space by allowing more separation between the transit passengers and the freeway and by still maintaining ample distance between the train and the traveled way.

Examples of rail transit combined with freeway. Figure VIII-42 shows an eight-lane depressed freeway with rail rapid transit in the median of minimum width. A single station is provided between two major cross streets. Pedestrian rampways connect to the cross streets, the shelters being at the cross-street level. The traffic density on the near cross street shows the general undesirability of pedestrian access on a cross street that functions as a part of freeway ramps.

Figure Figure VIII-43 shows a six-lane major street overpassing an eight-lane depressed freeway with rail transit in the median. A sheltered rampway at the right leads to a rail transit station in the median of the freeway.

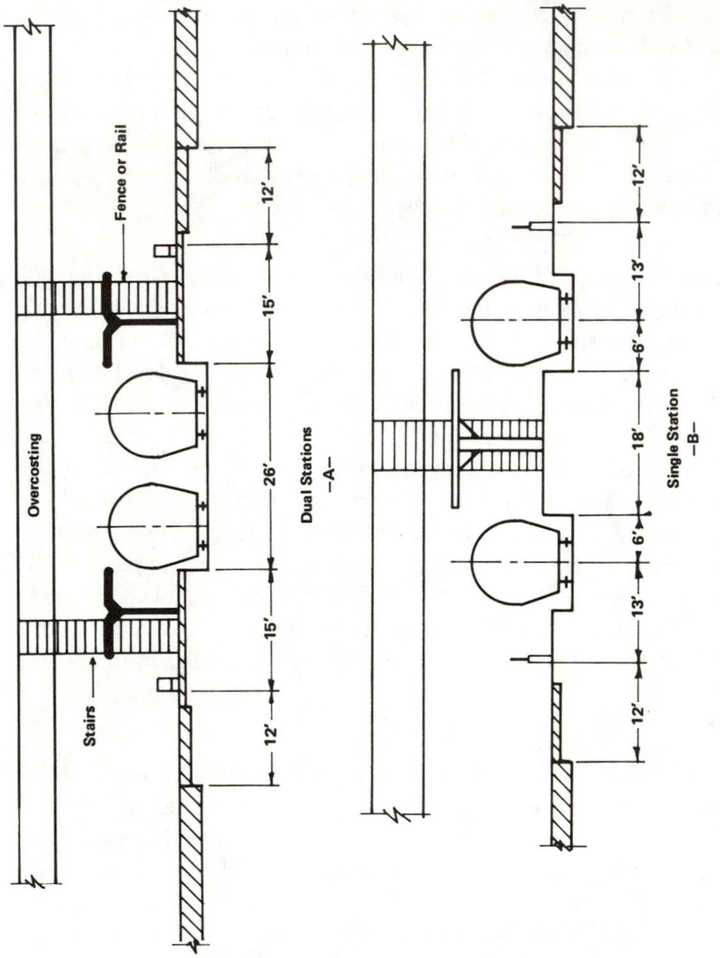

Note: Rail transit dimensions and clearances are not considered as standard.

Figure VIII-41. Transit station layout.

Figure VIII-42. Depressed freeway with rail rapid transit in the median.

Figure VIII-43. Depressed freeway with rail rapid transit in the median.

RURAL FREEWAYS

General Design Considerations

Rural freeways are similar in concept to urban ground-level freeways, but the alinement and cross-sectional elements are more liberal in design, which is commensurate with the higher design speed and, generally, with the greater availability of right-of-way.

Where terrain permits, a design speed of 70 mph should be used on rural freeways. A design speed of 70 mph will result in a safer freeway than one designed for a lower speed. In some cases it will also require fewer lanes to accommodate a given volume of traffic at the desired level of service, which for rural freeways is usually level-of-service B. In addition to the increased safety and capacity, for a given level of service higher design speeds result in a more comfortable driving environment and reduce both fuel consumption and operating costs. Freeways are initially designed to accommodate anticipated traffic for about 20 years and remain in service for a much longer time. Any cost-savings realized by initially constructing for lower design speeds may be outweighed by the high costs, disruption to the environment, and inconvenience to traffic that accompany the reconstruction of major facilities.

Although level-of-service B is desirable for rural freeways, level-of-service C may be necessary on auxiliary facilities where volumes are unusually high. Rural freeways generally have four through-traffic lanes except on approaches to metropolitan areas where six or more lanes may be provided. Interchanges are usually provided with intersecting highways classified collectors and higher. Local roads may be terminated at the freeway, connected to frontage roads or other local roads for continuity of travel, or carried over or under the freeway by grade separation with or without interchange.

Alinement and Profile

Being generally designed for high-volume and high-speed operation, rural freeways should have smooth flowing horizontal and vertical alinements. Proper combinations of flat curvature, shorter tangents, gentle grades, variable median widths, and separate roadway elevations all combine to enhance the safety and esthetic aspects of freeways. Advantage should be taken of favorable topographic conditions to achieve the desired goals. Changing median widths on tangent should be avoided where possible to avoid a distorted appearance.

Because the rural road network has a wider spacing than its urban counterpart, rural freeways are usually constructed near ground level with smooth and relatively flat profiles controlled by drainage and earthwork considerations and less by the need for frequent grade separations and interchanges. If the need for elevated or depressed sections should occur, the guidelines for urban freeways are appropriate.

Even though the profile may satisfy all the design controls, the finished vertical alinement may appear forced and angular if minimum criteria are used. Checking of profile designs in long continuous plottings, in which the foreshortened aspect can be simulated, can help avoid an undesirable roller coaster alinement in rolling terrain.

The relation of horizontal and vertical alinement should be studied simultaneously to obtain a desirable combination.

Medians

Median widths of about 50 to 90 ft are common on rural freeways. The 50 ft dimension shown in Figure VIII-44A provides for 6-ft graded shoulders and 6:1 foreslopes with a 3-ft median ditch depth. Adequate space is provided for vehicle recovery; however, median piers may require shielding for heavier traffic volumes in accordance with Chapter IV. The 90-ft dimension shown in Figure VIII-44B permits the designer to use independent profiles in rolling terrain to blend the freeway more appropriately with the environment while maintaining flat slopes for vehicle recovery. In flat terrain the 90-ft median is also suitable when stage construction requires the addition of two future 12-ft traffic lanes.

Where the terrain is extremely rolling or the land is not suitable for cultivation or grazing, or both, a wide variable median having an average of 150 ft or more, as shown in Figure VIII-44C, may be attainable. Such a width permits the use of independent roadway alinement, both horizontally and vertically, to its best advantage in blending the freeway into the natural topography. Foreslopes and backslopes used within the clear zone should provide for vehicle recovery. The remaining median width may be left in its natural state of vegetation, trees, and rock outcroppings to reduce maintenance costs and add scenic interest to passing motorists. The combination of independent alinement and a natural parklike median is most pleasing to motorists. For driver reassurance, the opposing roadway should be in view at frequent intervals.

In suburban areas where right-of-way restrictions dictate or in mountainous terrain where location is difficult, median widths in the range of 10 to 30 ft may be necessary, as shown in Figure VIII-44D. In such instances the median is usually paved and where the roadway pavements are crowned, underground drainage should be provided. Considering the usual developing-area traffic volumes as well as operational characteristics in mountainous areas, a median barrier is usually warranted as a safety measure.

Emergency crossovers on rural freeways are normally provided where interchange spacing exceeds 5 mi to avoid extreme adverse travel for emergency and law enforcement vehicles. Between interchanges, emergency crossovers are spaced at 3- to 4-mi intervals. Maintenance crossovers may be required at one or both ends of interchange facilities, depending on inter-

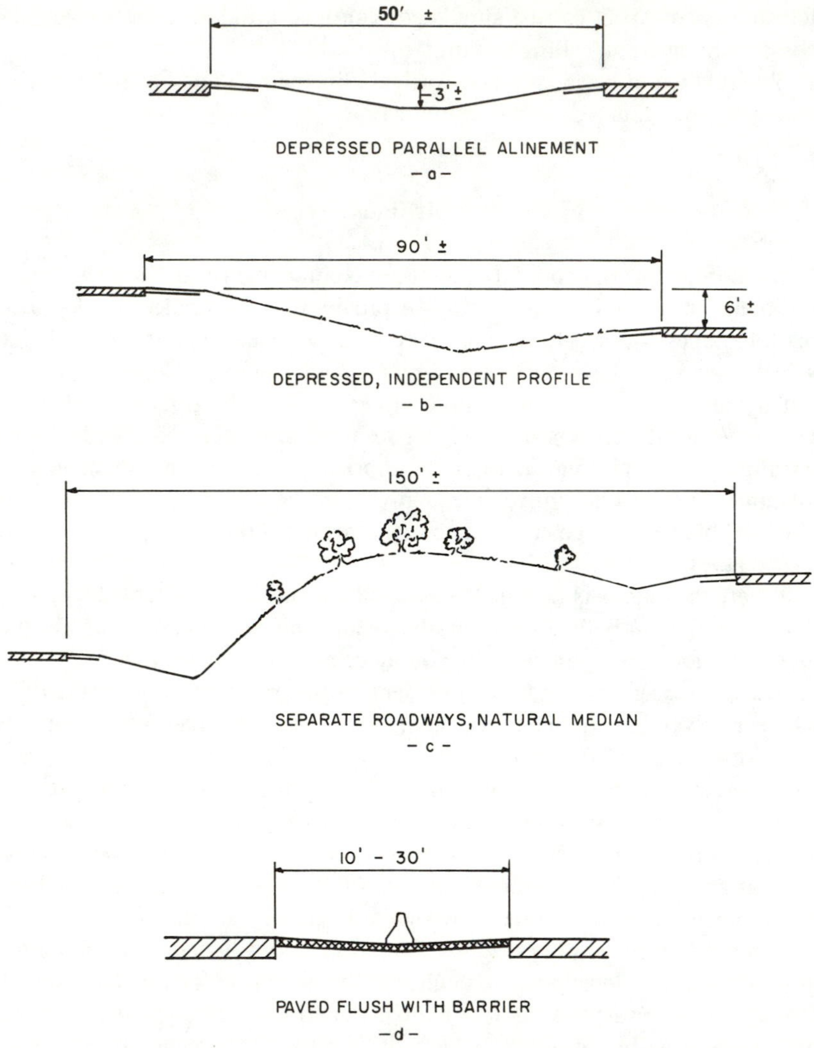

Figure VIII-44. Typical rural medians.

change type, for the purpose of snow removal and at other locations to facilitate maintenance operations. Maintenance or emergency crossovers generally should not be located closer than 1,500 ft to the end of a speed-change taper of a ramp or to any structure. Crossovers should be located only where above-minimum stopping sight distance is provided and preferably should not be located within curves requiring superelevation.

The width of the crossover should be sufficient to provide safe turning movements and should have a surface capable of supporting the maintenance equipment used on it. The crossover should be depressed below shoulder level to be inconspicuous to traffic and should have 10:1 or flatter sideslopes to minimize its effect as an obstacle to uncontrolled vehicles. Crossovers should not be placed in restricted-width medians unless the median width is sufficient to accommodate the vehicle length, say 25 ft or more. Where median barriers are employed, each end of the barrier at the opening may require a crashworthy terminal. Reference should be made to the AASHTO *Roadside Design Guide* (**4**).

Examples of Rural Medians

Figure VIII-45 illustrates a variable-width median with independently designed roadways.

Figure VIII-46 illustrates an 88-ft median in flat terrain with parallel alinement of the roadways.

Figure VIII-47 illustrates a narrow, flush, paved median with barrier used on an approach to an urbanized area where right-of-way is restricted.

Sideslopes

Flat, rounded sideslopes, fitting with the topography and consistent with available right-of-way, should be provided. Foreslopes of 6:1 or flatter are recommended in cut sections and for fills of moderate height as discussed in Chapter IV. Where fill heights are intermediate, broken back foreslopes may be used to provide a vehicle recovery area. For high fills, steeper slopes protected by guardrail may be necessary. Backslopes of 3:1 or flatter permit normal landscaping and erosion control practices and ease maintenance operations. In highly productive agricultural areas, steeper slopes may be used, but the combination of foreslope, backslope, and ditch configuration should permit vehicle recovery. Where rock or loess deposits are encountered, backslopes may be nearly vertical, but should be located where practical to provide an adequate recovery area for errant vehicles.

Figure VIII-45. Independently designed roadways with variable width median.

Figure VIII-46. Rural freeway with 88-ft median through flat terrain.

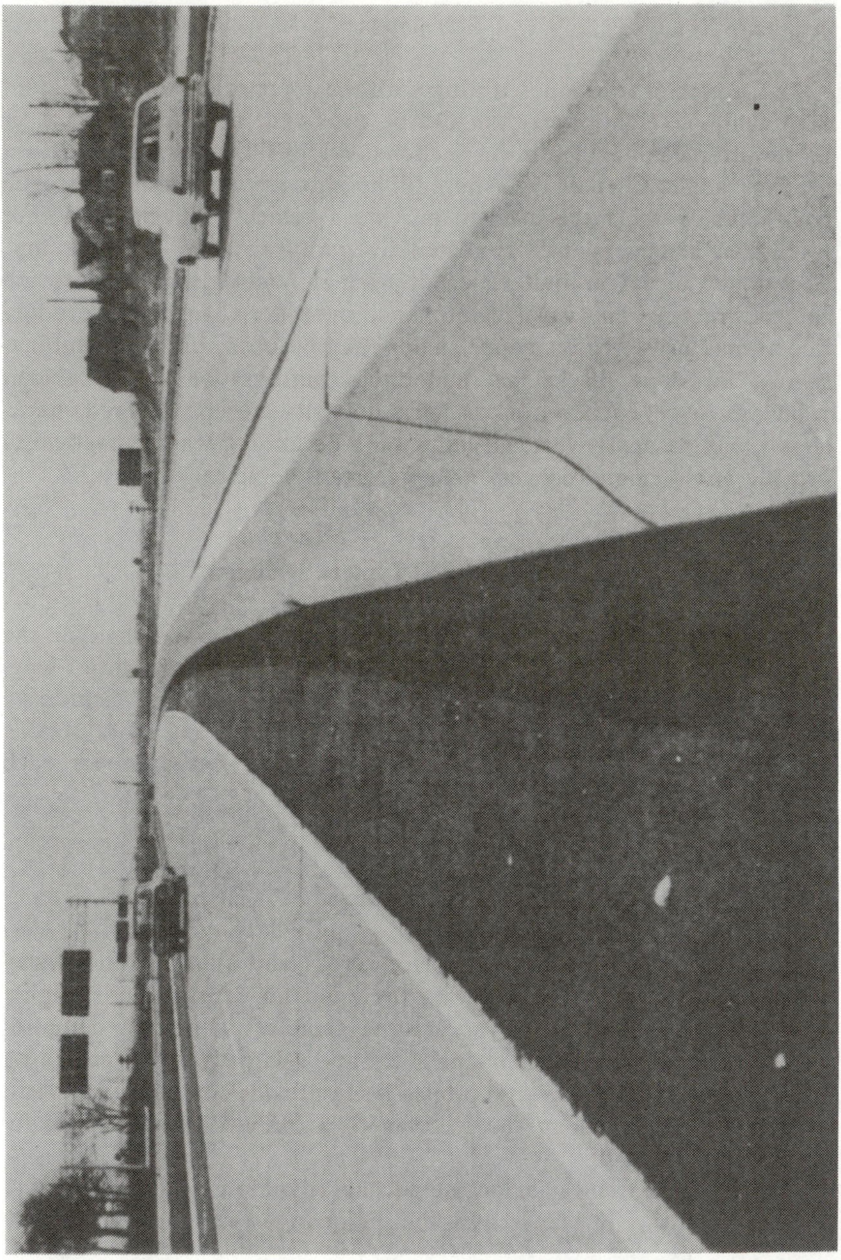

Figure VIII-47. Freeway approaching urban area with narrow paved median and barrier.

Typical Cross Section and Right-of-Way

Cross sections of rural freeways vary considerably and depend on the number of lanes, the availability and use of land taken for right-of-way, and the type of terrain. The general right-of-way widths to include all elements except frontage roads in a composite highway cross section are shown in Figure VIII-48. About 50 ft should be added for each frontage road included.

Frontage Roads

Although the need for local service across and along rural freeway corridors is considerably less than the need for service along highly developed urban freeways, there is nevertheless a certain demand. Usually, frontage roads are intermittent and relatively short, providing access to one or more severed properties to provide continuity of a local road by relocation along the freeway to a connection with a grade-separated crossroad.

Where a rural freeway is located parallel to and in close proximity to a major highway, in many cases the major highway is converted to a continuous two-way frontage road and serves as a collector facility.

Because of the lack of continuity and the type of service being provided, newly constructed frontage roads are normally two-way facilities in rural areas. Since traffic operations at two-way frontage road intersections with grade-separated crossroads are more complex, such intersections are generally located as far as practicable from grade-separation structures and interchange ramp terminals.

Rural frontage roads are generally outside the control-of-access line but within the right-of-way limits.

Design details for rural frontage roads are similar to those used for local roads, as discussed in Chapters III and V.

Examples of Rural Frontage Roads

Figure VIII-49 illustrates a simple, short, two-way frontage road used to furnish access to a farm unit severed by the freeway location.

Figure VIII-50 shows, in the upper left quadrant, an example of a frontage road used for the dual purpose of providing access to a farm unit and maintaining the continuity of a local road. The frontage road in the upper right quadrant provides access to residential, commercial, and farm property.

Figure VIII-51 shows the freeway in the corridor of a major route that has been converted to a continuous two-way frontage road providing access to properties and maintaining local road continuity between widely spaced freeway grade separations.

Figure VIII-52 illustrates two-way frontage roads along both sides of the freeway in the vicinity of a sparsely developed rural community. The frontage roads collect and distribute traffic from private entrances and local roads and

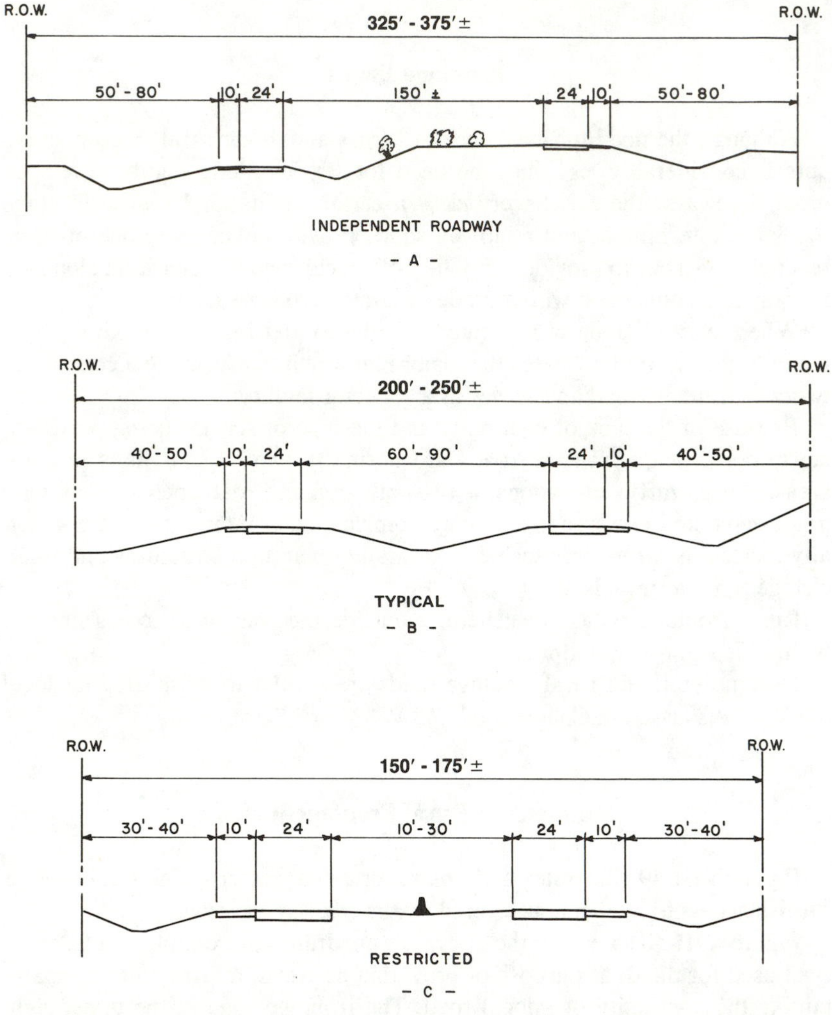

Figure VIII-48. Cross sections and right-of-way widths for four-lane rural freeways.

Figure VIII-49. Short frontage road for farm access.

Figure VIII-50. Frontage road serving residential, commercial, and farm property.

Figure VIII-51. Existing road converted to frontage road.

Figure VIII-52. Two-way frontage roads on both sides of freeway.

streets. They also provide access to the freeway via a connection to a crossroad and interchange not shown but located beyond the upper left corner of the photograph.

REFERENCES

1. *Highway Capacity Manual, Special Report 209*. Washington, D.C.: Transportation Research Board, 1985.
2. *Standard Specifications for Highway Bridges*. Washington, D.C.: AASHTO, 1984.
3. *Guide for the Design of High-Occupancy Vehicle and Public Transfer Facilities*. Washington, D.C.: AASHTO, 1983.
4. *Roadside Design Guide*. Washington, D.C.: AASHTO, 1989.

Chapter IX
AT-GRADE INTERSECTIONS

INTRODUCTION

An intersection is defined as the general area where two or more highways join or cross, including the roadway and roadside facilities for traffic movements within it. Each highway radiating from an intersection and forming part of it is an intersection leg. The common intersection of two highways crossing each other has four legs. It is not recommended that an intersection have more than four legs.

An intersection is an important part of a highway because, to a great extent, the efficiency, safety, speed, cost of operation, and capacity depend on its design. Each intersection involves through-or cross-traffic movements on one or more of the highways concerned and may involve turning movements between these highways. These movements may be handled by various means, depending on the type of intersection.

There are three general types of intersections: intersections at grade, grade separations without ramps, and interchanges. The last two are discussed separately in other chapters. Certain intersection design elements, primarily those concerning the accommodation of turning movements, are common and applicable to intersections at grade and interchanges. The design elements in the following discussions apply to at-grade intersections and their appurtenant features.

GENERAL DESIGN CONSIDERATIONS AND OBJECTIVES

The main objective of intersection design is to reduce the severity of potential conflicts between motor vehicles, buses, trucks, bicycles, pedestrians, and facilities while facilitating the convenience, ease, and comfort of people traversing the intersections. The design should be fitted closely to the natural transitional paths and operating characteristics of the users.

Four basic elements enter into design considerations of at-grade intersections.

A. Human Factors

1. Driving habits
2. Ability to make decisions
3. Driver expectancy
4. Decision and reaction time
5. Conformance to natural paths of movement
6. Pedestrian use and habits

B. Traffic considerations

1. Design and actual capacities
2. Design-hour turning movements
3. Size and operating characteristics of vehicle
4. Variety of movements (diverging, merging, weaving, and crossing)
5. Vehicle speeds
6. Transit involvement
7. Accident experience

C. Physical Elements

1. Character and use of abutting property
2. Vertical alinements at the intersection
3. Sight distance
4. Angle of the intersection
5. Conflict area
6. Speed-change lanes
7. Geometric features
8. Traffic control devices
9. Lighting equipment
10. Safety features
11. Bicycle traffic

D. Economic Factors

1. Cost of improvements
2. Effects of controlling or limiting rights-of-way on abutting residential or commercial properties where channelization restricts or prohibits vehicular movements
3. Energy consumption

TYPES AND EXAMPLES OF AT-GRADE INTERSECTIONS

General Considerations

The basic types of at-grade intersections are the T intersection (with multiple variations of angular approach), the four-leg intersection, and the multileg intersections. In each particular case the type is determined primarily by the number of intersecting legs, the topography, the traffic pattern, and the desired type of operation.

A basic intersection type can vary greatly in scope, shape, and degree of channelization. Once the type is established, the design controls and criteria covered in Chapter II and the elements of intersection design given in Chapter III as well as in this chapter must be applied to arrive at a suitable geometric plan. In this section each type of intersection is discussed separately, and likely variations of each are demonstrated. It is not practical to show all possible variations, but the types demonstrated are sufficient to cover the general application of at-grade intersection design. Many other variations of types and treatment may be found in the *NCHRP 279* (**1**), which shows examples in detail that are not included in this policy manual.

Although many of these examples are in urban areas, the principles involved apply equally to design in rural areas. Some minor variations occur with different kinds of traffic control, but all of the types shown lend themselves to cautionary or nonstop control, stop control for minor traffic volumes, four-way stop control, and signal control, both fixed time and traffic actuated.

Simple intersections are first presented, followed by the more complex types, some of which are special adaptations. Traffic volume, speed, and the character of highway or highways are the principal factors considered in the determination of the intersection type. The condition for which each intersection type may be suited is discussed in the following section.

Three-Leg Intersections

Basic Types of Intersections

Basic forms of T intersections are illustrated in Figures IX-1 and IX-2 as well as in Figures IX-53 through IX-56.

The most common type of T intersection is shown in Figure IX-53A. The normal pavement widths of both highways should be maintained except for the paved returns or where widening is required for the type of design vehicle anticipated. This type of unchannelized intersection fits junctions of minor or local roads and generally junctions of minor roads with more important highways where the roadway intersection is not generally more than 30° from normal. This would allow the angle of intersection to range from approximately 60° to 120°. In rural areas this type is usually used in conjunction with two-lane highways carrying light traffic. In suburban or urban areas it may be satisfactory for higher volumes and multilane roads.

Where speeds are high and turning movements in sufficient number increase the hazard, an additional area of surfacing or flaring may be provided for maneuverability, as shown in Figures IX-53 and IX-54.

The use of auxiliary lanes increases capacity and reduces the hazard caused by turning vehicles. Left turns from the through highways are particularly hazardous because vehicles must slow down and perhaps stop before completing the turn. Intersections with auxiliary lanes enable following through vehicles to maneuver around these slower vehicles. Existing intersections can have an auxiliary lane added with minimal difficulties to provide the types shown in Figures IX-53, and IX-54.

Figure IX-53B shows an added lane, on the side of the through highway adjacent to the intercepted road, acting as a continuous speed-change lane for right offturns. This arrangement is applicable where the right-turning movement from the through highway (movement 1) is substantial and the left-turning movement from the through highway (movement 2) is minor.

Figure IX-54A shows an added lane on the side opposite the intercepted road. This type of facility is commonly referred to as a "left turn lane" or a "right-hand passing lane." This arrangement is applicable where the left-turning movement from the through highway (movement 2) and the through movement (movement 5) are substantial, and the right-turning movement (movement 1) is minor. Here the added lane affords the opportunity for the following through drivers to pass on the right of the slower moving or stopped vehicles preparing to turn left. A driver turning left from the through highway naturally edges toward the center, and through traffic is encouraged to pass on the right of the vehicle slowing down or stopping to turn left.

Another flared arrangement, Figure IX-54B, may be effected by adding auxiliary lanes on each side of the through highway approaching the intersection. Such an arrangement may be appropriate where the capacity of the two-lane highway at the intersection is taxed by the volume and where signal control may be required, usually in developing areas. For such conditions in rural areas, the two-lane highway, preferably, would be converted to a divided section through the intersection, as shown in Figure IX-1. In addition to adding auxiliary lanes to the through highway, the intercepted road may be widened

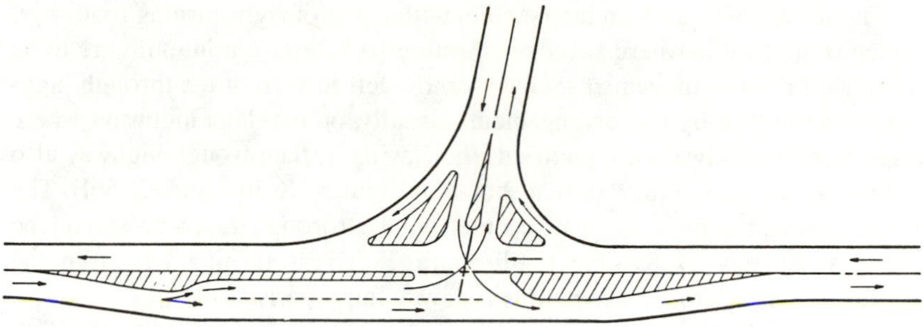

Figure IX-1. Channelized high-type "T" intersections.

on one or both sides, as shown at the left in Figure IX-55A, for better maneuverability and increased capacity of the intercepted road.

Channelized Three-Leg Intersections

Figures IX-55 and IX-56 depict channelized intersections, supported with discussion of general application and functional characteristics.

The figures discussed in the following paragraphs indicate various geometric designs relative to the use of channelization islands at three-leg intersections. When island design requires a convex section, there should be sufficient cross-sectional area of the island to ensure its proper function. That is, to channelize and separate turning movements, the total cross-sectional area should not be considered in the design when it is less than 75 ft^2. The undesirable turning paths described in conjunction with some of the intersections should be given particular attention. Unusual or sophisticated design geometrics should be avoided where possible. Figure IX-55A shows an intersection with a right-turn lane from the through roadway. This is accomplished by increasing the return radius between the two roadways a sufficient dimension to permit an individual turning roadway which is separated from the normal traveled ways by an island. The approach roadway may include a section of an auxiliary lane for the accommodation of right-turn traffic. Such an addition is a function of the number of vehicles desiring to perform a right turn movement as well as the approach speeds and the number of vehicles desiring to continue in the through direction.

Figure IX-55B shows an intersection with a pair of right-turning roadways, which is applicable where speeds or turning paths above minimum are to be provided for these movements. However, the left turn from the through highway is not helped by this arrangement. Usually, on two-lane highways where right-turning roadways are justified, the flaring of the through highway also is in order, along the lines of that shown in Figures IX-56A and IX-56B. The right-turning roadway for traffic entering the through highway should be made as narrow as feasible to discourage drivers turning left from the through highway from entering this roadway improperly.

Figure IX-56A shows a channelized intersection consisting of one divisional island on the intercepted road. Space for the island is made by flaring the pavement edges of the intercepted road and by using larger than minimum edge radii for right-turning movements. To fit the paths of left-turning vehicles the end of the island is generally located about 8 to 12 ft from the pavement edge of the through highway. This design is adaptable to two-lane highways over a wide range of volumes, particularly where space is not available for turning roadways and where simplicity is desired. For intermediate-to-heavy volumes (relative to the capacity of the highways), the through highway preferably should be flared.

Figure IX-56B shows an intersection with a divisional island and right-turning roadways, a desirable plan on important two-lane highways carrying intermediate-to-heavy traffic volumes, e.g., peak-hour volumes greater than 500 vehicles on the through highway with substantial turning movements. All movements through the intersection are accommodated on separate lanes. The divisional islands shown in Figure IX-56 differ in location with respect to the centerline. Either one may be used on each plan.

Figure IX-2 is a three-leg rural intersection. The two-lane major highway has been converted to a divided highway through the intersection. This intersection is well designed, with liberal use of painted raised bars in the median. A right turn lane in the upper right quadrant accommodates a non-restricted exit from the major route, and a separate left-turn storage lane serves higher turning volumes than can be accommodated by the intersection treatment shown in Figure IX-54B. A stop sign or signal normally controls the traffic intersecting the through highway.

Four-Leg Intersections

Basic Types

Basic types of four-leg intersections are shown in Figures IX-3 through IX-7. The overall principles of design, island arrangements, use of auxiliary

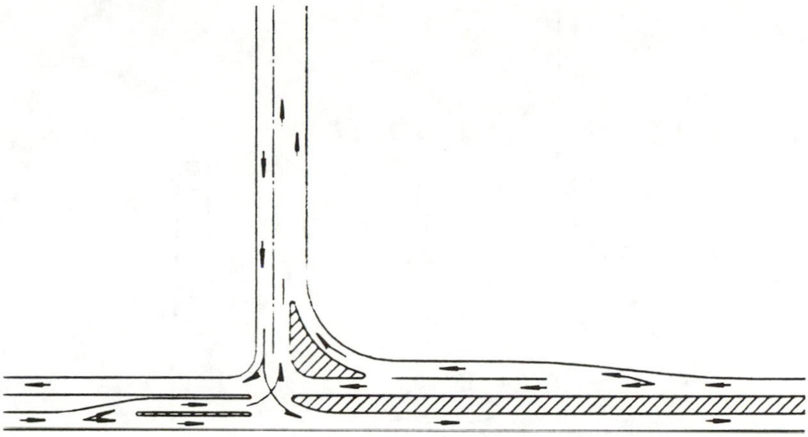

Figure IX-2. Existing three-leg rural intersection, channelized "T".

pavements, and many of the previous discussions for T intersections also apply to four-leg intersections.

Figure IX-3A illustrates the simplest form of an unchannelized four-leg intersection suitable for crossings of minor or local roads and often suitable for minor road crossings of major highways. The roadway intersection is not generally more than 30° from normal and the volume of turning traffic is low. This would allow the angle of intersection to range from approximately 60° to 120°. Approach pavements are continued through the intersection, and the corners are rounded to accommodate turning vehicles.

Figure IX-3B illustrates a flared intersection having additional capacity at the intersection for through and turning movements. An auxiliary lane on each side of the normal pavement at the intersection enables through vehicles to pass slow-moving and standing vehicles preparing to turn.

Depending on the relative volumes of traffic and the type of traffic control used, flaring is accomplished by parallel auxiliary lanes, as on the highway shown horizontally, or by pavement tapers, as shown on the crossroad. Flaring generally is similar on opposite legs. Parallel auxiliary lanes are essential where traffic volume on the major highway is near the uninterrupted flow capacity of the highway or where through and cross traffic is sufficiently high to require signal control. Such lanes are desirable also for lower volume conditions. The length of added pavement should be computed as it is for speed-

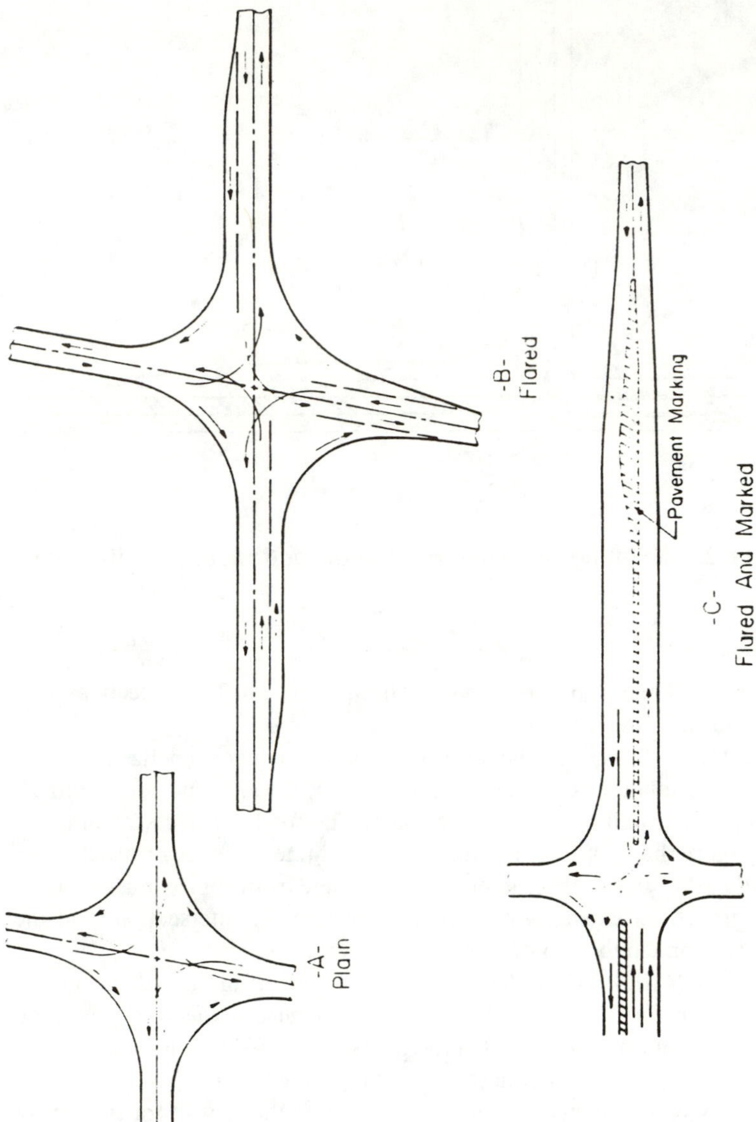

Figure IX-3. Unchannelized four-leg intersections, plain and flared.

change lanes, and the length of uniform lane width, exclusive of taper, normally should be greater than 150 ft on the approach side and 200 ft on the exit side of the intersection.

Figure IX-3C shows a flared intersection with a pavement marking area that divides traffic approaching the intersection and with provision for a median lane suitable for two-lane highways where speeds are high, intersections are infrequent, and the left-turning movements from the highway are hazardous. Pavement widening should be effected gradually, preferably with edge-reverse curves of 1° or flatter. The pavement marking at the widest point should be at least 12 ft, and the through-traffic lane on each side of it should be 2 to 3 ft wider than the normal lane width on the approaches. Near the cross road, where the full widening is attained, the overall width of pavement is about 40 ft. This plan affords better protection for vehicles turning left from the major highway than does the arrangement in Figure IX-3B, which is better suited for intersections requiring signal control. Pavement marking is not as positive a separator as a curbed divisional island, but it is advantageous where sand or snow is a maintenance problem and where any introduced curbed island may be hazardous, as on high-speed rural highways.

Channelized Four-Leg Intersections

The usual arrangement of these intersection types with simple channelization is shown in Figures IX-4 and IX-5. At other than minor intersections, right-turning roadways are often provided, as shown in Figure IX-4A, for the more important turning movement or where large vehicles are to be accommodated and at minor intersections in quadrants where the angle of turn greatly exceeds 90°.

Figure IX-4B shows an oblique-angle intersection with a skew angle of 45° or more and with separate two-way turning roadways in the acute-angle quadrants. Vehicles can turn readily to the right or left, and awkward maneuvers and encroachments at the intersection proper are eliminated. The multiple points of intersection and the large skew angle may combine to make this type of intersection undesirable. Preferably, one or both highways should be realined to reduce the skew angle. When realinement cannot be obtained, extensive application of proper signing and signal controls is recommended.

Figure IX-5A shows a crossing with right-turning roadways in all four quadrants that is suitable where space is available and turning movements are important, particularly in suburban areas where pedestrians are present. Where the intersecting highways are only two lanes wide, this arrangement is

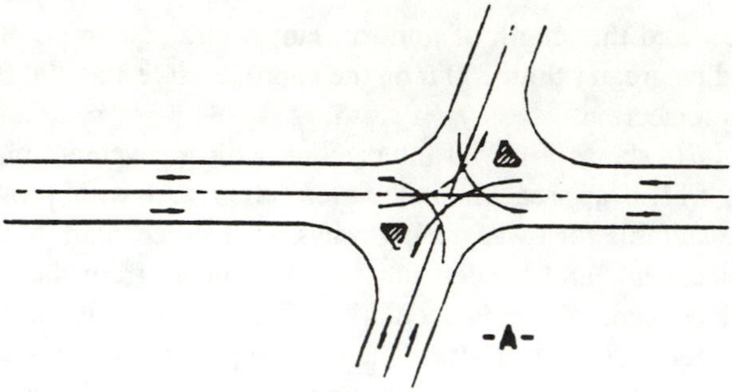

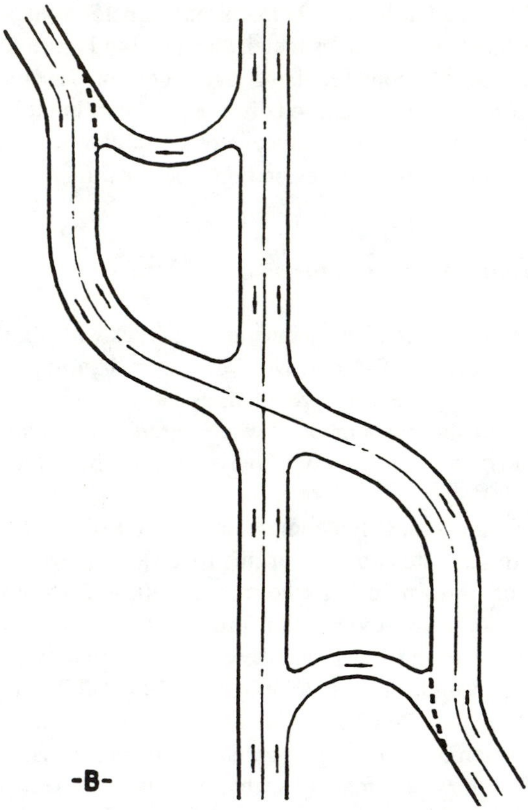

Figure IX-4. Channelized four-leg intersections.

not common. Where one or more of the right-turning movements require separate turning roadways, additional lanes generally are needed for the complementary left-turning movements. In the latter case the highway is widened, as shown in Figures IX-3B and IX-5C.

Figure IX-5B illustrates an intersection with divisional islands on the crossroad. This plan fits a wide range of volumes, its capacity being governed by the pavement widths provided through the intersection. The simplicity of the plan, in many cases, will make it preferable to that of Figure IX-5C.

Figure IX-5C is an appropriate plan, other than at a minor crossroad, for a two-lane highway operating near capacity or carrying moderate volumes at high speeds. The two-lane approach pavement on the major highway is converted to a four-lane section with a divisional island. The additional pavement areas are used for speed change, maneuvering, and storage of turning vehicles. Cross and turning volumes and sizes of vehicles to be accommodated determine the form of channelization on the crossroad.

The simplest form of intersection on a divided highway has minimum all-paved areas for right turns and a median opening conforming to designs given in later discussions in this chapter. Often the speed and volume of through and turning traffic justify a higher type of channelization and details to fit predominant movements.

Figure IX-6A shows an intersection where the turning movements in the lower right quadrant are significant, and a median lane and right-turning roadway are provided for these movements. Other turning movements are minor. The left-turning movement from the left to the top of the plan is accommodated by a tapered median end that permits at least one passenger vehicle to pause off the through pavement while waiting to turn.

Figure IX-6B illustrates a design for a high-speed divided highway intersecting a major crossroad. Right-turning roadways with speed change lanes and median lanes for left turns afford a high degree of efficiency in operation and high capacity and permit through traffic on the highway to operate at reasonable speed. Proper use of signal controls is required.

Figure IX-7A shows a design with double (two-lane) median lane for one left-turning movement. It requires traffic signal control with a separate signal phase for the left turn and is particularly suitable in urban areas where there is a heavy turning movement in one quadrant. The two-lane median lane should be separated from the through lanes by either an elongated island, as shown, or by pavement markings. Pavement markings, contrasting pavements, and signs should be used to discourage the through driver from inadvertently entering the median lane. Left-turning vehicles leave the through pavement to enter the median lane in single file but, once within it, store in two lanes. On receiving the green indication, turning is accomplished simulta-

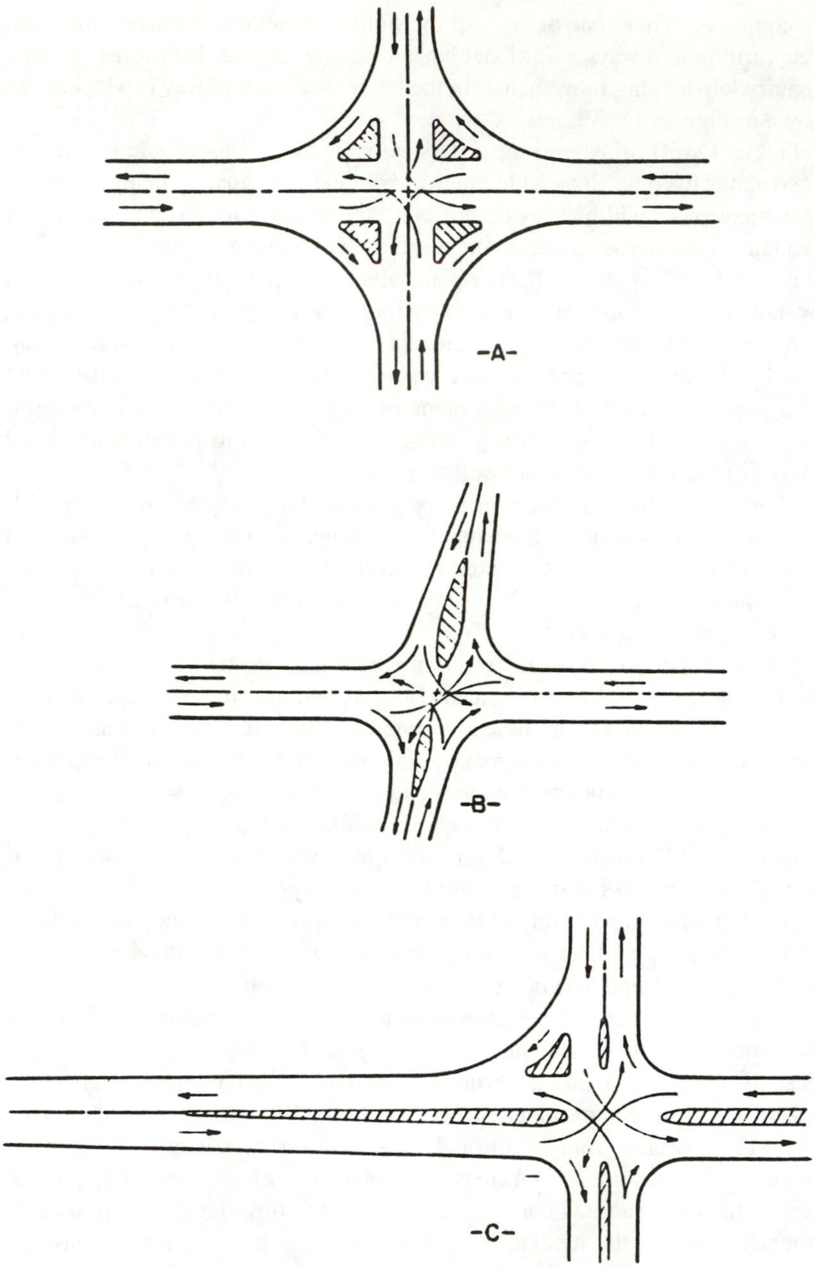

Figure IX-5. Four-leg intersections, channelized "T".

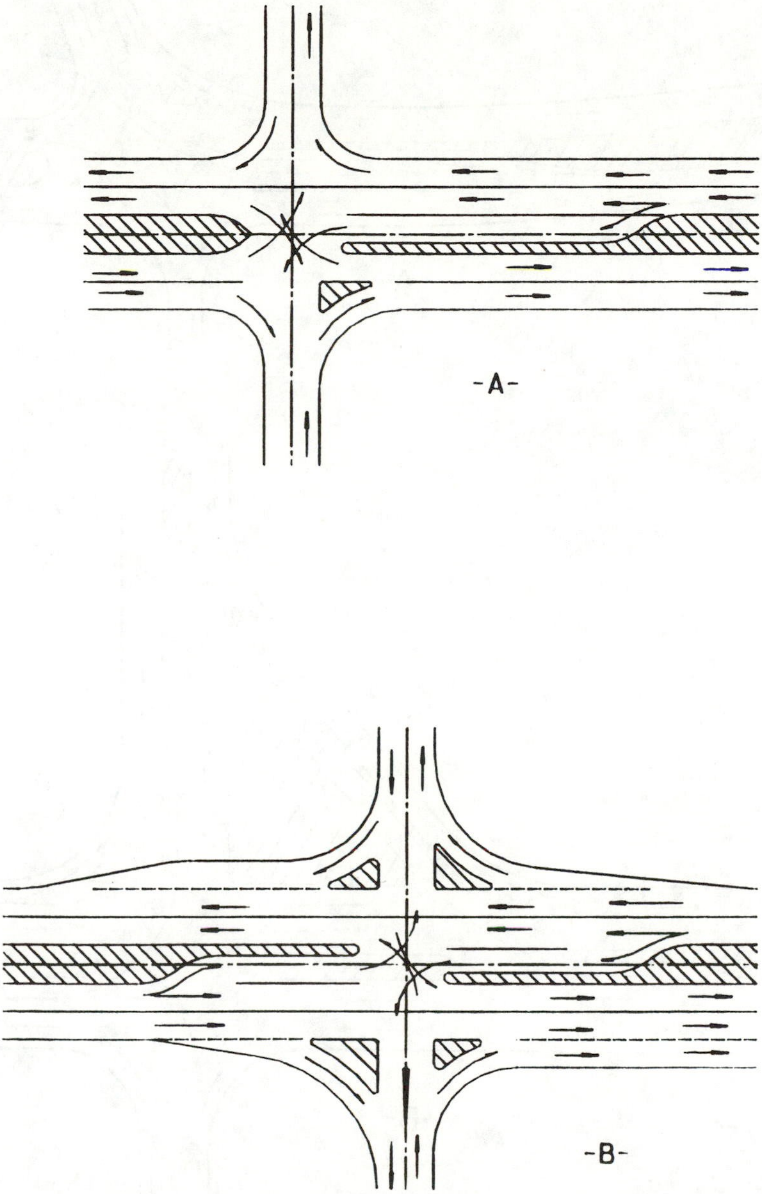

Figure IX-6. Four-leg interchange (channelized high-type).

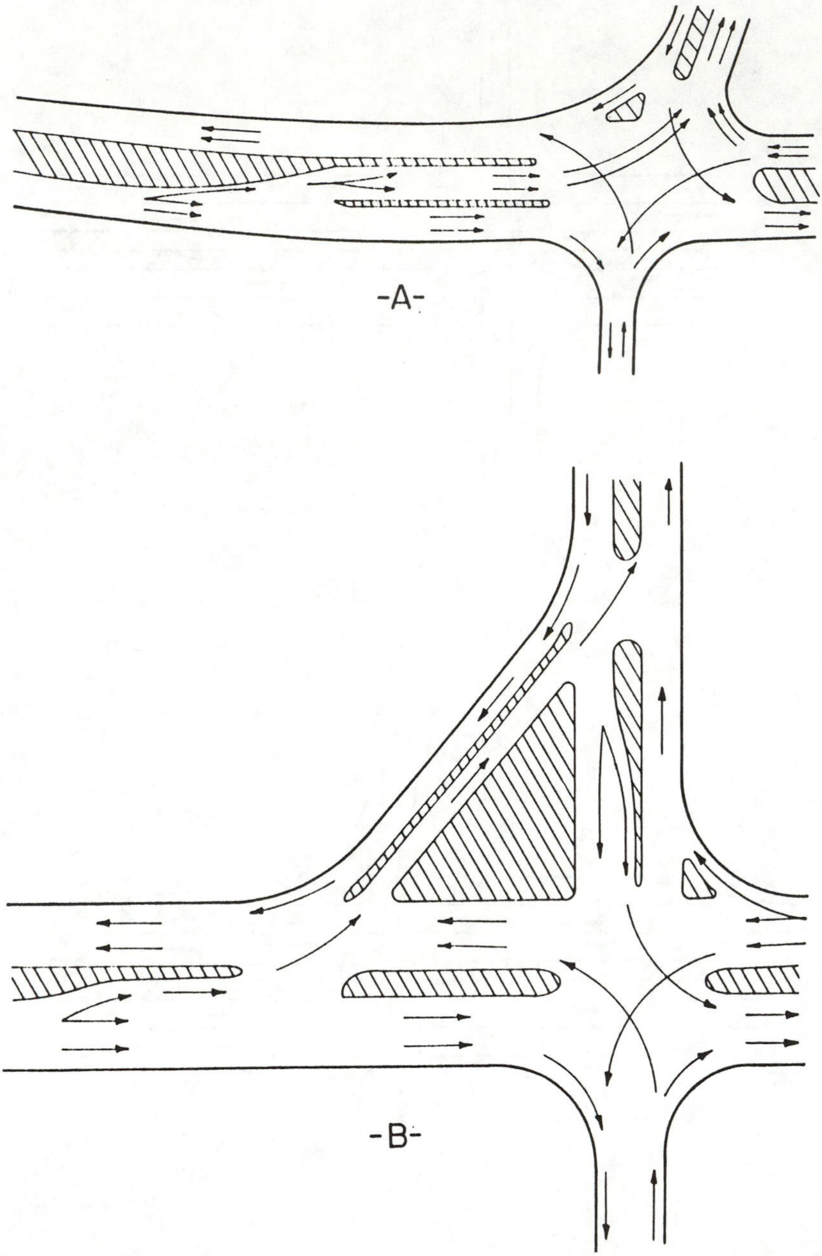

Figure IX-7. Four-leg intersections (channelized high-type).

neously from both lanes. The median opening and the crossroad pavement should be sufficiently wide to receive two lanes of vehicles.

Figure IX-7B is a design for unusually heavy through volumes and a high turning volume in one quadrant. The heavy left-turning movement is removed from the main point of intersection by providing a separate diagonal roadway and creating two additional intersections. A high degree of efficiency is obtained by a system of progressively synchronized traffic signals and proper signal timing in relation to the distance and pavement widths between the three points of conflict. The three intersections should be at least 200 ft, and preferably 300 ft or more, apart. A median lane for the left-turning movement onto the diagonal roadway should be two lanes wide. The right-turning movement using the diagonal roadway may flow continuously, and an auxiliary lane along each of the major roadways may be desirable. This design may be used where a grade separation is not feasible, as in flat terrain with traffic having a high volume of heavy trucks, or where it is necessary to defer the construction of a grade separation. Where movements in the other quadrants reach the proportions of through movements, additional diagonal roadways might be provided to advantage, but with major turning movements in more than one quadrant, a grade separation generally is imperative. Before using the design shown in Figure IX-7B, careful consideration should be given to operational characteristics since this design, in effect, creates two additional intersections.

Multileg Intersections

Multileg intersections are those with five or more intersection legs. They should be avoided wherever possible. Where volumes are light and stop control is used, it may be satisfactory to have all intersection legs intersect at a common, all-paved area. At other than minor intersections, safety and efficiency are improved by rearrangements that remove some conflicting movements from the major intersection. These rearrangements are accomplished by realining one or more of the intersecting legs and combining some of the traffic movements at adjacent subsidiary intersections, as shown in Figure IX-8, or in some cases, making one or more legs one way away from the intersection.

Figure IX-8A shows the simplest use of this principle on an intersection with five approach legs. The diagonal leg is realined to join the upper road at sufficient distance from the main point of the intersection to form two distinct intersections, each of which can be simply operated. The left to lower right is likely to be the major highway, and the diagonal leg is realined to locate the extra intersection on the less important road.

Figure IX-8B illustrates an intersection with six approach legs, two of which are realined in adjacent quadrants to form a simple crossing several hundred feet to the right of the major intersection, which is thereby converted to a four-leg crossing. This pattern applies where the top-to-bottom highway at the left is the more important route. If the left-to-right highway is more important, it may be preferable to bend the diagonal legs toward the other highway and thereby create three separate intersections along the minor highway. The designs in Figure IX-8 are shown in the simplest form. Separate turning lanes and divisional islands, as are necessary to fit the particular situation, may be used.

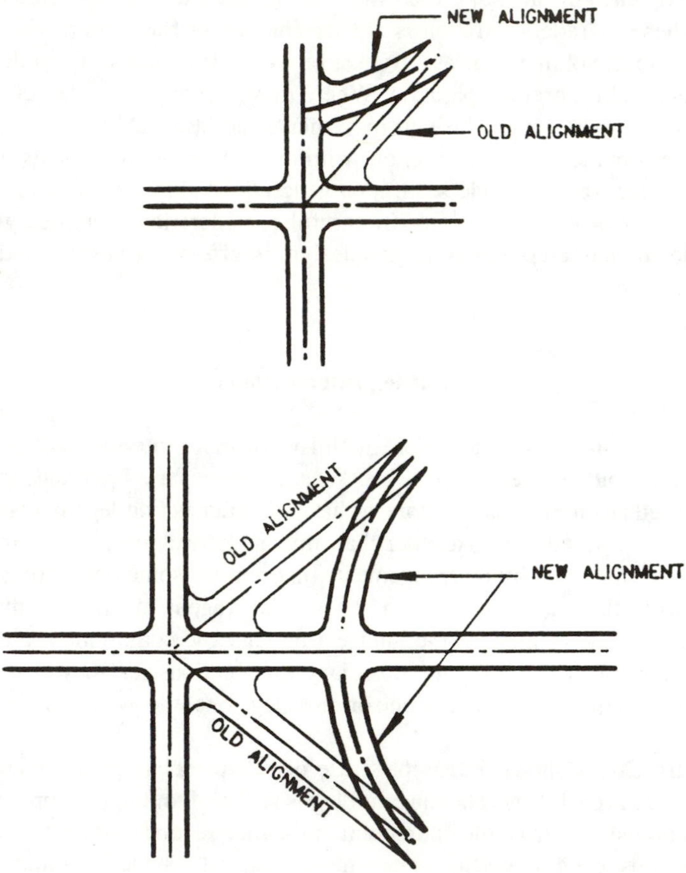

Figure IX-8. Realigning multileg intersections.

CAPACITY ANALYSIS

Capacity analysis is one of the most important considerations in the design of intersections. This subject is discussed at length in Chapter II and is discussed throughout this chapter as it relates to the various elements of intersection design. Optimum capacities can be obtained when at-grade intersections include auxiliary lanes, proper use of channelization, and traffic control devices. For more complete coverage of capacity of intersections, including procedures for making capacity computations, reference should be made to the *Highway Capacity Manual (HCM)* (**2**), and guidance for its use is given in Chapter II.

ALINEMENT AND PROFILE

General Considerations

Intersections are points of conflict between vehicles, pedestrians, and facilities and hence are potentially hazardous. The alinement and grade of the intersecting roads, therefore, should permit users to discern and perform readily the maneuvers necessary to pass through the intersection safely and with a minimum of interference by other users. To these ends the alinement should be as straight and the gradients as flat as practical. The sight distance should be equal to or greater than the minimum values for specific intersection conditions, as derived and discussed later. If it is not, users have difficulty in discerning the actions of other users, in reading and discerning the messages of traffic control devices, and in controlling their operations.

Site conditions generally establish definite alinement and grade limitations on intersecting roads. It is often possible to modify the alinement and grades, however, and thereby improve traffic conditions and reduce hazards, particularly on rural highways.

Alinement

Regardless of the type of intersection, for safety and economy, intersecting roads should generally meet at or nearly at right angles. Roads intersecting at acute angles require extensive turning roadway areas and tend to limit visibility, particularly for drivers of trucks. When a truck turns on an obtuse angle, the driver has blind areas on the right of the vehicle. Acute-angle intersections increase the exposure time of the vehicles crossing the main traffic flow and may increase the accident potential. The practice of realining roads intersecting at acute angles in the manner shown in Figure IX-9 (A and B) has proved to be beneficial. If the realinement curves fail to permit safe operations at speeds not much below that on the approaches of the major highway, however, they may prove to be as great a hazard as the acute-angle crossing itself. Warning signs for reduction of speed should be provided well in advance of such curves.

The practice of constructing short radii horizontal curves on side road approaches to achieve right angle intersections should be avoided whenever possible. These curves result in increased lane infringements because vehicles tend to smooth out the curves by encroaching on a portion of the opposite lane. Also, the traffic control devices at the intersection may be obscured resulting in the need for the installation of advanced signing.

Another method of realining an intersecting road originally at an acute angle is to make a staggered intersection, as shown in Figure IX-9 (C and D). Only a single curve is introduced on each crossroad leg, but crossing vehicles must turn onto the major highway and then reenter the minor highway. (The terms "major highway" and "minor highway" are used here to indicate the relative importance of the highways through the intersection rather than the functional character of the highways).

Where the realinement of the minor road is as shown in Figure IX-9C, the access continuity is poor because a crossing vehicle must reenter the minor road by a left turn off the major highway. This design arrangement should not be used unless traffic on the minor highway is very moderate and it is anticipated that the destinations on the minor highway are local and that almost all the traffic is turning onto the major highway and not proceeding through the intersection on the minor highway.

Where the alinement of the minor road is as shown in Figure IX-9D, the access continuity is good because a crossing vehicle first turns left onto the major highway (a maneuver that can be done safely by awaiting an opening in the through-traffic stream) and then turns right to reenter the minor road, thus interfering little with through traffic.

Although a right angle crossing normally is desired some deviation is permissible. Reconstructing an intersection to provide an angle of at least 60 degrees provides most of the benefits while reducing the right of way takings and construction costs often associated with providing a right angle intersection. The width of the roadway on the approach curves should be consistent with Table IX-3 in order to reduce the potential for lane infringement.

Where a large portion of the traffic from the minor highway turns onto the major highway, rather than continuing across the major highway, the double-intersection design may be advantageous regardless of the right or left entry.

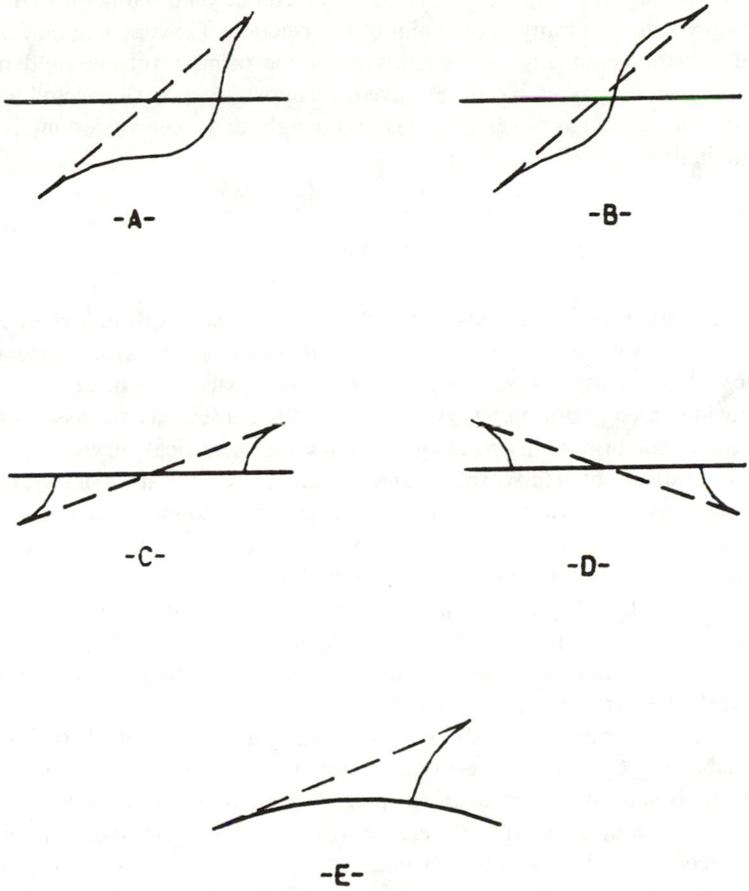

Figure IX-9. Realinement variations at intersections.

An alinement that intersects two roads at their junction to form a five-or-more-leg intersection should also be avoided.

Intersections on sharp curves should be avoided wherever possible because the superelevation and widening of pavements on curves complicate the intersection design.

Where the major highway is curving and a subordinate highway constitutes an extension of one tangent, realining the subordinate highway is advantageous, as shown in Figure IX-9E, to guide traffic onto the main highway and improve the visibility at the point of intersection. This practice may have the disadvantage of adverse superelevation for turning vehicles and may require further study when curves have high superelevation slopes and when the approach road has adverse grades and a sight distance restriction due to the grade line.

Profile

Combinations of grade lines that make vehicle control difficult should be avoided at intersections. Substantial grade changes should be avoided at intersections, but it is not always feasible to do so. Adequate sight distance should be provided along both highways and across the corners, as discussed later, even where one or both intersecting highways are on vertical curves.

The gradients of intersecting highways should be as flat as practical on those sections that are to be used for storage space for stopped vehicles, sometimes referred to as storage platforms. Most vehicles with either manual or automatic shifts must have the brakes applied to stand still while the motors are running unless they are stopped on a gradient flatter than 1 percent.

The calculated stopping and accelerating distances for passenger cars on grades of 3 percent or less differ little from the distances at the level. Grades steeper than 3 percent require correction of the several design factors to produce conditions equivalent to those on level highways. Most vehicle operators are unable to judge the increase or decrease in stopping or accelerating distance that is necessary because of steep grades. Their normal deductions and reactions thus may be in error at a critical time. Accordingly, grades in excess of 3 percent should be avoided on intersecting highways. Where conditions make such design unduly expensive, grades should not exceed about 6 percent, with a corresponding adjustment in design factors.

The profile gradelines and cross sections on the intersection legs should be adjusted for a distance back from the intersection proper to provide a smooth junction and proper drainage. Normally, the gradeline of the major highway should be carried through the intersection, and that of the crossroad should be adjusted to it. This design requires transition of the crown of the minor highway to an inclined cross section at its junction with the major highway. For simple unchannelized intersections involving low design speed and stop control signals or signs, it may be desirable to warp the crowns of both roads into a plane at the intersection, the particular plane depending on direction of drainage and other conditions. Changes from one cross slope to another should be gradual. Intersections with a minor road crossing a multilane divided highway with a narrow median and superelevated curve should be avoided whenever possible because of the difficulty in adjusting grades to provide a suitable crossing. Gradelines of separate turning roadways should be designed to fit the cross slopes and longitudinal grades of the intersection legs.

As a rule, the alinement and grades are subject to greater restriction at or near intersecting roads than on the open road. Their combination at or near the intersection must produce traffic lanes that are clearly visible to the operators at all times and clearly understandable for any desired direction of travel, free from sudden appearance of potential hazards, and consistent with the portions of the highway just traveled.

The combination of vertical and horizontal curvature should allow adequate sight distance at an intersection. A sharp horizontal curve following a crest vertical curve is undesirable in the intersection area.

INTERSECTION CURVES

Widths for Turning Roadways at Intersections

The pavement and roadway widths of turning roadways at intersections are governed by the volumes of turning traffic and the types of vehicles to be accommodated and may be designed for one-way or two-way operation, depending on the geometric pattern of the intersection. Widths determined for turning roadways are also applied on all roadways within an intersection. The turning radii and the pavement cross slopes are functions of design speed and type of vehicles. For an in-depth study of the criteria involved, see Chapter III.

Clearance Outside Pavement Edges

Minimum Designs for Sharpest Turns

Where it is necessary to provide for turning vehicles within minimum space, as at unchannelized intersections, the minimum turning paths of the design vehicles apply. Sharpest turns of these vehicles are shown in this chapter and in Chapter II, and the paths of the inner rear wheel and the front overhang are shown. The paths indicated, which are slightly greater than the minimum paths of nearly all vehicles in each class, are the minimums attainable at speeds less than 10 mph, and consequently offer some leeway in driver behavior. Layouts patterned to fit these paths of design vehicles are considered satisfactory as minimum designs where turning speeds of 10 mph or less are appropriate. Tables IX-1 and IX-2 summarize minimum edge of pavement designs for various design vehicles.

The dimensions and turning radii of design vehicles are identified in Chapter II. In this chapter, the following design vehicles are addressed: passenger car (P), single unit truck and bus (SU), medium tractor-semitrailer combinations (WB-40), larger tractor-semitrailer combination (WB-50), tractor-semitrailer-full trailer combinations (W-60), tractor-semitrailer combinations (WB-62 & WB-67), tractor-semitrailer-full trailer combinations (WB-96) and

Angle of Turn (degrees)	Design Vehicle	Simple Curve Radius	Simple Curve Radius with Taper		
			Radius (ft)	Offset (ft)	Taper (ft:ft)
30	P	60	—	—	—
	SU	100	—	—	—
	WB-40	150	—	—	—
	WB-50	200	—	—	—
	WB-62	360	220	3.0	15:1
45	P	50	—	—	—
	SU	75	—	—	—
	WB-40	120	—	—	—
	WB-50	—	120	2.0	15:1
	WB-62	—	140	4.0	15:1
60	P	40	—	—	—
	SU	60	—	—	—
	WB-40	90	—	—	—
	WB-50	—	95	3.0	15:1
	WB-62	—	140	4.0	15:1
75	P	35	25	2.0	10:1
	SU	55	45	2.0	10:1
	WB-40	—	60	2.0	15:1
	WB-50	—	65	3.0	15:1
	WB-62	—	140	4.0	20:1

Table IX-1. Minimum edge-of-pavement designs for turns at intersections.

Angle of Turn (degrees)	Design Vehicle	Simple Curve Radius	Simple Curve Radius with Taper		
			Radius (ft)	Offset (ft)	Taper (ft:ft)
90	P	30	20	2.5	10:1
	SU	50	40	2.0	10:1
	WB-40	—	45	4.0	10:1
	WB-50	—	60	4.0	15:1
	WB-62	—	120	4.0	30:1
105	P	—	20	2.5	8:1
	SU	—	35	3.0	10:1
	WB-40	—	40	4.0	10:1
	WB-50	—	55	4.0	15:1
	WB-62	—	115	3.0	30:1
120	P	—	20	2.0	10:1
	SU	—	30	3.0	10:1
	WB-40	—	35	5.0	8:1
	WB-50	—	45	4.0	15:1
	WB-62	—	100	5.0	25:1
135	P	—	20	1.5	15:1
	SU	—	30	4.0	8:1
	WB-40	—	30	8.0	6:1
	WB-50	—	40	6.0	10:1
	WB-62	—	80	5.0	20:1
150	P	—	18	2.0	10:1
	SU	—	30	4.0	8:1
	WB-40	—	30	6.0	8:1
	WB-50	—	35	7.0	6:1
	WB-62	—	60	10.0	10:1
180	P	—	15	0.5	20:1
	SU	—	30	1.5	10:1
	WB-40	—	20	9.5	5:1
	WB-50	—	25	9.5	5:1
	WB-62	—	55	10.0	15:1

Table IX-1. Continued.

Angle of Turn (degrees)	Design Vehicle	3-Centered Compound		3-Centered Compound	
		Curve Radii (ft)	Symmetric Offset (ft)	Curve Radii (ft)	Asymmetric Offset (ft)
30	P	—	—	—	—
	SU	—	—	—	—
	WB-40	—	—	—	—
	WB-50	—	—	—	—
	WB-62	—	—	—	—
45	P	—	—	—	—
	SU	—	—	—	—
	WB-40	—	—	—	—
	WB-50	200-100-200	3.0	—	—
	WB-62	460-240-460	2.0	120-140-500	3.0-8.5
60	P	—	—	—	—
	SU	—	—	—	—
	WB-40	—	—	—	—
	WB-50	200- 75-200	5.5	200- 75-275	2.0-6.0
	WB-62	400-100-400	15.0	110-100-220	10.0-12.0
75	P	100- 75-200	2.0	—	—
	SU	120- 45-120	2.0	—	—
	WB-40	120- 45-120	5.0	120- 45-200	2.0-6.5
	WB-50	150- 50-150	6.0	150- 50-225	2.0-10.0
	WB-62	440- 75-440	15.0	140-100-540	5.0-12.0
90	P	100- 20-100	2.5	—	—
	SU	120- 40-120	2.0	—	—
	WB-40	120- 40-120	5.0	120- 40-200	2.0-6.0
	WB-50	180- 60-180	6.0	120- 40-200	2.0-10.0
	WB-62	400- 70-400	10.0	160- 70-360	6′-10′0″
105	P	100- 20-100	2.5	—	—
	SU	100- 35-100	3.0	—	—
	WB-40	100- 35-100	5.0	100- 55-200	2.0-8.0
	WB-50	180- 45-180	8.0	150- 40-210	2.0-10.0
	WB-62	520- 50-520	15.0	360- 75-600	4.0-10.5

Table IX-2. Minimum edge-of-pavement designs for turns at intersections.

Angle of Turn (degrees)	Design Vehicle	3-Centered Compound Curve Radii (ft)	3-Centered Compound Symmetric Offset (ft)	3-Centered Compound Curve Radii (ft)	3-Centered Compound Asymmetric Offset (ft)
120	P	100- 20-100	2.0	—	—
	SU	100- 30-100	3.0	—	—
	WB-40	120- 30-120	6.0	100- 30-180	2.0-9.0
	WB-50	180- 40-180	8.5	150- 35-220	2.0-12.0
	WB-62	520- 70-520	10.0	80- 55-520	24.0-17.0
135	P	100- 20-100	1.5	—	—
	SU	100- 30-100	4.0	—	—
	WB-40	120- 30-120	6.5	100- 25-180	3.0-13.0
	WB-50	160- 35-160	9.0	130- 30-185	3.0-14.0
	WB-62	600- 60-600	12.0	100- 60-640	14.0-7.0
150	P	75- 85-75	2.0	—	—
	SU	100- 30-100	4.0	—	—
	WB-40	100- 30-100	6.0	90- 25-160	1.0-12.0
	WB-50	160- 35-160	7.0	120- 30-180	3.0-14.0
	WB-62	480- 55-480	15.0	140- 60-560	8.0-10.0
180	P	50- 15-50	0.5	—	—
	SU	100- 30-100	1.5	—	—
	WB-40	100- 20-100	9.5	85- 20-150	6.0-13.0
	WB-50	130- 25-130	9.5	100- 25-180	6.0-13.0
	WB-62	800- 45-800	20.0	100- 55-900	15.0-15.0

Table IX-2. Continued.

tractor-semitrailer-full trailer-full trailer combination (WB-114). The remaining design vehicles including articulated buses and motor homes and passenger cars pulling trailers or boats are not addressed in this Chapter since these vehicles are generally selected to control the design. Should any of these vehicles be established as the design vehicle, refer to Chapter II for information relating to the turning templates. Additional information on design characteristics of large trucks can be found in References 3 and 4.

In the design of the edge of pavement for the minimum path of a given design vehicle, it is assumed that the vehicle is properly positioned within the traffic lane at the beginning and end of the turn, i.e., 2 ft from the edge of pavement on the tangents approaching and leaving the intersection curve. Curve designs for edge of pavement conforming to this assumption are shown beginning with Figure IX-10. They closely fit the inner wheel paths of the various design vehicles, clearing them by 2 ft or more throughout most of the turn, and at no point by less than 9 in. Differences in the inner path of vehicles turning left and right are not sufficient to be significant in design. Although not shown separately in the figures, these edge designs also apply for left-turn layouts, such as a left turn to leave a divided highway at very low speed.

The design should be modified where alinement conditions such as curvature prior to or at the end of the turn provides the assumed positioning. Superimposing the appropriate design vehicle turning template is the most expeditious way to customize a design for special conditions.

At an intersection with low right-turn volumes, the designer may determine that the need for a deceleration and right-turn lane is not warranted. In this instance, the composition of the shoulder may be improved for greater load capacities to permit right-turn vehicles to utilize the shoulder with no detrimental effects to the shoulder. Where right-turn volumes are high, consideration should be given to providing a right turn lane which should have a surface similar to the mainline pavement surface. Appropriate provisions should be made to provide deceleration lanes for the right turning vehicles. Full or partial shoulders should be considered where deceleration lanes for right turns are utilized.

Passenger Vehicles

Three minimum designs for the inner edge of pavement for a 90° right turn to accommodate the P design vehicle are shown in Figure IX-10. A 25-ft radius on the inner edge of pavement (the solid line in Figure IX-10A) is the sharpest simple arc that clears the inner wheel path and does so by about 9 in. near the end of the arc. A simple curve of 30-ft radius, shown dotted in the same figure, provides a 1.2-ft clearance at the ends of the curve but has a clearance of about 5.4 ft at the middle of the curve. On a radius of more than 30 ft most drivers would naturally use a turning radius flatter than the minimum for their passenger vehicles and more or less follow the edge of pavement.

The edge design shown in Figure IX-10C is a practical equivalent to a curve transitional for most or all of its length. It consists of a three-centered curve with radii of 100, 20 and 100 ft the center of the middle curve being located 22.5 ft from the extension of the tangent edges (which measurement includes a 2.5 ft offset). This design involves little extra pavement as compared with the simple curve of 30-ft radius, the pavement areas in a single quadrant between the tangent edges of pavement produced being 24 yd^2 for the edge design and 21.5 yd^2 for the simple curve. However, the edge design is preferred because it fits the design vehicle path more closely. A design that closely approximates the three-centered curve layout in the field is shown in Figure IX-10B. It consists of a simple offset curve and connecting tapers.

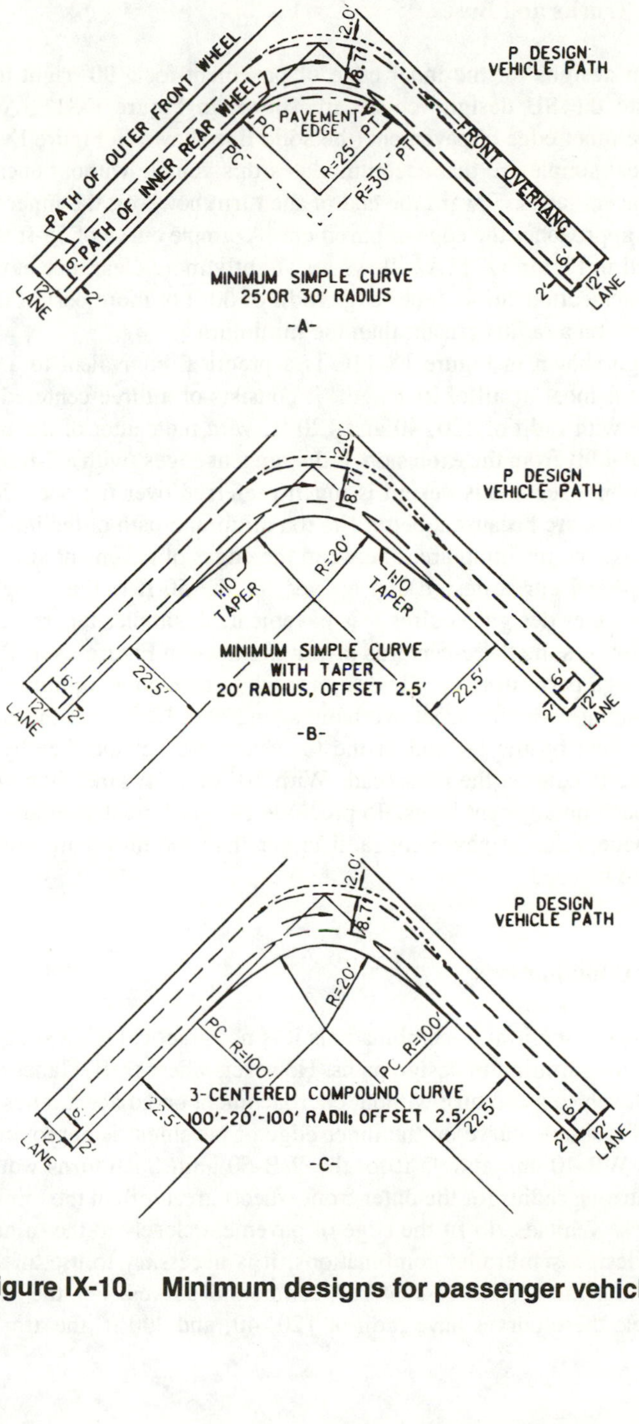

Figure IX-10. Minimum designs for passenger vehicles.

Single-Unit Trucks and Buses

Minimum designs for the inner edge of pavement for a 90° right turn to accommodate the SU design vehicle are shown in Figure IX-11. A 50-ft radius on the inner edge of pavement (the solid line shown in Figure IX-11A) is the sharpest simple arc that accommodates this vehicle without encroachment on adjacent lanes. Toward the end of the turn, however, the inner wheel path closely approaches the edge of pavement. A simple curve of 55-ft radius, shown dotted in Figure IX-11A, allows for slightly more clearance at the far end of the intersection curve. Inner-edge radii of 60 ft or more permit the SU vehicle to turn on a radius greater than the minimum.

The design shown in Figure IX-11C is a practical equivalent to a curve transitional for most or all of its length. It consists of a three-centered compound curve with radii of 120, 40 and 120 ft, with the center of the middle curve located 42 ft from the extension of the tangent edges (with a 2-ft offset). In an operational sense this design is much preferred over the one with the minimum simple arc because it better fits the minimum path of the inner rear wheel. Because the resulting areas between the edges of pavement are 53 yd^2 for the compound curve design and 60 yd^2 for the 50-ft radius, single-arc design, the former design requires less pavement. A simpler tapered design that closely follows the three-centered layout is shown in Figure IX-11B.

In any design permitting the SU design truck to turn on its minimum path without swinging wide, the front overhang swings out 12 ft from the edge of tangent pavement on the far end of the turn, and the vehicle thereby fully occupies a 12-ft lane on the crossroad. With 10- or 11-ft lanes, the vehicle would encroach on adjacent lanes. To preclude encroachment with lanes less than 12 ft wide, edge of pavement radii larger than the minimum indicated would have to be used.

Semitrailer Combinations

For the design semitrailer combinations it is not practical to fit simple circular arcs to their minimum design paths. However, where traffic lanes are 12 ft wide, such vehicles can turn without encroachment on adjacent lanes when the radius of a simple curve on the inner edge of pavement is approximately 75 ft for the WB-40 unit and 95 ft for the WB-50 unit. Such turns would be made on a turning radius (of the outer front wheel) greater than the minimum shown of these vehicles. To fit the edge of pavement closely to the minimum path of the design semitrailer combinations, it is necessary to use an asymmetrical arrangement of three-centered compound curves. For the WB-40 design vehicle these curves have radii of 120, 40, and 200 ft, the arc of the

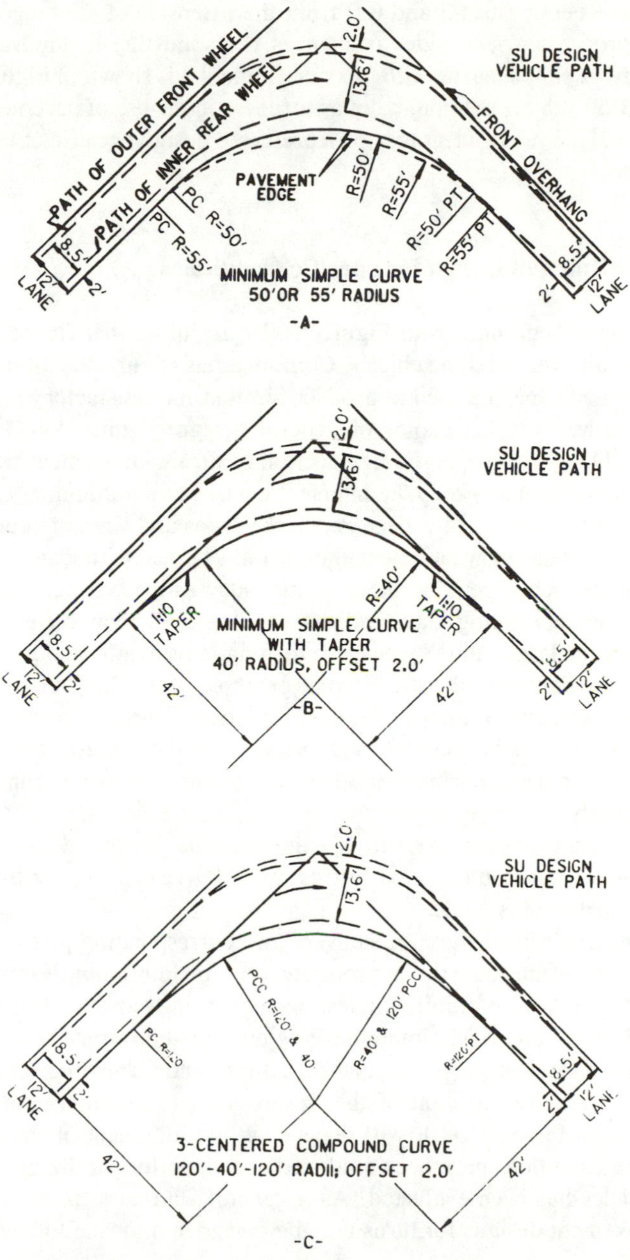

Figure IX-11. Minimum designs for single unit trucks and buses.

middle curve being offset 2 and 6 ft from the extension of the tangent edges on the approach and exit sides (shown as the solid line in Figure IX-12), respectively. A simple curve with taper combination is shown in Figure IX-13 for the WB-50 vehicle. Although not as efficient in the use of pavement as the asymmetrical curve layout, it may be a preferred design because of its ease of construction.

Choice of Minimum Design for Specific Conditions

The designs beginning with Figure IX-10 are those that fit the sharpest turns of the different design vehicles. Combinations of curves with radii other than those shown may be used to produce similar and satisfactory results. Of particular concern is the choice of general design (Figures IX-11, IX-12, IX-13, and IX-14) for a specific intersection or turning movement where it is desirable or necessary to keep the intersection area to a minimum. The selection of any one of the designs depends on the types and sizes of vehicles that will be turning and to what extent they should be accommodated. In turn, these elements may depend on other factors such as the type, character, and location of the intersecting roads and traffic volumes thereon, the number and frequency of the larger units involved in turning movements, and the effect of these larger vehicles on other traffic. For example, if turning traffic is nearly all passenger vehicles, it may not be cost-effective to design for large trucks, provided that it remains possible for an occasional large truck to turn by swinging wide and encroaching on other traffic lanes without disrupting traffic significantly.

It is necessary, therefore, for the designer to analyze the likely paths and encroachments that result when a turn is made by vehicles larger than those for which the design is made.

From the analysis of these maneuvers and corresponding paths, together with other pertinent data, the appropriate type of minimum design can be selected. Applications of minimum designs for turning movements are numerous, even in rural areas. Minimum designs may be appropriate where speeds are generally very low, property values are high, and turning movements are low in volume. The selection of the design vehicle for minimum design, beginning with Figure IX-10, will depend on the judgment of the designer after all the conditions have been analyzed and the effect of the operation of larger vehicles has been evaluated. As a general summary, three minimum edge-of-pavement designs for turns may be considered for the following conditions:

P design (Figure IX-10). This design is used at intersections in conjunction with parkways where absolute minimum turns are stipulated, at local road

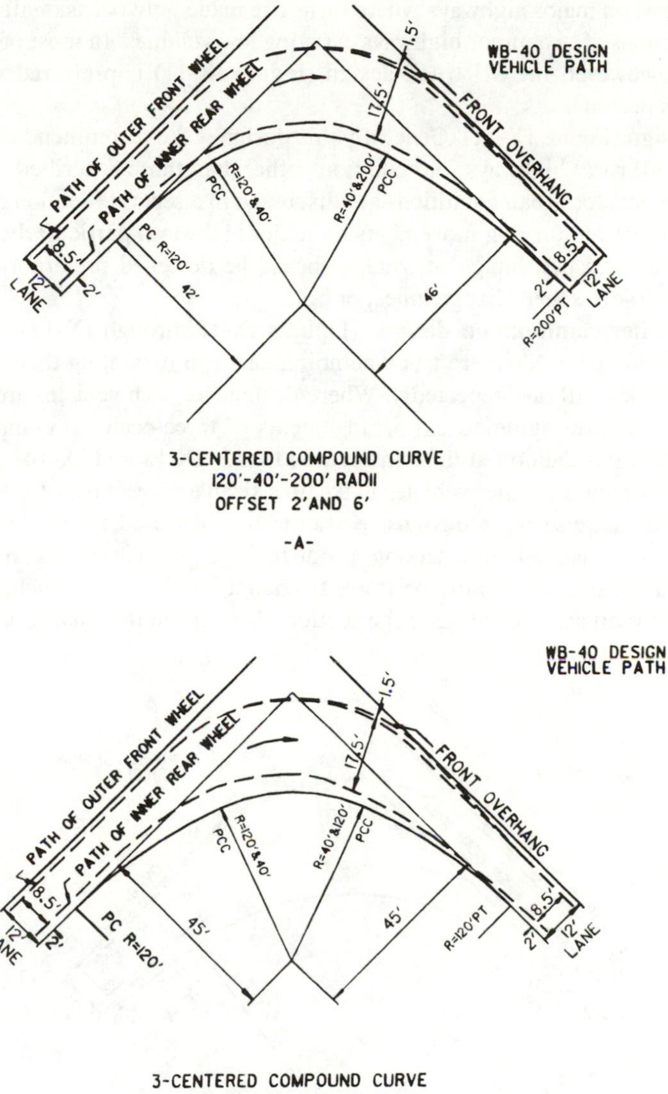

Figure IX-12. Minimum designs for semitrailer combinations (WB-40 design vehicle path).

intersections on major highways where turns are made only occasionally, and at intersections of two minor highways carrying low volumes. In most of these instances, however, the SU truck design (Figure IX-11) is preferred where conditions permit.

SU design (Figure IX-11). Generally this design is the recommended minimum for all rural highways for conditions other than those described. Turning movements for urban conditions are discussed in a separate section of this chapter. Important turning movements on major highways, particularly those involving a large percentage of trucks, should be designed preferably with even larger radii, speed change lanes, or both.

Semitrailer combination designs (Figures IX-12 through IX-17). These designs should be used where truck combinations approximating this size of design vehicles will turn repeatedly. Where designs for such vehicles are warranted, the simpler symmetrical arrangements of three-centered compound curves (shown as the dotted lines in Figures IX-12, IX-13 and IX-14) generally are preferred if smaller vehicles make up a sizable percentage of the turning volume. Because these designs, particularly when used in two or more quadrants of an at-grade intersection, produce large paved areas that may be difficult to control, it is usually desirable to channelize them, for which somewhat larger radii are needed. (See the section "Islands" in this chapter).

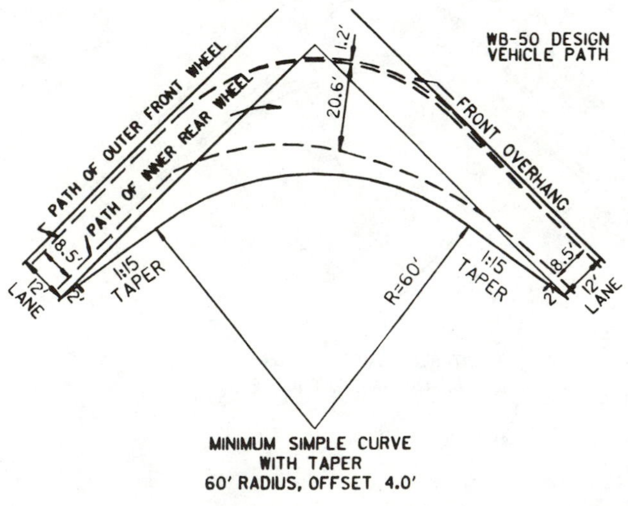

Figure IX-13. Minimum designs for semitrailer combinations (WB-50 design vehicle path).

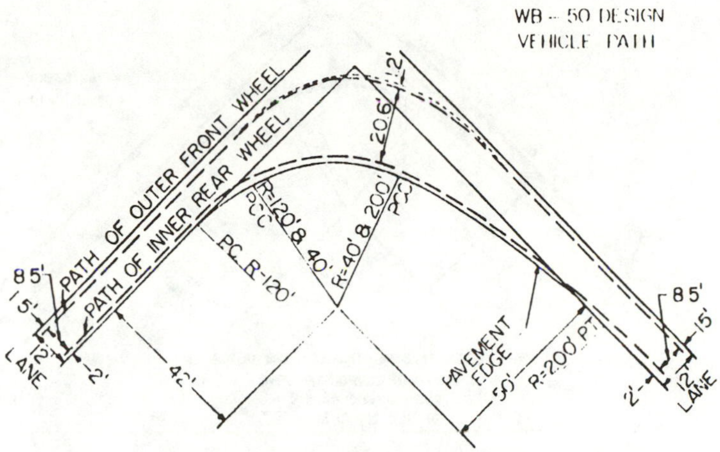

WB-50 SEMITRAILER COMBINATION
3 CENTERED COMPOUND CURVE;
120'-40'-200' RADII, OFFSET 2' AND 10'

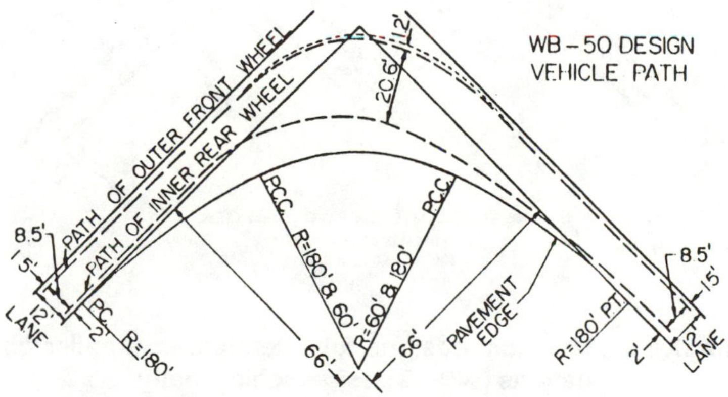

WB-50 SEMITRAILER COMBINATION
3 CENTERED COMPOUND CURVE;
180'-60'-180', OFFSET 6'

Figure IX-14. Minimum designs for semitrailer combinations (WB-50 design vehicle path).

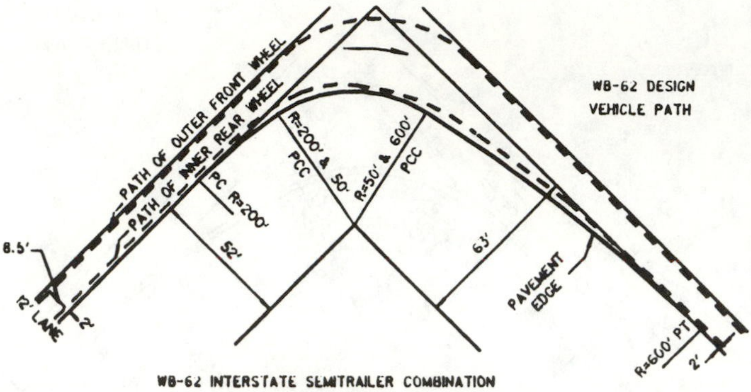

WB-62 INTERSTATE SEMITRAILER COMBINATION
3-CENTERED COMPOUND CURVE
200'-50'-600' RADII
OFFSET 2' AND 13'

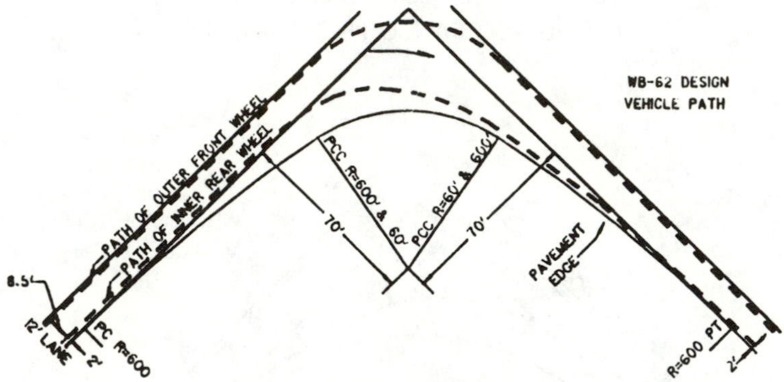

WB-62 INTERSTATE SEMITRAILER COMBINATION
3 CENTERED COMPOUND CURVE
600'-60'-600', OFFSET 10'

Figure IX-15. Minimum designs for interstate semitrailer combinations (WB-62 design vehicle path).

Other design vehicle combinations are shown in (Figures IX-18, IX-19 and IX-20). These designs should be used where these design vehicles will turn repeatedly.

Minimum or near-minimum designs for turning movements often are necessary at channelized intersections, particularly where various conflicting

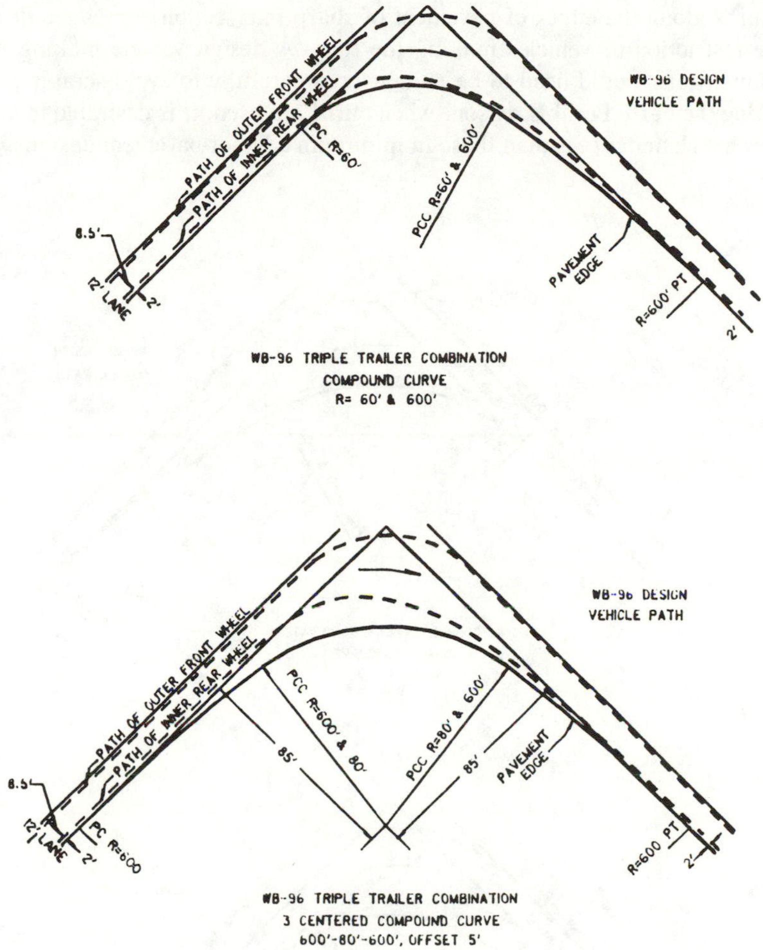

Figure IX-16. Minimum designs for triple trailer combinations (WB-96 design vehicle path).

movements must be controlled, where signal control is required, and at minor intersections where only occasional vehicles turn. Minimum designs also may be used on high-speed, high-volume highways at locations where rights-of-way are limited, but in such instances they should be used in conjunction with speed-change lanes.

Curbs along the edges of pavement of sharp intersection curves result in some restriction of vehicles making the turn. A design vehicle making its minimum turn would need to be maneuvered carefully to avoid scraping or jumping the curb. For this reason, when curbs are used, it is desirable to use somewhat flatter curves than those in minimum edge-of-pavement designs.

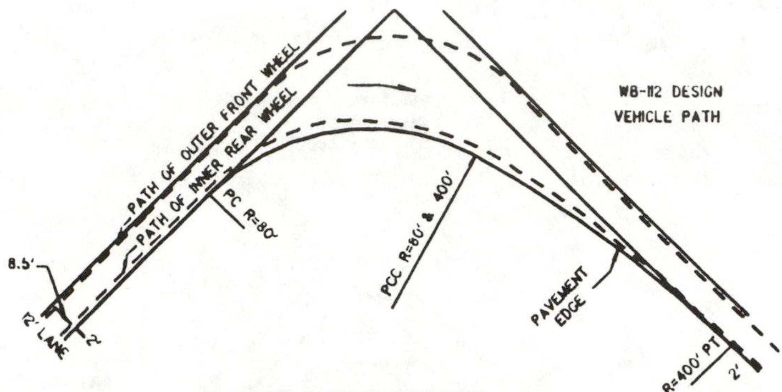

WB-112 DOUBLE TRAILER COMBINATION
COMPOUND CURVE
R= 80' & 400'

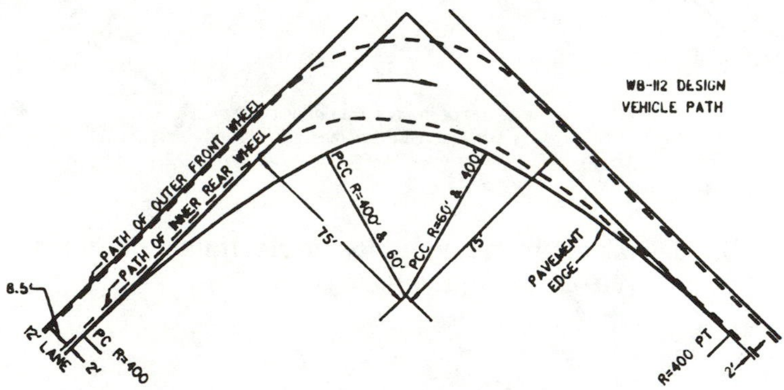

WB-112 DOUBLE TRAILER COMBINATION
3 CENTERED COMPOUND CURVE
400'-60'-400', OFFSET 15'

Figure IX-17. Minimum designs for double trailer combinations (WB-112 design vehicle path).

Oblique-Angle Turns

Minimum designs for the edge of pavement at oblique-angle intersections were developed in the same manner as those for right angle intersections by plotting the paths of the design vehicles on the sharpest turns and fitting curves or combinations of curves to the paths of inner rear wheels. Suggested minimum designs in which three-centered compound curves are used for each design vehicle when making the sharpest turn are given in Table IX-2 for various angles of turn.

For convenience, the intersection angle condition is indicated by the angle of turn, which is the angle through which a vehicle travels in making a turn. It is measured from the extension of the tangent on which a vehicle approaches to the corresponding tangent on the intersecting road to which the vehicle turns. This angle is the same as that commonly called the delta or central angle in surveying terminology. With angles of turn less than 90° the radii required to fit the minimum paths of vehicles are longer than those suggested for right angle turns. With angles of turn more than 90° the radii are decreased and larger offsets of the central arc are necessary.

The designs given in Table IX-2 are those suggested to fit the sharpest turns of the different design vehicles. Some other combinations of curves also may be used with satisfactory results. The use of tapers with simple curves is another method for design of edge of pavement for turns at intersections, and dimensions for such combinations are given in Table IX-1. Tapers are necessary to keep the intersection area to a minimum, and any design given in Tables IX-1 or IX-2 may be chosen, in accordance with the types and sizes of vehicles that will be turning and the extent to which these vehicles should be accommodated. The previous general discussion in the section "Choice of Minimum Design for Specific Conditions" also applies to these values.

At 90° intersections with inner edges of pavement designed for passenger vehicles, all trucks can turn by making a wide swing onto adjacent traffic lanes, which on two-lane roads are opposing traffic lanes. For angles of turn less than 90°, trucks also can turn on an inner edge of pavement designed for passenger vehicles with even less encroachment than that for the 90° turns. For turning angles of more than 90°, the minimum design for the P design vehicle must be adjusted to ensure that all turning trucks remain within two lanes of pavement on each highway. In this regard, for turning angles of 120° or more, the same dimensions of three-centered curves, as required by the P design vehicle (100, 20 and 100 ft) may be used, but the offset of the middle curve should be increased from 2.5 ft to as much as 10 ft for the 180° turn. Where space is available, even for minor roads, a design based on the SU design vehicle would be preferred generally. With the SU truck design, the WB-40 design vehicle will encroach only slightly, if at all, on adjacent traffic lanes, and the WB-50 design vehicle will encroach only partially on other lanes.

Design for angles of turn more than 90° may result in unnecessarily large paved intersections, portions of which are often unused. This situation may lead to confusion among drivers and hazard to pedestrians. These conditions may be alleviated to a considerable extent by using three-centered asymmetric compound curves or by using larger radii coupled with corner islands, a step in channelization as discussed in a following section. On major highways intersecting at oblique angles, separate roadways for right-turning traffic should be provided, if feasible, in the quadrants where vehicles turn more than about 120°.

Curvature for Turning Movements, Urban Streets

Because of space limitations and generally lower operating speeds in urban areas, curve radii for turning movements may be smaller than those normally used in rural areas. Curvature (corner radius) to accommodate turning movements depends largely on the number and types of turning vehicles. Minimum turning paths for passenger vehicles and all other design vehicles are included in Chapter II.

The street development standards of most cities provide curb radii of 5 to 30 ft, most of which are between 10 and 15 ft. Most passenger cars operating at very low speed on lanes 10 ft or more in width are able to make a right turn with a curb radius of about 15 ft with little encroachment on other lanes. However, operation of these vehicles at increased speeds or of larger vehicles even at very low speed generally results in substantial encroachment on adjacent lanes at either the beginning or the end of the turn, or both. Where there are curb parking lanes on both of the intersecting streets and parking is restricted thereon or for some distance from the corner, the extra width provided by the restriction serves to increase the usable radius. On most streets, curb radii of 10 to 15 ft are reasonable because streets and sidewalks are generally confined within the public right-of-way, and larger radii can be obtained only by narrowing sidewalks at corners and increasing the length of pedestrian crosswalks. However, to ensure efficient traffic operation on arterial streets carrying heavy traffic volumes, it is desirable to provide corner radii of 15 to 25 ft for passenger vehicles and of 30 to 50 ft for most trucks and buses to expedite turns to and from the through lanes. Where large truck combinations turn frequently, somewhat larger radii for turns at intersections are required.

The WB-62 truck is generally not engaged in local distribution but is used principally for "over-the-road" transportation between trucking terminals. Ideally, these terminals are located near highway facilities designed to accommodate the vehicles serving them.

If it is necessary to route WB-62 trucks over local streets to reach the terminals, careful consideration should be given to the network to be used. Generally, this network should not include narrow streets, streets with relatively small curb radii at intersections with cross streets or streets with parking.

Effect of Curb Radii on Turning Paths

The effect of curb radii on the paths of various design vehicles turning through an angle of 90° on streets without parking lanes is shown in Figures IX-18 and IX-19. Figure IX-18A shows the effects of a 15-ft radius. With 12-ft lanes, a design passenger vehicle can turn with no encroachment on an adjacent lane at the exit, but the SU, BUS, and WB-40 design vehicles must swing wide on both streets and occupy two lanes on each street. To turn within two lanes of the cross street, the WB-50 design vehicle occupies a part of the third lane on the approach pavement. The path of a WB-62 design vehicle is not shown in this figure or the figures that follow because it has a narrower turning path than a WB-50 design vehicle.

Figure IX-18B shows operation around a 25-ft curb radius. The passenger vehicle can turn within a 12-ft lane, and except for the WB-62 or larger truck, trucks can turn (by swinging wide) within the confines of two lanes on each street.

Figure IX-19A shows operation around a 30-ft radius. The SU design vehicle can approach the turn within the right lane and turn into the second lane on the cross street. A WB-40 design vehicle encroaches slightly on the third lane. The BUS and WB-50 design vehicles must occupy a portion of the third lane at the exit.

Figure IX-19B shows the trucks that can turn around a 40-ft radius from the outer lane without encroaching on the third lane of the cross street. This type of maneuver is practicable for turns from an arterial street where the cross street is four lanes or wider, because the second lane on the cross street normally would be free of traffic as a result of signal or stop sign control on the cross street. Turns from the cross street to the arterial about such radii also would not be disturbing where signal control is used, but without such control, drivers of trucks must wait for a proper gap in traffic to turn into the second lane on the major street.

Figure IX-19 shows the path of the WB-62 truck turning around a 40 ft curb radius. This vehicle must occupy most of the third lane of the cross street.

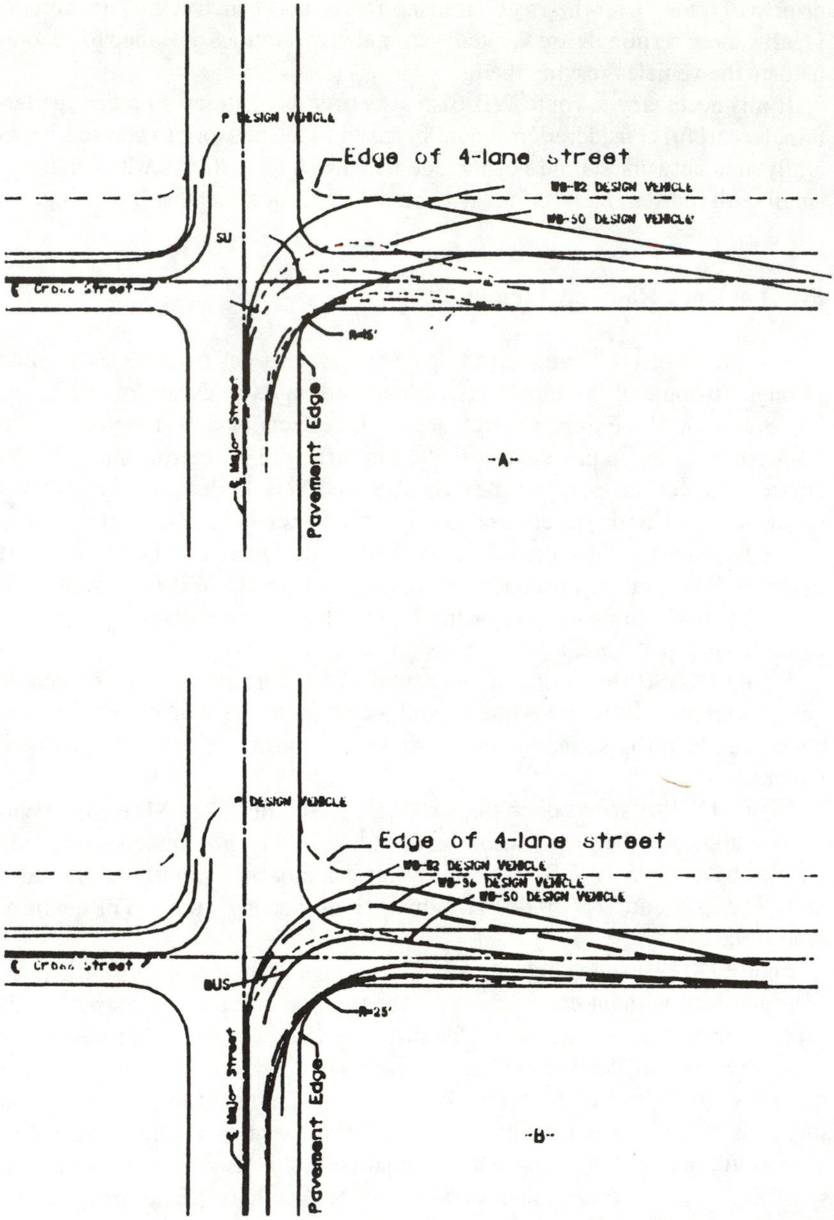

Figure IX-18. Effect of curb radii on turning paths of various design vehicles.

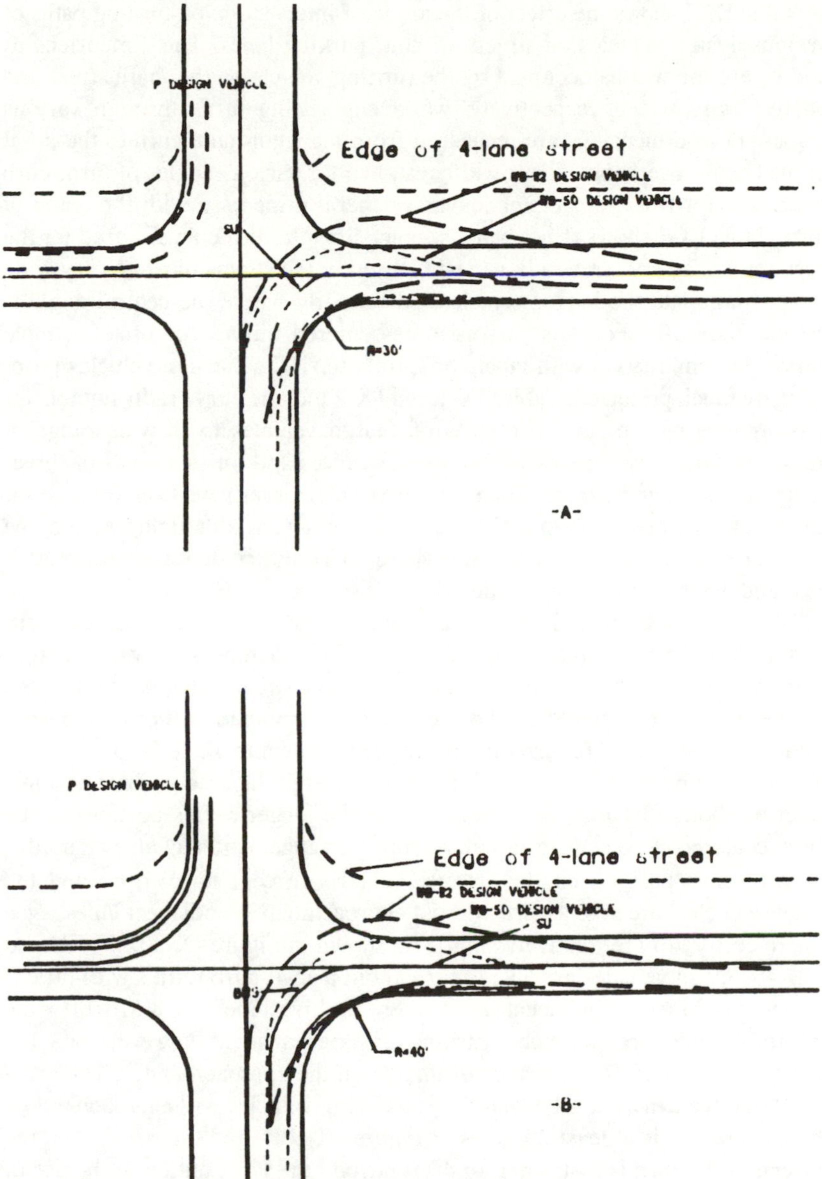

Figure IX-19. Effect of curb radii on turning paths of various design vehicles.

Table IX-3 shows the effect of the angle of intersection on turning paths of various design vehicles on streets without parking lanes. The dimensions d_1 and d_2 are the widths occupied by the turning vehicle on the main street and on the cross street, respectively, while negotiating turns through various angles. Both dimensions are measured from the right-hand curb to the point of maximum overhang. These widths, shown for various angles of turn, curb radii, and for two types of maneuvers, generally increase with the angle of turn. Table IX-3 shows that when a simple, single-radius curve is used for the curb return, a very large radius must be used or the streets must be very wide to accommodate the longer vehicles, particularly where the central angle is greater than 90°. For this reason three-centered curves (or offset, simple curves in combination with tapers or spirals to fit the paths of vehicles properly) are much preferred. Tables IX-1 and IX-2 include curve radii suitable for accommodating the several classes of design vehicles for a wide range of angles of turn. Data are shown for simple curves and for two types of three-centered compound curves. The radii for simple curves have been omitted for angles of turn greater than 90° for the reasons given, although they may be used for separate turning roadways where sufficient rights-of-way are available and where there is little pedestrian traffic.

As stated above, turning around a 40 ft curb radius the WB-62 truck occupies most of the third lane of the cross street. Encroachment on adjacent lanes is more severe with the smaller curb radii and street networks with these smaller radii curbs should not be used by this combination. For this reason, values for only the 40 ft curb radius have been shown in Table IX-3.

As shown by the footnote for Table IX-4, straight tapered wedges having a taper of about 15:1 may be substituted for the longer radius portions of the three-centered curves. A spiral-curve-spiral arrangement may also be used.

With curb parking on the streets, vehicles (except for WB-62 and the larger vehicles) are able to turn without encroachment on adjacent lanes, even where curb radii are relatively small. As shown in Figure IX-20, the SU and WB-40 design vehicles are able to turn about a 15-ft curb radius with little if any encroachment on adjacent lanes where parking lanes are 8 to 10 ft wide and traffic lanes are 12 ft wide. Parking must be restricted, however, for a distance of at least 15 ft in advance of the PC on the approach and 30 ft beyond the PT on the exit. The BUS and WB-50 design vehicles will encroach unless the curb radius is at least 25 ft, as in Figure IX-20B, and the parking on the far end of the turn is restricted for 40 ft beyond the PT. Caution in the use of radii of 15 or 25 ft is advised where parking is permitted, because traffic volumes may increase to the point where all parking will be prohibited either during rush hours or throughout the day, and the same turning conditions will prevail as shown in Figures IX-18 and IX-19 and Table IX-3.

△	Design Vehicle	d₂ (ft) for Cases A and B Where: R = 15′ A	B	R = 20 A	B	R = 25 A	B	R = 30 A	B	R = 40 A	B
30°	SU	14	13	14	13	13	13	13	13	13	13
	BUS	22	17	19	17	19	17	19	17	18	17
	WB-40	14	14	14	14	14	14	14	14	14	14
	WB-50	20	17	20	17	20	17	19	16	18	16
	WB-62	—	—	—	—	—	—	—	—	27	17
60°	SU	19	16	19	16	17	15	16	15	14	14
	BUS	28	21	26	20	24	20	23	19	22	18
	WB-40	24	19	22	19	21	19	19	18	17	16
	WB-50	31	22	27	21	28	20	25	19	22	18
	WB-62	—	—	—	—	—	—	—	—	30	22
90°	SU	26	20	23	18	19	16	17	15	13	13
	BUS	38	23	33	22	30	22	25	21	22	18
	WB-40	31	22	27	21	23	21	19	18	17	16
	WB-50	42	22	37	24	32	22	29	21	22	18
	WB-62	—	—	—	—	—	—	—	—	39	23
120°	SU	34	22	27	19	21	18	17	16	13	13
	BUS	46	28	40	25	32	23	26	19	19	18
	WB-40	37	23	29	22	24	22	19	18	17	16
	WB-50	50	29	43	28	36	27	30	26	22	18
	WB-62	—	—	—	—	—	—	—	—	26	22
150°	SU	40	25	32	21	22	19	17	16	12	12
	BUS	48	28	40	25	32	23	22	18	17	16
	WB-40	39	24	29	22	23	22	19	18	17	16
	WB-50	53	31	46	28	36	27	28	26	22	18
	WB-62	—	—	—	—	—	—	—	—	20	18

NOTES: p design vehicle turns within 12 ft width where R = 15 ft or more. No parking on either street.

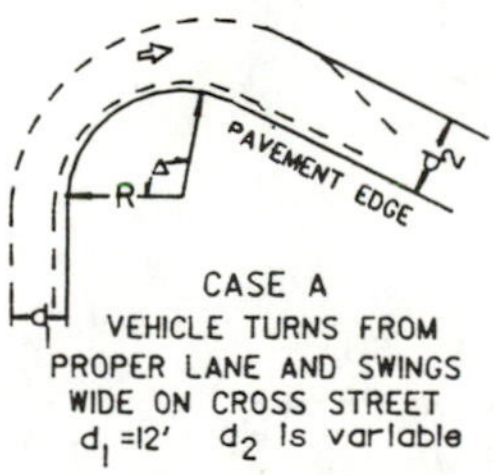

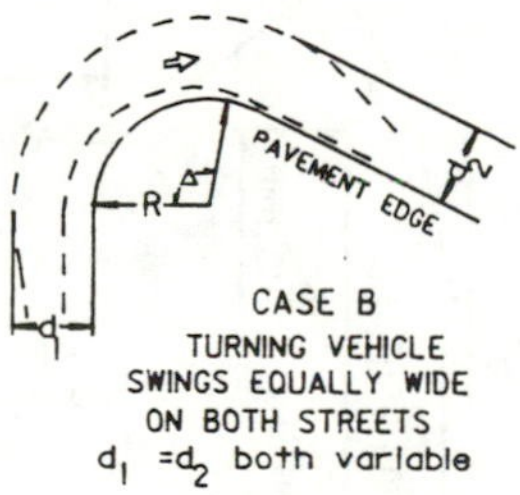

Table IX-3. Cross-street width occupied by turning vehicle for various angles of intersection and curb radii.

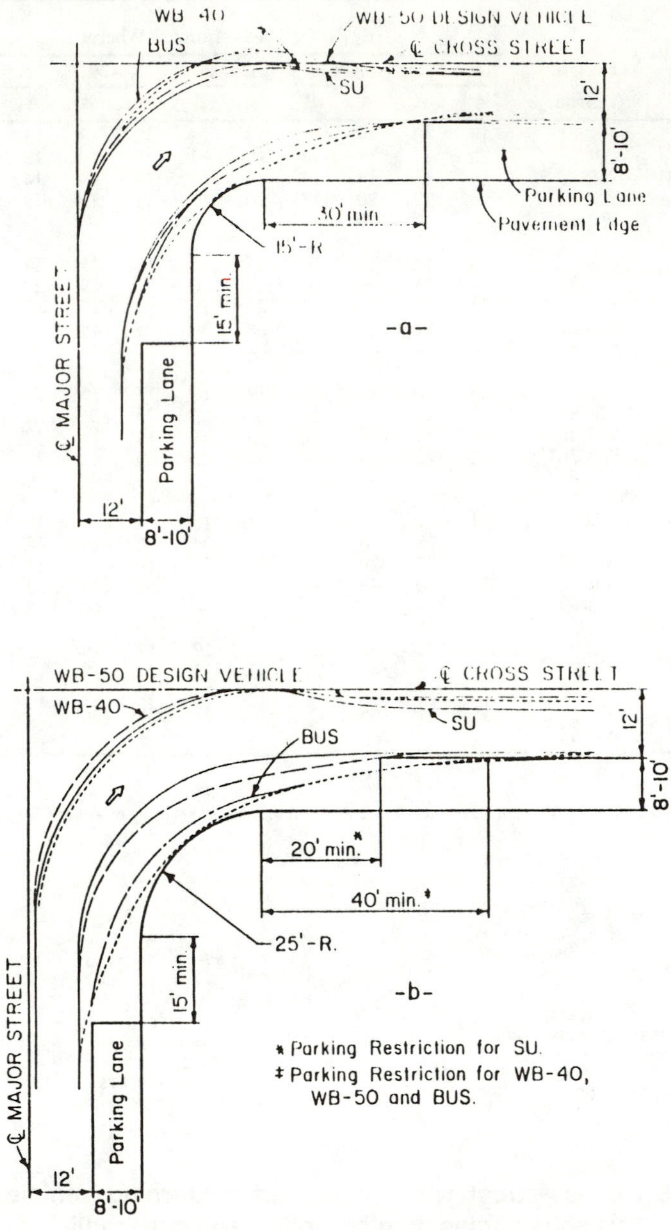

Figure IX-20. Effect of curb radii and parking on turning paths.

For arterial street design, adequate radii for vehicles must be balanced against the needs of pedestrians and the difficulty of acquiring additional right-of-way or corner setbacks. Because the corner radius often is a compromise, its effect on pedestrians in combination with vehicular movements should be examined.

Crosswalk distances and the correspondingly required additional right-of-way or corner setback increase with the radius of curb return. On the basis of the assumptions that the sidewalk centerline at a right-angle intersection is in line with the middle of a border and that the same curb radius is used on all four corners, the added crosswalk distances between curbs as compared with the normal curb-to-curb street widths are shown in Figure IX-21.

The additional right-of-way or corner setback resulting from various curb radii for border widths of 10 and 20 ft is shown in Figure IX-22. The dimensions shown in Figures IX-21 and IX-22 vary somewhat with intersection angles that are either more or less than 90°.

The dimensions presented in Figures IX-21 and IX-22 demonstrate why curb radii of only 10 to 15 ft have been used in most cities. Where larger radii are used, an intermediate refuge or median island is desirable or crosswalks may need to be offset so that crosswalk distances are not objectionable.

In summary, corner radii at intersection on arterial streets should satisfy the requirements of the drivers using them to the extent practical and in consideration of the amount of right-of-way available, the angle of the intersection, numbers of pedestrians, width and number of lanes on the intersecting streets, and amounts of speed reductions. The following summary is offered as a guide:

1. Radii of 15 to 25 ft are adequate for passenger vehicles. These radii may be provided at minor cross streets where there is little occasion for trucks to turn or at major intersections where there are parking lanes. Where the street has sufficient capacity to retain the curb lane as a parking lane for the foreseeable future, parking should be restricted for appropriate distances from the crossing.
2. Radii of 25 ft or more at minor cross streets should be provided on new construction and on reconstruction where space permits.
3. Radii of 30 ft or more at major cross streets should be provided where feasible so that an occasional truck can turn without too much encroachment.
4. Radii of 40 ft or more, and preferably three-centered compound curves or simple curves with tapers to fit the paths of appropriate design vehicles, should be provided where large truck combinations and buses turn frequently. Larger radii are also desirable where speed reductions would cause problems.

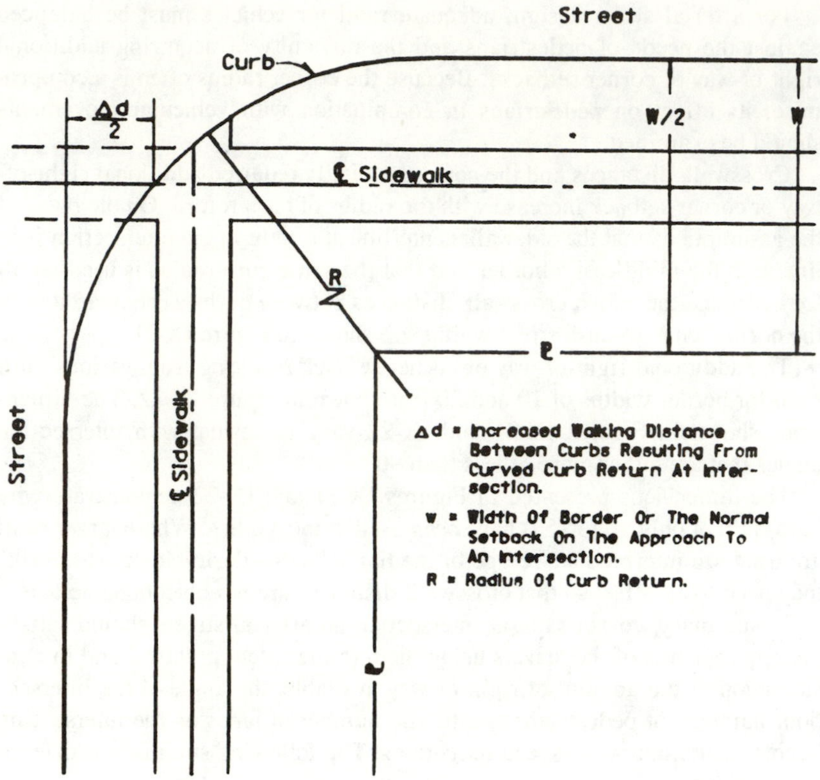

Curb Radius, R	Added Crosswalk Distance Δd	
	W=10'	W=20'
Feet	Feet	Feet
10	3	0
20	14	5
30	27	15
40	42	27
50	57	40

Figure IX-21. Variations in length of crosswalk with corner-curb radius and width of border.

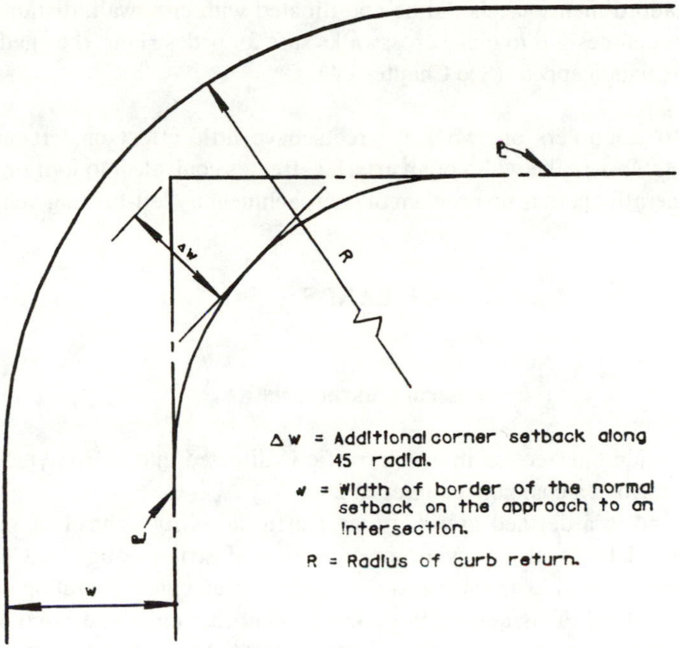

Curb Radius, R	Additional Corner Setback, ΔW	
	W=10′	W=20′
feet	feet	feet
10	0	0
20	4	0
30	8	4
40	13	8
50	17	13

Figure IX-22. Variations in corner setback with corner-curb radius and width of border.

5. Radii dimensions should be coordinated with crosswalk distances or special designs to make crosswalks safe for pedestrians, the aged, and the handicapped. (See Chapter IV.)

Curb radii at corners on two-way streets have little effect on left-turning movements. Where the width of an arterial street is equivalent to four or more lanes, generally there is no problem of encroachment by left-turning vehicles.

ISLANDS

General Characteristics

An at-grade intersection in which traffic is directed into definite paths by islands is termed a channelized intersection.

An island is a defined area between traffic lanes for control of vehicle movements. Islands also provide an area for pedestrian refuge and traffic control devices. Within an intersection a median or an outer separation is considered an island. This definition makes evident that an island is no single physical type—it may range from an area delineated by a curb to a pavement area marked by paint.

Islands generally are included in intersection design (channelization) for one or more of the following purposes:

1. Separation of conflicts.
2. Control of angle of conflict.
3. Reduction in excessive pavement areas.
4. Regulation of traffic and indication of proper use of intersection.
5. Arrangements to favor a predominant turning movement.
6. Protection of pedestrians.
7. Protection and storage of turning and crossing vehicles.
8. Location of traffic control devices.

MINIMUM DESIGN FOR TURNING ROADWAYS

Where the inner edges of pavement for right turns at intersections are designed to accommodate semitrailer combinations or where the design permits passenger vehicles to turn at speeds of 15 mph or more, the pavement area at the intersection may become excessively large for proper control of traffic. To avoid this condition, an island may be provided to form a separate turning roadway (a connecting roadway for traffic turning between two intersection legs).

At-grade intersections which have large paved areas, such as those with large corner radii and those at oblique angle crossings, permit and encourage uncontrolled vehicle movements, require long pedestrian crossings, and have unused pavement areas. Even at a simple intersection, appreciable areas may exist on which some vehicles can wander from natural and expected paths. Conflicts may be reduced in extent and intensity by layout design to include islands.

Islands generally are either elongated or triangular in shape and are situated in areas normally unused as vehicle paths. The dimensions depend on the particular intersection design. Islands should be located and designed to offer little hazard to vehicles, be relatively inexpensive to build and maintain, and occupy a minimum of roadway space but yet be commanding enough that motorists will not drive over them. Island details depend on particular conditions and should be designed to conform to the general principles that follow.

Curbed islands are sometimes difficult to see at night because of the glare from oncoming headlights or from distant luminairies or roadside businesses. Accordingly, where curbed islands are used, the intersection should have fixed-source lighting or appropriate delineation.

Where various intersections are involved in a given project and the warrants are sufficiently similar, in order to enhance driver expectancy it is desirable to provide a common geometric design for each intersection. The designer should also refer to part V of the *Manual of Uniform Traffic Control Devices for Streets and Highways (MUTCD)* (**5**) for design guidance.

Painted, flush medians and islands may be preferred to the curbed type under certain conditions, including the following: in lightly developed areas; at intersections where approach speeds are relatively high; where there is little pedestrian traffic; where fixed-source lighting is not provided; and where signals, signs, or lighting standards are not needed on the median or island. Painted islands may also be placed at the pavement edge. At some intersections, both curbed and painted islands may be desirable. All pavement markings should be reflectorized. The use of thermoplastic striping, raised dots, spaced and raised retroreflective markers, and other forms of long-life markings also may be desirable. This subject is discussed in the section "Traffic Control" in Chapter VI.

Islands serve three primary functions: (1) channelization—to control and direct traffic movement, usually turning, (2) division—to divide opposing or same direction traffic streams, usually through movements, and (3) refuge—to provide refuge for pedestrians. Most islands combine two or all of these functions.

Channelized Islands

Islands to control and direct traffic movement should guide motorists into the proper channel for their intended route. The confusing traffic movements resulting from too spacious areas may be eliminated by conversion of the normally unused areas into islands that leave little to the discretion of drivers. Channelizing islands may be of many shapes and sizes, depending on conditions and dimensions of the intersection, some of which are illustrated in Figure IX-23. A common form is one of triangular shape, island a, to separate right-turning traffic from through traffic. Central islands such as c and d may serve as a guide around which turning vehicles operate.

Channelizing islands should be placed so that the proper course of travel is immediately obvious, easy to follow, and of unquestionable continuity. When designing an island, attention should be given to the fact that the driver's eye view is different from the plan view. Particular care must be taken where the channelization is on or beyond a crest of a vertical curve, however slight, or where there is substantial horizontal curvature on the approach to or through the channelized area. The outlines of islands should be easily flowing curved lines or straight lines nearly parallel to the line of travel. Where islands separate turning traffic from through traffic, the radii of curved portions should equal or exceed the minimum for the turning speeds expected. Drivers should not be confronted suddenly with an unusable area in the normal vehicle path. Islands first approached by traffic should be indicated by a gradually widening marking or a conspicuously roughened strip that directs traffic to each side.

Intersections with multiple turning lanes may require three or more islands to channelize the various movements. There is a practical limitation to the use of multiple islands for channelizing traffic. A group of islands outlining several one-way lanes may cause confusion and result in inadvertent entrance of a one-way lane by opposing traffic. Such layouts may be confusing only to drivers using them for the first time, and on later passages the same drivers may use them as desired. However, the hazard of multiplicity suggests that a few large islands are preferable to a greater number of smaller islands. At intersections where the area for multiple-lane channelization is restricted, it may be advisable to try temporary layouts of movable stanchions or sandbags and observe traffic flow with several variations of sizes or shapes of islands before designing and constructing the permanent islands.

Properly placed islands are advantageous where through and turning movements are heavy. At minor intersections on two-lane highways, however, channelization may be of questionable value, especially in rural areas where small curbed islands were provided. The use of curbed islands generally should be reserved for multilane highways or streets and for the more important intersections on two-lane highways. In or near urban areas where speeds are low and drivers are accustomed to confined facilities, channelization can be expected to work well. Curbed islands generally should not be used in rural areas and at isolated locations.

Marked channelization (painting or striping) can be made to increase efficiency and safety and has the advantage of easy modification when warranted by driver behavior. If a more positive barrier is required, curbed islands may be constructed, but the marked channelization may well serve initially to establish the best layout arrangement before permanent construction is established. It should be noted, however, that inclement weather decreases the effectiveness of marked channelization.

Divisional Islands

Divisional islands often are introduced on undivided highways at intersections. They alert the drivers to the crossroad ahead and regulate traffic through the intersection. These islands are particularly advantageous in controlling left turns at skewed intersections and at locations where separate channels are provided for right-turning traffic. A variety of divisional islands that divide opposing traffic are illustrated in Figure IX-23.

Where an island is introduced at an intersection to divide opposing traffic on a four-lane road or on a major two-lane highway carrying high volumes, particularly where future conversion to a wider pavement is likely, two full lanes on each side of the dividing island should be provided. In other instances narrower pavements may be used. For moderate volumes, pavement widths shown under case II (one-lane one-way operation with provision for passing a stalled vehicle) in Chapter III and in Table III-21 are appropriate. For light volumes and where small islands are necessary, widths on each side of the island corresponding to case I in Table III-21 may be used.

Widening a roadway to include a divisional island (Figure IX-24) should be done in such a manner that the proper paths to follow are unmistakably evident to drivers. The alinement should require no appreciable conscious effort in vehicle steering. Often the highway is on a tangent, and to introduce dividing islands, reverse curve alinement must be used. In rural areas where speeds are generally high, necessary reversals in curvature should preferably be 1° or less. Sharper curves may be used on intermediate-speed roads, but preferably should not be greater than 2°. Usually, the pavement in each direction of travel is bowed out, more or less symmetrically about the centerline, as shown in Figure IX-24. Widening may also be effected on one side only, one of the pavements continuing through the intersection on a straight course. When this arrangement is used for a two-lane road that is planned for future

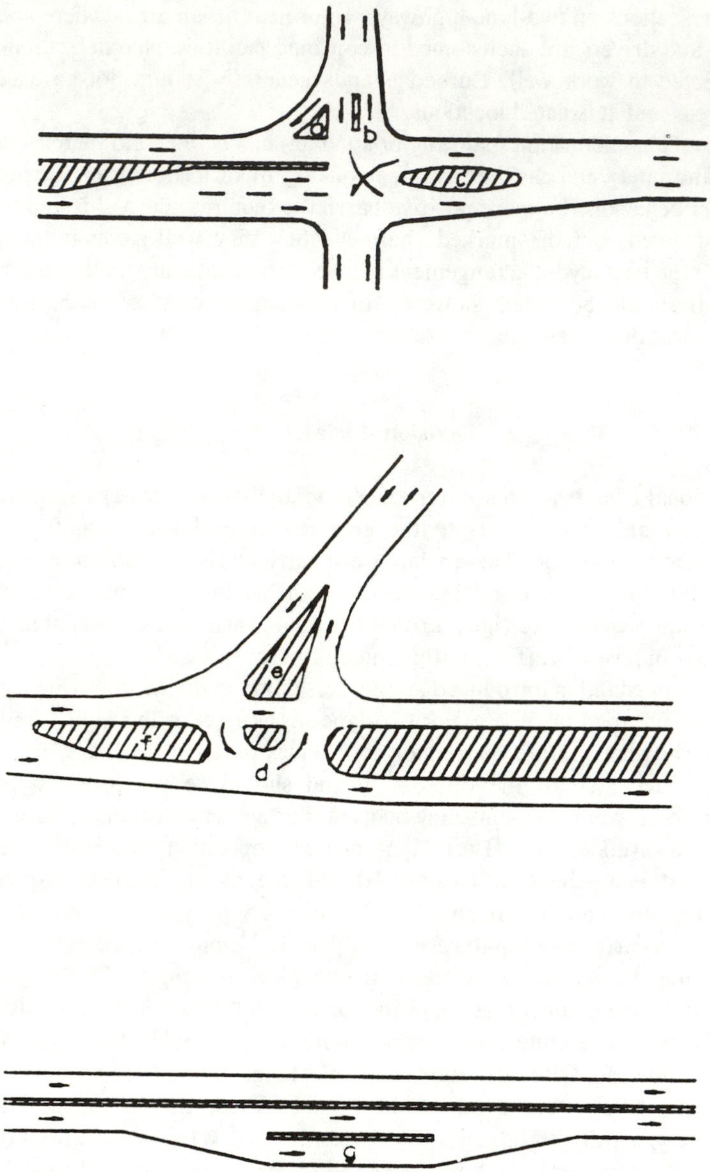

Figure IX-23. General types and shapes of islands.

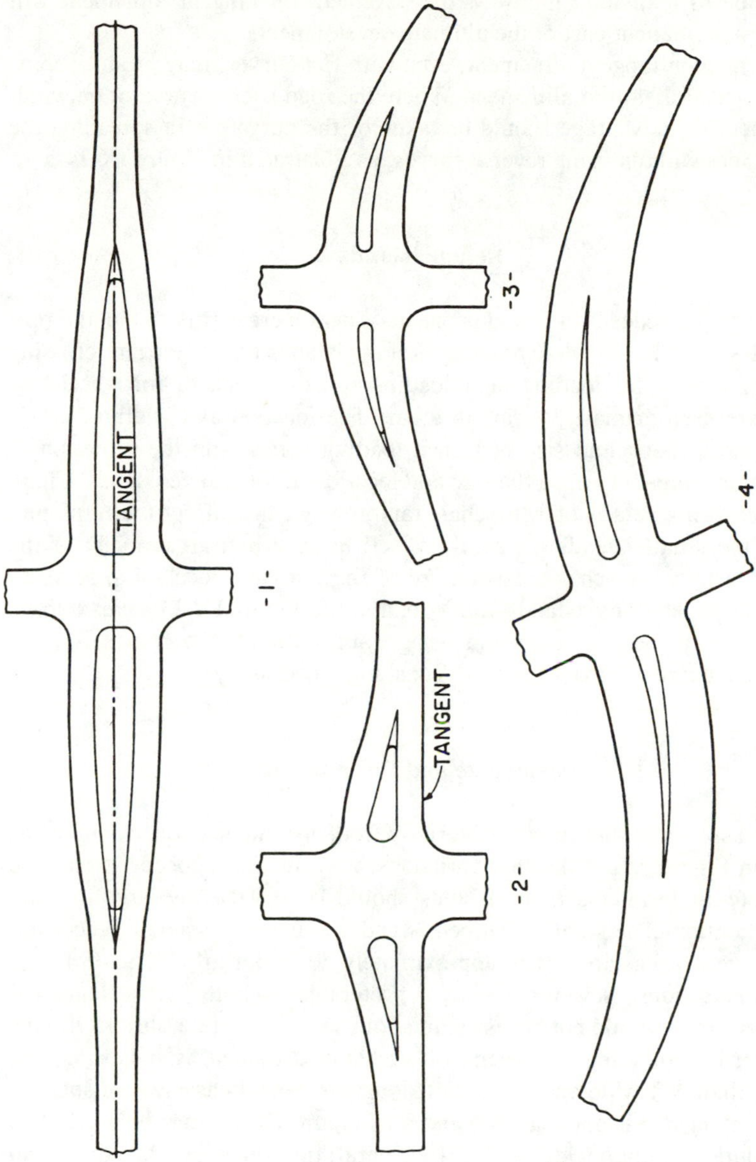

Figure IX-24. Alinement for addition of divisional islands at intersections.

conversion to a divided highway, the pavement on tangent alinement will become a permanent part of the ultimate development.

Widening on tangent alinement, even with flat curves, may produce some appearance of distorted alinement. Where the road is on a curve or on winding alinement, advantage should be taken of the curvature in spreading the traffic lanes without using reverse curves, as illustrated in Figure IX-24.

Refuge Islands

A refuge or pedestrian island is one at or near a crosswalk to aid and protect pedestrians crossing the roadway. Refuge islands for pedestrians crossing a wide highway, for loading or unloading transit riders, or for wheelchair ramps are used primarily in urban areas. The location and width of crosswalks, the location and size of transit loading zones, and the provision of wheelchair ramps influence the size and location of the refuge islands. Chapter IV contains details of wheelchair ramp design that affect the minimum size of the small islands. In rural as well as in urban areas many of the islands designed for channelization are of the type and location to serve as refuge for pedestrians. Islands a, b, c, e, and f in Figure IX-23 are examples. The general principles for island design apply directly to refuge islands, except that barrier curbs usually are considered necessary.

Island Size and Designation

Island sizes and shapes vary materially from one intersection to another, as shown in Figure IX-23. Further variations, not illustrated, occur at multiple and acute-angle intersections. Islands should be sufficiently large to command attention. The smallest curbed island that normally should be considered is one that has an area of approximately 50 ft^2 for urban and 75 ft^2 for rural intersections. However, 100 ft^2 is preferable for both. Accordingly, triangular islands should not be less than about 12 ft, and preferably 15 ft, on a side after the rounding of corners. Elongated or divisional islands should be not less than 4 ft wide and 20 to 25 ft long. In special cases where space is limited, elongated islands, as in b and g of Figure IX-23, may be reduced to an absolute minimum width of 2 ft. In general, introducing curbed divisional islands at isolated intersections on high-speed highways is undesirable unless special attention is directed to providing high visibility for the islands. Curbed divisional islands introduced at isolated intersections on high-speed highways should be at least 100 ft and preferably several hundred feet in length. When they cannot be long, they may be preceded by visibly roughened pavement, jiggle bars, or marking. When situated in the vicinity of a high point in the roadway profile or at or near the beginning of a horizontal curve, the approach end of the curbed island should be extended to be clearly visible to approaching drivers.

Islands should be delineated or outlined by a variety of treatments, depending on their size, location, and function. The type of area in which the intersection is located, rural versus urban, also governs the design. In a physical sense, islands can be divided into three groups: (1) raised islands outlined by curbs, (2) islands delineated by pavement markings, buttons, or raised (jiggle) bars placed on all-paved areas; and (3) nonpaved areas formed by the pavement edges, possibly supplemented by delineators on posts or other guideposts, or a mounded-earth treatment beyond and adjacent to the pavement edges.

Curbed island treatment, group 1, is universal and is the most positive. In rural areas where curbs are not common, this treatment often is limited to islands of small to intermediate size. Conversely, in urban areas, use of this type of island is generally standard practice.

Island delineation on paved areas, group 2, is used in urban districts where speeds are low and space is limited. In rural areas this type is used where maintenance problems and snow removal make curbs undesirable or where high approach speeds make any curb a potential hazard. Group 2 islands also are applicable on low-volume highways where the added expense of curbs may not be warranted and where the islands are not large enough for delineation by pavement edges alone.

The group 3 treatment necessarily applies to other than small islands and is logical primarily at rural intersections where there is space for large-radius intersection curves.

The central area of large islands in most cases has a turf or other vegetative cover. As space and the overall character of the highway determine, low landscape planting may be included, but it must not obstruct sight distance. Ground cover or plant growth, such as turf, vines, and shrubs, is desirable for an island and provides excellent contrast with the paved areas assuming that the ground cover is cost justified and can be properly maintained. Small curbed islands may be mounded, but where pavement cross slopes are outward, large islands should be depressed to avoid draining across the pavement. This feature is especially desirable where alternate freezing and thawing of stored snow may occur. For small curbed islands and in areas where growing conditions are not favorable, some form of paved surface is used on the island. In many respects the curbed-island cross section design is similar to that discussed in the section "Medians—Width and Cross Section" in Chapter IV.

Delineation and Approach-End Treatment

Delineation of small curbed islands is effected primarily by curbs. Large curbed islands may be sufficiently delineated by color and texture contrast of vegetative cover, mounded earth, shrubs, guard posts, signs, or any combination of these. In rural areas island curbs should nearly always be a mountable type, except where there is a definite need for a non-mountable curb, as at structures or pedestrian crossings. In special cases, curbs, as shown in Chapter IV are suitable, preferably of heights not more than 6 to 9 in. Either type of curb could be appropriate in urban areas, depending on the condition. High-visibility curbs are advantageous at critical locations or on islands and roadway forks approached by high-speed traffic.

The outline of a curbed island is determined by the edge of through-traffic lanes and turning roadways, with lateral clearance, if any, to the curbed island sides. The points at the intersections of the curbed island are rounded or beveled for visibility and construction simplicity. The amount that a curbed island is offset from the through-traffic lane is influenced by the type of edge treatment and other factors such as island contrast, length of taper or auxiliary pavement preceding the curbed island, and traffic speed. Island curbs are introduced rather suddenly and should be offset from the edge of through-traffic lanes even if they are mountable. A mountable curb at an island need not be offset from the edge of a turning roadway, except to reduce its vulnerability. Barrier curbs should be offset from edges of through and turning roadway pavements.

Details of triangular curbed island design are shown in Figures IX-25 and IX-26. The lower right corner of each curbed island is designed for approach-end treatment. Three curbed island sizes, small, intermediate, and large, are shown for two general cases of through-traffic lane edge: (1) The curbed island edge is determined by offset clearance from the through pavement, and (2) the curbed island edge is at the outside of a shoulder carried through the intersection. Small curbed islands are those of minimum or near-minimum size, as previously discussed. Large curbed islands are those with side dimensions at least 100 ft long. All curbed islands in Figures IX-25 and IX-26 are shown with approach noses rounded on appropriate radii of 2 to 3 ft and with minimum rounding (about 1-ft radius) at merging ends. The lower left corner is rounded with a radius of 2 to 5 ft.

Figure IX-25 shows curbed islands adjacent to through-traffic lanes. Where there are no curbs on the approach pavement, the minimum offset of the edge of the curbed island should be 2 to 3 ft. Where the approach pavement has a mountable curb, a similar curb on the curbed island could be located at the edge of the through lane where there is sufficient length of curbed island to effect a gradual taper from the nose offset. Non-mountable

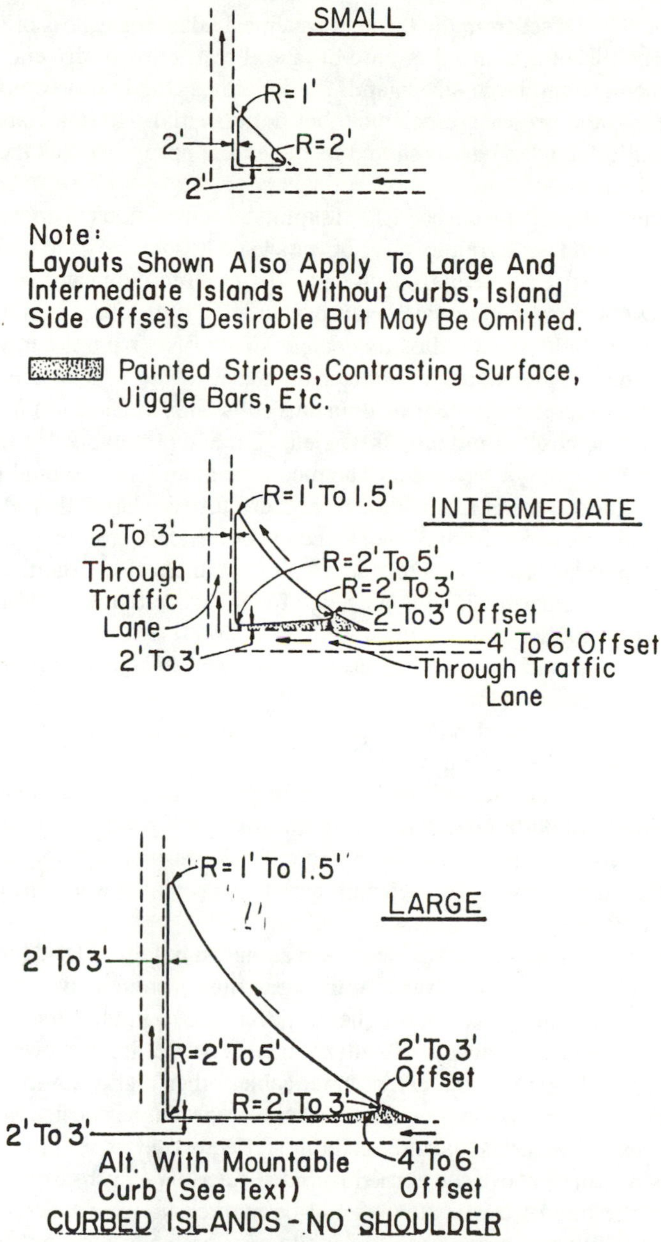

Figure IX-25. Details of triangular island design (curbed islands, no shoulder).

curbs should be offset from the through pavement edge, regardless of the size of the curbed island, to avoid a sense of lateral restriction to drivers. Where the intermediate and large-size islands are uncurbed, the indicated offsets of the curbed island proper are desirable but not essential. However, any fixed objects within the island areas should be offset an appropriate distance from the through lanes.

The approach end of a curbed island should be conspicuous to approaching drivers and should be definitely clear of vehicle paths, physically and visually, so that drivers will not veer from the island. A raised delineator (nonrigid) may be desirable on the approach end, particularly if the curbed island is so narrow that it could be straddled by vehicles. The offset from the travel lane to the approach nose should be greater than that to the side of the curbed island, normally about 2 ft. For median curbed islands the face of curb at the approach island nose should be offset at least 2 ft and preferably 4 ft from the normal median edge of pavement. The median curbed island should then be gradually widened to its full width. For other curbed islands the total nose offset should be 4 to 6 ft from the normal edge of through pavement and 2 to 3 ft from the pavement edge of a turning roadway. Large offsets should be provided where the curbed island is preceded by a deceleration lane or a gradually widening auxiliary pavement.

When an approach shoulder is used, the curbed island should be offset from the through travel lane by an amount equal to the shoulder width except where a deceleration or turn lane is provided. Where speeds are intermediate or high and the curbed island is preceded by a deceleration lane or a gradually widened auxiliary pavement, it may be desirable to offset the nose of large curbed islands an additional 2 to 4 ft. In heavily developed areas an offset as small as 1 ft may be appropriate for non-mountable curbs where the approach nose of the curb island is offset and tapered, particularly where pedestrian traffic is a factor.

Curbed islands should be ramped down as shown in Figure IX-27 and provided with devices to give advance warning to approaching drivers both day and night. Pavement markings, roughened pavement, or jiggle bars in front of the approach nose are particularly advantageous on the areas shown as stippled in Figure IX-25. To the extent practicable, other high-visibility indications should be used, such as reflectorized curbs, signs located near the approach ends of curbed islands suitably reflectorized or illuminated, or reflectors mounted above the curbed island surface. The curbs of all islands located in the line of traffic flow should be marked in accordance with the MUTCD (**2**). Where approaches to islands may be made in more than one direction, caution should be exercised in determining whether to ramp the curbed island down so that adequate protection is provided for traffic signal equipment and signs located in the island.

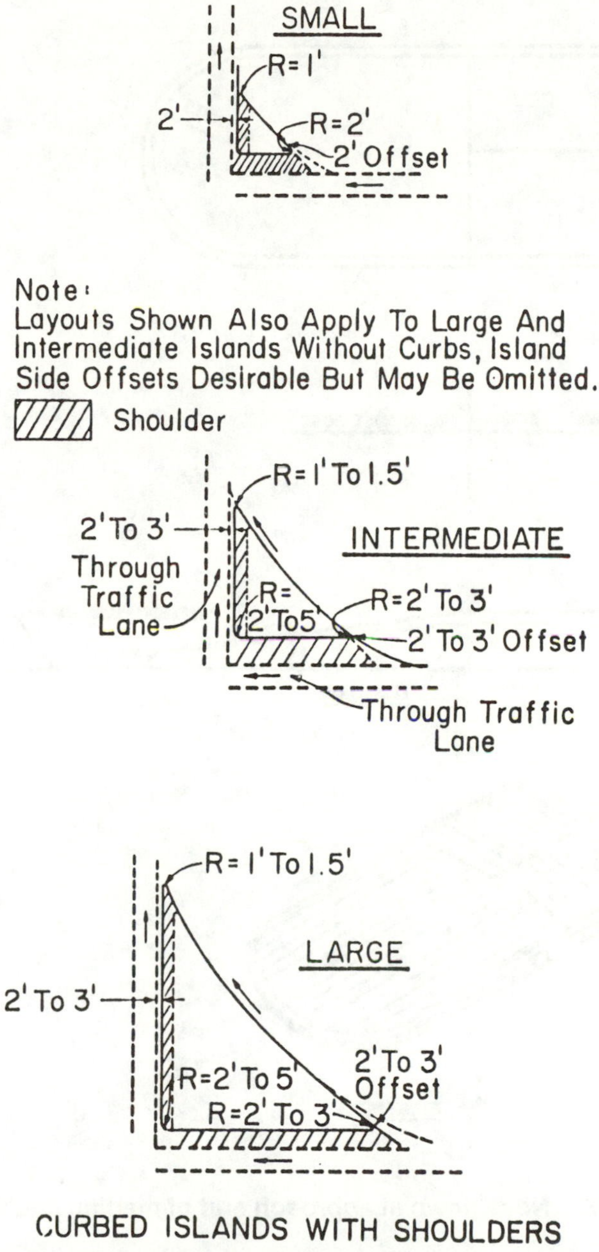

Figure IX-26. Details of triangular island design (curbed island with shoulders).

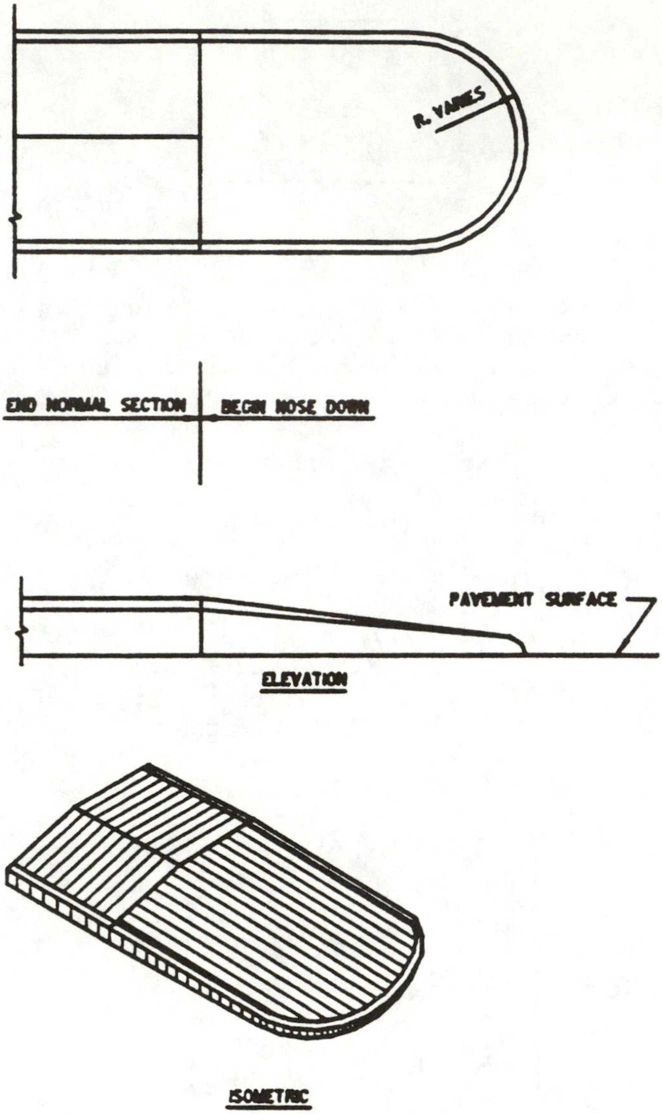

Figure IX-27. Nose down at approach end of median and island.

Delineation and warning devices are especially pertinent at approach ends of median-curbed islands, which usually are in the direct line of approaching traffic. In rural areas the approach should consist of a gradually widening center stripe (Figure IX-28). Although not as frequently obtainable, this approach should be strived for in urban areas also. Preferably, it should gradually change to a raised marking of color and texture contrasting with that of the traffic lanes or to jiggle bars that may be crossed readily even at considerable speed. This section should be as long as practicable. The cross sections in Figure IX-28 demonstrate the transition. The face of curb at the approach island nose should be offset at least 2 ft and preferably 4 ft from the normal (median) edge of pavement, and the widened pavement gradually should be transitioned to the normal width toward the crossroad. The end of the curbed island at the crossroad is designed as a median end. The delineation and approach-end treatment described for triangular and median-curbed islands generally apply to other forms of curbed island. However, pavement markings or raised pavement areas at approach island ends may not be necessary at secondary curbed islands situated within a multiple-island intersection. The approach-end treatment is primarily pertinent to curbed islands first approached by traffic entering an intersection.

Right-Angle Turns With Corner Islands

The principal controls for the design of turning roadways are the alinement of the inner pavement edge and the pavement width, so that the design vehicle can be accommodated while turning at a speed consistent with the operation of the intersection. With radii greater than minimum, these controls result in an area large enough for an island, generally triangular in shape, between the inner edge of the turning roadway and the pavement edges (extended) of the two through highways. Such an island is desirable for delineating the path of through and turning traffic, for the placement of signs, and for providing refuge for pedestrians. Larger islands may be necessary to locate signs and to facilitate snow-removal operations. The inner edge of pavement on the turning roadway should be designed to provide at least the minimum island and the minimum width of turning roadway pavement. The turning roadway pavement should be wide enough to permit the outer and the inner wheel tracks of a selected vehicle to be within the edges of the pavement by about 2 ft on each side. Generally, the turning roadway pavement width should not be less than 14 ft. However, pavement widths designed for larger vehicles may be reduced with paint or color contrasting treatments to channelize passenger cars and discourage the formation of two turning lanes.

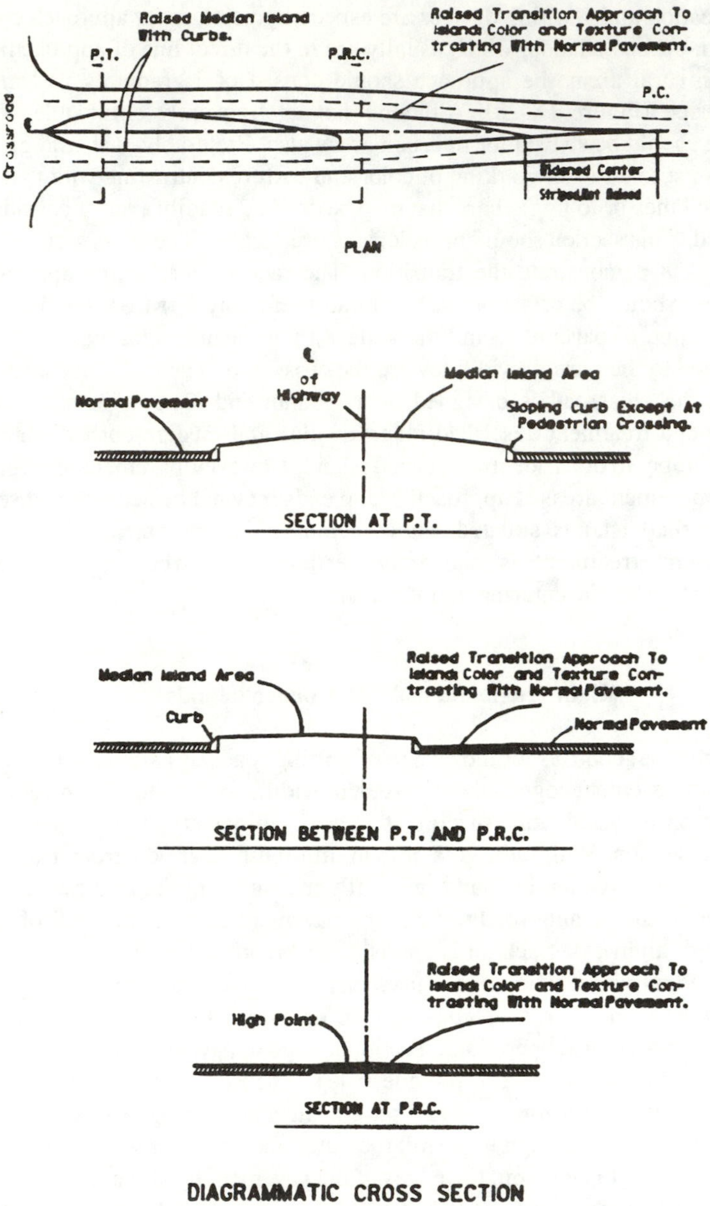

Figure IX-28. Details of divisional island design.

Figure IX-29 shows minimum designs for turning roadways for a 90° right turn to fit these controls. A design based on a minimum island and a minimum width of channel of 14 ft (Figure IX-29A) results in a circular arc of 60-ft radius (not shown) on the inner pavement edge of the turning roadway or in a three-centered curve (as shown) with radii of 150, 50, and 150 ft with the middle curve offset 3 ft from the tangent edges extended. This design not only permits passenger vehicles to turn at a speed of 15 mph but also enables SU design vehicles to turn on a radius (outer front wheel) of approximately 65 ft and still clear the pavement edges of the channel by about 1 ft on each side, as shown by the vehicle paths in the figure.

By increasing the pavement width to 18 ft and using the same combination of curves but with the middle curve offset 5 ft from the tangent edges extended, a more desirable arrangement results (Figure IX-29B). This design enables the SU design vehicle to use a 70-ft turning radius with clearances and makes it possible for the WB-50 vehicle to negotiate the turn with only slight encroachment on adjacent through-traffic lanes.

At locations where a significant number of semi-trailer combinations, particularly the longer units, will be turning, the arrangement shown in Figure IX-29C should be used. This design, consisting of a minimum curve of 65-ft radius, offset 6 ft, and terminal curves of 180 ft radii, generally provides for a WB-50 design vehicle passing through a 20-ft turning roadway pavement and greatly benefits the operation of smaller vehicles.

However, the designer should be aware of larger semi-trailer combinations on designated roadways and the effects these vehicles will have on turning roadway pavements. The designer should reference the truck turning templates in Chapter II to meet their design needs. As previously stated, turning roadway pavement widths can be reduced with paint or color contrasting treatments to channelize passenger cars and discourage the formation of two turning lanes.

The island in all instances should be located about 2 ft outside the pavement edges extended, as shown in Figure IX-29C. For high-speed highways the offset to the through lanes would be desirably greater. When of minimum or near-minimum size, it should be delineated by curbs. In rural areas the use of the painted islands should be given prior consideration. If curbed islands are necessary, the curbs should be mountable to make them less hazardous to through traffic and to allow for greater latitude in the maneuvering of large vehicles. (See Figures IX-25 and IX-26 and accompanying discussion.)

For each minimum design shown in Figure IX-29, a three-centered symmetric compound curve is recommended; however, asymmetric compound curves could also be used, particularly where the design provides for the turning of trucks. Although an equivalent simple curve of a given radius is noted

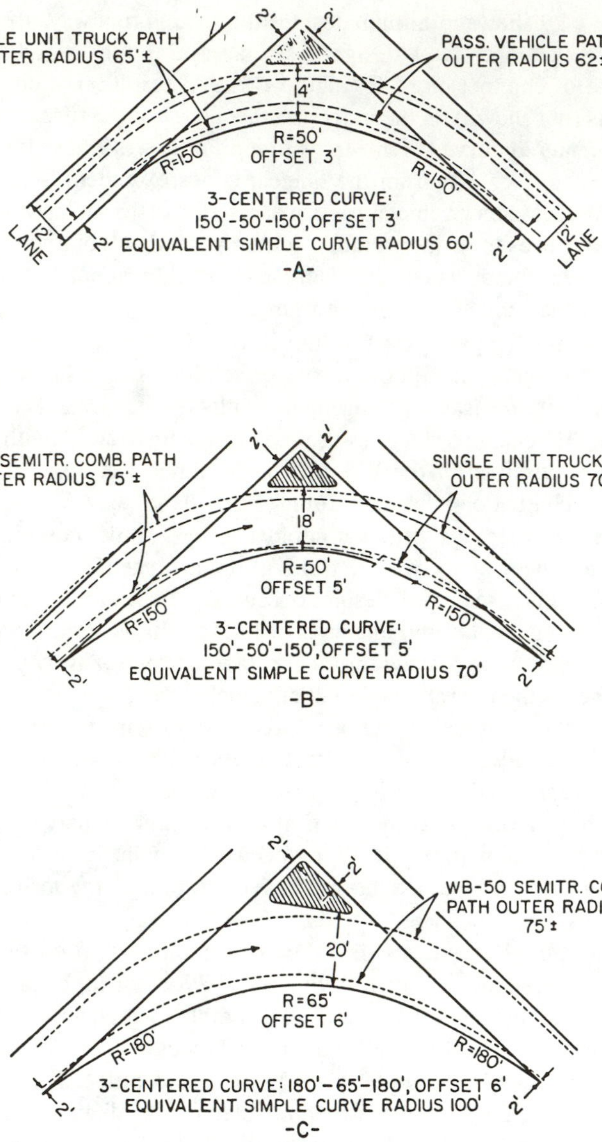

Figure IX-29. Designs for turning roadways with minimum corner islands.

in the figure in each case, its use in the two latter designs may result in design vehicle encroachments on the shoulder or the island.

Oblique-Angle Turns with Corner Islands

Minimum design dimensions for oblique-angle turns, determined on a basis similar to that for right-angle turns, are given in Table IX-4. Curve design for the inner edge of pavement, turning roadway pavement width, and the approximate island size are indicated for the three chosen design classifications described at the bottom of the table. For a particular intersection the designer may choose from the three minimum designs shown in accordance with the size of vehicles, the volume of traffic anticipated, and the physical controls at the site.

In Table IX-4 no design values are given for angles of turn less than 75°. Turning roadways for flat-angle turns involve relatively large radii and are not considered in the minimum class. Such arrangements require individual design to fit site controls and traffic conditions. For angles of turn between 75° and 120° the designs are governed by a minimum island, which provides for turns on more than minimum turning radii. For angles of 120° or more, the sharpest turning paths of the selected vehicles and arrangements of curves on the inner edge of pavement to fit these paths generally control the design, the resulting island size being greater than the minimum. The inner edge of pavement arrangements for designs B and C for turning angles of 120° to 150° are the same as those given in Table IX-2 for single-unit trucks and semitrailer combinations, respectively. The size of islands for the large turning angles given in the last column of Table IX-4 indicates the otherwise unused and uncontrolled areas of pavement that are eliminated by the use of islands.

APPLICATION AT TURNING ROADWAY TERMINALS

An important part of intersection design is the provision of a suitable pavement edge alinement where a turning roadway departs from or joins the through highway pavement. Ease and smoothness of operation result when the pavement edge is designed with spirals or compound curves and is of a shape and length to avoid abrupt deceleration by drivers before they leave the through pavement, to permit development of superelevation in advance of the maximum curvature, and to enable vehicles to follow natural turning paths.

Various degrees of transitional treatments for an exit from a highway are illustrated in Figures IX-30 and IX-31 for design (turning) speeds of 20 and 30 mph, respectively. As p, the offset from the edge of the through pavement

Angle of Turn (degrees)	Design Classification	Three-Centered Compound Curve Radii (ft)	Three-Centered Compound Curve Offset (ft)	Width of Lane (ft)	Approx. Island Size (sq ft)
75	A	150-75-150	3.5	14	60
	B	150-75-150	5.0	18	50
	C	180-90-180	3.5	20	50
90[a]	A	150-50-150	3.0	14	50
	B	150-50-150	5.0	18	80
	C	180-65-180	6.0	20	125
105	A	120-40-120	2.0	15	70
	B	100-35-100	5.0	22	50
	C	180-45-180	8.0	30	60
120	A	100-30-100	2.5	16	120
	B	100-30-100	5.0	24	90
	C	180-40-180	8.5	34	220
135	A	100-30-100	2.5	16	460
	B	100-30-100	5.0	26	370
	C	160-35-160	9.0	35	640
150	A	100-30-100	2.5	16	1400
	B	100-30-100	6.0	30	1170
	C	160-35-160	7.1	38	1720

[a] Illustrated in Figure IX-29.

NOTES: Asymmetric three-centered compound curves and straight tapers with a simple curve can also be used without significantly altering the width of pavement or corner island size.

Painted island delineation is recommended for islands less than 75 ft^2 in size.

Design classification:

A—Primarily passenger vehicles; permits occasional design single-unit truck to turn with restricted clearances.

B—Provides adequately for SU; permits occasional WB-50 to turn with slight encroachment on adjacent traffic lanes.

C—Provides fully for WB-50.

Table IX-4. Minimum designs for turning roadways.

to the minimum radius curve produced, is increased, progressively smoother and more adequate turnout facilities are provided. A spiral curve of minimum length (Table IX-5) joining the edge of through pavement with the intersection curve (Figure IX-30B) is a substantial improvement over the simple curve (Figure IX-30A) in easing the turning path. This short spiral is not satisfactory for development of superelevation on the turnout because the wedge of auxiliary pavement, areas a-b-c, is too small for this purpose. By doubling the minimum length of spiral (Figure IX-30C), p is increased to nearly 9 ft, an increase that provides for a more directional path and a substantially greater pavement area (a-b-c-d) for development of superelevation than in the previous example. Similar, although generally not as satisfactory, results can be obtained by compounding the intersection curve with an arc of double the minimum radius (Figure IX-30D).

A more desirable edge alinement is obtained by using a three step compound curve (Figure IX-30E). In this case the 90-ft radius curve is preceded by 180- and 360-ft radii curves of approximately the minimum lengths given in Table IX-6. This arrangement requires more space as indicated by the offset p of approximately 23 ft. The design is superior to the preceding examples in providing a more gradual turnout with some space furnished for deceleration off the through traffic lane, and an area of transitional pavement on which a relatively flat circular curve is used and a combining spiral between it and the minimum radius curve. The radius selected for the initial curve corresponds to a design speed of about 45 mph, 25 mph more than the speed for the minimum (90 ft radius) intersection curve. The end of the flatter curve, or the beginning of the spiral, is positioned where the widening obtained approximates a lane width.

Minimum lengths of spiral for intersection curves (Table IX-5) and lengths of circular arcs for compound intersection curves (Table IX-6) are also discussed in Chapter III under these headings.

Figures IX-30E and IX-31F indicate the desirable type of alinement where the right turn movements are to be channelized, particularly where the turning movements are relatively large or large trucks need to be accommodated. Where these designs are not feasible, arrangements similar to those shown in Figures IX-30C and IX-30D should be used. A design with a single circular arc, as in Figure IX-30A, generally should be avoided.

The same general explanation applies to Figure IX-31, which demonstrates designs with a minimum radius of 230 ft.

Figures IX-30 and IX-31 illustrate turning roadway exits, but similar arrangements are applicable to turning roadway entrances, except that the approach nose becomes a merging end without offset from either edge of pavement.

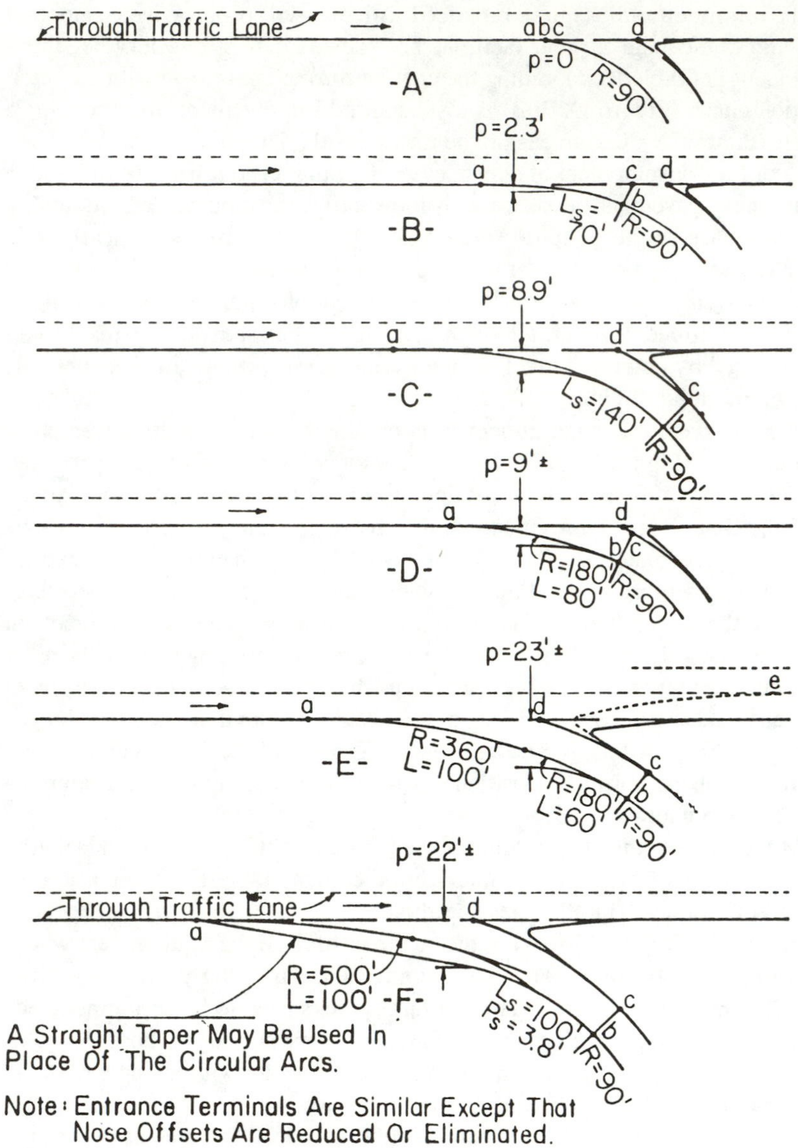

Figure IX-30. Use of transition and compound curves at turning roadway terminal designs for 20 mph turns.

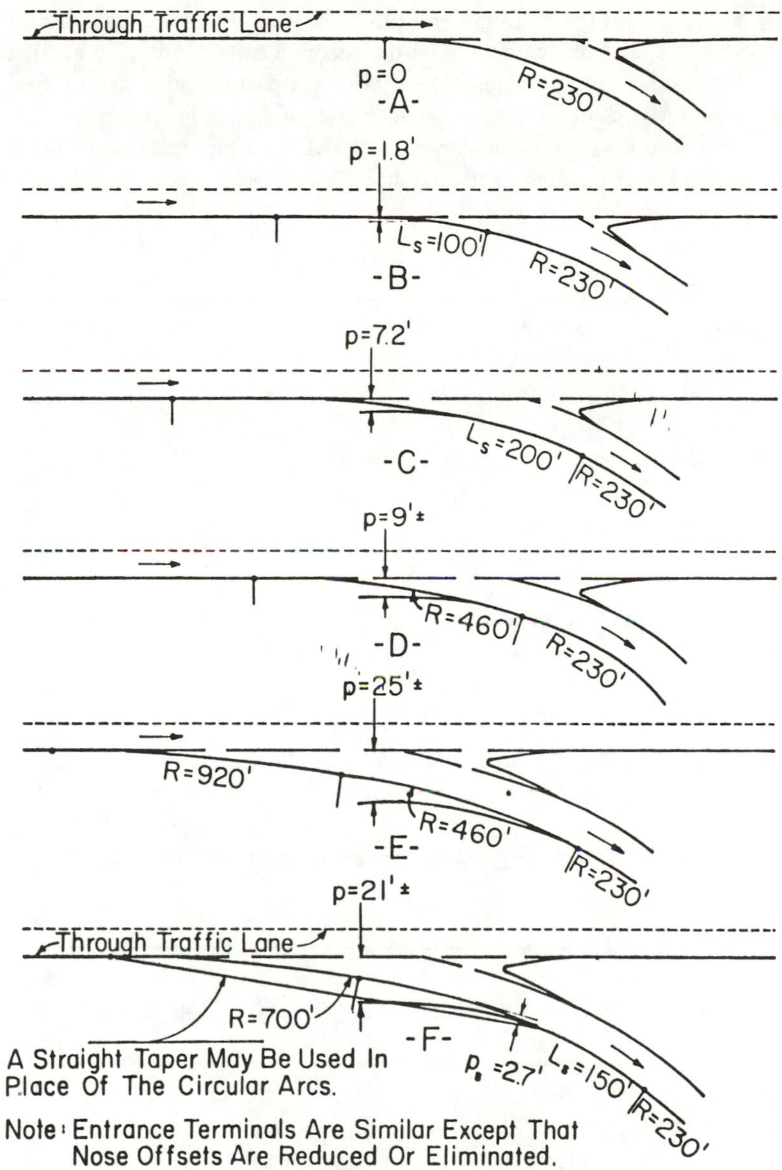

Figure IX-31. Use of transition and compound curves at turning roadway terminal designs for 30 mph turns.

In the design of Figures IX-30 and IX-31, it is assumed that part or all of the necessary speed change occurs on the through traffic lane. The arrangements shown also are applicable when joining a parallel auxiliary or speed change lane. If the outer lane is a deceleration lane, the pavement taper beyond the nose would be that shown by the dashed line in Figure IX-30E, joining the edge of through pavement at e. Designs with speed change lanes are discussed later in this chapter.

Design (turning) speed (mph)	20	25	30	35	40	45
Minimum radius (ft)	90	150	230	310	430	550
Assumed C (ft/sec)	4.0	3.75	3.5	3.25	3.0	2.75
Calculated length of Spiral (ft)	70	87	105	134	156	190
Suggested minimum length of spiral (ft)	70	90	110	130	160	200
Corresponding circular curve offset from tangent (ft)	2.2	2.2	2.2	2.3	2.5	3.0

Table IX-5. Minimum lengths of spiral for intersection curves.

Radius (ft)	**100**	**150**	**200**	**250**	**300**	**400**	**500 or more**
Length of circular arc (ft)							
Minimum	40	50	60	80	100	120	140
Desirable	60	70	90	120	140	180	200

Table IX-6. Lengths of circular arc for a compound intersection curve when followed by a curve of one-half radius or preceded by a curve of double radius.

In an alternate method of designing turning roadway terminals, a straight taper to an offset circular arc may be used instead of a transition or the longer radius arc of a compound curve. This arrangement (Figure IX-30F) requires more paved area, but it provides a gradual turnout and deceleration off the through-traffic lanes. Tapers are discussed later in this chapter.

SIGHT DISTANCE

General Considerations

Each intersection contains several potential vehicle conflicts. The possibility of these conflicts actually occurring can be greatly reduced through the provisions of proper sight distances and appropriate traffic controls. The avoidance of accidents and the efficiency of traffic operations still depend on the judgement, capabilities, and the response of the individual driver.

The operator of a vehicle approaching an intersection at-grade should have an unobstructed view of the entire intersection and sufficient lengths of the intersecting highway to permit control of the vehicle to avoid collisions. When traffic at the intersection is controlled by signals or signs, the unobstructed views may be limited to the area of control. The sight distance considered safe under various assumptions of physical conditions and driver behavior is directly related to vehicle speeds and to the resultant distances traversed during perception and reaction time and braking.

Arterial highways in urban areas are preferential facilities with stop signs or traffic signal control of crossing or entering traffic. At intersections, greater stopping sight distance should be available along the arterial because more hazard is involved than on a section of highway free of intersections and because of the time required by vehicles, particularly trucks, to cross the relatively wide pavement of an arterial from the stop position. This principle applies to signalized intersections as well as to those with only stop signs, as there may be a malfunction of the signals, or it may be desirable to place the signals periodically on a flashing operation. The latter action is the same as stop sign control.

Sight distance at a crossroad or street should be sufficient along the predominant highway to avoid the hazard of collision between a vehicle starting to cross the highway from this stop position and a vehicle on the through highway operating at the design speed and appearing after the crossing movement has begun.

Approaches—Sight Triangle

There must be an unobstructed sight distance along both approaches of both roads at an intersection and across their included corners for a distance sufficient to allow the operators of vehicles, approaching simultaneously, to see each other in time to prevent collisions at the intersection. Each driver has three possible actions: to accelerate, to slow down, or to stop. At each intersection, assumptions are made about the physical layout and the actions of the vehicle operators on both intersecting roads. For each case, the space-time-velocity relationship indicates the sight triangle required to be free of obstructions or to establish the necessary modification in approach speeds when a sight triangle with less than the required dimensions is used. The sight triangle is shown on Figure IX-32A. If feasible, any object within the sight triangle high enough above the elevation of the adjacent roadway to constitute a sight obstruction, should be removed or lowered. Such objects may include: buildings, cut slopes, hedges, trees, bushes, unmowed grass or tall crops.

Departures—Sight Triangle

After a vehicle has stopped at an intersection, the driver must have sufficient sight distance to make a safe departure through the intersection area. As shown on Figure IX-32B, the intersection design should provide adequate sight distances for any of the various vehicular maneuvers such as crossing the intersecting roadway or turning into the intersecting roadway. These maneuvers must be provided for as well as the sight distances which must be provided for vehicles approaching on the major highway either from the right or the left. Distance "d" is the distance traveled by the respective vehicle on the major roadway traveling at design speed during the time required for the stopped vehicle to depart from its stopped position and either cross the intersection or to turn into its desired leg of the major roadway.

Intersection Control

There are four types of controls that apply to at-grade intersections. These types are:

I. No Control, but allowing vehicles to adjust speed.
II. Yield Control where vehicles on the minor intersecting roadway must yield to vehicles on the major intersecting roadway.

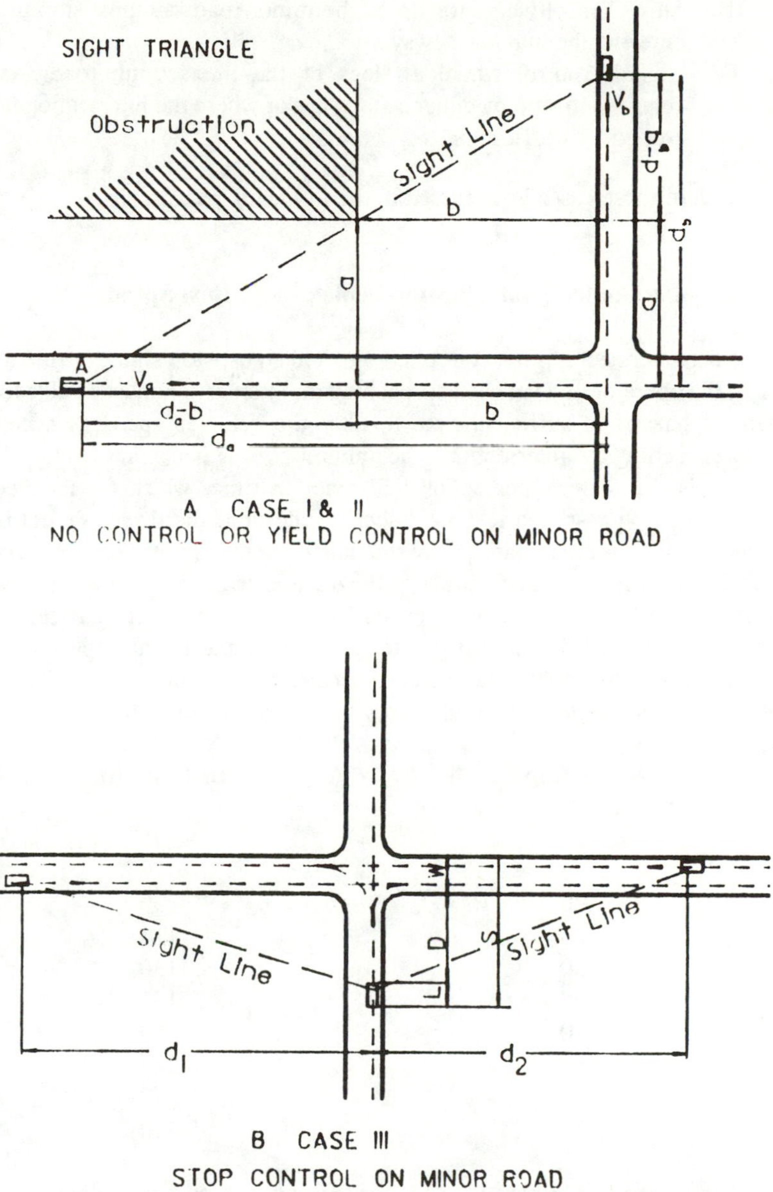

Figure IX-32. Sight distance at intersections, minimum sight triangle.

III. Stop Control where traffic on the minor roadway must stop prior to entering the major roadway.
IV. Signal Control where all legs of the intersecting roadways are required to stop by either a stop sign or where the intersection is controlled by traffic signals.

Each of these cases is discussed in the following paragraphs.

Case I—No Control, but Allowing Vehicles to Adjust Speed

Where a crossing is not controlled by yield signs, stop signs, or traffic signals, the operator of a vehicle approaching an intersection must be able to perceive a hazard in sufficient time to alter the vehicle's speed as necessary before reaching the intersection. The amount of time necessary to start deceleration is the driver's perception and reaction time, which for intersection design, can be assumed to be 2.0 sec. In addition, the driver should begin actual braking some distance from the intersection to accomplish deceleration and avoid collision. A lower limit for the distance from an intersection, where a driver can first see a vehicle approaching on the intersecting road, is that which is traversed during 2.0 sec. for perception and reaction plus an additional 1.0 sec. to actuate braking or to accelerate to regulate speed. The average distances traveled by a vehicle in 3.0 sec. are the following:

Speed (mph)	Distance (ft)
10	45
15	70
20	90
25	110
30	130
35	155
40	180
50	220
60	260
70	310

Under Case I, in Figure IX-32A, for the given speeds, the sight triangle is determined by the above minimum distances along the road.

Referring to Figure IX-32A, highway A with a 50 mph speed and highway B with a 30 mph speed, require an unobstructed sight triangle between points on highways A and B with legs extending at least 220 ft and 130 ft, respectively, from the intersection. These, or greater distances, will permit a vehicle on either road to change speed before reaching the intersection.

Intersections with sight triangles having dimensions approximately equal to those indicated are not necessarily safe. There is potential confusion to operators with the possibility of a driver on one highway being confronted with a succession of vehicles on the intersecting highway. Even where only one vehicle on each of the adjacent legs approaches an intersection, both vehicles may begin to slow down and reach the intersection at the same time. One important factor is that a vehicle approaching a noncontrolled intersection must yield the right-of-way to vehicles approaching the intersection on the right. Noncontrolled intersections should be used only in the design of rural intersections on lightly traveled two lane roads where the cost of achieving greater sight distance is prohibitive. Where this minimum triangle cannot be provided, traffic control devices should be used to slow down or stop vehicles on one road even if both roads are lightly traveled.

The preferred design for "noncontrolled" crossings allows the operators of vehicles on both highways to see the intersection and the traffic on the intersecting highway in sufficient time to stop the vehicle before reaching the intersection. The safe stopping distances for intersection design are the same as those used for design in any other section of highway. Chapter III, section under "Stopping Sight Distance," contains stopping sight distance criteria.

For two highways of known design speeds, the sight triangle determined by the safe stopping sight distance is a safer design than that with the adjusted speed distances. An operator sighting a vehicle on the intersecting road may stop, if necessary, or otherwise change speed to avoid a hazard. Confusion of the drivers is possible because both may slow down and reach the intersection at the same time. However, this danger is slight because of the great number of speed change possibilities, the time available, and the normal decrease in speed as an intersection is approached under such conditions.

Each vehicle approaching an intersection must have the appropriate clear sight distance for vehicles approaching the intersection along both the right and left legs of the cross road. All of the above data are based on an intersection with flat grades on the approach legs. Appropriate corrections for approach legs on grades must be made in accordance with data in Chapter III, section under "Effects of Grades."

Case II—Yield Control for Minor Roads

This type of design requires that the minor roadway be posted with yield signs as its legs intersect the major roadway. Reference is made to Figure IX-32A.

The sight distance for the vehicle operator on the minor road must be sufficient to allow the operator to observe a vehicle on the major roadway, approaching from either the left or the right, and then through perception, reaction and braking time, bring the vehicle to a stop prior to reaching the intersecting roadway. Where a proper sight distance cannot be achieved for the driver of the vehicle on the minor road, it may be necessary to have a posted speed reduction as the minor road approaches the major cross road.

In the event that the minor roadway vehicle must stop, the data for enabling stopped vehicles to cross or enter the major roadway, as explained under Case III, are applicable.

Proper adjustments for all distances must be made when any of the approach legs are on other than level grades. Such adjustments are stated in Chapter III, section under "Effects of Grades."

When an obstruction, which cannot be removed except at a prohibitive cost, fixes the vertices of a sight triangle to points that are less than the design speed safe stopping sight distance from the intersection, appropriate adjustments must be made for the design speed and for the safe stopping sight distance for at least one of the roadways concerned.

If the vehicles on one of the roads are permitted to travel at the design speed of that roadway, the critical corresponding speed on the intersecting road may be evaluated in terms of the known design speed and the dimensions to the known obstructions.

In a typical case, as in Figure IX-32A, speed V_a is known and a and b are the known distances to the sight obstruction from the respective paths of vehicles A and B. The critical speed V_b of vehicle B can then be evaluated in terms of known factors. Distance d_a is the stopping distance for vehicle A. When vehicle A is at a distance d_a from the intersection and the drivers of vehicles A and B first sight each other, vehicle B is a distance d_b from the intersection. By the similar triangle proportion:

$$d_b = \frac{a \quad d_a}{d_a - b}$$

and the critical speed V_b is that for which the stopping distance is d_b.

The signs on road B showing the safe speed with which to approach the intersection should be located so that a driver can reduce speed to V_b by the time the vehicle reaches the point that is the distance d_b from the intersection. Similar calculations may be used to determine how far back an obstruction should be moved to provide sufficient sight distance for safe driving at desired vehicle speeds on the respective roads. Proper adjustments to all distances must be made where any of the intersection approach legs are on other than a level grade. Reference is made to Chapter III, section under "Effects of Grades."

Case III—Stop Control on Minor Roads

Where traffic on the minor road of an intersection is controlled by stop signs, the driver of the vehicle on the minor road must have sufficient sight distance for a safe departure from the stopped position even though the approaching vehicle comes in view as the stopped vehicle begins its departure movements. Reference is made to the sight triangle on Figure IX-32b. The driver of the vehicle entering the intersection from the major road must have an unobstructed view within the sight triangle. The intersection sight distance value (d_1 or d_2) will establish one leg of the sight triangle. The leg of the stop-controlled road or street will be determined by the assumed location of the driver's eye. This is established as 20 feet behind the edge of pavement or curb line. There are three basic maneuvers that occur at the average intersection. These maneuvers are:

A. To travel across the intersecting roadway by clearing traffic on both the left and the right of the crossing vehicle;
B. To turn left into the crossing roadway by first clearing traffic on the left and then to enter the traffic stream with vehicles from the right; and,
C. To turn right into the intersecting roadway by entering the traffic stream with vehicles from the left.

These conditions are shown by the Cases A, B, and C, respectively, on Figure IX-35. To assist the user, the figures necessary in the analysis for all Case III conditions are outlined below and designated as General (G), Case IIIA (A), Case IIIB (B) and Case IIIC (C):

(G) Figure IX-32—illustration of sight lines for Cases I, II and III.
(G) Figure IX-33—to determine t_a for P, SU and WB-50 vehicles.
(G) Figure IX-34—to determine the time-distance relationship for P vehicles and for SU and WB-50 trucks.
(G) Figure IX-34A—to determine the time-speed relationship for P vehicles.
(G) Figure IX-35—to determine distance d and d_2 for two-lane highways.
(A) Figure IX-35A—to determine distance S for Case IIIA conditions for two-lane and multiland situations.

(B) Figure IX-36—illustrates Case IIB—in calculating the necessary sight distance to the right on a two-lane highway.

(B) Figure IX-37—illustrates Case IIIB for a multilane highway in calculating the necessary sight distance to the right for left turning vehicles.

(C) Figure IX-38—illustrates Case IIIC conditions in establishing stopping sight distances.

(A) Figure IX-39—shows the sight distance d necessary for safe crossing from a stopped position.

(B,C) Figure IX-40—illustrates for Cases IIIB and IIIC the sight distances required for the maneuvers.

The stop condition criterion is applicable to two-lane, two-directional roadways through multilane divided highways. Where the principal roadway is either undivided or divided with a narrow median (the median is not of sufficient width to store the design vehicle), the departure maneuvers are treated as a single operation. Where the major roadway is divided and has a wide median (wide enough to safely store the design vehicles), the departure maneuvers are considered as two operations. The first operation concerns the traffic approaching from the left for all three maneuvers; that is, crossing the entire roadway, crossing a part of the roadway and turning left into the cross road or turning right into the cross road. The second phase concerns traffic from the right for the first two operations; that is, continuing to cross the major roadway or turning left and merging with traffic from the right. Details are shown in Figures IX-35 and IX-35A.

Case IIIA—Crossing Maneuver

The sight distance for a crossing maneuver is based on the time it takes for the stopped vehicle to clear the intersection and the distance that a vehicle will travel along the major road at its design speed in that amount of time. See Figure IX-35, and Figure IX-35A Case IIIA. This distance may be calculated from the equation:

$$d = 1.47V\,(J + t_a)$$

where: d = sight distance along the major highway from the intersection, ft;

V = design speed on the major highway, mph;

J = sum of the perception time and the time required to actuate the clutch or actuate an automatic shift, sec; and

t_a = time required to accelerate and traverse the distance S to clear the major highway pavement, sec.

The term J represents the time necessary for the vehicle's operator to look in both directions of the roadway, to perceive that there is sufficient time to cross the road safely, and to shift gears, if necessary, preparatory to starting. It is the time from the driver's first look for possible oncoming traffic to the instant the car begins to move. Some of these operations are done simultaneously by many drivers, and some operations, such as shifting of gears, may be done before looking up or down the road. Even though most drivers may require only a fraction of a second, a value of J should be used in design to represent the time taken by the slower driver. A value of 2.0 seconds is assumed.

The time t_a required to cover a given distance during acceleration depends on the vehicle acceleration, which, for passenger cars, seldom equals the rate the vehicle is capable of. Rather, this vehicle acceleration rate is considerably less, as governed by the temperament and other characteristics of the individual driver and by the immediate conditions. Few drivers operate at the maximum acceleration of their vehicles in crossing a major highway, most drivers accelerate somewhat more rapidly than normal but less than the full vehicle acceleration rate.

The solid line curve in Figure IX-33 labeled 'P', is the recommended time-distance relationship of a typical passenger vehicle to be used in computing t_a, the time required to cross the major highway in Case IIIA conditions, and also to clear the sight distance to the left under Case IIIB. The time-distance data for the 'P' design vehicle shown in Figure IX-33 was developed from studies performed by the University of Michigan Transportation Research Institute (UMTRI) in Reference (7). The time-distance data in Figure IX-33 for the SU and WB-50 design vehicles is the same as in the 1984 edition of this AASHTO book, as updated information was not available at the time of publication.

The acceleration of buses and trucks is substantially lower than that of passenger vehicles, particularly for heavily loaded trucks and truck combinations. The high gear ratios (in low gear) necessary in moving the larger units result in very low accelerations. From truck operation studies, the time-distance relationship for acceleration of SU and WB-50 design vehicles has been determined resulting in the assumed relationship shown in Figure IX-34. On flat grades the accelerating time for the SU and WB-50 vehicles is about 140 to 170 percent, respectively, of that for passenger vehicles. Figure IX-34A illustrates the time-speed relationship between time and speed for passenger vehicles.

The value of t_a can be read directly from Figure IX-33 for nearly level con-

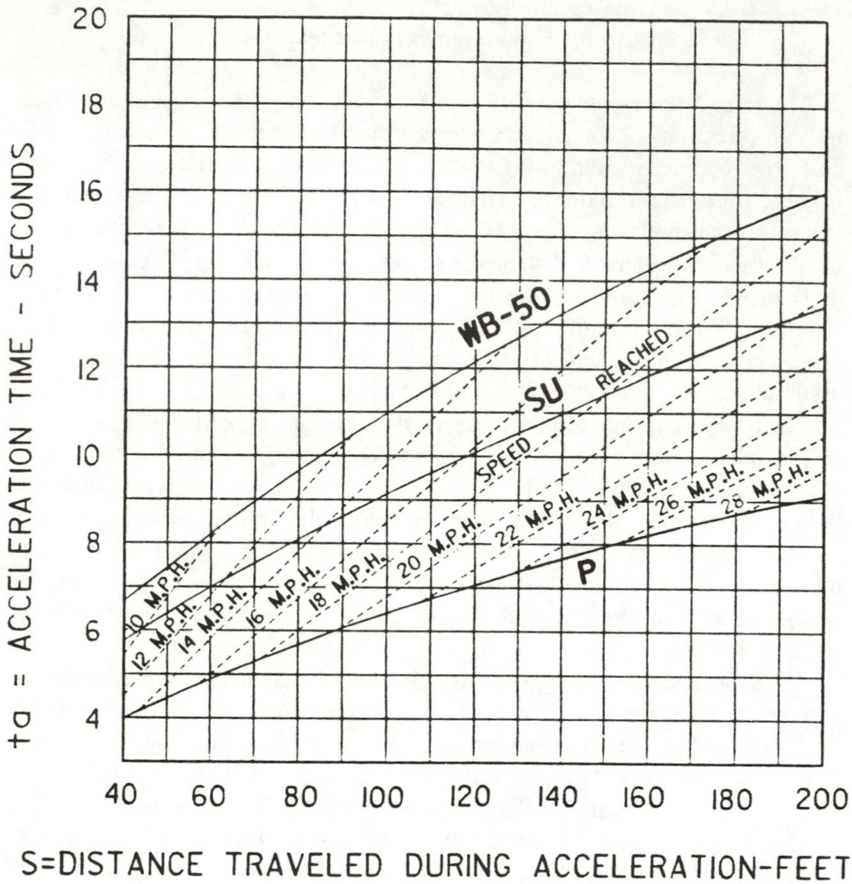

Figure IX-33. Sight distance at intersections (Case III, acceleration from stop).

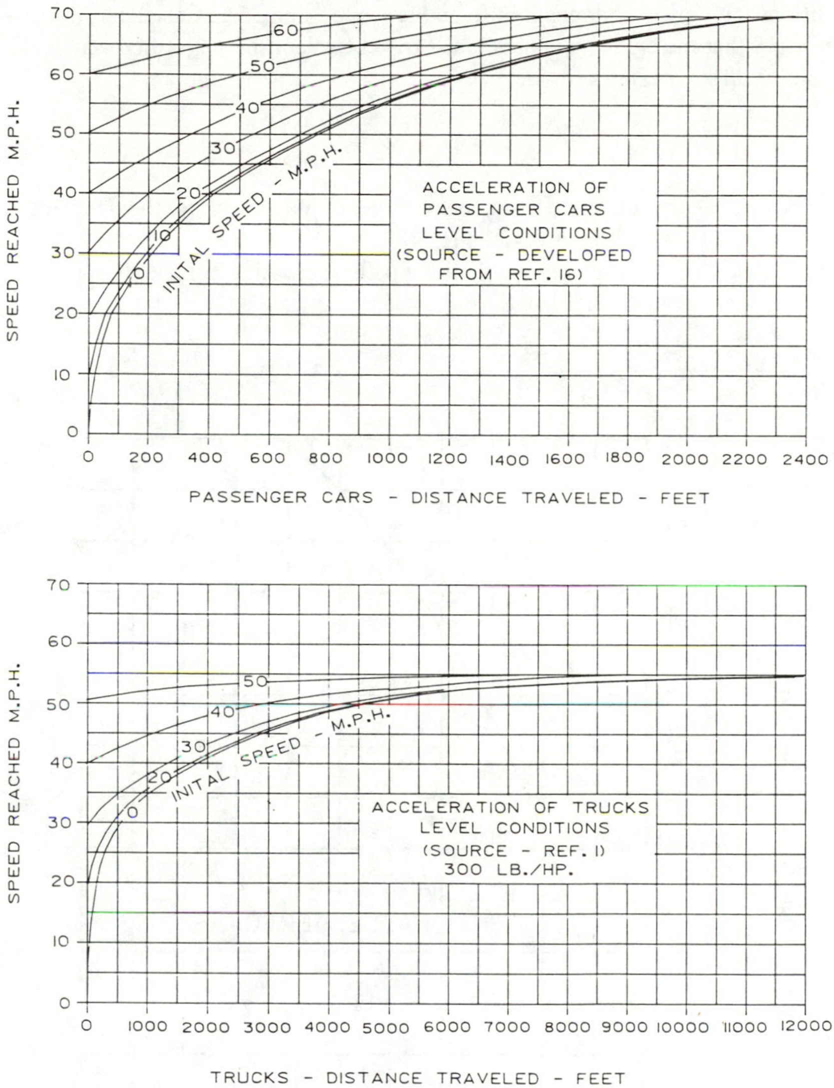

Figure IX-34. Acceleration curves.

ditions for a given distance S in ft. In Figures 35 and 35A, Case IIIA, the distance S that the crossing vehicle must travel to clear the major highway is the sum of three distances (in feet):

$$S = D + W + L$$

where: D = distance from near edge of pavement to the front of a stopped vehicle, ft;
W = pavement width along path of crossing vehicle, ft; and
L = overall length of vehicle, ft.

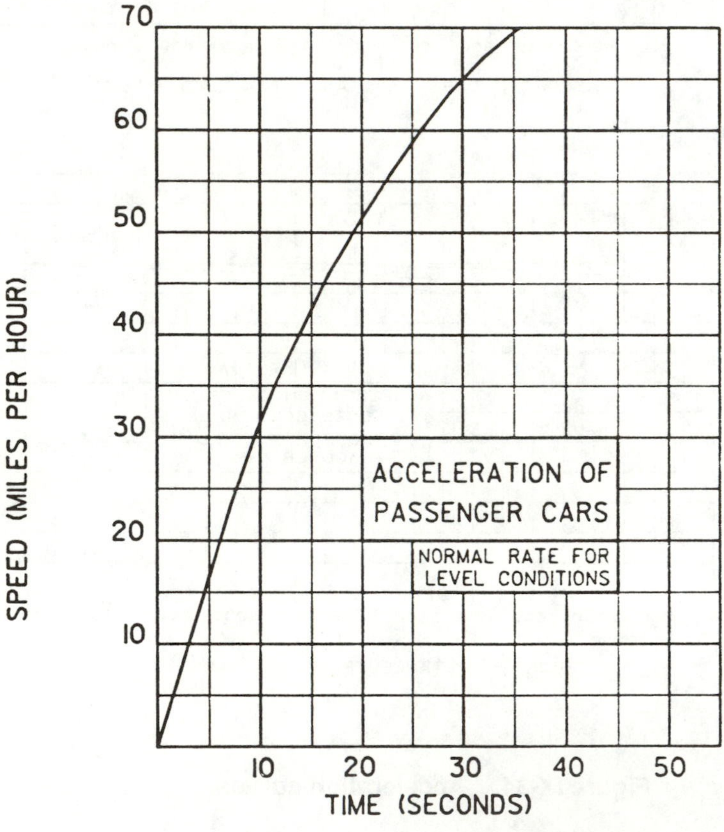

Figure IX-34A. Acceleration curve time—speed relationship for passenger vehicles.

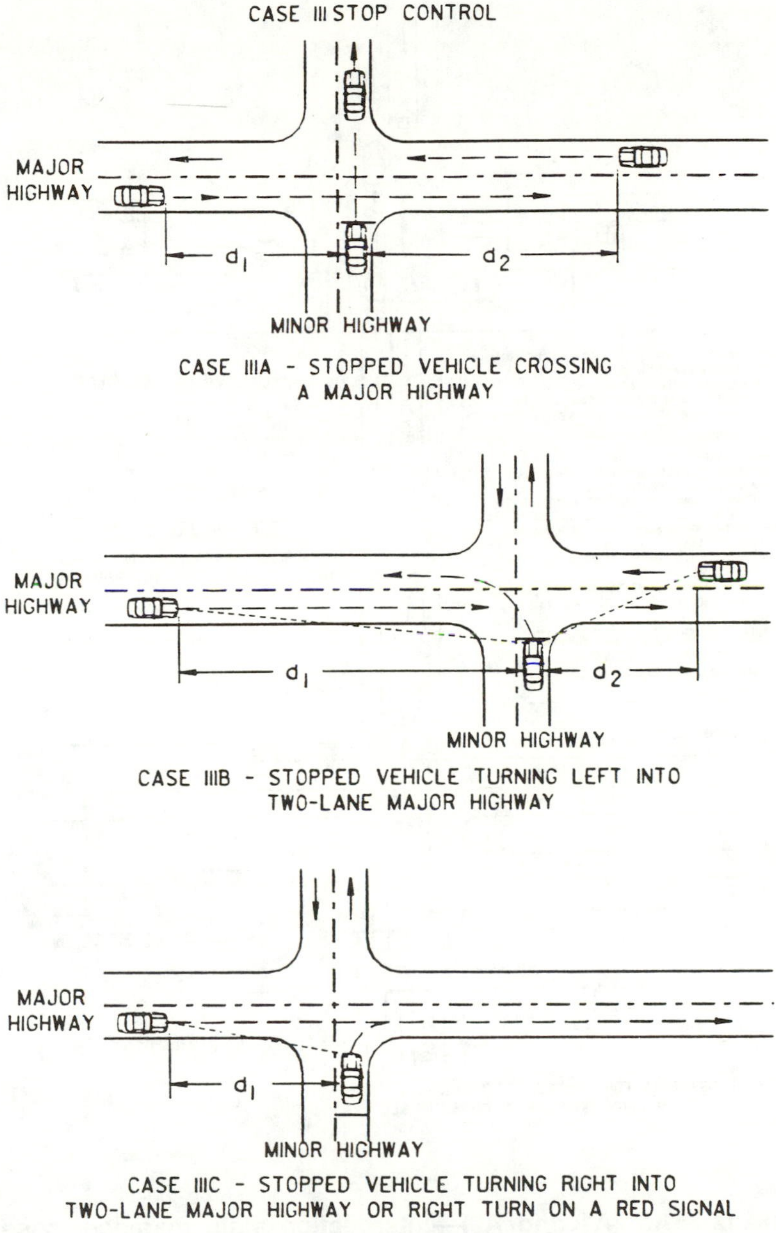

Figure IX-35. Intersection sight distance at at-grade intersections.

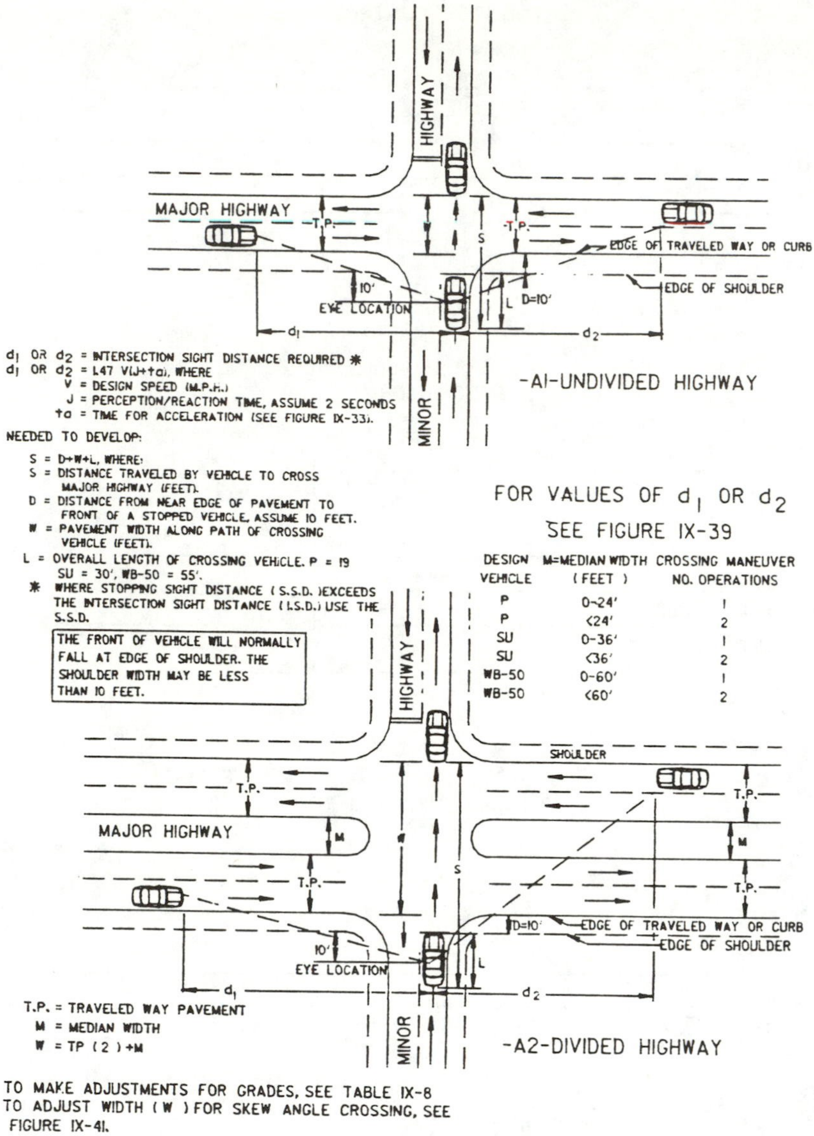

Figure IX-35A. (A1 and A2)—Intersection sight distance (case IIIA—level—90 degree crossings).

Most drivers stop their vehicles as close to the edge of the intersection pavement as convenient; however, many leave several feet. For general design purpose, a D value of 10 ft is assumed (measured from pavement edge to front of vehicle) for the Case III conditions.

The value of W is measured along the path of the accelerating vehicle. The principal variation in W is the number of lanes on the major highway. Lane widths of 12 ft are used.

The values of L, as shown in Chapter II, are 19, 30, 50, and 55 ft for the P, SU, WB-40, and WB-50 vehicles, respectively.

Figure IX-39 shows the sight distance d necessary for safe crossing from a stopped position with pavements of 12-ft lanes at nearly right-angle intersections. The sets of sloping lines give the values for the three types of crossing vehicles, each with a reference to two, four, and six-lane widths. Values for other pavement widths can be calculated readily by the use of proper t_a, as determined from Figure IX-33. These values are for level conditions. For grade adjustments, refer to Chapter III.

In testing whether a sight distance along a major highway is adequate at an intersection as given in Figure IX-39, the distance should be measured from a height of eye of 3.50 ft (for passenger cars) to the top of an object, 4.25 ft above the pavement. Stopping sight distance values for each design speed as given in Chapter III, in the Section "Stopping Sight Distance", is also shown on Figure IX-39. Comparing the d values contained in Chapter III, indicates that in certain instances for 'P' design vehicles above 50 mph, the stopping sight distance will exceed the intersection sight distance (d). Where this occurs, to permit a crossing from a stopped position, the stopping sight distance will govern.

Divided highways, where the median width is equal to or greater than the vehicle length, allow the crossing to be made in two steps. The vehicle crosses the first pavement, stops within the protected area of the median opening, and there awaits an opportunity to complete the second crossing step. For divided highways with median widths less than L, the length of the design vehicle (Figures IX-32 and IX-35A), the median width should be included as part of the W value. Figure IX-35A, the divided highway, shows each design vehicle and the median width that determines the number of steps for each crossing maneuver.

Where the stopping sight distance along a major highway is less than the Case IIIA distance at an intersection, vehicles on the major highway may have to adjust their speeds. Intersection warning signs indicating the appropriate approach speed may be provided. When the sight distance along the major highway and the pavement width along the path of the crossing vehicle are known, the appropriate approach speed can be computed or obtained directly from Figure IX-39.

The Case IIIA distances required for highways of high design speed (especially the multilane type) are large, and the average driver's keenness of perception may be inadequate for proper use of such distances. Drivers may be able to cross the openings in the through-traffic stream that are less than the Case IIIA distances. To cross such openings, either the crossing driver must accelerate at a higher rate or the through driver must be traveling at a lower speed or must slow down to permit the crossing. Because these operations may be hazardous, it is desirable to use the indicated Case IIIA sight distance for the control-type vehicle.

Case IIIB—Turning Left into a Major Highway

Figures IX-35 and IX-36, Case IIIB, illustrate a P vehicle entering a cross road from a stopped position by clearing vehicles approaching from the left and then by turning left and entering the traffic stream approaching from the right.

For a vehicle to safely complete this maneuver it is necessary for the driver to have sufficient sight distance to the left to cross the near lane(s) without interfering with oncoming traffic in this lane. The stopped vehicle must also have sufficient sight distance to the right to allow the operator of the vehicle to turn left and accelerate to a speed where he does not significantly interfere with the vehicles coming from the right.

The sight distance to the left may be calculated from the equation.

$$d = 1.47\,V\,(J + ta)$$

where d = sight distance to the left along the major highway from the intersection, ft,

V = design speed on the major highway, mph;

J = 2 sec.

ta = time required to accelerate and traverse the distance to clear the traffic in the lane approaching from the left. This distance is the sum of the ten foot setback to the stop line, the length of the vehicle, and the distance travelled to cross the opposing lane(s) (approximately 1.5 X lane width).

The necessary sight distance to the right to make a left turn from a stop sign is calculated making the assumption that the mainline vehicle will slow down to a speed 85 percent of the design speed and will always be at least 1.9 $\times$ speed in feet away from the turning vehicle.

Figure IX-36 should be referred to when calculating the necessary sight distance to the right on a two lane highway with no median. To calculate the necessary sight distance first determine from Figure IX-34 the distance required for the turning vehicle to reach a speed of 85 percent of the mainline design speed—this is P. Next determine the time required to travel this distance from Table IX-7. To this time, add J (2 sec.). Multiply this time by 1.47 $\times$ (85 percent of the mainline design speed). This will give Q; see Figure IX-36. The letter h is equal to P-16- (1.9 $\times$ 85 percent of the mainline design speed)-vehicle length. Subtracting h from Q gives the necessary stopping sight distance.

In the case of undivided highways or divided highways with narrow medians (the median width is less than the length of the design vehicle), both clearances are performed simultaneously as one operation. For divided highways with wide medians (the median is wider than the length of the design vehicle), the maneuvers are performed as two operations. The stopped vehicle must first have a proper sight distance to depart from a stopped position and cross traffic approaching from the left. The crossing vehicle may then stop and store in the median prior to performing the second operation. This operation requires the necessary sight distance for vehicles approaching on the right to allow the crossing vehicle to depart from the median, to turn left into the cross road and to accelerate without being overtaken by vehicles approaching from the right.

As is shown in Figure IX-37 Case IIIB multilane, adding additional lanes or a median results in the accelerating left turning vehicle achieving average running speed closer to the intersection. The distance that n and h are decreased by is the width of the extra lanes from the left plus the width of the median. The necessary sight distance is therefore decreased by the same distance. This distance is minor when compared with the required sight distance, especially at higher speeds. In the case of undivided highways or divided highways with narrow medians (median width less than the length of the design vehicle), the driver of the entering vehicle will check for adequate sight distance for vehicles approaching from both the left and right before entering the major highway in one maneuver.

For divided highways with wide medians (median wider than the length of the design vehicle), the maneuver is performed as two-operations. The stopped vehicle must first have proper sight distance to depart from a stopped position and cross traffic approaching from the left. The crossing vehicle may then stop and store in the median prior to performing the second operation. This operation requires the necessary sight distance for vehicles approaching from the right to allow the crossing vehicle to depart for the median, turn left into the major highway and then accelerate without being overtaken by mainline vehicles approaching from the right.

Figure IX-40 contains data for a P design vehicle turning left into a cross highway. Curve B-1 indicates the sight distances required for the turning

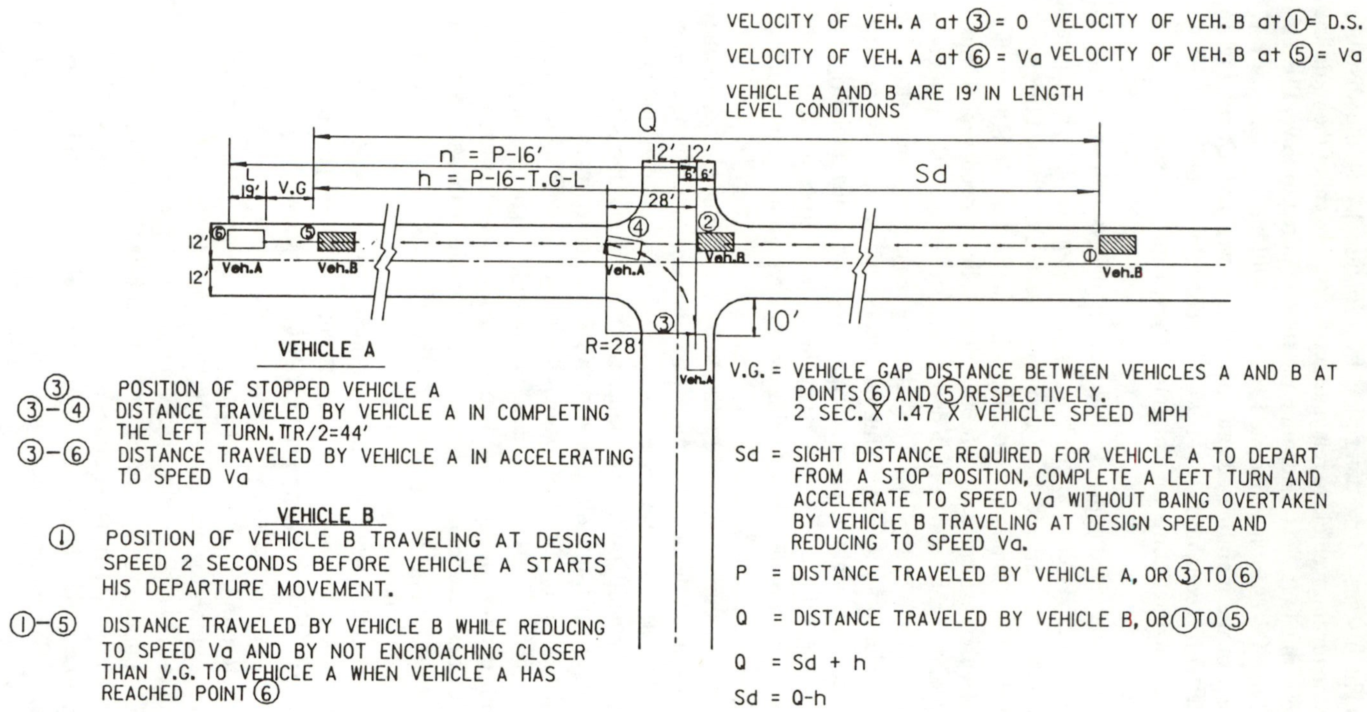

Figure IX-36. Case IIIB, stopped vehicle turning left onto two-lane major highway.

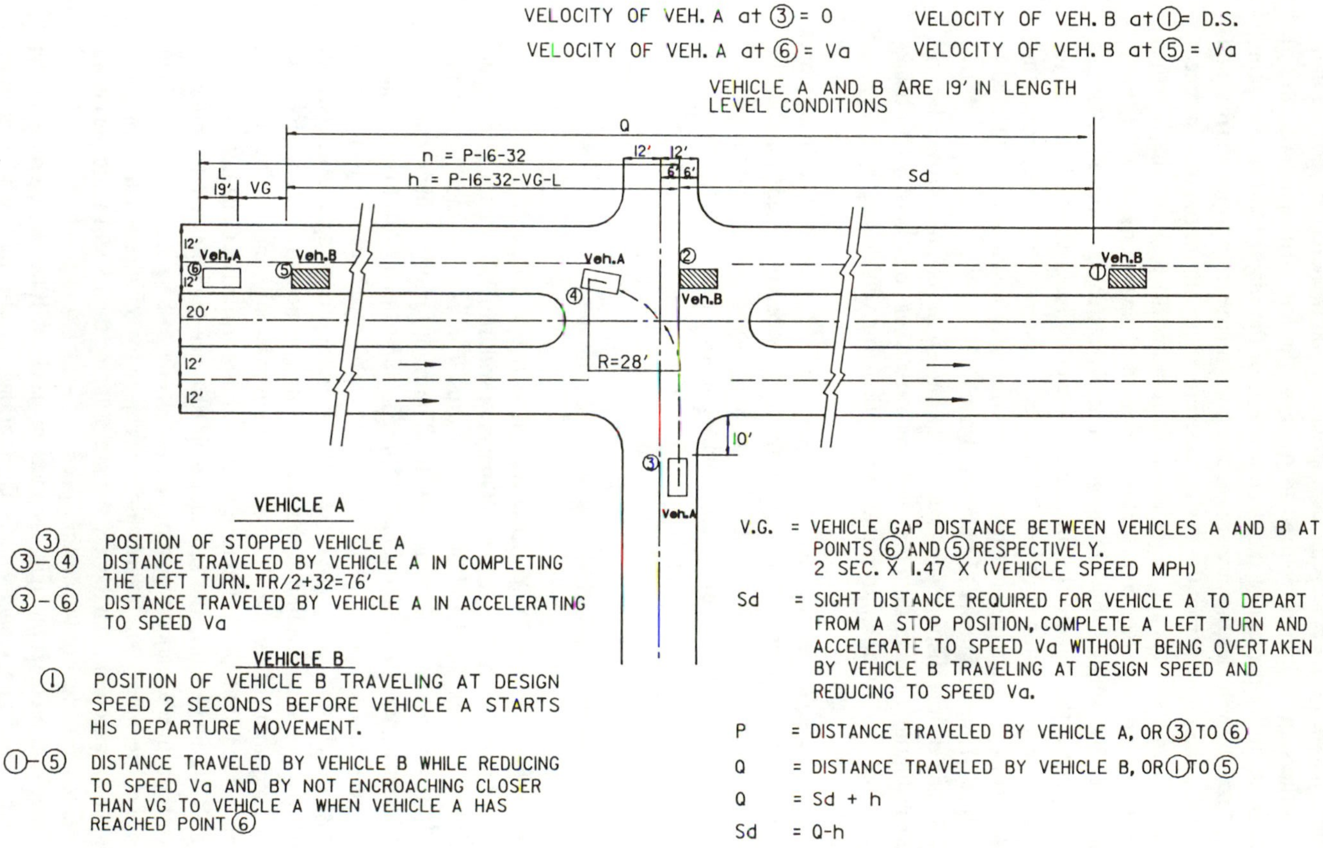

Figure IX-37. Case IIIB, stopped vehicle turning left onto multilane highway.

maneuver with respect to vehicles approaching on the left. Curve B-2b indicates the sight distance required to perform the left turn maneuver and to accelerate to 85 percent of the design speed of the major road before being overtaken by vehicles that are approaching the intersection from the right and reducing their speed from the design speed to 85 percent of the design speed. Curve B-2a indicates the sight distance required to perform the left turn maneuver and to accelerate to the design speed of the major road before being overtaken by vehicles that are approaching from the right and maintaining design speed throughout the turning maneuvers. The last condition requires the longest sight distance. Where the grades of the intersection legs are other than flat, corrections should be made to the sight distances as outlined in Chapter III, section under "Effects of Grades."

Required sight distances for trucks making left turns into a cross road will be substantially longer than for passenger cars. These relationships for trucks can be derived using appropriate assumptions for vehicle acceleration rates and turning paths.

Required sight distance for a stopped vehicle turning left from a major highway into a minor highway (or entrance) may be computed from the formula $D = 1.47V (J + t_a)$ = the time required to accelerate and traverse the distance to clear traffic in the approaching lane. t_a may be obtained directly from Figure IX-33.

Case IIIC—Turning Right into a Major Highway

Figures IX-35 and IX-38, Case IIIC, illustrate a vehicle departing from a stopped position, turning right and merging with traffic from the left. Figure IX-38 illustrates the detail of the criteria used in establishing the stopping sight distances shown in Figure IX-40. The right-turning vehicle must have sufficient sight distance for vehicles approaching from the left to complete its right turn and to accelerate to a predetermined speed relating to the major cross road before being overtaken by traffic approaching the intersection from the left and traveling at the same predetermined speed. The sight distance requirement for the right turn maneuver is only one to three feet less than that required for the left turn maneuver as outlined in Case IIIB. On Figure IX-40 curve Cb is the safe stopping distance for a passenger vehicle to turn right into a major roadway and accelerate to 85 percent of the design speed of the major road before being overtaken by vehicles that are approaching the intersection from the left and are reducing their speed from the design speed to 85 percent of the design speed.

Required intersection sight distance for trucks will be considerably longer than that for the passenger vehicle. Corrections must be made where the inter-

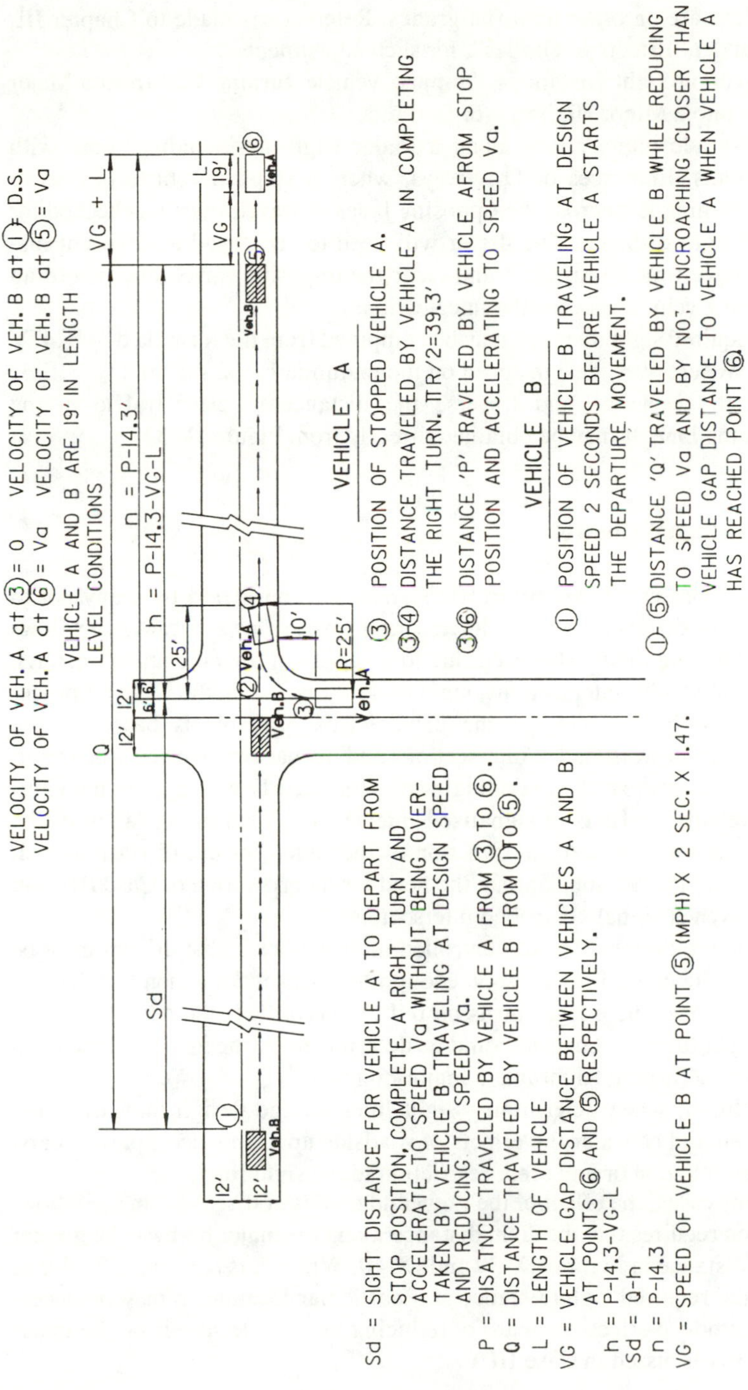

Figure IX-38. Case IIIC, stopped vehicle turning right onto two-lane major highway.

secting legs are on other than flat grades. Reference is made to Chapter III, section under "Effects of Grades", for such adjustments.

Intersection Sight Distance—Stopped Vehicle Turning Left from a Major Highway into a Minor Highway (or Entrance).

All at-grade intersections along a Major Highway including those with Minor Roads, Entrances or Driveways, where a stopped vehicle desires to enter by turning left across the opposing lanes of travel, must be checked for adequate sight distance. The driver will need to see ahead a sufficient distance to have time to turn left and clear the opposing travel lane before an approaching vehicle reaches the intersection.

The required sight distance may be computed from the formula $d = 1.47V(J + t_a)$. Where V = design speed on the mainline J = 2 sec. and t_a = time required to accelerate and traverse the distance to clear traffic in the approaching lane. It may be obtained directly from Figure IX-33.

Case IV—Signal Control

Intersections controlled by traffic signals are considered by many not to require sight distance between intersecting traffic flows because the flows move at separate times. However, due to a variety of operational characteristics associated with intersections, sight distance based on the Case III procedures should be available to the driver. This principle is based on the increased driver workload at intersections and the hazard involved when vehicles turn onto or cross the major highway. The hazard associated with unanticipated vehicle conflicts at signalized intersection, such as, violation of the signal, right turns on red, malfunction of the signal, or use of flashing red/yellow mode, further substantiate the need for incorporation of Case III sight distance even at signal-controlled intersections.

A basic requirement for all controlled intersections is that drivers must be able to see the control device soon enough to perform the action it indicates. At intersections where right turns on red are permitted, the departure sight line for right turning vehicles should be determined by the methods discussed in Case IIIC, Turning Right into a major highway.

In addition, when determining sight lines for the design maneuver, the designer should consider the effects of roadside appurtenances, parked cars, snow accumulation or any other restriction to the sight line.

In many cases, the effect of the skew and grades as discussed in the following section requires that the sight distance along the major highway be greater than the distance in Figures IX-39 and IX-40. Where it is not cost-effective to provide the required sight distance at a particular location, it may be necessary to introduce effective means of reducing the vehicle speeds on the major highway as discussed in Case IIIA.

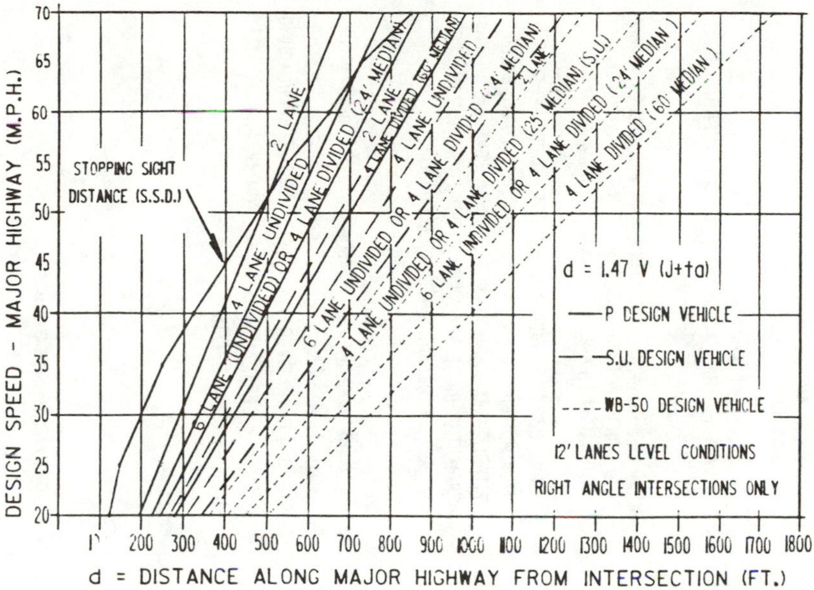

Figure IX-39. Sight distance at intersections (Case IIIA, required sight distance along a major highway).

Effect of Skew

When two highways intersect at an angle considerably less than a right angle, e.g., less than 60°, and when realinement to increase the angle of intersection is not justified, some of the factors for determination of corner sight distance may need adjustment.

Figure IX-41 shows the sight triangles at an oblique-angle intersection, for Case I, II and III conditions. For Case I and II intersections, the sight lines A-B and B-C are smaller or larger than they would be for a right-angle intersection, but because the derivations are based on the relations of speed and distances along the roads, the actual distance A-B is immaterial as long as the area within the triangle, having legs d_a and d_b measured along the road, is free of sight obstructions.

Where obstructions at oblique intersections limit sight distances, the distances a and b to be used in the calculations (equations under heading of Case II, Yield Control for Minor Roads) should be measured parallel to the roads, as shown in Figure IX-41.

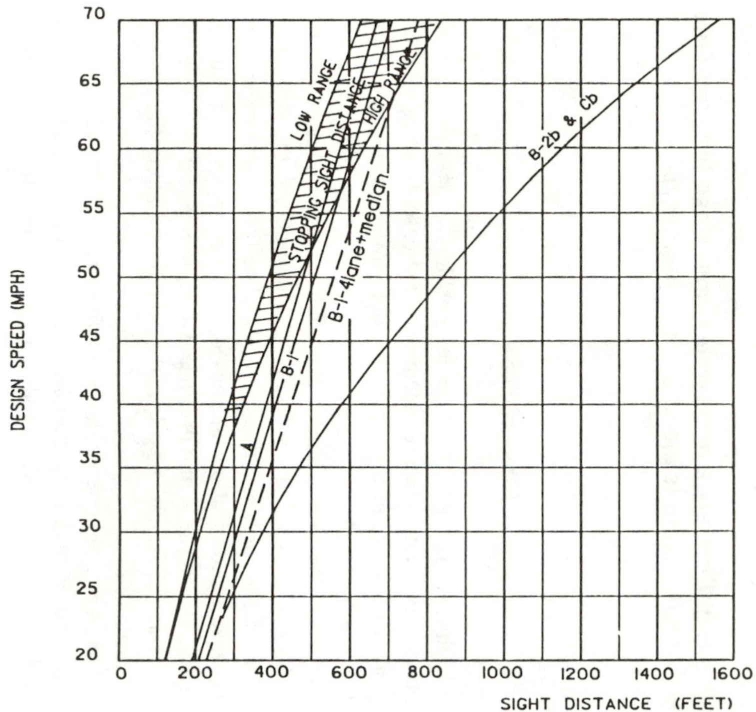

A- SIGHT DISTANCE FOR P VEHICLE CROSSING 2-LANE HIGHWAY FROM STOP. (SEE DIAGRAM)

B-I- SIGHT DISTANCE FOR P VEHICLE TURNING LEFT INTO 2-LANE HIGHWAY ACCROSS P VEHICLE APPROACHING FROM LEFT. (SEE DIAGRAM)

B-I-4Lane+median SIGHT DISTANCE FOR P VEHICLE TURNING LEFT INTO 4-LANE HIGHWAY ACCROSS P VEHICLE APPROACHING FROM LEFT. (SEE DIAGRAM)

B-2b- SIGHT DISTANCE FOR P VEHICLE TO TURN LEFT INTO 2-LANE HIGHWAY AND ATTAIN 85% OF DESIGN SPEED WITHOUT BEING OVERTAKEN BY A VEHICLE APPROACHING FROM THE RIGHT REDUCING SPEED FROM DESIGN SPEED TO 85% OF DESIGN SPEED. (SEE DIAGRAM)

Cb- SIGHT DISTANCE FOR P VEHICLE TO TURN RIGHT INTO 2-LANE HIGHWAY AND ATTAIN 85% OF DESIGN SPEED WITHOUT BEING OVERTAKEN BY A VEHICLE APPROACHING FROM THE LEFT AND REDUCING FROM DESIGN SPEED TO 85% OF DESIGN SPEED.

Figure IX-40. Intersection sight distance at at-grade intersection (Case IIIB and Case IIIC).

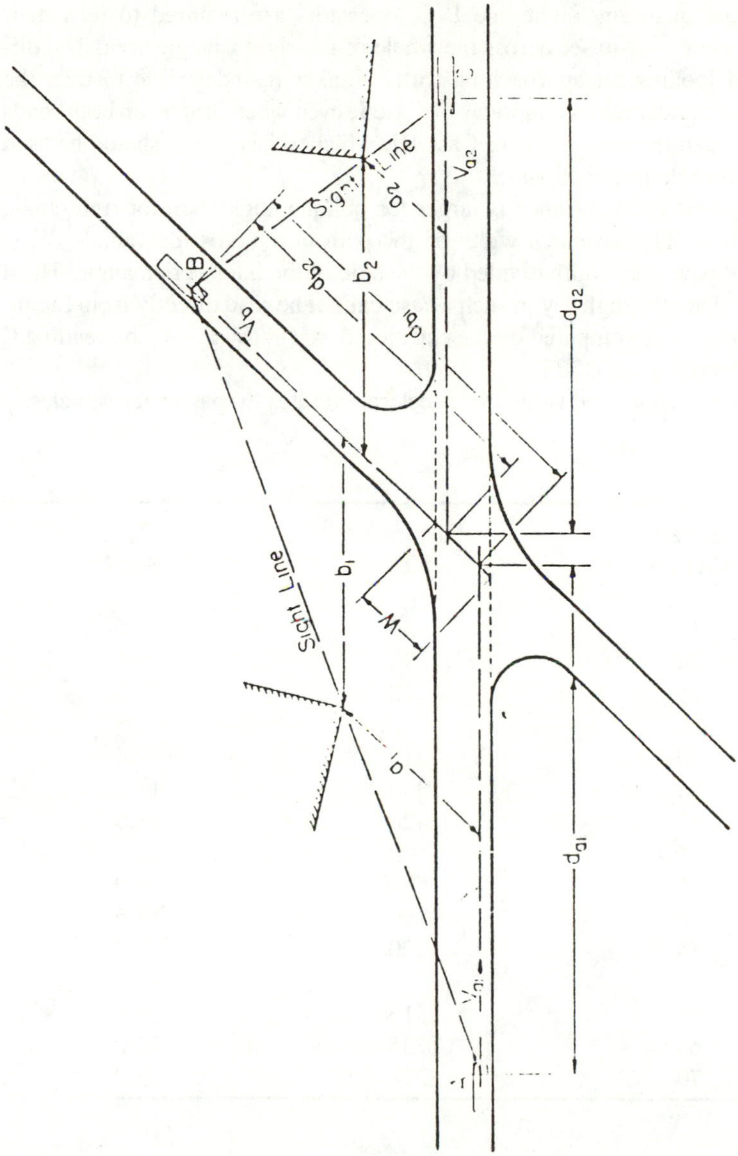

Figure IX-41. Sight distance at intersections, effect of skew.

For an obtuse-angle quadrant the angle between the sight lane A-B and the path of either vehicle is small, and vehicle operators can look across the full sight triangle area with but a little side glance from the vehicle path. For an acute-angle quadrant, sight line B-C, operators are required to turn their heads considerably to see across the whole of the sight triangle area. The difficulty of looking for approaching traffic makes it undesirable to treat the intersection under the assumptions of Case I, even where traffic on both roads is light. Treatment by Case II or Case III, whichever is larger, should be used at oblique-angle intersections.

In Case III the S distance is larger for oblique-angle than for right-angle intersections. The pavement width on the path of the crossing vehicle, W, is the actual pavement width divided by the side of the intersection angle. The d distance along the highway in such a case cannot be read directly from Figure IX-39, but can be computed by the equation $d = 1.47V (J + t_a)$ by reading t_a directly from Figure IX-33.

Table IX-7 shows the values for acceleration rates for passenger vehicles.

Speed (MPH)	Distance (FT)	T_a (Sec.)
15	50	4.5
20	90	6.1
25	140	7.3
30	215	9.4
35	305	11.3
40	420	13.5
45	570	15.9
50	760	18.6
55	1000	21.7
60	1315	25.4
65	1735	30.0
70	2320	35.9

Table IX-7. Acceleration rates for passenger vehicles.

Effect of Vertical Profiles

The evaluation of sight distance at intersections in Case II is based on the safe stopping distance of a vehicle traveling at a stated speed on level highways. One or more of the roads approaching the intersection may not be level. A vehicle descending a grade requires a somewhat greater distance to stop than does one on a level grade; also, a vehicle ascending a grade requires less distance in which to stop.

The differences in stopping distances on various grades at intersections are the same as those given in Chapter III. The differences indicate that grades up to 3 percent have little effect on stopping sight distances. Grades on an intersection leg should be limited to 3 percent unless the sight distances are greater than the lower limits or stopping on a level grade, in which case the grades should not be greater than 6 percent.

In the Case III derivation, the time required to cross the major highway is materially affected by the grade of crossing on the minor road. Normally, the grade across an intersection is so small that it need not be considered, but when curvature on the major road requires the use of superelevation, the grade across it may be significant, in which case the sight distance along the major road should be increased.

The effect of grade on acceleration can be expressed as a multiplicand to be used with the time, t_a , as determined for level conditions for a given distance. The approximate ratios of the accelerating time on the grade to that on the level, for the likely range of crossing distances, are given in Table IX-8.

Ratio, accelerating time on grade to accelerating time on level section					
	Crossroad Grade (percent)				
Design Vehicle	−4	−2	0	+2	+4
P	0.7	0.9	1.0	1.1	1.3
SU	0.8	0.9	1.0	1.1	1.3
WB-50	0.8	0.9	1.0	1.2	1.7

Table IX-8. Effect of gradient on accelerating time at intersections.

The value of t_a from Figure IX-33, adjusted by the appropriate ratio, can be used in the equation $d = 1.47V (J + t_a)$ to determine the proper sight distance d for the crossroad grade condition.

Sight Distance at Ramp Terminals

Although ramp terminals may be considered part of interchange design, where ramps for interchanges intersect the secondary facility, the terminals should be planned in accordance with design principles for at-grade intersections and with particular attention to sight distance characteristics. An added sight distance element, not discussed previously, involves a bridge structure supporting the secondary facility over or under the major highway.

The distances traveled by vehicles entering the crossroad by turning left from the ramp, before clearing the lane used by a vehicle approaching from the left, are about 60 ft for passenger cars, 90 ft for SU vehicles, and 120 ft for WB-50 vehicles. These distances are based on the assumptions that the turning vehicles accelerate from a stopped position on the ramp with the front of the stopped vehicle 10 ft from the edge of through pavement (D = 10 ft and, in Case III in Figure IX-32, that the turning vehicles follow the minimum turning paths for the respective design vehicles and that the vehicles enter a two-lane two-way highway.

The time for acceleration, t_a , may be obtained from Figures IX-33 and IX-34 with an assumed perception and preparation time of 2.0 sec.

The resulting sight distance requirements for various design speeds of the various design vehicles are given in Table IX-9. The speed on the crossroad through the interchange is given in this table. The design speed used should be that of the faster vehicles expected on the section of the highway in the interchange area.

Both the horizontal sight triangle (Figure IX-42A) and the vertical curvature (Figure IX-42B) should be checked to ensure that the sight distance shown in Table IX-9 is provided. The check may be done graphically or mathematically. The design must also account for stopping sight distance for the privileged vehicle (the vehicle having legal right-of-way). The ability of the driver of the privileged vehicle to observe other vehicles at the ramp terminal or proceeding onto the secondary facility is a highly significant element of the total design.

Figure IX-42A shows the check made at a typical interchange to ensure sufficient sight distance past the bridge parapet. The sight line is from the ramp terminal to the privileged vehicle. However, the preceding comments with respect to the privileged vehicle must be taken into account with the height of eye being 3.50 ft (for passenger cars). A similar check should be

Assumed Design Speed On Crossroad Through the Interchange	Sight distance Required to Permit Design Vehicle to Turn Left from Ramp to Crossroad (ft) and not Interfere with oncoming traffic from left.[a] Design Vehicle Assumed at Ramp Terminal			Sight Distance Available to Entering Vehicle When Vertical Curve on Crossroad is Designed for Stopping Sight Distance (ft)[b]	
	P	SU	WB-50	P	SU or WB-50
70	710	1060	1430	1295	1610
60	610	910	1230	980	1220
50	510	760	1030	705	875
40	410	610	820	500	620
30	310	460	620	305	380

[a] Sight distance measured from height of eye of 3.5 ft. for P, and 8 ft for SU and WB-50 design vehicles to an object height of 4.25 ft. high.
[b] Minimum available stopping sight distance based on the assumption that there is no horizontal sight obstruction and that $S<L$.
Where stopping sight distance exceeds intersection sight distance, S.S.D. shall be used. (See Chaper III.)

Table IX-9. Required sight distance to the left along the crossroad at terminals of ramps at interchanges.

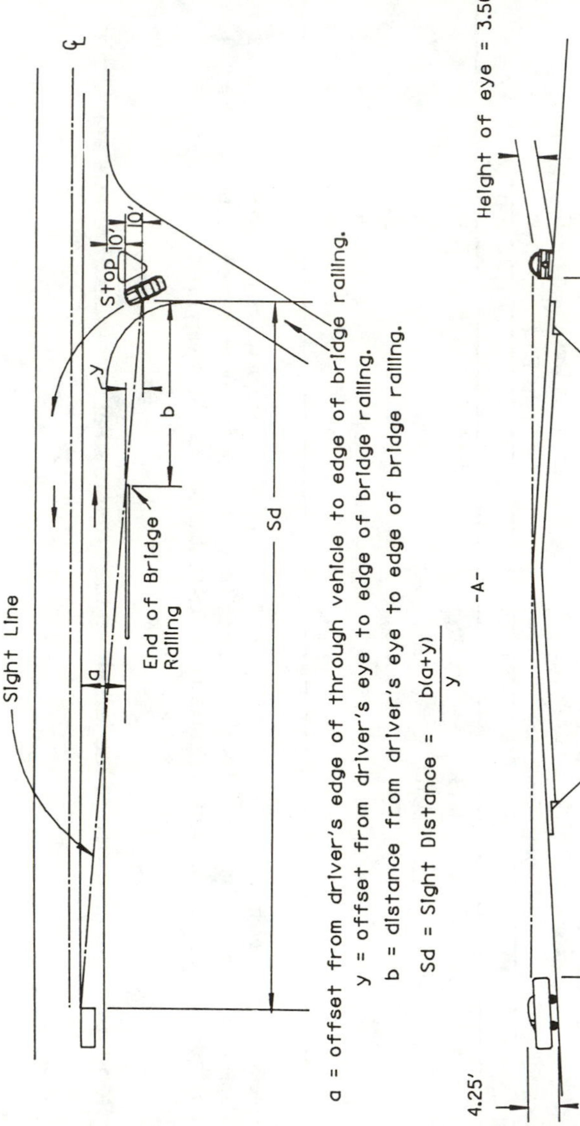

a = offset from driver's edge of through vehicle to edge of bridge railing.

y = offset from driver's eye to edge of bridge railing.

b = distance from driver's eye to edge of bridge railing.

$$Sd = \text{Sight Distance} = \frac{b(a+y)}{y}$$

Figure IX-42. Measurement of sight distance at ramp terminals.

made where the crossroad underpasses the major highway, but in this case the abutment or columns instead of the parapet end would have to be cleared and set back. Cut slope intersections should also be daylighted to ensure safe sight distance at the intersection.

In some cases vertical curves longer than those necessary to provide safe stopping distance for the crossroad may have to be provided. In other cases the ramp terminals may have to be relocated farther from the structure. Where this adjustment cannot be made in the design to provide sufficient sight distance, the structure may have to be designed to increase the lateral clearance, or traffic signals may have to be considered.

STOPPING SIGHT DISTANCE AT INTERSECTIONS FOR TURNING ROADWAYS

General Considerations

The values for stopping sight distance as computed in Chapter III for open highway conditions are applicable to turning roadway intersections of the same design speed. The same method of derivation is suitable for turning roadways for design speeds of 10, 15, 20, 25, and 35 mph. For a brake reaction time of 2.5 sec. for each design speed and coefficients of friction varying from 0.44 to 0.37 for design speeds of 10 to 25 mph, respectively, and 0.34 for 35 mph, the stopping sight distances in Table IX-10 result. These distances have been rounded upward to provide an increased factor of safety.

These sight distances are the lower limits that should be available at all points along a turning roadway; wherever feasible, longer sight distances should be provided. They apply as controls in design of both vertical and horizontal alinement.

Vertical Control

The length of vertical curve is predicated, as it is for open highway conditions, on sight distance measured from the height of eye of 3.50 ft to the height of object of 6 in. Formulas shown in the Section "Crest Vertical Curves" in Chapter III apply directly. Figure IX-43 and Table IX-11 show the relation between design speed, algebraic difference in gradient, and length of crest vertical curve to provide safe stopping distance. Figure IX-43 is based on the rounded values for K. The method of rounding is the same as that explained for Figure III-39. The lower left portions of the heavy lines, for the

Design speed (mph)	10	15	20	25	30	35	40
Stopping Sight Distance (ft)	50	75	125	250	200	225-250	275-325

Table IX-10. Stopping sight distance for turning roadways.

low algebraic differences in grades, are terminated at selected practical lengths of vertical curve, equal in feet to 3V in miles per hour. The factor K is constant for each design speed and the length of vertical curve in feet is found by multiplying A, the algebraic differences in percent of grades, by K.

For design speeds of less than 35 mph, sag vertical curves, as governed by headlight sight distances, theoretically should be longer than crest vertical curves. Lengths of sag vertical curves, found by substituting the stopping sight distances from Table IX-10, in the formulas in the section "Sag Vertical

Design Speed (mph)	Stopping Sight Distance (ft)	K = Rate of Vertical Curvature Length (ft) per Percent of A	
		Calculated	Rounded
20	125-125	8.6	10- 10
25	150-150	14.4- 16.1	20- 20
30	200-200	23.7- 28.8	30- 30
35	225-250	35.7- 46.4	40- 50
40	275-325	53.6- 73.9	60- 80
45	325-400	76.4-110.2	80-120
50	400-475	106.6-160.0	110-160
55	450-550	140.4-217.6	150-220
60	525-650	189.2-302.2	190-310
65	550-725	227.1-394.3	230-400
70	625-850	282.8-530.9	290-540

Table IX-11. Design controls for crest vertical curves based on stopping sight distance.

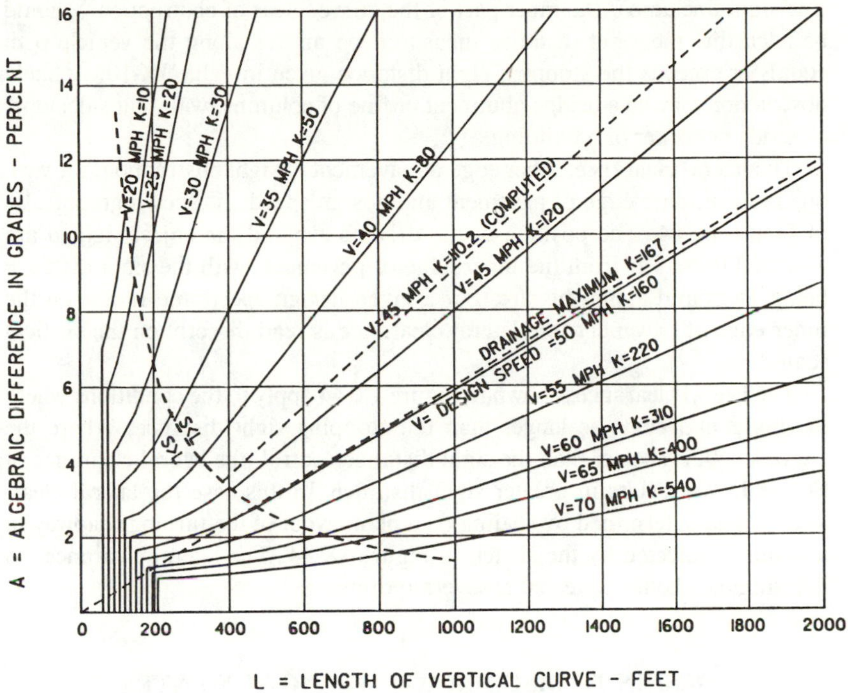

Figure IX-43. Design controls for crest vertical curves (stopping sight distance, open road conditions).

Curves" in Chapter III, are 25 to 60 percent greater than those for crest vertical curves in Figure IX-43. Because the design speed of most turning roadways is governed by the horizontal curvature and the curvature is relatively sharp, a headlight beam parallel to the longitudinal axis of the vehicle ceases to be a control. Thus, for turning roadways the lower limit of lengths of vertical curves in sags may be considered to be the same as those on crests (Figure IX-43). Where feasible, longer lengths for both crest and sag vertical curves should be used.

Horizontal Control

The sight distance control as applied to horizontal alinement has an equal, if not greater, effect on design of turning roadways than the vertical control.

The sight line across the inner part of the curve, clear of obstructions, should be such that the sight distance measured on an arc along the vehicle path equals or exceeds the stopping sight distance given in Table IX-10. A likely obstruction may be a bridge abutment or line of columns, wall, cut sideslope, or a side or corner of a building.

The lateral clearance, inner edge of pavement to sight obstruction, for various radii of inner edge of pavement and design speeds, is shown graphically in Figure IX-44. The position of the driver's eye and the object sighted are assumed to be 6 ft from the inner edge of pavement, with the sight distance being measured along this arc. For a given design speed and radius on the inner edge of pavement, the lateral clearance is read directly on the vertical scale.

The lateral clearances shown in Figure IX-44 apply to the conditions where the horizontal curve is longer than the stopping sight distance. Where the curve length is shorter than the sight distance control, the lateral clearance of Figure IX-44 results in greater sight distance. In this case the lateral clearance is best determined by scaling on a plan layout of the turning roadway in a manner indicated by the sketch in Figure IX-44. The lateral clearance, so determined, should be tested at several points.

DESIGN TO DISCOURAGE WRONG-WAY ENTRY

An inherent problem of interchanges is the possibility of a driver entering one of the exit terminals from the crossroad and proceeding along the major highway in the wrong direction in spite of signing. This wrong way entrance is becoming more of a problem with the increased number of interchanges. Attention to several details of design at the intersection can discourage this hazardous maneuver.

As shown in Figures IX-45 and IX-46, a sharp or angular intersection is provided at the junction of the left edge of the ramp entering the crossroad and the right edge of the through pavement. The control radius should be tangent to the crossroad centerline, not the edge. This type of design discourages the improper right turn onto the one-way ramp.

As shown in the same figures, islands can be used in the terminal areas where ramps intersect the crossroads. The islands provide a means of channelizing the traffic into proper paths and can be effectively used for sign placement. Design of the islands should take into consideration initial or future signal installations at the ramp terminals.

Provision of a median as a deterrent to wrong-way movement, as illustrated in Figure IX-46, is a very effective treatment. The median makes the left-turn movement onto the exit ramp terminal very difficult, and a short-ra-

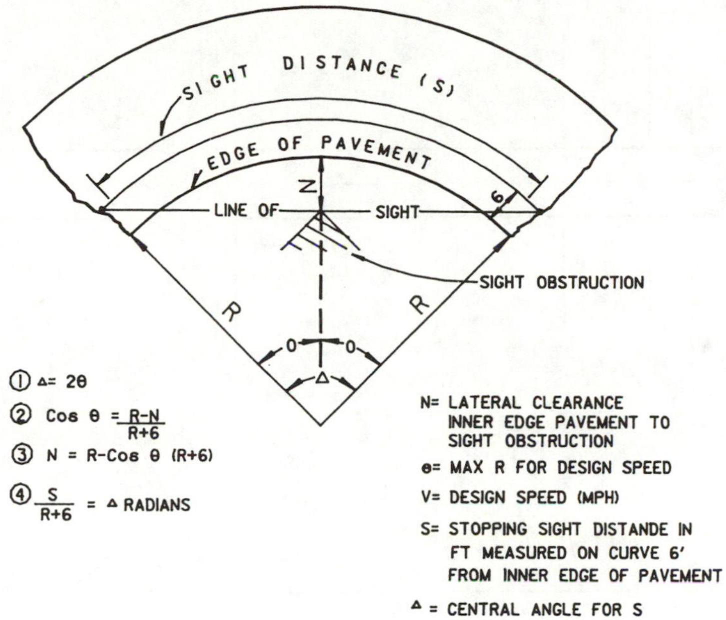

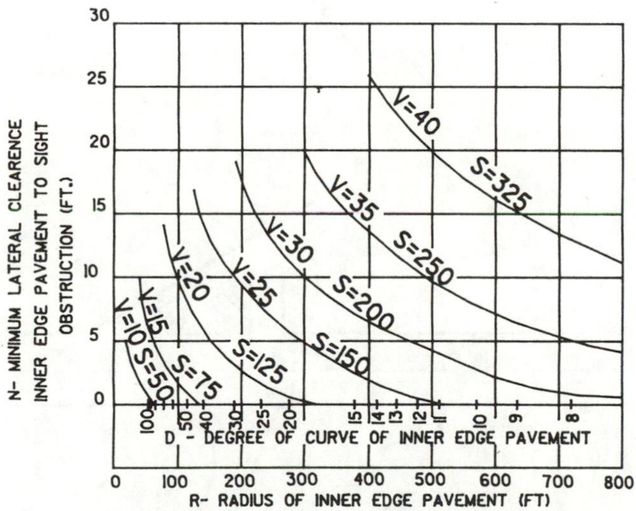

Figure IX-44. **Lateral clearance to sight obstruction inside of horizontal curves providing stopping distance for turning roadways.**

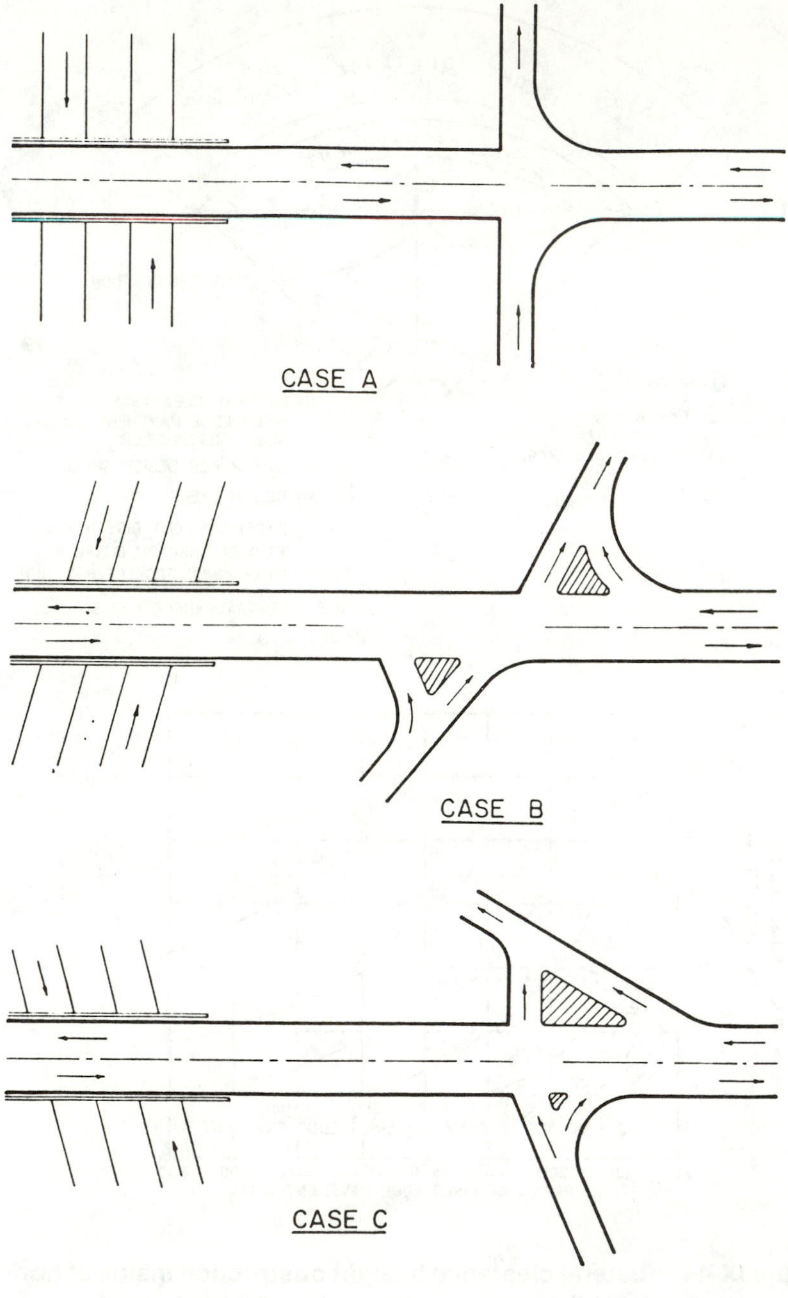

Figure IX-45. Two-lane crossroad designs to discourage wrong-way entry.

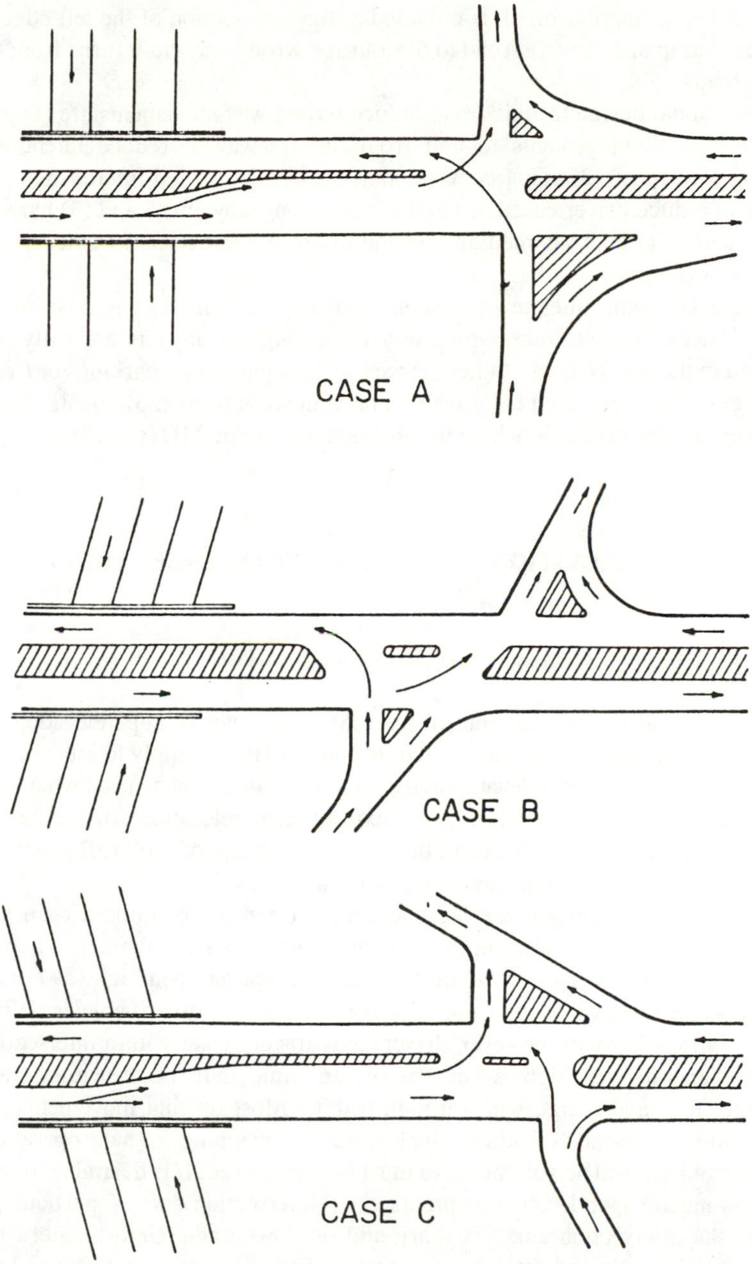

Figure IX-46. Divided crossroad designs to discourage wrong-way entry.

dius curve or angular break is provided at the intersection of the left edge of the exit ramp and the crossroad to discourage wrong-way right turns from the crossroads.

Additional design techniques to reduce wrong-way movements are (1) providing for all movements to and from the freeway to reduce intentional wrong-way entry, (2) using conventional, easily recognized interchange patterns to reduce driver confusion and hence wrong-way entry, and (3) narrowing the arterial highway median opening to prevent left-turn movements onto freeway off-ramps.

Signs and supplementary pavement markings are among the most important devices to discourage wrong-way turns. Signing layouts are fully discussed in the MUTCD (**5**). Other devices such as pavement markings or flashing lights have been used effectively to prevent wrong-way movements. These devices are discussed elsewhere in this book and in the MUTCD (**5**).

SUPERELEVATION FOR CURVES AT INTERSECTIONS

General Design Considerations

The general factors that control the maximum rates of superelevation for open highway conditions as discussed in Chapter III also apply to intersection curves. Maximum superelevation rates up to 0.12 ft/ft may be used where climatic conditions are favorable. The maximum superelevation rate for curves at intersections is 0.10 ft/ft except that a maximum rate of 0.08 ft/ft generally should be used where snow and icing conditions prevail.

In intersection design, the feasible curves often are of limited radii and length. Within an intersection, drivers anticipate the sharp curves and accept operation with higher side friction than they accept on open highway curves of the same radii when their speed is not affected by other vehicles. When other traffic is present, however, drivers may travel slower within intersection areas than on open highway curves of the same radii because they must emerge from and merge with through traffic. Most turning movements are made in the presence of other vehicles, but in designing for safe operation, periods of light traffic volumes also must be considered. It is desirable to provide as much superelevation as practical on intersection curves, particularly where the intersection curve is sharp and on downgrade. Unfortunately, the practical difficulty of attaining superelevation without abrupt cross-slope change at the turning roadway terminals, primarily because of sharp curvature and short lengths of turning roadway, often prevents the development of a desirable rate of superelevation. This fact has been recognized in use of low rates of superelevation for sharp curvature in the development of minimum radii for given intersection design speeds. Designs with gradually changing curvature, effected by the use of compound curves or spirals, or both, permit desirable development of superelevation.

Where radii greater than the minimum for a given design speed are used, they should be superelevated at less than the maximum rate to effect a balance in design, as that between sharper and flatter curves. Such greater radii are encountered where (1) the speed is established by one or both terminal curves of a turning roadway, and the central portion is on a considerably flatter curvature, (2) the turning roadway as a whole is on relatively flat curvature, but there is stop sign control in front or in back of the curve, or (3) the speed is low with respect to the curvature as a result of other speed limitations on the connecting roadway or on one or both of the intersecting highways. Table IX-12 shows suggested superelevation rates in relation to design speed, derived in much the same manner as those for open highway curves. The wide variation in likely speeds on intersection curves, as that between periods when speed is affected by traffic volume and periods when traffic is light, precludes need for precision, so a range in superelevation rate is given for each combination of design speed and radius of intersection curve. A maximum rate of 0.10 is shown, but any other maximum rate can be used. Rates in the upper half or third of the indicated range are preferred. The rate of 0.02 is considered a practical minimum for effective drainage across the surface.

Radius (ft)	Degree of Curve	Range in Superelevation Rate for Intersection Curves with Design Speed (mph) of					
		15	20	25	30	35	40
50	—	.02-.10	—	—	—	—	—
90	63.6	.02-.07	.02-.10	—	—	—	—
150	38.2	.02-.05	.02-.08	.04-.10	—	—	—
230	24.8	.02-.04	.02-.06	.03-.08	.06-.10	—	—
310	18.5	.02-.03	.02-.04	.03-.06	.05-.09	.08-.10	—
430	13.3	.02-.03	.02-.03	.03-.05	.04-.07	.06-.09	.09-.10
600	9.6	.02	.02-.03	.02-.04	.03-.05	.05-.07	.07-.09
1000	5.7	.02	.02-.03	.02-.03	.03-.04	.04-.05	.05-.06
1500	3.8	.02	.02	.02	.02-.03	.03-.04	.04-.05
2000	2.9	.02	.02	.02	.02	.02-.03	.03-.04
3000	1.9	.02	.02	.02	.02	.02	.02-.03

NOTE: Preferably use superelevation rate in upper half or third of indicated range. In areas where snow or ice is frequent, use maximum rate of 0.08.

Table IX-12. Superelevation rates for curves at intersections.

Superelevation Runoff

The principles of superelevation runoff, discussed in Chapter III for open highway conditions, apply generally to intersection curves. Primarily, the control of the rate of change of cross slope is that of riding comfort and appearance. In many instances the value of the cross slope is determined by the rate of change in elevation of a point on the pavement (generally the pavement edge) with respect to the point about which the pavement is being rotated. For through highway profiles the gradient of one edge of the pavement with respect to the centerline should be no greater than 0.50 percent for a design speed of 50 mph and 0.67 percent for a design speed of 30 mph, as discussed in Chapter III.

Logically, the pavement edge grades should vary with the design speed. A rate of 0.71 percent is applied for a 15 or 20 mph design speed with a decrease in rate to 0.45 percent for 60 mph, the latter being that used for open highway design. These values are shown in Table IX-13. For a 24 ft pavement as a base width, the change in pavement edge grade may be varied up to 25 percent above or below the tabulated values, the lower grades being applicable to wider pavements and the higher grades to narrower pavements.

Usually, the profile of one edge of turning roadway pavement is established first, and the profile on the other edge is developed by stepping up or down from the first edge by the amount of desired superelevation. This step is done by plotting a few control points on the second edge by using the values from Table IX-13 to approximate the change in superelevation rate from one point to the next, and then plotting a smooth profile for the second edge of pavement as close to the plotted points as is feasible. Drainage may be an additional control, particularly for curbed pavements.

Design Speed (mph)	15-25	30	40	50	55	60	65	70
Maximum relative gradients for profiles between the edge of two lane pavement and the centerline (percent)	0.71	0.67	0.58	0.50	0.47	0.45	0.41	0.40

NOTE: The above table is recommended for curves through intersections and coincides with the Table III-14. The word "MAXIMUM" in Table IX-13 is used in place of "DESIGN" to allow more flexibility in selected 'e' max. through the intersection curves. Change in pavement edges divided by the roadway width (24 ft) should not exceed those rates set in the table.

Table IX-13. Maximum rate of change in pavement edge elevation for curves at intersections.

Development of Superelevation at Turning Roadway Terminals

Superelevation commensurate with curvature and speed seldom is practicable at terminals where: (1) a flat intersection curve results in little more than a widening of the through-traffic pavement, (2) it is desirable to retain the cross slope of the through pavement, and (3) there is a practical limit to the difference between the cross slope on the through pavement and that on the intersection curve. Too great a difference in cross slope may cause vehicles traveling over the cross over crown line to sway sideways. When vehicles, particularly high-bodied trucks, cross the crown line at other than low speed and at an angle of about 10° to 40°, the body throw may make vehicle control difficult.

General Procedure

For design of an exit from a highway, the through traffic lanes may be considered fixed in profile and cross slope. As the exit curve diverges from the through traveled way, the curved (or tangent) edge of the widening pavement section can only gradually vary in elevation from the edge of through lane. Shortly beyond the point where the full width of the turning roadway is attained, an approach nose separates the two pavements. Where the exit curve is relatively sharp and without taper or transition, little superelevation in advance of the nose can be developed in the short distance available. Beyond the nose substantial superelevation usually can be attained, the amount depending on the length of the turning roadway curve. Where this curve deviates gradually from a traveled way, a desirable treatment of superelevation may be effected.

The method of developing superelevation at turning roadway terminals is illustrated diagrammatically in Figures IX-47 through IX-50. Figure IX-47 illustrates the variation in cross slope where a turning roadway leaves a through road that is on tangent. From point A to B, the normal cross slope on the through-traffic lane is extended to the outer edge of auxiliary pavement. The width at B is nominal, 1 to 3 ft, and projecting the cross slope of the through pavement across this width simplifies construction. Beyond point B, the width is sufficient that the cross slope on the auxiliary pavement can begin to be steeper than the cross slope on the adjacent through-traffic lane, as at C. At D where the full width of the turning roadway is attained, a still greater slope can be used. Superelevation is further increased adjacent to the nose at E and is facilitated somewhat by sloping downward the pavement wedge formed

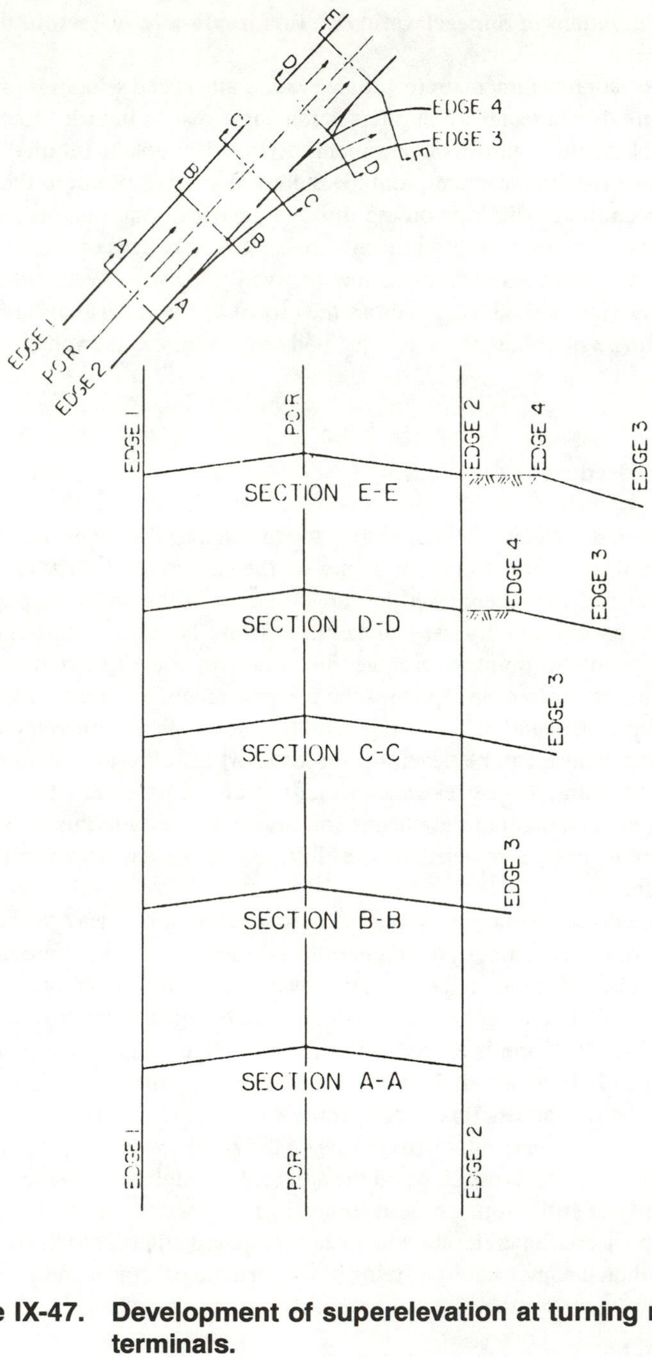

Figure IX-47. Development of superelevation at turning roadway terminals.

between the right edge of the travel way and the extended left pavement edge of the turning roadway. Beyond the nose, as at E, the pavement is transitioned as rapidly as conditions permit until the full desired superelevation is attained.

Figure IX-48 is a similar illustration for the condition where the through lanes and the turning roadway curve in the same direction. The desired superelevation on the exit pavement, which generally is steeper than that on the through lanes, can be attained in a relatively short distance. At B the cross slope of the through lane is extended over the widened pavement. At C and D somewhat steeper cross slopes are introduced, the full superelevation being reached in the vicinity of E.

A less favorable situation occurs when the joining facilities curve in opposite directions, as in Figure IX-49. Because of the rate of superelevation on the through pavement, it may be impracticable to slope the auxiliary pavement in a direction opposite to that of the through lanes for reasons of appearance, riding quality, and safety. In a typical treatment for a moderate rate of superelevation, the rate of cross slope on the through pavement is extended onto the auxiliary pavement, as at B. At C it may still continue upward, but at a lesser rate. The break between the two slopes becomes more pronounced at D, the added pavement being nearly horizontal. Some superelevation is introduced at the nose, either by a single crown line centering on the nose or by a double break in the cross slope over the pavement wedge in front of the nose. Most of the superelevation must be gained beyond the nose.

On designs with a parallel speed change lane, as in Figure IX-50, part of the cross slope change may be made over the length of this lane. Usually, more than half of the total superelevation rate can be attained at D, and the full desired superelevation can be reached at or just beyond the nose.

The discussion and arrangements illustrated in Figures IX-47 through IX-50 for exit terminals are also directly applicable to entrance terminals, except that the details at the merging end are different from those of an approach nose. The merging end of an entrance terminal would be located in proximity of D.

Turn-Lane Cross-Slope Rollover

The design control at the crossover line (not to be confused with the crown line normally provided at the centerline of a pavement) is the algebraic difference in cross slope rates of the two adjacent pavements. Where both pavements slope down and away from the crossover crown line, the algebraic difference is the sum of their cross slope rates; where they slope in the same direction, it is the difference of their cross slope rates. A desirable maximum

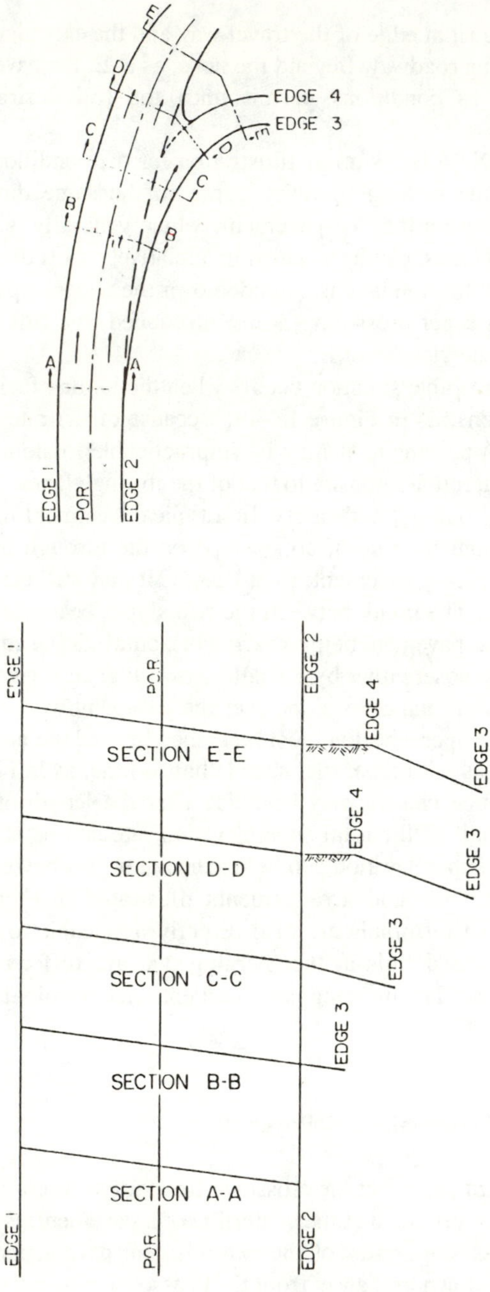

Figure IX-48. Development of superelevation at turning roadway terminals.

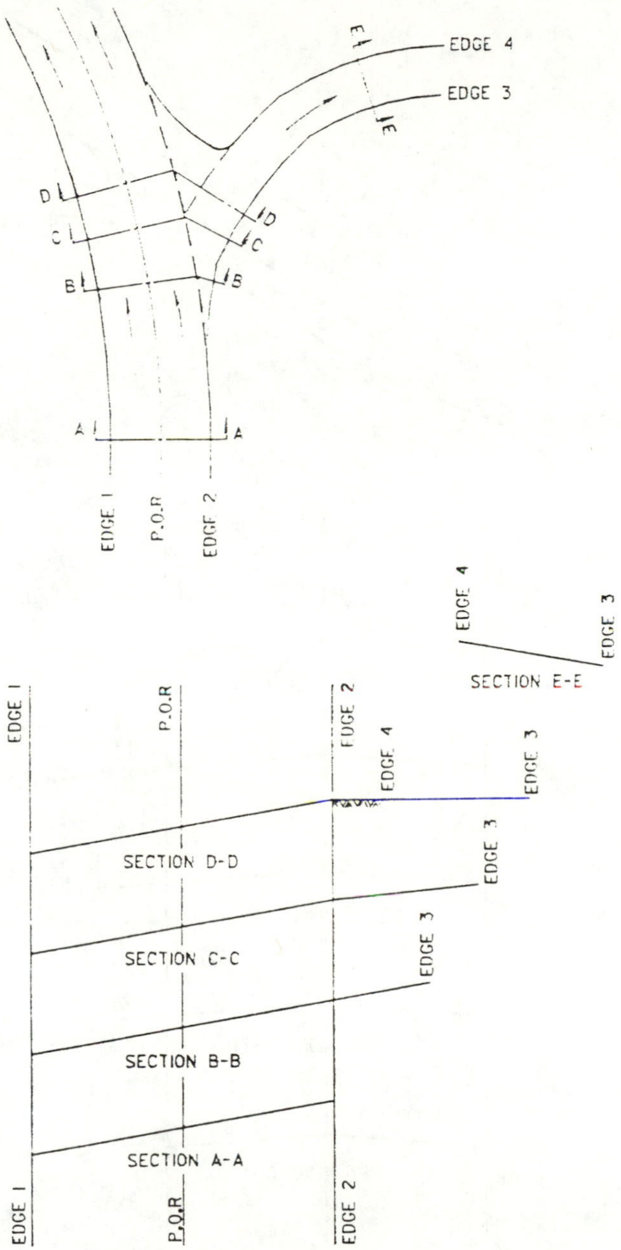

Figure IX-49. Development of superelevation at turning roadway terminals.

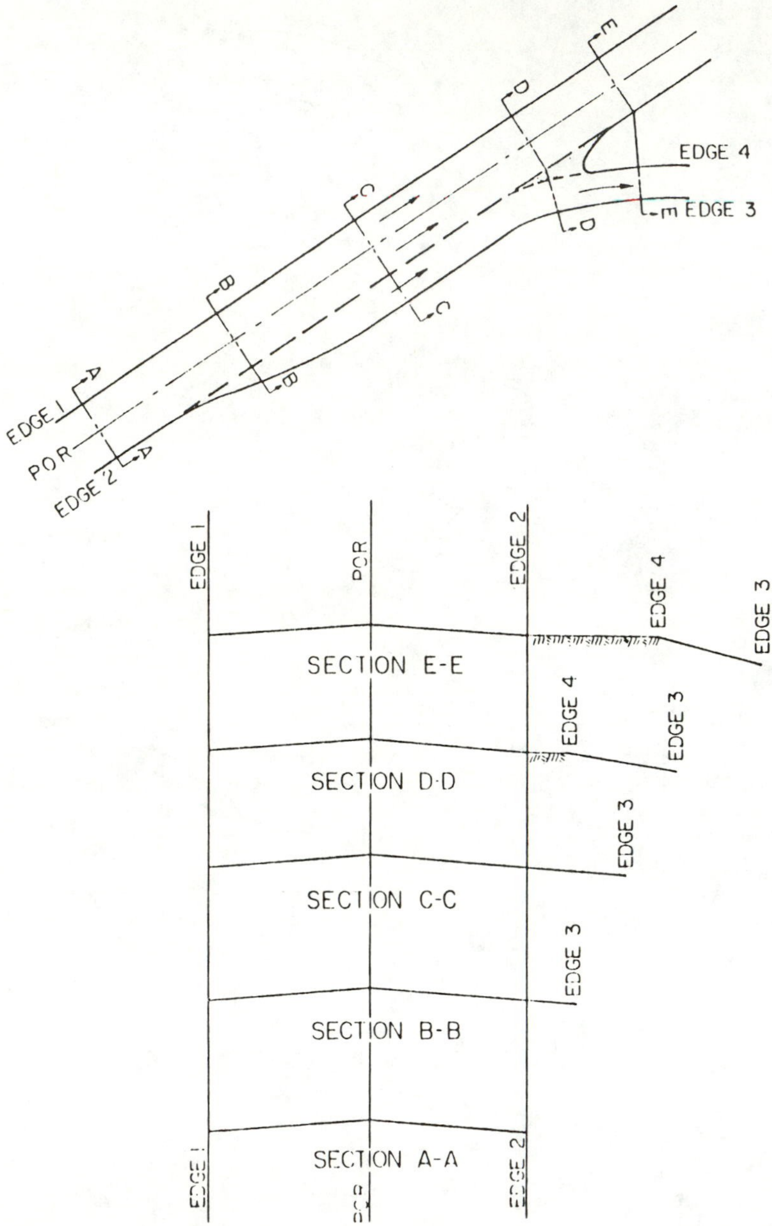

Figure IX-50. Development of superelevation at turning roadway terminals.

algebraic difference at a crossover crown line is 4 or 5 percent, but it may be as high as 8 percent at low speeds and where there are few trucks. The suggested maximum differences in cross slope rates at a crown line, related to the speed of turning traffic, are given in Table IX-14.

Superelevation Transition and Gradeline Control

The attainment of superelevation over the gradually widening auxiliary pavement and over the whole of the turning roadway terminals should not be abrupt. The design should be in keeping with the superelevation and cross slope controls, previously determined and given in Tables IX-12 through IX-14.

As an example, consider an arrangement as in Figure IX-47, in which the limiting curve of the turning roadway has a radius of 230 ft, corresponding to a design speed of 30 mph. From Table IX-12 the maximum superelevation rate of this curve may be between 6 and 10 percent and preferably not less than 8 percent. Because the pavement width is variable, the transition of cross slope change should be developed by using the method of pavement edge change in grade with respect to the point of rotation for a full-width auxiliary lane. Elevations developed by this method should then be converted to a change in elevation between the edge of the pavement of the through lane and the edge of the full-width pavement of the auxiliary lane and then be prorated for the actual partial widths of the auxiliary lane. In this example, the pavement edge change in grade should be no greater than 0.67 percent, as shown in Table IX-13.

Design Speed of Exit or Entrance Curve (mph)	Maximum Algebraic Difference in Cross Slope at Crossover Line (percent)
15 and 20	5.0 to 8.0
25 and 30	5.0 to 6.0
35 and over	4.0 to 5.0

Table IX-14. Maximum algebraic difference in pavement cross slope at turning roadway terminals.

An alternate method, which has been noted with respect to ridability, comfort, and appearance of the roadway in cross slope transition areas, is to establish a rate of change in the pavement cross slope. The rate of pavement cross slope is a function of pavement width and the change in grade of the pavement edge with respect to the point of pavement rotation. This method results in the pavement edge grade being equal to the pavement width, which is rotated, times the rate of change in pavement cross slope. Thus, if the pavement edge grade change is 0.67 percent and the width of pavement being rotated (the assumption being that the full width of the auxiliary lane is applied for calculating the grade change of the pavement edge) is 12 ft, the rate of change in pavement cross slope is 5.58 percent per 100-ft station.

In Figure IX-47, if the cross slope on the through pavement is 1 percent and the distance from A to B and also from B to C is 50 ft, trial cross slope rates would be 1 percent at A, 3.79 percent at B, and 6.58 percent at C. Here the cross over crown line control (Table IX-14) is barely satisfied, because at the critical section C, the algebraic difference in cross slope rates is 5.58 percent. If the remaining lengths of C to D and D to E are 25 ft apart, the cross slope rate would be 7.97 percent at D and 9.35 percent at E.

The cross slope of the pavement edge in front of the nose at E could be some intermediate rate, say 4 percent. On second trial a better adjustment of superelevation transition results by using a lower change in the cross slope rate for the turning roadway, such as 4 percent per 100-ft length.

This procedure of establishing superelevation cross slopes at given points is a preliminary step in design. Elevations on the pavement edges resolved from these cross slopes serve as control points for drawing the edge of pavement profiles on the turning roadway. Excellent practical results are obtained by plotting to large vertical scale the profiles for both pavement edges of the turning roadway and the edge and centerline of through pavement in juxtaposition on a single profile drawing. Important points such as approach noses or merging ends also are located. Only one profile, either the stationed centerline or edge of pavement, is depicted in true length, but the inaccuracy in length of the other profiles is small and it is easy to locate points thereon in the field by radial measurement from the stationed line. The three-dimensional condition can be readily visualized.

Mathematically derived vertical curves, as used for open highways, are not always practicable at intersections, but the profile curves can be developed readily with a spline or irregular curve templates. All necessary elevations can be read directly from the profiles when they are drawn to large enough vertical scale. The final profile may not precisely produce the selected cross slope at all of the control points, but this problem is not serious as long as the cross slope change is progressive and within the design control limits. The principal criterion is the development of smooth-edge profiles that do not appear distorted to the driver. Another method of obtaining a three-dimensional presentation is to plot contour lines on a layout of the intersection area. A 50-scale or in complex areas a 20-scale drawing will provide an accurate picture with the additional advantage of showing drainage patterns, sumps, and irregular slope conditions.

TRAFFIC CONTROL DEVICES

Traffic control devices are necessary for regulating, warning, and guiding traffic and are a primary determinant in the safe and efficient operation of intersections. It is essential that intersection design be accomplished simultaneously with the development of traffic control plans to ensure that sufficient space is provided for proper installation of traffic control devices. Geometric design should not be considered complete nor should it be implemented until it has been determined that needed traffic devices will have the desired effect in controlling traffic.

Most of the intersection types illustrated and described in the following discussions are adaptable to either signing control, signal control, or a combination of both. At intersections not requiring signal control, the normal pavement widths of the approach highways are carried through the intersection with the possible addition of speed change lanes, median lanes, auxiliary lanes, or pavement tapers. Where volumes are sufficient to require signal control, the number of lanes for through movements may also need to be increased. Where the volume approaches the uninterrupted flow capacity of the intersection leg, the number of lanes in each direction may have to be doubled at the intersection to accommodate the volume under stop-and-go control. Other geometric features that may be affected by signalization are length and width of storage areas, location and position of turning roadways, spacing of other subsidiary intersections, access connections, and the possible location and size of islands to accommodate signal posts or standards.

At high-volume intersections-at-grade, the design of the signals should be sophisticated enough to respond to the varying traffic demands, the objective being to keep the vehicles moving through the intersection. Factors affecting capacity and computation procedures for signalized intersections are covered in the HCM (**2**).

An intersection requiring traffic signal control is best designed by considering jointly the geometric design, capacity, analysis, design hour volumes, and physical controls. Details on the design and location of most forms of traffic control signals, including the general warrants, are given in Part IV of the MUTCD (**5**).

GENERAL INTERSECTION TYPES

General Design Considerations

General types of at-grade intersections and terminology are indicated in Figures IX-51 and IX-52. The geometric forms are the three-leg, four-leg, and multileg intersections. Further classification includes such variations as unchannelized, flared, and channelized intersections. Details and specific adaptations of each general type are demonstrated under the heading "Types and Examples of At-Grade Intersections."

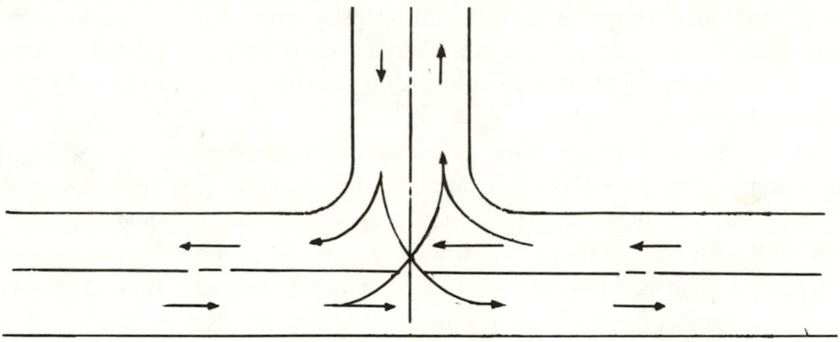

UNCHANNELIZED - T

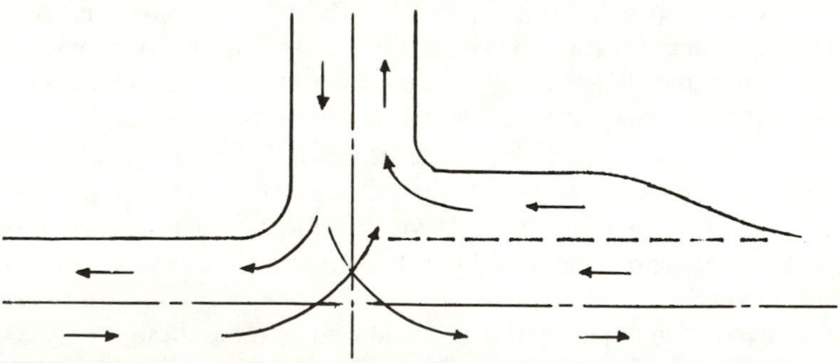

T-INTERSECTION WITH RIGHT TURN LANE

Figure IX-51. General types of at-grade intersections.

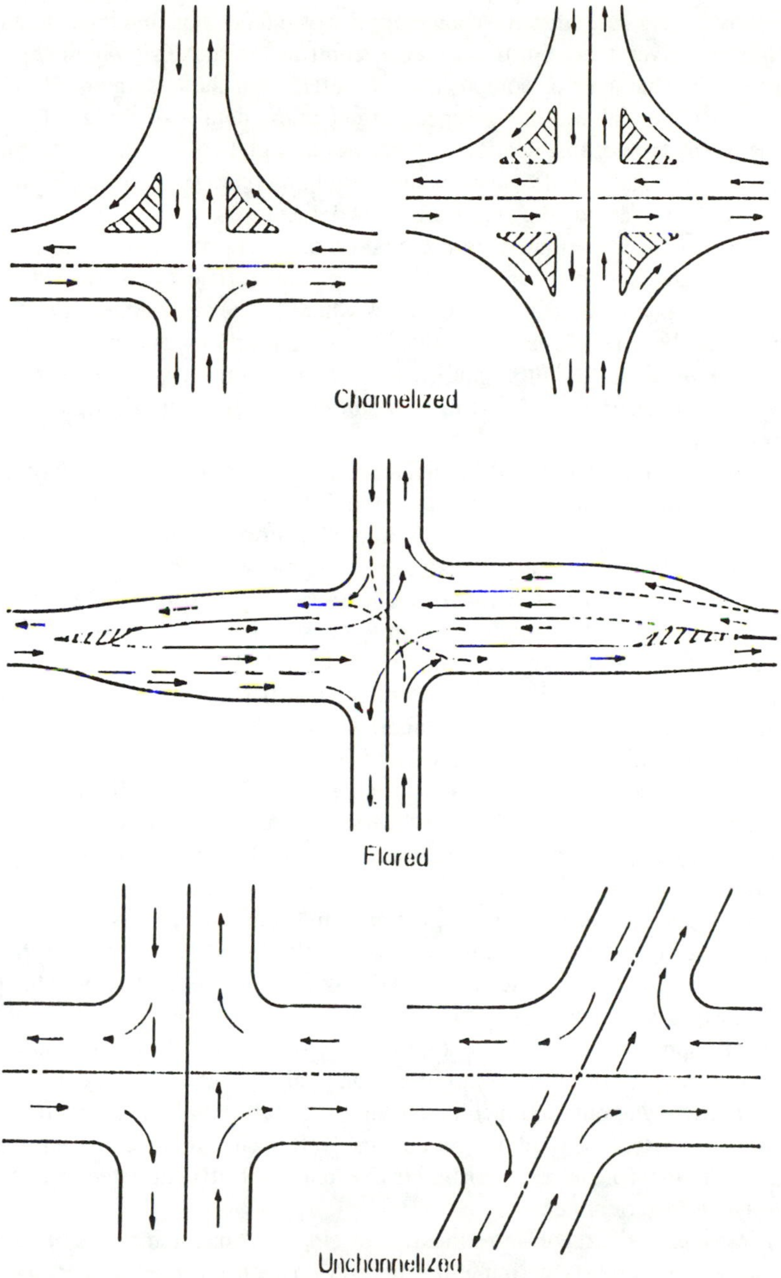

Figure IX-52. General types of at-grade intersections.

Many factors enter into the choice of type of intersection and the extent of design of a given type, but the principal controls are the design-hour traffic volume, the character or composition of traffic, and the design speed. The character of traffic and design speed affect many details of design, but in choosing the type of intersection they are not as significant as the traffic volume. Of particular significance are the actual and relative volumes of traffic involved in various turning and through movements.

Left-turn facilities should be established on roadways where traffic volumes are high enough or safety considerations are sufficient to warrant them. Table IX-15 is a guide to traffic volumes where left-turn facilities should be considered. For the volumes shown, left turns and right turns from the minor street can be equal to, but be not greater than, the left turns from the major street. Some examples of left-turn facilities are given in Figures IX-51 through IX-56.

Additional information on left turn lanes as well as suggested length of left turn lanes can be found in References 1 and 6. In the case of double left turns, a capacity analysis of the intersection should be performed to determine what traffic controls are necessary in order for it to function properly.

Local conditions and the cost of right-of-way often influence the type of intersection selected as well as many of the design details. Limited sight distance, for example, may make it necessary to control traffic by yield signs, stop signs, or traffic signals when the traffic densities are less than those ordinarily considered necessary for such control. The alinement and grade of the intersecting roads and the angle of intersection may make it advisable to channelize or use auxiliary pavement areas, regardless of the traffic densities. In general, traffic service, highway design designation, physical conditions, and cost of right-of-way are considered jointly in choosing the type of intersection.

For the general benefit of through-traffic movements the number of crossroads, intersecting roads, or intersecting streets should be minimal. Where intersections are closely spaced on a two-way facility, it is seldom possible to provide signals for completely coordinated traffic movements at reasonable speeds in opposing directions on that facility. At the same time the resultant road or street patterns must permit travel on roadways other than the predominant highway without too much inconvenience. Traffic analysis is required to determine whether the road or street pattern, left open across the predominate highway, is adequate to serve normal traffic plus the traffic diverted from any terminated road or street.

The patterns of traffic movement at the intersections and the volume of traffic on each approach, including pedestrians, during one or more peak periods of the day are indicative of the type of traffic control devices necessary, the pavement widths required (including auxiliary lanes), and where applicable, the degree of channelization needed to expedite the movement of all traffic. The differing arrangement of islands and the shape and length of auxiliary pavements depend on whether signal control is necessary.

	Advancing Volume			
Opposing Volume	**5% Left Turns**	**10% Left Turns**	**20% Left Turns**	**30% Left Turns**
	40-mph Operating Speed			
800	330	240	180	160
600	410	305	225	200
400	510	305	275	245
200	640	470	350	305
100	720	515	390	340
	50-mph Operating Speed			
800	280	210	165	135
600	350	260	195	170
400	430	320	240	210
200	550	400	300	270
100	615	445	335	295
	60-mph Operating Speed			
800	230	170	125	115
600	290	210	160	140
400	365	270	200	175
200	450	330	250	215
100	505	370	275	240

Table IX-15. Guide for left turn lanes on two lane highways (6).

The composition and character of traffic are a design control. Movements involving large trucks necessitate larger intersection areas and flatter approach grades than those required at intersections where traffic consists predominantly of passenger cars. Bus stops located near an intersection may further modify the arrangement. Approach speeds of traffic also have a bearing on the geometric design as well as on control devices and markings.

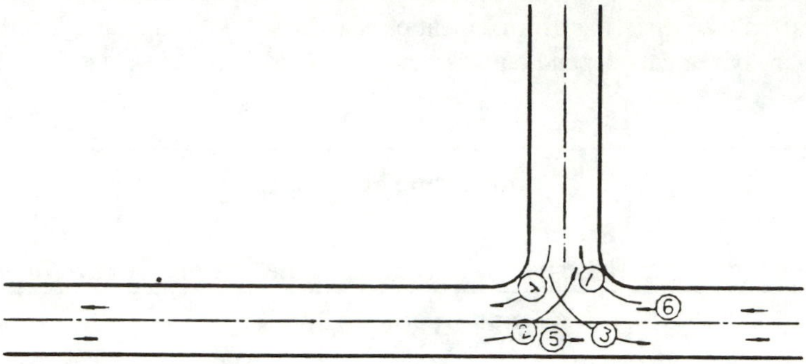

PLAIN 'T' INTERSECTION

-A-

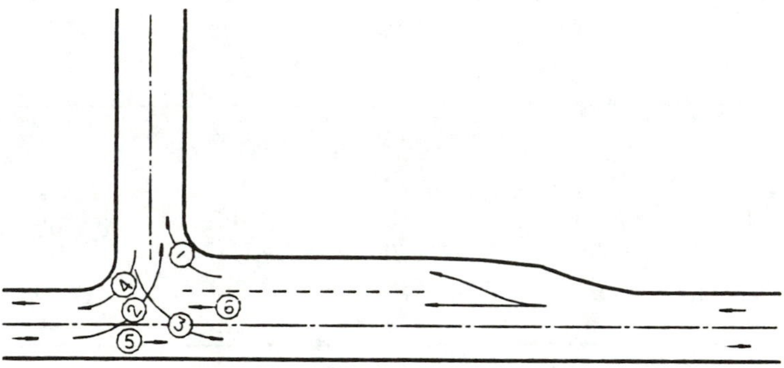

'T' INTERSECTION

(With Right Turn Lane)

-B-

Figure IX-53. "T" intersections.

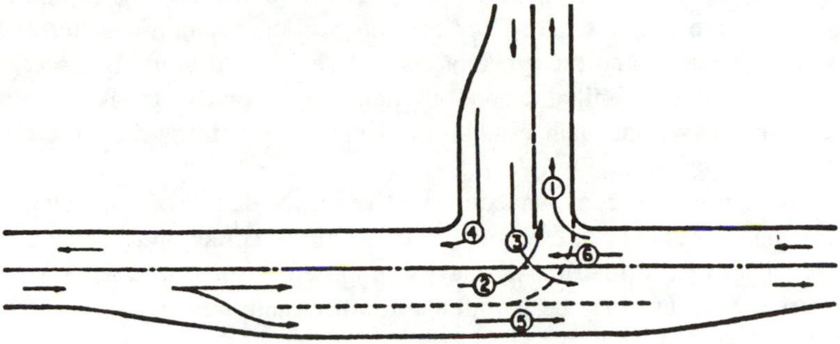

'T' INTERSECTION (WITH RIGHT HAND PASSING LANE)

-A-

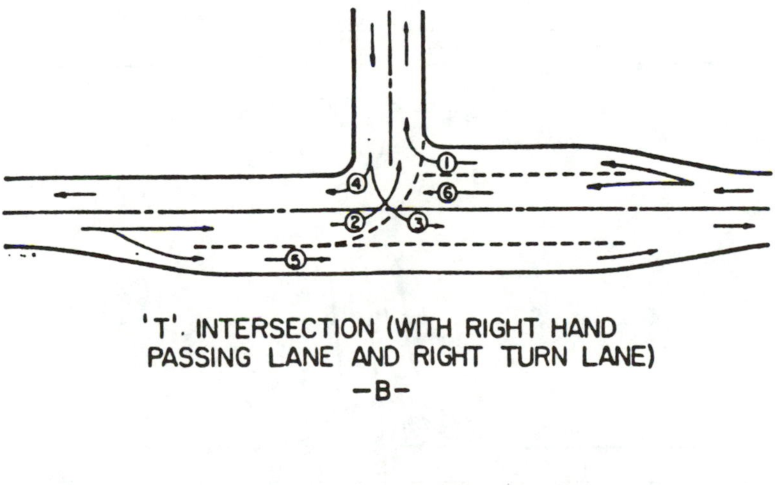

'T'. INTERSECTION (WITH RIGHT HAND PASSING LANE AND RIGHT TURN LANE)

-B-

Figure IX-54. "T" intersections.

The number and locations of the approach roadways and their angles of intersection are major controls for the intersection geometric pattern, the location of islands, and the types of control devices. At-grade intersections preferably should be limited to no more than four approach legs. Two or more crossroads intersecting an arterial highway in close proximity should be combined into a single crossing.

The distances between intersections influence the degree of channelization at any one particular intersection. For example, where intersections are closely spaced, turn restrictions may be imposed at some intersections and pedestrian crossings may be prohibited at others and thus make some chan-

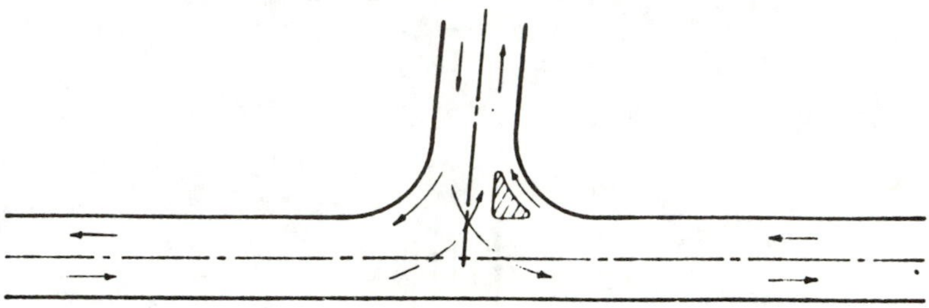

WITH SINGLE TURNING ROADWAY

-A-

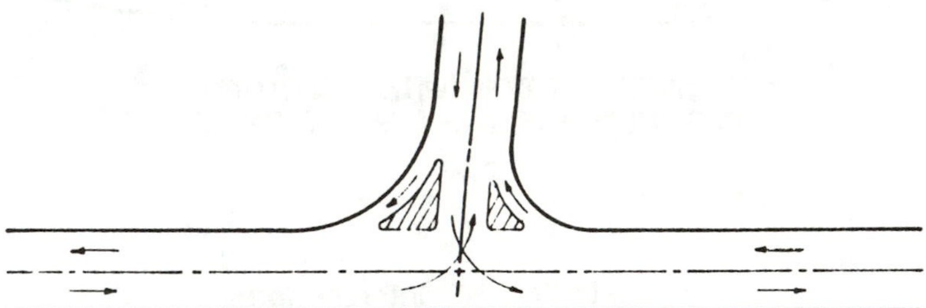

WITH A PAIR OF TURNING ROADWAYS

-B-

Figure IX-55. Channelized "T" intersections.

nelizing islands and auxiliary pavement areas unnecessary, or it may be necessary to introduce continuous auxiliary lanes between two or more intersecting roads or streets to handle a buildup and weaving of traffic. Where crossroads are widely spaced, each at-grade intersection must necessarily accommodate all crossing, turning, and pedestrian movements.

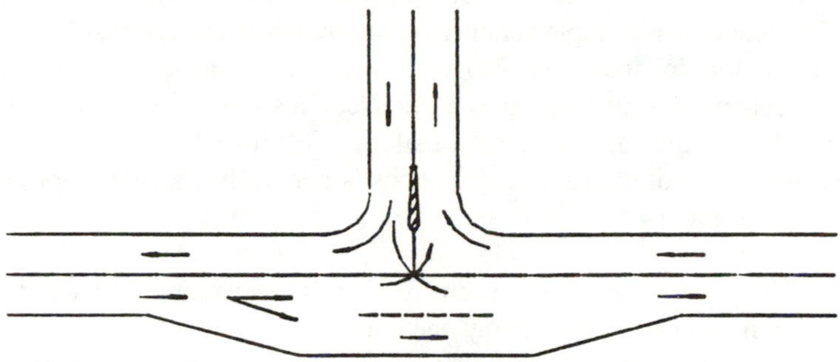

WITH DIVISIONAL ISLAND AND RIGHT PASSING LANE

-A-

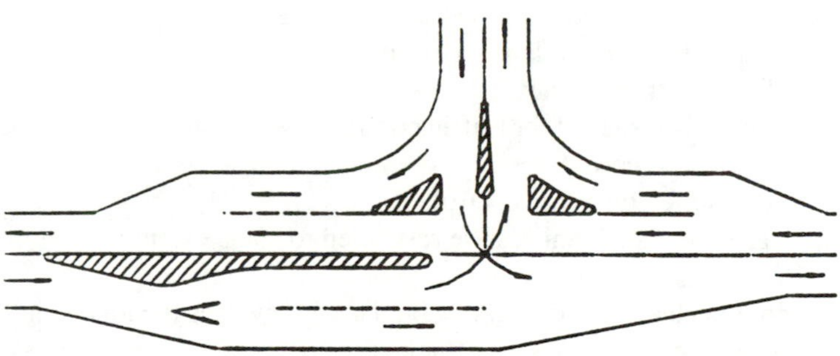

WITH DIVISIONAL ISLANDS AND TURNING ROADWAYS

-B-

Figure IX-56. Channelized "T" intersections.

CHANNELIZATION

Channelization is the separation or regulation of conflicting traffic movements into definite paths of travel by traffic islands or pavement marking to facilitate the safe and orderly movements of both vehicles and pedestrians. Proper channelization increases capacity, improves safety, provides maximum convenience, and instills driver confidence. Improper channelization has the opposite effect and may be worse than none at all. Over channelization should be avoided because it could create confusion and worsen operations. In some cases a simple channelization improvement can result in a dramatic reduction in accidents. Most of these cases involve some type of left-turn treatment. Left-turn lanes at intersections reduce rear-end exposure and provide a comfortable means for making a left turn.

Channelization of at-grade intersections is generally warranted for one or more of the following factors:

1. The paths of vehicles are confined by channelization so that not more than two paths cross at any one point.
2. The angle and location at which vehicles merge, diverge, or cross are controlled.
3. The amount of paved area is reduced and thereby decreases vehicle wander and narrows the area of conflict between vehicles.
4. Clearer indications are provided for the proper path in which movements are to be made.
5. The predominant movements are given priority.
6. Areas are provided for pedestrian refuge.
7. Separate storage lanes permit turning vehicles to wait clear of through-traffic lanes.
8. Space is provided for traffic control devices so that they can be more readily perceived.
9. Prohibited turns are controlled.
10. The speeds of vehicles are restricted to some extent.

Design of a channelized intersection usually involves the following significant controls: the type of design vehicle, the cross sections on the cross roads, the projected traffic volumes in relation to capacity, the number of pedestrians, the speed of vehicles, the location of any required bus stops, and the type and location of traffic control devices. Furthermore, the physical controls such as right-of-way and terrain have an effect on the extent of channelization that is economically feasible.

Certain principles should be followed in the design of a channelized intersection, but the extent to which they are applied will depend on the characteristics of the total design plan. These principles are:

1. Motorists should not be confronted with more than one decision at a time.
2. Unnatural paths that require turns greater than 90° or sudden and sharp reverse curves should be avoided.
3. Areas of vehicle conflict should be reduced as much as possible. However, merging and weaving areas should be as long as conditions permit. Channelization should be used to keep vehicles within well-defined paths that minimize the area of conflict.
4. Traffic streams that cross without merging and weaving should intersect desirably at right angles with a range of 60—120 degrees acceptable.
5. The angle of intersection between merging streams of traffic should be appropriate to provide adequate sight distance.
6. The points of crossing or conflict should be studied carefully to determine if such conditions would be better separated or consolidated to simplify design with appropriate control devices added to ensure safe operation.
7. Refuge areas for turning vehicles should be provided clear of through traffic.
8. Prohibited turns should be blocked wherever possible.
9. Location of essential control devices should be established as a part of the design of a channelized intersection.
10. Channelization may be desirable to separate the various traffic movements where multiple phase signals are used.

SPEED-CHANGE LANES AT INTERSECTIONS

Drivers leaving a highway at an intersection are usually required to reduce speed before turning. Drivers entering a highway from a turning roadway accelerate until the desired open-road speed is reached. When undue deceleration or acceleration by leaving or entering traffic takes place directly on the highway traveled way, it disrupts the flow of through traffic and often is hazardous. To preclude or minimize these undesirable aspects of operation at intersections, speed-change lanes are standard practice on highways having expressway characteristics and are frequently used on other main highway intersections.

A speed-change lane is an auxiliary lane, including tapered areas, primarily for the acceleration or deceleration of vehicles entering or leaving the through-traffic lanes. The terms "speed-change lane," "deceleration lane," or "acceleration lane," as used here, apply broadly to the added pavement joining the traveled way of the highway or street with that of the turning roadway and do not necessarily imply a definite lane of uniform width. A speed-change lane should be of sufficient width and length to enable a driver to maneuver a vehicle into it properly, and once into it, to make the necessary change between the speed of operation on the highway or street and the lower speed on the turning roadway. Deceleration and acceleration lanes may be designed in conjunction with each other, the relationship depending on the arrangement of the intersection and traffic requirements. They may be designed as parts of at-grade intersections but are especially important at ramp junctions where turning roadways meet high-speed traffic lanes.

Warrants for the use of speed-change lanes cannot be stated definitely. Many factors must be considered, such as speeds, traffic volumes, capacity, type of highway, service provided, the arrangement and frequency of intersections, and accident experience. Observations and considerable experience with speed-change lanes have led to the following general conclusions:

1. Speed-change lanes are warranted on high-speed and on high-volume highways where a change in speed is necessary for vehicles entering or leaving the through-traffic lanes.
2. All drivers do not use speed-change lanes in the same manner; some use little of the available facility. As a whole, however, these lanes are used sufficiently to improve the overall safety and operation of the highway.
3. Use of speed-change lanes varies with volume, the majority of drivers using them at high volumes.
4. The directional type of speed-change lane consisting of a long taper fits the behavior of most drivers and does not require maneuvering on a reverse-curve path.
5. Deceleration lanes on the approaches to at-grade intersections that also function as storage lanes for turning traffic are particularly advantageous, and experience with them generally has been favorable. Such lanes reduce hazard and increase capacity.

A median lane provides refuge for vehicles awaiting an opportunity to turn, and thereby keeps the highway traveled way clear for through traffic. The width, length, and general design of median lanes are similar to those of any other deceleration lane but their design includes some additional features discussed under the heading "Auxiliary Lanes" discussed later in this chapter.

Deceleration lanes always are advantageous, particularly on high speed roads, because the driver of a vehicle leaving the highway has no choice but to slow down on the through-traffic lane if a deceleration lane is not provided. The failure to brake by the following drivers because of a lack of alertness causes many rear-end collisions. Acceleration lanes are not always desirable at stop-controlled intersections where entering drivers can wait for an opportunity to merge without disrupting through traffic. Acceleration lanes are advantageous on roads without stop control and on all high-volume roads even with stop control where openings between vehicles in the peak-hour traffic streams are infrequent and short. (For additional design guidance relative to lengths of deceleration and acceleration auxiliary lanes, refer to Chapter X.)

MEDIAN OPENINGS

General Design Considerations

Medians are discussed in Chapter IV chiefly as an element of the cross section. General ranges in width are given, and the width of the median at intersections is treated briefly. For intersection conditions the median width, the length of the opening, and the design of the median end are developed in combination to fit the character and volume of through and turning traffic. Where the traffic pattern at an at-grade intersection shows that nearly all traffic travels through on the divided highway and the volume is well below capacity, a median opening of the simplest and least costly design may be sufficient. This type of opening permits vehicles to make cross and turning movements, but in doing so they may encroach on adjacent lanes and usually will not have a protected space clear of other traffic. A traffic pattern showing appreciable cross and turning movements or through traffic of high speed and high volume justifies a design in which the median is of a shape and scale to enable turning movements to be made, perhaps at above-minimum speed without encroachment on adjacent lanes and with little or no interference between traffic movements.

The design of a median opening and median ends should be based on traffic volumes and type of turning vehicles, as discussed in Chapter II. Cross and turning traffic must operate in conjunction with the through traffic on the divided highway. This requirement makes it necessary to know the volume and composition of all movements occurring simultaneously during the design hours. The design of a median opening becomes a matter of considering what traffic is to be accommodated, choosing the design vehicle to use for layout controls for each cross and turning movement, investigating whether larger vehicles can turn without undue encroachment on adjacent lanes, and finally checking the intersection for capacity. If the capacity is exceeded by the traffic load, the design must be expanded, possibly by widening or otherwise adjusting widths for certain movements. Traffic control devices such as yield signs, stop signs, or traffic signals may be required to regulate the various movements effectively and improve the effectiveness of operations.

Control Radii for Minimum Turning Paths

An important factor in designing median openings is the path of each design vehicle making a minimum left turn at 10 to 15 mph. Where the volume and type of vehicles making the left-turn movement call for higher than minimum speed, the design may be made by using a radius of turn corresponding to the speed deemed appropriate. However, the minimum turning path at low speed is needed for minimum design and for testing layouts developed for one design vehicle for use by an occasional larger vehicle.

The paths of design vehicles making right turns are given in Chapter II and are discussed in this chapter in the section "Intersection Curves." Any differences between the minimum turning radii for left turns and those for right turns are small and are insignificant in highway design. Minimum 90° left-turn paths for design vehicles are shown in Figure IX-57. Figure IX-57A shows these paths positioned as they would govern median end design for vehicles leaving a divided highway. Figure IX-57B shows them positioned for left turns to enter a divided highway. In both cases it is assumed that the inner wheel of each design vehicle clears the median edge and centerline of the crossroad by 2 ft at the beginning and end of the turn. For comparison, circular arcs of 40-, 50-, and 75-ft radii and tangent to the crossroad centerline and the median edge are also shown. The transition paths of the inner rear wheels are long, particularly for the semitrailer vehicles when completing the turn. Where the controlling circular arc is sharper than these long transition paths, drivers can, and habitually do, swing wide and turn on a reverse or S-curve path instead of turning directly to traverse the minimum paths shown.

The edges of pavement that most closely fit the paths of turning vehicles are transitional; however, for sharp turns at intersections, designs closely fitting these paths are three-centered curves. Design guidance covering three-centered curves is discussed in the section "Intersection Curves" in this chapter. The same curves are applicable to left turns and should be used where there is a physical edge of pavement for left turns, as in a channelized intersection and on ramps for the predominant highway.

The customary at-grade intersection on a divided highway does not have a continuous physical edge of pavement delineating the left-turn path. Instead, the driver has guides at the beginning and at the end of the left-turn operation:

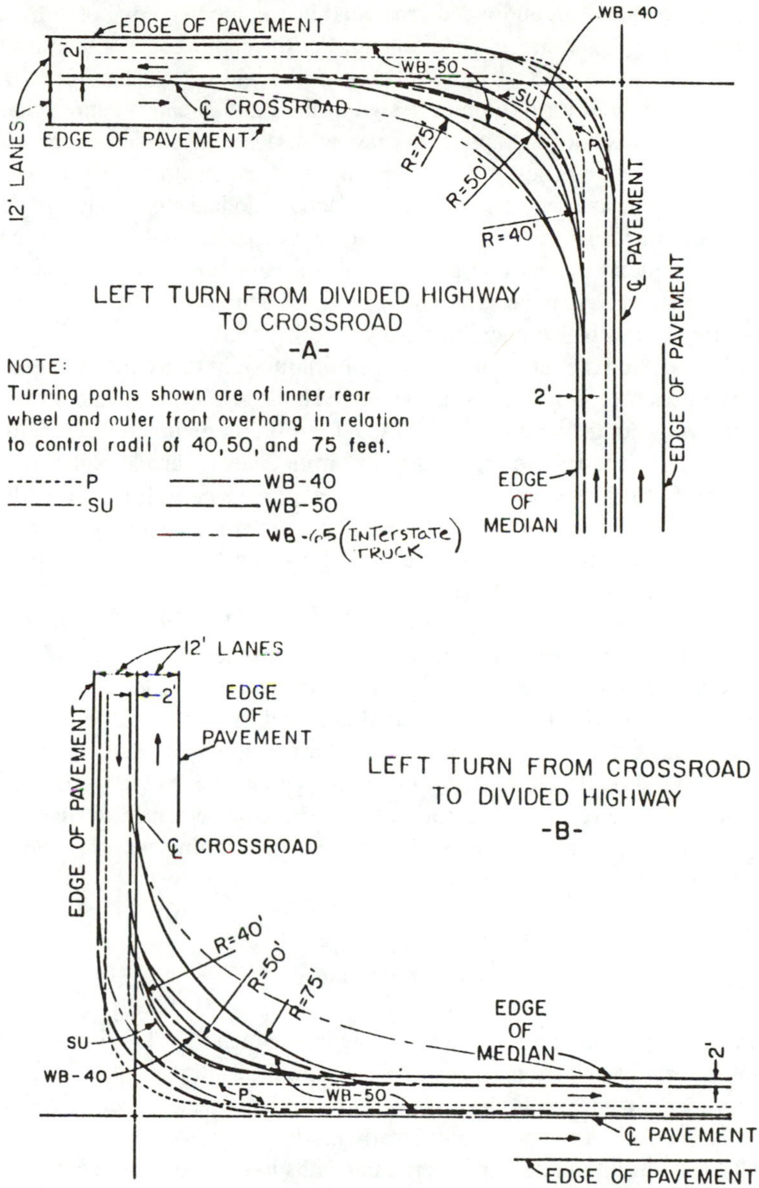

Figure IX-57. Control radii at intersections for 90° left turns.

(1) the centerline of an undivided crossroad or the median edge of a divided crossroad, and (2) the curved median end. For the central part of the turn the driver has the open central intersection area in which to maneuver. Under these circumstances for minimum design of the median end, the precision of compound curves does not appear necessary, and simple curves for the minimum assumed edge of left turn have been found satisfactory. The larger the simple curve radius used, the better it will accommodate a given design vehicle, but the resulting layout for the larger curve radius will have a greater length of median opening and greater paved areas than one for a minimum radius. These areas may be sufficiently large to result in erratic maneuvering by small vehicles, which may interfere with other traffic.

By considering the range of radii for minimum right turns and the need for accommodation of more than one type of vehicle at the usual intersections, the following control radii can be used for minimum practical design of median ends: a control radius of 40 ft accommodates P vehicles suitably and occasional SU vehicles with some swinging wide; one of 50 ft accommodates SU vehicles and occasional WB-40 vehicles with some swinging wide; and one of 75 ft accommodates WB-40 and WB-50 vehicles with only minor swinging wide at the end of the turn.

These relations are shown generally in Figures IX-58 through IX-60 and the accompanying Tables IX-16 through IX-18. In the following explanation, each control radius design is tested for use by larger vehicles and for occasional movements other than those for which the design is developed. The figures indicate how the design may be tested for protection of cross traffic. This test is followed by development of median opening designs for traffic conditions in which the volumes of through and turning movements are such that it is desirable to provide space in the median for turning vehicles clear of through traffic.

Shape of Median End

One form of a median end at an opening is a semicircle. This simple design is satisfactory for narrow medians. The several disadvantages of semicircular ends for medians greater than about 10 ft in width are widely recognized, and other more desirable shapes are generally used.

Alternate minimum designs for median ends to fit the design control radii of 40, 50, and 75 ft are shown in Figures IX-58 through IX-63. The paths of design vehicles are shown on the basis of the inner rear wheel beginning and ending the left-turn maneuver 2 ft from the edge of the median and the centerline of the undivided crossroad. The alternate minimum designs are a semicircular end and a bullet nose form. The indicated PC of the control

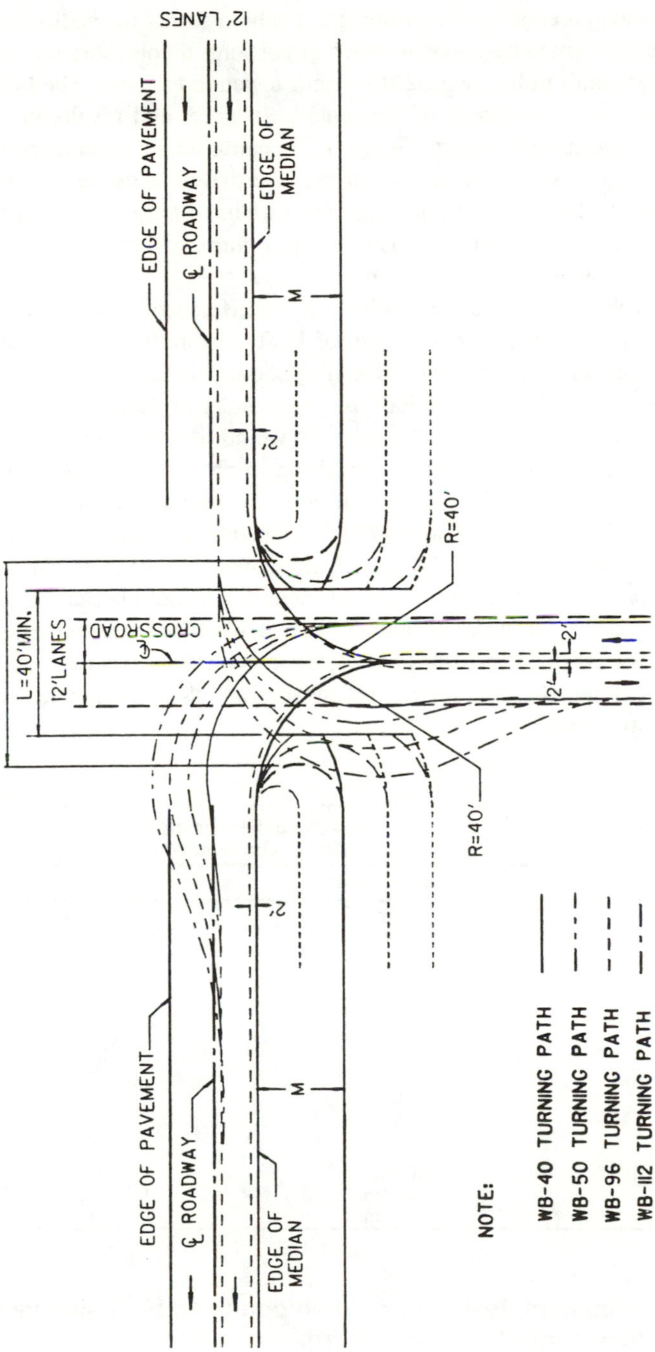

Figure IX-58. Minimum design of median openings (P design vehicle, control radius of 40 ft).

radius on the median edge is a common PC for both forms of median end. The bullet nose is formed by two symmetrical portions of control radius arcs and an assumed small radius, e.g., 2 ft is used, to round the nose. The bullet nose design closely fits the path of the inner rear wheel and results in less intersection pavement and a shorter length of opening than the semicircular end. These advantages are operational in that the driver of the left-turning vehicle channelized for a greater portion of the path has a better guide for the maneuver, and the elongated median is better positioned to serve as a refuge for pedestrians crossing the divided highway.

For medians about 4 ft wide there is little or no difference between the two forms of median end. For a median width of 10 ft or more the bullet nose is superior to the semicircular end and preferably should be used in design. On successively wider medians the bullet nose end results in shorter lengths of openings. For median widths greater than 14 ft and a 40-ft control radius (Figure IX-58), the minimum length of opening to provide for cross traffic becomes a positive control. The minimum ends for medians about 14 ft or more wide also take the shape of squared or flattened bullet ends, the flat end being parallel to the crossroad centerline. This shape retains the advantages over semicircular median ends regardless of the median width because of the channelizing control. The bullet nose curves are such as to position the left-turning vehicles to turn to or from the crossroad centerline; whereas as the semicircular end tends to direct the left off movement onto the opposing traffic lane of the crossroad.

Width Median M	L= Minimum Length of Median Opening, Feet	
	Semicircular	Bullet Nose
4	76	76
6	74	60
8	72	53
10	70	47
12	68	43
14	66	40′ Min.
16	64	40′ Min.
20	60	40′ Min.
24	56	40′ Min.
28	52	40′ Min.
32	48	40′ Min.
36	44	40′ Min.
40	40 Min.	40′ Min.
50	40 Min.	40′ Min.
60	40 Min.	40′ Min.

Table IX-16. Minimum design of median openings (P design vehicle, control radius of 40 ft).

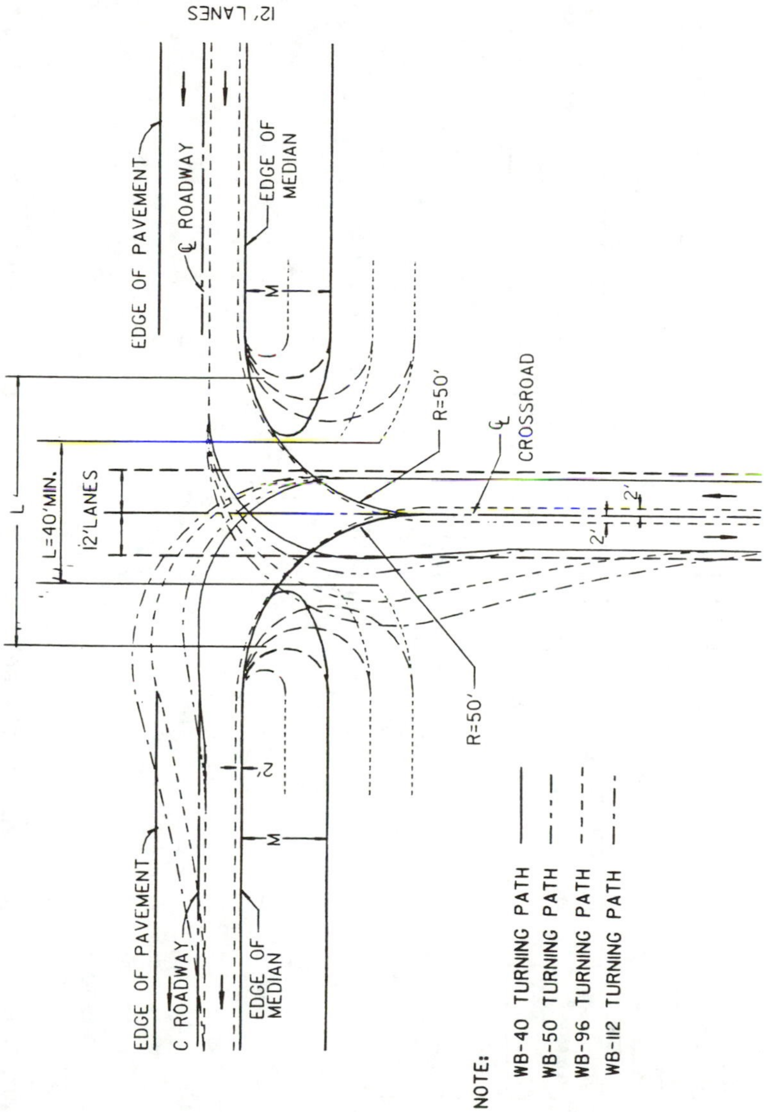

Figure IX-59. Minimum design of median openings (SU design vehicle, control radius of 50 ft).

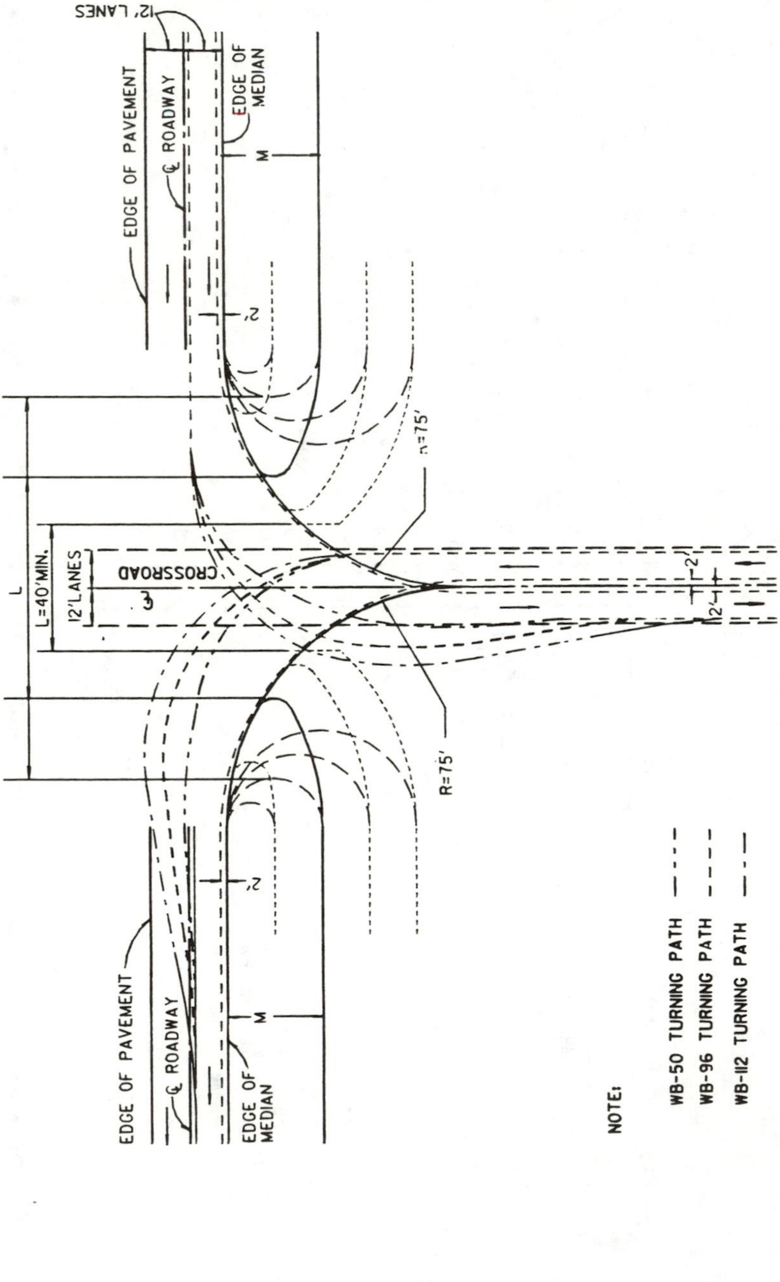

Figure IX-60. Minimum design of median opening (WB-40 design vehicle, control radius of 75 ft).

Width Median M	L = Minimum Length of Median Opening, Feet	
	Semicircular	Bullet Nose
4	96	96
6	94	76
8	92	68
10	90	62
12	88	58
14	86	53
16	84	50
20	80	44
24	76	40′ Min.
28	72	40′ Min.
32	68	40′ Min.
36	64	40′ Min.
40	60	40′ Min.
50	50	40′ Min.
60	40′ Min.	40′ Min.
70	40′ Min.	40′ Min.

Table IX-17. Minimum design of median openings (SU design vehicle, control radius of 50 ft).

Width Median M	L = Minimum Length of Median Opening, Feet	
	Semicircular	Bullet Nose
4	146	122
6	144	115
8	142	110
10	140	105
12	138	100
14	136	96
16	134	92
20	130	85
24	126	78
28	122	73
32	118	67
36	114	62
40	100	57
60	90	40′ Min.
80	70	40′ Min.
100	50	40′ Min.
110	40′ Min.	40′ Min.
120	40′ Min.	40′ Min.

Table IX-18. Minimum design of median openings (WB-40 design vehicle, control radius of 75 ft).

Minimum Length of Median Opening

For any three- or four-leg intersection on a divided highway the length of median opening should be as great as the width of crossroad roadway pavement plus shoulders and in no case less than 40 ft (or less than the crossroad pavement plus 8 ft). Where the crossroad is a divided highway, the length of opening should be at least equal to the width of the crossroad roadways plus that of the median plus 8 ft.

The use of a 40-ft minimum length of opening without regard to the width of median or the control radius should not be considered except at very minor crossroads. The 40-ft minimum length of opening does not apply to openings for U-turns, as discussed elsewhere.

Median Openings Based on Control Radii for Design Vehicles Noted

Passenger Vehicles

Figure IX-58 shows minimum median opening designs based on a control radius of 40 ft for a 90° intersection. The control radius is made tangent to the upper median edge and to the centerline of the undivided crossroad, thereby locating the semicircular median end or forming a portion of a bullet nose end. The resulting lengths of opening vary with the width of median, as shown in the tabulation on the figure. For each of the median widths indicated, the channelizing and area differences between the semicircular and bullet nose ends are apparent.

The control radius of 40 ft accommodates P design vehicles making turns somewhat above minimum, the actual path of which is not shown. (See Figure IX-57 instead.)

The paths of the WB-40, and WB-50 design vehicles making minimum left turns both off and onto the divided highway are shown in Figure IX-58 to indicate how these large vehicles can turn at an intersection designed for passenger cars. Only the inner wheel track and outer front overhang paths are indicated. The paths are depicted from a position parallel to the median edge or centerline of the crossroad and at the beginning of the turn, and they indicate that swinging wide and reversing are required at the end of the turn. Drivers of large vehicles making sharp left turns also may swing right before turning left. However, paths might be a combination of these two extremes, swinging out before beginning the left turn with infringement on the outer lane of the divided highway and also swinging wide and reversing at the end of the turn. The path with the parallel movement at the beginning of the turn is shown because it indicates the maximum encroachment.

In Figure IX-58, the WB-40 design vehicle turning from the divided highway encroaches about 3 ft beyond the two-lane (projected) crossroad edge of pavement, and the WB-50 vehicle encroaches about 11 ft. With wide crossroads this encroachment is within the median opening, but with two-lane crossroads, as shown in Figure IX-58, the encroachment may be beyond the median end, particularly with wide medians having a minimum length of opening. As the left turn is completed, the encroachment may be beyond the edge of pavement for right turns located diagonally opposite the beginning of the left-turn movement off the divided highway. With wide crossroads this encroachment does not extend beyond the right-turn edge of pavement (not shown in Figure IX-58), but with two-lane crossroads and narrow medians it may extend beyond. By swinging over a few feet on the divided highway before beginning the turn, most drivers could pass through these openings and remain on the paved areas. Although this procedure is used extensively, it should be discouraged through a more expansive design where possible.

For turns onto the divided highway the paths show varying degrees of encroachment on the right lane of the divided highway. The SU design vehicle encroaches about 1 ft on the right lane of a four-lane divided highway, the WB-40 vehicle about 5 ft, and the WB-50 vehicle about 10 ft. These distances can be lessened by the drivers anticipating the turn and swinging right before turning left, if space is available. This space depends on the median width, the length of opening as governed by the number of lanes on the crossroad, and other limitations such as triangular islands for channelizing right-turn movements.

Figure IX-58 indicates that minimum median openings based on a control radius of 40 ft are not well suited for lengths of opening for two-lane crossroads because trucks cannot turn left without difficult maneuvering and encroachment on median ends or outer shoulders, or both, depending on the median width. It may be suitable for wide crossroad pavements, but for these cases it is advantageous to use a control radius greater than 40 ft, which enables all vehicles to turn at a little greater speed and enables trucks to maneuver and turn with less encroachment. Figures IX-58 through IX-60 show the squared or truncated bullet nose design in conjunction with the 40-ft minimum length of opening. Provision of longer tapers not only avoids this somewhat awkward-looking design but also provides for other important objectives as well. This topic is discussed further in the section "Above-Minimum Designs for Direct Left Turns."

Single-Unit Trucks or Buses

Figure IX-59 shows minimum median opening designs for a 90° intersection, based on a control radius of 50 ft. The basis of development, median ends, and turning paths shown are similar to those of Figure IX-58. As indicated in Figure IX-57, the control radius of 50 ft accommodates the SU design vehicle making minimum left turns without encroachment on adjacent lanes. The paths of the WB-40 and WB-50 design vehicles making left turns both off and onto the divided highway are shown in Figure IX-59 to indicate how these large vehicles can turn at an intersection designed for the SU vehicles.

The WB-50 design vehicle would encroach about 4 ft beyond a 24-ft crossroad in turning off the divided highway, but encroachment could be reduced by swinging wide at the beginning of the turn. In turning onto the divided highway it would encroach about 7 ft on the adjacent lane, a distance that could be reduced but not eliminated by swinging wide at the beginning of the turn; but, to do so, the length of opening would have to be greater than the 40-ft minimum.

Figure IX-59 indicates that minimum lengths of median openings based on a control radius of 50 ft are suited for truck operation, except that WB-50 vehicles will encroach on adjacent lanes. For these cases additional advantage is gained by using a control radius greater than 50 ft where turnings by WB-50 semitrailers are expected.

Semitrailer Combinations

Figure IX-60 shows minimum median opening designs for a 90° intersection, which are based on a control radius of 75 ft while Figure IX-60A is based on a control radius of 100 ft. The 75-ft control radius is sufficiently large to accommodate a WB-40 design vehicle, and the minimum path of the WB-50 vehicle indicates that it can also use this design without undue encroachments. The left turn to leave the divided highway can be made within a two-lane crossroad. In the left turn to enter the divided highway the WB-50 vehicle would encroach on the adjacent lane about 2 ft. This encroachment can be avoided by swinging wide at the beginning of the turn.

Effect of Skew

A control radius for design vehicles as the basis for minimum design of median openings results in lengths of openings that increase with the skew

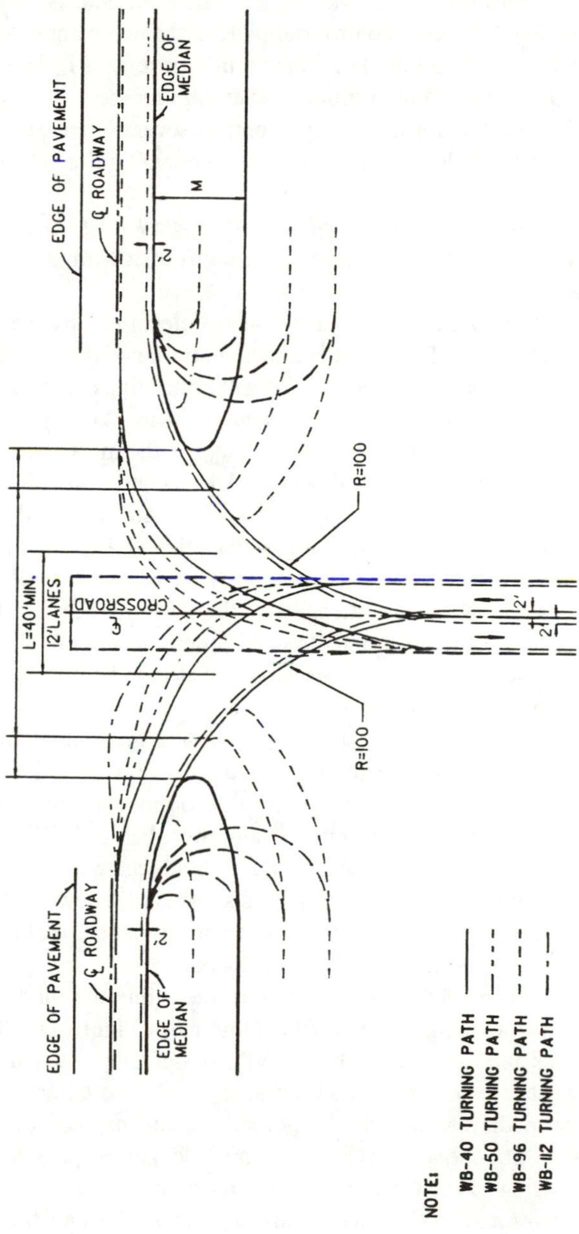

Figure IX-60A. Minimum design of median openings (radius of 100 ft).

angle of the intersection. Although the bullet nose end remains preferable, the skew introduces other variations in the shape of median end. At a skewed crossing it is necessary to use the control radius R in the acute angle to locate the PT on the median edge, point (1) as shown in Figure IX-61. The arc for this radius is the equivalent of the minimum inner path for the vehicle in turning more than 90°. With this PT as a design control, several alternate designs that depend on the skew angle, median width, and control radius may be considered.

Semicircular ends (A in Figure IX-61) result in very long openings and minor channelizing control for vehicles making a left turn with less than 90° in the turning angle.

A symmetrical bullet nose B with curved sides determined by the control radius and tangent at points (1) and (2) is a layout similar to those in Figures IX-58 through IX-60. This design also has little channelizing control for vehicles turning left less than 90° from the divided highway. An asymmetrical bullet nose (C in Figure IX-61) with radii R and R_2 has the most positive control and less paved area than designs A and B. The second radius R_2, which exceeds R, is that determined to be tangent at point (2) and also to the crossroad centerline. In this design the nose is offset from the median center paths.

An asymmetrical bullet nose (D in Figure IX-61) also can be developed by use of the control radius R for both turns, with the PT values on the median edge, and points (1) and (3) not opposite each other. Because of the rather extreme asymmetry of this nose, this design is not deemed desirable, and it may not provide a sufficient length of opening.

The length of the opening of these alternates for a given median width decreases in the order discussed, A to D. For wide medians and a large skew the length of openings may not be sufficient to accommodate the crossroad and flattened ends, or an above-minimum design must be used. With design ends B, C, or D, a flattened bullet nose treatment for adequate length of opening would be made parallel to the crossroad, as shown at the lower left, C-E, in Figure IX-61 for the length of opening L . In such cases the use of a control radius larger than the minimum should be considered.

Table IX-19 shows typical values obtained for the minimum median ends designed with a control radius of 50 ft (the same as in Figure IX-59) for a range of skew angles and median widths. Lengths of openings, measured normal to the crossroad, are shown for median ends A, B, and C, as in Figure IX-61. For any one median width with semicircular ends and skew of 20° and 40°, the openings are about one-third and two-thirds longer, respectively, than that for a 90° crossing. Likewise, the symmetrical bullet nose ends give lengths of openings for a 20° skew that are about one and one-half times, and for a 40° skew about two times, the lengths for a 90° crossing.

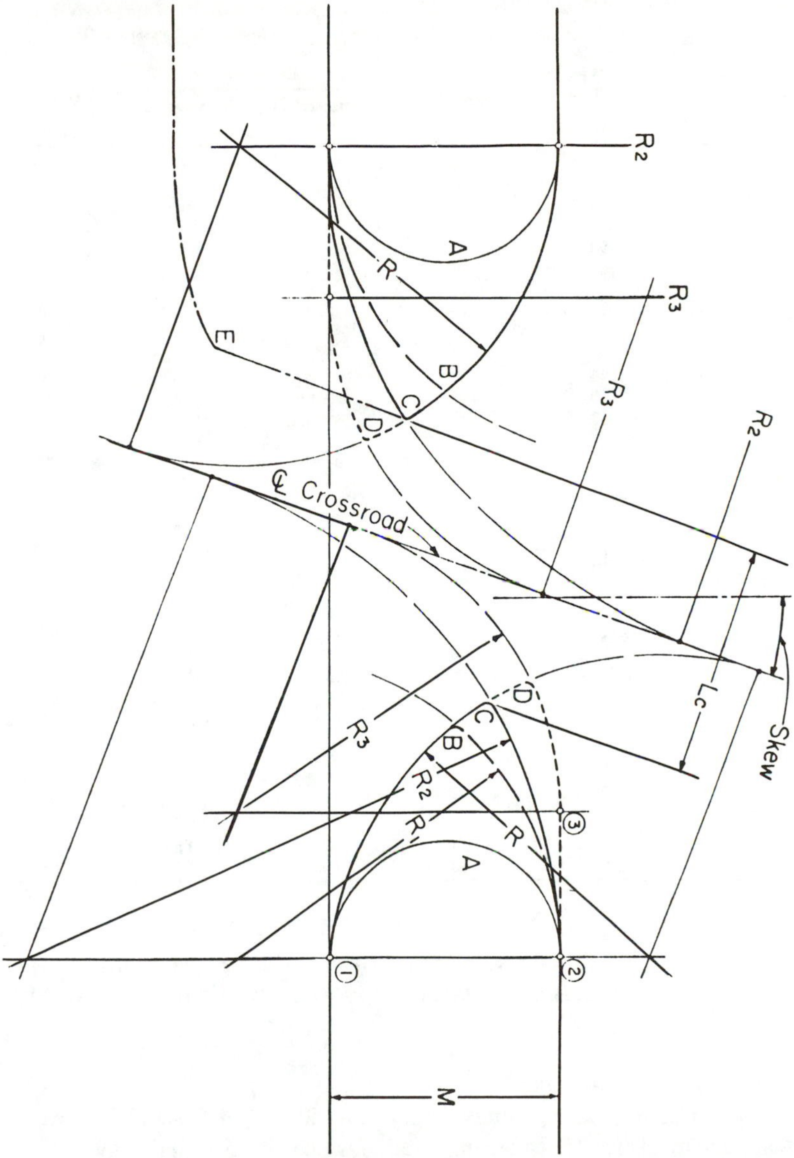

Figure IX-61. Minimum design of median openings (effect of skew).

Skew Angle (degrees)	Width of Median (ft)	Semi-circular A	Bullet Nose Symmetrical B	Bullet Nose Asymmetrical C	R for Design C (ft)
		Length of Median Opening Measured Normal to the Crossroad (ft)			
0	10	90	62	—	—
	20	80	44	—	—
	30	70	40 min.	—	—
	40	60	40 min.	—	—
	50	50	40 min.	—	—
	60	40	40 min.	—	—
10	10	106	80	77	70
	20	94	58	54	68
	30	82	45	40 min.	65
	40	71	40 min.	40 min.	63
	50	60	40 min.	40 min.	61
	60	47	40 min.	40 min.	59
20	10	121	97	90	97
	20	107	74	65	92
	30	94	59	48	86
	40	81	48	40 min.	82
	50	68	40 min.	40 min.	76
	60	54	40 min.	40 min.	71
30	10	135	114	105	140
	20	120	91	77	130
	30	104	75	58	120
	40	90	62	42	110
	50	76	51	40 min.	100
	60	60	40	40 min.	90
40	10	148	127	118	210
	20	131	106	90	193
	30	115	90	68	174
	40	98	77	51	156
	50	81	64	40 min.	139
	60	64	51	40 min.	121

Table IX-19. Effect of skew on minimum design for median openings (typical values based on control radius of 50 ft).

In general, median openings longer than 80 or 100 ft should be avoided, regardless of skew. This plan may call for special channelization, left-turn lanes, or adjustment to reduce the crossroad skew, all of which result in above-minimum designs.

Preferably, each skew crossing should be studied separately with trial graphical solutions on a scale of 1 in. = 20 ft to 1 in. = 50 ft to permit the designer to make comparisons and choose the preferred layout. In general the asymmetrical bullet nose end (C in Figure IX-61) is preferable. Where end B is not greatly different, the practical aspects of symmetry may make it preferable. In some cases ends D and C-E may well be considered in the alternate studies.

For the preceding discussion the design controls for minimum median openings for left turns are summarized in Table IX-20.

Design Vehicles Accommodated	Control Radius (ft)		
	40	50	75
Predominant	P	SU	WB-40
Occasional	SU	WB-40	WB-50

Table IX-20. Design controls for minimum median openings.

ABOVE-MINIMUM DESIGNS FOR DIRECT LEFT TURNS

Median openings that enable vehicles to turn on minimum paths, and at 10 to 15 mph, are adequate for intersections where traffic for the most part proceeds straight through the intersection. Where through-traffic volumes and speeds are high and left-turning movements are important, undue interference with through traffic should be avoided by providing median openings that permit turns without encroachment on adjacent lanes. This arrangement would enable turns to be made at speeds above that for the minimum vehicle paths and provide space for vehicle protection while turning or stopping. The general pattern for minimum design can be used with larger dimensions.

A variety of median opening arrangements may be considered that depend on the control dimensions (width of median and width of crossroad or street, or other) and the size of vehicle to be used for design control.

Median openings having above-minimum control radii and bullet nose median ends are shown in Figure IX-62. The design controls are the three radii R, R_1, and R_2. Radius R is the control radius for the sharpest portion of the turn, R_1 defines the turnoff curve at the median edge, and R_2 is the radius of the tip. When a sufficiently large R_1 is used, an acceptable turning speed for vehicles leaving the major road is ensured and a sizable area inside the inner edge of through-traffic lane between points (1) and (2) may be available for speed change and protection from turning vehicles. Radius R_1 may vary from about 80 to 400 ft, or more.

The tabulated values shown, 90, 150, and 230 ft, are established minimum radii for turning speeds of 20, 25, and 30 mph, respectively. In this case the ease of turning probably is more significant than the turning speeds, because

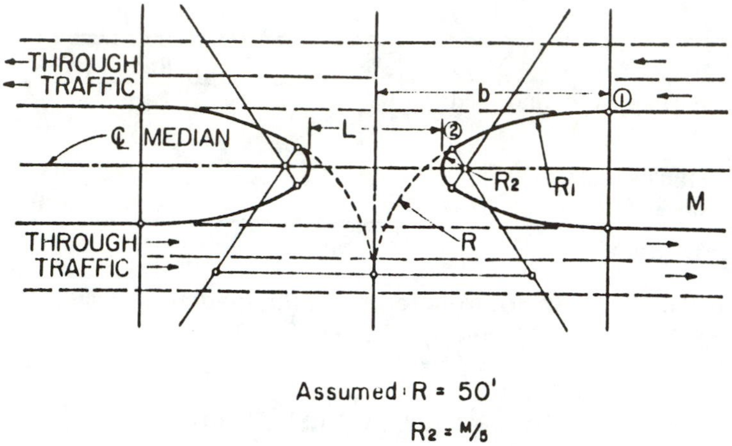

M WIDTH OF MEDIAN, FEET	DIMENSIONS IN FEET, WHEN					
	R1 = 90'		R1 = 150'		R1 = 230'	
	L	b	L	b	L	b
20	58	65	66	78	71	90
30	48	68	57	85	63	101
40	40	71	50	90	57	109
50	—	—	44	95	51	115
60	—	—	—	—	46	122
70	—	—	—	—	41	128

Figure IX-62. Above minimum design of median openings (typical bullet nose ends).

the vehicle will need to slow down to about 10 to 15 mph at the sharp part of the turn or may need to stop at the crossroad. Radius R_2 can vary considerably, but is pleasing in proportion and appearance when it is about one-fifth of the median width. Radius R is tangent to the crossroad centerline (or edge of crossroad median). Radii R and R_1 comprise the two-centered curve between the terminals of the left turn. For simplicity, the PC is established at point (2). Radius R cannot be smaller than the minimum control radius for the design vehicle, or these vehicles will be unable to turn to or from the intended lane even at low speed. To avoid a large opening, R should be held to a reasonable minimum, e.g., 50 ft, as used in Figure IX-62.

The length of median opening is governed by the radii. For medians wider than about 30 ft coupled with a crossroad of four or more lanes, the control radius R generally will need to be greater than 50 ft or the median opening will be too short. A rounded value can be chosen for the length of opening, e.g., 50 or 60 ft, and that dimension can be used to locate the center for R. Then R becomes a check dimension to ensure the workability of the layout. The tabulation of values in Figure IX-62 shows the resultant lengths of median openings over a range of median widths for three assumed values of R_1 and for R assumed to be 50 ft. Dimension b is included as a general design control and for comparison with other above-minimum designs.

The median end designs in Figure IX-62 do not positively provide protection areas within the limits of the median width. A design using $R_1 = 100$ ft or more provides space for at least a single passenger vehicle to pause in an area clear of both the through-traffic lanes and the crossroad lanes; with wide medians such radii may provide enough protection space for larger design vehicles. At skew intersections, above-minimum designs with bullet nose median ends can be applied directly. Where the skew is 10° or more, adjustments in R and R_2 from the values shown are required to provide the necessary length of opening.

INDIRECT LEFT TURNS AND INDIRECT U-TURNS

General Design Considerations

Divided highways require median openings to provide access for crossing traffic in addition to left-turning and U-turning movements. The discussions to follow deal with the various design methods that accommodate these movements predicated on median width.

At intersections where the median is too narrow to provide a lane for left-turning vehicles and the traffic volumes or speeds, or both, are relatively high, safe, efficient operation is particularly troublesome. Vehicles that slow down or stop in a lane primarily used by through traffic to turn left greatly increase the potential for rear-end collision.

Other factors that require special design considerations for left- and U-turning movements are the required turning paths of the various design vehicles in conjunction with narrow medians. The necessity to turn left or U-turn in the urban or heavily developed residential or commercial sectors also presents serious problems with respect to safety and efficient operation.

The design plans shown in Figures IX-63 and IX-64 offer several options with respect to indirect left turns and also provide for indirect U-turning movements. Figure IX-63 involves a jug-handle-type ramp or diagonal roadway that intersects a secondary crossing roadway. The motorist exits via the jug-handle-type ramp, makes a left turn onto the crossroad, and then makes a left turn onto the divided highway to complete the left-turn or U-turning maneuver.

Figure IX-64 shows an at-grade loop that can serve as an alternate to the jug-handle-type ramp. The loop design might be considered when the jug-handle-type ramps necessitate costly right-of-way, the opposite quadrant being less costly. There might be other justifications in selecting the loop instead of the ramp, such as improved vertical alinement and comparative grading costs.

An additional advantage of the loop occurs where the intersecting road is on one side of the divided highway. The accommodation for an indirect left turn or indirect U-turn could be accomplished in placing the loop on the side opposite the intersecting roadway.

Figure IX-65 illustrates a design that provides for indirect left turns to be made from the right, via separate turning roadways connected to a crossroad. Such arrangements have the advantage of eliminating left turns from the through lanes and providing storage for left-turning vehicles not available on the highway itself. The left-turning vehicles with little extra travel distance are able to cross the main highway safely with appropriate traffic control devices. Figure IX-65 illustrates three design options that might be adaptable to various roadway patterns. The turn from bottom to left is accomplished via the added left-turn slip ramp at the lower right (similar to previous discussions). This arrangement permits safe left turns onto the minor road under traffic signal protection and prevents cars making left turns from blocking the lane adjacent to the medians. Where there is a parallel roadway nearby, the added ramp may connect to it, as shown in the upper left or alternately as shown by the dashed-line connection. However, this design is less desirable because the vehicles must pass through the crossroad intersection and create delays by reducing speed in turning right. This problem might be overcome by the introduction of auxiliary lanes if space is available.

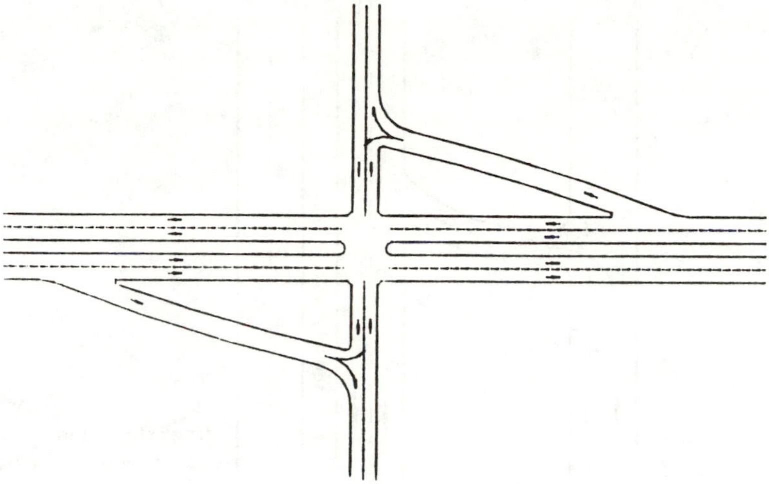

Figure IX-63. Jug-handle-type ramp with crossroad.

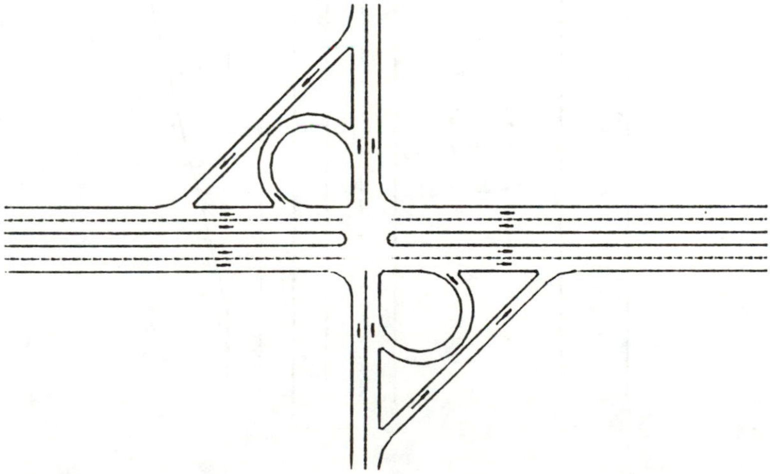

Figure IX-64. At-grade loop (surface loop) with crossroad.

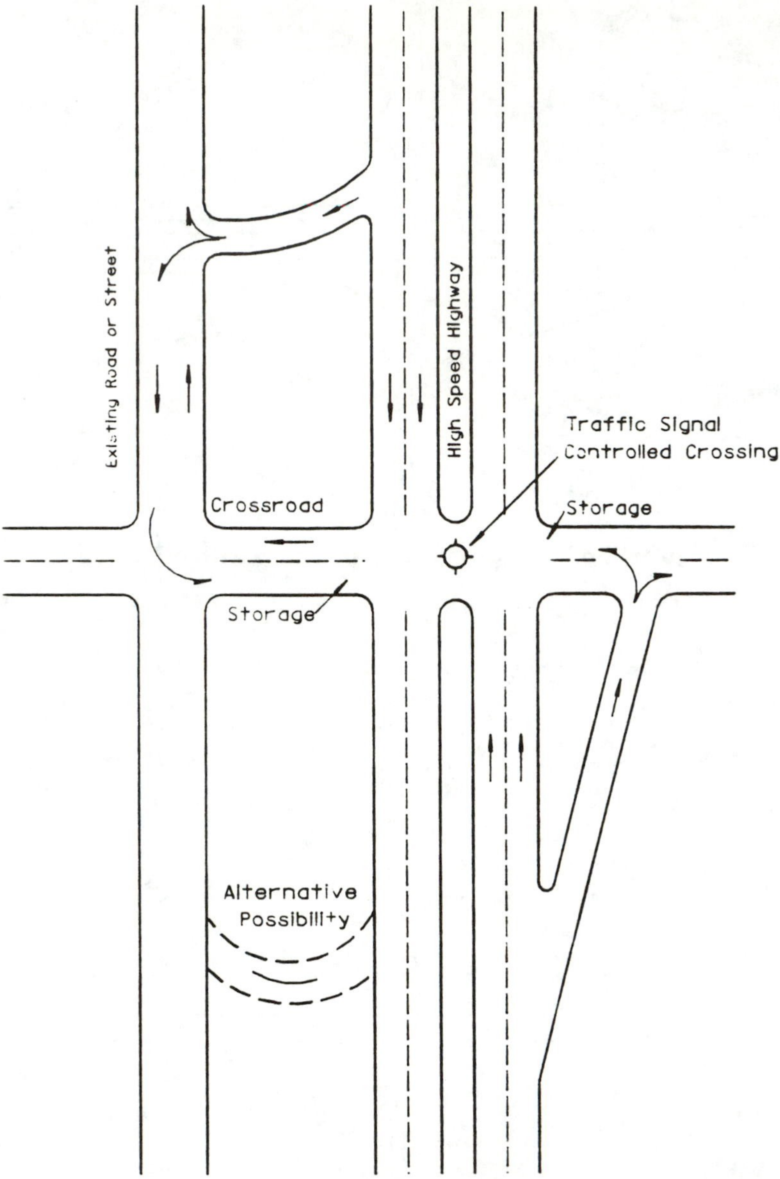

Figure IX-65. Special indirect left-turn designs for traffic leaving highway with narrow median.

Indirect Left Turn or Indirect U-Turn—Using Local Streets

Highways without control of access that involve narrow nontraversable medians and where the adjacent property owners enter the divided highway by right turn only must gain access to the opposite traveled way by one of three types of operation and control. The first option is to use the interconnecting street patterns. This operation requires making the initial right turn, proceeding by continuous right turns around the block to the median opening that services the secondary crossroads, and then turning left. Variations of access to the divided highway would also prevail for the property owners on the adjacent street patterns. However, the around-the-block principle would still control movements with respect to exit and return trips. The around-the-block option would require careful examination of existing turning radii to accommodate SU vehicles and the WB vehicles that may be required to use this method of indirect left turns or indirect U-turns. This system requires careful design attention with respect to restrictive parking, regulatory signs, and signal control devices in the proximity of each intersection.

The second alternative, which would only benefit the property owners abutting the divided highway, is to provide median openings for the individual properties. This option would defeat a major purpose of the median and would lead to complete erosion of this control feature.

The around-the-block alternative is not always feasible, especially where radial routes traverse areas that are only partially developed and where there is no established pattern of adjacent roadways, often with no existing roads or streets running parallel to the through highway. Even where there is a suitable network of adjacent roadways, the adverse travel is objectionable. The necessity of passing through four intersections and the left turn to the arterial might be as much of a hindrance to the free flow of traffic as a left turn might be.

The third alternative is use of the design principles previously described with respect to constructing jug-handle-type ramps or at-grade intersecting loops.

From the preceding discussion it is apparent that wherever possible, a newly designed divided highway should have a median width that can accommodate normal left turns and U-turns by using a median storage lane that will protect and store the design-hour turning volume.

Indirect Left Turn or Indirect U-Turn-Wide Medians

Figure IX-66 illustrates an indirect left turn for two arterials where left turns are heavy on both roads. The north-south roadway is undivided and the east-west roadway is divided with a wide median. Because left turns from the north-south road would cause congestion because of the lack of storage, left turns from the north-south road are prohibited at the main intersection. Left-turning traffic turns right onto the divided road and then makes a U-turn at a one-way crossover located in the median of the divided road. Auxiliary lanes are highly desirable on each side of the median between the crossovers for storage of turning vehicles.

The crossover should be 400 to 600 ft away from the intersection to allow the left-turn traffic to approach the intersection on a green signal. This scheme provides a slight increase in capacity at very little cost with no additional acquisition of right-of-way. The main disadvantage is that the left-turn traffic has to pass through the same intersection twice. This maneuver also may be confusing to motorists unfamiliar with the design and thus requires special signing.

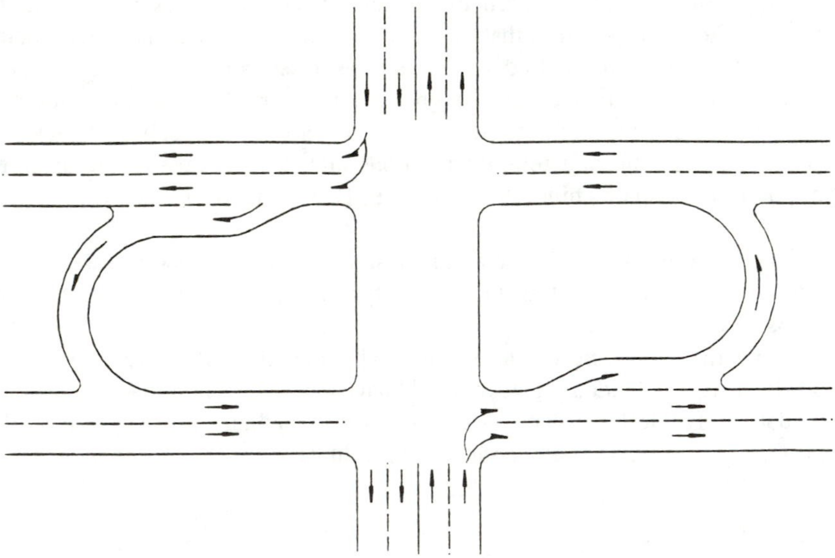

Figure IX-66. Indirect left turn through a crossover.

Special left-turn considerations may also be required at major crossroads, particularly at isolated intersections in suburban areas, where one or more of the left-turning movements are so large that they cannot be handled by the conventional median lanes and where there is insufficient width to install two median lanes. In such instances the jug-handle-type ramp or the at-grade loop pattern that provides an exit or entrance ramp in two of the crossing quadrants would be required.

Location and Design of U-Turn Median Openings

Median openings designed to accommodate vehicles making U-turns only are needed on some divided highways in addition to openings provided for cross and left-turning movements. Separate U-turn median openings may fit at the following locations:

1. Locations beyond intersections to accommodate minor turning movements not otherwise provided in the intersection or interchange area. The major intersection area is kept free for the important turning movements, in some cases obviating expensive ramps or additional structures.
2. Locations just ahead of an intersection to accommodate U-turn movements that would interfere with through and other turning movements at the intersection. Where a fairly wide median on the approach highway has few openings, U-turning is necessary to reach roadside areas. Advance separate openings to accommodate them outside the intersection proper will reduce interference.
3. Locations occurring in conjunction with minor crossroads where traffic is not permitted to cross the major highway but instead is required to turn right, enter the through traffic stream, weave to the left, U-turn, then return. On high-speed or high-volume highways the difficulty and long lengths required for weaving with safety usually make this design pattern undesirable unless the volumes intercepted are light and the median is of adequate width. This condition may occur where a crossroad with high volume traffic, a shopping area, or other traffic generator that requires a median opening nearby and additional median openings would not be practical.
4. Locations occurring where regularly spaced openings facilitate maintenance operations, policing, repair service of stalled vehicles, or other highway-related activities. Openings for this purpose may be needed on controlled-access highways and on divided highways through undeveloped areas.

5. Locations occurring on highways without control of access where median openings at optimum spacing are provided to serve existing frontage developments and at the same time minimize pressure for future median openings. A preferred spacing at 0.25 to 0.50 mi is suitable in most instances. Fixed spacing is not necessary, nor is it fitting in all cases because of variations in terrain and local service requirements.

One requirement for a satisfactory design for a U-turn is that the width of the highway, including the median, be sufficient to permit the turn to be made without encroachment beyond the outer edges of the pavements. Desirably, the median should be wide enough to permit the turn to be accomplished from the lane next to the median onto the lane nearest the shoulder on the roadway of opposing traffic.

Medians of 20 ft and wider are needed for passenger and truck traffic, to permit vehicles to turn from the inner lane (next to the median) on one pavement to the outer lane of a two-lane opposing pavement. A median left-turn lane is highly desirable in advance of the U-turn opening to eliminate stopping on the through lanes. This scheme would increase the required median width by approximately 12 ft.

Wide medians are uncommon in highly developed sectors. Consequently, it is necessary to consider U-turn designs of a lower standard. Where right-of-way is restricted, speeds are low, and signal control is used to provide sufficient gaps in the traffic stream, medians 10 to 15 ft wide may be used for U-turn openings. This design permits passenger vehicles to turn from the inner lane to the shoulder on a four-lane divided highway and from the inner to the outer lane on a six-lane facility. Preferably, openings for U-turns should be located in advance of an intersecting road or street, at least 100 ft from the median end, to keep the entrance onto the U-turn free of vehicles stopped by traffic signals.

Normally, U-turns should not be permitted from the through lanes. However, where medians have adequate width to shield a vehicle standing in the median opening, it may be permissible. Minimum widths of median required for U-turns by different design vehicles turning from the lane adjacent to the median are given in Figure IX-67. These dimensions are for a four-lane divided facility.

Figure IX-68 illustrates special U-turn designs with narrow medians. In Figure IX-68A the U-turning vehicle swings right from the outer lane, loops around to the left, stops clear of the divided highway until a suitable gap in the traffic stream develops, and then makes a normal left turn onto the divided highway. In Figure IX-68B the U-turning vehicle begins on the inner lane of the divided highway, crosses the through-traffic lanes, loops around to the left, and then merges with the traffic. To deter vehicles from stopping on through lanes, a left-turn lane with proper storage capacity should be provided to accommodate turning vehicles.

CONTINUOUS LEFT-TURN LANES (TWO-WAY)

The foregoing discussion on design for indirect left turns and indirect U-turns brings into focus the difficulties involved in providing access to abutting property, especially where such access is by commercial vehicles. These conditions are very common in commercial and industrial areas where property values are high and rights-of-way for wide medians are difficult to acquire. Under such conditions, a paved, flush, traversable median 10 to 16 ft wide, which can be used for left-turn storage by traffic in either direction, is the optimum type of design. Figure IX-69 illustrates two kinds of typical multilane two-way marking applications, which are very effective in sectors

Type of Maneuver		M - Min. width of median - feet for design vehicle						
		P	WB-40	SU	BUS	WB-50	WB-60	TDT
		Length of design vehicle						
		19'	50'	30'	40'	55'	65'	118'
Inner Lane to Inner Lane	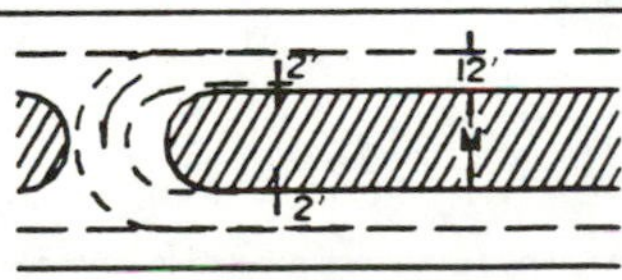	30	61	63	63	71	71	101
Inner Lane to Outer Lane	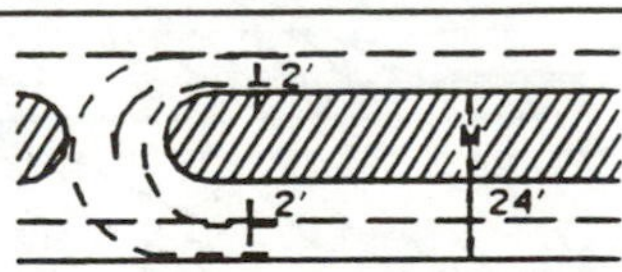	18	49	51	51	59	59	89
Inner Lane to Shoulder	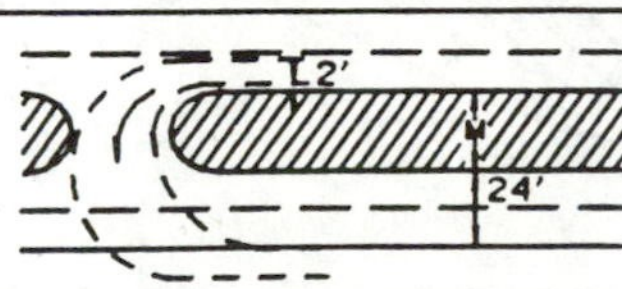	8	39	41	41	49	49	79

Figure IX-67. Minimum designs for U-turns.

requiring opposing left turns to individual properties. In general, continuous left-turn lanes should be used only in an urban setting where operating speeds are relatively low and where there are no more than two through lanes in each direction. The operational characteristics of left turning lanes with more than two through lanes in each direction is the subject of ongoing research and caution is recommended at this time when considering more than a five lane cross section. This subject is discussed in greater detail in Chapter IV and in the MUTCD (**5**).

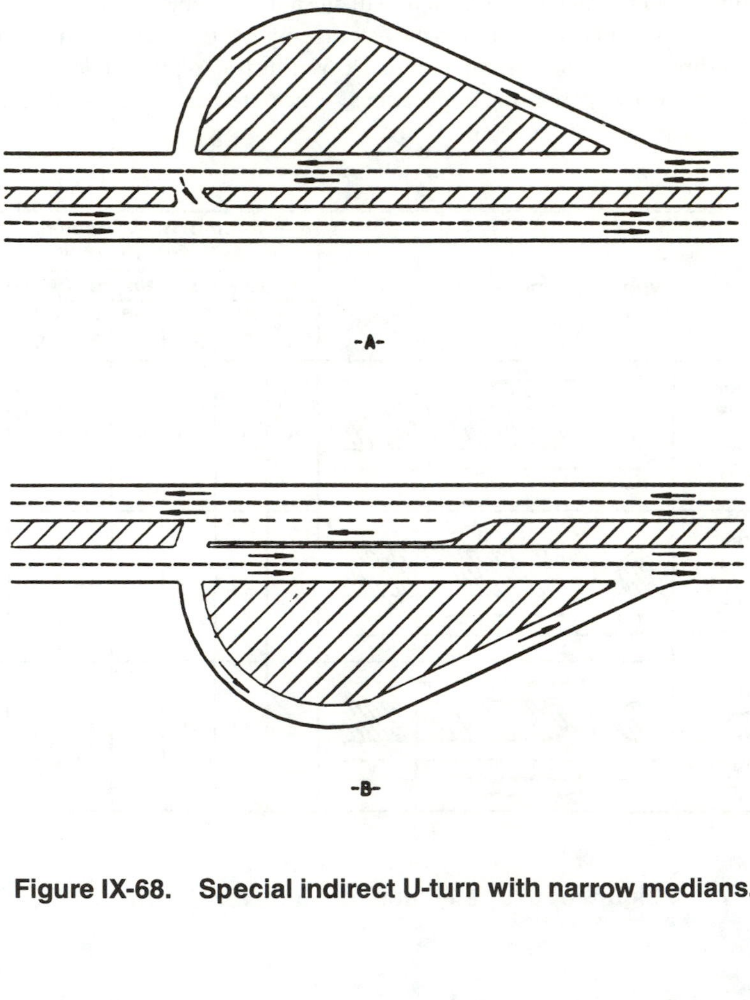

Figure IX-68. Special indirect U-turn with narrow medians.

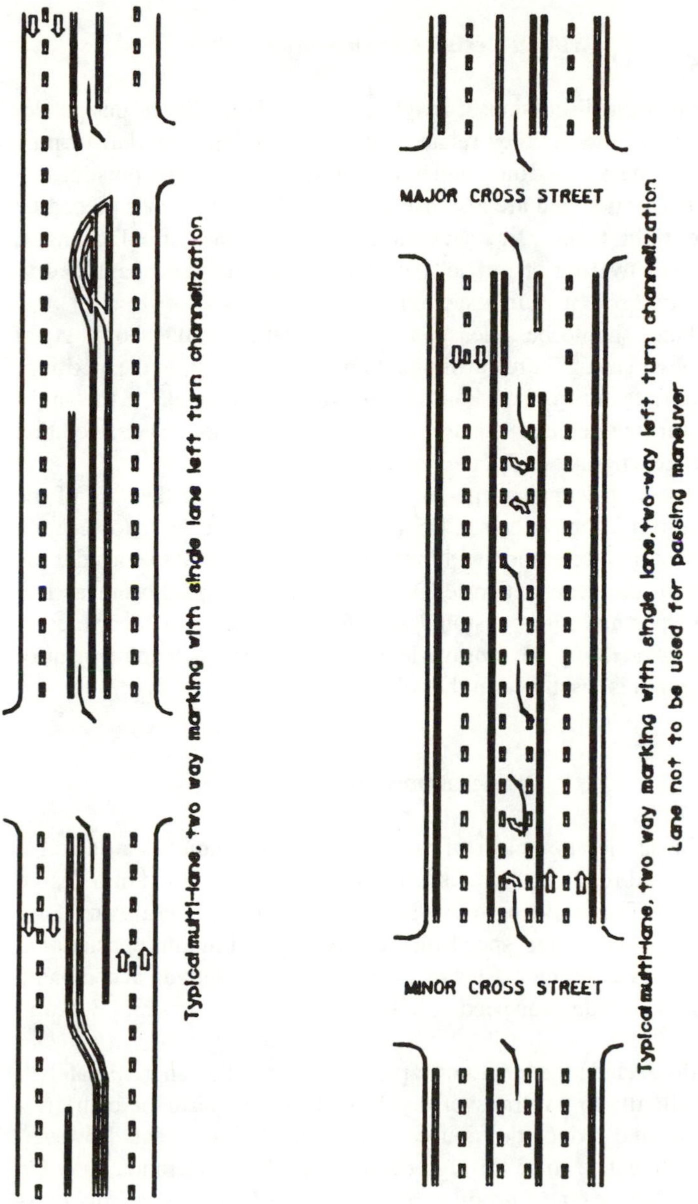

Figure IX-69. Multilane two-lane marking.

AUXILIARY LANES

General Design Considerations

From the foregoing discussions it is appropriate to deal with the design elements of auxiliary lanes as they relate to median openings and left-turning movements. In general, auxiliary lanes are used preceding the openings as described in this section and are also used at at-grade intersections preceding and following right turning movements to accommodate speed changes, weaving and maneuvering of traffic. Auxiliary lanes may also be added to increase capacity through an intersection.

Auxiliary lanes should be at least 10 ft wide and desirably should equal that of the through lane. Where curbing is to be used adjacent to the auxiliary lane, an appropriate curb offset should be provided. The length of the auxiliary lanes for turning vehicles consists of three components: (1) deceleration length (2) storage length and (3) entering taper.

Desirably, the total length of the auxiliary lane should be the sum of the length for these three components. Common practice, however, is to accept a moderate amount of deceleration within the through lanes and to consider the taper as a part of the deceleration length. Where intersections occur as frequently as four per mile, it is customary to forego most of the deceleration length and to provide only the storage length plus taper. Each component of the auxiliary length is discussed in the following section.

Deceleration Length

Provision for deceleration clear of the through-traffic lanes is a desirable objective on arterial roads and streets and should be incorporated into design whenever feasible. The total length required is that needed for a safe and comfortable stop from the design speed of the highway. Minimum deceleration lengths for auxiliary lanes on grades of 2 percent or less, with an accompanying stop condition, for design speeds of 30, 40 and 50 mph are 235, 315 and 435 ft respectively.

These lengths exclude the length of taper which should be approximately 8 to 15 ft longitudinally to 1 ft transversely. Long tapers indicate the path drivers are to follow and reduce the unused portion of the pavement. However, long tapers tend to entice some through drivers into the deceleration lane. On the other hand, shorter tapers produce better "targets" for approaching drivers and give positive visibility of the added auxiliary lane.

On many urban facilities it will not be feasible to provide full length for deceleration. In such cases at least a part of the deceleration must be accomplished before entering the auxiliary lane. However, the lengths given should be accepted as a desirable goal and should be provided where practical and feasible. Deceleration lengths shown are applicable to both left and right-turning lanes, but speed is usually lower in the right lane than in the left lane.

Storage Length

The auxiliary lane should be sufficiently long to store the number of vehicles likely to accumulate during a critical period. The storage length should be sufficient to avoid the possibility of left-turning vehicles stopping in the through lanes. The storage length should be sufficiently long so that the entrance to the auxiliary lane is not blocked by vehicles standing in the through lanes waiting for a signal change or for a gap in the opposing traffic flow.

At unsignalized intersections the storage length, exclusive of taper, may be based on the number of turning vehicles likely to arrive in an average 2-min period within the peak hour. As a minimum requirement, space for at least two passenger cars should be provided; with over 10 percent truck traffic, provisions should be made for at least one car and one truck. The 2-min waiting time may need to be changed to some other interval that depends largely on the opportunities for completing the left-turn maneuver. These intervals, in turn, depend on the volume of opposing traffic. Where the volume of turning traffic is high, a traffic signal will usually be required.

At signalized intersections the required storage length depends on the signal cycle length, the signal phasing arrangement, and the rate of arrivals and departures of left-turning vehicles. The storage length is a function of the probability of occurrence of events and should usually be based on one and one-half to two times the average number of vehicles that would store per cycle, which is predicated on the design volume. This length will be sufficient to serve heavy surges that occur from time to time. As in the case of unsignalized intersections, provision should be made for storing at least two vehicles.

Traffic signal design fundamentals are discussed further in MUTCD (**5**).

Where turning lanes are designed for two-lane operation, the storage length is reduced to approximately one-half of that required for single-lane operation. Reference should be made to the *Highway Capacity Manual* (**2**).

Taper

On high-speed highways the taper of an auxiliary lane should generally conform to that discussed under "Ramp Terminals" in the section "Speed-Change Lanes" in Chapter X. However, on most urban streets approaching at-grade intersections, shorter tapers are satisfactory because of lower operating speeds.

Some States permit the tapered section of deceleration auxiliary lanes to be constructed in a "squared-off" section at full paving width and depth. This construction requires a painted delineation of the taper and is only applicable to noncurbed sections. The design requires transition of the outer or median shoulders around the squared-off beginning of the deceleration lane.

The squared-off design principle can be applied to median deceleration lanes, and it can also be used at the beginning of deceleration right-turn exit terminals when there is a single exit lane. When two or more exit lanes are used, the tapered designs discussed in the section "Speed-Change lanes" in Chapter X is recommended.

Additional guidance for lengths of tapers may be found in the "Manual of Uniform Traffic Control Devices" (**5**) which recommends the following for design speeds equal to or greater than 45 mph: length of taper (L) equals the design speed (S) multiplied by the offset (W), L = SXW. For design speeds equal to or less than 40 mph; the length of taper (L) equals the offset (W) multiplied by the design speed (S) squared, then divided by 60, or $L = WS^2/60$.

Some State highway agencies believe that the abrupt squared-off beginning of deceleration exits offers improved driver commitment to the exit maneuver and also contributes to driver security and safety because of the elimination of the unused portion of long tapers. For example, a 15:1 taper ratio is recommended for design speeds up to 50 mph. If the auxiliary lane was 12 ft wide, the taper length would be 180 ft. This approach section could be 12 ft wide and constructed for its full length. The painted delineation taper could be 12 ft wide and constructed for its full length. The painted delineation taper could be established on a 4:1 ratio that would produce a stripe approximately 50 ft long. This design principle could be applied to all deceleration design speeds and should operate with relatively quick exit maneuvers to the full-width lane under low design speeds and still provide for the longer exit maneuvers at the higher design speeds.

The longitudinal location along the highway, where a vehicle will move from the through lane to a full-width deceleration lane, will vary depending on many factors. These factors include the type of vehicle, the driving characteristics of the vehicle operator, the speed of the vehicle, weather conditions, and lighting conditions.

This design principle could also improve capacity by offering an early exit to the daily users that would diminish slow-down interference to the through movements.

The discussions to follow and figure references do not relate to the "squared-off" design section. However, the design principles described in the preceding can be applied.

Straight-line tapers are frequently used, as shown in Figure IX-70A. The taper rate may be 8:1 for design speeds up to 30 mph and 15:1 for design speeds of 50 mph. A short curve is desirable at either end, as shown in Figure IX-70B, but may be omitted for ease of construction. Where curves are used at the ends, the tangent section should be about one-third to one-half of the total length. Figure IX-70C shows a taper with a symmetrical reverse curve; a length of about 90 ft or more is necessary. At low speeds, taper (2) will operate as desired and taper (1) probably will not. A more desirable reverse-curve taper has the turnoff curve radius about twice that of the second curve (Figure IX-70D). When 100 ft or more in length is provided, tapers (1) and (2) are suitable for low-speed operation. All the dimensions and configurations shown in Figure IX-70 are applicable to right-turn lanes as well as left-turn lanes.

Median Left-Turn Lanes

A median left-turn lane is an auxiliary lane for storage or speed change of left-turning vehicles located at the left of one-directional pavement within a median or divisional island. Accident potential, inconvenience, and considerable loss in efficiency of operation are evident on divided highways where such lanes are not provided. Median lanes, therefore, should be provided at intersections and at other median openings where there is a high volume of left turns or where the vehicular speeds are high. Minimum designs of median openings are shown on Figures IX-58 through IX-60 and in Tables IX-16 through IX-18. Median lane designs for various widths of median are shown in Figures IX-71 and IX-72.

Median widths of 20 ft or more are desirable at intersections with single median lanes, but widths of 16 to 18 ft permit reasonably adequate arrangements. Where two median lanes are used, a median width of at least 28 ft is desirable to permit the installation of two 12-ft lanes and a 4-ft separator. Although not equal in width to a normal traveled lane, a 10-ft lane with a 2-ft curbed separator or with traffic buttons or paint lines, or both, separating the median lane from the opposing through lane may be acceptable where speeds are low and the intersection is controlled by traffic signals.

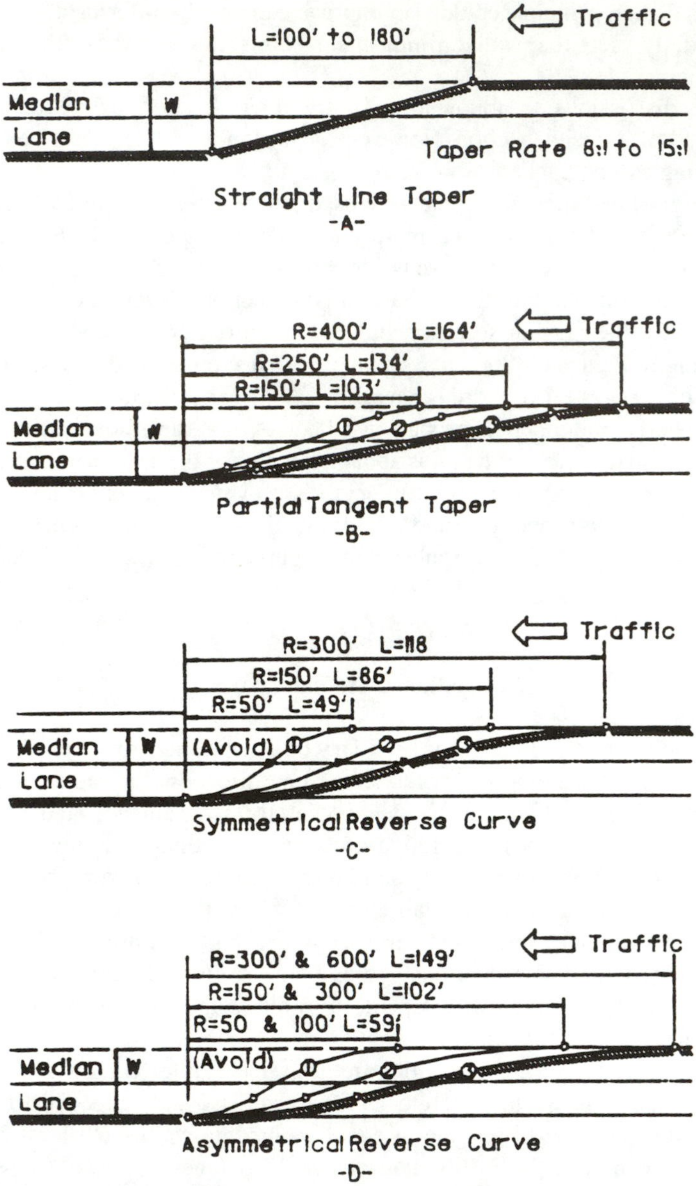

Dimensions are also applicable for right turn flares.

Figure IX-70. Taper design for auxiliary lanes.

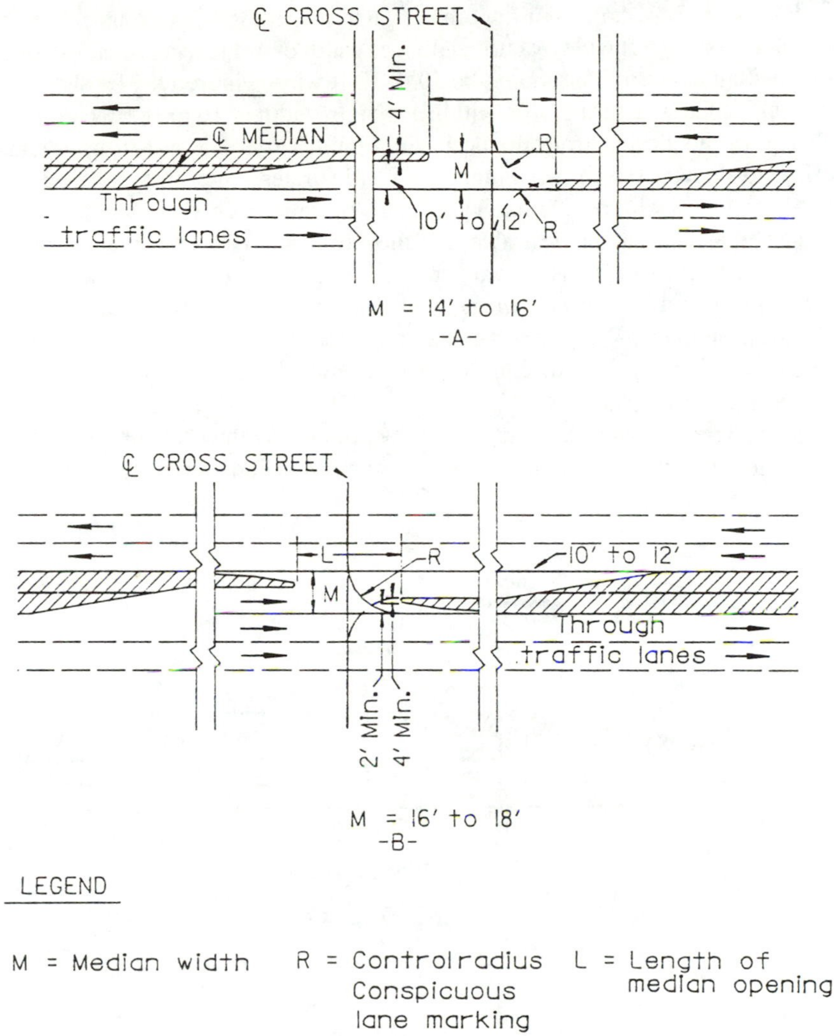

Figure IX-71. Median left-turn design.

Figure IX-71A shows a minimum design for a median left-turn lane within a median 14 to 16 ft wide. A curbed divider width of 4 ft is recommended and the median left-turn lane should be 10 to 12 ft wide. Figure IX-71B shows a typical median left-turn design within a median width of 16 to 18 ft. The only change in this design from the design in Figure IX-71A is a 2-ft minimum offset to the approach nose. Figure IX-72 illustrates a more liberal median left-turn design within a median width of 18 ft or more. On these medians the elongated tapers may be desirable. For medians 18 ft wide or more, a flush, color-contrasted divider is recommended to delineate the area between the turning lane and the adjacent through lane in the same direction of travel.

Pavement markings, contrasting pavement texture, signs and physical separators may be used to discourage the through driver from inadvertently entering the wrong lane.

Left turning vehicles leave the through pavement to enter the median lanes in single file, but once within it, store in two lanes and, on receiving the green

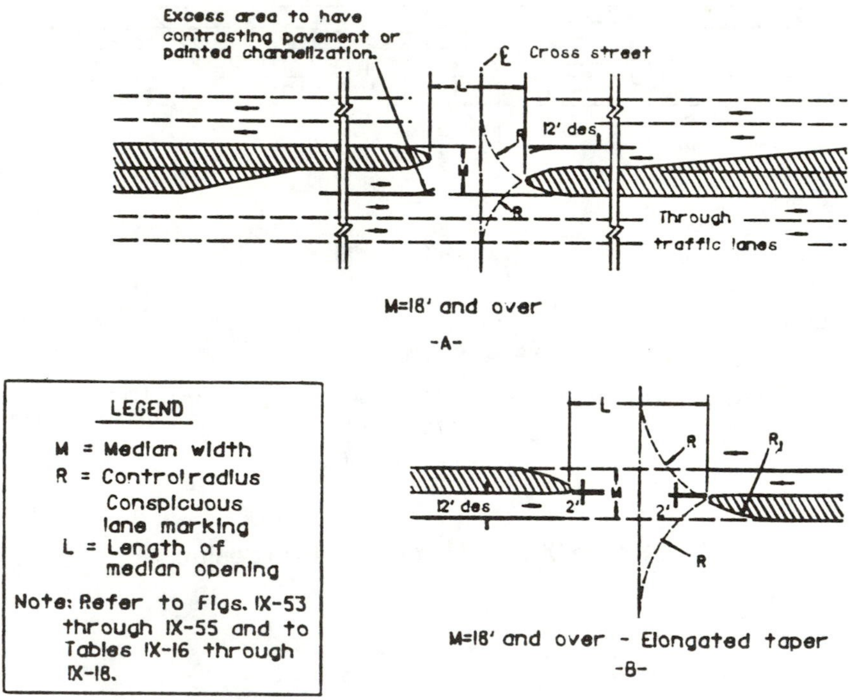

Figure IX-72. Median left-turn design.

indication, turn simultaneously from both lanes. With three-phase signal control, such an arrangement results in an increase in capacity of approximately 180 percent of that of a single median lane. Occasionally, there are operational problems as a result of the two-abreast turns, especially the problems of sideswipe accidents. These usually result from too sharp a turning radius or a roadway that is too narrow. The receiving leg of the intersection must have adequate width to accommodate two lanes of turning traffic. A width of 30 ft is used by several highway agencies. Double turning lanes should only be used with signalization providing a separate turning phase.

Median End Treatment

The form of treatment given the end of the narrowed median adjacent to lanes of opposing traffic depends largely on the available width. The narrowed median may be curbed to delineate the lane edge, to separate opposing movements, to provide space for necessary signs, markers, and lighting standards and to protect pedestrians. For a discussion on "ramped down" approach ends used for curbed medians, refer to the Section in this chapter relating to approach-end treatment. To serve these purposes satisfactorily the minimum narrowed median width of no less than 4 ft is recommended and is preferably 6 to 8 ft wide. These dimensions can be provided within a median 16 to 18 ft wide and a turning lane width of 12 ft.

For medians wider than about 18 ft as shown in Figure IX-72, it is usually preferable to aline the left-turn lane in a manner that will reduce the width of the divider to 6 to 8 ft immediately in advance of the intersection, rather than to aline it exactly parallel with and adjacent to the through lane. This alinement will place the vehicle waiting to make the turn as far to the left as practical and thus provide appropriate visibility of opposing through traffic. The advantages of aligning the left-turn lanes are: (1) better visibility of opposing through traffic; (2) decreased possibility of conflict between opposing left-turn movements within the intersection; and (3) more left-turn vehicles served in a given period of time, particularly at a signalized intersection.

Any excess area within the median to the right of the left-turn lane should be paved, preferably with contrasting pavement, or provided with painted channelization and maintained flush with the roadway surface as described in the preceding section.

For curbed dividers 4 ft or more in width at the narrowed end, the curbed nose can be offset from the opposing through-traffic lane 2 ft or more, with gradual taper beyond to make it less vulnerable to contact by through traffic, as shown in Figure IX-71B. The shape of the nose for curbed dividers 4 ft wide usually is semicircular, but for wider width the ends are normally shaped to a bullet nose pattern to conform better with the paths of turning vehicles.

SIMULTANEOUS LEFT TURNS

Simultaneous left turns may be considered on an intersection of two major at-grade highways. Figure IX-73 indicates traffic patterns that should be considered in the design. Marking details are given in Part III of the MUTCD (**5**).

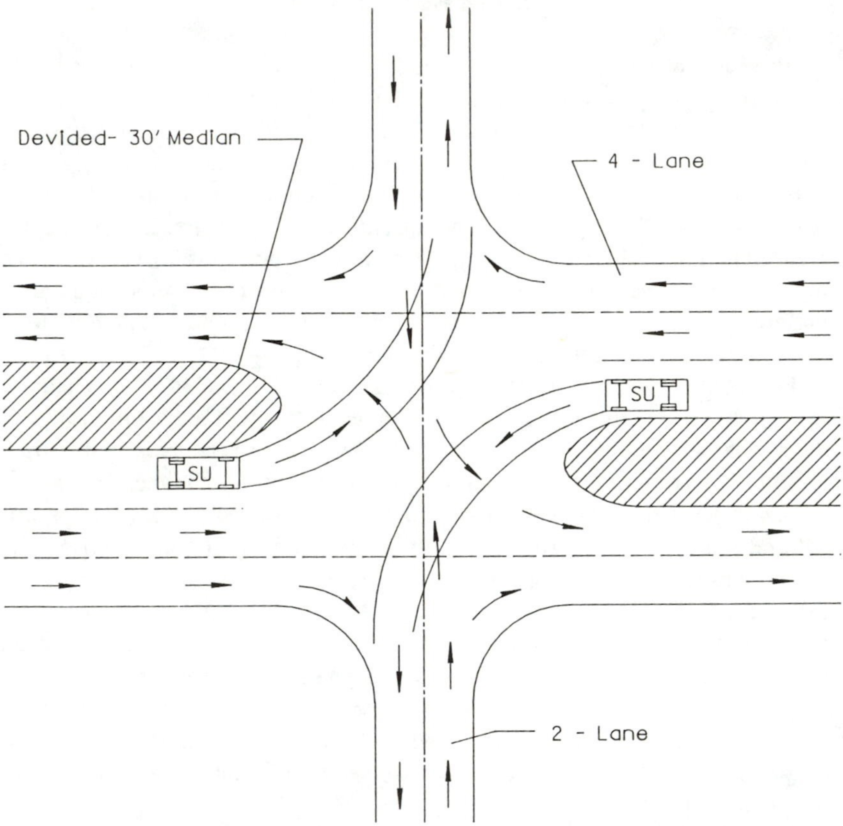

Figure IX-73. Four-leg intersection providing simultaneous left turns.

INTERSECTION DESIGN ELEMENTS WITH FRONTAGE ROADS

Frontage road cross-sectional elements, functional characteristics, and service value as collectors are discussed in Chapters IV, VI and VII. The discussion to follow concerns frontage road design elements with respect to the operational features where the frontage road intersects the major highway. Frontage roads are generally required contiguous to arterials or freeways where adjacent property owners are not permitted direct access to the major facility. Short lengths of frontage roads may be desirable along urban arterials to preserve the capacity and safety of the arterial through control of access. Much of the improvement in capacity and safety may be offset by the added hazard introduced where the frontage road and arterial intersect the at-grade crossroad. The added hazard results in part from the increase in the number of conflicting movements and from the confusing pattern of roadways and separations, which lead to wrong-way entry. Inevitably, where an arterial is flanked by frontage roads, the problems of design and traffic control at intersections are far more complex than where the arterial consists of a single roadway. Three intersections (two, if there is only one frontage road) actually exist at each cross street.

In lightly developed areas, such as through single-family residential neighborhoods, an intersection designed to fit minimum turning paths of passenger vehicles may operate satisfactorily. In heavily developed areas, however, particularly through commercial districts where frontage roads receive heavy use, an intersection designed to meet minimum geometric requirements will seldom operate satisfactorily unless certain traffic movements are prohibited. Separate signal indications can be used to relieve some of the conflicts between the various movements but only at the expense of delay to most of the traffic.

The preferred alternative to restricting turns is to design the intersection with expanded dimensions, particularly the width of outer separation. This design permits the intersections between the crossroad and the frontage roads to be well removed from the crossroad intersection with the main lanes.

For satisfactory operation with moderate-to-heavy traffic volumes on the frontage roads, the outer separation should preferably be 150 ft or more in width at the intersection. The 150-ft dimension is derived on the basis of the following considerations:

1. This dimension is about the shortest acceptable length needed for placing signs and other traffic control devices to give proper direction to traffic on the crossroad.

2. It usually affords acceptable storage space on the crossroad in advance of the main intersection to avoid blocking the frontage road.
3. It enables turning movements to be made from the main lanes onto the frontage roads without seriously disrupting the orderly movement of traffic.
4. It facilitates U-turns between the main lanes and two-way frontage road. (Such a maneuver is geometrically possible with a somewhat narrower separation but is extremely difficult with commercial vehicles).
5. It alleviates the potential of wrong-way entry onto the through lanes of the predominant highway.

Accordingly, outer separations at intersections should be 150 ft or more in width wherever practical and feasible. Narrower separations are acceptable where frontage road traffic is very light, where the frontage road operates one way only, or where some movements can be prohibited. Turning movements that are affected most by the width of outer separation are (1) left turns from the frontage road onto the crossroad, (2) U-turns from the through lanes of the predominant highway onto a two-way frontage road, and (3) right turns from the through lanes of the predominant highway onto the crossroad. By imposing the restrictions, as may be appropriate on some or all of these movements, outer separations as narrow as 8 ft may operate satisfactorily. With such narrow separations caution must be exercised in assessing the risk of wrong way entry onto the through lanes.

Except for the width of the outer separation, the design elements for intersections involving frontage roads are much the same as those for conventional intersections-at-grade. Figure IX-74 shows two arrangements of highways with frontage roads intersecting cross streets at grade. Turning movements are shown on the assumption that frontage road volumes are very light and that all movements will be under traffic signal control.

As illustrated, deceleration or storage lanes may be provided for the right-turn movements adjacent to the through roadway. Because traffic turning right must cross the path of traffic on the frontage road, the need for such storage lanes is usually greater in this case than in the case of conventional intersections. Storage and speed-change lanes are clearly delineated by pavement markings. Contrasting surfaces are also desirable.

Figure IX-74A shows a simple intersection design with an outer separation of 150 ft or more in width. The intersections of the two-way frontage roads and the crossroad are sufficiently removed from the through pavements that they might operate as separate intersections. The major elements in the design of the outer intersections are adequate width, adequate radii for right turns,

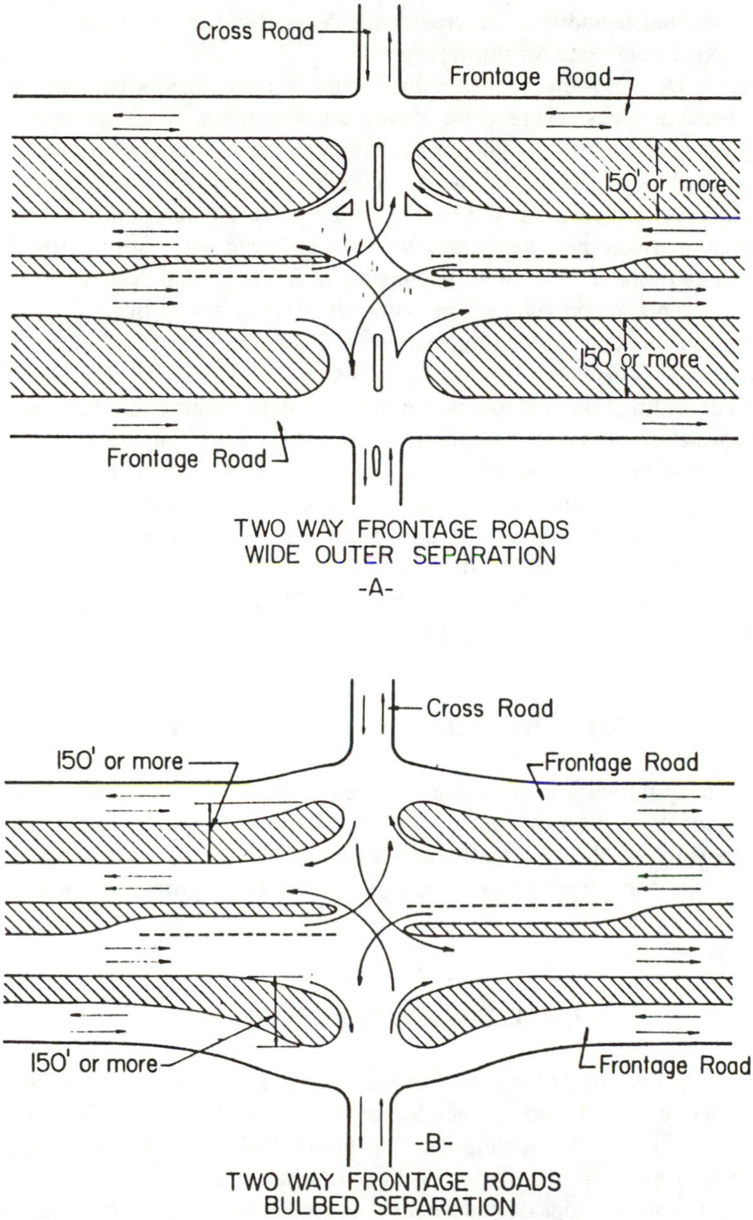

Figure IX-74. Intersections with frontage roads.

and divisional islands on the crossroad. A right-of-way of about 600 ft or more would be needed for this design.

Figure IX-74B shows a design that would be adaptable for two-way frontage roads in areas where right-of-way considerations would preclude the design shown in Figure IX-74A. With narrow outer separations between intersections a bulb treatment of the outer separations, as shown, formed by a reverse-curve alinement of the frontage road on each side of the crossroad is required to widen the outer separation to a desirable width at the crossroad. The length of the reverse curve is a matter of frontage road design, governed by design speeds and right-of-way controls. The widths of the outer separation bulbs should be based on the pattern and volumes of traffic, but the right-of-way controls may also govern because extra area is required at the intersection. The width of outer separation at the crossroad opening should be at least 60 ft, which might be acceptable for light-to-moderate frontage road traffic, but preferably it should be 150 ft or more. A width of 32 ft for the outer separation is the minimum that will permit a U-turn by a passenger car from the through lanes onto the frontage road. Widths of 70 ft or more are necessary for trucks and buses. Where such movements are likely to occur frequently, the width of separations should be considerably greater, desirably 150 ft or more.

WHEELCHAIR RAMPS AT INTERSECTIONS

When designing a project that requires curbs and adjacent sidewalks to accommodate pedestrian traffic, proper attention should be given to the requirements of the physically handicapped whose mobility depends on wheelchairs and other devices. Design criteria and figure illustrations are given in Chapter IV.

LIGHTING AT INTERSECTIONS

Lighting may affect the safety of highway and street intersections as well as efficiency of traffic operations. Statistics indicate that the evening accident rate is higher than that during daylight hours. This fact, to a large degree, may be attributed to impaired visibility. In urban and suburban areas where there are concentrations of pedestrians and roadside and intersectional interferences, fixed-source lighting tends to reduce accidents. Whether or not rural at-grade intersections should be lighted depends on the planned geometrics and the turning volumes involved. Intersections that generally do not require channelization are seldom lighted. However, for the benefit of nonlocal highway users, lighting at rural intersections is desirable to aid the driver in ascertaining sign messages during nondaylight periods.

Intersections with channelization, particularly multiple-road geometrics, should include lighting. Large channelized intersections especially need illumination because of the higher range of turning radii that are not within the lateral range of vehicular headlight beams. Vehicles approaching the intersection also must reduce speed. The indication of this need should be definite and visible at a distance from the intersection that may be beyond the range of headlights. Illumination of at-grade intersections with fixed-source lighting accomplishes this need.

The planned location of intersection luminaire supports should be designed to present the least possible hazard to out-of-control vehicles. The breakaway support base should not be used within limits of an at-grade intersection, particularly in densely developed areas with adjacent sidewalks. When struck, these light standards could be a problem for pedestrians and compound damage to adjacent property and other vehicles.

Additional discussions and design guidance are given in Reference (**8**).

DRIVEWAYS

Driveways are, in effect, at-grade intersections and should be designed consistent with the intended use. For further discussion of driveways refer to Chapter IV. The number of accidents is disproportionately higher at driveways than at other intersections; thus their design and location merit special consideration.

Driveways should not be situated within the functional boundary of at-grade intersections. This boundary would include the longitudinal limits of auxiliary lanes. The regulation and design of driveways are intimately linked with the right-of-way and zoning of the roadside. On new highways the necessary right-of-way can be obtained to provide the desired degree of driveway regulation and control. In many cases additional right-of-way can be acquired on existing highways or agreements can be made to improve existing undesirable access conditions. Often the desired degree of driveway control must be effected through the use of police powers to require permits for all new driveways and through adjustments of those in existence.

The objective of driveway regulation is desirable spacing and a proper layout plan. Its attainment depends on the type and extent of legislative authority granted the highway agency. Many States and cities have developed standards for driveways and have separate units to handle the details incidental to checking requests and issuing permits for new driveways or changes in driveway connections to their road systems. Major features of design and controls are suggested in References (**9**) (**10**) and (**11**).

RAILROAD GRADE CROSSINGS

A railroad highway crossing, like any highway-highway intersection, involves either a separation of grades or a crossing at-grade. The geometrics of a highway and structure that entails the overcrossing or undercrossing of a railroad are substantially the same as those for a highway grade separation without ramps.

The horizontal and vertical geometrics of a highway approaching an at-grade railroad crossing should be constructed in a manner that does not necessitate a driver divert attention to roadway conditions.

Horizontal Alinement

Desirably, the highway should intersect the tracks at a right angle with no nearby intersections or driveways. This layout enhances the driver's view of the crossing and tracks and reduces conflicting vehicular movements from crossroads and driveways. To the extent practical, crossings should not be located on either highway or railroad curves. Roadway curvature inhibits a driver's view of a crossing ahead and a driver's attention may be directed toward negotiating the curve rather than looking for a train. Railroad curvature may inhibit a driver's view down the tracks from both a stopped position at the crossing and on the approach to the crossings. Those crossings that are located on both highway and railroad curves present maintenance problems and poor rideability for highway traffic due to conflicting superelevations.

Vertical Alinement

It is desirable that the intersection of highway and railroad be made as level as possible from the standpoint of sight distance, rideability, braking and acceleration distances. Vertical curves should be of sufficient length to insure an adequate view of the crossing.

In some instances the roadway vertical alinement may not meet acceptable geometrics for a given design speed because of restrictive topography or limitations of right of way. Acceptable geometrics necessary to prevent drivers of low-clearance vehicles from becoming caught on the tracks would provide the crossing surface at the same plane as the top of the rails for a distance of 2 ft outside of the rails. The surface of the highway should also not be more than 3 in higher nor 6 in lower than the top of nearest rail at a point 30 ft from the rail unless track superelevation dictates otherwise as shown on Figure IX-75.

General

The geometric design of railroad-highway grade crossings must be made jointly with the determination of the warning devices to be used. When only passive warning devices, such as signs and pavement markings are used, the highway drivers are warned of thc crossing location, but must determine whether or not there are train movements for which they should stop. On the other hand, when active warning devices such as flashing light signals or automatic gates are used, the driver is given a positive indication of the presence or the approach of a train at the crossing. A large number of significant variables must be considered in determining the type of warning device to be installed at a railroad grade crossing. For certain low-volume highway crossings where adequate sight distance is not available, it may be necessary to install additional signing to provide a safe crossing.

Traffic control devices for railroad-highway grade crossings consist primarily of signs, pavement markings, flashing light signals, and automatic gates. Standards for design, placement, installment, and operation of these devices are covered in the MUTCD (**5**) as well as the use of various passive warning devices. Some of the considerations for evaluating the need for active warning devices at a grade crossing include the type of highway, volume of vehicular traffic, volume of railroad traffic, maximum speed of the railroad trains, permissible speed of vehicular traffic, the volume of pedestrian traffic, accident record, sight distance, and the geometrics of the crossing. The potential for complete elimination of grade crossings without active traffic

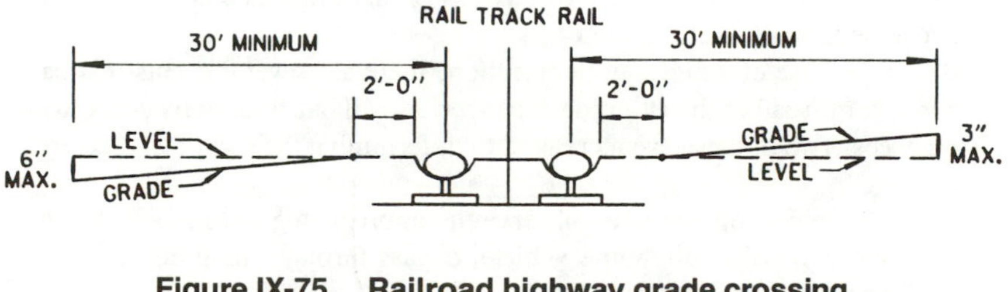

Figure IX-75. Railroad highway grade crossing.

control devices, for example, closing lightly used crossings and installing active devices at other more heavily used crossings, should be given prime consideration.

If it is established that active grade crossing traffic control devices are needed, the basic active device, flashing light signals, is used. When additional warning is desirable, the criteria or warrants recommended for evaluating the need for automatic gates at a grade crossing in addition to the above, include the existence of multiple main line tracks; multiple tracks at or in the vicinity of the crossing which may be occupied by a train or locomotive so as to be sure the movement of another train approaching the crossing; a combination of high speeds and moderately high volumes of highway and railroad traffic; and a substantial number of school buses or trucks carrying hazardous materials using the crossing.

These guidelines are not all inclusive. There will always be situations that are not covered by these guidelines and must be evaluated using good engineering judgment. Additional information on railroad-highway grade crossings can be found in References (**12**) (**13**) (**14**) (**15**) and (**16**).

Numerous hazard index formulas have been developed to assess the relative potential hazard at a railroad grade crossing on the basis of various combinations of its characteristics. Although no single formula has universal acceptance, each has its own values in establishing an index, that when used with sound engineering judgment, provides a basis for a selection of the type of warning devices to be installed at a given crossing.

The geometric design of a railroad-highway grade crossing involves the elements of alinement, profile, sight distance, and cross section. The requirements may vary with the type of warning devices used. Where signs and pavement markings are the only means of warning, the highway should cross the railroad at or nearly at right angles. Even when flashing lights or automatic gates are used, small intersection angles should be avoided. Regardless of the type of control, the roadway gradient should be flat at and adjacent to the railroad crossing to permit vehicles to stop when necessary and then proceed across the tracks without difficulty.

Sight distance is a primary consideration at crossings without train-activated warning devices. A complete discussion of sight distance at-grade crossings can be found in Reference (**14**).

As in the case of a highway intersection, there are several events that can occur at a railroad-highway grade intersection without train-activated warning devices. Two of these events related to determining the sight distance are:

1. The vehicle operator can observe the approaching train in a sight line that will safely allow the vehicle to pass through the grade crossing prior to the train's arrival at the crossing.

2. The vehicle operator can observe the approaching train in a sight line that will permit the vehicle to be brought to a stop prior to encroachment in the crossing area.

Both of these maneuvers are shown as Case I on Figure IX-76. The sight triangle consists of the two major legs, that is, the sight distance, d_H, along the highway and the sight distance, d_T, along the railroad tracks. Case I of Table IX-21 indicates values of the sight distances for various speeds of the vehicle and the train. These distances are developed from two basic formulas:

$$d_H = 1.47\, V_v t + \frac{V_v^2}{30f} + D + d_e$$

and

$$d_T = \frac{V_T}{V_v}\left[(1.47)\, V_v t + \frac{V_v^2}{30f} + 2D + L + W)\right]$$

where:

- d_H = sight distance leg along the highway allows a vehicle proceeding to speed V_v to cross tracks safely even though a train is observed at a distance d_T from the crossing or to safely stop the vehicle without encroachment of the crossing area, ft;
- d_T = sight distance leg along the railroad tracks to permit the maneuvers described as for d_h, ft;
- V_v = velocity of the vehicle, mph;
- V_T = velocity of the train, mph;
- t = perception/reaction time, which is assumed to be 2.5 sec; this is the same value used in Chapter III to develop the minimum safe stopping distance;
- f = coefficient of friction, which is assumed to be same values used and shown in Table III-1 for the development of the minimum safe stopping distance;
- D = distance from the stop line or front of the vehicle to the nearest rail, which is assumed to be 15 ft;
- d_e = distance from the driver to the front of the vehicle, which is assumed to be 10 ft;
- L = length of vehicle, which is assumed to be 65 ft; and
- W = distance between outer rails; for a single track, this value is 5 ft.

Corrections must be made for skew crossings and other than flat highway grades.

When a vehicle has stopped at a railroad crossing, the next maneuver is to depart from the stopped position. It is necessary that the vehicle operator have

$$d_H = 1.47\ V_v\ t + \frac{V_v^2}{30f} + D + de$$

$$d_T = \frac{V_T}{V_v}\left(1.47\ V_v t + \frac{V_v^2}{30f} + 2D + L + W\right)$$

d_H = Sight Distance Along Highway
d_T = Sight Distance Along Railroad Tracks
V_v = Velocity of Vehicle
t = Perception/Reaction Time (assumed 2.5 Sec.)
f = Coefficient Of Friction (See Table III-1)
D = Distance From Stop Line To Near Rail (Assumed 15 ft.)
W = Distance Between Outer Rails (Single Track W=5ft.)
L = Length Of Vehicle (Assumed 65ft.)
V_T = Velocity Of Train
de = Distance From Driver To Front Of Vehicle (Assumed 10ft.)

Adjustments Must Be Made For Skew Crossings.
Assumed Flat Highway Grades Adjacent To And At Crossings.

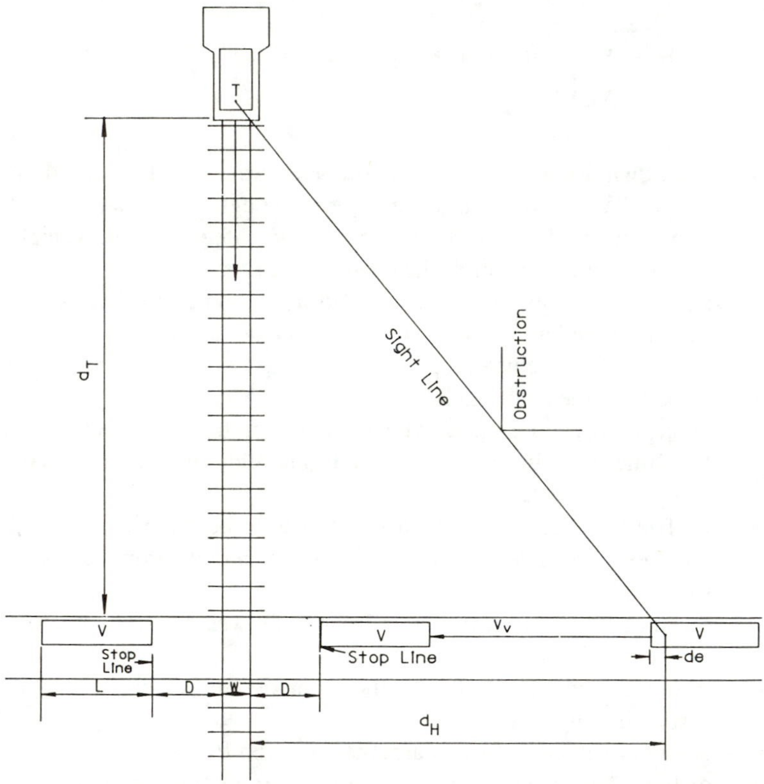

Figure IX-76. Case I: moving vehicle to safely cross or stop at railroad crossing.

Train Speed (mph)	Case II Departure from Stop	Case I Moving Vehicle						
		Vehicle Speed (mph)						
	0	10	20	30	40	50	60	70
	Distance Along Railroad from Crossing, d_T (ft)							
10	240	145	103	99	103	112	122	134
20	480	290	207	197	207	224	245	269
30	719	435	310	296	310	337	367	403
40	959	580	413	394	413	449	489	537
50	1,200	725	517	493	517	561	611	671
60	1,439	870	620	591	620	673	734	806
70	1,679	1,015	723	690	723	786	856	940
80	1,918	1,160	827	789	827	898	978	1,074
90	2,158	1,305	930	887	930	1,010	1,101	1,209
	Distance Along Highway from Crossing, d_H (ft)							
		69	132	221	338	486	659	865

Table IX-21. Required design sight distance for combination of highway and train vehicle speeds; 65-ft truck crossing a single set of tracks at 90°.

a sight distance along the tracks that will permit sufficient time to accelerate the vehicle and clear the crossing prior to the arrival of a train even though the train might come into view as the vehicle is beginning its departure process. Figure IX-77 indicates this maneuver. Case II on Table IX-21 contains various values of departure sight distance for a range of train speeds. These values are obtained from the formula:

$$d_T = 1.47\, V_T \left[\frac{V_G}{a_1} + \frac{L + 2D + W + -da + J}{V_G} \right]$$

where:

d_T = sight distance along railroad tracks, ft;

V_T = velocity of train, mph;

V_G = maximum speed of vehicle in first gear, which is assumed to be 8.8 fps;

a_1 = acceleration of vehicle in first gear, which is assumed to be 1.47 ft/sec^2;

L = length of vehicle, which is assumed to be 65 ft;

D = distance from stop line to nearest rail, which is assumed to be 15 ft;

J = perception/reaction time, which is assumed to be 2.0 sec;

W = distance between outer rails; for a single track, this value is 5 ft; and

d_a = $\frac{V_G^2}{2a_1}$ or distance vehicle travels while accelerating to maximum speed in first gear;

$$\frac{V_G^2}{2a_1} = \frac{(8.82)^2}{(2)(1.47)} = 26.4 \text{ ft}$$

Adjustments for skew crossings and for highway grades other than flat are necessary.

Sight distances of the order shown in Table IX-21 are desirable at any railroad grade crossing not controlled by active warning devices. Their attainment, however, is difficult and often impracticable, except in flat, open terrain.

In other than flat terrain, it may be necessary to rely on speed control signs and devices and to predicate sight distance on a reduced vehicle speed of operation. Where sight obstructions are present, it may be necessary to install active traffic control devices that will bring all highway traffic to a stop before crossing the tracks and will warn drivers automatically in time for an approaching train.

For safety the driver of a stopped vehicle at a crossing should see enough of the railroad track to be able to cross it before a train reaches the crossing, even though the train may come into view immediately after the vehicle starts to cross. The length of the railroad track in view on each side of the crossing must be greater than the product of the train speed and the time necessary for the stopped vehicle to start and cross the railroad. The required sight distance along the railroad track may be determined in the same manner as it is for a stopped vehicle crossing a preference highway, which is covered previously in

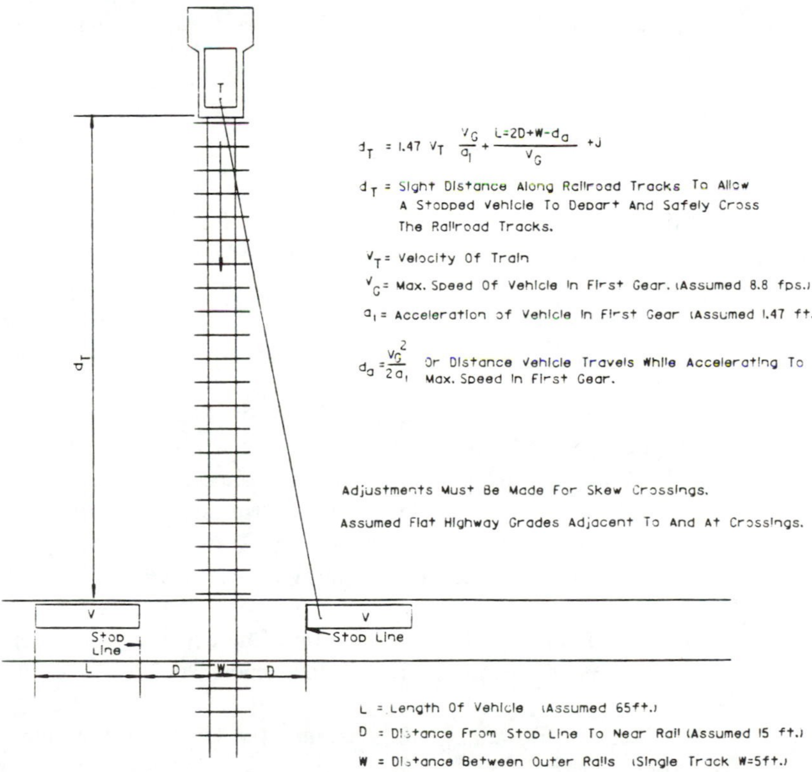

Figure IX-77. Case II: departure of vehicle from stopped position to cross single railroad track.

this Chapter. In order for vehicles to cross two tracks from a stopped position, with the front of the vehicle 15 ft from the closest rail, sight distances along the railroad, in feet, should be determined by the formula with a proper adjustment for the W value.

The highway traveled way at a railroad crossing should be constructed for a suitable length with all-weather surfacing. A roadway section equivalent to the current or proposed cross section of the approach roadway should be carried across the crossing. The crossing surface itself should have a riding quality equivalent to that of the approach roadway. If the crossing surface is in poor condition, the driver's attention may be devoted to choosing the smoothest path over the crossing. This effort may well reduce the attention given to observance of the warning devices or to the primary hazard of the crossing, which is the approaching train. Information regarding various surface types that may be used can be found in "Railroad-Highway Grade Crossing Surfaces" (**16**).

REFERENCES

1. Intersection Channelization Design Guide, NCHRP 279.
2. *Highway Capacity Manual, Special Report No. 209*, Washington, D.C.: Transportation Research Board, 1985.
3. "Symposium on Geometric Design for Large Trucks: Transportation Research Record 1052," National Research Council, Transportation Research Board, 1986.
4. "Twin-Trailer Trucks" Special Report 211, National Research Council, Transportation Research Board, 1986.
5. Federal Highway Administration, *Manual on Uniform Traffic Control Devices*. Washington, D.C.: U.S. Government Printing Office, 1988.
6. Harmelink, M.D. "Volume Warrants for Left-Turn Storage Lanes at Unsignalized Grade Intersections." Highway Research Record No. 211, 1967.
7. Fancher, P.S., UMTRI, "Vehicle Acceleration Characteristics Influencing Highway Design", Appendix F, NCHRP Report No. 270, 1984, pp. 146-169.
8. Walton, N.E. and Rowan, N.J. "Warrants for Highway Lighting." NCHRP Report 152, 1974.
9. "General Framework for Implementing Access Control Techniques", Vol. I. Technical guidelines for the Control of Direct Access to Arterial Highways, Report No. FHWA-RD-76-87.
10. "Detailed Description of Access Control Techniques," Vol. II. Technical

guidelines for the Control of Direct Access to Arterial Highways, Report No. FHWA-RD-76-87.
11. "Guidelines for Driveway Location and Design," Institute of Transportation Engineers, 1986.
12. "Traffic Control Devices and Rail-Highway Grade Crossings," National Research Council, Transportation Research Board, Transportation Research Record 1114, 1987.
13. Taggart, R.C. et al "Evaluating Grade-Separated Rail and Highway Crossing Alternatives" NCHRP Report 288, 1987.
14. "Railroad-Highway Grade Crossing Handbook." Report No. FHWA-TS-86-215, September, 1986.
15. Richards H.A. and Bridges, G.S. "Railroad Grade Crossings." Chapter 1 in Traffic Control and Roadway Elements-Their Relationship to Highway Safety. Automotive Safety Foundation, 1968.
16. "Railroad-Highway Grade Crossing Surfaces." Implementation Package 79-8, Federal Highway Administration, August, 1979.

Chapter X
GRADE SEPARATIONS AND INTERCHANGES

INTRODUCTION AND GENERAL TYPES OF INTERCHANGES

The ability to accommodate high volumes of traffic safely and efficiently through intersections depends largely on what arrangement is provided for handling intersecting traffic. The greatest efficiency, safety, and capacity are attained when the intersecting through-traffic lanes are separated in grades. An interchange is a system of interconnecting roadways in conjunction with one or more grade separations that provides for the movement of traffic between two or more roadways or highways on different levels.

The type of grade separation and interchange, along with its design, is influenced by many factors, such as highway classification, character and composition of traffic, design speed, and degree of access control. These controls plus signing requirements, economics, terrain, and right-of-way are of great importance in designing facilities with adequate capacity to safely accommodate the traffic demands.

Each interchange site should be studied and alternate designs made to determine the most fitting arrangement of structures and ramps and accommodation of bicycle and pedestrian traffic through the interchange area. Interchanges vary from single ramps connecting local streets to complex and comprehensive layouts involving two or more highways.

The basic types of interchanges are shown in Figure X-1. Any one type can vary extensively in shape and scope, and there are numerous combined types that are difficult to designate by separate names. An important element of interchange design is the assembly of one or more of the basic types of ramps, which are discussed elsewhere in this chapter. The layout for any specific ramp and type of traffic movement will reflect site conditions of topography and culture, cost, and the degree of flexibility in traffic operation desired. The last factor should predominate in design, but the practical aspects of cost and site conditions frequently are deciding factors in the type and scope of ramps.

Figure X-1 (A and B) shows typical three-leg interchanges. Figure X-1A is a trumpet-type interchange, named for the trumpet or jug-handle ramp configuration; Figure X-1B is a three-level, directional, three-leg interchange. Figure X-1C is an interchange with ramps in one quadrant that is not suitable for freeway systems but becomes very practical for an interchange between a

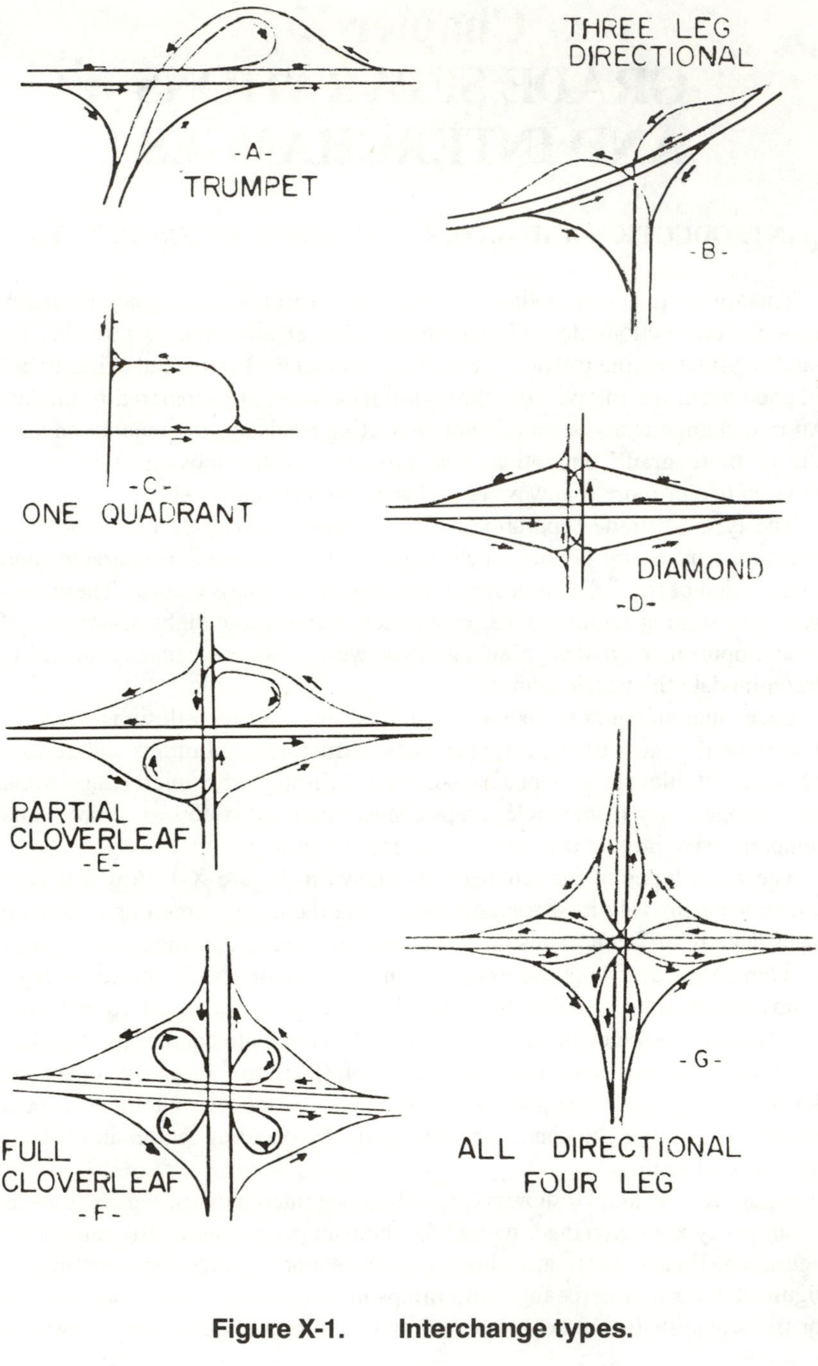

Figure X-1. Interchange types.

major highway and a parkway. This design is appropriate for parkways because design speed is usually in the low ranges, large trucks are prohibited, and turning movements are light. A typical diamond interchange is illustrated in Figure X-1D. Diamond interchanges have numerous other configurations incorporating frontage roads and continuous collector or distributor roads. Figure X-1E is a partial cloverleaf that contains two cloverleaf-type loops and two diagonal ramps. Varying configurations favor heavier traffic movements. A full cloverleaf, as shown in Figure X-1F, gives each interchanging movement an independent ramp; however, it generates weaving maneuvers that must occur either in the area adjacent to the through lanes or on collector-distributor roads. Figure X-1G is a fully directional interchange.

WARRANTS FOR INTERCHANGES AND GRADE SEPARATIONS

An interchange is a useful and an adaptable solution for many intersection problems, but because of the high initial cost, its use to eliminate existing traffic bottlenecks or to correct existing hazardous conditions is limited to those cases where the required expenditure can be justified. An enumeration of the specific conditions or warrants justifying an interchange at a given intersection is difficult and, in some instances, cannot be conclusively stated. Because of the wide variety of site conditions, traffic volumes, highway types, and interchange layouts, the warrants that justify an interchange may differ at each location. The conditions that should be considered to reach a rational decision are the available warrants. These are shown as the following six items:

1. Design designation. The conclusion to develop a highway with full control of access between selected terminals becomes the warrant for providing highway grade separations or interchanges for all intersecting highways. Although access control, provision of medians, and elimination of parking and pedestrian traffic are important, the separation of grades on freeways provides the greatest increment of safety. Once it has been decided to develop a route as a freeway, it must be determined whether each intersecting highway should be terminated, rerouted, or provided with a grade separation or interchange. For those crossroads that cannot be terminated, the individual warrant for a separation or interchange is absorbed in the decision to develop the freeway. The chief concern is the continuous flow on the major road. If traffic on the minor road will cross the freeway, a grade separation or interchange is provided. Thus, an intersection that might warrant only traffic signal control, if considered as an isolated case, will warrant a grade separation or interchange when considered as a part of a freeway.

2. Elimination of bottlenecks or spot congestion. Insufficient capacity at the intersection of heavily traveled routes results in intolerable congestion on one or all approaches. Inability to provide essential capacity with an at-grade facility provides a warrant for an interchange where development and available right-of-way permit. Even on facilities with partial control of access, the elimination of random signalization contributes greatly to improvement of free-flow characteristics.

3. Elimination of hazard. Some at-grade intersections have a disproportionate rate of serious accidents. Lacking inexpensive methods of eliminating hazards, a highway grade separation or interchange may be warranted. Accident-prone intersections frequently are found at the junction of comparatively lightly traveled highways in sparsely settled rural areas where speeds are high. In such areas, structures usually can be constructed at little cost compared with urban areas, right-of-way is not expensive, and these lower cost developments can be justified by the elimination of only a few serious accidents. Serious accidents at heavily traveled intersections, of course, also provide a warrant for interchange facilities. In addition to greater safety, the interchange also expedites all movements.

4. Site topography. At some sites grade-separation designs are the only type that can be made economically. The topography at the site may be such that any other type of intersection, to meet required standards, is physically impossible to develop or is equal or greater in cost.

5. Road-user benefits. The road-user costs due to delays at congested at-grade intersections are large. Such items as fuel, tires, oil, repairs, time, and accidents that require speed changes, stops, and waiting generate expenses well in excess of those for intersections permitting uninterrupted or continuous operation. In general, interchanges require somewhat more total travel distance than direct crossings at grade, but the added cost of the extra travel distance is less than the saving in cost effected by the reduction in stopping and delay costs. The relation of road-user benefits to the cost of improvement indicates an economic warrant for that improvement. For convenience the relation is expressed as a ratio, the annual benefit divided by the annual capital cost of the improvement. Annual benefit is the difference in road-user costs for the existing condition and in the condition after improvement. Annual capital cost is the sum of interest and amortization for the cost of the improvement. The larger the ratio, the greater the justification insofar as road-user benefits are concerned. Comparison of these ratios for design alternates is an important factor in determining the type and extent of improvement to be made. If used for justifying a single project or design, a ratio in excess of one is necessary for minimum economic justification. Furthermore, interchanges usually are adaptable to stage construction, and initial stages may produce incremental benefits that compare even more favorably with incremental costs.

6. Traffic volume warrant. A traffic volume warrant for interchange treatment would be the most tangible of any interchange warrant. Although a specific volume of traffic at an intersection cannot be completely rationalized as the warrant for an interchange, it is an important guide, particularly when combined with the traffic distribution pattern and the effect of traffic behavior. However, volumes in excess of the capacity of an at-grade intersection would certainly be a warrant. Interchanges are desirable at cross streets with heavy traffic volumes but may be difficult to provide because of costly right-of-way. The elimination of conflicts due to high crossing volume greatly improves the movement of traffic.

Some warrants for separations are given in warrants for interchanges; additional warrants are as follows:

1. Local roads and streets that cannot feasibly be terminated outside the right-of-way limits of freeways.
2. Access to areas not served by frontage roads or other means of access.
3. Railroad grade separations.
4. Unusual concentrations of pedestrian traffic (for instance, a city park developed on both sides of a major arterial).
5. Bikeways and routine pedestrian crossings.
6. Access to mass transit stations within the confines of a major arterial.
7. Free-flow aspects of certain ramp configurations and completing the geometry of interchanges.

ADAPTABILITY OF HIGHWAY GRADE SEPARATIONS AND INTERCHANGES

There are three general types of intersections—at-grade intersections, highway grade separations without ramps, and interchanges. Each has a field of usage in which it is practical, but the limits are not sharply defined. There is much overlapping and the final selection of intersection type frequently is a compromise after joint consideration of design traffic volume and pattern, cost, topography, and availability of right-of-way.

Traffic and Operation

Through traffic is accommodated to varying degrees of efficiency on all types of intersections. Where traffic on minor crossroads is small and subordinated to that on a major road, through traffic on the major road is little inconvenienced on at-grade intersections, particularly where topography is flat. Through traffic on the minor crossroads is delayed, but that on the major road proceeds without difficulty or delay. Where the minor crossroad traffic volume is sufficient to justify traffic signal controls to enable it to proceed through the intersection, difficulty and delay are experienced by all through traffic. Where through and crossroad volumes are nearly equal, about 50 percent of the traffic is required to stop.

Through traffic has no difficulty or delays at highway grade separations except where approach gradients are long and steep and many heavy trucks are included in the traffic stream. Ramps at interchanges have no severe effect on through traffic except where the capacity is not adequate, the merging or speed-change lanes are not of adequate length, or a full complement of turning roadways is not provided.

Turning traffic can affect operation and is accommodated to varying degrees on at-grade intersections and interchanges. At interchanges, ramps are provided for the turning movements. Where turning movements are light and some provision must be made for all turning movements, a one-quadrant ramp design may suffice. However, the left-turning movements on both through highways may provide no more freedom of movement than an intersection at grade would at that site. Ramps provided in two quadrants may be located so that crossings of through movements occur only at the crossroad and the major highway is free of such interference. An interchange with a ramp for every turning movement is suited for heavy volumes of through traffic and for any volume of turning traffic, provided the ramps and terminals are designed with sufficient capacity.

Right-turning movements at interchanges follow simple direct or nearly direct paths on which there is little possibility of driver confusion. Cloverleaf interchanges require loop paths for the left-turning movements, maneuvers that confuse many drivers, require added travel distance, and in some cases induce weaving movements. The diamond pattern of ramps is simple and is more adaptable than a cloverleaf in cases where direct left turns are fitting on the minor road. However, where traffic on the minor road is sufficient to justify the expenditure to eliminate the at-grade left turns, a cloverleaf or higher type interchange should be considered.

Except on freeways, interchanges usually are provided only where crossing and turning traffic cannot readily be accommodated on a less costly at-grade intersection. Confusion by some drivers appears unavoidable on interchanges, but such difficulties are minor compared with the benefits in the reduction of delays, stops, and accidents. As interchanges become more frequent and drivers gain experience in operating on them, as interchange designs are improved, and as the quality and use of signing and other control devices are increased, confusion is minimized. In areas where interchanges are infrequent the publicity, education, and enforcement—regarding proper usage of the ramp patterns provided—are valuable in ensuring efficient operation.

Interchanges are adaptable to all kinds of traffic. The presence in the traffic stream of a high proportion of heavy trucks makes provision of interchanges especially desirable. Avoidance of slowdowns and stops for heavy vehicles, which have much less accelerating ability than passenger cars, will do much toward retention of the capacity of the intersecting highways.

Site Conditions

In rolling or hilly topography, interchanges usually can be well fitted to the existing ground, and the through roads often can be designed to higher standards with separated gradelines than would be the case if an at-grade intersection is provided. Such terrain may also serve to simplify the design of some ramps. Other ramps, however, may require steep grades, be unduly long, or both, depending on the site terrain.

Interchange design is simple in flat terrain, but it may be necessary to introduce grades that may not favor vehicle operation. Interchanges in flat terrain generally are not as pleasing in appearance as those fitted to rolling terrain. When it is possible to regrade the whole of the interchange area and to landscape it properly, most of the deficiency in appearance can be overcome.

The right-of-way required for an interchange is dependent largely on the number of turning movements for which separate ramps need to be provided. The actual area required for any particular interchange also depends on the types of highway, topography, and the overall standards of interchange development. Definitely related to right-of-way requirements is the impact on property access that may occur with provision of an interchange. The separation of grades may require adjustment in the existing highway profiles and complicate local access, or the provision of ramps may require circuitous travel paths.

Type of Highway and Intersecting Facility

Interchanges are practical for all types of intersecting highways and for any range of design speeds. The hazard from stopping and direct turns at an intersection increases with the design speed so that high-design-speed highways warrant interchange treatment earlier than low-design-speed roads with similar traffic volumes. The ramps provided for high-design-speed highways should permit suitably high turning speeds, for which built-in facilities for changing speed assume greater importance.

Interchanges provide areas suitable for landscape development. For some conditions the necessary two-level development is a disadvantage in regard to appearance and may block the view. On the other hand, due regard to the architectural features in the structural design and to the flattening and rounding of slopes for erosion control and landscape treatment can provide a development that will almost always prove esthetically pleasing. Attention to such aspects may require the use of above-minimum layouts rather than the use of the simplest or least costly type of structure or the development of ramps with a minimum of grading.

Interchanges are essential components of freeways. With full control of access, grade separations are required at all crossroads of sufficient importance to prohibit their termination. The type of interchange will vary with the terrain, development along the highway, and right-of-way conditions, but in general it will be based on ramp layouts to expedite entrance to or exit from the freeway. Ramp connections also may involve frontage roads.

The extent or degree to which local service must be maintained or provided also is of concern in the selection of the type of intersection. Local service can be provided readily on certain types of at-grade intersections, whereas considerable additional facilities may be necessary on some types of interchanges.

Safety

Elimination or minimization of crossing and turning conflicts is essential in providing a reasonable degree of safety, especially at intersections. Regardless of design, signing and signalization, at-grade intersections have an ever present potential for vehicle-contact type accidents. This is due, in part, to conflicting crossing and turning movements occurring within a limited area.

By separating the grades of the intersecting roadways, accidents caused by crossing and turning movements can be reduced. The grade separation structure, itself, is somewhat a hazard; however, this can be minimized by the use of adequate clear roadway widths and protective devices at bridge abutments and piers. Where access between the intersecting roadways must be provided, an interchange is appropriate for providing the maximum degree of safety. Depending on the type of interchange used, left turns may be entirely eliminated or confined to the lower classified crossroad. Right-turning traffic can be accommodated on high design type ramps that provide operation approaching the equivalent of free flow. Conflicts caused by crossing traffic will be eliminated or minimized.

Stage Development

Where the ultimate development consists of a single grade-separation structure, stage construction may be uneconomical unless provisions are made in the original design for a future stage of construction. Ramps, however, are well adapted to stage development.

Economic Factors

Initial Costs

The interchange is the most costly type of intersection. The combined cost of the structure, ramps, through roadways, grading and landscaping of large areas, and possible adjustments in existing roadways and utilities generally exceeds the cost of an at-grade intersection. Directional interchanges involve more than one structure, and their cost usually is greater than any simple interchange.

Maintenance Costs

Each type of intersection has appreciable and distinct maintenance costs. Interchanges have large pavement and variable slope areas, the maintenance of which, together with that of the structure, signs, and landscaping, exceeds that of an at-grade intersection. Interchanges often involve maintenance and operation costs for lighting.

Vehicular Operating Costs

In a complete analysis of the adaptability of interchanges, it is necessary to compare vehicular operating costs of all traffic with those for other intersections. The values are so dependent on traffic, site, and design that a general comparison cannot be cited. Through traffic at an interchange usually follows a direct path with only a minor speed reduction. The added vehicular costs for the rise and fall in passing over and under the structure may need to be considered only when grades are steep, which condition usually is limited to the minor intersecting roads. Right-turning traffic is subject to added vehicular costs of deceleration and acceleration and may also be subject to the costs of operation on a grade, but travel distance is usually shorter than that on an at-grade intersection. Left-turning traffic is subject to the added costs of acceleration and deceleration and usually to the added travel distance compared with direct left turns at grade. Directional ramps may eliminate large speed changes and save travel distance, as compared with at-grade intersections. For any one vehicle these differences in operating costs may appear insignificant, but when considered in cumulative totals, they may indicate a pronounced benefit to traffic as a whole in the use of one particular type of intersection. For intermediate-to-heavy traffic the total vehicle-operating costs at an intersection usually will be lower with an interchange than with an at-grade design, especially if the through movements predominate.

GRADE SEPARATION STRUCTURES

Introduction

Various types of structures are employed to separate the grades of two intersecting roadways or a highway and a railroad. Although many phases of structural design also must be considered, this discussion is confined to the geometric features of grade separation structures. Some phases of structural design necessarily are referred to because of their effect on geometric design. This discussion largely concerns highway grade separations, but most of the geometric design features also apply to railroad grade separations.

Types of Separation Structures

Grade separation structures are identified by three general types: (1) through, (2) partial through, and (3) deck type. The deck type is most common for grade separations; however, the through or partial through are appropriate for railroad structures. In special cases where the spans are long and it is necessary to severely limit the difference in elevation between the roadways, truss bridges may be used.

Through girder bridges, in comparison to through deck-type bridges, will decrease vertical restrictions. In the case where the upper roadway must span from hilltop to hilltop and vertical clearance is not a problem, deck-type structures such as trusses, arches, girders, etc., could be appropriate. A through plate girder bridge is often used for railroad separations when the railroad overpasses the highway or street. The through plate girder and the through truss bridges produce a greater sense of visual restriction than deck-type structures; therefore, lateral clearances from the normal pavement edge should be as great as is feasible.

In any single separation structure care should be exercised in maintaining a constant clear roadway width and a uniform protective railing or parapet.

The type of structure best suited to grade separations is one that gives drivers little sense of restriction. Where drivers take practically no notice of a structure over which they are crossing, their behavior is the same or nearly the same as at other points on the highway, and sudden, erratic changes in speed and direction are unlikely. On the other hand, it is virtually impossible not to notice a structure overpassing the roadway being used. For this reason every effort should be made to design the structure so that it fits the environment in a pleasing and functional manner without drawing unnecessary or distracting attention. Collaboration between the bridge engineer and highway engineer throughout the various stages of planning and design can provide excellent results in this regard. Structures of this character should have liberal lateral clearances on the roadways at each level. All piers and abutment walls should be suitably offset from the traveled way. The finished underpass roadway median and off-shoulder slopes should be rounded and there should be a transition to backslopes to redirect errant vehicles away from protected or unprotected structural elements.

A grade separation structure should conform to the natural lines of the highway approaches in alinement, profile, and cross section. The structure should be designed to fit the highway. Fitting the structures to the highway may result in variable structural widths, flared pavements, flared parapets or bridge railing, and nonsymmetrical substructure units. Such dimensional variations are recognized as essential by both highway and bridge engineers and result in individual designs for each separate structure. In addition to the above geometric considerations, other conditions such as span lengths, depths of structure, foundation material at the site, esthetics, safety, and especially skew may substantially influence the engineering and cost feasibility of the structure being considered. The bridge engineer should be consulted during alinement (horizontal and vertical) studies, and close coordination should be maintained throughout the design phase so that the most prudent design from the standpoint of safety, functionality, and economics of the total highway (including the bridge) can be selected. Many times a minor adjustment in alinement can substantially reduce serious structural problems, especially with wide structures.

For the overpass highway the deck-type structure is most suitable. Although they may present both lateral and vertical clearance problems on the lower roadway, the supports relative to the upper roadway are underneath and out of sight. The deck-type bridge at the upper roadway has unlimited vertical clearance; lateral clearance is controlled only by location of the protective barrier. The parapet system should provide a freedom of view from the passing vehicles insofar as practicable; however, redirection capability should have precedence. The parapet and railing should have an appearance of strength and the ability to safely redirect the design vehicle(s) under the design impact conditions. Consideration must also be given to containing and redirecting larger vehicles crossing the structure. The end posts of through trusses shall be protected by a suitable approach traffic barrier and transition section. Spans at highway grade separations should never be great enough to require through trusses. In special cases where spans are long and it is necessary to limit the difference in elevation between the two roadways, all feasible designs are compared by the bridge engineer for suitability including economic and esthetic considerations.

For the underpass highway the most desirable structure from the standpoint of vehicular operation is one that will span the entire highway cross section and provide a lateral clearance of structural supports from the edge of pavement that is consistent with design requirements. The lateral space between the edge of pavement and the structural supports is built as wide and flat as practicable to provide usable recovery space for errant vehicles and to prevent distraction in the peripheral field of vision. In the case of depressed roadways, lateral clearances may be reduced, as discussed in the section "Lateral Clearances" later in this chapter. On divided highways center supports should be used only where the median is wide enough to provide sufficient lateral clearance or narrow enough to require protective barriers. The standard lateral clearance of an underpass at piers or abutments may allow sufficient room to construct additional lanes under the structure if necessary, but at a sacrifice of recovery space. In anticipation of future widening, the piers or abutment design must provide footings with sufficient cover after widening. The bridge engineer must be advised when future widening is contemplated. A greater sense of openness results with end spans than with full depth abutments. Perched stub or semistub abutments can also provide necessary visual clearance.

In urban areas not all of the streets that cross the main roadway are of sufficient prominence to warrant interchange ramps with the main line. A sufficient number of cross streets must be separated in grade to preserve the continuity of traffic flow on the local street system. As a matter of economics, however, it is seldom practical to continue all cross streets across the main line. Most streets that cross the major roadway, whether or not they connect with it, experience a rapid increase in traffic following construction of the major roadway as a result of intensified land development and local street closures within the main-line corridor. Terminated and through streets may be intercepted by one-way frontage roads on each side of the main facility. Access between the main roadway and frontage roads can be provided by slip ramps at prescribed intervals to meet traffic requirements.

On elevated facilities with viaduct construction the cross streets are relatively undisturbed; however, on all other types of roadways considerable savings can be effected by terminating some of the less important cross streets. Special consideration is needed relative to the spacing and treatment of cross streets on these roadways. Arterials and other major channels of cross traffic should continue across the main line without interruption or deviation. Grade separations should be of sufficient number and capacity to handle adequately not only the normal cross traffic but also the traffic diverted to cross streets from the other streets terminated by the main facility and the traffic generated by access connections to and from the main line. Thus, determination of the number and location of cross streets to be separated in grade requires a thorough analysis of traffic on the street system, in addition to that on the main line and its interchanges.

Insofar as throughway operation is concerned, there is no minimum spacing or limit to the number of grade-separated cross streets. The number and their location along one corridor are governed by the local street system, existing or planned. Depending on features of the city street network such as the length of block, the presence or absence of frontage roads, and degree of adjacent urban development, it may be necessary to provide more crossings than otherwise required for the principal cross streets. Where frontage roads are not provided or where they are used only intermittently, more crossings may be necessary than would otherwise be required in order that all areas may be conveniently accessible. Other factors that may affect the number and spacing of cross streets are the location of schools, recreational areas, and other public facilities, school bus routes, and fire fighting equipment routes. Whereas in and near downtown districts, cross streets continuing across the throughway may be located at intervals of two or three blocks, and sometimes every block, in intermediate areas they are likely to be three to five blocks apart, and in residential or outlying districts they should be at greater intervals.

Cross streets must also fit the existing, revised, or expected future pattern of transit operation and the needs of pedestrians. For the most part, pedestrians are accommodated on structures that also serve vehicular traffic. Because extra travel distance is more acceptable for vehicular travel than for pedestrians, it is appropriate to add separate pedestrian crossings, particularly where there are large numbers of pedestrians, as would be the case near schools, churches, and factories.

Although the streets that are to cross the major roadway should be selected during the planning stage, it does not necessarily follow that all crossings need to be constructed initially. Normally, structures carrying the major roadway should be constructed initially, as it is impractical to disrupt the main line after it is open to traffic. However, some of the planned structures to carry cross streets over the major roadway may be deferred until fully justified by traffic growth or other planned developments. The system of overcrossing streets should be coordinated with and shown in the design of the major roadway, and a plan should be developed showing those that are to be constructed initially and those that are to be provided later. Such a plan should show the traffic circulation scheme at initial and later stages, and it should be checked periodically against traffic needs of the major roadway, the interchanges, and the street system.

The new cross street structure and approaches are usually designed for projected traffic 10 to 20 years in the future. In many cases the existing cross street is of a lower standard at either side of the separation structure than at the newly designed separation. Improvement of the lower standard cross street may not be scheduled for several years. Therefore, there should be a suitable transition of the new work to the existing facility in a manner that will promote the safe, orderly movement of traffic.

In many instances the existing street approaching the major roadway requires some improvements to facilitate traffic and increase capacity to ensure free movement to and from the major roadway. Typical improvements are pavement widening, control of parking and pedestrians, improvements of intersections with traffic signals, marking, channelization, and one-way operation where appropriate.

Where a city street underpasses a major roadway, the underside of the structure is a design feature that deserves special treatment for esthetics. Because of numerous pedestrians and slower moving traffic, the underside of a structure as viewed from the cross street is especially noticeable to the local citizens. It should be as open as feasible to allow the maximum amount of light and air below. An open-type structural design is also needed to improve the sight distance, especially if there are intersections adjacent to the structure.

On elevated viaduct sections of roadway the local street system may be left relatively undisturbed unless it is necessary to realine the cross street or widen it for additional capacity. Structural openings should allow for future approach width and vertical clearance.

Cross street overcrossings and undercrossings have many features in common. These include such items as pavement widths, corner curb radii, storage for turning vehicles, horizontal clearances, curbs, and sidewalks.

A variety of typical highway separation structures are depicted in Figures X-2 and X-3. The bridge span arrangement is dictated principally by the required clear roadside recovery area, although sight distance is important on all roadways and diamond-type interchanges.

A single simple-span girder bridge may be used with spans of up to approximately 150 ft. The single simple-span girder bridge accommodates conditions of severe skew and horizontal curvature. Spans of greater length

Figure X-2. Typical grade separation structures with closed abutments.

Figure X-3. Typical grade separation structures with open-end spans.

require greater structure depth and higher approach embankments. The structure depth for single-span girder bridges is approximately 1/15 to 1/30 of the span.

The conventional type of overpass structure over divided highways is currently a two-span deck-type bridge. When bridging with two or more spans, the deck-girder-type bridge, either steel or concrete, is usually continuous in design for reasons of economy, some saving in structure depth, and the elimination of deck joints over the piers.

As an alternative to the girder bridge, a deck-type single-span rigid frame or a three-span rigid frame slant leg bridge may be used for esthetic purposes in favorable circumstances. A spandrel arch bridge at special geographic locations where excess vertical clearance is available and the skew is not severe may be economically and esthetically desirable when foundation support is adequate. The length of arch span may be of the order of several hundred feet if the arch shape is not too flat. The spandrel arch bridge is inherently of pleasing appearance.

Two or more structures are not uncommon at interchanges with direct connections for left-turning movements. In special cases with direct connections, several structures may be combined to form one multilevel structure. Two variations of roadways crossing at three or four levels are shown in Figure X-4. Designs that include three- and four-level structures may not be more costly than the equivalent number of conventional structures to provide the same traffic service, particularly in urban areas where right-of-way costs are high.

Over Versus Under

General Design Considerations

A detailed study should be made at each proposed highway grade separation to determine whether the main road should be carried over or under the structure. Often the choice is dictated by features such as topography or highway classification. It may be necessary to make several nearly complete preliminary layout plans before a decision regarding the most desirable general plan can be reached. Some statements of over-versus-under preference follow, but such general guides should be subordinated to detailed studies of the separation as a whole.

At any one site the governing conditions as to which road should be carried over usually fall into one of three groups: (1) the influence of topography predominates, and the design must be closely fitted to it; (2) the topography does not favor any one arrangement; and (3) the alinement and gradeline controls

—A—

—B—

Figure X-4. Multilevel grade separation structures.

of one highway are sufficiently important to subordinate those of the other and, possibly, to adopt an arrangement contrary to one that fits the site topography. As a rule, a design that best fits the existing topography is the most pleasing and economical to construct and maintain, and this factor becomes the first consideration in design. The chief exception to this is the case where a major road is sufficiently predominant in design to overweigh topographic and crossroad controls. Where topography does not govern, as is common in the case of flat topography, it may be necessary to study secondary factors, and the following additional points of general consideration need to be examined:

1. For the most part designers are governed by the need for economy, which is obtained by designs fitting existing topography, not only along the intersecting highways but for the whole of the area to be used in ramps and slopes. Thus, it is necessary to consider alternates in the interchange area as a whole to decide the over-versus-under question.
2. There is a certain traffic warning advantage on an undercrossing highway. As a driver approaches, the structure looms ahead, makes obvious the upper level crossroad, and gives advance warning of likely interchange connections. In level topography where the overpass highway is built up to cross over a highway at normal ground level, there also is similar warning advantage. Where an undercrossing highway dips beneath a crossroad to horizon level, this advantage is minimized.
3. Through traffic is given esthetic preference by a layout in which the more important road is the overpass. A wide overlook is possible from the structure and its approaches, giving drivers a minimum feeling of restriction.
4. Where turning traffic is significant, the ramp profiles are best fitted when the major road is at the lower level. The ramp grades then assist turning vehicles to decelerate as they leave the major highway and to accelerate as they approach it, rather than the reverse. In addition, on diamond-type interchanges the ramp terminal is visible to drivers as they leave the major highway.
5. In rolling topography or in rugged terrain, major road overcrossings may be attainable only by a forced alinement and rolling gradeline. Where there otherwise is no pronounced advantage to the selection of either an underpass or an overpass, the type that provides the better sight distance on the major road (desirably safe passing distance if the road is two-lane) should be preferred.
6. An overpass offers the best possibility for stage construction, both in the highway and the structure, with minimum impairment of the

original investment. The initial development of only part of the ultimate width is a complete structure and roadway in itself. By lateral extension of both or construction of a separate structure and roadway for a divided highway, the ultimate development is reached without loss of the initial facility.

7. Troublesome drainage problems may be reduced by carrying the major highway over without altering the grade of the crossroad. In some cases the drainage problem alone may be sufficient reason for choosing to carry the major highway over rather than under the crossroad.
8. Where topography control is secondary, the cost of bridges and approaches may determine whether the major roadway underpasses or overpasses the minor facility. A cost analysis that takes into account the bridge type, span length, roadway cross section, angle of skew, soil conditions, and cost of approaches will determine which of the two intersecting roadways should be placed on structure.
9. An underpass may be more advantageous where the major road can be built close to the existing ground, with continuous gradient, and with no pronounced grade changes. Where the widths of the roads are greatly different, the quantity of earthwork makes this arrangement more economical. Because the minor road usually is built to lower standards than the major road, grades on it may be steeper and sight distances shorter, with resultant economy in grading volume and pavement area on the shorter length of road to be rebuilt above the general level of the surrounding country.
10. Frequently, the choice of an underpass at a particular location is determined not by conditions at that location but by the design of the highway as a whole. Grade separations near urban areas constructed as parts of a depressed expressway, or as one raised above the general level of adjoining streets, are good examples of cases where decisions regarding individual grade separations are subordinated to the general development.
11. Where a new highway crosses an existing route carrying a large volume of traffic, an overcrossing by the new highway causes less disturbance to the existing route and a detour is usually not required.
12. The overcrossing structure has no limitation as to vertical clearance, which can be a significant advantage in the case of oversized loads requiring special permits on a major highway or route.
13. Desirably, the roadway carrying the highest traffic volume should have the fewest number of bridges for better ridability and fewer conflicts when repair and reconstruction are necessary.
14. In some instances it may be necessary to have the higher volume

facility depressed and crossing under the lower volume facility to reduce noise impact.

15. In some instances the lower volume facility should be carried over if there is a pronounced economic advantage.

Structure Widths

Roads with wide shoulders, wide gutters, and flat slopes are the safest and give the driver a sense of freedom. Poles, walks, bridge columns, bridge railing, and parapets located close to the pavement are potential hazards and cause drivers to shy away from them. For this reason the clear width on bridges should preferably be as wide as the approach pavement plus shoulders in order to give drivers a sense of openness. On bridges that are long and costly, some compromise from the desirable may be necessary.

On long bridges, particularly on long-span structures where cost per square foot is greater than the cost on short-span structures, widths less than ideal may be acceptable; however, economy alone should not be the governing factor in determining structure widths. The analysis of traffic characteristics, safety features, emergency contingencies, and cost/benefit ratios should be fully considered before the desirable structure width is compromised.

In arriving at a decision as to the width of the roadway over or under a grade separation, in determining the dimensions, location, and design of the structure as a whole, and in detailing features adjacent to the road, the designer should aim to provide a facility on which driver reaction and vehicle placement will be essentially the same as elsewhere on the intersecting roads. The width should not be so great as to result in the high cost of structure without proportionate value in usefulness and safety.

Underpass Roadway

The type of structure for an underpass should be determined for the dimensional, load, foundation, and general site requirements for each case. Only the dimensional details are reviewed herein.

Although it is an expensive element, an underpass is only one component of the total facility. It is desirable that the entire roadway section including the median, pavements, shoulders, and clear roadside areas be carried through the structure without change. However, there are structural design limitations, vertical roadway control limitations, and limitations due to skewed crossings, appearance or esthetic dimension relations, and cost factors, such as those encountered in lengthy depressed sections, that may require some reduction in the basic roadway cross section. On the other hand, where conditions permit a substantial length of freeway to be developed with desirable lateral dimensions, an isolated overpass along the section should not be designed to be a restrictive segment. In such case the additional structural costs are considered to be warranted as proper parts of the total facility.

Lateral Clearances

Minimum lateral clearances at underpasses are illustrated in Figure X-5. For a two-lane undercrossing roadway or for an undivided multilane roadway, the cross section width will vary, depending on the design standards for the particular functional classification and traffic volume. The minimum lateral clearance from the edge of the traveled way to the face of the protective barrier should be the normal shoulder width.

On divided highways the left clearances are usually governed by the median width; a median of 10-ft minimum width may be used on a four-lane roadway to provide 4-ft shoulders and rigid median barrier, but for six or more lanes the minimum median width should be 22 ft to provide 10-ft shoulders and a rigid median barrier. Figure X-5A shows the minimum lateral clearances to a continuous median barrier, either concrete or metal, for the basic roadway section and through an underpass where there is no center support. The same clearance dimensions would be applicable for a continuous wall on the left. Where a concrete median barrier is used, its base should be alined with respect to the traffic lane, as shown in Figure X-5A.

Figure X-5B shows the minimum right clearance as applicable to a continuous wall section. Figure X-5B shows the use of a concrete barrier constructed integrally with the wall. For this situation the right clearance should be measured to the base of the barrier. For design with a continuous concrete barrier on the right, usually a section similar to a median barrier, Figure X-5B is applicable. The same type of barrier may be used as an introduced feature where conditions lead to structure design with full-depth abutments.

As discussed in earlier chapters, high-speed highways should be designed with surfaced shoulders flush with the surface of the through-traffic lane. Continuous curbs should be limited to special drainage systems and slope walls on the outside of shoulders, either left or right. Such curbs should be carried through the underpass. Where it is necessary to provide walkways, the full shoulder section should be maintained and the span increased by the width of the walk. When highway agencies believe that a curb is necessary along solid abutments or walls, a concrete barrier may be used. See Chapter IV for other types of barriers, warrants, and placements of same.

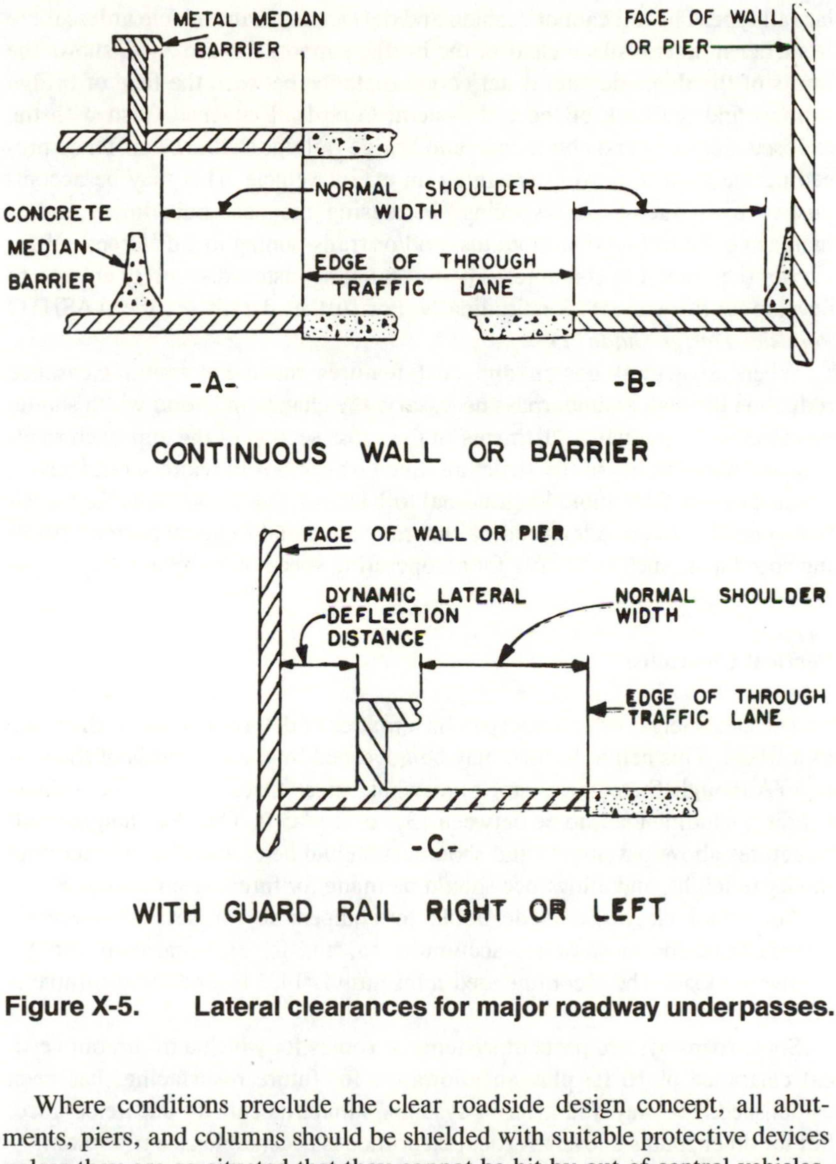

Figure X-5. Lateral clearances for major roadway underpasses.

Where conditions preclude the clear roadside design concept, all abutments, piers, and columns should be shielded with suitable protective devices unless they are so situated that they cannot be hit by out-of-control vehicles. Protective devices are usually not required along continuously walled sections.

Guardrail installed along the face of an exposed pier or abutment should have a clearance appropriate to the dynamic lateral deflection of the particular rail type. The rail cannot cushion and deflect an errant vehicle unless there is sufficient lateral space clear of the bridge support. Figure X-5C shows the limits of the dynamic lateral deflection distance between the face of bridge support and the back of the rail system. Guardrail attached flush with the exposed faces of piers, abutments, and bridge railings should be stiffened preceding the hazard to avoid snagging an errant vehicle. This may be accomplished by reducing post spacing, increasing the post imbedment and/or increasing the rail section modulus, and/or transitioning to a different, stiffer barrier (i.e., metal to concrete). The rail must be fastened securely enough to develop its full strength longitudinally. For further details see the AASHTO *Roadside Design Guide* (**1**).

Where structural design and cost features make horizontal clearance reduction through an underpass necessary, the change in lateral width should be effected as gradual adjustments in the cross section of the approach roadway and not abruptly at the structure. Such transitions in width should have a gradual rate of 50 or more longitudinal to 1 lateral. One good guide for transition control is to use a longitudinal/lateral ratio equal to the expected operating speed to 1, such as 55 to 1 for an operating speed of 55 mph.

Vertical Clearance

The clear height of an underpass is a matter of determination for the route as a whole. This height in turn may be governed by the standards of the system. Although State laws vary somewhat, most States permit the vehicle height, including load, to be between 13.5 and 14.5 ft. The clear height of all structures above pavements and shoulders should be at least 1 ft greater than the legal height, and allowance should be made for future resurfacing.

Additional clearance is desirable to compensate for several pavement resurfacings, for snow or ice accumulation, and for an occasional slightly overheight load. The recommended minimum is 14.5 ft, and the desirable is 16.5 ft.

Some roadways are parts of systems or routes for which a minimum vertical clearance of 16 ft, plus an allowance for future resurfacing, has been established. Freeway and arterial systems standards call for such clearance, but for other routes a lower vertical clearance is acceptable as a minimum.

In order to permit the movement of exceptionally high loads through an urban area, it is desirable to have at least one route with structures designed so that the movement can be readily handled. This design could entail the use of deck-type bridges, mast-arm-supported traffic signals, which can be swung to one side, street lights mounted higher than normal, underground utilities, etc.

Where a depressed facility is a parkway with traffic restricted to passenger vehicles, the vertical clearance at structures desirably should be 15 ft and in no case less than 12.5 ft. The minimum clearance should be obtained within all portions of the roadway.

Overpass Roadway

The roadway dimensional design of an overpass or other bridge should be the same as that of the basic roadway. The bridge is a small part of the continuous roadway and should be designed without change in cross section dimensions, except for very costly conditions.

The following paragraphs cover the general dimensional features for single structures of the type and size used at a grade separation, a stream crossing, or a single-structure interchange. Overpasses usually are deck structures. Their major dimensional features are the parapet rail system, lateral clearances, and the median treatment, where applicable. Typical overpass structures are shown in Figure X-6. For further discussion see also the sections under "Curbs," "Walkways," and "Traffic Barriers" in Chapter IV.

Bridge Railings

The typical rail has some form of concrete base or parapet on which metal or concrete rail or rails are mounted on structurally adequate posts. The bridge railing should be designed to accommodate the design vehicle(s) on the structure under the design impact conditions. Under these conditions the design vehicle should be safely redirected, without penetration or vaulting over the railing. Likewise, the railing should not pocket or snag the design vehicle, causing abrupt deceleration or spinout, and it should not cause the design vehicle to roll over.

Most bridge railings in service are of a rigid, nonyielding type. Several railings incorporate energy-absorbing features in their design to reduce vehicle impact severity. Where noise is a factor, solid rails should be considered for their added value in noise attenuation.

At certain locations there may be a need to provide a pedestrian walkway or bicycle path on the freeway overpass. In these situations a barrier-type bridge rail of adequate height should be installed between the pedestrian walkway and the roadway. A pedestrian rail or screen would be necessary on the outer edge of the walkway.

Figure X-6. Typical overpass structures.

Lateral Clearances

On overpass structures it is desirable to carry the full width of the approach roadway across all structures. For facilities other than freeways, exception is made on major structures on which, because of their high unit cost, the selection of cross section dimensions should be subject to individual economic studies. Refer to previous chapters on arterials, collectors, and local roads and streets for permissible deviations from providing full approach roadway width across bridges.

In the case of a curbed roadway the minimum structure width should match the curbed approach roadway.

When the full approach roadway width is continued across the structure, the parapet rail, both left and right, should aline with the guardrail on the approach roadway. For example, where the standard practice of the highway agency is to place the longitudinal barrier 2 ft from the outer edge of the surfaced shoulder, the bridge rail usually is placed 2 ft outside the effective edge of the shoulder. This width increment provides additional clearance for high-speed operation and door-opening space for vehicles stopped on the shoulder of the structure. Some agencies prefer to place the roadway longitudinal barrier 2 ft from the outer edge of the shoulder and the bridge rail at the shoulder edge. In this case a transition rate of about 20:1 is appropriate to taper the longitudinal barrier into the bridge rail.

At some interchanges extra space for speed-change lanes or weaving sections is required across overpass structures. Where the auxiliary lane is a continuation of a ramp, the horizontal clearance to the bridge rail should be at least equal to the width of shoulder on the approach ramp. Where the auxiliary lane is a weaving lane connecting entrance and exit ramps or is a parallel-type speed-change lane across the entire structure, the clearance to the parapet should be of uniform width, at least equal to the shoulder width on the ramp.

Medians

On a divided highway with a wide median or one being developed in stages, the overpass will likely be built as two parallel structures. The approach width of each roadway should be carried across each individual structure. If separate parallel structures are used, the width of opening between structures is unimportant.

Where the approach is a multilane undivided pavement or one with a flush median less than 4 ft wide, a raised median is considered unnecessary on short bridges of about 100 ft in length and desirable on bridges of 400 ft or more in length. On bridges between 100 and 400 ft in length, local conditions such as traffic volume, speed, sight distance, need for roadway lighting standards, future improvement, approach cross section, number of lanes, and whether the roadway is to be divided determine whether or not medians are warranted.

Where there are narrow or moderate widths on approaches to long single structures, the structure should be wide enough to accommodate the same type of median barrier as is used in the narrow median of the approach roadway.

Horizontal Distance to Effect Grade Separation

The distance required for adequate design of a grade separation depends on the design speed, the roadway gradient, and the amount of rise or fall necessary to effect the separation. Figure X-7 shows the horizontal distances required in flat terrain. It may be used as a guide for preliminary design to determine quickly whether or not a grade separation is feasible for given conditions, what gradients may be involved, and what profile adjustments, if any, may be necessary on the cross street. These data also may serve as a general guide in other than flat terrain, and adjustments can be made in the length of the terminal vertical curves. The chart is useful where the profile is rolled to overpass some cross streets and to underpass others, and it is useful for design of an occasional grade separation on a facility located at ground level, such as a major street or an at-grade expressway.

The horizontal distance required to effect a grade separation can be determined from Figure X-7 for gradients ranging from 2 to 7 percent and for design speeds ranging from 30 to 70 mph. Design speeds of 50 to 70 mph are applicable to urban freeways, and 40 mph (30 mph in special cases) is used on major arterials. The curves are derived with the same approach gradient on each side of the structure. However, values of D from Figure X-7 also are applicable to combinations of unequal gradients. Distance D is equal to the length of the initial vertical curve, plus one-half the central vertical curve, plus the length of tangent between the curves. Lengths of vertical curves, both sag and crest, are minimums based on the minimum stopping sight distance. Longer curves are desirable. Length D applies equally to an overpass and an underpass despite the fact that the central crest vertical curve may be longer than the central sag vertical curve for comparable values of H and G.

Certain characteristics and relations in Figure X-7 are worthy of note:

1. For the usual profile rise (or fall) required for a grade separation (H of 25 ft or less), gradients greater than 3 percent for a design speed of 70

mph, 4 percent for 60 mph, 5 percent for 50 mph, and 6 percent for 40 mph cannot be used. For values of H less than 25 ft, flatter gradients than cited above generally must be used. The lower terminal of the gradient lines on the chart, marked by a small circle, indicates the point where the tangent between curves is zero and below which a design for the given grade is not feasible; i.e., a profile condition where the minimum central and end curves for the gradient would overlap.

2. For given H and design speed, distance D is shortened a negligible amount by increasing the gradient above 4 percent for a design speed of 50 mph, and above 5 percent for 40 and 30 mph. Distance D varies to a greater extent, for given H and G, with changes in design speed.

A 20- to 22-ft difference in elevation is usually required at a grade separation of two highways for essential vertical clearance and structural thickness. The same dimension generally applies to a highway undercrossing a railroad, but about 28 ft is required for a highway overcrossing a main line railroad. In level terrain these vertical dimensions correspond to H, the rise or fall needed to effect a grade separation. In practice, however, H may vary over a wide range because of topography. Where a relatively short distance is available for a grade separation, it may be necessary to reduce H to keep D within the distance available. This reduction is accomplished by raising or lowering the intersecting street or railroad.

For example, assume a highway undercrossing is at a cross street whose elevation is 5 ft above the highway approach point (A) (Figure X-7). The required H is 15 ft. The design speed is 60 mph and only 820 ft are available to effect the separation of grades. From Figure X-7 it is seen that a gradient steeper than 3 percent cannot be used and in a length of 820 ft, a fall (H distance) of only 12.5 ft can be effected and still have adequate vertical curves. Therefore, it is necessary to raise the elevation of the cross street 2.5 ft. Another alternative may be an overcrossing, in which case the cross street would have to be lowered to an elevation of 7.5 ft below point (A) or 12.5 ft from its existing elevation.

Grade Separations Without Ramps

There are many situations where grade separations are provided without ramps. Major arterials often intersect existing highways that must be kept open for service but on which traffic is minor. Lacking a suitable relocation plan for the crossroad, a highway grade separation without ramps may be provided. All drivers desiring to turn to or from that road are required to use other existing routes and enter or leave the throughway at other locations. In

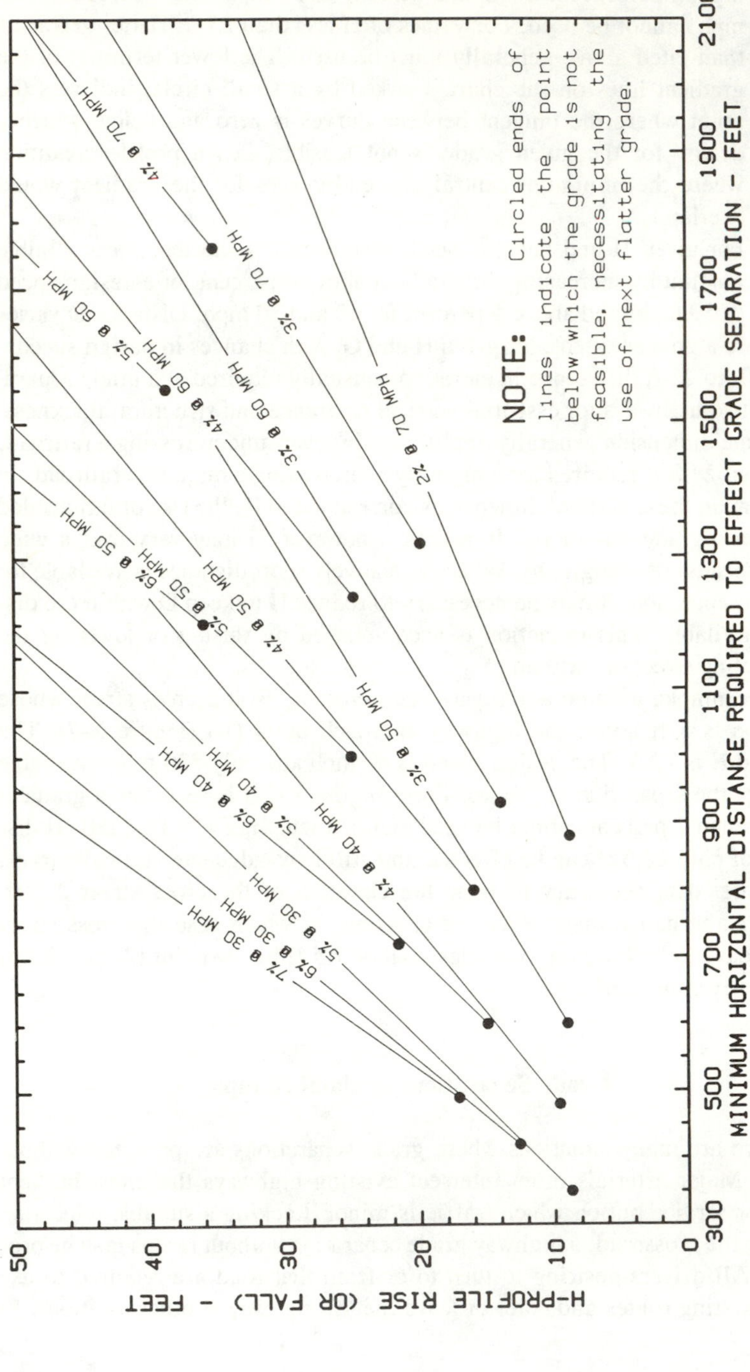

Figure X-7. Flat terrain, distance required to effect grade separation.

NOTE: "MINIMUM VERTICAL CLEARANCE SHOULD BE CHECKED UNDER THE OUTSIDE EDGE OF THE OVERCROSSING STRUCTURE"

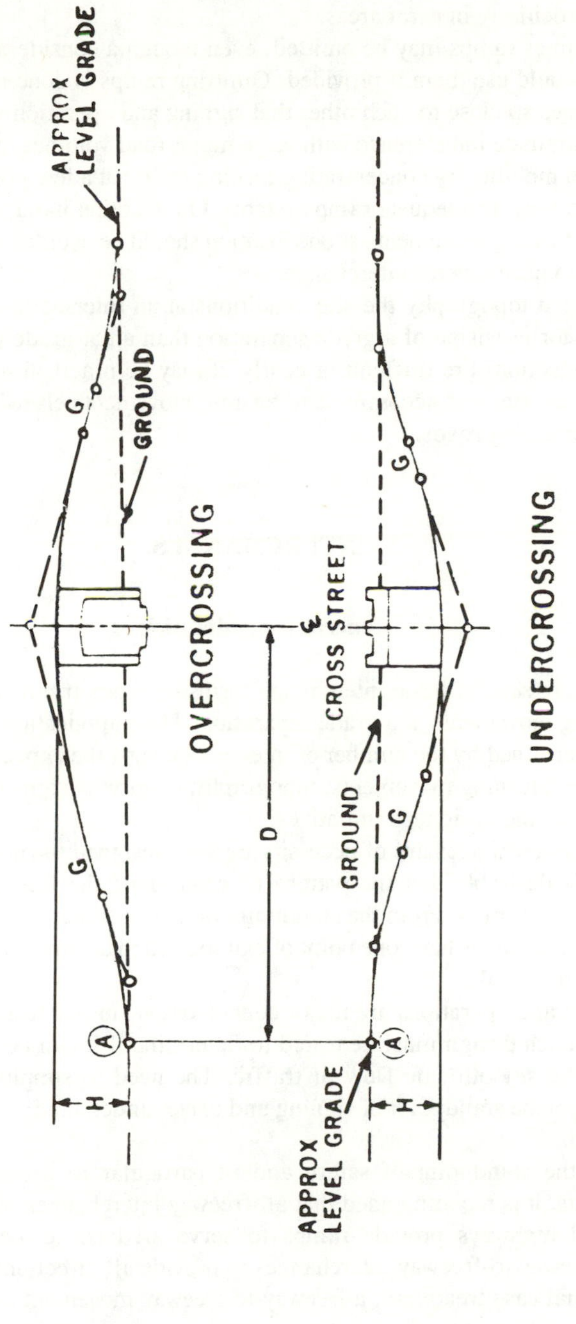

Figure X-7. Continued.

some instances these vehicles may have to travel a considerable extra distance, particularly in rural areas.

Sometimes ramps may be omitted, even though a considerable number of vehicles would use them if provided. Omitting ramps is done to avoid having interchanges so close to each other that signing and operation would be difficult, to eliminate interference with large major road volumes, and to increase safety and mobility by concentrating turning traffic at a few points where it is feasible to provide adequate ramp systems. On the other hand, undue concentration of turning movements at one location should be avoided where it would be better to have several interchanges.

In rugged topography the site conditions at an intersection may be more favorable for provision of a grade separation than an at-grade intersection. If ramp connections are difficult or costly, it may be practical to omit them at the structure site and accommodate turning movements elsewhere by way of other intersecting roads.

INTERCHANGES

General Considerations

There are several basic interchange forms or geometric patterns of ramps for turning movements at a grade separation. Their application at a particular site is determined by the number of intersection legs, the expected volumes of through and turning movements, topography, culture, design controls, proper signing, and the designer's initiative.

While interchanges are of necessity custom designed to fit specific conditions, it is desirable that the pattern of exits along the freeway have some degree of uniformity. From the standpoint of driver expectancy, it is desirable that all interchanges have one point of exit located in advance of the crossroad wherever practical.

Signing and operations are major considerations in the design of the interchanges. Each design must be tested to determine if it can be signed to provide for the smooth safe flow of traffic. The need to simplify interchange design from the standpoint of signing and driver understanding can hardly be overstated.

From the standpoint of safety and in particular to prevent wrong-way movements, it is recommended that all freeway interchanges with nonaccess-controlled highways provide ramps to serve all basic directions. Drivers expect freeway-to-freeway interchanges to provide all directional movements. As a special case treatment, a freeway-to-freeway movement may be omitted if the turning traffic is minor and can be accommodated and given the same route signing over other freeway facilities.

For convenience, the types and examples of interchanges are shown in general terms of three- and four-leg intersections and special designs requiring two or more structures. The general types of interchanges are shown either schematically or as examples of existing facilities.

Three-Leg Designs

An interchange at an intersection with three intersecting legs consists of one or more highway grade separations and one-way roadways for all traffic movements. When two of the three intersection legs form a through road and the angle of intersection is not acute, the term T interchange applies. When all three intersection legs have a through character or the intersection angle with the third intersection leg is small, the interchange may be considered a Y type. A clear distinction between the T and Y types is not necessary or important. Regardless of such factors as the intersection angle and through-road character, any one basic interchange pattern may apply for widely variant conditions.

Figure X-8 illustrates patterns of three-leg interchanges with one grade separation. Figures X-8A and X-8B show the widely used trumpet pattern. Through-traffic movements, from points a to c, are on direct alinement. A criterion for selection of either design is the relative volumes of the left-turning movements, the more direct alinement favoring the heavier volume and the loop the lesser volume. Skewed crossings are more desirable than right-angle crossings because the skewed crossing has a somewhat shorter travel distance and flatter turning radius for the heavier left-turning volume, and there is less angle of turn for both left turns. In Figure X-8A, the curvature of the loop b-a begins before the structure, warning the driver to anticipate a major break in curvature. The transition spirals provide for a smooth speed change and steering maneuver both into the loop and onto the high-speed facility. The oblong shape of the loop allows the curvature of the high-volume left turn, c to b, to be flattened, allowing higher operating speeds to be attained. The loop exit of Figure X-8B is placed well in advance of the structure to provide sufficient deceleration length in the approach to the break in curvature. Curves with spiral transitions are effective in developing the desired shape of ramps. The curvature of the left turn, b-a, is initiated in advance of the structure for driver anticipation.

The other type of three-leg single-structure interchange shown in Figure X-8C is less common. Figure X-8C is a plan with loops for both left-turning movements. With respect to traffic, it is inferior to Figures X-8A and X-8B

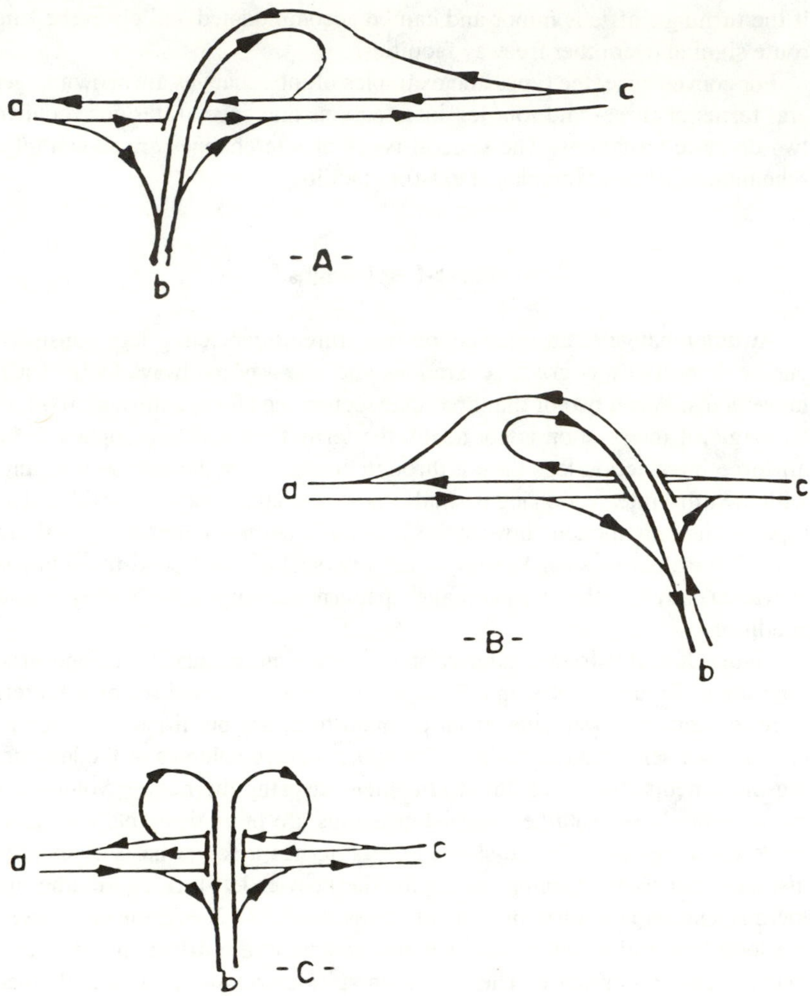

Figure X-8. Three-leg interchanges with single structures.

because both left turns use loops and must weave across each other. Although the pattern is adaptable to interchanges where the left-turning volumes are not great, there appears to be little reason for its selection if Figures X-8A and X-8B are equally adaptable to the site conditions. For comparable conditions, construction costs for Figures X-8A and X-8B should be about the same. Figure X-8C has an excellent field of usage as the initial stage of an ultimate cloverleaf. A collector-distributor road is provided to eliminate weaving on the main road. In the second stage the roadway forming the fourth leg opposite the stem of the T is developed and the remaining ramps are added.

The small radii loop ramps are not considered to be an adequate method of terminating a freeway-type roadway.

Figure X-9 illustrates high-type T and Y interchanges, each with more than one structure or with one three-level structure, that provide for all of the movements without loops. These types are more costly than single structure types and are justified only where all movements are large.

In Figure X-9A, all movements are directional. Three structures are required. Weaving is avoided. This plan is suitable for the intersection of a through freeway with the terminal of another major freeway. Some or all of the interchanging movements will require at least two-lane roadways. All entrances and exits are designed as branch connections or major forks, as discussed later in this chapter. The alinement of this scheme may be adjusted to reduce the right-of-way requirements, forming an interchange with only one three-level structure, as illustrated in Figure X-9B.

Operationally, Figure X-9A might be superior to Figure X-9B because of the inherent sharp curvature on movement c-b in Figure X-9B. A complete cost comparison requires a special analysis, but there usually is little difference. In some cases the more complex three-level structure has been found to be less costly.

Figure X-9C illustrates a three-leg interchange with a double jug-handle pattern. This pattern applies where it is necessary to carry a freeway through with a minimum of deviation but where the intersecting road is of considerable importance also. Interchanging traffic leaves the freeway on the right or enters on the right. Ramps are usually only single-lane roadways. This pattern requires the use of three structures, at least two of which must span double roadways. As shown in Figure X-9D, the basic pattern can be arranged so that the two left-turn ramps and the through road meet at a common point where a three-level structure replaces the three structures shown.

Figure X-9E is another variation of the general plan of Figures X-9C and X-9D. Separate roadways are provided for each left-turning movement with two 2-level structures separating the ramps from the through movements. The grade separation structures must be spaced sufficiently far apart to permit the placement of the separate ramp, b-a, between them, thus avoiding the third structure of Figure X-9C. This design may be altered, as shown in Figure X-9F. This arrangement provides smoother alinement on the ramps, but successful operation depends on provision of a suitable-length weaving section for these two movements.

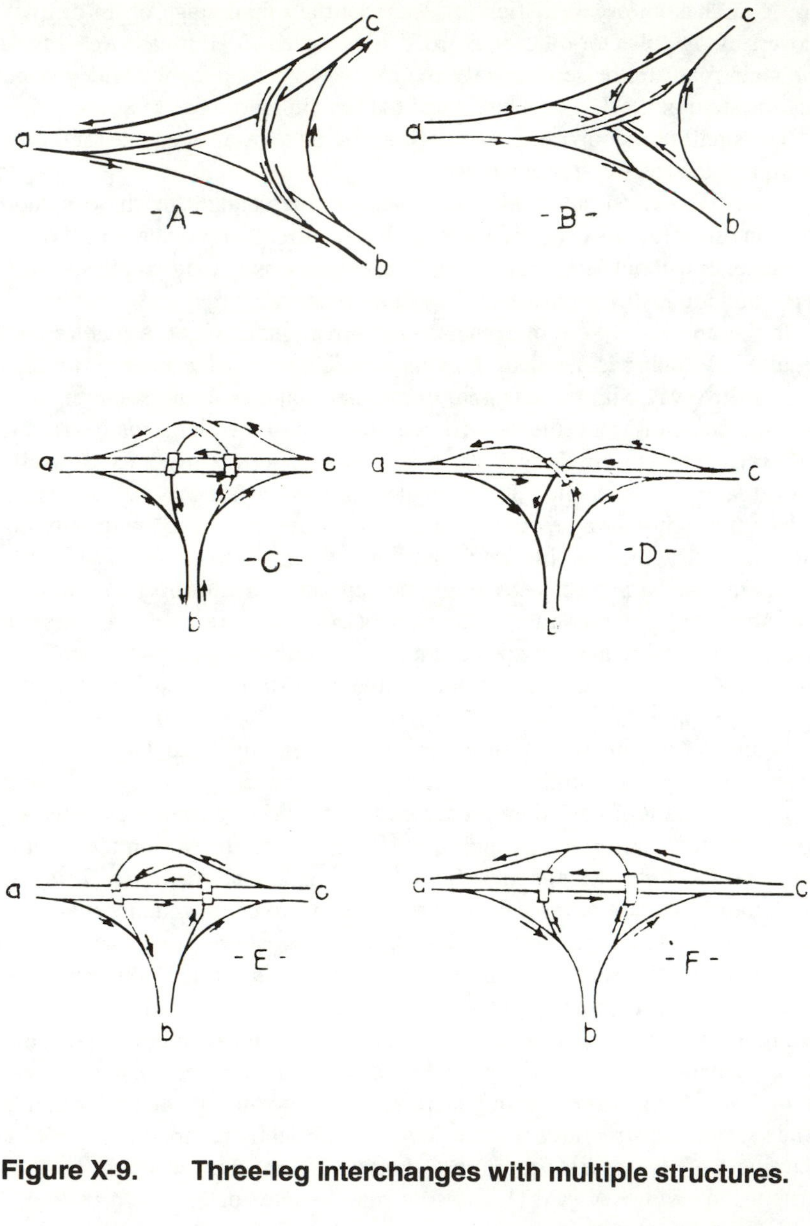

Figure X-9. Three-leg interchanges with multiple structures.

Examples of Unique Three-Leg Designs

Figure X-10 shows a trumpet interchange at the junction of a freeway and a major local road in a rural area. A unique feature is that the local road overpasses one roadway of the freeway and underpasses the other because of the steep slope on the left. This pattern also explains the relatively sharp radius on the loop. The design favors the heavier traffic movement that is provided the semidirect connection, and the loop handles the lighter volume.

Figure X-11 shows an interchange between two freeways in a rural area. The directional design with large radii permit high speed operation for all movements. The frontage roads are desirably one-way operations with connections from the interchange roadways being as long as practical. Major forks and following ramp terminals should have sufficient separation distance to provide for smooth traffic operations.

There are three separate structures. Turning movements are made from and to the right side of the bottom-to-top roadway despite the higher volume in the left-to-top movements.

Figure X-12 illustrates a directional, three-leg interchange between two freeways. This liberal design permits turns at full freeway speed. A full interchange to a local street is provided within the major interchange. The long spacing between successive exits and entrances ensures good operation.

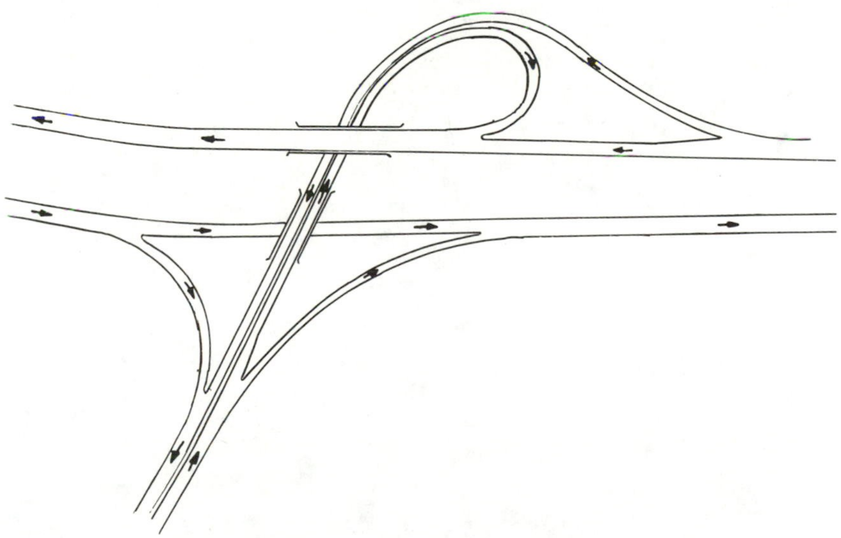

Figure X-10. Three-leg interchange (T-type or trumpet).

Figure X-13 shows a directional, three-leg interchange between two freeways at a river crossing. The turning roadways are liberally designed to permit high-speed operation. Note the major fork and branch connection on the river crossings. A much more expansive gore area is provided on the divergence than on the convergence. The expansive gore area on the divergence is provided for a recovery area and possible installation of an attenuator.

The two-lane exit and entrance at the bridge in the foreground of Figure X-14 serve both the local street system and the freeway-to-freeway movements. All interchange movements are usually provided and the exits on the curves are properly designed to discourage inadvertent exits. The exit in the

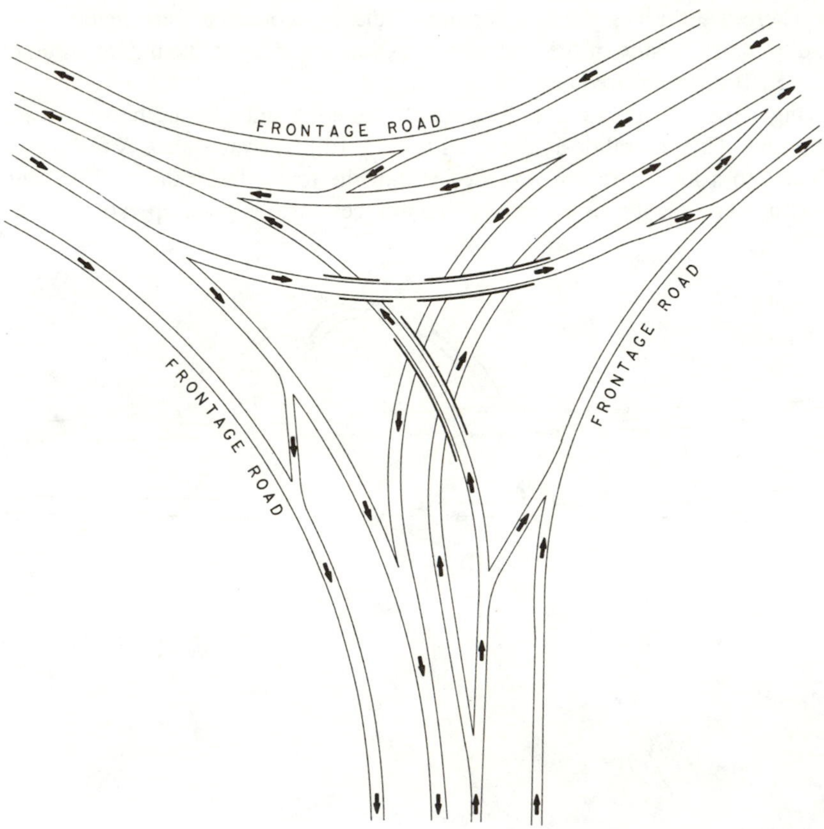

Figure X-11. Three-leg interchange semidirectional design.

Figure X-12. **Directional three-leg interchange with minor interchange provided internally.**

Figure X-13. **Directional three-leg interchange of a river crossing.**

bottom of the foreground is placed so that it commences in advance of the main-line curve. The gores are liberally designed with good delineation.

Figure X-15 shows a modified trumpet-type interchange for an urban facility intersecting with a freeway. The approach curvature of the urban facility and the lack of available right-of-way are common problems at this type of interchange. The main traffic movement, which in this case is the urban facility entering the freeway, usually receives the highest and most direct priority.

Four-Leg Designs

Interchanges with four intersection legs may be grouped under five general types: (1) ramps in one quadrant (ramps in two or three quadrants are discussed as partial cloverleafs), (2) diamond interchanges, (3) partial cloverleafs, (4) full cloverleafs, and (5) interchanges with direct and semidirect connections. Operational characteristics and adaptations of each type are discussed separately. Actual examples of existing or planned interchanges are demonstrated under each type.

Figure X-14. Trumpet-type freeway-to-freeway interchange.

Figure X-15. Modified trumpet interchange at an urban facility and a freeway.

Ramps in One Quadrant

Interchanges with ramps in only one quadrant have application for intersections of roads with low volumes. Where a grade separation is provided at an intersection because of topography, even though volumes do not justify the structure, a single two-way ramp of near-minimum design usually will suffice for all turning traffic. The ramp terminals may be plain T intersections.

Locations where designs of this type are applicable are very limited. A typical location would be at the intersection of a scenic parkway and a State or county two-lane highway. Turning movements would probably be light and without trucks. Terrain and the preservation of natural environment would take precedence over providing additional ramps.

At some interchanges it may be necessary to limit ramp development to one quadrant because of topography, culture, or other controls, even though the traffic volume justifies more extensive turning facilities. With ramps in one quadrant only, a high degree of channelization at the terminals and the median, together with left-turn lanes on the through facilities, normally is required to control turning movements properly.

In some instances a one-quadrant interchange may be constructed as the first step in a stage construction program. In this case the initial ramps should be designed as a part of the ultimate development.

Figure X-16A illustrates a one-quadrant interchange at the intersection of a State highway and a scenic parkway located in a rural mountainous area. The elongated shape of the ramp was determined largely by topography. Traffic entering both through roadways is under stop sign control. Although traffic volumes are low, the turning traffic is a substantial proportion of the total volume.

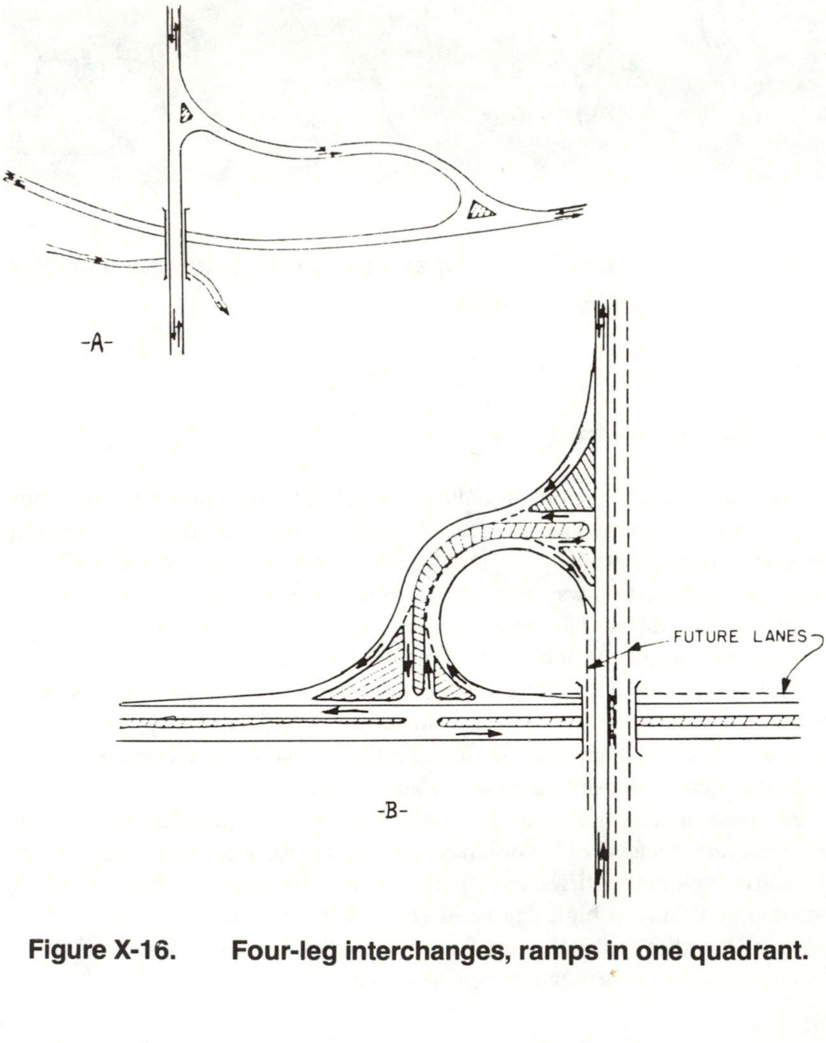

Figure X-16. Four-leg interchanges, ramps in one quadrant.

Figure X-16B is a one-quadrant interchange designed to function as an early phase of stage construction. On future construction it is readily adaptable to become a part of a full or partial cloverleaf interchange without major renovation. The channelization, although elaborate, is conducive to safety and attractive landscaping.

Diamond Interchanges

The simplest and perhaps most common type of interchange is the diamond. A full diamond interchange is formed when a one-way diagonal-type ramp is provided in each quadrant. The ramps are alined with free-flow terminals on the major highway, and the left turns at grade are confined to the crossroad. The diamond interchange has several advantages over a comparable partial cloverleaf; all traffic can enter and leave the major road at relatively high speeds, left-turning maneuvers entail little extra travel, and a relatively narrow band of right-of-way is required, sometimes no more than that for the highways themselves.

Diamond interchanges have application in both rural and urban areas. They are particularly adaptable to major-minor crossings where left turns at grade on the minor road are fitting and can be handled without particular hazard or difficulty. The intersection on the crossroad formed by the terminals functions as any other T intersections at grade and should be designed as outlined in Chapter IX. However, because these intersections have four legs, two of which are one-way, they present a problem in traffic control to prevent wrong-way entry from the crossroad. For this reason a median should be provided on the crossroad to facilitate proper channelization. This median should at least be painted, but a depressed median or a raised median with a mountable curb is preferred. In most cases additional signing to help prevent improper use of the ramps will have to be incorporated in the design of the interchange. This problem is further discussed in Chapter IX in the section "Design to Discourage Wrong-Way Entry." Diamond interchanges usually require signalization where cross streets are carrying moderate-to-large traffic volumes. The capacity of the ramps and that of the cross street may be determined by the signal-controlled ramp terminals. In such a case pavement widening may be required on the ramps or on the cross street through the interchange area, or both. A single-lane ramp usually serves traffic from the freeway, but it may have to be widened to two or three lanes or channelized for storage at the cross street, or both, in order to develop the necessary capacity for the at-grade condition. This design would preclude stored vehicles from extending too far along the ramps or onto the freeway. Left-turning movements in the most common plans, Figure X-17, usually necessitate multiphase control.

Diamond interchanges can assume a variety of patterns as illustrated in Figures X-17 and X-18. Figure X-17 shows those most commonly used. They may be designed with or without frontage roads. Designs with frontage roads are likely in built-up areas, often as part of a series of such interchanges along a freeway. Ramps should connect to the frontage road a minimum of 350 ft from the crossroad. Desirable lengths will be several hundred feet longer to provide adequate weaving length, space for vehicle storage, and turn lanes at the crossroad.

Figure X-17C is a spread diamond rural interchange with the potential for conversion to a cloverleaf.

In a diamond interchange the greatest impediment to smooth operations is left-turning traffic at the crossroad terminal. Arrangements that may be suitable to reducing traffic conflicts are shown in Figures X-18 and X-19.

By using a split diamond (i.e., each pair of ramps connected to a separate crossroad about a block apart), as shown in Figure X-18A, conflicts are minimized by handling the same traffic at four rather than two crossroad intersections, reducing the left-turn movements at each intersection from two to one. An undesirable feature is that traffic leaving the freeway cannot return at the same interchange and continue in the same direction. Frontage roads (shown dotted) are optional.

Figure X-18B shows a split diamond in conjunction with a pair of one-way cross streets and one-way frontage roads. Simplicity of layout and operation of the crossroad and the at-grade terminals result. Traffic leaving the freeway for services or other reasons is afforded ready access to return and continue the journey in the same direction.

Figure X-18C shows a diamond interchange with frontage roads and separate turnaround provisions. These are highly desirable if the cross street has heavy traffic volumes and there is considerable demand for the U-turning movement. The turnaround roadways are adjacent to the cross street with additional width provided beneath the structure or, if the cross street overpasses the freeway, on top. As an alternative, separate structures may be provided for the U-turn movements.

Figure X-19 shows diamond interchanges with more than one structure. The plan in Figure X-19A and the "criss-cross" arrangement in Figure X-19B are sometimes dictated by topographic conditions or right-of-way restrictions. The results operationally are the same as those shown in Figure X-18A. The layout of Figure X-19B also may be used to eliminate weaving between two closely spaced interchanges. These layouts may be further modified by the use of one-way operation on the cross streets, this operation having the same

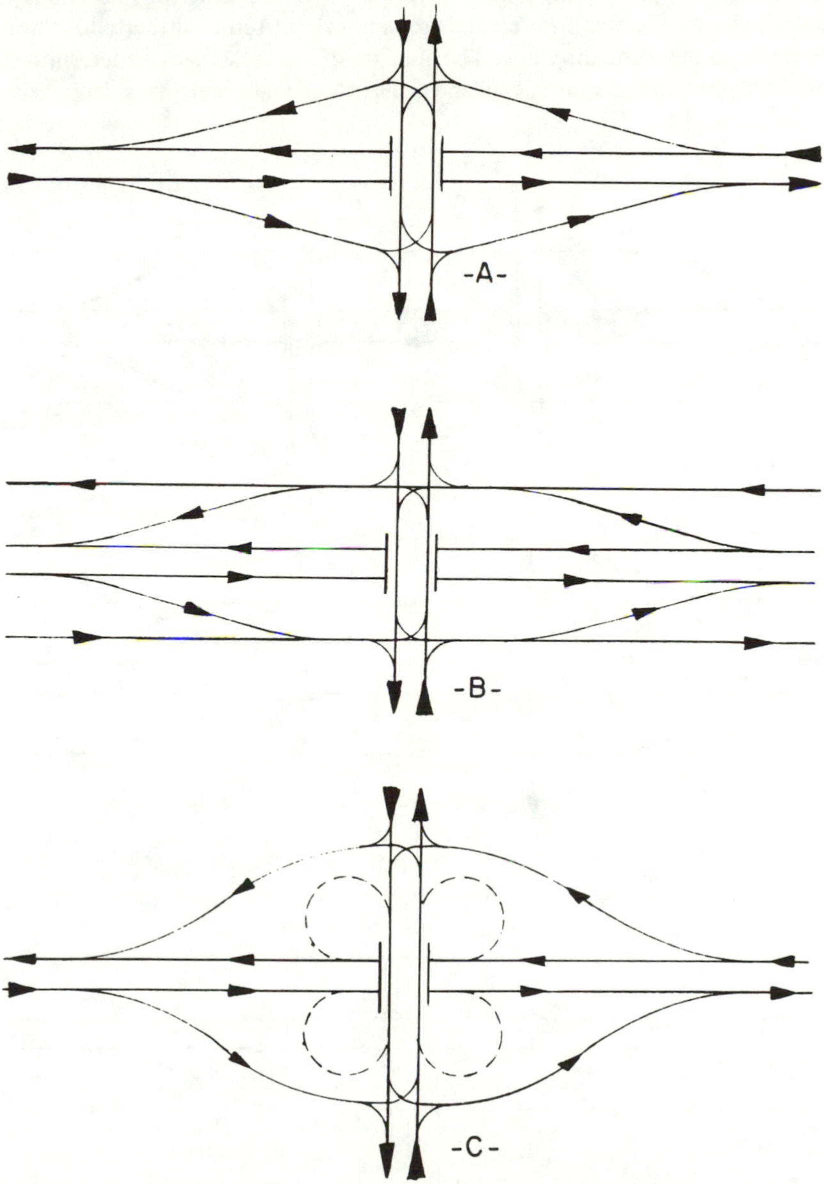

Figure X-17. Diamond interchanges, conventional arrangements.

advantages as shown in the plan of Figure X-19B. The deficiency of both layouts is that traffic that has left the freeway cannot return directly to it and continue in the same direction. The spacing of the crossroads is determined primarily by grade requirements and acceleration and deceleration lengths.

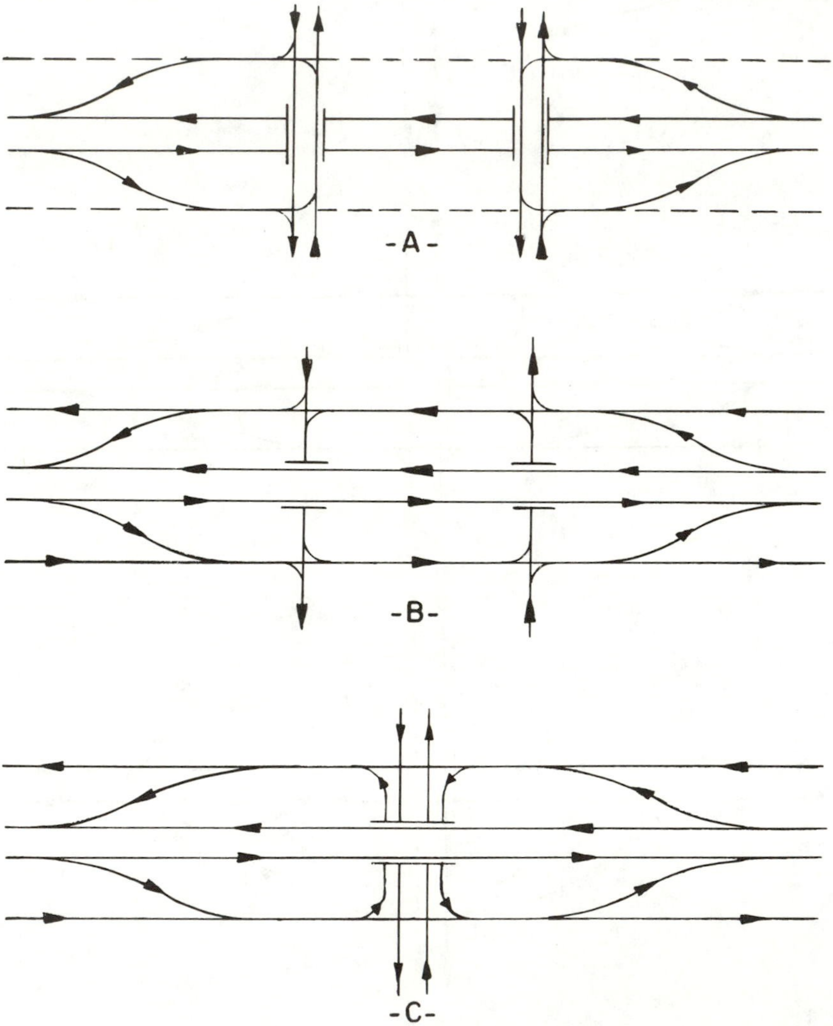

Figure X-18. Diamond interchange arrangements to reduce traffic conflicts.

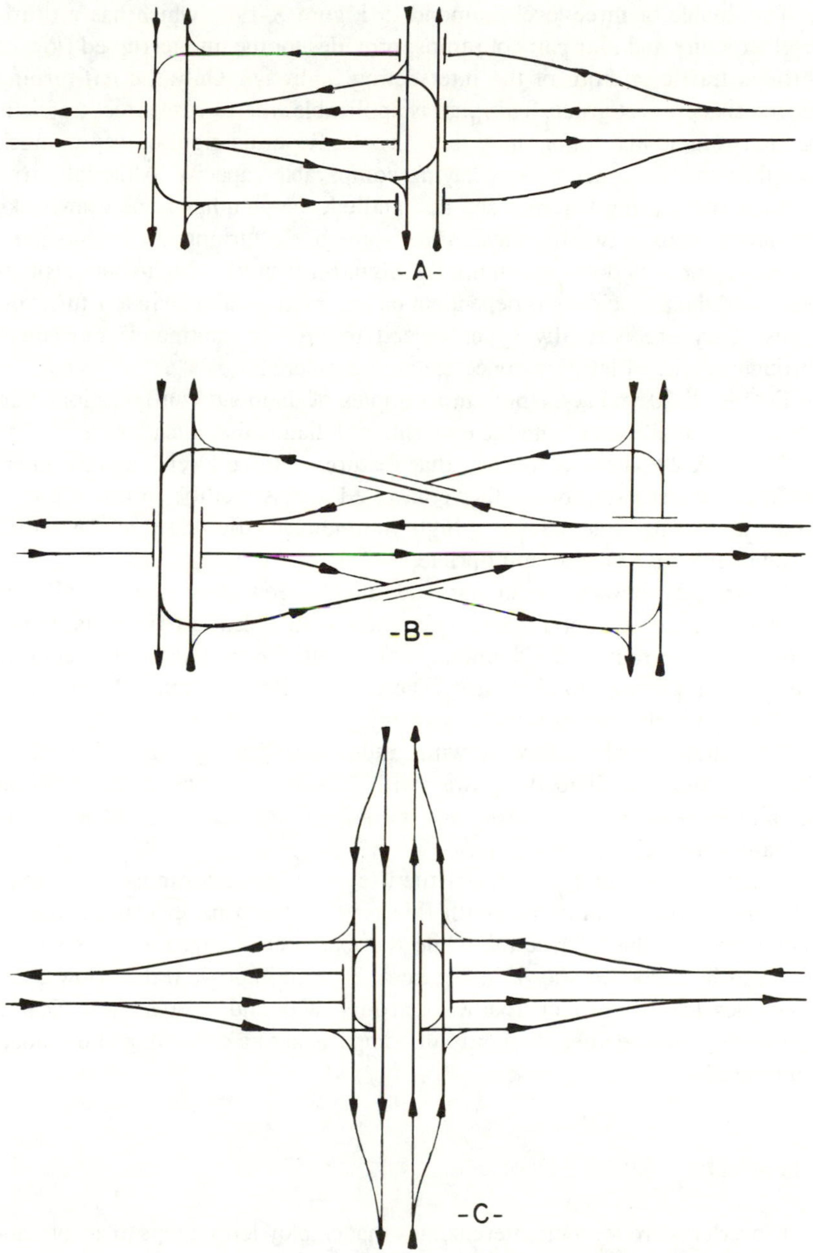

Figure X-19. Diamond interchanges with additional structures.

The double or three-level diamond in Figure X-19C, which has a third-level structure and four pairs of ramps, provides for the uninterrupted flow of through traffic on both of the intersecting highways. Only the left-turning movements cross at grade. This plan is applicable where cross-street traffic is of high volume and topography is favorable. Required right-of-way is much less than that for other layouts having comparable capacity. Although large through and turning volumes can be handled, the plan has some drawbacks for intersections of two freeways in that some of the turning movements must either stop or slow down substantially. Signals are used in high-volume situations, and their efficiency is dependent on the relative balance in left-turn volumes. They are normally synchronized to provide continuous movement through a series of left turns once the area is entered.

Figures X-20 and X-21 present examples of diamond configurations that are somewhat different from the conventional diamond application.

Figure X-20 shows a freeway that features a three-level diamond interchange at the crossing of another controlled access facility. In urban areas, where a crossing street carries a high volume of traffic, the three-level diamond interchange may be appropriate.

Figure X-21 shows a spread diamond interchange with provision in all four quadrants for future loop ramps. When there is sufficient traffic to justify the additional loop ramps, the diamond ramps will also be modified to convert the interchange to a full cloverleaf. The interchange is currently located in a rural area, but the area is rapidly developing.

The ramps are all 14 ft wide with grades less than 1 percent. The minor road is a four-lane divided highway with left-turning lanes at the diamond ramp terminals. Traffic control consists only of stop signs for traffic on the off-ramps from the freeway.

It may be beneficial to consider the use of 'X' pattern ramps at diamond interchanges in urban areas. With this type of ramp pattern, the entrance occurs prior to the intersection while the exit occurs after the cross street. This configuration, as shown in Figure X-22, can improve traffic flow characteristics for the through roadways around diamond interchanges. Driver expectancy may be altered slightly in comparison to a conventional diamond configuration.

Cloverleafs

Cloverleafs are four-leg interchanges that employ loop ramps to accommodate left-turning movements. Interchanges with loops in all quadrants are referred to as "full cloverleafs" and all others as "partial cloverleafs." A full

Figure X-20. Freeway with a three-level diamond interchange.

cloverleaf may not be warranted at major-minor crossings where, with the provision of only two loops, freedom of movement for traffic on the major road can be maintained by confining the direct at-grade left turns to the minor road. The principal disadvantages of the cloverleaf are the extra travel distance required for left-turning traffic, the weaving maneuver generated, the very short weaving length typically available, and the relatively large right-of-way areas required for it. When collector-distributor roads are not used, further disadvantages are weaving on the main line, the double exit on the main line, and problems associated with signing for the second exit.

Figure X-21. Existing four-leg interchange with diamond stage construction.

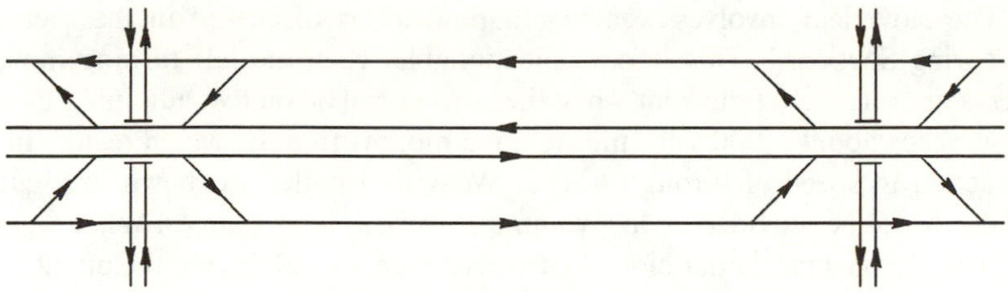

Figure X-22. X-pattern ramp arrangement.

Because cloverleafs are so expansive, considerably more so than diamond interchanges, they are less common in urban areas, their being better adapted to suburban or rural areas where space is available and there is a need to avoid restrictive at-grade left turns.

The travel distance on a loop, as compared with that of a direct left turn at grade, increases rapidly with an increase in design speed. On a loop designed for 20 mph (90-ft radius) the extra travel distance is approximately 600 ft; for a design speed of 25 mph (150-ft radius) it is about 1,000 ft; and for a design speed of 30 mph (230-ft radius) it is around 1,500 ft. Thus, for a 5-mph increase in design speed, extra travel distance increases over 50 percent; required area of right-of-way increases about 130 percent. Travel time on loops varies almost directly with the design speed, the increased speed being more than balanced by increased distance. For an increase of 5 mph in loop design speed, travel time increases 20 to 30 percent or approximately 7 seconds. This travel time increase actually is somewhat less when the overall maneuver is considered because of deceleration and acceleration outside the limits of the loop proper. (In any case the travel time via a loop may be much less than it is when turning left directly if there is a delay at the crossing at grade.) The advantages of increased speed must be weighed against the disadvantages of extra travel time and distance and increased right-of-way. It should also be noted that large trucks may not be able to operate efficiently on smaller radii curves. Considering all factors, experience shows that the practical size of loops resolves into approximate radii of 100 to 150 ft for minor movements on highways with design speeds of 50 mph or less and 150 to 250 ft for more important movements on highways with higher design speeds. A continuous additional lane is required to provide for deceleration, acceleration, and weaving between the on-and off-loop ramps. Additional structure width or length is usually required for this lane. The lateral placement of bridge abutments or columns, or main-line bridge widths may have to be reviewed due to the effects of offtracking at the exit end of the ramp curve.

The cloverleaf involves weaving maneuvers as discussed in the section "Weaving Sections." This is not objectionable when the left turning movements are relatively light, but when the sum of traffic on two adjoining loops approaches about 1,000 vph, interference mounts rapidly, which results in a reduction in speed of through traffic. Weaving lengths, as shown in Figure X-68, should be provided on low-volume cloverleaf interchanges. When weaving volume on a particular cloverleaf weave exceeds 1,000 vph, the quality of service on the main facility deteriorates rapidly, thus generating a need to transfer the weaving section from the through lanes to a collector-distributor road. A loop rarely operates with other than a single line of vehicles, regardless of pavement width, and it thus has a design capacity limit of 800 to 1,200 vph, the higher figure being applicable only where there are no trucks and where the design speed for the ramp is 30 mph or higher. Loop ramp capacity is, therefore, a major control in cloverleaf designs. Loops may be made to operate with two lanes abreast, but only by careful attention to design of the terminals and to design for weaving requirements, which would necessitate widening by at least two additional lanes through the separation structure. To accomplish this type of design the terminals must be separated by such great distances and the loop radii must be made so large that cloverleafs with two-lane loops are not economical from the standpoint of right-of-way, construction cost, and indirection of travel. Loops to operate for two lanes, therefore, are considered exceptional cases.

Where left turns on both the main facility and the crossroad are prohibited and all turning movements must be accommodated, a four-quadrant cloverleaf interchange is the minimum type interchange that will suffice. When a full cloverleaf interchange is used in conjunction with a freeway and the sum of the traffic on two adjoining cloverleaf loops approaches about 1000 vph, collector-distributor roads should be considered. However, on high-type facilities and freeways, full cloverleaf interchanges without collector-distributor roads are acceptable where weaving is less than 1000 vph. Collector-distributor roads are generally not cost effective where the ramp volumes are low and not expected to increase significantly. The use of acceleration or deceleration lanes with cloverleaf interchanges is one possible alternative to collector-distributor roads.

Various arrangements, shapes, and features of cloverleaf interchanges are demonstrated by actual examples that follow.

Figure X-23 shows an existing cloverleaf interchange between a freeway and an expressway with partial control of access, located on the edge of a rapidly expanding suburban area. Collector-distributor roads have been provided along the freeway in expectation of heavy weaving volumes. Because of the high unit right-of-way cost, this design with the collector-distributor roads and loop ramps with smaller radii was more economical than a design with

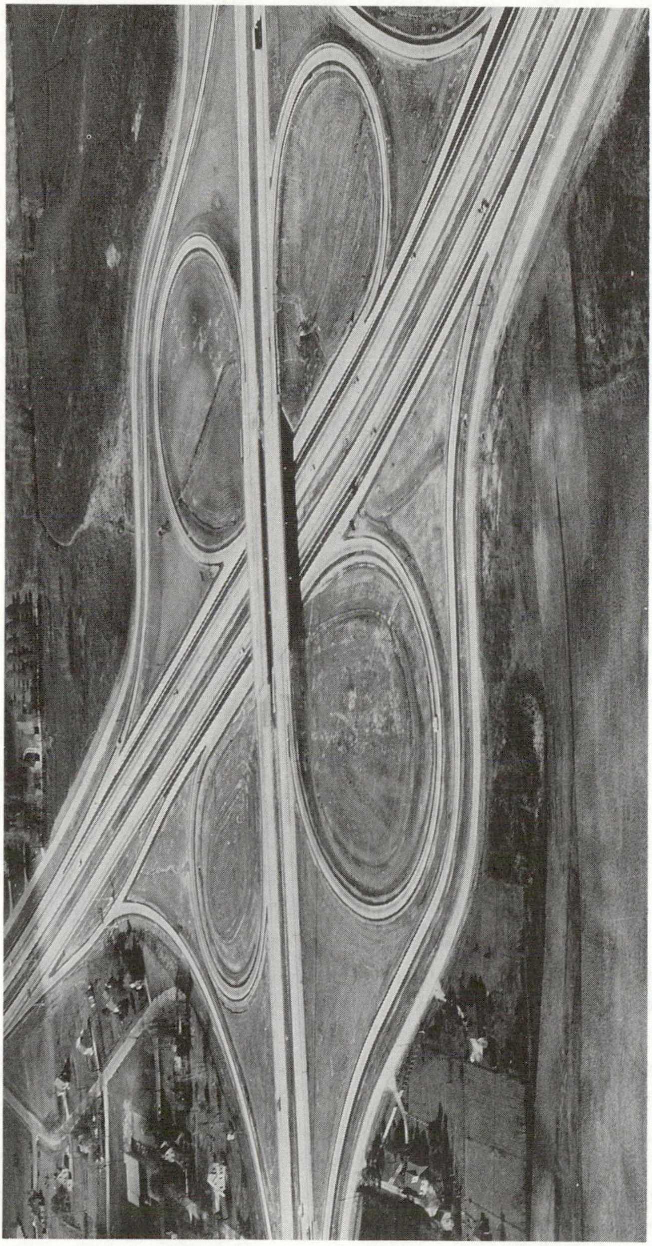

Figure X-23. **Four-leg interchange, full cloverleaf with collector-distributor roads.**

ramps of larger radii that would have been used if the collector-distributor roads had not been provided. The grades are relatively flat, 3 percent being the maximum.

Figure X-24 shows a cloverleaf interchange between a freeway and a divided arterial street. Collector-distributor roads serve all the ramp movements on the freeway.

Partial Cloverleaf Ramp Arrangements

In the design of partial cloverleafs the site conditions may offer a choice of quadrants to use. There is considerable operational advantage in certain arrangements of ramps. These are discussed and summarized in the following analysis. At a particular interchange site, topography and culture may be the factors that determine the quadrants in which the ramps and loops can be developed.

Ramps should be arranged so that the entrance and exit turns create the least impediment to the traffic flow on the major highway. The following guidelines should be considered in the arrangement of the ramps at partial cloverleafs:

1. The ramp arrangement should enable major turning movements to be made by right-turn exits and entrances.
2. Where through-traffic volume on a major highway is decidedly greater than that on the intersecting minor road, preference should be given to an arrangement placing the right turns, either exit or entrance, on the major highway even though it results in a direct left turn off the crossroad.

These controls do not always lead to the most direct turns. Instead, drivers frequently are required to first turn away from or drive beyond the road of their intended direction. Such arrangements cannot be avoided if the through-traffic movements, for which the separation is provided, are to be facilitated to the extent possible.

Figure X-25 diagrammatically illustrates the manner in which the necessary turning movements are made for various two- and three-quadrant cloverleaf arrangements.

When ramps in two quadrants are adjacent and on the same side of the minor road, as shown in Figure 25-A, or diagonally opposite each other, as shown in Figures X-25E and X-25F, all turning movements to and from the major road are accomplished by right turns. Any choice between the arrangement in Figure X-25A and its alternate arrangement (ramps in the other two quadrants) will depend on the predominant turning movements or the availa-

Figure X-24. Cloverleaf interchange with collector-distributor roads.

bility of right-of-way, or both. When the ramps in two quadrants are adjacent but on the same side of the major road (Figures X-25B and X-25D), four direct left turns fall on the major road. This arrangement and its alternate are the least desirable of the six possible arrangements of the two-quadrant developments, and their use should be avoided.

The arrangement with ramps in diagonally opposite quadrants has an advantage in that the turning movements in both directions in the quadrants of the ramps are made by desirable right-turn exits and entrances. At interchanges where turning movements in one quadrant predominate, the best two-quadrant arrangement has ramps in that quadrant and in the diagonally opposite one. Where turning movements in two adjacent quadrants are of nearly the same importance, arrangements of Figures X-25A, X-25E, and X-25F are applicable in that all turns to and from the major road are on the right, but the arrangement in Figure X-25E is preferable because the ramps are on the near side of the structure as drivers approach on the major road. With this plan it is feasible to provide for high-speed exit turns from the major road, and drivers desiring to turn are not confused by improper near-side ramps on the left, as shown in Figure X-25F, or by hidden right-turn ramps on the far side of the structure.

There are four possibilities in arrangement of ramps in three quadrants, the arrangement in Figure X-25G and the alternate arrangements in which ramps in two quadrants are on the upper side of the major road. In an arrangement with ramps in three quadrants, six of the eight turning movements can be made by right-turn exits and entrances, and the other two are made by right turns on the major road and corresponding left turns on the minor road. The determination of which quadrant is to be without ramps is usually dependent on the availability of right-of-way and the predominant turning movements to be handled.

In some cases it is desirable to provide diagonal ramps in all four quadrants, but with loops in one, two, or three of the quadrants. Figure X-25F shows a design with loops in diagonally opposite quadrants. This design has the advantage of providing all right exits. Storage of vehicles waiting to make the left turn at the at-grade intersections occurs on the ramp and not on either of the through highways. There is no weaving required on the major highway.

Examples of Partial Cloverleaf Designs

Figure X-26 shows an existing partial cloverleaf design with ramps in diagonally opposite quadrants in a suburban area, arranged to require minimum frontage on the crossroad, a two-lane highway that overpasses a six-lane freeway. The ramps are 22 ft wide.

Directional islands and merge lanes at the ramp terminals permit free-moving right turns to and from the secondary road. The only traffic control required is stop signs at the secondary road for the left turns from the off-ramps. Protected left-turn bays on the secondary road are desirable.

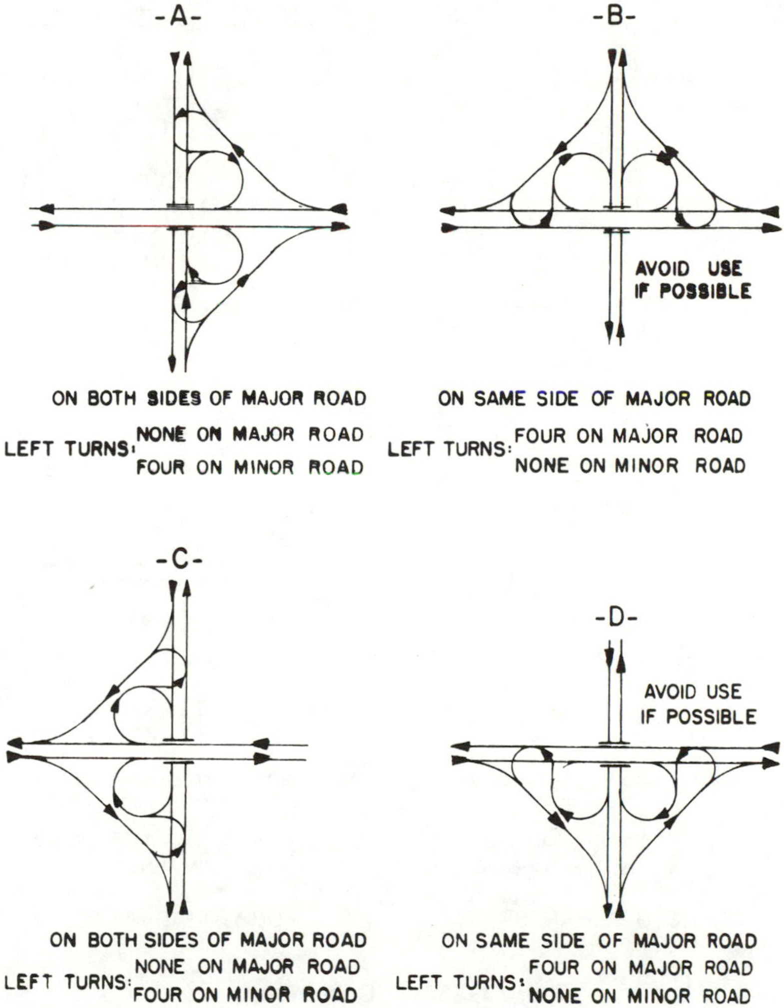

Figure X-25. Schematic of partial cloverleaf ramp arrangements, exit and entrance turns.

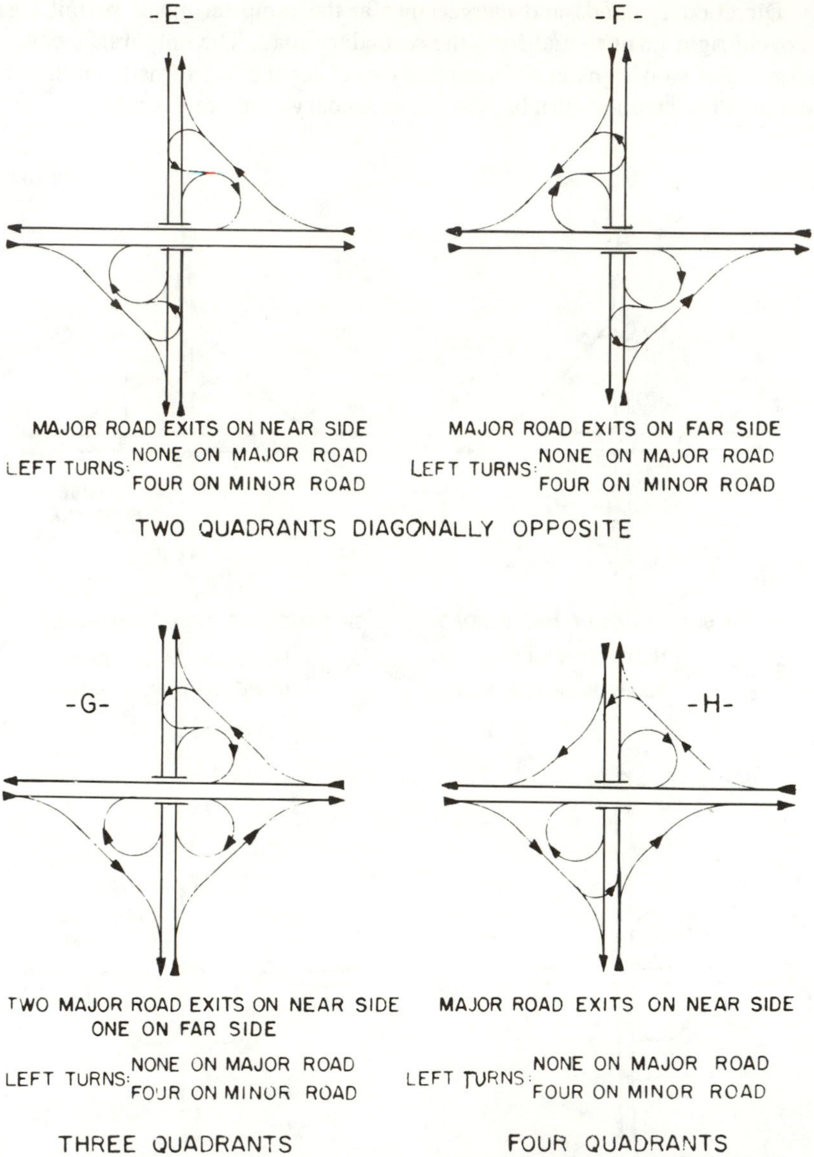

Figure X-25. Continued.

Figure X-26. Four-leg interchange (two-quadrant cloverleaf).

Minimum radii of 180 ft are provided for the two loops, and existing vegetation has been retained within the loops.

Figure X-27 shows an existing partial cloverleaf with ramps in diagonally opposite quadrants. In relation to the major highway the ramps are in opposite quadrants from those of the two previous examples. A major highway crosses over a four-lane freeway. Ramps are located to avoid the two quadrants with heavy commercial and residential development. Direct left turns are confined to the minor road where the terminals are channelized by divisional islands that permit control of turning traffic. The intersection of the ramps with the minor highway in the lower right quadrant is signalized. The outer connections are designed to encourage high-speed merging with the freeway traffic. The loops have slightly larger radii than the previous example and are designed for a speed of 30 mph.

Figure X-28 shows an existing partial cloverleaf interchange in a rural area with diagonal ramps in all four quadrants. Loop ramps are provided in two diagonally opposite quadrants. This arrangement has the advantage that left turns are not required from the crossroad to the ramps.

Any other arrangement of two loop ramps and four diagonal ramps does require left turns to be made from the minor road onto a ramp. Note the triangular island channelization at the ends of the two ramps from which left turns must be made in order to travel from the freeway onto the two-lane high-volume crossroad. The design provides adequate left-turn storage and a free-flow right-turn movement from the freeway to the crossroad.

The loop ramps are carried on separate roadways under structure spans, so the loop traffic merges with the outer connection traffic before entering the freeway. This design entails a considerable extra cost but reduces the points of conflict on the freeway.

Directional and Semidirectional Interchanges

Direct or semidirect connections are used for important turning movements to reduce travel distance, increase speed and capacity, eliminate weaving, and avoid the indirection in driving on a loop. Higher levels of service can be realized on direct connections, and in some instances on semidirect ramps, than on loops because of relatively high speed and the likelihood of more adequate terminal design. Often a direct connection is designed with two lanes. In such cases the ramp capacity may approach the capacity of an equivalent number of lanes on the through highway.

In rural areas there rarely is a volume justification for provision of direct

Figure X-27. Four-leg interchange (two-quadrant cloverleaf).

connections in more than one or two quadrants. The remaining left-turning movements usually are handled satisfactorily by loops or at-grade intersections. At least two structures are required for such an interchange. There are many possible arrangements with direct and semidirect connections, but only the more basic forms are discussed herein.

A direct connection is defined as a one-way roadway that does not deviate greatly from the intended direction of travel. Interchanges that use direct connections for the major left-turn movements are termed directional interchanges. Direct connections for one or all left-turn movements would qualify an interchange to be termed directional even if the minor left-turn movements are accommodated on loops.

When one or more interchange connections are indirect in alinement yet more direct than loops, the interchange is described as semidirectional. All left-turn connections or only those that accommodate major left-turn movements may be semidirect in alinement.

Figure X-28. Four-leg interchange (partial cloverleaf).

On direct or semidirect interchanges usually more than one highway grade separation is involved.

Semidirect or direct connections for one or more left-turning movements are often required at major interchanges in urban areas. Interchanges involving two freeways nearly always call for directional layouts. In such cases turning movements in one or two quadrants often are comparable in volume to through movements. Direct or semidirect connections, as compared with loops, have shorter travel distance, higher speeds of operation, and a higher level of service, and they often avoid the necessity for weaving.

There are many schemes for directional interchanges that use various combinations of directional, semidirectional, and loop ramps. Any one of them may be appropriate for a certain set of conditions, but only a limited number of patterns are used generally. These layouts fill the least space, have the fewest or least complex structures, minimize internal weaving, and fit the common terrain and traffic conditions. Basic patterns of selected semidirectional and all directional interchanges are illustrated in Figures X-29 and X-30, with distinctions made as to types with and without weaving.

With Loops and Weaving

Common arrangements where turning movements in one quadrant predominate are shown in Figures X-29A and X-29B. The predominant turning movement bypasses the central portion of the interchange, and does so via two-lane terminals. The minor turning movements pass through weaving sections between loops on each highway. In both figures a semidirect connection is used without affecting the alinement of the intersecting highways. Both schemes involve three structures, and the area occupied is about the same as or somewhat more than a full cloverleaf.

The efficiency and capacity of all the schemes shown in Figure X-29 may be improved by eliminating weaving on the main roadways through the use of a collector-distributor road, as shown dashed in Figure X-29.

With Loops and No Weaving

Semidirectional interchanges with loops but that entail no weaving are shown in Figure X-30. Spread of the through lanes is not required for any of the schemes; however, four or more structures are required. Single exits on the right along with right-hand entrances enhance the operational characteristics of these designs.

Fully Directional

These interchanges are generally required at the intersections of high-volume freeways. They are costly in construction but fill comparatively little area and have high capacity for both through and turning traffic. They require detailed, time-consuming study, and each interchange generally costs several

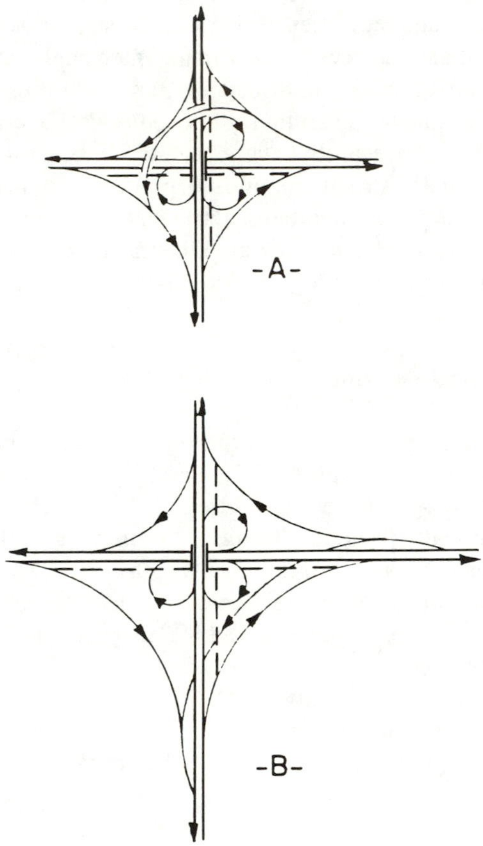

NOTE: WEAVING ADJACENT TO THE THROUGH LANES IS ELIMINATED BY PROVIDING COLLECTOR-DISTRIBUTOR ROADS AS SHOWN BY DOTTED LINES.

Figure X-29. Semidirect interchanges with weaving.

million dollars, justifying study of all likely alternatives; even minor adjustments often prove to be important. A detailed discussion is, therefore, not within the purview of this policy, but Figures X-31A through X-31C are included to show diagrammatic layouts.

Weaving, left side exits, and left side entrances are undesirable within directional interchanges; however, there may be instances where they cannot

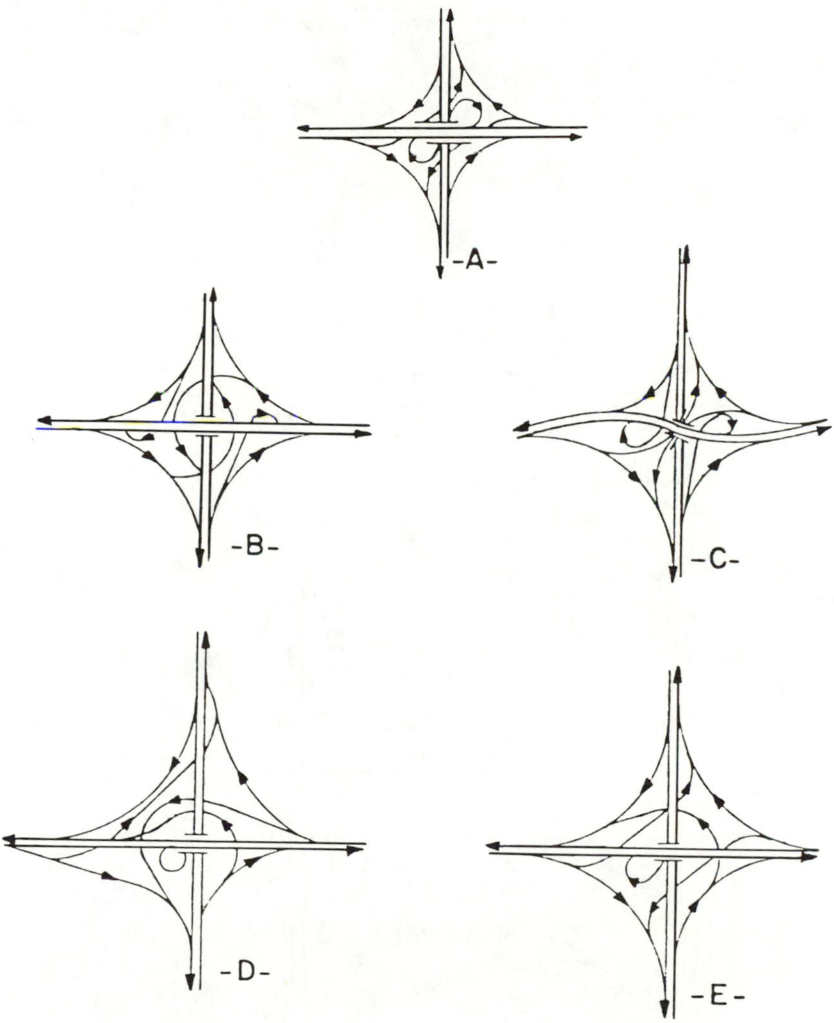

Figure X-30. Semidirect interchanges with no weaving.

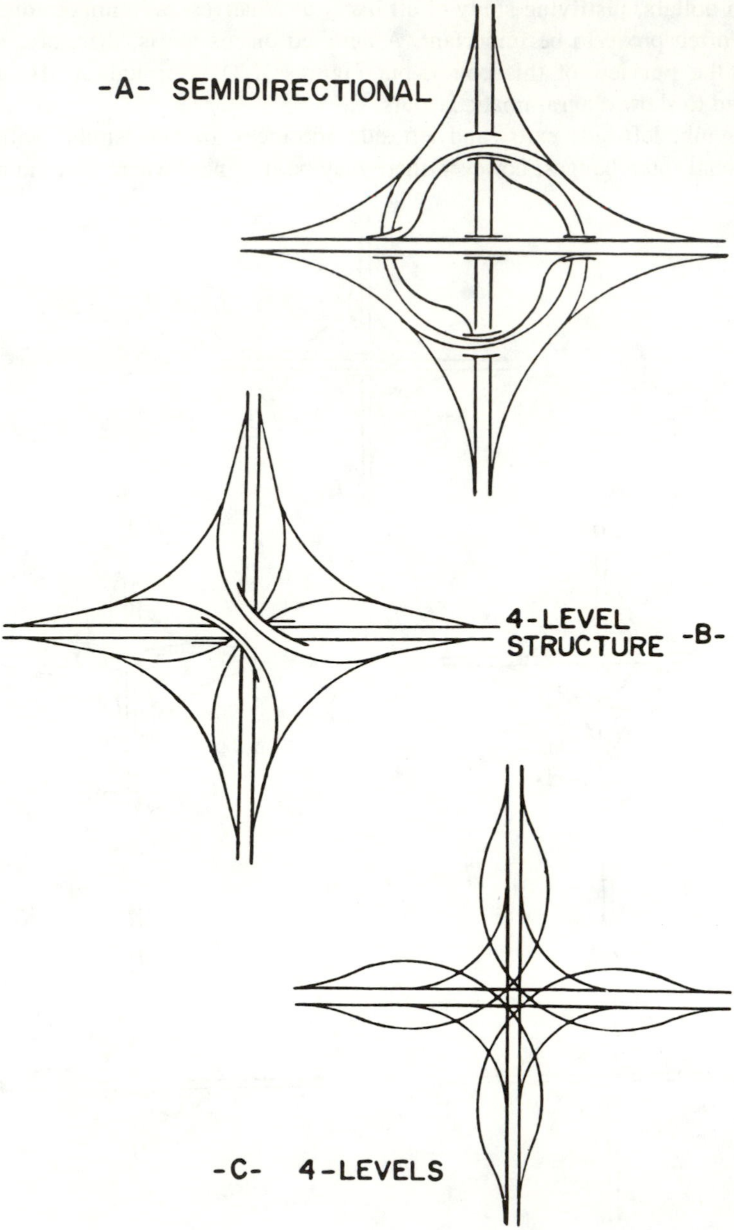

Figure X-31. Semidirectional and directional interchanges—multilevel structures.

be reasonably avoided because of site restrictions or other considerations. With heavy left-turn movements the terminals should be designed as major forks and branch connections, as covered later in this chapter.

The most widely used type of directional interchange is the four-level layout system shown in Figure X-31B. A variation of this type is the four-level interchange with two exits from both major roadways, as shown in Figure X-31C.

Figure X-32 shows an existing interchange in a suburban area at the crossing of two high-volume freeways.

Other examples of semidirectional or directional interchanges are shown in Figures X-33 through X-37.

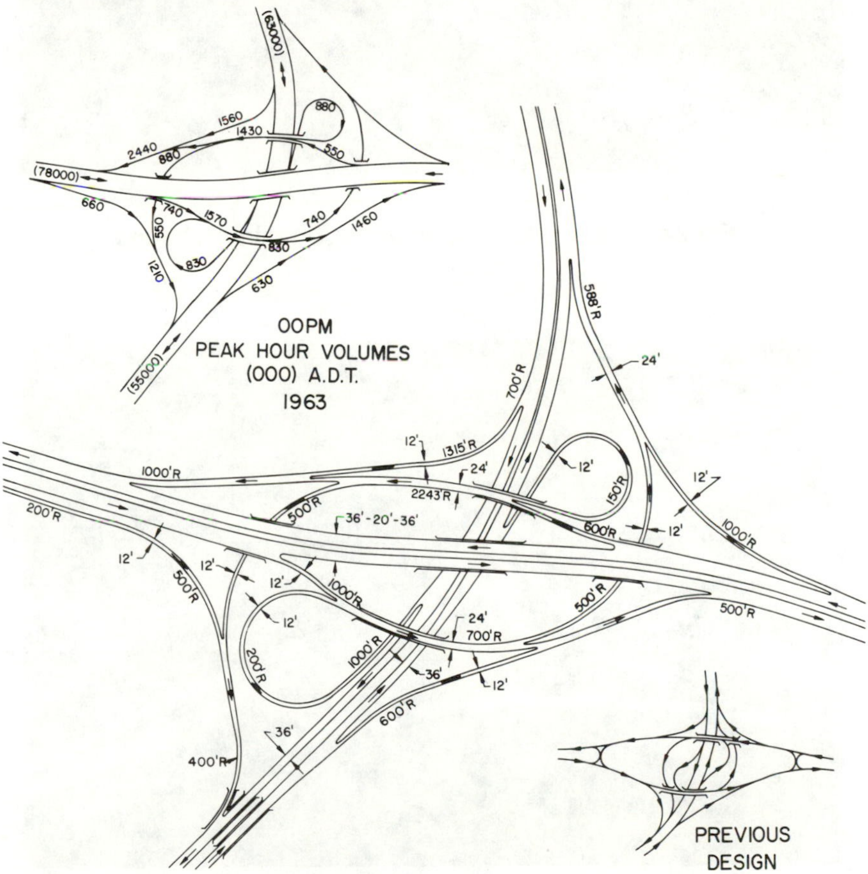

Figure X-32. Directional interchange, two semidirect connections.

Figure X-33. Four-level directional interchange.

Figure X-34. Four-level directional interchange.

Figure X-35. Semidirectional interchange with loops.

Figure X-36. Semidirectional interchange with a single loop.

Figure X-37. Four-level semidirectional interchange with two exits and two entrances on each leg.

Other Interchange Types

Offset Interchanges

Figure X-38 illustrates an offset interchange arrangement between freeways that may have application where there are major buildings or other developments near the crossing of the freeways. It consists of a pair of trumpet interchanges, one on each highway, which are connected to each other with a ramp highway. The length of the connecting roadway depends on the distances between the trumpet interchanges and the crossing of the freeways. As illustrated in Figure X-38, the ramp highway may be developed in a manner to include local service connections, in this case a diamond interchange.

A disadvantage of this type of interchange is the substantial indirection for six of the eight turning movements between the freeways. However, when one pair of these movements is predominant, location of the ramp highway in the quadrant to favor them makes the plan well fitted. When considered from a city street system, the overall configuration may be confusing to nonrepeat drivers. As a part of the continuing flow on one-way roadways with adequate signing, however, most motorists would be unaware of the unusual pattern.

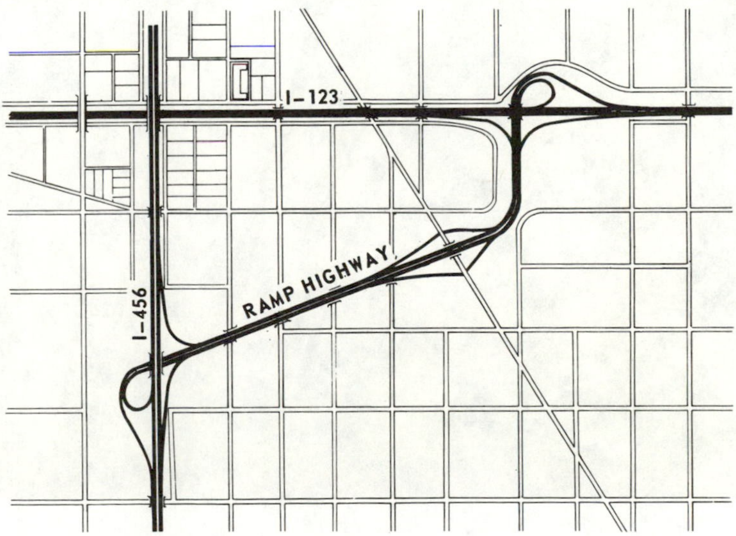

Figure X-38. Offset interchange via ramp highway.

Combination Interchanges

When one or two turning movements have very high volumes with respect to the other turning movements, analysis may indicate the need for a combination of two or more of the previously discussed interchanges.

Figure X-39 shows an existing diamond interchange in which a semidirectional ramp has been added to accommodate the high-volume left-turning traffic. The complementary high-volume right-turning movement is provided with a liberal radius to facilitate high speeds also. Because the cross street connects a city on the left with a four-lane freeway, relatively high volumes

Figure X-39. Four-leg interchange, diamond with a semidirect connection.

result in that direction. This design requires two more structures than a diamond interchange.

Three of the crossroad terminals are channelized with separate right- and left-turning roadways.

Figure X-40 shows an existing cloverleaf interchange between two freeways in which a semidirect connection has been substituted for the loop ramp in the upper left quadrant. The interchange is located at the edge of a suburban area that is rapidly developing both industrially and residentially, and considerably higher volumes are expected in the future. The ramps along the left-to-right roadway have been combined so that the traffic merges with the freeway traffic at a single point at the upper left. The semidirectional turning roadway permits traffic to travel at operating speeds approaching that on the main roadways. The complement of this movement is provided with a high-type two-lane ramp also with more liberal radii than provided for the remaining movements. The design was complicated by an adjacent railroad and two important local roads, one of which was relocated from the center of the interchange area. An additional structure and spans were required for the local road crossings to provide the two high-speed directional ramps.

Figure X-41 shows an existing four-leg interchange, one-half diamond and one-quadrant cloverleaf. The interchange located near the boundary of an urban area is between a freeway designed for 70 mph and a highway designed for 45 mph on which access is not controlled. The design speed of the loop ramp is 25 mph. The land use and topography, which include a railroad, stream, country road, and existing commercial development, influenced the

Figure X-40. Four-leg interchange, cloverleaf with a semidirect connection.

design. Operation is reported as generally satisfactory, with the exception that the accident rate for the intersections on the local road is greater than anticipated. This occurrence may make it necessary to provide traffic signal control on the crossroad.

Unusual ramp configurations are shown in Figures X-42 through X-45.

Figure X-42 shows a complex interchange arrangement at a crossing of two major routes in a downtown district. A braided type design is used in the center of the photograph. The upper part is a good example of a modified three-leg, all-directional interchange. Local connections are also provided. There has been a minimum of disruption to the city street system.

Figure X-43 shows an elevated freeway section with connection to a downtown area. The spurs in the center of the photo help disperse the traffic destined for the downtown area.

Figure X-44 shows a freeway interchange with varied ramp arrangements. The upper quadrants are elongated partial cloverleafs. The elongated arrangement may be used to save certain buildings and to provide a liberal weaving distance between the entrance and exit on the loops. The lower quadrants

Figure X-41. Four-leg interchange, one-half diamond and one-quadrant cloverleaf.

have lateral-type ramps that connect to an arterial running parallel to the freeway.

Figure X-45 shows the unusual arrangement of a three-level cloverleaf. In this case, environmental constraints and other site restrictions necessitated the use of this configuration. Advance traffic studies were carefully prepared to ensure that the loop ramps would continue to function properly as the traffic volumes increased. Signing was also critical to the proper operation of the facility. Another interesting aspect of this interchange is the park-and-ride public transportation facility provided underneath the structures on the lowest

Figure X-42. Complex interchange arrangement.

Figure X-43. Elevated freeway section with connections to downtown area.

Figure X-44. Varied ramp arrangements.

Figure X-45. Freeway with a three-level cloverleaf interchange.

level. While this interchange does not have widespread applicability, it exemplifies the type of innovative design which is sometimes needed to provide a functional facility under restrictive conditions.

General Design Considerations

Interchange-Type Determination

The need to use interchanges may occur in the design of all functionally classified roadways, as discussed previously under "Warrants for Interchanges" in this chapter.

Interchange types are covered in two categories, "systems interchanges" and "service interchanges." The term systems interchanges is used to identify interchanges that connect freeway to freeway. The term service interchange applies to interchanges that connect the freeway to lesser facilities.

In rural areas the problem of interchange-type selection is solved on the basis of service demand. When the intersecting roadways are freeways, all-directional interchanges may be in order for high turning volumes.

A combination of directional, semidirectional, and loop ramps may be appropriate where turning volumes are high for some movements and low for others. When loop ramps are used in combination with direct and semidirect ramp designs, it is desirable that the loops be arranged so that weaving sections will not be created.

A cloverleaf interchange is the minimum design that can be used at the intersection of two fully controlled access facilities or where left turns at grade are prohibited. A cloverleaf interchange is adaptable in a rural environment where right-of-way is not prohibitive and weaving is minimal.

In the decision process to use cloverleaf interchanges, careful attention should be given to the potential improvement in operational quality that would be realized if the design included collector-distributor roads on the major roadway.

Simple diamond interchanges are the most common type of interchange for the intersection of a major roadway with a minor facility. The capacity of a diamond interchange is limited by the capacity of the at-grade terminals of the ramps at the crossroad. High through and turning volumes could preclude the use of a simple diamond unless signalization is used.

Partial cloverleaf designs may be appropriate where rights-of-way are not available in one or two quadrants or where one or two movements in the interchange are disproportionate to the others, especially when they require left turns across traffic. In the latter case, loop ramps may be utilized to accommodate the heavy left-turn volume.

Generally, interchanges in rural areas are widely spaced and can be designed on an individual basis without any appreciable effect from other interchanges within the system. The final configuration of an interchange may be determined by the need for route continuity, uniformity of exit patterns, single exits in advance of the separation structure, elimination of weaving on the main facility, signing potential, and availability of right-of-way. Sight distance on the highways through a grade separation should be at least as long as that required for stopping and preferably longer. Where exits are involved, decision sight distance is preferred, although not always practical.

Interchange-type determination in an urban environment requires considerable analysis of prevailing conditions so that the most practicable interchange configurations can be developed.

At a new location it is desirable that interchanges be planned into the location studies so that the final alinement is compatible, both horizontally and vertically, with interchange sites. Generally, in urban areas, interchanges are so closely spaced that each interchange may be influenced directly by the preceding or following interchange to the extent that additional traffic lanes may be required to satisfy capacity, weaving, and lane balance.

On a continuous urban route all the interchanges should be integrated into a system design rather than considered on an individual basis. Line sketches for the entire urban corridor can be developed and several alternate interchange combinations developed for analysis and comparisons.

During the analysis procedure a thorough study of the crossroad should be made to determine its potential for handling the heavier volume of traffic that an interchange would discharge. The ability of the crossroad to receive and discharge traffic from the main roadway has considerable bearing on the interchange geometry. For example, loop ramps may be needed to eliminate heavy left turns on a conventional diamond interchange.

In the process of developing preliminary line-sketch studies, systems interchanges may be inserted at freeway-to-freeway crossings and varying combinations of service interchanges developed for lesser crossroads. Generally, cloverleaf interchanges with or without collector-distributor roads are not practicable for urban construction because of the excessive right-of-way requirements.

Once several alternates have been prepared for the system design, they can be compared on the following principles: (1) capacity, (2) route continuity, (3) uniformity of exit patterns, (4) single exits in advance of the separation structure, (5) with or without weaving, (6) potential for signing, (7) cost, (8) availability of right-of-way, (9) potential for stage construction, and (10) compatibility with the environment. The most desirable alternatives can be retained for plan development.

In the case of an isolated interchange well removed from the influence of other interchanges, the criteria set forth for rural interchange determination apply.

Figure X-46 depicts interchanges that are adaptable on freeways as related to classifications of intersecting facilities in rural, suburban, and urban environments.

Approaches to the Structure

Alinement, profile, and cross section. Traffic passing through an interchange should be afforded the same degree of utility and safety as that given on the approaching highways. The standards for design speed, alinement, profile, and cross section in the intersection area, therefore, should be as high as those on the approaching highways, even though they may be difficult to attain. The presence of the structure itself is somewhat of a hazard, which should not be augmented by low geometric standards that tend to encourage unsafe driver behavior. Preferably, the geometric standards at the highway grade separation should be higher than those for the highway to counterbalance any possible sense of restriction caused by abutments, piers, curbs, and rails. Desirably, the alinement and profile of the through highways at an interchange should be relatively flat with high visibility. Sometimes it will be possible to design only one of the intersecting roads on a tangent with flat grades. Preferably, the major highway should be so treated.

The general controls for horizontal and vertical alinement and their combination, as stated in Chapter III, should be adhered to closely. In particular, any relatively sharp horizontal or vertical curves should be avoided. Horizontal curvature that begins at or near a pronounced crest or sag should be avoided. The gradients on intersecting roads at an interchange should be kept to a minimum and in no case should they exceed the maximums established for open-highway conditions. Gradients that may slow down commercial vehicles or that may be difficult to negotiate when icy should be avoided. Reduction of vehicle speeds by long upgrades encourage passing, which is hazardous in the vicinity of ramp terminals. Slow-moving through vehicles also encourage abrupt cutting in by vehicles leaving and entering the highways.

The alinement and cross section of the approaches to a grade separation without ramps involve no special problems except where a change in width is made to include a middle pier or where the median is narrowed for structure economy. With ramps, changes in alinement and cross section may be

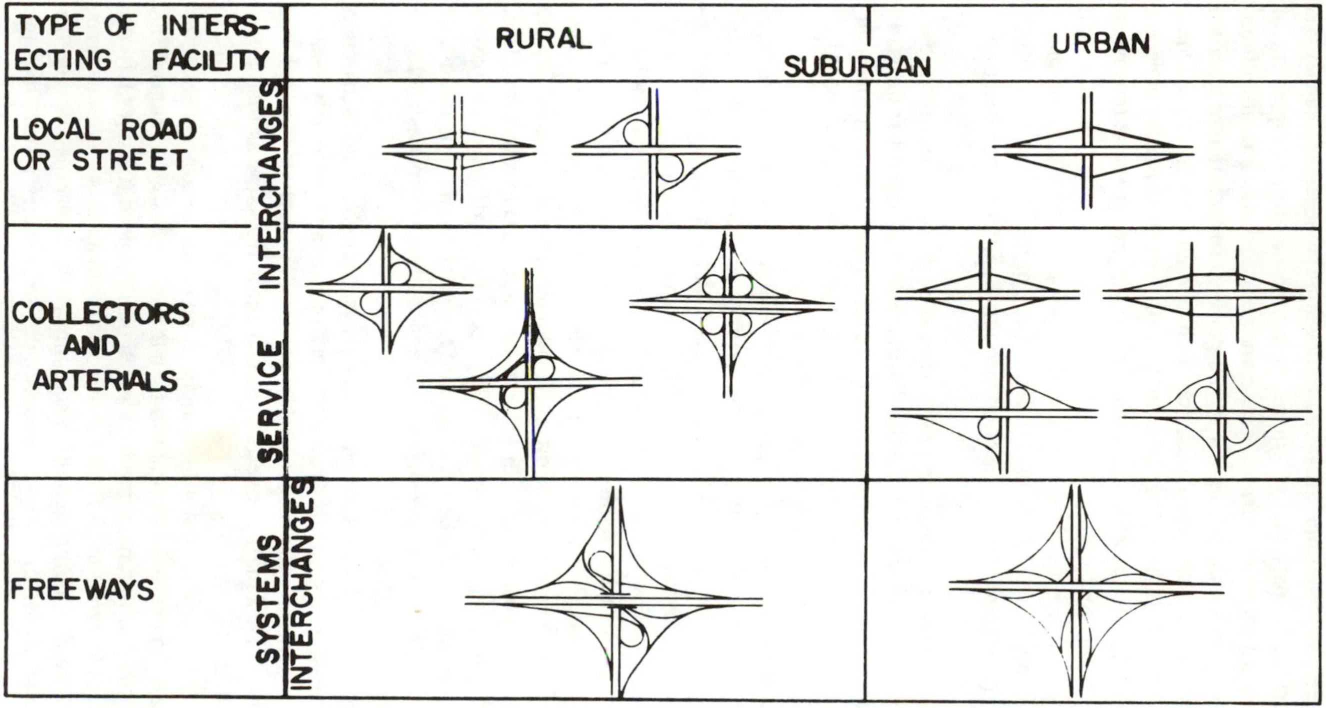

Figure X-46. Adaptability of interchanges on freeways as related to types of intersecting facilities.

required to ensure proper operation and to develop the necessary capacity at the ramp terminals, particularly where there is not a full complement of ramps and where some left turns at grade are necessary. On a divided highway, direct left turns may call for a widening of the cross section to ensure a suitably wide median for a combined speed-change and storage lane. On an undivided multilane highway the introduction of a median usually is necessary to ensure that the direct left turn is made to the proper ramp, thus minimizing hazard and confusion. Where a two-lane highway is carried through an interchange, wrong-way left turns are likely to occur, even with the provision of a full complement of ramps. For high-speed or high-volume conditions this factor may warrant a divided section through the interchange area to prevent such turns.

At an interchange a four-lane highway should be divided. Four-lane roads may carry enough traffic to justify the elimination of at-grade left turns. To ensure that drivers desiring to turn left will use the proper ramps, a nontraversable median should be provided. At-grade left turns preferably should be accommodated within a suitably wide median.

Widening or narrowing to gain the desired width for a divisional island in an interchange area is done in the same manner as that done at any other intersection. Many alinement possibilities occur, and some of the more typical situations are illustrated in Figure X-47. Figure X-47A shows the customary symmetrical development of a divisional island on a four-lane undivided highway. Traffic in each direction is required to traverse two reverse curves. Figure X-47B shows a divisional island developed on a four-lane undivided highway in which the centerline is offset through the interchange area. Traffic in each direction is enabled to enter without curvature and to effect the offset with but one reverse curve, which desirably occurs after drivers have passed beyond possible hazards at the structure and ramp terminals. It is not usable on existing four-lane highways unless the approaches are reconstructed to obtain the centerline offset.

Sight distance. Sight distance on the highways through a grade separation should be at least as long as that required for stopping and preferably longer. Where exits are involved, decision sight distance is preferred, although not always practical. Design of the vertical alinement is the same as that at any other point on the highway.

The horizontal sight distance limitations of piers and abutments at curves usually present a more difficult problem than that of vertical limitations. With curvature of the maximum degree for a given design speed (Chapter III), the normal lateral clearance at piers and abutments of underpasses does not provide the minimum stopping sight distance. Similarly, on overpasses with the sharpest curvature for the design speed, sight deficiencies result from the

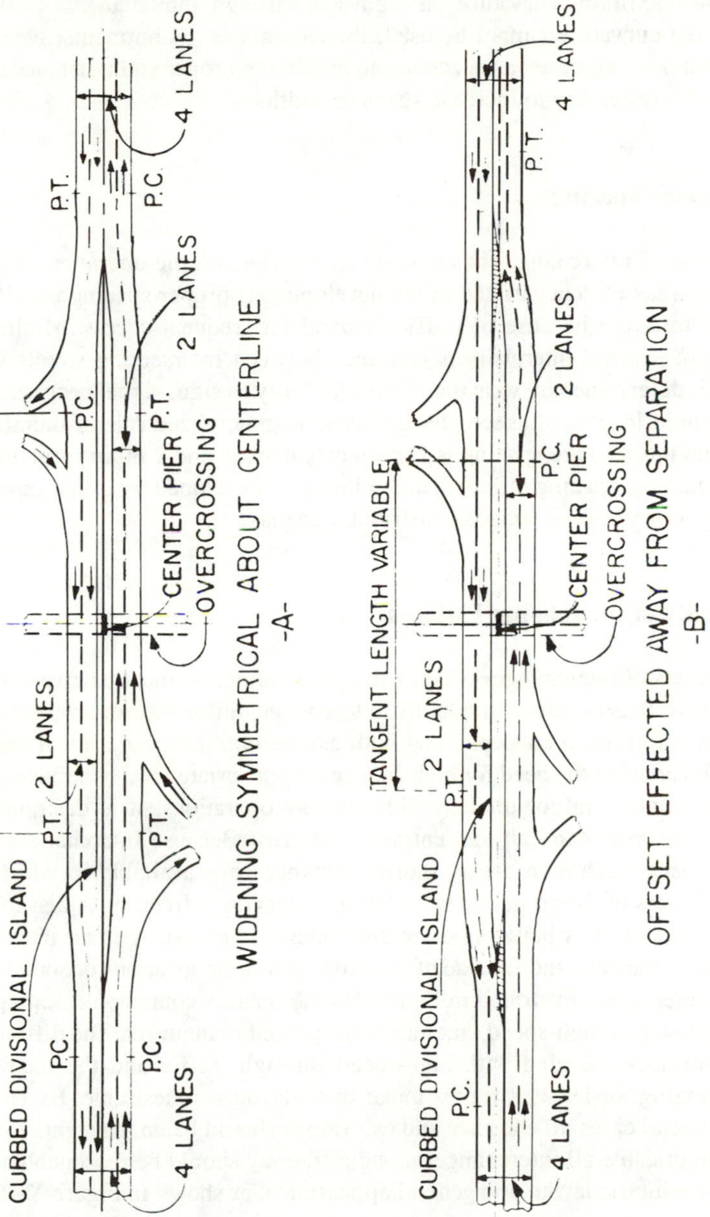

Figure X-47. Widening for divisional island at interchanges.

usual practical offset to bridge rails. This factor emphasizes the need for use of below-maximum curvature on highways through interchanges. If sufficiently flat curvature cannot be used, the clearances to abutments, piers, or rails should be increased as necessary to obtain the proper sight distance even though it is necessary to increase spans or widths.

Interchange Spacing

Spacing of interchanges has a pronounced effect on the operation of freeways. In areas of concentrated urban development, proper spacing usually is difficult to attain because of traffic demand for frequent access. Minimum spacing of arterial interchanges (distance between intersecting streets with ramps) is determined by weaving volumes, ability to sign, signal progression, and required lengths of speed-change lanes. A generalized rule of thumb for minimum interchange spacing is 1 mi in urban areas and 2 mi in rural areas. In urban areas, spacing of less than 1 mi may be developed by grade-separated ramps or by adding collector-distributor roads.

Uniformity of Interchange Patterns

If a series of interchanges is being designed, attention must be given to the group, as well as to each individually. Interchange uniformity and route continuity are interrelated concepts, and both can be obtained under ideal conditions. Considering the need for high capacity, appropriate level of service, and maximum safety in conjunction with freeway operations, it is desirable to provide uniformity in exit and entrance patterns. Because interchanges are closely spaced in urban areas, shorter distances are available in which to inform drivers of the course to be followed in leaving a freeway. A dissimilar arrangement of exits between successive interchanges, such as an irregular sequence of the near and far side of structure exit ramp locations or some left-off movements, as illustrated in Figure X-48A, causes confusion resulting in slowing down on high-speed lanes and unexpected maneuvers. The difficulty of left-entrance merging with high-speed through traffic and the requisite lane changing for left-exit ramps make these layouts undesirable. Except in highly special cases all entrance and exit ramps should be on the right. To the extent practicable all interchanges along a freeway should be reasonably uniform in geometric layout and general appearance, as shown in Figure X-48B.

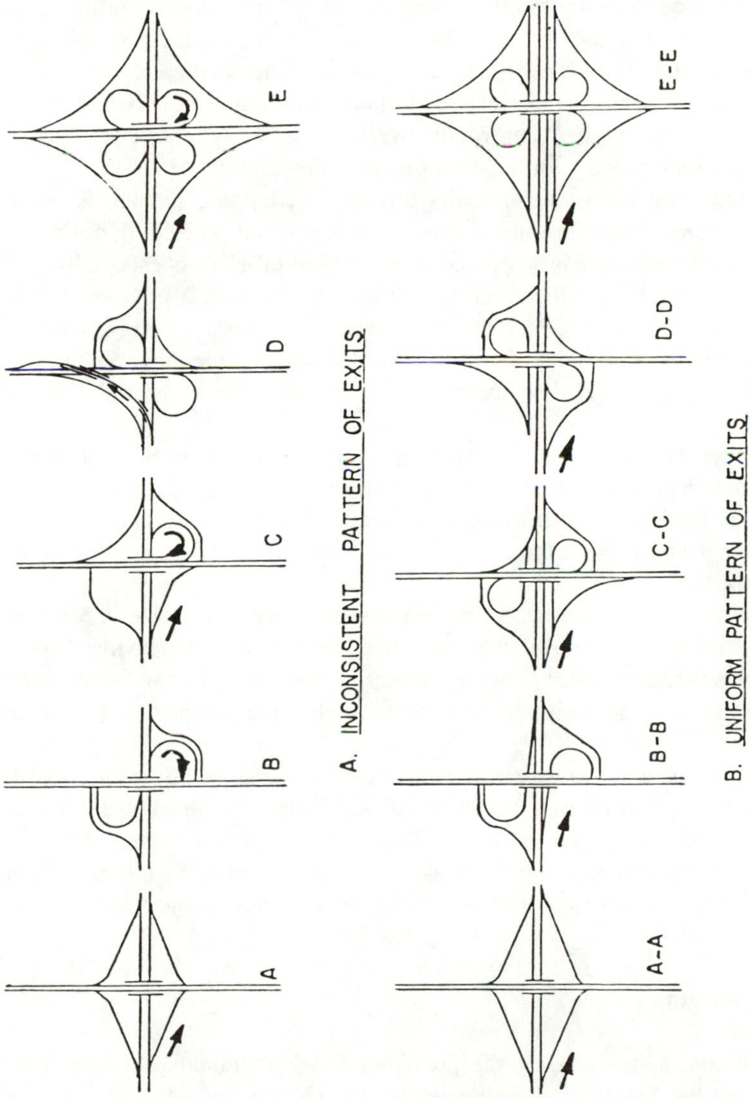

Figure X-48. Arrangement of exits between successive interchanges.

Overlapping Routes

In some situations it is necessary that two or more routes occupy a single alinement within a corridor. Generally, in rural areas the problem of handling overlapping routes is limited to signing and maintaining route continuity. In urban areas the complexity of the problem increases with the probability of weaving and the need for additional capacity and lane balance.

In urban areas it is preferable not to have overlapping routes, especially the short-distance type. When routes overlap, signing is more complicated, and the decision process for the driver is more demanding.

The provision for route continuity through overlapping sections is essential. However, in some instances this provision poses a problem in determining which route should have precedence, and the problem is especially true when both routes have the same classification. Through a process of subclassification, i.e., U.S., State, city, or county route, a priority may be established for one of the overlapping roadways. All other factors being equal, priority should be assigned to the route that handles the highest volume of through traffic.

Once priority for one of the overlapping roadways has been established, basic lanes, lane balance, and other principles of interchange design can be applied to the design of the overlapping section. The lower classified facility should enter and exit on the right, thus conforming to the concept of route continuity.

On overlapping roadways, weaving is usually involved. However, on long overlaps the longer area available minimizes the problem of weaving. Where the overlap is short, as that between successive interchanges, careful attention must be given to the design of weaving sections, lane balance, and optional lanes.

In a situation where a major arterial would be overlapped by a lesser roadway, the minor facility may be designed as a collector-distributor road with transfer roads connecting the two facilities, as shown in Figure X-49. This design removes weaving from the major roadway and transfers it to the minor facility. See the discussion of collector-distributor roads in this chapter.

Route Continuity

Route continuity refers to the provision of a directional path along and throughout the length of a designated route. The designation pertains to a route number or a name of a major highway.

Route continuity is an extension of the principle of operational uniformity

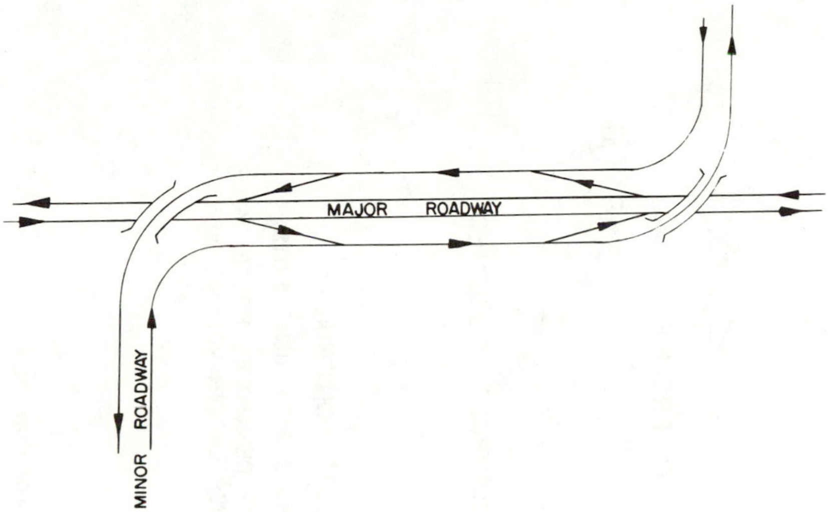

Figure X-49. Collector-distributor road on major-minor roadway overlap.

coupled with the application of proper lane balance and the principle of maintaining basic number of lanes.

The principle of route continuity simplifies the driving task in that it reduces lane changes, simplifies signing, delineates the through route, and reduces the driver's search for directional signing.

Desirably, the through driver, especially the unfamiliar driver, should be provided a continuous through route on which it is not necessary to change lanes and vehicular operation occurs on the left of all other traffic.

In the process of maintaining route continuity, particularly through cities and bypasses, interchange configurations need not always favor the heavy movement but rather the through route. In this connection the heavy movements can be designed on flat curves with reasonably direct connections and auxiliary lanes, equivalent operationally to through movements.

Figure X-50 illustrates the principle of route continuity as applied to a hypothetical route, Interstate 15, as it intersects other major high-volume routes (service interchanges not shown).

In Figure X-50B route continuity is disrupted by other routes exiting or entering on the left, except for the northbound direction of the last interchange.

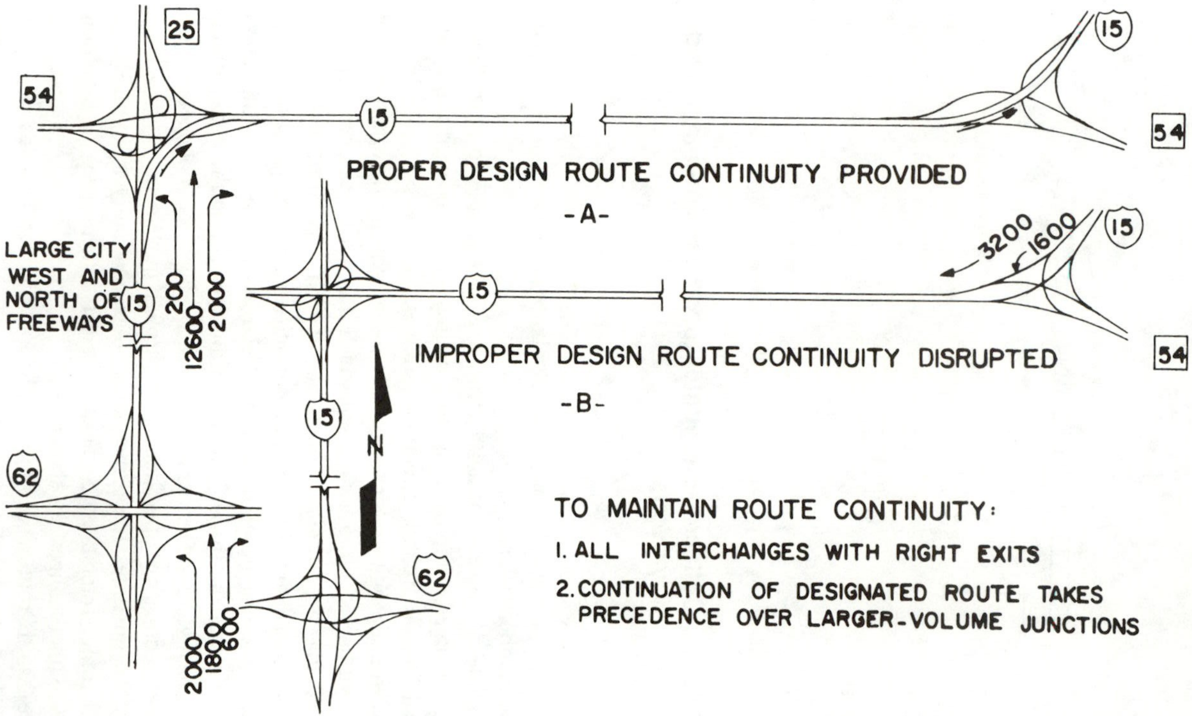

Figure X-50. Interchange forms to maintain route continuity.

In Figure X-50A route continuity is maintained on the designated route by keeping it on the left of all other entering or exiting routes.

Signing and Marking

Ease of operation at interchanges, that is, clarity of paths to be followed, safety, and efficiency, depends largely on their relative spacing, the geometric layout, and effective signing. The location of and minimum distances between ramp junctions depend to a large degree on whether or not effective signing can be provided to inform, warn, and control drivers. Location and design of interchanges, individually and as a group, should be tested for proper signing. Signs should conform to the *Manual on Uniform Traffic Control Devices* (MUTCD) (**2**).

Pavement striping, delineators, and other markings also are important parts of driver communication at interchanges. These should be uniform and consistent, following the national standards of the MUTCD ((**2**).

Basic Number of Lanes

Fundamental to establishing the number and arrangement of lanes on a freeway is the designation of the basic number of lanes. Any route of arterial character should maintain a certain consistency in the number of lanes provided along it. Thus, the basic number of lanes is defined as a minimum number of lanes designated and maintained over a significant length of a route, irrespective of changes in traffic volume and requirements for lane balance. Stating it another way, the basic number of lanes is a constant number of lanes assigned to a route, exclusive of auxiliary lanes.

As illustrated in Figure X-51, the basic number of lanes on freeways is maintained over significant lengths of the routes, as A to B or C to D. The number of lanes is predicated on the general volume level of traffic over a substantial length of the facility. The volume considered here is the DHV (normally, representative of the morning or evening weekday peak). Localized variations are ignored, so volumes on short sections below the general level would theoretically have reserve capacity, and volumes on short sections somewhat above the general level would be compensated for by the addition of auxiliary lanes introduced within these sections. Typical examples of lane balance are shown in Figure X-52.

An increase in the basic number of lanes is required where traffic builds up sufficiently to justify an extra lane and where such buildup raises the volume level over a substantial length of the following facility.

The basic number of lanes may be decreased where traffic is reduced sufficiently to drop a basic lane, provided there is a general lowering of the volume level on the freeway route as a whole. Lane reductions are discussed later in this chapter.

Coordination of Lane Balance and Basic Number of Lanes

To realize efficient traffic operation through and beyond an interchange, there should be a balance in the number of traffic lanes on the freeway and

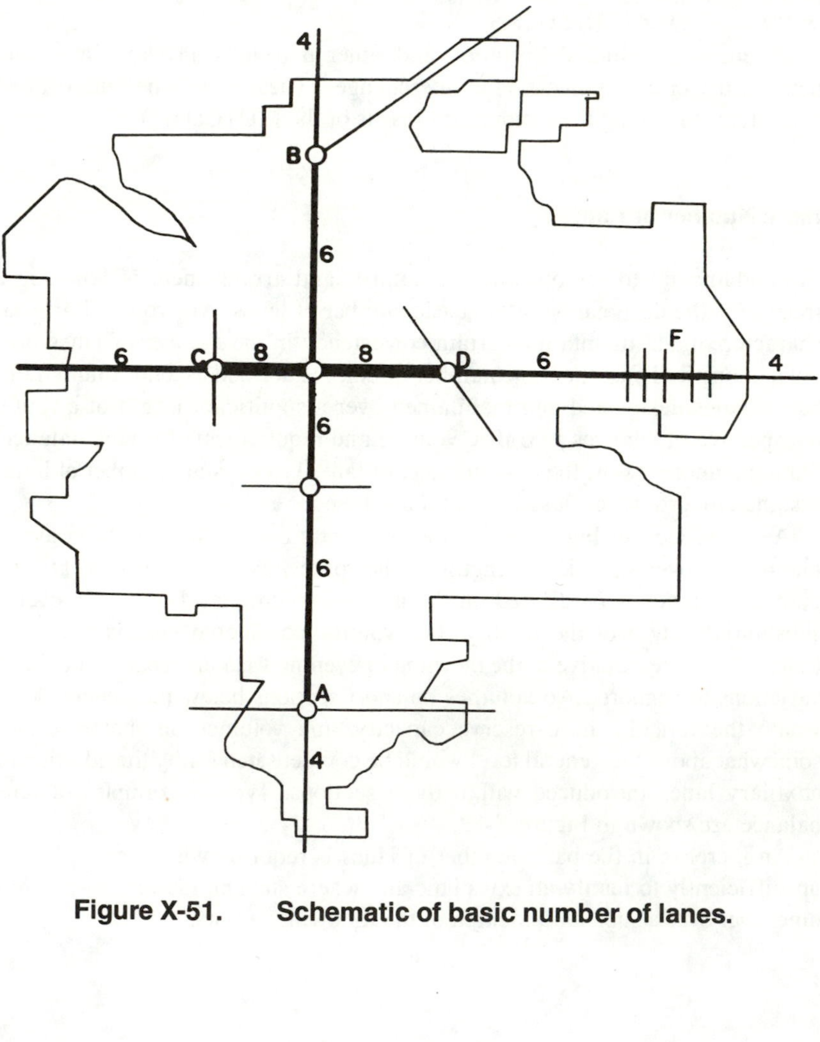

Figure X-51. Schematic of basic number of lanes.

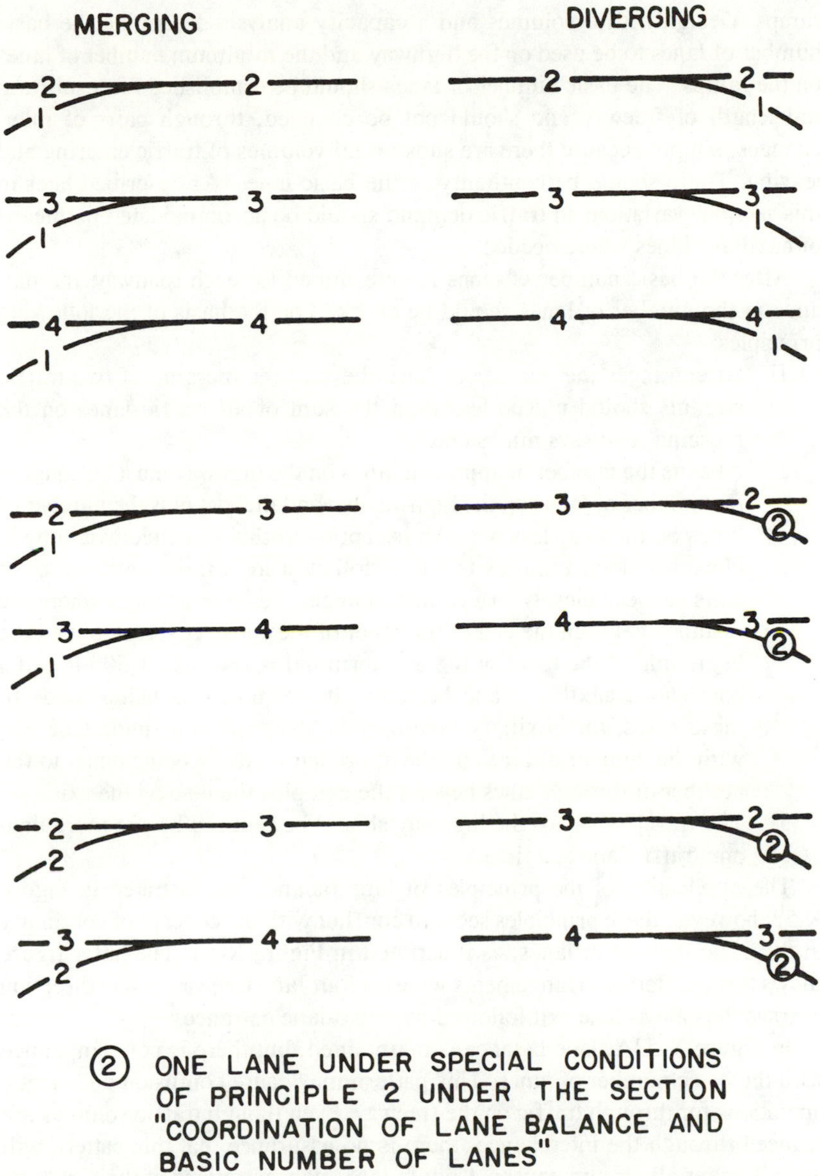

Figure X-52. Typical examples of lane balance.

ramps. Design traffic volumes and a capacity analysis determine the basic number of lanes to be used on the highway and the minimum number of lanes on the ramps. The basic number of lanes should be established for a substantial length of freeway and should not be changed, through pairs of interchanges, simply because there are substantial volumes of traffic entering and leaving. There should be continuity in the basic lanes. As described later in this section, variations in traffic demand should be accommodated by means of auxiliary lanes where needed.

After the basic number of lanes is determined for each roadway, the balance in the number of lanes should be checked on the basis of the following principles:

1. At entrances the number of lanes beyond the merging of two traffic streams should not be less than the sum of all traffic lanes on the merging roadways minus one.
2. At exits the number of approach lanes on the highway must be equal to the number of lanes on the highway beyond the exit plus the number of lanes on the exit, less one. An exception to this principle would be at cloverleaf loop ramp exits which follow a loop ramp entrance or at exits between closely spaced interchanges; i.e., interchanges where the distance between the end of the taper of the entrance terminal and the beginning of the taper of the exit terminal is less than 1,500 ft and a continuous auxiliary lane between the terminals is being used. In these cases, the auxiliary lane may be dropped in a single-lane exit with the number of lanes on the approach roadway being equal to the number of through lanes beyond the exit plus the lane on the exit.
3. The traveled way of the highway should be reduced by not more than one traffic lane at a time.

The application of the principles of lane balance is illustrated in Figure X-52; however, these principles seem to conflict with the concept of continuity in the basic number of lanes, as illustrated in Figure X-53. The latter figure shows three different arrangements where a four-lane freeway in one direction of travel has a two-lane exit followed by a two-lane entrance.

In Figure X-53A, lane balance is maintained, but there is not compliance with the basic number of lanes. This pattern may cause confusion and erratic operations for through traffic on the freeway. Even though traffic volumes are reduced through the interchange, there is no assurance that this pattern will prevail under all circumstances. Unduly large concentrations of through traffic may be caused by special events or by closures or reduction in capacity of other parallel facilities because of accidents or maintenance operations. Under such circumstances any lanes that have been dropped on a freeway between some interchanges (based on capacity and lane-balance requirements as dictated by the normal DHV) produce definite bottlenecks.

The arrangement shown in Figure X-53B provides continuity in the basic number of lanes but does not conform with the principles of lane balance. With this arrangement the large exiting or entering traffic volume requiring two lanes would have difficulty in diverging from or merging with the mainline flow.

Figure X-53C illustrates an arrangement in which the concepts of lane balance and basic number of lanes are brought into harmony by means of build-

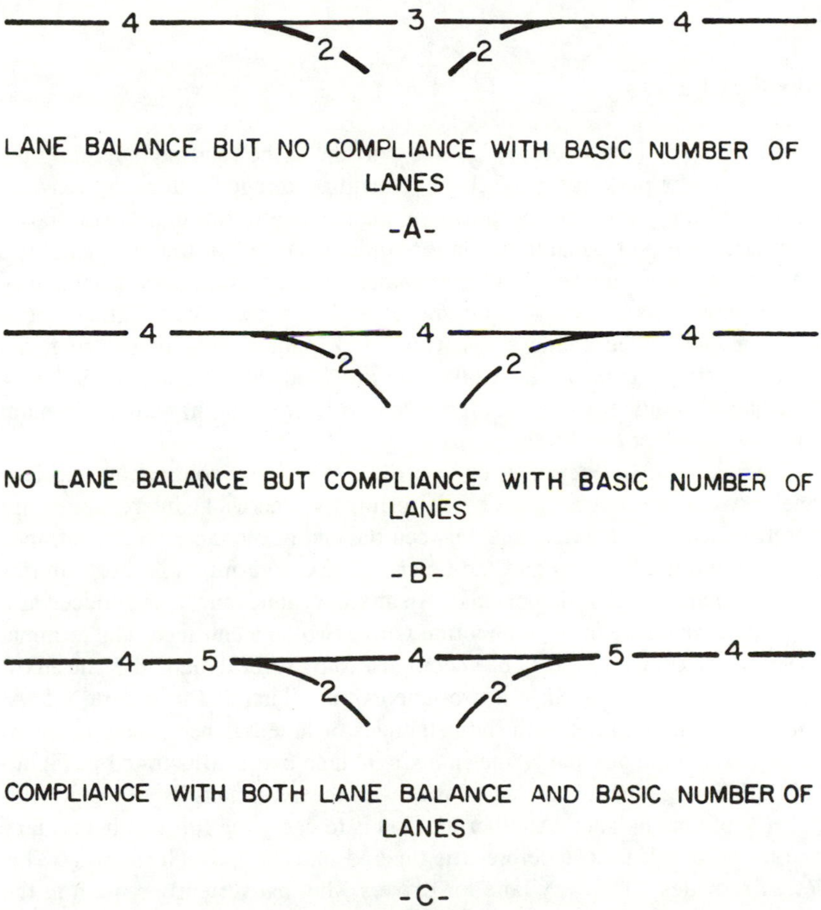

Figure X-53. Coordination of lane balance and basic number of lanes.

ing on the basic number of lanes, that is, by adding auxiliary lanes or removing them from the basic width of the traveled way. Auxiliary lanes may be added to satisfy capacity and weaving requirements between interchanges, to accommodate traffic pattern variations at interchanges, and for simplification of operations (as reducing lane changing). The principles of lane balance must always be applied in the use of auxiliary lanes. In this manner the necessary balance between traffic load and capacity is provided, and lane balance and needed operational flexibility are realized.

Design details of multilane ramp terminals with auxiliary lanes are covered later in this chapter.

Auxiliary Lanes

An auxiliary lane is defined as the portion of the roadway adjoining the traveled way for parking, speed change, turning, storage for turning, weaving, truck climbing, and other purposes supplementary to through-traffic movement. The width of an auxiliary lane should equal that of the through lanes. An auxiliary lane may be provided to comply with the concept of lane balance or to comply with capacity requirements in the case of adverse grades or to accommodate speed changes, weaving, and maneuvering of entering and leaving traffic. Where auxiliary lanes are provided along freeway main lanes, the adjacent shoulder would desirably be 8 to 12 ft in width, with a minimum 6-ft-wide shoulder considered.

Operational efficiency may be improved by using a continuous auxiliary lane between the entrance and exit terminals where (1) interchanges are closely spaced, (2) the distance between the end of the taper on the entrance terminal and the beginning of the taper on the exit terminal is short, and/or (3) local frontage roads do not exist. An auxiliary lane may be introduced as a single exclusive lane or in conjunction with a two-lane entrance. The termination of the auxiliary lane may be accomplished by several methods. The auxiliary lane may be dropped in a two-lane exit as illustrated in Figure X-54A. This treatment complies with the principles of lane balance. Some agencies prefer to drop the auxiliary lane in a single lane exit as illustrated in Figure X-54B. This treatment is in accordance with the exceptions listed under principle (2) of lane balance. Another method is to carry the full-width auxiliary lane to the physical nose before it is tapered into the through roadway. This design provides a recovery lane for drivers who inadvertently remain in the discontinued lane. (See Figure X-54C.) When these methods of terminating the auxiliary lane (Figures X-54B and X-54C) are used, the exit gore should be visible throughout the length of the auxiliary lane.

If local experience with single-exit design indicates problems with turbulence in the traffic flow caused by vehicles attempting to recover and proceed on the through lanes, the recovery lane should be extended 500 to 1,000 ft before being tapered into the through lanes. (See Figure X-54D.) Within large interchanges, this distance should be increased to 1,500 ft. When an auxiliary lane is carried through one or more interchanges, it may be dropped as indicated above, or it may be merged into the through roadway approximately 2,500 ft beyond the influence of the last interchange. (See Figure X-54E.)

When interchanges are widely spaced, it might not be feasible or necessary to extend the auxiliary lane from one interchange to the next. In such cases, the auxiliary lane originating at a two-lane entrance should be carried along the freeway for an effective distance beyond the merging point, as shown in Figure X-55A2, or an auxiliary lane introduced for a two-lane exit should be carried along the freeway for an effective distance in advance of the exit and extended onto the ramp, as shown in Figure X-55B2. Auxiliary lanes that do not extend to the next interchange may be terminated or introduced as indicated in Figures X-55A1, X-55A2, X-55B1, and X-55B2. Figures X-55A1 and X-55B1 utilize taper designs, whereas Figures X-55A2 and X-55B2 are parallel designs.

Generally, the parallel designs are preferred. While the tapered designs are acceptable, some agencies are apprehensive concerning the inside merge on the tapered entrance ramps. Lane should not be less in length than those shown for single-lane ramps in Tables X-4 and X-6 with adjustments for grades as suggested in Table X-5. What the effective length of the introduced auxiliary lane should be under these circumstances is not known precisely. Experience indicates that minimum distances of about 2,500 ft are required to produce the necessary operational effect and to develop the full capacity of two-lane entrances and exits.

For those instances where an auxiliary lane extends for a long distance from an entrance at one interchange to an exit at the next interchange, unfamiliar motorists may perceive the auxiliary lane as an additional through lane. For these situations, an auxiliary lane may be terminated as discussed in the subsequent section entitled "Lane Reductions" or by providing a two-lane exit.

Auxiliary lanes are used to balance the traffic load and maintain a more uniform level of service on the highway. Such added lanes facilitate the positioning of drivers at exits and bringing them onto the highway at entrances. Thus, the concept is very much related to signing and route continuity. Careful consideration should be given to the design treatment of an auxiliary lane because it may have the potential for trapping a driver at its termination point or the point where it is continued onto a ramp or turning roadway.

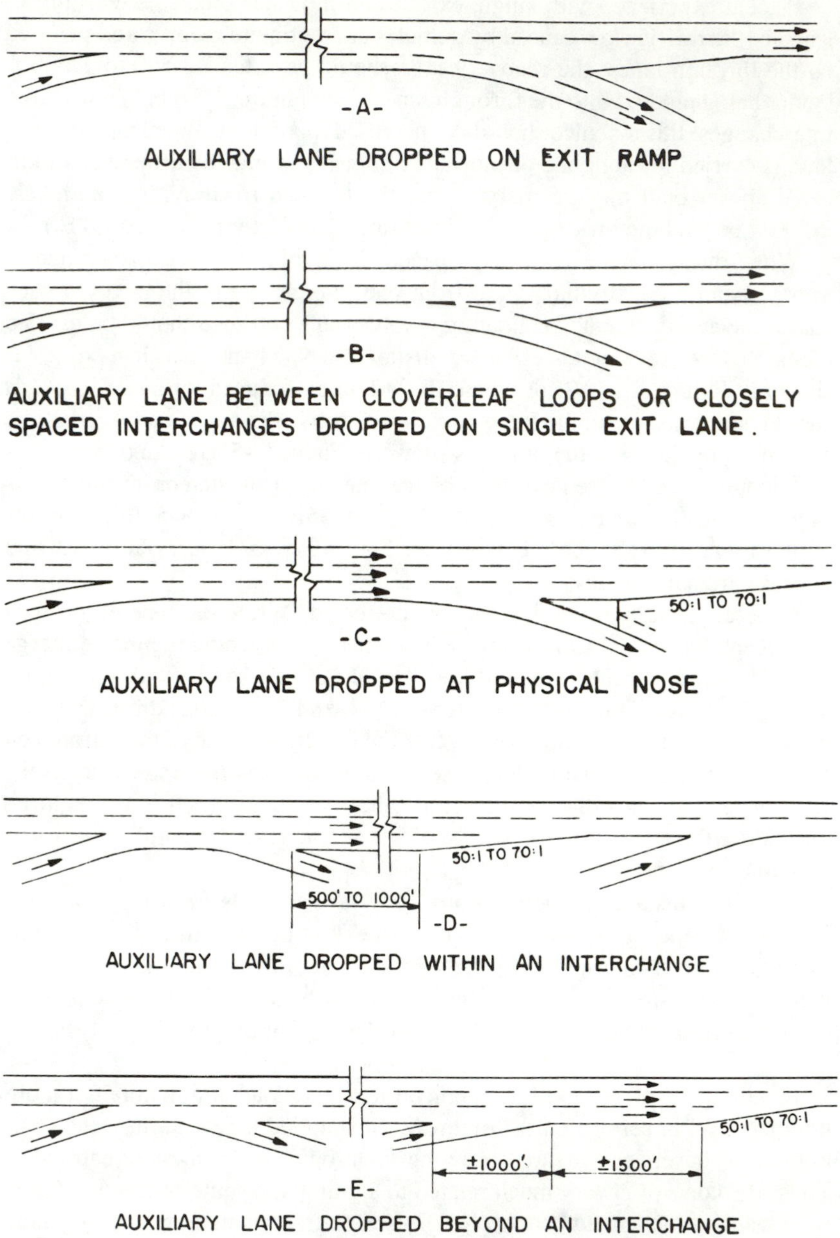

Figure X-54. Alternatives in dropping auxiliary lanes.

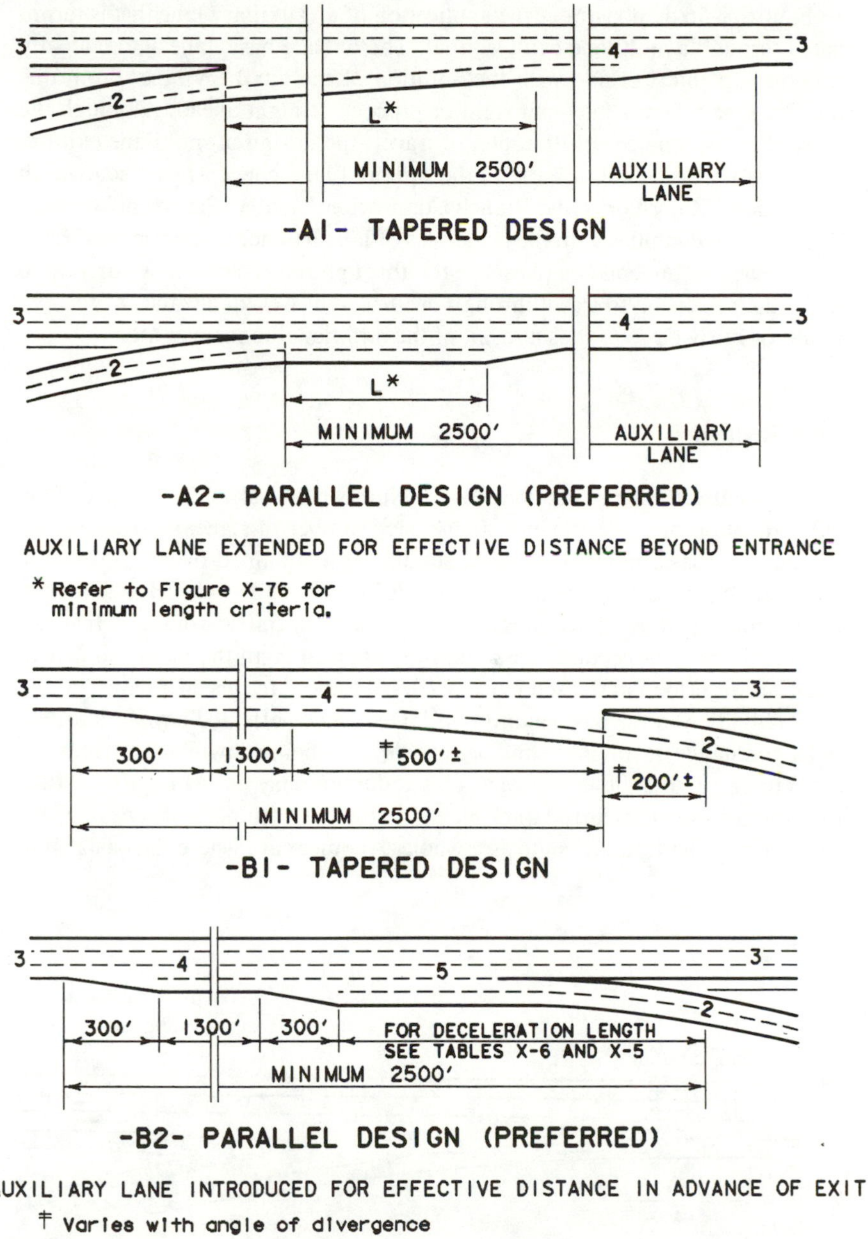

Figure X-55. **Coordination of lane balance and basic number of lanes through application of auxiliary lanes.**

Figure X-56 demonstrates the application of an auxiliary lane that is terminated through a multilane exit terminal. The outside basic lane automatically becomes an interior lane with the addition of the auxiliary lane. From this interior lane a driver may exit right or proceed straight ahead. Although the driver has two choices of direction of travel, the design of multilane exit terminals is not to be confused with the optional lane concept, as discussed in the section "Major Forks and Branch Connections" under "Pavement Widths." The example complies with the principles of lane balance and basic number of lanes. The design lends emphasis to the through route, and allows drivers to make their decision to travel through or turn right well in advance of the exit point, or fairly close to it as a result of the additional maneuver area.

Lane Reduction

As discussed in the section "Basic Number of Lanes" and the section "Coordination of Lane Balance and Basic Number of Lanes" earlier in this chapter, the basic number of lanes should be maintained over a significant length of freeway. Lane reductions should not be made between and within interchanges simply to accommodate variations in traffic volumes. Instead, auxiliary lanes, as needed, are added or removed from the basic number of lanes, as described in the section "Auxiliary Lanes" in this chapter.

A reduction in the basic number of lanes may be effected beyond a principal interchange involving a major fork or at a point downstream from the interchange with another freeway. This reduction may be made provided the exit volume is sufficiently large to change the basic number of lanes beyond this point on the freeway route as a whole. Another case where the basic num-

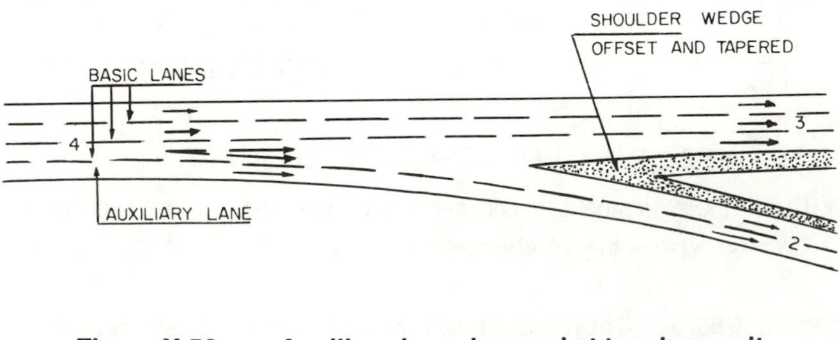

Figure X-56. Auxiliary lane dropped at two-lane exit.

ber of lanes may be reduced is where a series of exits, as in outlying areas of the city, causes the traffic load on the freeway to drop sufficiently to justify the lesser basic number of lanes. The lane reduction of a basic lane or an auxiliary lane may be effected at a two-lane exit ramp or between interchanges.

If a basic lane or an auxiliary lane is to be dropped between interchanges, it should be accomplished at a distance of 2,000 to 3,000 ft from the previous interchange to allow for adequate signing.

The reduction should not be made so far downstream that motorists become accustomed to a number of lanes and are surprised by the reduction. (See Figure X-54C.) Desirably, the lane-drop transition should be effected on tangent horizontal alinement and on the approach side of any crest vertical curve. A sag vertical curve is also a good location for a lane drop because it provides good visibility. Preferably, the lane reduction should be made on the right following an exit ramp because there is likely to be less traffic in that lane. A right-side lane reduction has advantages in that speeds are generally lower and the merging maneuver from the right is more familiar to most motorists, it being similar to a merge at an entrance ramp. Left-side lane reductions may not function as well because of generally higher speeds and the less familiar left-side merge.

The end of the lane drop should be tapered into the highway in a manner similar to that at a ramp entrance. Preferably, the rate of taper should be longer than that for a ramp to provide an additional factor of safety. The minimum taper rate should be 50:1, and 70:1 is desirable.

If there is a lane reduction of a basic lane or an auxiliary lane within an interchange, it should be effected in conjunction with a two-lane exit, as shown in Figure X-54A, or in a single-lane exit with an adequate recovery lane, as discussed in "Auxiliary Lanes."

Weaving Sections

Weaving sections are highway segments where the pattern of traffic entering and leaving at contiguous points of access results in vehicle paths crossing each other.

Weaving sections may occur within an interchange, between entrance ramps followed by exit ramps of successive interchanges, and on segments of overlapping roadways.

Because considerable turbulence occurs throughout weaving sections, interchange designs that eliminate weaving or remove it from the main facility are desirable. Weaving sections may be eliminated from the main facility by the selection of interchange forms that do not have weaving or by the incorporation of collector-distributor roads.

Although interchanges that do not involve weaving operate better than those that do, interchanges with weaving areas nearly always are less costly than those without. Designs to avoid weaving movements require a greater number of structures or larger and more complex structures, with some direct connections. Joint evaluation of the total interchange cost and the specific volumes to be handled is required to reach a sound decision between design alternates.

Where cloverleaf interchanges are used, consideration should be given to the inclusion of collector-distributor roads on the main facility, or possibly both facilities where warranted.

The capacity of weaving sections may be seriously restricted unless adequate length and width are provided through the weaving section along with lane balance. See Chapter II for procedures for determining weaving lengths and widths. Refer to the *Highway Capacity Manual* (HCM) (**3**) for capacity analysis of weaving sections.

Collector-Distributor Roads

Collector-distributor roads between two interchanges and continuous collector-distributor roads are discussed in Chapter VIII. Collector-distributor roads within an interchange are discussed in the following.

A full cloverleaf interchange in an urban or suburban area is a typical example of a single interchange that should be analyzed for the need for collector-distributor roads within the interchange.

Collector-distributor roads may be one or two lanes in width, the determining factor being capacity requirements. Lane balance should be maintained at entrances and exits to and from the main line, but strict adherence is not mandatory on the collector-distributor road proper because weaving is handled at reduced speed. The design speed usually ranges from 40 to 50 mph, but should not be less than 10 mph below the design speed of the main roadway. Operational problems will occur if collector-distributor roads are not properly signed, especially those servicing more than one interchange.

Outer separations between the main line and the collector-distributor roads should be as wide as practicable; however, minimum widths are tolerable. The minimum width should be great enough for shoulder widths on the collector-distributor roads equal to that on the main line and for a suitable barrier to prevent indiscriminate crossovers.

The advantages of using collector-distributor roads within an interchange are that weaving is transferred from the main roadway, single entrances and exits are developed, all main-line exits occur in advance of the structure, and a uniform pattern of exits can be maintained. See the "Two-Exit Versus Single-Exit Interchange Design" discussion following.

Two-Exit Versus Single-Exit Interchange Design

In general, interchanges that are designed with single exits are superior to those with two exits, especially if one of the exits is a loop ramp or the second exit is a loop ramp preceded by a loop entrance ramp. Whether used in conjunction with a full cloverleaf or with a partial cloverleaf interchange, the single-exit design may improve operational efficiency of the entire facility.

The purposes for developing single exits, where applicable, are as follows:

1. To remove weaving from the main facility and transfer it to a slower speed facility.
2. To provide a high-speed exit from the main roadway for all exiting traffic.
3. To simplify signing and the decision process.
4. To satisfy driver expectancy by placing the exit in advance of the separation structure.
5. To provide uniformity of exit patterns.
6. To provide decision sight distance for all traffic exiting from the main roadway.

The full cloverleaf interchange, where a weaving section exceeds 1000 vph, is an example where operational efficiency may be improved by the development of single exits and entrances.

The loop ramps of a full cloverleaf interchange create a weaving section adjacent to the outside through lane, and considerable deceleration-acceleration must occur in the through lane. By using collector-distributor roads, as shown in Figure X-57, a single exit is provided and weaving is transferred to the collector-distributor road. The second exit of a cloverleaf interchange occurs beyond the separation structure, and in many cases it is hidden behind a crest vertical curve. The single-exit design places the exit from the main line in advance of the structure and is conducive to a uniform pattern of exits. Where the through roadway overpasses the crossroad in a vertical curve, it may be more difficult to develop full decision sight distance for the loop ramp exit of a conventional cloverleaf interchange. The use of the single-exit design may make it easier to obtain the desired decision sight distance due to the exit occurring on the upgrade.

Some arrangements of partial cloverleaf loop ramps may feature single exits, as shown in Figure X-25F, and still be inferior because they do not provide any of the desirable purposes previously discussed.

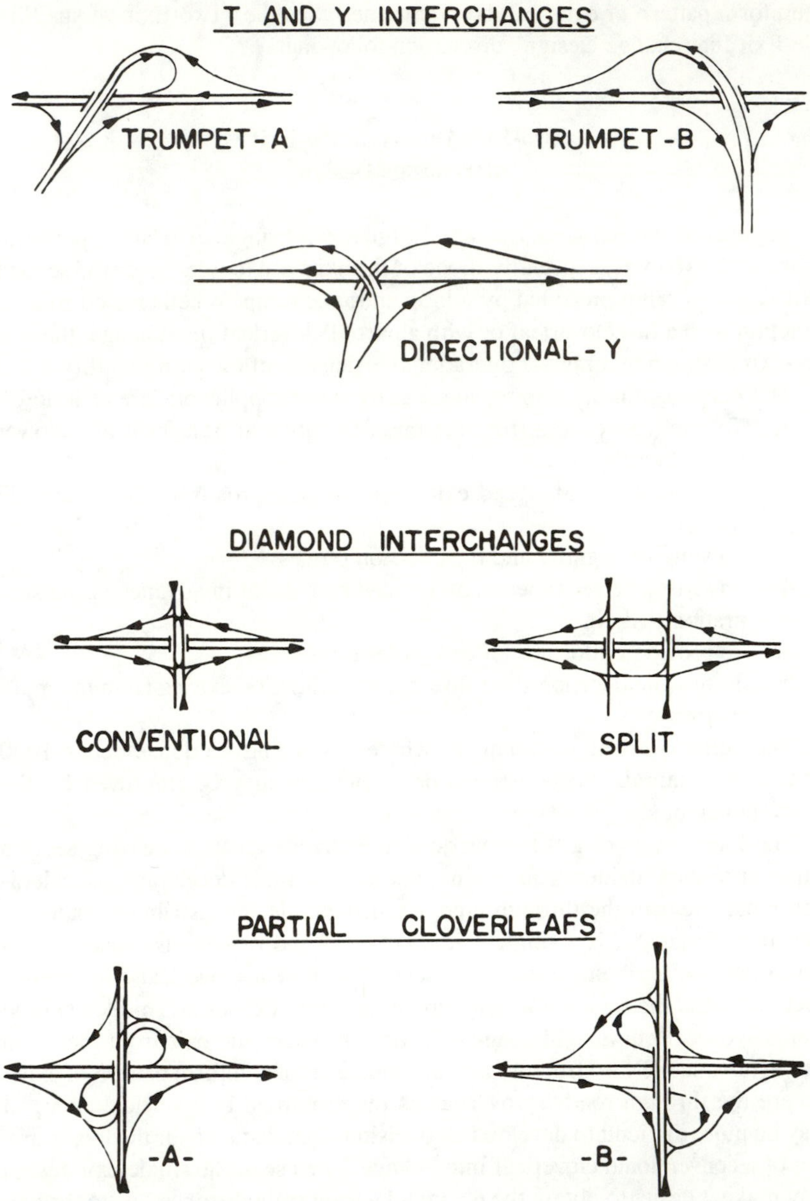

Figure X-57. Interchange forms with one and two exits.

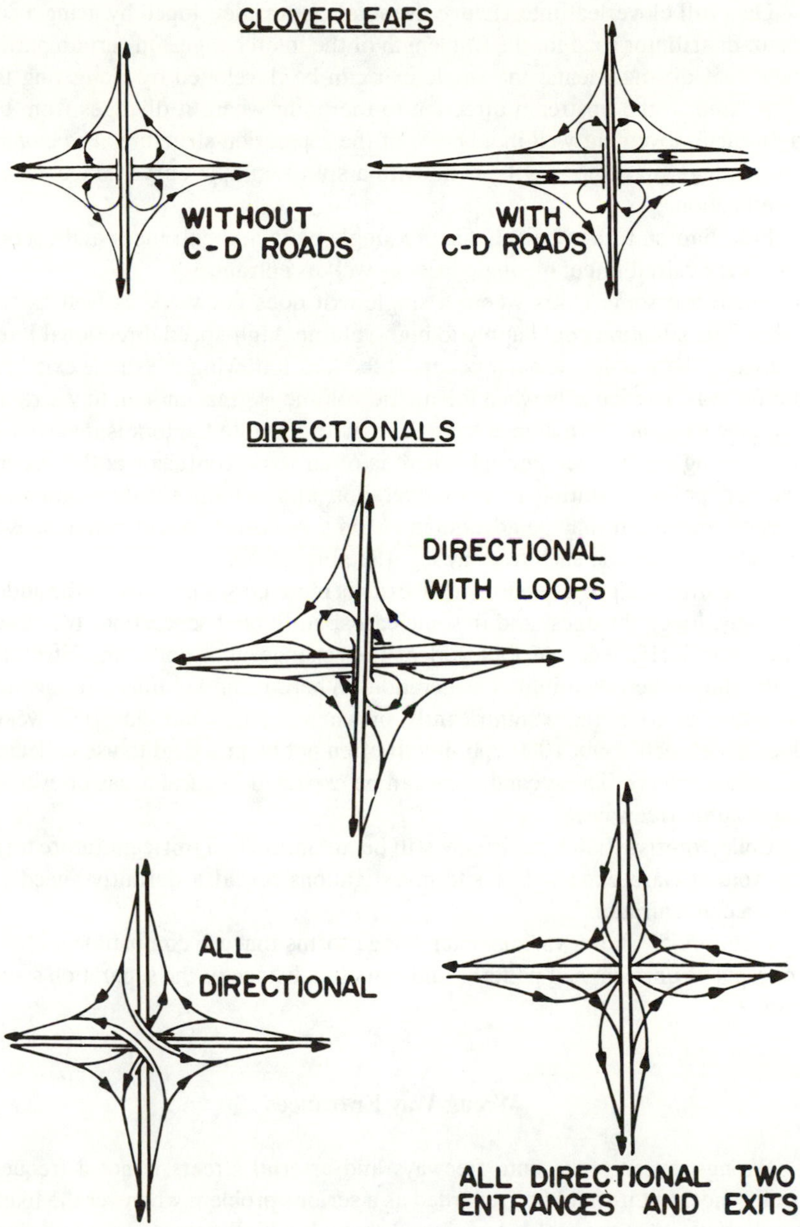

Figure X-57. Continued.

On a full cloverleaf interchange the single exit is developed by using a collector-distributor road for the full length of the interchange. On certain partial cloverleaf arrangements, the single exit can be developed by elongating the loop ramp in the upstream direction to the point where it diverges from the right-turn movement well in advance of the separation structure. The elongation of the loop ramp may be done with a spiral, simple curve, tangent, or a combination.

Full diamond interchanges feature single exits and entrances and accomplish the desired intent of single exits as well as entrances.

There are some cases where a single exit does not work as well as two exits. This situation could apply to high-volume, high-speed directional interchanges. The problem usually occurs at the fork following the single exit from the freeway, particularly when the traffic volume is great enough to warrant a two-lane exit and the distance from the exit terminal to the fork is insufficient for weaving and proper signing. There is often some confusion at this second decision point, resulting in poor operation and a high accident potential. Because of this, it may be advantageous on some directional interchanges to provide two exits on each freeway leg. (See Figure X-37.)

Generally, the provision for single exits is more costly because of the added roadway, longer bridges, and in some cases, additional separation structures. The overall efficiency of a cloverleaf interchange with collector-distributor roads should be taken into consideration. Where ramp volumes are low and not expected to increase significantly, or where a particular cloverleaf weave does not exceed about 1000 vph, it will often not be practical to use collector-distributor roads. These conditions can be expected in rural areas or even on low volume freeways.

Collector-distributor roads may still be an option if significant future turning volumes are expected or site investigations reveal a definitive need for such a configuration.

Figure X-57 shows various interchange forms that are compatible with the concepts of uniform exit patterns and exits in advance of the separation structure.

Wrong-Way Entrances

Wrong-way entrance onto freeways and arterial streets is not a frequent occurrence, but it should be regarded as a serious problem wherever the likelihood exists, because each occurrence has such a high potential for culminating in a serious accident. This problem should be given special consideration at all stages of design. Most wrong-way entrances occur at freeway offramps, at intersections at-grade along divided arterial streets, and at transitions from undivided to divided highways. Several factors that contribute to wrong-way entrances are related to interchange design. These factors concern the type of interchange and more particularly the local road terminal of the exit ramps. They are discussed below.

Partial interchanges are particularly troublesome. Where provision is not made for any one or more of the movements at an interchange, wrong-way entry may occur. Exit ramps that connect to two-way frontage roads are conducive to wrong-way entry. Without channelization on the frontage road, they appear as open entries. Some of the "scissors" channelization has proved to be confusing, resulting in wrong-way use.

Exit ramps with a sweeping connection to the street (outer connection, loop, and some diamond ramps) have a low rate of wrong-way entry. One-way ramps that connect as an unchannelized "T" intersection are most likely to lead to wrong-way entry.

Unusual or odd arrangements of exit ramps are confusing and conducive to wrong-way entry. An example is the button-hook or J-shaped ramp that connects to a parallel or diagonal street or frontage road, which is often well removed from the interchange structure and other ramps. Another case is a pair of right-turn connections to a lateral or parallel street (frontage road) offset from the separation structure.

On undivided crossroads, a nontraversable median (except at turn points) introduced within the interchange limits helps prevent wrong-way entry on diamond, partial cloverleaf, and full cloverleaf interchanges. Where adjacent off-and-on ramps join a minor road, the ramp pavements should be separated. The ramp-crossroad intersection at a diamond interchange should be well removed from any other nearby intersection, such as frontage road-crossroad intersection. Local road connections within the length of any exit ramp should be avoided. Temporary ramp terminals warrant special attention in layout details to avoid easy direct wrong-way entry paths.

Left-side main roadway exit ramps should be avoided because they may appear to be a right-side entrance ramp to a confused motorist. Many wrong-way entries occur at night when volumes are low and control devices are less effective. This occurrence suggests consideration of lighting for difficult conditions.

Open sight distances throughout the entire length of the ramp help prevent wrong-way use. Especially important is the driver's view of the ramp terminal when approaching from the cross street.

In the design of any interchange, consideration should be given to the likelihood of wrong-way travel and to the possible measures that may be taken in the design and traffic control for preventing or discouraging such usage. Signing to prevent wrong-way entry should be in accordance with the MUTCD (**2**).

Ramps

Types and Examples

The term "ramp" includes all types, arrangements, and sizes of turning roadways that connect two or more legs at an interchange. The components of a ramp are a terminal at each leg and a connecting road, usually with some curvature, and on a grade. Generally, the horizontal and vertical alinement standard of ramps is below that of the intersecting highway, but in some cases it may be equal.

Figure X-58 illustrates several types of ramps and their characteristic shapes. Numerous shape variations are in use but each can be classed broadly as one of the types shown. Each ramp generally is a one-way roadway. Diagonal ramps (Figure X-58A) almost always are one-way but usually have both a left- and right-turning movement at the terminal on the minor intersecting road. Although shown as a continuous curve in Figure X-58A, a diagonal ramp may be largely tangent or wishbone in shape with a reverse curve. Diamond-type interchanges generally have four diagonal ramps.

The loop may have single turning movements (left or right) or double turning movements (left and right) at either or both ends. Figure X-58D shows the case where there are only single turns made at both ends of the ramp. With this loop pattern a left-turning movement is made without an at-grade crossing of the opposing through traffic. Instead, drivers making a left turn travel beyond the highway separation, swing right, and turn through approximately 270° to enter the other highway. The loop usually involves more indirect travel distance than that for other type ramps.

With a semidirect connection (Figure X-58C), the driver makes a right turn first, swinging away from the intended direction, gradually reversing, and then completing the movement by following directly around and entering the other road. This semidirect connection may also be used for right turns, but there is little reason for its use if the conventional diagonal can be provided. A descriptive term frequently associated with this type of ramp is "jughandle," the obvious plan shape. Travel distance on this ramp is less than that for a comparable loop and more than that for a direct connection. At least three structures or a three-level structure is needed. Figure X-58D is termed an outer connection, while Figure X-58E is referred to as a direct connection.

The different ramp patterns of an intersection, i.e., the different types of interchanges, are made up of various combinations of these types of ramps. For example, the trumpet has one loop, one semidirectional ramp, and two right directional or diagonal ramps.

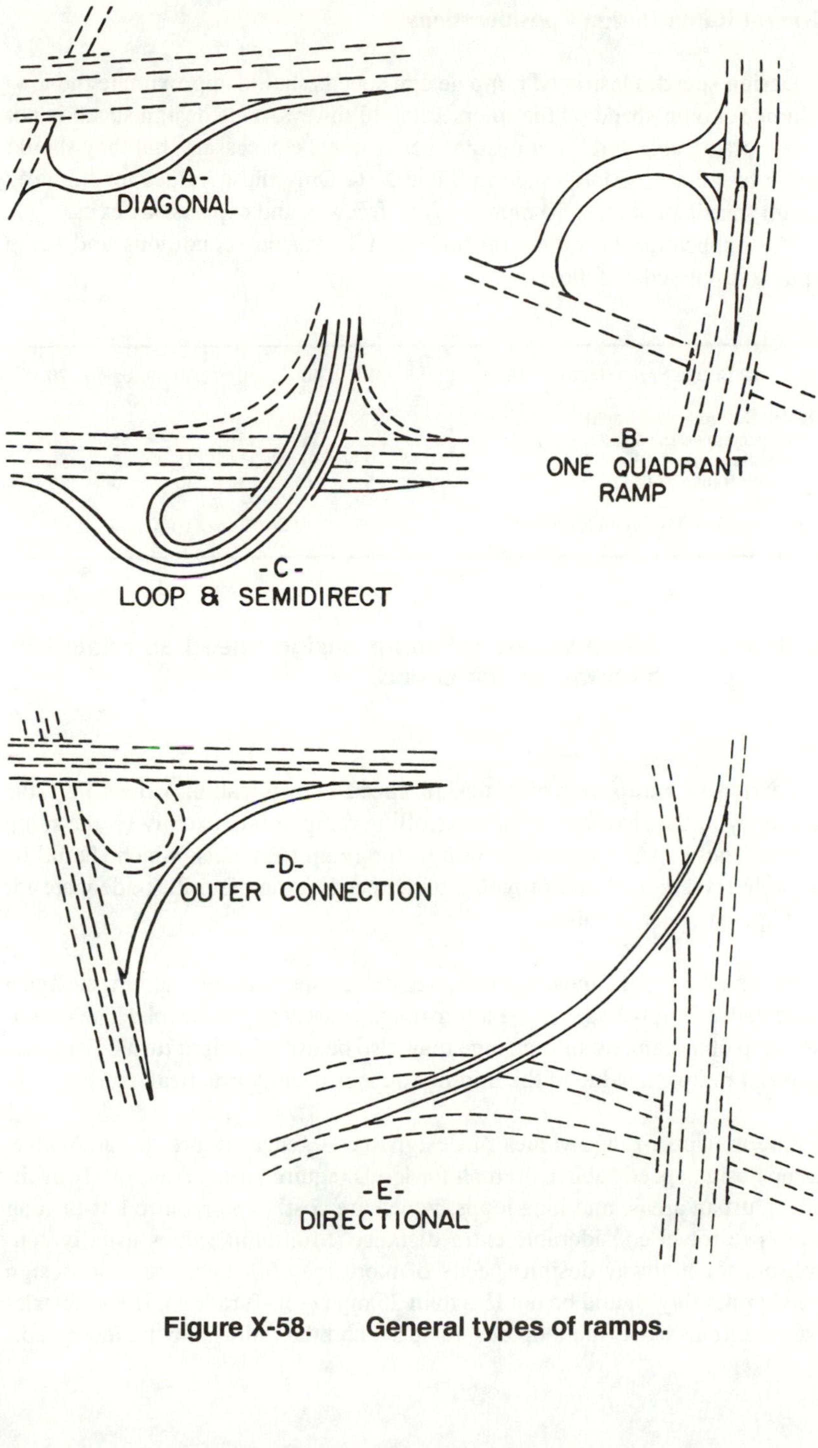

Figure X-58. General types of ramps.

General Ramp Design Considerations

Design speed. Desirably, ramp design speeds should approximate the low-volume running speed on the intersecting highways. This design speed is not always practicable and lower design speeds may be necessary, but they should not be less than the low range in Table X-1. Only those values for highway design speeds of at least 50 mph apply to freeway and expressway exits.

The application of values in Table X-1 to various conditions and ramp types is discussed as follows.

Highway Design Speed (mph)	30	40	50	60	65	70
Ramp Design Speed (mph)						
Upper Range (85%)	25	35	45	50	55	60
Middle Range (70%)	20	30	35	45	45	50
Lower Range (50%)	15	20	25	30	30	35
Corresponding Minimum Radius (ft)	See Table III-6.					

Table X-1. Guide values for ramp design speed as related to highway design speed.

Portion of ramp to which design speed is applicable. Values in Table X-1 apply to the sharpest or the controlling ramp curve, usually on the ramp proper. These speeds do not pertain to the ramp terminals, which should be properly transitioned and provided with speed-change facilities adequate for the highway speed involved.

Ramps for right turns. Upper-range design speed often is attainable and a value between upper range and lower range usually is practicable. The diagonal ramp of a diamond interchange may also be used for right turns. For these diagonal ramps, a value in the middle range is usually practical.

Loops. Upper-range values of design speed generally are not attainable. Ramp design speeds above 30 mph for loops require large areas, rarely available in urban areas, and long loops, which are costly and require left-turning drivers to travel considerable extra distance. Minimum values usually control, but for highway design speeds of more than 50 mph, the loop design speed preferably should be not less than 25 mph (150-ft radius). If less restrictive conditions exist, the loop design speed and the radius may be increased.

Semidirect connections. Design speeds between the middle and upper ranges shown in Table X-1 should be used. A design speed less than 30 mph should not be used. Generally, for short, single lane ramps, a design speed greater than 50 mph is not practical. For two-lane ramps, values in the middle and upper ranges are appropriate.

Direct connections. Design speeds between the middle and upper ranges shown in Table X-1 should be used. The minimum preferably should be 40 mph and in no case less than 35 mph.

Different design speeds on intersecting highways. The highway with the greater design speed should be the control in selecting the design speed for the whole of a ramp. However, the ramp design speed may vary, the portion of the ramp closer to the lower speed highway being designed for the lower speed. The latter point is particularly applicable where the ramp is on an upgrade from the higher speed highway to the lower speed highway.

At-grade terminals. Where a ramp joins a major crossroad or street, forming an intersection at-grade, Table X-1 is not applicable to that portion of the ramp near the intersection because a stop sign or signal control is normally employed. This terminal design should be predicated on near-minimum turning conditions, as given in Chapter IX.

Curvature. The factors and assumptions of minimum-turning roadway curves for various speeds are discussed in Chapter III. They apply directly to design of ramp curves. Compound curves and spiral transitions are desirable to effect the desired shape of ramps, to meet site conditions and other controls, and to fit the natural paths of vehicles. Caution should be exercised in the use of compound curvature to prevent unexpected and abrupt speed adjustments. Additional design information regarding the use of spiral transitions and compound curves is contained in Chapter III.

The general shape of a ramp evolves from the type of ramp selected, as previously described and shown in Figure X-58. The specific shape of a ramp may be influenced by such factors as the traffic pattern, traffic volume, design speeds, topography, culture, intersection angle, and type of ramp terminal.

Several forms may be used for the loop and outer connection of a semidirectional interchange, as shown diagrammatically in Figure X-59A. The loop, except for its terminals, may be a circular arc or some other symmetrical or asymmetrical curve, formed with spiral transitions. The asymmetrical arrangement may fit where the intersecting roads are not of the same importance and the ramp terminals are designed for different speeds, the ramp in part functioning as a speed-change area. Similar shapes may be dictated by right-of-way controls, profile and sight distance conditions, and terminal location. The freeway terminal should normally be placed in advance of the structure. The most desirable alinement for an outer connection is one on a continuous curve (line A). This arrangement, however, may involve questionably extensive right-of-way. Another desirable arrangement has a central tangent and terminal curves (lines B-B and C-C). Where the loop is more important than the outer connection, reverse alinement on the outer connection may be used to reduce the area of right-of-way, as shown by line D-D. Any combination of lines B, C, and D may be used for a feasible shape.

In Figure X-59A the loop and the outer connection are separated, as is generally desirable. However, where the movements are minor and economy is desired, a portion of the two ramps may be combined into a single two-way pavement. Where this design is used, a barrier should separate the traffic in the two directions. This design is generally discouraged.

Diagonal ramps may assume a variety of shapes, depending on the pattern of turning traffic and right-of-way limitations. As shown in Figure X-59B, the ramp may be a diagonal tangent with connecting curves (solid line). To favor a right-turning movement the ramp may be on a continuous curve to the right with a spur to the left for left turns. On restricted right-of-way along the major highway, it may be necessary to use reverse alinement with a portion of the ramp being parallel to the through roadway.

Diagonal ramps, of a type usually called "slip ramps", connect with a parallel frontage road as shown in Figure X-59C. Where this design is used, it is desirable to have one-way frontage roads. Ramps to two-way frontage roads introduce the possibility of wrong-way entry onto the through lanes. If used, special attention should be given in the design and signing of ramps to two-way frontage roads to discourage the possibility of wrong-way entry.

The shape of a semidirect connection (Figure X-59D) is influenced by the location of the terminals with respect to the structures, the extent to which the structure pavements are widened, and the curve radii necessary to maintain a desired turning speed for an important left-turning movement. The angular position or the curvature may be dictated somewhat by the relative design speeds of the intersection legs and by the proximity of other roadways.

Sight distance. Sight distance along a ramp should be at least as great as the safe stopping sight distance. Sight distance for passing is not required. There should be a clear view of the entire exit terminal, including the exit nose and a section of the ramp pavement beyond the gore.

The sight distance on a freeway preceding the approach nose of an exit ramp should exceed the minimum stopping sight distance for the through traffic design speed, desirably by 25 percent or more. Decision sight distance, as discussed in Chapter III, is a desirable goal. There should be a clear view of the entire exit terminal, including the exit nose.

See Chapter III for ranges in design values for stopping sight distance on horizontal and vertical curves for turning roadways and open road conditions.

Grade and profile design. Ramp grades should be as flat as feasible to minimize the driving effort required in maneuvering from one road to another. Most ramps are curved and steep grades on them hamper the flow of

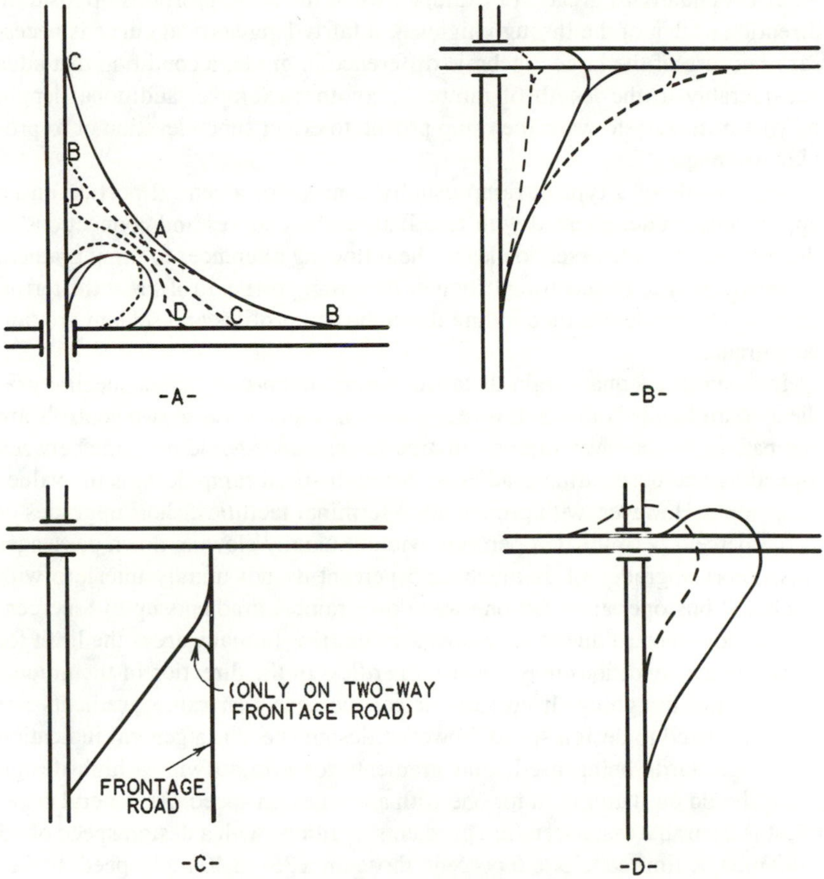

Figure X-59. Ramp shapes.

traffic. The slowing down of vehicles on an ascending ramp is not as serious as on a through road, provided the speed is not decreased sufficiently to result in a peak-hour backup onto the through road. Most diamond ramps are only 400 to 1,200 ft long, and the short central portion with steepest gradient has only moderate operational effect. Accordingly, gradients on ramps may be steeper than those on the intersecting highways, but a precise relation cannot be set. General values of limiting gradient can be shown, but for any one ramp the gradient to be used is dependent on a number of factors peculiar to that site and quadrant alone. The flatter the gradient on a ramp, the longer it will be, but the effect of gradient on the length of the ramp is less than generally thought. The conditions and designs at ramp terminals frequently have an effect equally as great. For example, when the ramp profile is opposite in direction to that of the through highway, a fairly long vertical curve is necessary because of the large algebraic difference in grade, a condition that adds considerably to the length of ramp. As another example, additional length may be necessary to warp the ramp profile to effect superelevation or to provide drainage.

The profile of a typical ramp usually consists of a central portion on an appreciable grade, coupled with terminal vertical curves and connections to the profiles of the intersection legs. The following references to ramp gradient pertain largely to the central portion of the ramp profile. Profiles at the terminals largely are determined by the through-road profiles and seldom are tangent grades.

In general, adequate sight distance is more important than a specific gradient control and should be favored in design. Usually, these two controls are compatible. On one-way ramps a distinction can and should be made between ascending and descending gradients. For high-speed ramp designs the values cited apply. However, with proper ramp terminal facilities, short upgrades of 7 to 8 percent permit safe operation without unduly slowing down passenger cars. Short upgrades of as much as 5 percent do not unduly interfere with truck and bus operation. On one-way down ramps, gradients up to 8 percent do not cause hazard due to excessive acceleration. In many areas the limit for snow and ice conditions may govern regardless of the direction of the grade.

From the foregoing discussion, it can be seen that ramp gradients are directly related to design speed; however, design speed is a general indication of the standards being used, and gradients for a ramp with a high design speed should be flatter than for one with a low design speed. As general criteria, it is desirable that ascending gradients on ramps with a design speed of 45 to 50 mph be limited to 3 to 5 percent; those for a 35- to 40-mph speed, to 4 to 6 percent; those for a 25- to 30-mph speed, to 5 to 7 percent; and those for a 15- to 25-mph speed, to 6 to 8 percent. Where topographic conditions dictate, grades steeper than desirable may be used. One-way descending gradients on ramps should be held to the same general maximums, but in special cases they may be 2 percent greater. Where ramp terminals are properly located and fit other design requirements and where the curvature conforms to a reasonable design speed, the ramp generally is long enough to effect the difference in levels with flat or, at the worst, not excessive gradients. The cases in which gradient is a determining factor in the length of the ramp are as follows: (1) for intersection angles of 70 degrees or less it may be necessary to locate the ramp farther from the structure than required for minimum alinement to provide a ramp of sufficient length for reasonable gradient; (2) where the intersection legs are on appreciable grade, with the upper road ascending and the lower road descending from the structure, the ramp will have to effect a large difference in elevation that increases with the distance from the structure; (3) where a ramp leaves the lower road on a downgrade and meets the higher road on a downgrade, the longer-than-usual vertical curves at the terminals may make a long ramp necessary to meet gradient limitations. It is apparent that the alinement and grade of a ramp must be determined jointly.

Vertical curves. Usually, ramp profiles assume the shape of the letter S with a sag vertical curve at the lower end and a crest vertical curve at the upper end. Additional vertical curves may be necessary, particularly on ramps that overpass or underpass other roadways. Where a crest or sag vertical curve extends onto the ramp terminal, the length of curve should be determined by using a design speed between those on the ramp and the highway. See Chapter III for design values for open roadway and turning roadway conditions.

Superelevation and cross slope. The following values should be used for cross slope design on ramps:

1. Superelevation rates, as related to curvature and design speed on ramps, are given in Table IX-12. The highest rate practicable should be used, preferably in the upper half or third of the indicated range particularly on descending ramps.
2. The cross slope on portions of ramps on tangent normally are sloped one way at a practical rate that may range from 1.5 percent to 2 percent for high-type pavements.
3. Superelevation runoff, or the change in superelevation rate per unit length of ramp, should not be greater than that in Table IX-13. The superelevation development is started or ended along the auxiliary pavement or the ramp terminal. Alternate profile lines for both edges should be studied to determine that all profiles match the control

points and that no unsightly bumps and dips are inadvertently developed. Spline profiles are very useful in developing smooth pavement edges.

4. Another important control in developing superelevation along the ramp terminal is that of the crossover crown line at the edge of the through-traffic lane. The maximum algebraic difference in cross slope between the auxiliary pavement and the adjacent through lane is shown in Table IX-14.
5. Three segments of a ramp should be analyzed to determine superelevation rates that would be compatible with the design speed and the configuration of the ramp. The exit terminal, the ramp proper, and the entrance terminal should be studied in combination to ascertain the design speed and superelevation rate. Three ramp configurations are considered for this discussion.

The diamond ramp usually consists of a high-speed exit, tangent or curved alinement on the ramp proper, and stop or yield conditions at the forward terminal. Deceleration to controlling curve speed should occur on the auxiliary lane of the exit, and continued deceleration to stop or yield conditions would occur on the ramp proper. Superelevation comparable to open road conditions would not be appropriate on the ramp proper or the forward terminal.

The loop ramp, as at a cloverleaf interchange, consists of a moderate-speed exit connecting to a slow-speed ramp proper, which in turn connects to a moderate-speed acceleration lane. The curvature of the ramp proper could be a simple curve or combination of curves and transitions. The design speed and superelevation rate would be determined by the curvature of the ramp proper. Superelevation would have to be gradually developed into and out of the curve for the ramp proper, as detailed later in this discussion.

Direct and semidirect ramps generally are designed with high-speed exits, moderate or high-speed ramp propers, and high-speed entrances. Superelevation rates comparable to open-road conditions are appropriate for high-speed direct and semidirect ramps of relatively longer lengths.

On ramps designed for speeds of 40 mph or less, superelevation ranges given in Table IX-12 are appropriate for design of the ramp proper.

The method of developing superelevation at free-flow ramp terminals is demonstrated diagrammatically in Figure X-60.

Figure X-60A shows a tapered exit from a tangent section with the first ramp curve falling beyond the required deceleration length. The normal crown is projected onto the auxiliary pavement, and no superelevation is required until the first ramp proper curve is reached.

Figure X-60B shows a parallel-type exit from a tangent section that leads into a flat exiting curve. At point b the normal crown of the through roadway is projected onto the auxiliary pavement. At point c the crown line can be gradually changed to start the development of superelevation for the exiting curve. At point d two breaks in the crossover crown line would be conducive to developing a full superelevation in the vicinity of the physical nose.

Figures X-60C and X-60D show ramp terminals on which the superelevation of the through roadway would be projected onto the auxiliary pavement.

Figure X-60E shows a parallel entrance on the high side of a curve. At point d the ramp would possibly be flat, and full superelevation would be attained at point c.

Figure X-60F shows a parallel exit from a tangent section with sharp curvature developing in advance of the physical nose. This design is typical for cloverleaf loops. Part of the cross slope change can be accomplished over the length of the parallel lane with about half of the total superelevation being developed at point b. Full superelevation of the ramp proper is reached beyond the physical nose.

Care should be exercised to see that the rate of cross slope change does not exceed criteria in Table IX-12 and that algebraic difference in cross slope does not exceed values in Table IX-14.

Gores. The term "gore" indicates an area downstream from the shoulder intersection points as illustrated in Figure X-61. The physical nose is a point, having some dimensional width separating the pavements, upstream from the gore. The painted nose is a point, having no dimensional width, occurring at the separation of pavements. The neutral area refers to the triangular area between the gore nose and the physical nose and also to the triangular area between the physical nose and the painted nose. The geometric layout of these is an important part of exit ramp terminal design. It is the decision point area that must be clearly seen and understood by approaching drivers. Further, the separating ramp roadway not only must be clearly evident but also must be of geometric shape to fit the likely speeds at that point. In a series of interchanges along a freeway the gores should be uniform and have the same appearance to drivers.

As a general rule, the width at the gore nose is typically between 20 and 30 feet including paved shoulders, measured between the traveled way of the main line and that of the ramp. This dimension may be increased if the ramp roadway curves away from the freeway immediately beyond the gore nose or if speeds in excess of 60 mph are expected to be common.

The entire triangular area should be striped to delineate the proper paths on each side and to assist the driver in identifying the gore area. The MUTCD (**2**) may be referenced for guidance in channelization. Standard or snow plowable raised reflective markers can be employed for additional delineation.

Rumble strips may be placed in the gore area but should not be located too

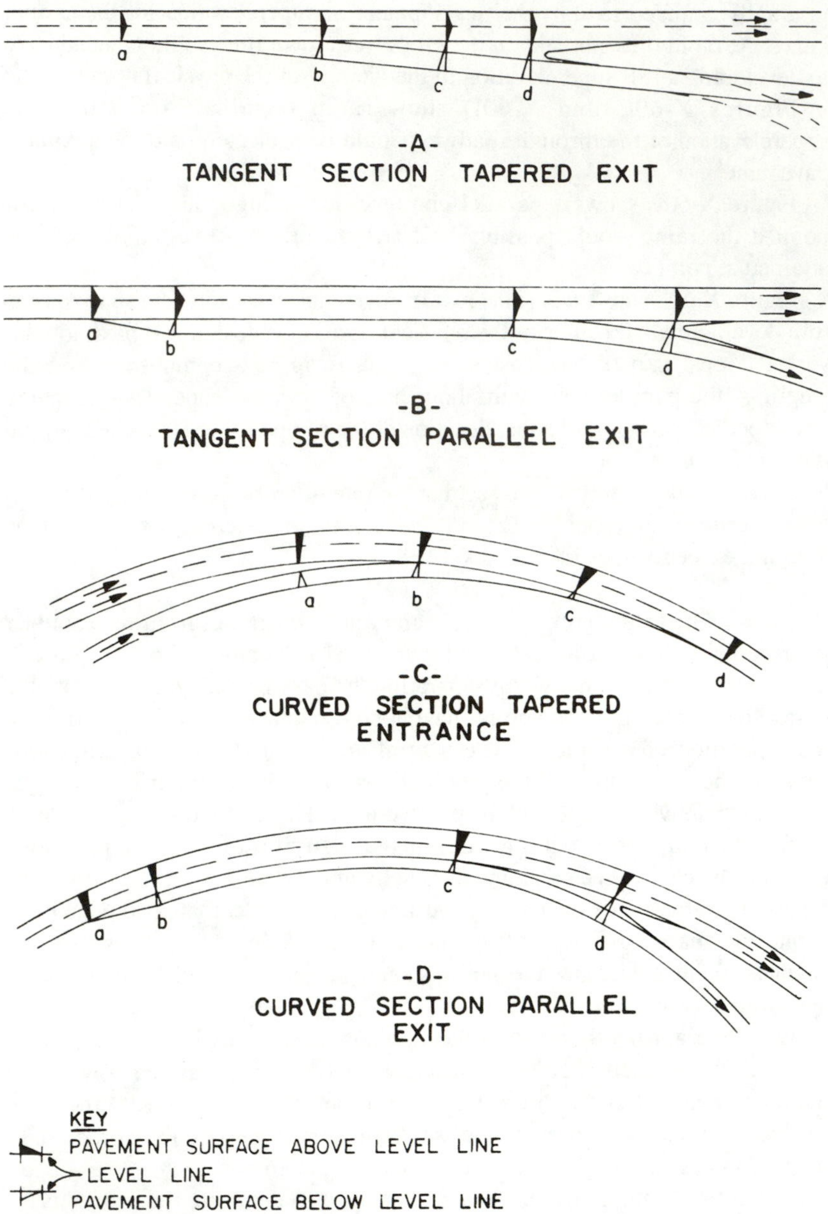

Figure X-60. Development of superelevation at free-flow ramp terminals.

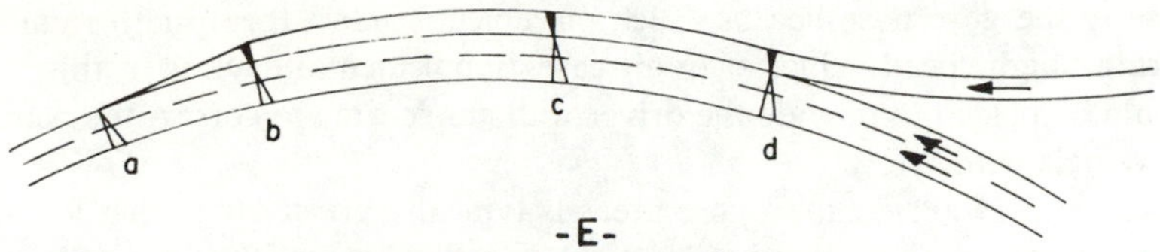

-E-
CURVED SECTION PARALLEL
ENTRANCE ON HIGH SIDE OF CURVE

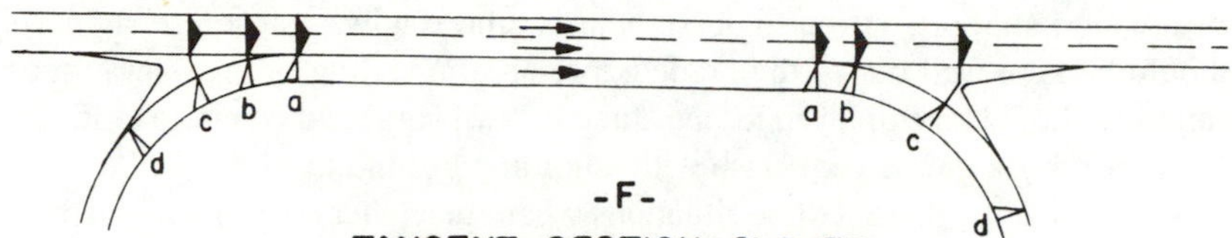

-F-
TANGENT SECTION CLOVER-
LEAF ENTRANCE AND EXIT

Figure X-60. Continued.

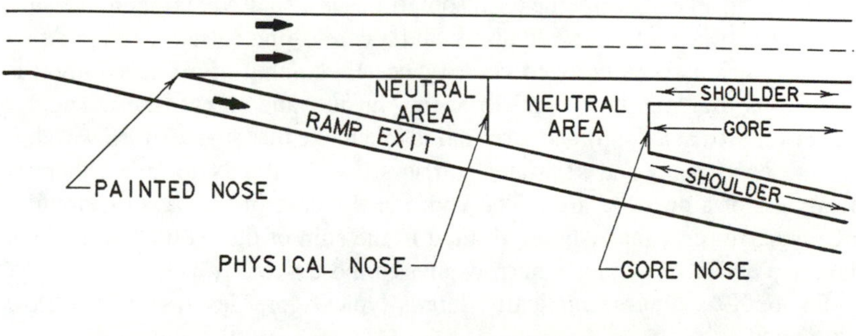

Figure X-61. Typical gore area characteristics.

close to the gore nose because such placement renders them ineffectual for warning high-speed vehicles. In all cases supplemental devices of this type should be placed to provide the driver with ample time to correct the path of his vehicle with safety.

The rate of accidents in gore areas is typically greater than that for run-off-the road accidents at other locations. For this reason, the gore area, and the unpaved area beyond, should be kept as free of obstructions as possible to provide a clear recovery area. The unpaved area beyond the nose should be graded as nearly level with the roadways as is practicable so that vehicles inadvertently entering will not be upset or abruptly stopped by steep slopes. Heavy sign supports, street light standards, and roadway structure supports should be kept well out of the graded gore area. Yielding or breakaway-type supports should be employed for the standard exit sign, and concrete footings, where used, should be kept flush with adjoining ground level.

Unfortunately, there will be situations where placement of a major obstruction in a gore is unavoidable. Gores that occur at exit ramp terminals on elevated structures are a prime example. Also, there are occasions when a bridge pier in a gore cannot be avoided. Guardrails and bridge rails are designed to handle angular impacts but are not effective in handling the kind of near head-on impacts that occur at these gores.

In recognition of this problem a considerable effort has been directed toward the development of cushioning or energy-dissipating devices for use in front of hazardous fixed objects. At present, several types and models of crash cushions are being used. These devices substantially reduce the severity of fixed-object accidents. In view of this reduction, adequate space should be provided for the installation of a crash-cushion device whenever it is found necessary to construct a major obstruction in a gore on a high-speed highway.

Reference may be made to Chapter IV and to the *Roadside Design Guide* **(1)** for details of installation.

Although the term "gore" generally refers to the area between a through roadway and an exit ramp, the term sometimes is also used to refer to the similar area between a through roadway and a converging entrance ramp. At an entrance terminal the point of convergence (beginning of all paved area) is defined as the "merging end." In shape, layout, and extent, the triangular maneuver area at an entrance terminal is much like that at an exit. However, it points downstream and separates traffic streams already in lanes, thereby being less of a decision area. The width at the base of the paved triangular area is narrower, usually being limited to the sum of the shoulder widths on the ramp and freeway plus a narrow physical nose 4 to 8 ft wide.

Figure X-62 diagrammatically details typical gore designs for free-flow exit ramps.

Figures X-62A and X-62B depict a recovery area adjacent to the outside through lane and moderate left offset of the ramp pavement.

Figure X-62C shows a major fork, with neither diverging roadway having priority. The offset is equal for each roadway, and striping or rumble strips are placed upstream from the physical nose. Desirably, curbs, utility poles, and sign supports should be omitted from the gore area, especially on high-

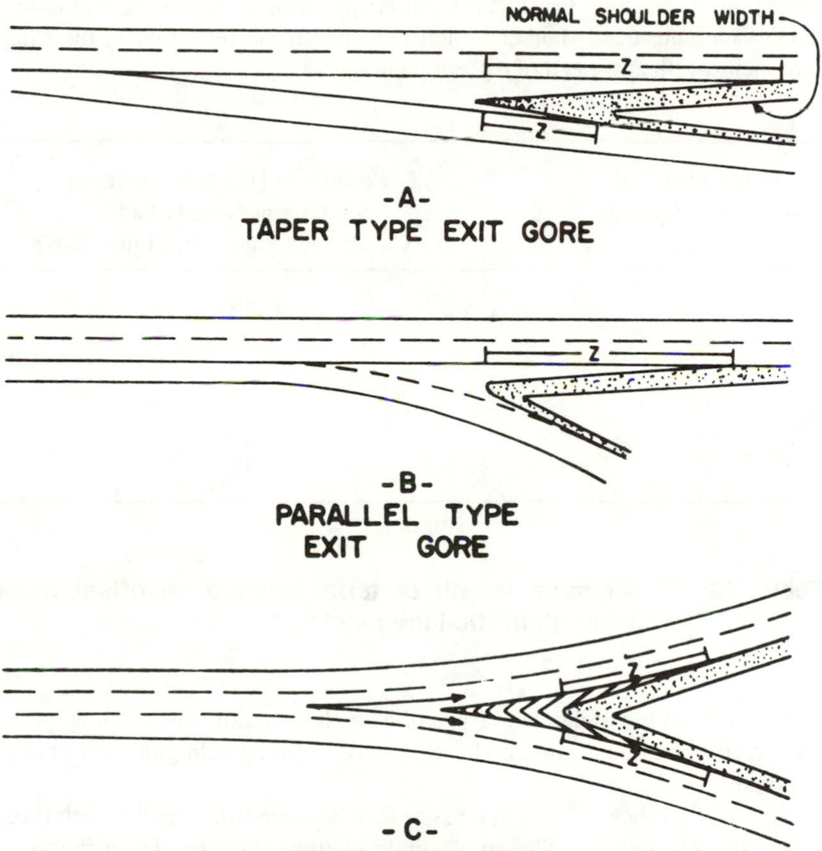

NOTE: FOR "Z" LENGTHS SEE TABLE X-2

Figure X-62. Typical gore details.

speed facilities. When curbs are used, they should be low-profile, mountable-type designs and the geometry of the gore area intersection points is usually curved. When curbs are not used, the geometry of the gore area intersection points can be squared or truncated.

Table X-2 gives minimum lengths for tapers beyond the offset nose. However, another alternative for allowing a recovery area is the use of the paved shoulder of the through lane for any necessary recovery maneuvers.

Figure X-63 shows an entrance ramp, as at a cloverleaf loop, where a reduction in the ramp pavement width is appropriate in order to maintain a single-lane entrance. Another option is to begin the reduction in the ramp pavement width at the end of the ramp curvature.

Design Speed of Approach Highway (mph)	Z–Length in Feet of Nose Taper per Foot of Nose Offset; for Conditions Shown in Figure X-62
30	15
40	20
50	25
60	30
70	35

Table X-2. Minimum length of taper beyond an offset nose (through-traffic-lane side).

Figure X-64 is a night photograph of a single-lane exit. The striping, pavement reflectors, delineators, and fixed-source lighting help guide the exiting motorist.

Figure X-65 shows a gore at a major fork between two freeways with three lanes on each roadway. The small angle of divergence results in the long, gradual split with a clear recovery area. Overhead signs are provided.

Figure X-66 shows a gore at a two-lane exit from an elevated freeway on a viaduct. There is sufficient space in the gore for an impact attenuator device.

Figure X-67 shows a typical gore and ramp terminal as it appears to traffic entering a freeway.

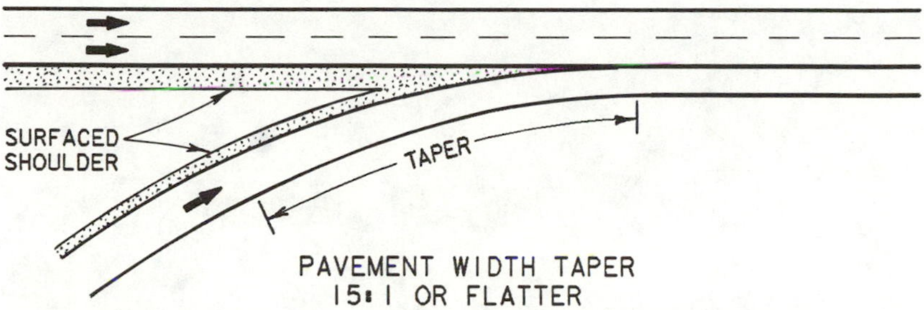

Figure X-63. Pavement narrowing on entrance ramps.

Figure X-64. Gore area, single-lane exit.

Figure X-65. Gore area, major fork.

Figure X-66. Gore area, two-lane exit on viaduct.

Figure X-67. Entrance terminal.

Pavement Widths

Width and cross section. Ramp pavement widths are governed by the type of operation, curvature, and volume and type of traffic. It should be noted that the roadway width for a turning roadway, as distinct from pavement width, includes the shoulders or equivalent clearance outside the edges of pavement. The section, "Widths for Turning Roadways", in Chapter III may be referenced for additional discussion on the treatments at the edge of pavement. Design widths of ramp pavements for various conditions are given in Table X-3. Values are shown for three general design traffic conditions, as follows:

Traffic Condition A—predominantly P vehicles, but some consideration for SU trucks.

Traffic Condition B—sufficient SU vehicles to govern design, but some consideration for semitrailer vehicles.

Traffic Condition C—sufficient bus and combination-type vehicles to govern design.

Traffic conditions A, B, and C are described in broad terms because design data regarding traffic in volume, or percentage of the total, for each type of vehicle are not available to define these traffic conditions with precision in relation to pavement width. In general, traffic condition A can be assumed as having a small volume of trucks or only an occasional large truck; traffic condition B as having a moderate volume of trucks, say in the range of 5 to 10 percent of the total traffic; and traffic condition C as having more and larger trucks.

Radius on Inner Edge of Pavement, R (ft)	Pavement Width (ft)								
	Case I One-Lane, One-Way Operation—No Provision for Passing a Stalled Vehicle			Case II One-Lane, One-Way Operation—With Provision for Passing a Stalled Vehicle			Case III Two-Lane Operation—Either One-Way or Two-Way		
	Design Traffic Condition								
	A	B	C	A	B	C	A	B	C
50	18	18	23	23	25	29	31	35	42
75	16	17	19	21	23	27	29	33	37
100	15	16	18	20	22	25	28	31	35
150	14	16	17	19	21	24	27	30	33
200	13	16	16	19	21	23	27	29	31
300	13	15	16	18	20	22	26	28	30
400	13	15	16	18	20	22	26	28	29
500	12	15	15	18	20	22	26	28	29
Tangent	12	15	15	17	19	21	25	27	27
Width Modification Regarding Edge of Pavement Treatment:									
No stabilized shoulder	None			None			None		
Mountable curb	None			None			None		
Barrier curb:									
one side	Add 1 ft			None			Add 1 ft		
two sides	Add 2 ft			Add 1 ft			Add 2 ft		
Stabilized shoulder, one or both sides	None			Deduct shoulder width; minimum Pavement width as under case I			Deduct 2 ft where shoulder is 4 ft or wider		

NOTE: [a]A = predominantly P vehicles, but some consideration for SU trucks.

B = sufficient SU vehicles to govern design, but some consideration for semitrailer vehicles.

C = sufficient bus and combination-types of vehicles to govern design.

Refer to Chapter III for additional information.

Table X-3. Design widths of pavements for turning roadways.

Shoulders and lateral clearances. Design values for shoulders and lateral clearances on the ramps are as follows:

1. The sum of one-way ramp pavement width and the paved shoulder width should not exceed the Case I width for the controlling traffic condition and the controlling ramp radius, plus 10 to 12 ft. Usually, the paved shoulder widths are made the same for the full ramp length and the controlling radius is that in effect for most of its length. For pavement support, a paved shoulder width of 2 to 4 ft is desirable on the left, with the remainder of the total paved width as the paved right shoulder. For example, with traffic condition C and a 400-ft radius the Case I shoulder-ramp-shoulder widths (using plus 10 ft) may be 2-16-8 or 4-16-6 or some other combination with a 26-ft total. While the 16 ft ramp pavement width may not allow for passing a stalled vehicle by itself, the use of a portion of the paved shoulder commonly allows traffic to get around the vehicle. For the same conditions but a Case II design, the width should be 2-22-2, for a total of 26 ft, since the ramp pavement itself includes a part of the shoulder width. A wider paved total width not only is unnecessary for the passing of a stalled vehicle but is undesirably hazardous in that passing of slow-moving vehicles or two-lane operation are encouraged, for which the ramp width and usually the length are deficient.
2. For Case III one-way operation, the sum of the right and left paved shoulder width should not exceed 10 to 12 ft.
3. Directional ramps with a design speed over 40 mph should have a paved right shoulder of 8 to 10 ft and a 2- to 6-ft paved left shoulder.
4. For freeway ramp terminals where the ramp shoulder is narrower than that on the freeway, the paved shoulder width of the through lane should be carried into the exit terminal, and should begin within the entrance terminal, with the transition to the narrower ramp shoulder effected gracefully on the ramp end of the terminal. Abrupt change should be avoided.
5. Ramps should have a lateral clearance on the right outside of the edge of the traveled way of at least 6 ft, and preferably 8 to 10 ft, and on the left a lateral clearance of at least 4 ft beyond the edge of paved shoulder.
6. Where ramps pass under structures, the total width including paved shoulders should be carried through without change. There should be a lateral clearance of at least 4 ft beyond the edge of paved shoulder.
7. Ramps on overpasses should have the full approach roadway width carried over the structure.
8. Edge lines and/or some type of color or texture differentiation between the traveled way and shoulder is desirable.

Shoulders and curbs. Shoulders are needed on ramps and terminals in interchange areas to provide stopping space clear of the traffic lanes, to minimize the effect of breakdowns, and to aid drivers who may be confused.

Highways preferably are designed without curbs, and ramps at interchanges are similarly treated. Curbs should be considered only to facilitate particularly difficult drainage situations as in urban areas where enclosed drainage is required because of restricted right-of-way. In some cases curbs are used at the ramp terminals but are omitted along the central ramp portions. The use of curbs on facilities designed for intermediate or higher speeds is not recommended except in special cases. Where curbs are not used, full-depth paving should be provided on shoulders because of the frequent use of shoulders for turning movements. On low-speed facilities curbs may be placed at the pavement edge. Barrier curbs are seldom used in conjunction with shoulders, except where pedestrian protection is required. On high-speed facilities, mountable curbs should desirably be placed at the outer edge of the shoulder. Because of fewer restrictions and more liberal designs, the need for curbs in rural areas seldom arises. See Chapter IV for a full discussion of shoulder cross-section elements.

Ramp Terminals

The terminal of a ramp is that portion adjacent to the through traveled way, including speed-change lanes, tapers, and islands. Ramp terminals may be the at-grade type, as at the crossroad terminal of diamond or partial cloverleaf interchanges, or the free-flow type where ramp traffic merges with or diverges from high-speed through traffic at flat angles. Design elements for the former type are discussed in Chapter IX, and those for the latter type are discussed in the following sections.

Terminals are further classified according to the number of lanes on the ramp at the terminal, either single or multilane, and according to the configuration of the speed-change lane, either taper or parallel type.

Left-hand entrances and exits. Left-hand entrances and exits are contrary to the concept of driver expectancy when intermixed with right-hand entrances and exits.

Extreme care should be exercised to avoid left-hand entrances and exits in the design of interchanges. Even in the case of major forks and branch connections, the less significant roadway should exit and enter on the right. (See the discussion of route continuity in this chapter.)

Left-side ramp terminals break up the uniformity of interchange patterns and in general create hesitant operation on the through roadways.

Left-hand entrances and exits are considered satisfactory for collector-distributor roads; however, their use on high-speed, free-flow ramp terminals is not recommended. Because left-hand entrances and exits are contrary to driver expectancy, special attention should be given to signing and the provision for decision sight distance in order to alert the driver that an unusual situation exists.

Terminal location and sight distance. Where diamond ramps and partial cloverleaf arrangements intersect the crossroad at grade, an at-grade intersection is formed. Desirably, this intersection should be located an adequate distance from the separation structure to provide sight distance that permits safe entry or exit on the crossroad. The criteria for entering sight distance along with stopping sight distance are detailed in Chapter III.

Drivers prefer and expect to exit in advance of the separation structure. The use of collector-distributor roads and single exits on partial cloverleafs and other types of interchanges automatically locates the main line exit in advance of the separation structures.

Designs that employ exits concealed behind crest vertical curves should be avoided, especially on high-speed facilities.

Desirably, high-speed entrance ramp terminals should be located on descending grades to aid truck acceleration. Adequate sight distance at entrance terminals should be available so that merging traffic on the ramp can adjust speed to merge into gaps on the main facility.

Loop ramps that are located beyond the structure, as in the conventional cloverleaf or in certain arrangements of partial cloverleafs, usually require a parallel deceleration lane. The actual exit from the auxiliary lane is difficult for drivers to locate even when sight distance is not restricted by a vertical curve. Locating the exit in advance of the structure via a single exit alleviates this problem. See the two-exit versus the single-exit discussion in this chapter.

Ramp terminal design. Profiles of ramp terminals should be designed in association with horizontal curves to avoid sight restrictions that will adversely affect operations. At an exit into a ramp on a descending grade, a horizontal curve ahead should not appear suddenly to a driver using the ramp. Instead the initial crest vertical curve should be made longer and sight distance over it increased so that the beginning and the direction of the horizontal curve are obvious to the driver in sufficient time for safe operation. At an entrance terminal from a ramp on an ascending grade, the portion of the ramp and its terminal intended for acceleration should closely parallel the through-lane profile to permit entering drivers to have a clear view ahead, to the side, and to the rear on the through road.

It is desirable that profiles of highway ramp terminals be designed with a platform on the ramp side of the approach nose or merging end. This platform should be a length of about 200 ft or more on which the profile does not greatly differ from that of the adjacent through-traffic lane.

A platform area should also be provided at the at-grade terminal of a ramp. The length of this platform should be determined from the type of traffic control involved at the terminal and the capacity requirements. For further discussion, see Chapter IX.

Traffic control. On major highways, ramps are arranged to facilitate all turning movements by merging or diverging maneuvers. On minor highways, some of the left-turning movements often are made directly at-grade. The left-turning movements leaving the crossing highway preferably should have median lanes. For low-volume crossroads the left-turning movements from the ramps normally should be controlled by stop signs. The right-turning movements from the ramps into multilane crossroads should be provided with an acceleration lane or generous taper or should be controlled by stop or yield signs. Ramps approaching stop signs should have close to 90° alinement and be nearly level for storage of several vehicles.

Traffic signal controls may be required at ramp terminals on the minor road where there is sufficient volume of through and turning traffic. In such cases the intersections formed at the terminals would be designed and operated in the same manner as any other traffic-signal-controlled intersections at grade. Signal controls should be avoided on express-type highways and confined to the minor highways on which other intersections are at-grade, and some include signal controls. In or near urban areas, signal control has considerable application at ramp terminals on streets crossing over or under an expressway. Here, the turning movements usually are sizable, and the cost of right-of-way and improvements thereon is high. As a result, appreciable saving may be effected by the use of diamond ramps with high-type terminals on the expressway and signalized terminals on the cross streets. Warrants for the installation of traffic signals that can be applied to diamond ramp terminals are given in Part IV of the MUTCD **(2)**.

Distance between a free-flow terminal and structure. The terminal of a ramp should not be near the grade-separation structure. If it is not possible to place the exit terminal in advance of the structure, the exiting terminal on the far side of the structure should be well removed in order that, when leaving, drivers have some distance, after passing the structure, in which to see the turnout and begin the turnoff maneuver. Sight distance comparable to guidelines set out under the decision sight distance discussion is recommended where practical. The distance between the structure and the approach nose at the ramp terminal should be sufficient for exiting drivers to leave the through lanes without undue hindrance to through traffic. Such distance also aids drivers entering from a far side terminal in having a clear view well back on the through road behind or to the left. These drivers may be able to see back along the road beyond the limits of the structure, but as a general rule, the crest of the profile at an overpass and the columns, abutments, and approach walls at an underpass obstruct or impair their view of the traffic stream with which they must merge.

Although there is no direct measure of the several variables involved, the conditions for determining the distance between a structure and far side approach nose are similar to those discussed for speed-change lanes. A reasonable minimum distance between the structure and an exit nose of about the length of speed-change taper is suggested. Decision sight distances are desirable but are not rigid controls for ramp design. The overall shape of the ramp to fit right-of-way or topographic controls may govern.

The safety and convenience of both through and turning traffic are enhanced by a long distance between structure and exit ramp terminal, but too long a distance for certain ramp arrangements such as cloverleaf loop ramps could result in unusually large additional right-of-way requirements as well as increased travel time and length on the loops. Where only one loop is required and it falls on the far side of the structure, a speed-change lane should be developed on the near side and carried across the structure if sight distance is a problem.

Ramp terminals on the near side of a grade separation need not be as far removed as those beyond the structure. Both the view of the terminal ahead for drivers approaching on the through road and the view back along the road for drivers on an entrance ramp are not affected by the structure. Where an entrance ramp curve on the near side of the structure requires an acceleration lane, the ramp terminal should be located to provide length for it between the terminal and the structure, or the acceleration lane could be continued through or over the structure. Where far-side terminals are located close to the structure, the sight distance should be checked because the horizontal sight line may be limited by the abutment or parapet.

Distance between successive ramp terminals. On urban freeways there are frequently two or more ramp terminals in close succession along the through lanes. To provide sufficient maneuvering length and adequate space for signing, a reasonable distance is required between terminals.

Spacing between successive outer ramp terminals is dependent on the classification of the interchanges involved, the function of the ramp pairs (entrance or exit), and weaving potential, when applicable.

The term "systems interchange" is used to identify an interchange that transfers traffic freeway to freeway. A "service interchange" is the designation for an interchange between a freeway and a local arterial.

The ramp-pair combinations are entrance followed by entrance (EN-EN), exit followed by exit (EX-EX), exit followed by entrance (EX-EN), entrance followed by exit (EN-EX) (weaving), and turning roadways.

Figure X-68 shows the minimum values for spacing of ramp terminals for the various ramp-pair combinations as they are applicable to the interchange classifications.

Where an entrance ramp is followed by an exit ramp, the absolute minimum distance between the successive noses is governed by weaving requirements. A notable exception to this length policy for EN-EX ramp combinations is the distance between loop ramps of cloverleaf interchanges. For these interchanges the distance between EN-EX ramp noses is primarily dependent on loop ramp radii and roadway and median widths. A recovery lane beyond the nose of the loop ramp exit is desirable.

When the distance between the successive noses is less than 1,500 ft, the speed-change lanes should be connected to provide an auxiliary lane. This auxiliary lane is provided for improved traffic operation over relatively short sections of the freeway route and is not considered as an addition to the basic number of lanes. See the section "Auxiliary Lanes" in this chapter for alternate methods of dropping these lanes.

Speed-change lanes. Drivers leaving a highway at an interchange are required to reduce speed as they exit on a ramp. Drivers entering a highway from a turning roadway accelerate until the desired open-road speed is reached. Because the necessary change in speed is usually substantial, provision should be made for acceleration and deceleration to be accomplished on auxiliary lanes to minimize interference with through traffic and to reduce accident potential. Such an auxiliary lane, including tapered areas, primarily for the acceleration or deceleration of vehicles entering or leaving the through-traffic lanes, is termed a speed-change lane. The term speed-change lane, deceleration lane, or acceleration lane as used herein applies broadly to the added pavement joining the traveled way of the highway with that of the turning roadway and does not necessarily imply a definite lane of uniform width. It is a part of the elongated ramp terminal area.

A speed-change lane should, as a minimum requirement, have sufficient length to enable a driver to make the necessary change between the speed of operation on the highway and the speed on the turning roadway in a safe and comfortable manner. Moreover, in the case of an acceleration lane, there should be additional length sufficient to permit adjustments in speeds of both through vehicles and entering vehicles so that the driver of the entering vehi-

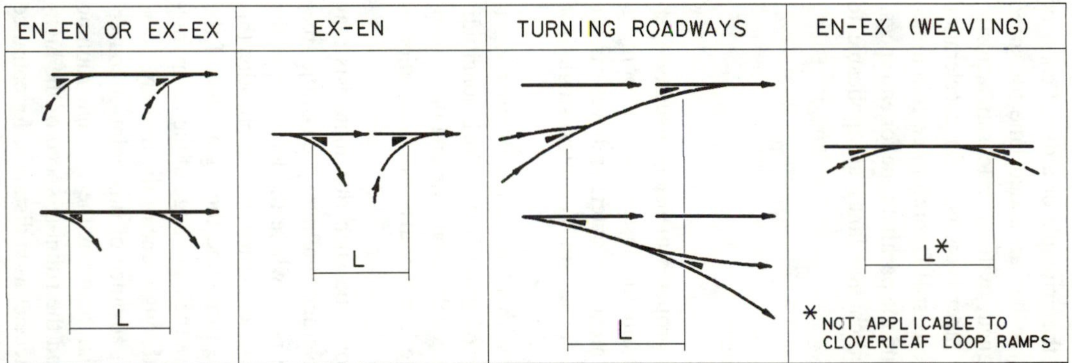

EN-EN OR EX-EX		EX-EN		TURNING ROADWAYS		EN-EX (WEAVING)			
						SYSTEM TO SERVICE INTERCHANGE		SERVICE TO SERVICE INTERCHANGE	
FULL FREEWAY	CDR OR FDR	FULL FREEWAY	CDR OR FDR	SYSTEM INTER-CHANGE	SERVICE INTER-CHANGE	FULL FWY.	CDR OR FDR	FULL FWY.	CDR OR FDR
MINIMUM LENGTHS MEASURED BETWEEN SUCCESSIVE RAMP TERMINALS									
1000	800	500	400	800	600	2000	1600	1600	1000

NOTE: FDR - FREEWAY DISTRIBUTOR ROAD EN - ENTRANCE
CDR - COLLECTOR DISTRIBUTOR ROAD EX - EXIT

THE RECOMMENDATIONS ARE BASED ON OPERATIONAL EXPERIENCE AND NEED FOR FLEXIBILITY AND ADEQUATE SIGNING. THEY SHOULD BE CHECKED IN ACCORDANCE WITH THE PROCEDURE OUTLINED IN THE HIGHWAY CAPACITY MANUAL (3) AND THE LARGER OF THE VALUES IS SUGGESTED FOR USE. ALSO, A PROCEDURE FOR MEASURING THE LENGTH OF THE WEAVING SECTION IS GIVEN IN CHAPTER 4 OF THE HIGHWAY CAPACITY MANUAL (3).

Figure X-68. Recommended minimum ramp terminal spacing.

cle can position himself opposite a gap in the through-traffic stream and maneuver into it before reaching the end of the acceleration lane. This latter requirement has much to do with both the configuration and length of an acceleration lane.

Speed-change lanes are designed in two general forms, the taper type and the parallel type. The taper type works on the principle of a direct entry or exit at a flat angle; whereas the parallel type has an added lane for speed change. Either type, when properly designed, will operate satisfactorily. There is a decided trend toward use of the taper type, both for deceleration and acceleration. However, the parallel type is still favored in certain areas. Some agencies use the taper type for exits and the parallel type for entrances.

See Chapter IX for discussion of speed-change lanes as applicable to at-grade intersections.

Single-Lane Free-Flow Terminals, Entrances

Taper type. The taper-type entrance of proper dimensions usually operates smoothly at all volumes up to and including the design capacity of merging areas. By relatively minor speed adjustment, the entering driver can see and use an available gap in the through-traffic stream. A typical single-lane taper-type entrance terminal is shown in Figure X-69A.

The entrance is brought into the freeway with a long, uniform taper. Operational studies show a desirable rate of taper of about 50:1 to 70:1 (longitudinal to lateral) between the outer edge of the acceleration lane and the edge of the through-traffic lane. The gap acceptance length is also a consideration as illustrated in Figure X-69.

The geometrics of the ramp proper should be such that motorists may attain a speed approximately equal to the average running speed of the freeway less 5 mph by the time they reach the point where the left edge of the ramp joins the traveled way of the freeway. For consistency of application this point of convergence of the left edge of the ramp and the right edge of the through lane may be assumed to occur where the right edge of the ramp traveled way is 12 ft from the right edge of the through lane of the freeway.

The distance required for acceleration in advance of this point of convergence is governed by the speed differential between the average running speed on the entrance curve of the ramp and the running speed of the highway. Table X-4 shows minimum lengths of acceleration distances for entrance terminals and their derivation. Figure X-69 shows the minimum lengths for gap acceptance. Where ramps occur in grades, the lengths should be adjusted in accordance with Table X-5.

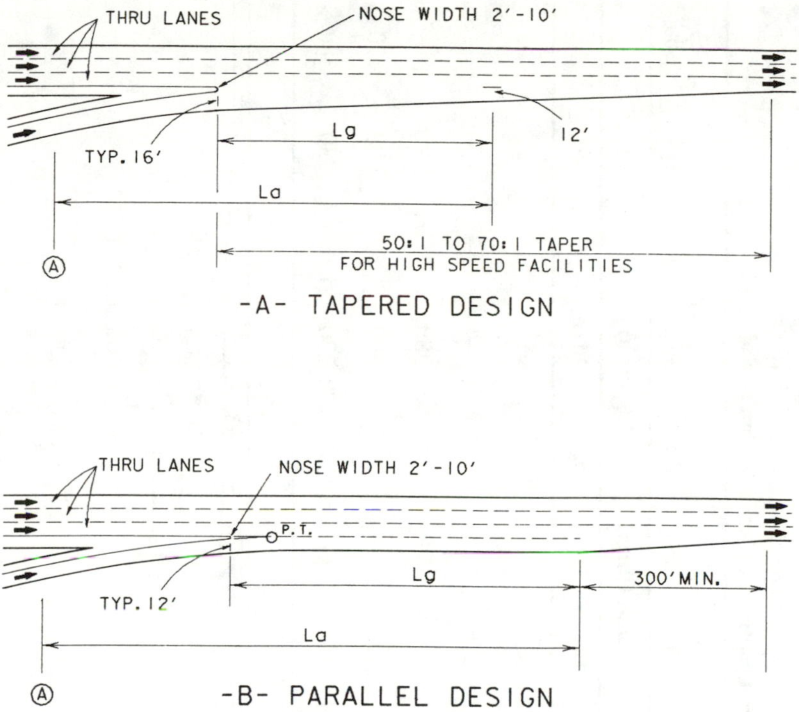

NOTES:

1. La IS THE REQUIRED ACCELERATION LENGTH AS SHOWN IN TABLE X-4 OR X-5.

2. POINT Ⓐ CONTROLS SAFE SPEED ON THE RAMP. La SHOULD NOT START BACK ON THE CURVATURE OF THE RAMP UNLESS THE RADIUS EQUALS 1000′ OR MORE.

3. Lg IS REQUIRED GAP ACCEPTANCE LENGTH. Lg SHOULD BE A MINIMUM OF 300′ TO 500′ DEPENDING ON THE NOSE WIDTH.

4. THE VALUE OF La OR Lg, WHICHEVER PRODUCES THE GREATEST DISTANCE DOWNSTREAM FROM WHERE THE NOSE WIDTH EQUALS TWO FEET, IS SUGGESTED FOR USE IN THE DESIGN OF THE RAMP ENTRANCE.

Figure X-69. Typical single-lane entrance ramps.

Highway		Acceleration Length, L (ft) For Entrance Curve Design Speed (mph)								
		Stop Condition	15	20	25	30	35	40	45	50
Design Speed (mph)	Speed Reached, V_a (mph)	and Initial Speed, V'_a (mph)								
		0	14	18	22	26	30	36	40	44
30	23	190	—	—	—	—	—	—	—	—
40	31	380	320	250	220	140	—	—	—	—
50	39	760	700	630	580	500	380	160	—	—
60	47	1,170	1,120	1,070	1,000	910	800	590	400	170
70	53	1,590	1,540	1,500	1,410	1,330	1,230	1,010	830	580

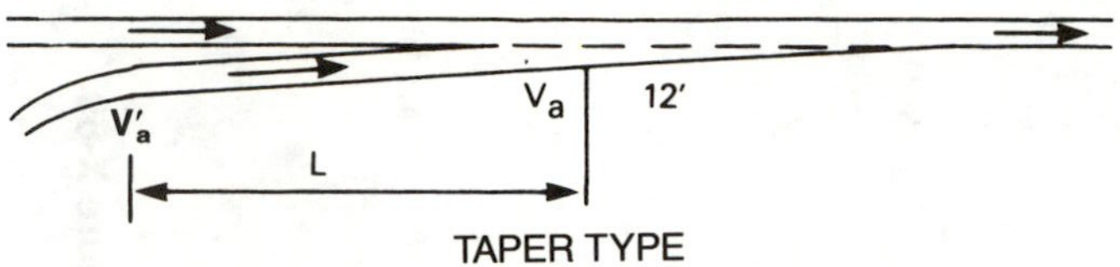

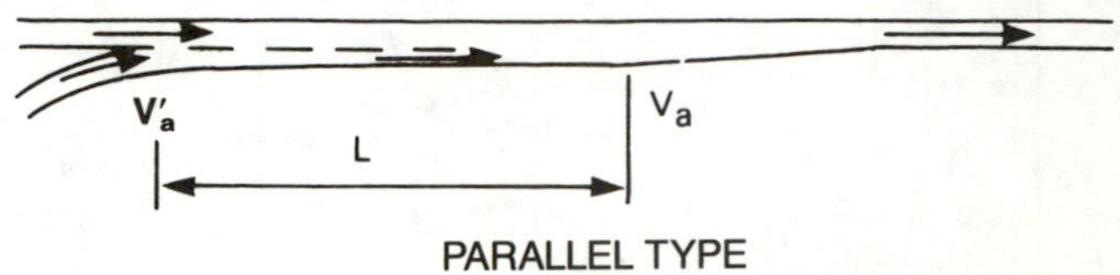

Note: Uniform 50:1 to 70:1 tapers are recommended where lengths of acceleration lanes exceed 1,300 feet.

Table X-4. Minimum acceleration lengths for entrance terminals with flat grades of 2 percent or less.

Parallel type. The parallel-type entrance provides an added lane of sufficient length to enable a vehicle to accelerate to near-freeway speed prior to merging. A taper is provided at the end of the added lane. The process of entering the freeway is similar to a lane change to the left. The driver is able to use the side-view and rear-view mirrors to monitor following and merging traffic.

A typical design of a parallel-type entrance is shown in Figure X-69B.

The taper at the downstream end of a parallel-type acceleration lane should be a suitable length to guide the vehicle gradually onto the through lane of the freeway. A length of approximately 300 ft is suitable for design speeds up to 70 mph.

The length of a parallel-type acceleration lane is generally measured from the point where the left-edge of the traveled way of the ramp joins the traveled way of the freeway to the beginning of the taper. It is noteworthy that, in the case of the taper type, acceleration is accomplished on the ramp upstream of the point of convergence of the two roadways; whereas, in the case of the parallel type, acceleration usually takes place downstream from this point. However, as in the case of the taper type, a part of the ramp proper may also be considered in the acceleration length, provided the curve approaching the acceleration lane has a long radius of say 1,000 ft or more, and the motorist on the ramp has an unobstructed view of traffic on the freeway to his left. The minimum acceleration lengths for entrance terminals are given in Table X-4, and the adjustments for grades are given in Table X-5.

The operational and safety benefits of long acceleration lanes are well recognized, particularly where both the freeway and ramp carry high traffic volumes. A long acceleration lane provides more time for the merging vehicles to find an opening in the through-traffic stream. An acceleration lane length of at least 1,200 ft, plus the taper, is desirable wherever it is anticipated that the ramp and freeway will frequently carry traffic volumes approximately equal to the design capacity of the merging area.

Single Lane Free-Flow Terminals, Exits

Taper type. The taper-type exit fits well the direct path preferred by most drivers, permitting them to follow an easy path within the diverging area. The taper-type exit terminal beginning with an outer edge alinement break usually provides a clear indication of the point of departure from the through lane and in general has been found to operate smoothly on high-volume freeways. The divergence angle is usually between 2° and 5°.

Studies of this type of terminal show that most vehicles leave the through lane at relatively high speeds, thereby reducing the possibilities of rear-end collisions because of the start of deceleration on the through lane. Once off the lane, the needed speed change can be effected as the exiting vehicle moves along the taper onto the ramp proper. Figure X-70A shows a typical design for a taper-type exit.

Deceleration should be effected after the vehicle has cleared the through-traffic lane and before reaching the point limiting design speed for the ramp proper. The length available for deceleration may be assumed to extend from a point where the right edge of the tapered wedge is about 12 ft from the right edge of the through lane, to the point controlling the safe speed for the ramp. The length provided between these points should be at least as great as the distance required to accomplish the necessary deceleration, which is governed by the speed of traffic on the through lane and the speed to be attained on the ramp. Deceleration may end in a complete stop, as at a crossroad terminal for a diamond interchange, or the critical speed may be governed by the curvature of the ramp roadway. Minimum deceleration lengths for various combinations of design speeds for the highway and for the ramp roadway are given in Table X-6. Grade adjustments are given in Table X-5.

The taper-type exit terminal design can be used advantageously in developing the desired long, narrow, triangular emergency maneuver area just upstream from the exit nose located in proper offset position from both the through lane and separate ramp lane. The taper configuration also works well in the length-width superelevation adjustments to effect a ramp cross slope different from that of the through lane.

The width of the recovery area or the distance between the inner edges of the diverging pavements at the ramp nose is usually 20 to 30 ft. This entire area should be paved to provide a safe maneuver and recovery area, but the desired travel path for the ramp roadway should be clearly delineated by pavement markings.

Parallel type. A parallel-type exit terminal usually begins with a taper, followed by a derived length of added full lane. A typical parallel-type exit terminal is shown in Figure X-70C. This type of terminal provides an inviting exit area, because the foreshortened view of the taper and the added width are very apparent. However, this design assumes that drivers will exit near the beginning of the added lane, and effect speed change thereafter. It requires a reverse-curve maneuver that is somewhat unnatural. Under low-volume conditions a driver may choose to avoid the reverse-curve exit path and turn directly off the through lane in the vicinity of the exit nose. Such a maneuver may result in undesirable deceleration on the through lane, in undesirable conflict on the deceleration lane, or in excessive speed in the exit-nose area.

The length of a parallel-type deceleration lane is usually measured from the point where the added lane attains a 12-ft width to the point where the alinement of the ramp roadway departs from the alinement of the freeway. Where the ramp proper is curved, it is desirable to provide a transition at the end of the deceleration lane. A compound curve may be used with the initial curve desirably having a long radius of, say, 1,000 ft or more. A transition or a long radius curve is also desirable if the deceleration lane connects with a relatively straight ramp. In such cases a portion of the ramp may be considered as a part of the deceleration length, thus shortening somewhat the required

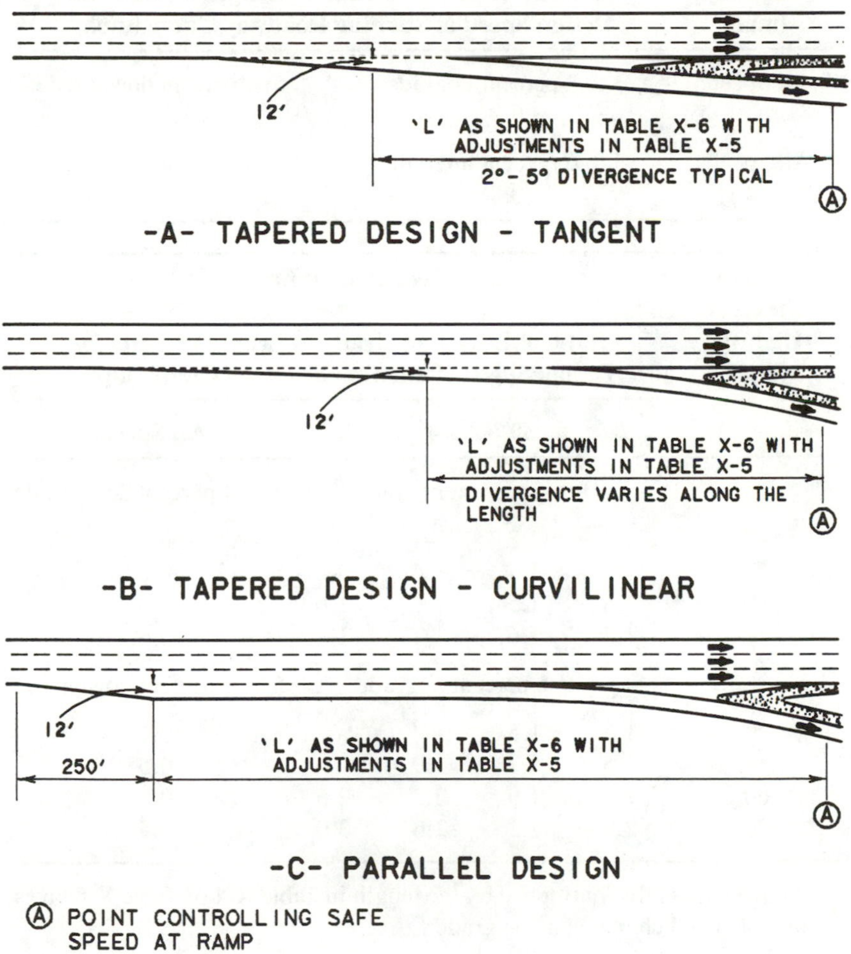

Figure X-70. Exit ramps—single lane.

length of contiguous parallel lane. Minimum lengths are given in Table X-6 and adjustments for grades in Table X-5. Longer parallel-type deceleration lanes are more likely to be used properly. Lengths of at least 800 ft are desirable.

Design Speed of Highway (mph)	Deceleration Lanes Ratio of Length on Grade to Length on Level for Design Speed of Turning Roadway Curve (mph)[a]	
All Speeds	3 to 4 percent upgrade 0.9	3 to 4 percent downgrade 1.2
All Speeds	5 to 6 percent upgrade 0.8	5 to 6 percent downgrade 1.35

Design Speed of Highway (mph)	Acceleration Lanes Ratio of Length of Grade to Length of Level for Design Speed of Turning Roadway Curve (mph)[a]				
	20	30	40	50	All Speeds
	3 to 4 percent upgrade				3 to 4 percent downgrade
40	1.3	1.3	—	—	0.7
50	1.3	1.4	1.4	—	0.65
60	1.4	1.5	1.5	1.6	0.6
70	1.5	1.6	1.7	1.8	0.6
	5 to 6 percent upgrade				5 to 6 percent downgrade
40	1.5	1.5	—	—	0.6
50	1.5	1.7	1.9	—	0.55
60	1.7	1.9	2.2	2.5	0.5
70	2.0	2.2	2.6	3.0	0.5

[a] Ratio from this table multiplied by the length in Table X-4 or Table X-6 gives length of speed change lane on grade.

Table X-5. Speed change lane adjustment factors as a function of grade.

Highway Design Speed, V (mph)	Average Running Speed, V_a (mph)	Deceleration Length, L (ft) For Design Speed of Exit Curve, V′ (mph)								
		Stop Condition	15	20	25	30	35	40	45	50
		For Average Running Speed on Exit Curve, V'_a (mph)								
		0	14	18	22	26	30	36	40	44
30	28	235	185	160	140	—	—	—	—	—
40	36	315	295	265	235	185	155	—	—	—
50	44	435	405	385	355	315	285	225	175	—
60	52	530	500	490	460	430	410	340	300	240
65	55	570	540	530	490	480	430	380	330	280
70	58	615	590	570	550	510	490	430	390	340

V = Design speed of highway
V_a = Average running speed on highway
V′ = Design speed of exit curve
V'_a = Average running speed on exit curve

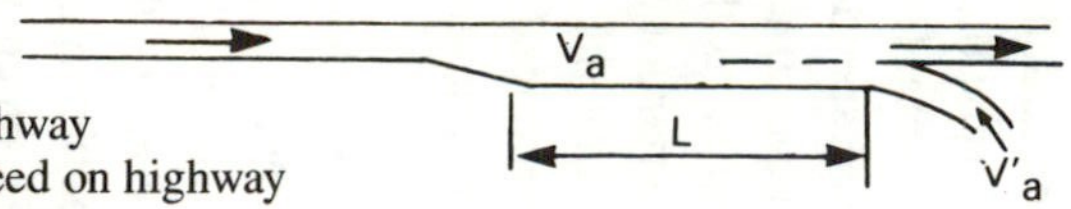

PARALLEL TYPE

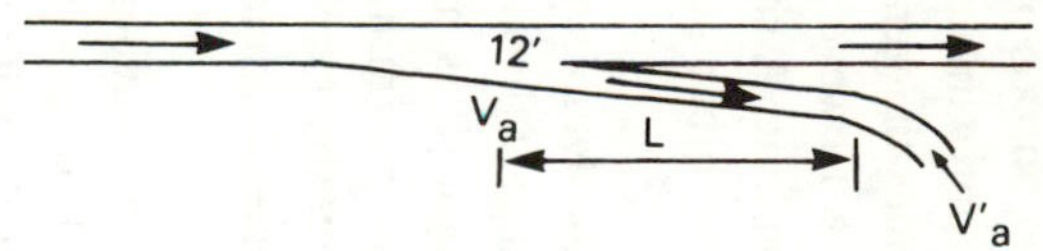

TAPER TYPE

Table X-6. Minimum deceleration lengths for exit terminals with flat grades of 2 percent or less.

The taper portion of a parallel-type deceleration lane desirably should have a taper of approximately 15 to 25 ft longitudinally to 1 ft transversely. A long taper indicates the general path to be followed and reduces the unused portion of pavement. However, a long taper tends to entice the through driver into the deceleration lane. A short taper produces a better "target" to the approaching driver, giving him a positive indication of the added lane ahead.

Free-flow terminals on curves. The previous discussion was stated in terms of tangent through-lane alinement. Because the curvature on most freeways is slight, it is usually not necessary to make any appreciable adjustments at ramp terminals on curves. However, where the curves on a freeway are relatively sharp and it is necessary to provide exits and entrances on these curves, some adjustments in design may be desirable to avoid operational difficulties.

On freeways having design speeds of 60 mph or more the curves are sufficiently gentle so that either the parallel type or the taper type of speed-change lane is suitable. With the parallel type the design is about the same as that on tangent and the added lane is usually on the same curvature as the main line. With the taper type the dimensions applicable to terminals located on tangent alinement are suitable for use on curves. A method for developing the alinement of tapered speed-change lanes on curves is illustrated in Figure X-71. The ramp is tapered at the same rate relative to the through-traffic lanes on the curved section as the tangent section.

Wherever a part of a tapered speed-change lane falls on curved alinement, it is desirable that the entire length be within the limits of the curve. Where the taper is introduced on tangent alinement just upstream from the beginning of the curve, the outer edge of the taper will appear as a kink at the point of curvature.

At ramp terminals on relatively sharp curves such as those that may occur on freeways having a design speed of 50 mph, the parallel type of speed-change lanes has an advantage over the taper type. At exits the parallel type is less likely to confuse through traffic, and at entrances this type will usually result in smoother merging operations. Parallel-type speed-change lanes at ramp terminals on curves are shown diagrammatically in Figure X-72.

Entrances on curved sections of highway are generally less of a problem than exits. Figures X-72A and X-72B show entrances with the highway curving to the left and right, respectively. It is important that the approach curve on the ramp has a very long radius as it joins the acceleration lane. This alines the entering vehicle with the acceleration lane and lessens the chances of motorists entering directly onto the through lanes. The taper at the end of the acceleration lane should be long, preferably about 300 ft. When reverse curve

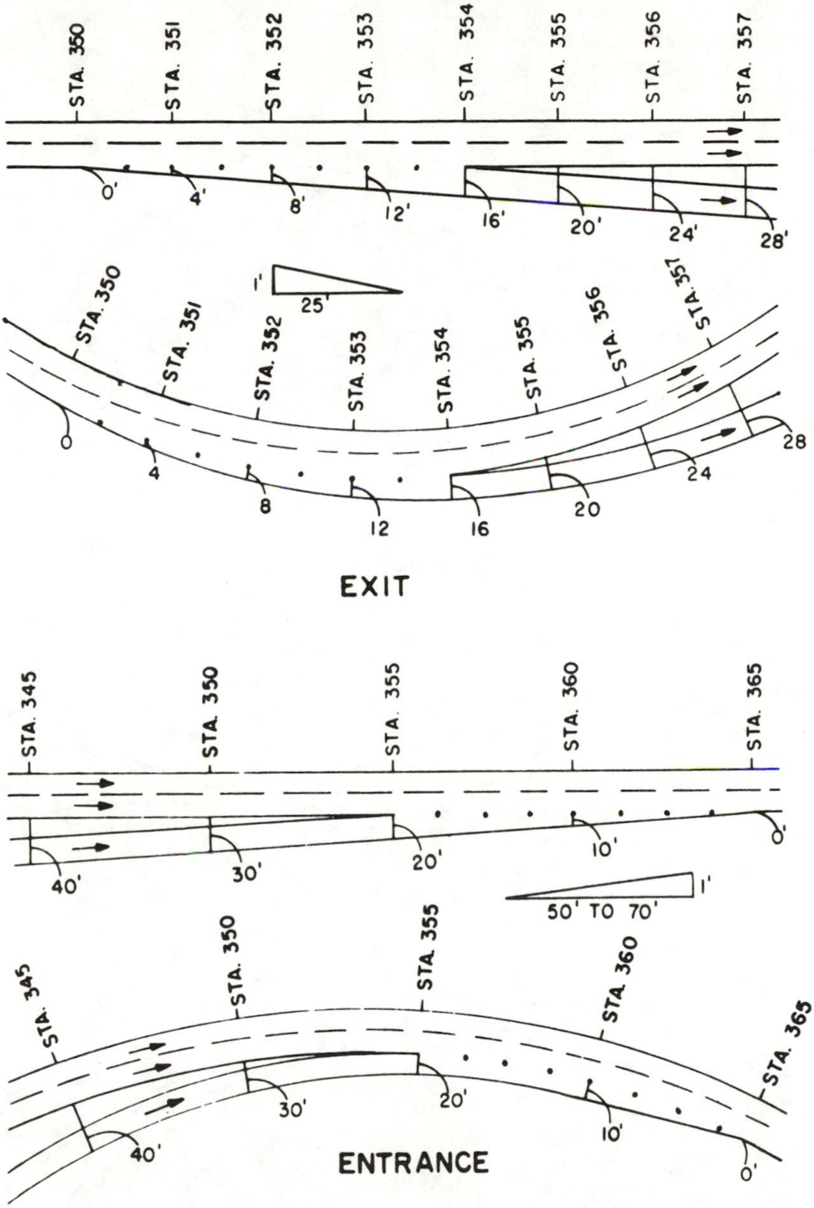

Figure X-71. Layout of taper-type terminals on curves.

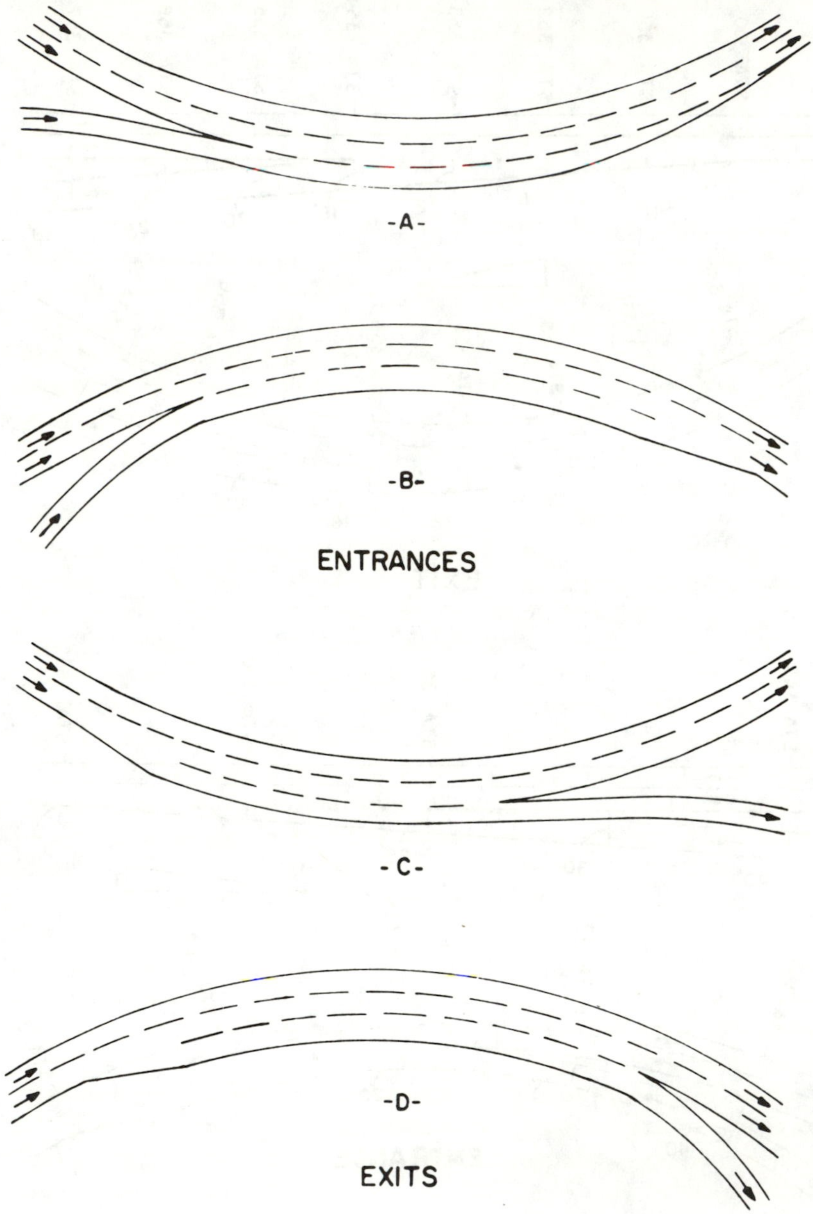

Figure X-72. Parallel-type ramp terminals on curves diagrammatic.

alinement occurs between the ramp and speed change lane, an intervening tangent should be used to aid in superelevation transition.

An exit may be particularly troublesome where the highway curves to the left (Figure X-72C) because traffic on the outside lane tends to follow the ramp. Exits on left-turning curves should be avoided, if possible. Caution must be used in positioning a taper-type deceleration lane on the outside of a left-turning main line curve. The design should provide a definite break in the right edge of pavement to provide a visual cue to the through driver so that he is not inadvertently led off the through roadway. In order to make the deceleration lane more apparent to approaching motorists, the taper should be shorter, preferably no more than 100 ft in length. The deceleration lane should begin either upstream or downstream from the PC. It should not begin right at the PC, as the deceleration lane appears to be an extension of the tangent, and motorists are more likely to be confused. The ramp proper should begin with a section of tangent or a long-radius curve to permit a long and gradual reversing of the superelevation.

An alternate design, which will usually avoid operational problems, is to locate the exit terminal a considerable distance upstream from the PC. A separate and parallel ramp roadway is provided to connect with the ramp proper.

With the highway curving to the right and the exit located on the right (Figure X-72D) there is also a tendency for vehicles to exit inadvertently. Again, the taper should be short to provide additional "target" value for the deceleration lane. With this configuration the superelevation of the deceleration lane is readily effected by continuing the rate from the through pavement and generally increasing it to the rate required on the ramp curve.

Figures X-73, X-74, and X-75 show several taper-type single-lane entrances and exits. In the lower roadway of Figure X-73 the first of the successive on-ramps is tapered into the freeway, whereas the second is carried ahead as an additional lane. Note that the ramps carry two lanes from the local street system, but there is a transition to a single lane. In the upper roadway of Figure X-73 there are successive exits, and a freeway lane is dropped after the first exit. There is a transition of the straight ramps to two lanes before joining the local street system. This transition provides additional storage capacity and facilitates turning movements.

Figure X-74 shows terminal treatments with curved ramps. At the exit on the upper roadway the taper-type deceleration lane is carried on tangent alinement well beyond the ramp nose before joining the long-radius curve on the ramp proper.

On the lower roadway shown in Figure X-75 the treatment of successive on-ramps is similar to that in Figure X-73. The first entrance is tapered into the freeway and the second entrance is carried ahead as an auxiliary lane. The auxiliary lane is of different color and texture than the through lanes and is

Figure X-73. Single-lane freeway terminals.

Figure X-74. Single-lane freeway terminals.

Figure X-75. Single-lane freeway terminals.

terminated at the next interchange downstream. On the upper roadway an auxiliary lane is provided for the exit. The auxiliary lane extends to the next interchange upstream. A recovery lane, which is tapered into the freeway at a ratio of 50:1, is provided beyond this exit.

Multilane free-flow terminals. Multilane terminals are required where traffic is too great for single-lane operation. Other considerations that may call for multilane terminals are through-route continuity, queuing on long ramps, lane balance, and design flexibility. The most common multilane terminals consist of two-lane entrances and exits at freeways. Other multilane terminals are sometimes termed "major forks" and "branch connections." The latter terms denote a separating and joining of two major routes.

Two-lane entrances. Two-lane entrances are warranted for two situations: either as branch connections or because of capacity requirements of the on-ramp. In order to satisfy lane-balance requirements, at least one additional lane must be provided downstream. This addition may be a basic lane if also required for capacity, or an auxiliary lane that may be dropped 2,500 to 3,000 ft downstream or at the next interchange. In some instances two additional lanes may be necessary because of capacity requirements.

If the two-lane entrance is preceded by a two-lane exit, there is probably no need to increase the basic number of lanes on the freeway from a capacity standpoint. In this case the added lane that results from the two-lane entrance is considered an auxiliary lane, and it may be dropped approximately 2,500 ft or more downstream from the entrance. Details of lane drops were discussed earlier in this chapter.

Figure X-76 illustrates simple two-lane entrance terminals where a lane has been added to the freeway. The number of lanes on the freeway has little or no effect on design of the terminal. Figure X-76A shows a taper-type entrance and Figure X-76B, a parallel-type entrance. Intermixing of the two designs is not recommended within a system route or an urban-area system.

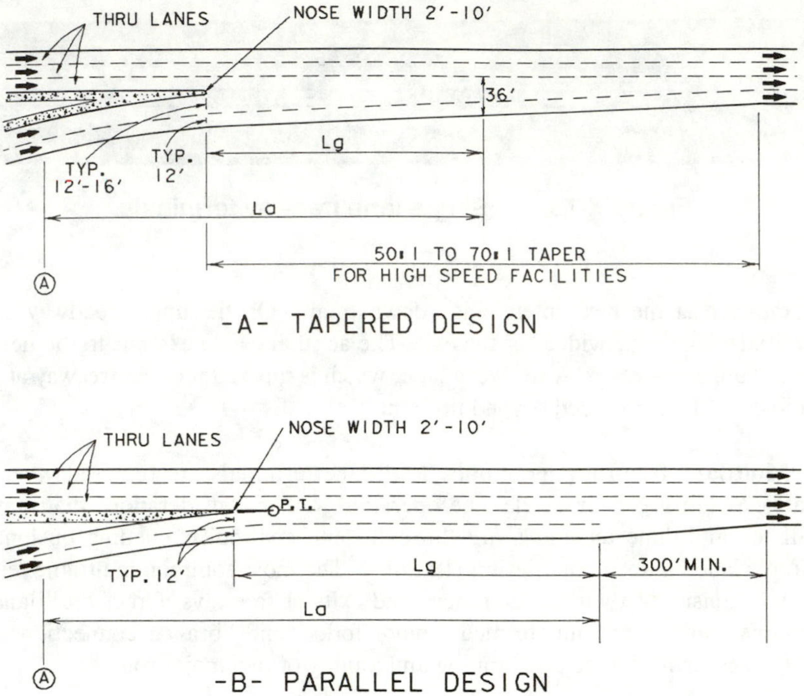

NOTES:

1. La IS THE REQUIRED ACCELERATION LENGTH AS SHOWN IN TABLE X-4 OR X-5.
2. POINT (A) CONTROLS SAFE SPEED ON THE RAMP. La SHOULD NOT START BACK ON THE CURVATURE OF THE RAMP UNLESS THE RADIUS EQUALS 1000' OR MORE.
3. Lg IS REQUIRED GAP ACCEPTANCE LENGTH. Lg SHOULD BE A MINIMUM OF 300' TO 500' DEPENDING ON THE NOSE WIDTH.
4. THE VALUE OF La OR Lg, WHICHEVER PRODUCES THE GREATEST DISTANCE DOWNSTREAM FROM WHERE THE NOSE WIDTH EQUALS TWO FEET, IS SUGGESTED FOR USE IN THE DESIGN OF THE RAMP ENTRANCE.

Figure X-76. Typical two-lane entrance ramps.

The basic form or layout of a two-lane taper-type entrance, as shown in Figure X-76A, is that of a single-lane taper, as described earlier in this chapter, with a second lane added to the right or outer side and continued as an added or auxiliary lane on the freeway. Table X-4 shows minimum lengths of acceleration distances for entrance ramps. The gap acceptance length is also a consideration as illustrated in Figure X-76A. Where ramp grades are involved, the lengths should be adjusted as shown in Table X-5. The length of a two-lane taper-type entrance is approximately the same as that for a single-lane entrance. As in the case of a single-lane entrance, it is most desirable that the geometrics of the ramp proper permit motorists to attain the approximate running speed of the freeway before reaching the tapered section.

With thc parallel type of two-lane entrance, as shown in Figure X-76B, the left lane of the ramp is continued onto the freeway as an added lane. The right lane of the ramp is carried as a parallel lane for several hundred feet and terminated with a tapered section which is a minimum of 300 ft long. The length of the right lane should, as a minimum, be determined from the acceleration length or gap acceptance length, as shown in Figure X-76B. Major factors in determining the needed length are the traffic volume on the ramp and the traffic volume on the freeway. Where traffic from the two lane entrance is only slightly in excess of the design capacity of a single lane ramp, a length equal to or greater than the values determined from the acceleration length or gap acceptance length, plus a taper, may suffice.

Where the predominant two-lane entrances in a particular State or locality are of the parallel type and drivers are accustomed to it, they are ill-prepared to use a taper type, and vice versa. Thus, a particular type of entrance terminal is sometimes condemned as being unsatisfactory when in fact the difficulty may be lack of uniformity. Either form of two-lane entrance is satisfactory if used exclusively within an area or a region, but they should not be intermixed along a given route.

Two-lane exits. Where traffic leaving the freeway at an exit terminal exceeds the design capacity of a single lane, it is necessary to provide a two-lane exit terminal. To satisfy lane balance requirements and not to reduce the basic number of through lanes, it is usually necessary to add an auxiliary lane upstream from the exit. A distance of approximately 1,500 ft is required to develop the full capacity of a two-lane exit. Typical designs for two-lane exit terminals are shown in Figure X-77; the taper is illustrated in Figure X-77A, and the parallel type in Figure X-77B.

In cases where the basic number of lanes is to be reduced beyond a two-lane exit, the basic number of lanes should be carried beyond the exit before the outer lane is dropped. This design provides a recovery area for any through vehicles that remain in that lane. This was discussed in the section "Lane Reduction" earlier in this chapter.

With the parallel type of two-lane exit, as shown in Figure X-77B, the operation is different from the taper type in that traffic in the outer through lane of the freeway must change lanes in order to exit. In fact, an exiting motorist is required to move two lanes to the right in order to use the right lane of the ramp. Thus, considerable lane changing is necessary in order for the exit to operate efficiently. This entire operation requires a substantial length of highway, which is dependent in part on the total traffic volume on the freeway and especially on the volume using the exit ramp. The total length from the beginning of the first taper to the point of departure of the ramp traveled way from the right-hand through lane of the freeway will range from 2,500 ft for turning volumes of 1,500 vph or less upward to 3,500 ft for turning volumes of 3,000 vph.

Figure X-78 features two-lane entrances and exits on a freeway. The two-lane entrance on the upper roadway is tapered into one additional lane on the freeway. Note the arrangement in advance of the freeway entrance. A one-lane ramp joins the two-lane ramp, and there is transition of these three lanes into two lanes before entering the freeway. On the bottom roadway a lane is dropped from the freeway following the two-lane exit. Note the subsequent split in the two-lane exit. The facility is constructed with sufficient distance between diverge points so that the second split operates smoothly.

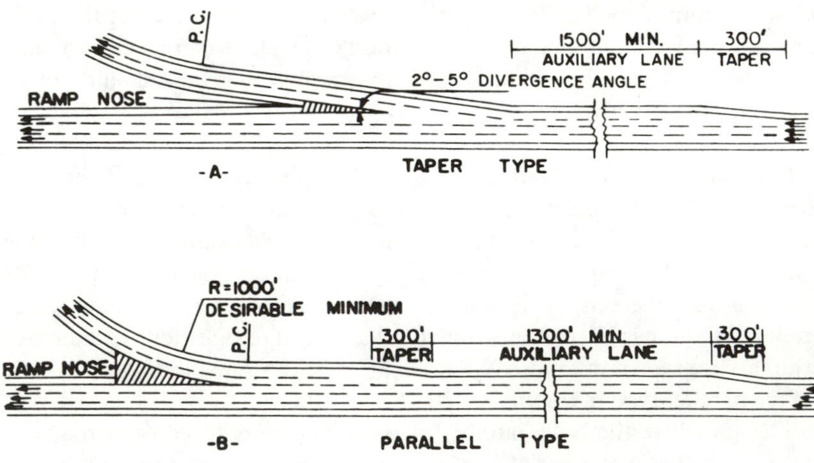

Figure X-77. Two-lane exit terminals.

In Figure X-79 the two-lane entrance and exit are shown on very flat angles, which encourage high-speed operation. The outside lanes of the entrance and exit are carried onto the freeway as auxiliary lanes of different color and texture. This design discourages through traffic from using the auxiliary lane. Note the subsequent split in the two-lane exit. The distance between the successive ramp terminals is well over 1,000 ft to provide adequate perception, reaction, and maneuvering time for the exiting motorists.

Figure X-80 illustrates a special arrangement of extremely high-volume entrances and exits. On the upper roadway, successive two-lane entrances are provided. A single lane is added to the freeway downstream from each of

Figure X-78. Multilane terminals.

Figure X-79. Multilane terminals.

these entrances. The flat entrance angles and long sight distances encourage smooth operations. The exit arrangement on the lower roadway is a major fork, the six freeway lanes splitting into four through lanes and three turning lanes. The three-lane exit subsequently splits into two and two.

The lower roadway of Figure X-81 illustrates a two-lane exit where the outside lane is an auxiliary lane that is introduced approximately 1,000 ft in advance of the gore area. The auxiliary lane is of different color and texture than the through lanes. Note that the ramp is transitioned from two to three lanes as it approaches the local street system.

Two-lane terminals on curved alinement. The design of ramp terminals where the freeway is on curved alinement is discussed under single-lane terminals. The same principles of design, in which offsets from the edge of roadway are used, may be used in the layout of two-lane terminals.

Major forks and branch connections. A major fork is defined as the bifurcation of a directional roadway, of a terminating freeway route into two

Figure X-80. Multilane terminals.

Figure X-81. Multilane terminals.

directional multilane ramps that connect to another freeway, or as the diverging area created by the separation of a freeway route into two separate freeway routes of about equal importance.

The design of major forks is subject to the same principles of lane balance as any other diverging area. The total number of lanes in the two roadways beyond the divergence should exceed the number of lanes approaching the diverging area by at least one. Desirably, the number of lanes should be increased by only one. Operational difficulties invariably develop unless traffic in one of the interior lanes has an option of taking either of the diverging roadways. Accordingly, the nose should be placed in direct alinement with the centerline of one of the interior lanes, as illustrated in Figure X-82A, B, or C, where the alinement of the two departing roadways are in curves. This interior lane is continued as a full-width lane, both left and right of the gore. Thus, the width of this interior lane will be at least 24 ft at the painted nose (prolongation of pavement-edge stripes) and preferably not over 28 ft. The length over which the widening from 12 to 24 ft takes place should be within the range of 1,000 or 1,800 ft. However, in the case where at least one of the approaches is on a tangent alinement and must continue on a tangent, a true optional interior lane cannot be physically developed. As such, the principles of the two lane exit facility must be used as shown in Figure X-82D.

In case of a two-lane roadway separating into two, two-lane routes there is no interior lane. In such cases it is advisable to widen the approach roadway to three lanes, thus creating an interior lane. The lane is added on the side of the fork that serves the lesser traffic volume or perhaps the non-preference freeway route. In the illustration, Figure X-82A, the right (lower) fork would be the more lightly traveled of the two. The widening from 36 ft for the approach roadway to about 48 or 50 ft at the painted nose should be accomplished in a continuous sweeping curve with no reverse curvature in the alinement of the pavement edges.

A branch connection is defined as the beginning of a directional roadway of a freeway formed by the convergence of two directional multilane ramps from another freeway or by the convergence of two freeway routes to form a single freeway route.

The number of lanes downstream from the point of convergence may be one lane fewer than the combined total on the two approach roadways. In some cases the traffic demand may require that the number of lanes going away from the merging area be equal to the sum of the number of lanes on the two roadways approaching it, and a design of this type will pose no operational problem. Such a design is illustrated in Figure X-83A.

Where a lane is to be dropped, which is the more common case, a means for accomplishing the reduction is discussed in the section "Lane Reduction" in this chapter. The lane that is terminated will ordinarily be the exterior lane

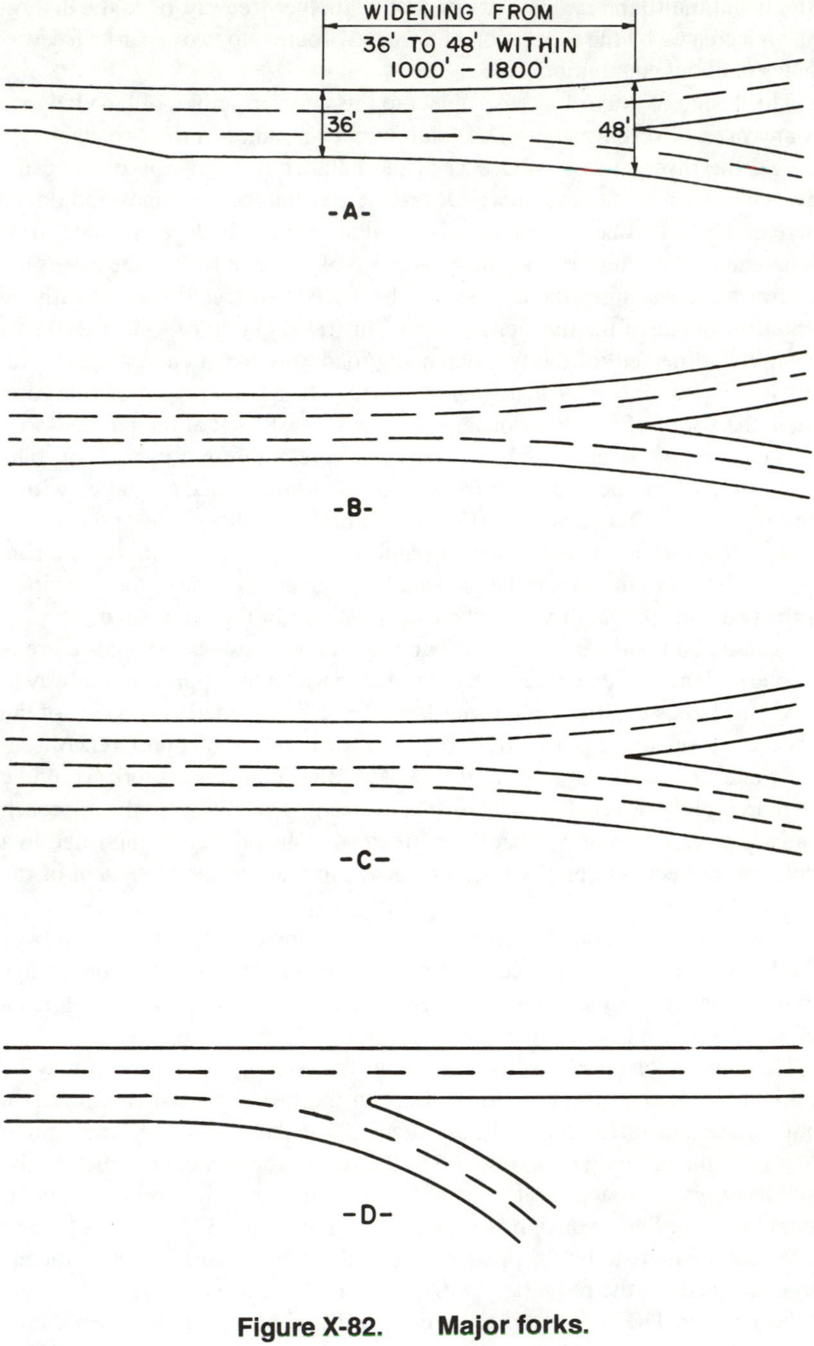

Figure X-82. Major forks.

from the roadway serving the lowest volume per lane. However, some considerations should also be given to the fact that the outer lane from the roadway entering from the right is the slow-speed lane for that roadway; whereas the opposite is true for the roadway entering from the left. If the traffic volumes per lane are about equal, it would be proper to terminate the lane on the right, as shown in Figure X-83B. In any case, consistency within an area or region is often more important than volume per lane since the latter may change with the specific design. The lane being terminated should be carried at full lane width for a distance of approximately 1,000 feet before being tapered out.

Another consideration is the possibility of a high-speed inside merge, as in Figure X-83C. This merge should be treated as any other high-speed merging situation by adding a lane on the right.

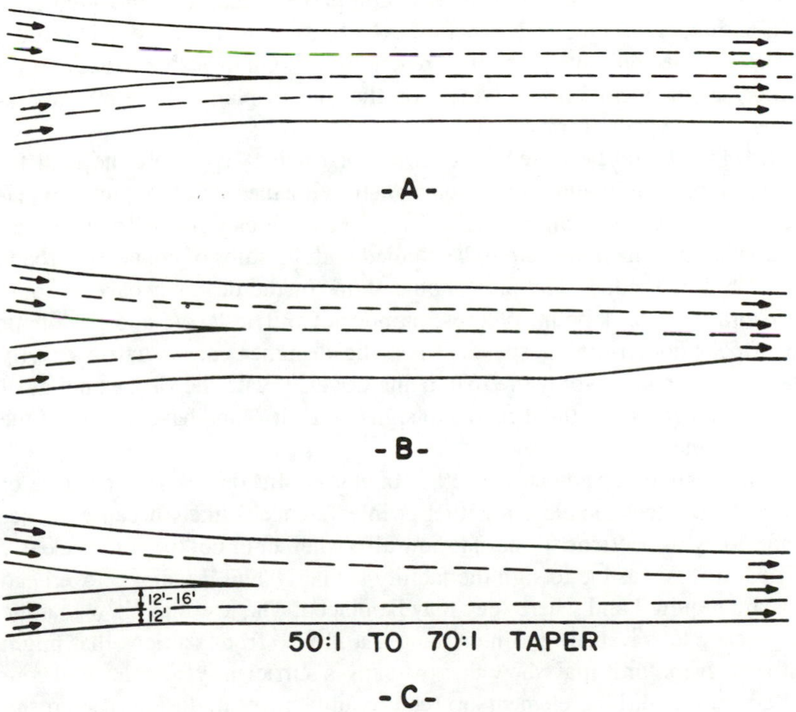

Figure X-83. Branch connections.

Other Interchange Design Features

Testing for Ease of Operation

Each section of freeway that includes a series of interchanges or a succession of exits and entrances should be tested for operational characteristics of the route after the preliminary design, including adaptability, capacity, and operational features, before each interchange is completed. The test is an evaluation of the section for ease of operation and for route continuity from a driver's point of view, both of which are affected by the location, proximity, and sequence of exits and entrances, the merging, diverging, and weaving movements necessary, practicability of signing, and clarity of paths to be followed.

A route may be tested by isolating that part of the plan, for each path, and examining it only with regard to other parts of the layout that will affect a driver on the path being tested. Viewing an entire plan, as it might be seen from the air, may give an impression of complexity because of the number of exits, entrances, ramps, and structures. Actually, it is not as complex to the drivers, who see only the path they are driving on. On the other hand, certain weaknesses of operation not evident on the overall plan will be revealed in testing a single path of travel.

The plan should be tested by drawing or tracing separately the path for each principal origin and destination and studying thereon those physical features that will be encountered by a driver. The test can also be made on an overall plan on which the path to be studied and the stubs of connecting roads are colored or shaded. The plan should show, or the designer have in some other form, the peak-hour volumes, number of traffic lanes, and peak-hour and offpeak-hour running speed. Thus, the designer can visualize exactly what the driver sees—only the road being traveled, with the various points of ingress and egress and the directional signs along it—and have a sense of the accompanying traffic.

Such an analysis indicates whether or not confusion is likely because of exits and entrances too close together or interference is likely because of successive weaving sections. It should show also whether or not the path is clearly defined, if it is feasible to sign the facility properly, and if major or overhead signs are required and where they may be placed. The test may show that the path is easy to travel, direct in character, and free from sections that might confuse drivers, or it may show that the path is sufficiently complex and confronted with disturbing elements to require adjustment in design. As a result it may be necessary to move or eliminate certain ramps. In an extreme case the test may show that it is necessary to change the overall pattern by eliminating an interchange, or to introduce collector-distributor roads in order to prevent interference with through traffic, or to make some other radical change in design.

Figure X-84 is a simple diagrammatic solution to a typical freeway operational problem. The freeway joins a major arterial at a branch connection and diverges at a major fork in a distance of approximately 1 to 3 miles. There may be other connections to and from the freeway between these points. The through freeway merges on the left at point X and diverges on the right at point Y. The desirable solution, as shown in this figure, does not require any lane changes on the through lanes of the freeway. Traffic on the local arterial enters and exits on the right, and there is no disruption of route continuity on either facility.

Grading and Landscape Development

Grading at an interchange is determined chiefly by the alinements, profiles, cross sections, and drainage requirements for the intersecting highways and ramps. Each through roadway or ramp should not be treated as a separate unit and graded to standard cross section without regard to its relation with adjacent roads and to the surrounding topography. Instead, the whole construction area should be designed as a single unit to keep construction and maintenance costs to a minimum, obtain maximum visibility, and enhance the appearance of the area. In some parts, such as at narrow sections between converging pavements, the slopes and grading controls may affect the alinement and profile design.

Contour grading design. An important and early step in interchange design is the initial bridge control study in which the preliminary alinement and profiles of the intersecting roads are developed to determine the controls for bridge design. Alternates in such elements as clearances, curbs, walks, and position and extent of walls should be examined in regard to general grading before conclusions are drawn for the bridge design, particularly for lengths of wing walls. Minor modifications in alinement and profile, in abutments and walls, and in related earthwork may produce a more desirable solution as a whole.

At interchanges, and elsewhere as feasible, steep roadside earth slopes should be avoided for all roads and ramps. Flat slopes should be used where feasible, for economical construction and maintenance, to increase safety, and

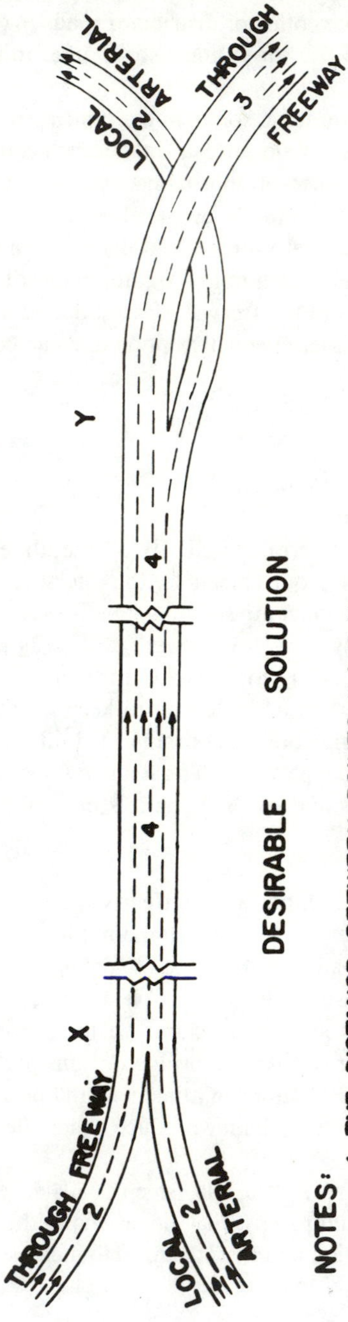

NOTES:

1. THE DISTANCE BETWEEN POINTS X AND Y MAY BE APPROXIMATELY 1 TO 3 MILES
2. NUMBER OF LANES ARE SHOWN ON EACH ROADWAY

Figure X-84. Diagram of freeway operational problem and solution.

to enhance the appearance of the area. Broad rounded drainageways or swalelike depressions should be used, where feasible, to encourage good turf and easy mower maintenance. V ditches and small ditches with steep side slopes should be avoided. Drainage channels and related structures should be as inconspicuous and maintenance-free as feasible. They should not be an eyesore, or a hazard to traffic. Transition grading between cut and fill slopes should be long and natural in appearance. The slopes should be well rounded and smooth to blend the highway into the adjacent terrain. The contours should have flowing continuity and be congruous with the form of the pavement ribbon and with the form of the adjacent topography as well.

Where good trees and other desirable features exist, the contour grading and drainage plan should be designed to protect and preserve these features, as feasible. This effort, however, must be consistent with the above stated objectives.

Seeding and planting. The seeding and planting of an area should be considered in the initial design. These items may be included in the grading or paving work or otherwise provided early to protect what usually is a large investment. Failure to establish turf or plantings early is costly because of erosion or structural failure. Present mechanized landscape methods are efficient and their application will go far toward thwarting destructive forces and raising the general esthetic level of the interchange. These methods need not be described here, but some unsafe practices are mentioned.

Proposed plantings should be selected with regard to their ultimate growth. Improperly located shrubs or trees may seriously shorten horizontal sight distance on curves and seriously interfere with lateral sight distance between adjacent roadways. Even low-lying ground covers can seriously shorten vertical sight distance on curving ramps.

Trees or shrubs may be used to outline travel paths or to give drivers a sense of an obstruction ahead. For example, the ends of a directional island or approach nose may be planted with low-growing shrubs that will be seen from a considerable distance and direct the driver's attention to the necessity for a turn. These shrubs should not be of the type that could cause vehicle damage on impact, nor should they ever obscure signs or warning devices.

The AASHTO *Roadside Design Guide* (**1**) may be referenced for guidance on minimum clear zones prior to planting trees which will mature to greater than 4 inches in diameter. Distances greater than the minimum are often necessary because of overhanging branches being a distraction, and leaves on pavements becoming a hazard, especially when wet. In areas where ice and snow are a problem, all trees should be planted an adequate distance from the traveled way to allow for snow drifting and to prevent icing in shaded areas.

Models

Three-dimensional models are helpful in the design of interchanges. Models are particularly useful in communicating the designer's ideas to lay-groups and others who are not trained to visualize three dimensions from the plans. Design concept teams and other officials find models helpful in analyzing proposed designs.

Highway models fall into two basic categories—design models and presentation models. Design models are simple and easily adjusted, thus permitting the designer to experiment with different concepts. Presentation models are more permanent and are valuable to highway officials in making a presentation to others who are not familiar with engineering terms and methods.

Design models. Design models are economical, skeletonized, scaled structures, which provide a rapid, practical means of showing the combination of roadway horizontal and vertical alinement in a proper scale relation. In simplest form, design models are made to examine the three-dimensional roadway relations. However, design models also may be more complete to show such things as earth slopes and grading plans.

A popular form for design models is the so-called peg model. Peg models are extremely easy to construct and to adjust to conform to indicated changes in roadway alinement and profile. Methods of using pegs to indicate roadway alinement and profile vary from simple techniques that merely indicate the approximate line and profile to those that show accurate horizontal and vertical curvature as well as roadway widths and superelevation.

In another technique the roadways are supported on continuous strips of material cut to the required profile. This method does not have the design adjustment flexibility of the peg method, but it is more suitable for contour grading design. Usually, the alinement and profile design should be practically completed before the contour-grading design begins.

When using cut profile strips, the profile is cut away at long structures. It is desirable to show such appurtenances as bridge piers and retaining walls. Bridge piers can be represented by dowels. Where a structure passes over another roadway, the dowels should be cut to show the bridge clearance. Dry sawdust or other suitable materials can be used to simulate earthwork for contour grading. Slopes can be adjusted to the correct grade, and ground levels can be finished off with reasonable accuracy. Where existing ground levels are to be retained, pins can be pushed into the base until their heads are at the required level shown on the base map. The sawdust is then graded between the pins.

Presentation models. A presentation model is one that is used primarily to illustrate the functional plan of an interchange or highway network. It may be extensive enough to include adjacent development or terrain conditions. The type of presentation model to be built can vary considerably, depending on its contemplated uses. The simplest type, the skeleton model, shows only the roadways on their correct alinement and their correct profiles and differs but little from some design models. No fill is shown, hence the term "skeleton." At the other extreme a fully finished model will show the surrounding land to the correct contour, land use, and even individual buildings, trees, watercourses, and other details. The more complete jobs are best handled by professional model builders.

Application of models. Development of contour grading drawings is sometimes difficult. The difficulty is particularly apparent in complicated interchange areas where variable roadways and numerous structures make it difficult to visualize from two-dimensional drawings the most desirable roadside grading arrangement. This problem is compounded by the flatter slopes employed for safety and ease of maintenance. The problem of removing surface drainage and locating drainage outlets becomes more difficult. The use of contour models with slopes permits the designer to quickly investigate various arrangements and determine which is the most suitable.

Profile and alinement design. Models can be used to advantage in investigating complicated pavement elevation and lane designs, particularly in merging and diverging areas of interchanges. For this purpose the type and scale model used should be such that the pavement elevations can be visually inspected and the elevations checked against those shown on the plans. It should also be easy to make small adjustments to the pavement elevations and to rearrange the through and auxiliary lanes.

Structure studies. Models may serve a variety of uses in bridge and retaining-wall studies. Compatibility of structures with site conditions and other adjacent structures can be examined. For esthetic purposes the driver should not be unnecessarily confronted with a variety of bridge designs at any one time. Further, he should not view a "forest" of turning roadway structure piers if it is possible to avoid it. The use of models provides a ready means to study these features.

Evaluation of alternative structure and walled embankment combinations is sometimes required for a particular area of a freeway Although it is quite easy to determine which arrangement is most economical by conventional methods, it is difficult to visualize which is best esthetically without a model. They can be developed with alternate set-in sections for the different designs under study.

Other elements. Models can be helpful in designing tree planting along freeway borders and in interchange areas. The landscape architect could experiment with alternate planting locations and arrangements to determine which areas would be suitable. Often the determination is related directly to grading plan development.

The use of models may be an aid for design and review of freeway signing, particularly in complicated interchange areas. Overpass structures, retaining walls, and excessive curvature may cut off or shorten the view of a sign, thus limiting the sign location. Study of such features also aids in determination of

Figure X-85. Design model.

the need for overhead signs, or if known, their best location. The position of some sign messages can be such that they are too easily interpreted for one traffic movement when actually they are intended for another. These as well as other signing problems may be avoided by an investigation of proposed sign locations with the use of a model. For study of signs, and landscaping also, the complete grading or presentation type of model usually is needed.

Models also can be very useful in the design location of lighting standards, guardrails, and other highway hardware items.

Examples of models are shown in Figures X-85 and X-86. Figure X-85 is a design model, and Figure X-86 is a presentation model.

Figure X-86. Presentation model.

REFERENCES

1. *Roadside Design Guide.* Washington, D.C.: AASHTO, 1989.
2. Federal Highway Administration, *Manual on Uniform Traffic Control Devices.* Washington, D.C.: U.S. Government Printing Office, 1988.
3. *Highway Capacity Manual*, Special Report No. 209. Washington, D.C.: Transportation Research Board, 1985.

INDEX